# STUDENT'S SOLUTIONS

## CINDY TRIMBLE & ASSOCIATES

# BEGINNING ALGEBRA
## SIXTH EDITION

## Elayn Martin-Gay
*University of New Orleans*

**PEARSON**

Boston   Columbus   Indianapolis   New York   San Francisco   Upper Saddle River
Amsterdam   Cape Town   Dubai   London   Madrid   Milan   Munich   Paris   Montreal   Toronto
Delhi   Mexico City   Sao Paulo   Sydney   Hong Kong   Seoul   Singapore   Taipei   Tokyo

The author and publisher of this book have used their best efforts in preparing this book. These efforts include the development, research, and testing of the theories and programs to determine their effectiveness. The author and publisher make no warranty of any kind, expressed or implied, with regard to these programs or the documentation contained in this book. The author and publisher shall not be liable in any event for incidental or consequential damages in connection with, or arising out of, the furnishing, performance, or use of these programs.

Reproduced by Pearson from electronic files supplied by the author.

ISBN-13: 978-0-321-78523-7
ISBN-10: 0-321-78523-1

1 2 3 4 5 6 EB 16 15 14 13 12

www.pearsonhighered.com

# Contents

# Chapter 1

1. a. $5 < 8$ since 5 is to the left of 8 on the number line.

   b. $6 > 4$ since 6 is to the right of 4 on the number line.

   c. $16 < 82$ since 16 is to the left of 82 on the number line.

2. a. $9 \geq 3$ is true, since $9 > 3$ is true.

   b. $3 \geq 8$ is false, since neither $3 > 8$ nor $3 = 8$ is true.

   c. $25 \leq 25$ is true, since $25 = 25$ is true.

   d. $4 \leq 14$ is true, since $4 < 14$ is true.

3. a. $3 < 8$

   b. $15 \geq 9$

   c. $6 \neq 7$

4. The integer $-10$ represents 10 meters below sea level.

5. a. The natural number is 25.

   b. The whole number is 25.

   c. The integers are $25, -15, -99$.

   d. The rational numbers are $25, \dfrac{7}{3}, -15, \dfrac{-3}{4}, -3.7, 8.8, -99$.

   e. The irrational number is $\sqrt{5}$.

   f. The real numbers are $25, \dfrac{7}{3}, -15, \dfrac{-3}{4}, \sqrt{5}, -3.7, 8.8, -99$.

6. a. $0 < 3$ since 0 is to the left of 3 on a number line.

   b. $15 > -5$ since 15 is to the right of $-5$ on a number line.

   c. $3 = \dfrac{12}{4}$ since $\dfrac{12}{4}$ simplifies to 3.

7. a. $|-8| = 8$ since $-8$ is 8 units from 0 on a number line.

   b. $|9| = 9$ since 9 is 9 units from 0 on a number line.

   c. $|-2.5| = 2.5$ since $-2.5$ is 2.5 units from 0 on a number line.

   d. $\left|\dfrac{5}{11}\right| = \dfrac{5}{11}$ since $\dfrac{5}{11}$ is $\dfrac{5}{11}$ unit from 0 on a number line.

   e. $\left|\sqrt{3}\right| = \sqrt{3}$ since $\sqrt{3}$ is $\sqrt{3}$ units from 0 on a number line.

8. a. $|8| = |-8|$ since $8 = 8$.

   b. $|-3| > 0$ since $3 > 0$.

   c. $|-7| < |-11|$ since $7 < 11$.

   d. $|3| > |2|$ since $3 > 2$.

   e. $|0| < |-4|$ since $0 < 4$.

## Vocabulary, Readiness & Video Check 1.2

1. The <u>whole</u> numbers are $\{0, 1, 2, 3, 4, ...\}$.

2. The <u>natural</u> numbers are $\{1, 2, 3, 4, 5, ...\}$.

3. The symbols $\neq$, $\leq$, and $>$ are called <u>inequality</u> symbols.

4. The <u>integers</u> are $\{..., -3, -2, -1, 0, 1, 2, 3, ...\}$.

5. The <u>real</u> numbers are {all numbers that correspond to points on the number line}.

6. The <u>rational</u> numbers are $\left\{\dfrac{a}{b}\,\middle|\, a \text{ and } b \text{ are integers, } b \neq 0\right\}$.

7. The <u>irrational</u> numbers are {nonrational numbers that correspond to points on the number line}.

8. The distance between a number $b$ and 0 on a number line is $\underline{|b|}$.

9. To form a true statement: $0 < 7$.

10. Five is greater than or equal to four; $5 \geq 4$

11. 0 belongs to the whole numbers, the integers, the rational numbers, and the real numbers; since 0 is a rational number, it cannot also be an irrational number.

12. absolute value

## Exercise Set 1.2

1. $7 > 3$ since 7 is to the right of 3 on a number line.

3. $6.26 = 6.26$

5. $0 < 7$ since 0 is to the left of 7 on a number line.

7. $-2 < 2$ since $-2$ is to the left of 2 on a number line.

9. $32 < 212$ since 32 is to the left of 212 on a number line.

11. $30 \leq 45$ since 30 is to the left of 45 on a number line.

13. $11 \leq 11$ is true, since $11 = 11$.

15. $10 > 11$ is false, since 10 is to the left of 11 on a number line.

17. $3 + 8 \geq 3(8)$ is false, since 11 is to the left of 24 on a number line.

19. $9 > 0$ is true, since 9 is to the right of 0 on a number line.

21. $-6 > -2$ is false, since $-6$ is to the left of $-2$ on a number line.

23. Eight is less than twelve is written as $8 < 12$.

25. Five is greater than or equal to four is written as $5 \geq 4$.

27. Fifteen is not equal to negative two is written as $15 \neq -2$.

29. The integer 14,494 represents 14,494 feet above sea level. The integer $-282$ represents 282 feet below sea level.

31. The integer $-28,000$ represents 28,000 fewer students.

33. 350 represents a deposit of $350. $-126$ represents a withdrawal of $126.

35. The number 0 belongs to the sets of: whole numbers, integers, rational numbers, and real numbers.

37. The number $-2$ belongs to the sets of: integers, rational numbers, and real numbers.

39. The number 6 belongs to the sets of: natural numbers, whole numbers, integers, rational numbers, and real numbers.

41. The number $\dfrac{2}{3}$ belongs to the sets of: rational numbers and real numbers.

43. The number $-\sqrt{5}$ belongs to the sets of: irrational numbers and real numbers.

45. False; rational numbers may be non-integers.

47. True

49. True

51. True

53. False; an irrational number may not be written as a fraction.

55. $-10 > -100$ since $-10$ is to the right of $-100$ on the number line.

57. $32 > 5.2$ since 32 is to the right of 5.2 on the number line.

59. $\dfrac{18}{3} < \dfrac{24}{3}$ since $6 < 8$.

61. $-51 < -50$ since $-51$ is to the left of $-50$ on the number line.

63. $|-5| > -4$ since $5 > -4$.

65. $|-1| = |1|$ since $1 = 1$.

67. $|-2| < |-3|$ since $2 < 3$.

69. $|0| < |-8|$ since $0 < 8$.

71. The tallest bar represents the greatest number of visitors; 2009

**73.** Look for the bars that have heights greater than 280; 2009, 2010

**75.** In 2001, there were 280 million visitors. In 2010, there were 281 million visitors.
280 million < 281 million

**77.** The 2009 cranberry production in Oregon was 49 million pounds, while the 2009 cranberry production in Washington was 16 million pounds.
49 million > 16 million

**79.** The 2009 cranberry production in Washington was 16 million pounds, while the 2009 cranberry production in New Jersey was 54 million pounds.
$54 - 16 = 38$
The production in Washington was 38 million pounds less or $-38$ million.

**81.** $-0.04 > -26.7$ since $-0.04$ is to the right of $-26.7$ on the number line.

**83.** The sun is brighter since $-26.7 < -0.04$.

**85.** The sun is the brightest since $-26.7$ is to the left of all other numbers listed.

**87.** $20 \le 25$ has the same meaning as $25 \ge 20$.

**89.** $6 > 0$ has the same meaning as $0 < 6$.

**91.** $-12 < -10$ has the same meaning as $-10 > -12$.

**93.** answers may vary

## Section 1.3 Practice

**1. a.** $36 = 4 \cdot 9 = 2 \cdot 2 \cdot 3 \cdot 3$

   **b.** $200 = 2 \cdot 100 = 2 \cdot 4 \cdot 25 = 2 \cdot 2 \cdot 2 \cdot 5 \cdot 5$

**2. a.** $\dfrac{63}{72} = \dfrac{3 \cdot 3 \cdot 7}{2 \cdot 2 \cdot 2 \cdot 3 \cdot 3} = \dfrac{7}{2 \cdot 2 \cdot 2} = \dfrac{7}{8}$

   **b.** $\dfrac{64}{12} = \dfrac{2 \cdot 2 \cdot 2 \cdot 2 \cdot 2 \cdot 2}{2 \cdot 2 \cdot 3} = \dfrac{2 \cdot 2 \cdot 2 \cdot 2}{3} = \dfrac{16}{3}$

   **c.** $\dfrac{7}{25} = \dfrac{7}{5 \cdot 5}$
There are no common factors other than 1, so $\dfrac{7}{25}$ is already in lowest terms.

**3.** $\dfrac{3}{8} \cdot \dfrac{7}{9} = \dfrac{3 \cdot 7}{8 \cdot 9} = \dfrac{3 \cdot 7}{2 \cdot 2 \cdot 2 \cdot 3 \cdot 3} = \dfrac{7}{2 \cdot 2 \cdot 2 \cdot 3} = \dfrac{7}{24}$

**4. a.** $\dfrac{3}{4} \div \dfrac{4}{9} = \dfrac{3}{4} \cdot \dfrac{9}{4} = \dfrac{3 \cdot 9}{4 \cdot 4} = \dfrac{27}{16}$

   **b.** $\dfrac{5}{12} \div 15 = \dfrac{5}{12} \cdot \dfrac{1}{15} = \dfrac{5 \cdot 1}{12 \cdot 15} = \dfrac{5}{12 \cdot 3 \cdot 5} = \dfrac{1}{36}$

   **c.** $\dfrac{7}{6} \div \dfrac{7}{15} = \dfrac{7}{6} \cdot \dfrac{15}{7} = \dfrac{7 \cdot 15}{6 \cdot 7} = \dfrac{15}{6} = \dfrac{3 \cdot 5}{2 \cdot 3} = \dfrac{5}{2}$

**5. a.** $\dfrac{8}{5} - \dfrac{3}{5} = \dfrac{8-3}{5} = \dfrac{5}{5} = 1$

   **b.** $\dfrac{8}{5} - \dfrac{2}{5} = \dfrac{8-2}{5} = \dfrac{6}{5}$

   **c.** $\dfrac{3}{5} + \dfrac{1}{5} = \dfrac{3+1}{5} = \dfrac{4}{5}$

   **d.** $\dfrac{5}{12} + \dfrac{1}{12} = \dfrac{5+1}{12} = \dfrac{6}{12} = \dfrac{1}{2}$

**6.** $\dfrac{2}{3} = \dfrac{2}{3} \cdot \dfrac{7}{7} = \dfrac{2 \cdot 7}{3 \cdot 7} = \dfrac{14}{21}$

**7. a.** $\dfrac{5}{11} + \dfrac{1}{7} = \dfrac{5 \cdot 7}{11 \cdot 7} + \dfrac{1 \cdot 11}{7 \cdot 11}$

$= \dfrac{35}{77} + \dfrac{11}{77}$

$= \dfrac{35+11}{77}$

$= \dfrac{46}{77}$

   **b.** $\begin{aligned} \dfrac{5}{21} &= \dfrac{10}{42} \\ -\dfrac{1}{6} &= -\dfrac{7}{42} \\ \hline &\phantom{=} \dfrac{3}{42} = \dfrac{1}{14} \end{aligned}$

c. $\dfrac{1}{3}+\dfrac{29}{30}-\dfrac{4}{5}=\dfrac{10}{30}+\dfrac{29}{30}-\dfrac{4\cdot 6}{5\cdot 6}$

$=\dfrac{10+29}{30}-\dfrac{24}{30}$

$=\dfrac{39-24}{30}$

$=\dfrac{15}{30}$

$=\dfrac{1}{2}$

8. $5\dfrac{1}{6}\cdot 4\dfrac{2}{5}=\dfrac{31}{6}\cdot\dfrac{22}{5}$

$=\dfrac{31\cdot 22}{6\cdot 5}$

$=\dfrac{31\cdot 2\cdot 11}{2\cdot 3\cdot 5}$

$=\dfrac{341}{15}$

$=22\dfrac{11}{15}$

9. $76\dfrac{1}{12}\ =\ \ 76\dfrac{1}{12}\ =\ \ 75\dfrac{13}{12}$

$\ \ \underline{-\ 35\dfrac{1}{4}}\ =\ \underline{-\ 35\dfrac{3}{12}}\ =\ \underline{-\ 35\dfrac{3}{12}}$

$\phantom{-\ 35\dfrac{1}{4}\ =\ -\ 35\dfrac{3}{12}\ =\ }40\dfrac{10}{12}=40\dfrac{5}{6}$

**Vocabulary, Readiness & Video Check 1.3**

1. A quotient of two numbers, such as $\dfrac{5}{8}$, is called a <u>fraction</u>.

2. In the fraction $\dfrac{3}{11}$, the number 3 is called the <u>numerator</u> and the number 11 is called the <u>denominator</u>.

3. To factor a number means to write it as a <u>product</u>.

4. A fraction is said to be <u>simplified</u> when the numerator and the denominator have no common factors other than 1.

5. In $7\cdot 3=21$, the numbers 7 and 3 are called <u>factors</u> and the number 21 is called the <u>product</u>.

6. The fractions $\dfrac{2}{9}$ and $\dfrac{9}{2}$ are called <u>reciprocals</u>.

7. Fractions that represent the same quantity are called <u>equivalent</u> fractions.

8. 5, Fundamental Principle of Fractions

9. The division operation changes to multiplication and the second fraction $\dfrac{1}{20}$ changes to its reciprocal $\dfrac{20}{1}$.

10. Find the LCD; two fractions must have the same or common denominator before you can subtract (or add).

11. The number $4\dfrac{7}{6}$ is not in proper mixed number form as the fraction part, $\dfrac{7}{6}$, should not be an improper fraction.

**Exercise Set 1.3**

1. 3 of the 8 equal parts are shaded; $\dfrac{3}{8}$

3. 5 of the 7 equal parts are shaded; $\dfrac{5}{7}$

5. $33=3\cdot 11$

7. $98=2\cdot 49=2\cdot 7\cdot 7$

9. $20=4\cdot 5=2\cdot 2\cdot 5$

11. $75=3\cdot 25=3\cdot 5\cdot 5$

13. $45=9\cdot 5=3\cdot 3\cdot 5$

15. $\dfrac{2}{4}=\dfrac{2}{2\cdot 2}=\dfrac{1}{2}$

17. $\dfrac{10}{15}=\dfrac{2\cdot 5}{3\cdot 5}=\dfrac{2}{3}$

19. $\dfrac{3}{7}=\dfrac{3}{7}$

**✓ 21.** $\dfrac{18}{30} = \dfrac{2 \cdot 3 \cdot 3}{2 \cdot 3 \cdot 5} = \dfrac{3}{5}$

**23.** $\dfrac{120}{244} = \dfrac{2 \cdot 2 \cdot 2 \cdot 3 \cdot 5}{2 \cdot 2 \cdot 61} = \dfrac{2 \cdot 3 \cdot 5}{61} = \dfrac{30}{61}$

**25.** $\dfrac{1}{2} \cdot \dfrac{3}{4} = \dfrac{1 \cdot 3}{2 \cdot 4} = \dfrac{1 \cdot 3}{2 \cdot 2 \cdot 2} = \dfrac{3}{8}$

**✓ 27.** $\dfrac{2}{3} \cdot \dfrac{3}{4} = \dfrac{2 \cdot 3}{3 \cdot 4} = \dfrac{2 \cdot 3}{3 \cdot 2 \cdot 2} = \dfrac{1}{2}$

**29.** $\dfrac{1}{2} \div \dfrac{7}{12} = \dfrac{1}{2} \cdot \dfrac{12}{7} = \dfrac{1 \cdot 12}{2 \cdot 7} = \dfrac{1 \cdot 2 \cdot 2 \cdot 3}{2 \cdot 7} = \dfrac{2 \cdot 3}{7} = \dfrac{6}{7}$

**✓ 31.** $\dfrac{3}{4} \div \dfrac{1}{20} = \dfrac{3}{4} \cdot \dfrac{20}{1} = \dfrac{3 \cdot 20}{4 \cdot 1} = \dfrac{3 \cdot 2 \cdot 2 \cdot 5}{2 \cdot 2} = \dfrac{3 \cdot 5}{1} = 15$

**33.** $\dfrac{7}{10} \cdot \dfrac{5}{21} = \dfrac{7 \cdot 5}{10 \cdot 21} = \dfrac{7 \cdot 5}{2 \cdot 5 \cdot 3 \cdot 7} = \dfrac{1}{2 \cdot 3} = \dfrac{1}{6}$

**✓ 35.** $\dfrac{25}{9} \cdot \dfrac{1}{3} = \dfrac{25 \cdot 1}{9 \cdot 3} = \dfrac{5 \cdot 5 \cdot 1}{3 \cdot 3 \cdot 3} = \dfrac{25}{27}$

**✓ 37.** $\text{Area} = l \cdot w = \dfrac{11}{12} \cdot \dfrac{3}{5}$

$= \dfrac{11 \cdot 3}{12 \cdot 5}$

$= \dfrac{11 \cdot 3}{2 \cdot 2 \cdot 3 \cdot 5}$

$= \dfrac{11}{2 \cdot 2 \cdot 5}$

$= \dfrac{11}{20}$ sq mi

**39.** $\text{Area} = \dfrac{1}{2}bh = \dfrac{1}{2} \cdot \dfrac{7}{8} \cdot \dfrac{4}{9} = \dfrac{7 \cdot 4}{2 \cdot 2 \cdot 4 \cdot 9} = \dfrac{7}{36}$ sq ft

**41.** $\dfrac{4}{5} - \dfrac{1}{5} = \dfrac{4-1}{5} = \dfrac{3}{5}$

**✓ 43.** $\dfrac{4}{5} + \dfrac{1}{5} = \dfrac{4+1}{5} = \dfrac{5}{5} = 1$

**✓ 45.** $\dfrac{17}{21} - \dfrac{10}{21} = \dfrac{17-10}{21} = \dfrac{7}{21} = \dfrac{7}{3 \cdot 7} = \dfrac{1}{3}$

**47.** $\dfrac{23}{105} + \dfrac{4}{105} = \dfrac{23+4}{105} = \dfrac{27}{105} = \dfrac{3 \cdot 3 \cdot 3}{3 \cdot 5 \cdot 7} = \dfrac{3 \cdot 3}{5 \cdot 7} = \dfrac{9}{35}$

**49.** $\dfrac{7}{10} = \dfrac{7 \cdot 3}{10 \cdot 3} = \dfrac{21}{30}$

**51.** $\dfrac{2}{9} = \dfrac{2 \cdot 2}{9 \cdot 2} = \dfrac{4}{18}$

**✓ 53.** $\dfrac{4}{5} = \dfrac{4 \cdot 4}{5 \cdot 4} = \dfrac{16}{20}$

**✓ 55.** $\dfrac{2}{3} + \dfrac{3}{7} = \dfrac{2 \cdot 7}{3 \cdot 7} + \dfrac{3 \cdot 3}{7 \cdot 3} = \dfrac{14}{21} + \dfrac{9}{21} = \dfrac{23}{21}$

**✓ 57.** $\dfrac{4}{15} - \dfrac{1}{12} = \dfrac{16}{60} - \dfrac{5}{60} = \dfrac{16-5}{60} = \dfrac{11}{60}$

**✓ 59.** $\dfrac{5}{22} - \dfrac{5}{33} = \dfrac{5 \cdot 3}{22 \cdot 3} - \dfrac{5 \cdot 2}{33 \cdot 2}$

$= \dfrac{15}{66} - \dfrac{10}{66}$

$= \dfrac{15-10}{66}$

$= \dfrac{5}{66}$

**61.** $\dfrac{12}{5} - 1 = \dfrac{12}{5} - \dfrac{5}{5} = \dfrac{12-5}{5} = \dfrac{7}{5}$

**63.** $1 - \dfrac{3}{10} - \dfrac{5}{10} = \dfrac{10}{10} - \dfrac{3}{10} - \dfrac{5}{10}$

$= \dfrac{10-3-5}{10}$

$= \dfrac{2}{10}$

$= \dfrac{2}{2 \cdot 5}$

$= \dfrac{1}{5}$

The unknown part is $\dfrac{1}{5}$.

**65.** $1 - \dfrac{1}{4} - \dfrac{3}{8} = \dfrac{8}{8} - \dfrac{1 \cdot 2}{4 \cdot 2} - \dfrac{3}{8} = \dfrac{8-2-3}{8} = \dfrac{3}{8}$

The unknown part is $\dfrac{3}{8}$.

**67.** $1 - \dfrac{1}{2} - \dfrac{1}{6} - \dfrac{2}{9} = \dfrac{18}{18} - \dfrac{1 \cdot 9}{2 \cdot 9} - \dfrac{1 \cdot 3}{6 \cdot 3} - \dfrac{2 \cdot 2}{9 \cdot 2}$

$\qquad\qquad = \dfrac{18 - 9 - 3 - 4}{18}$

$\qquad\qquad = \dfrac{2}{18}$

$\qquad\qquad = \dfrac{1}{9}$

The unknown part is $\dfrac{1}{9}$.

**69.** $5\dfrac{1}{9} \cdot 3\dfrac{2}{3} = \dfrac{46}{9} \cdot \dfrac{11}{3} = \dfrac{506}{27} = 18\dfrac{20}{27}$

**71.** $8\dfrac{3}{5} \div 2\dfrac{9}{10} = \dfrac{43}{5} \div \dfrac{29}{10}$

$\qquad\qquad = \dfrac{43}{5} \cdot \dfrac{10}{29}$

$\qquad\qquad = \dfrac{43 \cdot 5 \cdot 2}{5 \cdot 29}$

$\qquad\qquad = \dfrac{86}{29}$

$\qquad\qquad = 2\dfrac{28}{29}$

**73.** $17\dfrac{2}{5} + 30\dfrac{2}{3} = \dfrac{87}{5} + \dfrac{92}{3}$

$\qquad\qquad = \dfrac{87 \cdot 3}{5 \cdot 3} + \dfrac{92 \cdot 5}{3 \cdot 5}$

$\qquad\qquad = \dfrac{261}{15} + \dfrac{460}{15}$

$\qquad\qquad = \dfrac{721}{15}$

$\qquad\qquad = 48\dfrac{1}{15}$

**75.**
$$\begin{array}{r} 8\dfrac{11}{12} \\ -1\dfrac{5}{6} \\ \hline \end{array} \qquad \begin{array}{r} 8\dfrac{11}{12} \\ -1\dfrac{10}{12} \\ \hline 7\dfrac{1}{12} \end{array}$$

**77.** $\dfrac{10}{21} + \dfrac{5}{21} = \dfrac{10 + 5}{21} = \dfrac{15}{21} = \dfrac{3 \cdot 5}{3 \cdot 7} = \dfrac{5}{7}$

**79.** $\dfrac{10}{3} - \dfrac{5}{21} = \dfrac{10 \cdot 7}{3 \cdot 7} - \dfrac{5}{21} = \dfrac{70}{21} - \dfrac{5}{21} = \dfrac{65}{21}$

**81.** $\dfrac{2}{3} \cdot \dfrac{3}{5} = \dfrac{2 \cdot 3}{3 \cdot 5} = \dfrac{2}{5}$

**83.** $\dfrac{2}{3} \div \dfrac{3}{5} = \dfrac{2}{3} \cdot \dfrac{5}{3} = \dfrac{2 \cdot 5}{3 \cdot 3} = \dfrac{10}{9}$

**85.** $5 + \dfrac{2}{3} = \dfrac{15}{3} + \dfrac{2}{3} = \dfrac{15 + 2}{3} = \dfrac{17}{3}$

**87.** $7\dfrac{2}{5} \div \dfrac{1}{5} = \dfrac{37}{5} \div \dfrac{1}{5} = \dfrac{37}{5} \cdot \dfrac{5}{1} = \dfrac{37 \cdot 5}{5 \cdot 1} = \dfrac{37}{1} = 37$

**89.** $\dfrac{1}{2} - \dfrac{14}{33} = \dfrac{1 \cdot 33}{2 \cdot 33} - \dfrac{14 \cdot 2}{33 \cdot 2}$

$\qquad\qquad = \dfrac{33}{66} - \dfrac{28}{66}$

$\qquad\qquad = \dfrac{33 - 28}{66}$

$\qquad\qquad = \dfrac{5}{66}$

**91.** $\dfrac{23}{105} - \dfrac{2}{105} = \dfrac{23 - 2}{105} = \dfrac{21}{105} = \dfrac{21}{21 \cdot 5} = \dfrac{1}{5}$

**93.** $1\dfrac{1}{2} + 3\dfrac{2}{3} = \dfrac{3}{2} + \dfrac{11}{3}$

$\qquad\qquad = \dfrac{3 \cdot 3}{2 \cdot 3} + \dfrac{11 \cdot 2}{3 \cdot 2}$

$\qquad\qquad = \dfrac{9}{6} + \dfrac{22}{6}$

$\qquad\qquad = \dfrac{9 + 22}{6}$

$\qquad\qquad = \dfrac{31}{6}$

$\qquad\qquad = 5\dfrac{1}{6}$

**95.** $\dfrac{2}{3} - \dfrac{5}{9} + \dfrac{5}{6} = \dfrac{2 \cdot 2 \cdot 3}{3 \cdot 2 \cdot 3} - \dfrac{5 \cdot 2}{9 \cdot 2} + \dfrac{5 \cdot 3}{6 \cdot 3}$

$\qquad\qquad = \dfrac{12}{18} - \dfrac{10}{18} + \dfrac{15}{18}$

$\qquad\qquad = \dfrac{12 - 10 + 15}{18}$

$\qquad\qquad = \dfrac{17}{18}$

**97.** $5 + 4\dfrac{1}{8} + 4\dfrac{1}{8} + 15\dfrac{3}{4} + 15\dfrac{3}{4} + 10\dfrac{1}{2}$

$= \dfrac{40}{8} + \dfrac{33}{8} + \dfrac{33}{8} + \dfrac{126}{8} + \dfrac{126}{8} + \dfrac{84}{8}$

$= \dfrac{40 + 33 + 33 + 126 + 126 + 84}{8}$

$= \dfrac{442}{8}$

$= 55\dfrac{1}{4}$ feet

**99.** answers may vary

**101.** $5\dfrac{1}{2} - 2\dfrac{1}{8} = \dfrac{11}{2} - \dfrac{17}{8}$

$= \dfrac{11 \cdot 4}{2 \cdot 4} - \dfrac{17}{8}$

$= \dfrac{44}{8} - \dfrac{17}{8}$

$= \dfrac{44 - 17}{8}$

$= \dfrac{27}{8}$

$= 3\dfrac{3}{8}$ miles

**103.** The graph shows $\dfrac{21}{100}$ are miniplexes.

**105.** The largest sector of the graph corresponds to multiplexes, so multiplexes have the greatest number of screens.

**107.** The work is incorrect.

$\dfrac{12}{24} = \dfrac{2 \cdot 2 \cdot 3}{2 \cdot 2 \cdot 2 \cdot 3} = \dfrac{1}{2}$

**109.** The work is incorrect.

$\dfrac{2}{7} + \dfrac{9}{7} = \dfrac{2 + 9}{7} = \dfrac{11}{7}$

**Section 1.4 Practice**

**1. a.** $1^3 = 1 \cdot 1 \cdot 1 = 1$

**b.** $5^2 = 5 \cdot 5 = 25$

**c.** $\left(\dfrac{1}{10}\right)^2 = \left(\dfrac{1}{10}\right)\left(\dfrac{1}{10}\right) = \dfrac{1}{100}$

**d.** $9^1 = 9$

**e.** $\left(\dfrac{2}{5}\right)^3 = \left(\dfrac{2}{5}\right)\left(\dfrac{2}{5}\right)\left(\dfrac{2}{5}\right) = \dfrac{8}{125}$

**2. a.** $6 + 3 \cdot 9 = 6 + 27 = 33$

**b.** $4^3 \div 8 + 3 = 64 \div 8 + 3 = 8 + 3 = 11$

**c.** $\left(\dfrac{2}{3}\right)^2 \cdot |-8| = \dfrac{4}{9} \cdot 8 = \dfrac{32}{9}$ or $3\dfrac{5}{9}$

$75 + \div\ 3\left(5 - 3\right)^2$

$(2)$

**d.** $\dfrac{9(14 - 6)}{|-2|} = \dfrac{9(8)}{2} = \dfrac{72}{2} = 36$

**e.** $\dfrac{7}{4} \cdot \dfrac{1}{4} - \dfrac{1}{4} = \dfrac{7}{16} - \dfrac{4}{16} = \dfrac{3}{16}$

**3.** $\dfrac{6^2 - 5}{3 + |6 - 5| \cdot 8} = \dfrac{36 - 5}{3 + |1| \cdot 8} = \dfrac{31}{3 + 8} = \dfrac{31}{11}$

**4.** $4[25 - 3(5 + 3)] = 4[25 - 3(8)]$
$= 4[25 - 24]$
$= 4[1]$
$= 4$

**5.** $\dfrac{36 \div 9 + 5}{5^2 - 3} = \dfrac{4 + 5}{25 - 3} = \dfrac{9}{22}$

**6. a.** $2x + y = 2(2) + 5 = 4 + 5 = 9$

**b.** $\dfrac{4x}{3y} = \dfrac{4(2)}{3(5)} = \dfrac{8}{15}$

**c.** $\dfrac{3}{x} + \dfrac{x}{y} = \dfrac{3}{2} + \dfrac{2}{5} = \dfrac{15}{10} + \dfrac{4}{10} = \dfrac{19}{10}$

**d.** $x^3 + y^2 = 2^3 + 5^2 = 8 + 25 = 33$

**7.** $9x - 6 = 7x$
$9(4) - 6 \overset{?}{=} 7(4)$
$36 - 6 \overset{?}{=} 28$
$30 = 28$ False
$4$ is not a solution of $9x - 6 = 7x$.

**8. a.** Six times a number is $6x$, since $6x$ denotes the product of 6 and $x$.

   **b.** A number decreased by 8 is $x - 8$ because "decreased by" means subtract.

   **c.** The product of a number and 9 is $x \cdot 9$ or $9x$.

   **d.** Two times a number is $2x$, plus 3 is $2x + 3$.

   **e.** The sum of 7 and a number $x$ is $7 + x$.

**9. a.** A number $x$ increased by 7 is $x + 7$, so $x + 7 = 13$.

   **b.** Two less than a number $x$ is $x - 2$, so $x - 2 = 11$.

   **c.** Double a number $x$ is $2x$, added to 9 is $2x + 9$, so $2x + 9 \neq 25$.

   **d.** Five times 11 is $5(11)$, so $5(11) \geq x$, where $x$ is an unknown number.

## Calculator Explorations

**1.** $5^4 = 625$

**2.** $7^4 = 2401$

**3.** $9^5 = 59,049$

**4.** $8^6 = 262,144$

**5.** $2(20 - 5) = 30$

**6.** $3(14 - 7) + 21 = 3(7) + 21 = 21 + 21 = 42$

**7.** $24(862 - 455) + 89 = 9857$

**8.** $99 + (401 + 962) = 1462$

**9.** $\dfrac{4623 + 129}{36 - 34} = 2376$

**10.** $\dfrac{956 - 452}{89 - 86} = 168$

## Vocabulary, Readiness & Video Check 1.4

**1.** In the expression $5^2$, the 5 is called the base and the 2 is called the exponent.

**2.** The symbols ( ), [ ], and { } are examples of grouping symbols.

**3.** A symbol that is used to represent a number is called a variable.

**4.** A collection of numbers, variables, operation symbols, and grouping symbols is called an expression.

**5.** A mathematical statement that two expressions are equal is called an equation.

**6.** A value for the variable that makes an equation a true statement is called a solution.

**7.** Deciding what values of a variable make an equation a true statement is called solving the equation.

**8.** The order in which we perform operations does matter! We came up with an order of operations to avoid getting more than one answer when evaluating an expression.

**9.** The replacement value for $z$ is not used because it's not needed—there is no variable $z$ in the given algebraic expression.

**10.** No; the variable was replaced with 0 in the equation to see if a true statement occurred, and it did not.

**11.** We translate phrases to mathematical expressions and sentences to mathematical equations.

## Exercise Set 1.4

**1.** $3^5 = 3 \cdot 3 \cdot 3 \cdot 3 \cdot 3 = 243$

**3.** $3^3 = 3 \cdot 3 \cdot 3 = 27$

**5.** $1^5 = 1 \cdot 1 \cdot 1 \cdot 1 \cdot 1 = 1$

**7.** $5^1 = 5$

**9.** $7^2 = 7 \cdot 7 = 49$

**11.** $\left(\dfrac{2}{3}\right)^4 = \left(\dfrac{2}{3}\right)\left(\dfrac{2}{3}\right)\left(\dfrac{2}{3}\right)\left(\dfrac{2}{3}\right) = \dfrac{2 \cdot 2 \cdot 2 \cdot 2}{3 \cdot 3 \cdot 3 \cdot 3} = \dfrac{16}{81}$

**13.** $\left(\dfrac{1}{5}\right)^3 = \left(\dfrac{1}{5}\right)\left(\dfrac{1}{5}\right)\left(\dfrac{1}{5}\right) = \dfrac{1 \cdot 1 \cdot 1}{5 \cdot 5 \cdot 5} = \dfrac{1}{125}$

**15.** $(1.2)^2 = (1.2)(1.2) = 1.44$

**17.** $(0.04)^3 = (0.04)(0.04)(0.04) = 0.000064$

**19.** $5 + 6 \cdot 2 = 5 + 12 = 17$

**21.** $4 \cdot 8 - 6 \cdot 2 = 32 - 12 = 20$

**23.** $2(8 - 3) = 2(5) = 10$

**25.** $2 + (5 - 2) + 4^2 = 2 + 3 + 4^2 = 2 + 3 + 16 = 21$

**27.** $5 \cdot 3^2 = 5 \cdot 9 = 45$

**29.** $\dfrac{1}{4} \cdot \dfrac{2}{3} - \dfrac{1}{6} = \dfrac{2}{12} - \dfrac{1}{6} = \dfrac{1}{6} - \dfrac{1}{6} = 0$

**31.** $\begin{aligned} 2[5 + 2(8 - 3)] &= 2[5 + 2(5)] \\ &= 2[5 + 10] \\ &= 2[15] \\ &= 30 \end{aligned}$

**33.** $\dfrac{19 - 3 \cdot 5}{6 - 4} = \dfrac{19 - 15}{6 - 4} = \dfrac{4}{2} = 2$

**35.** $\dfrac{|6 - 2| + 3}{8 + 2 \cdot 5} = \dfrac{|4| + 3}{8 + 2 \cdot 5} = \dfrac{4 + 3}{8 + 2 \cdot 5} = \dfrac{4 + 3}{8 + 10} = \dfrac{7}{18}$

**37.** $\dfrac{3 + 3(5 + 3)}{3^2 + 1} = \dfrac{3 + 3(8)}{3^2 + 1} = \dfrac{3 + 3(8)}{9 + 1} = \dfrac{3 + 24}{9 + 1} = \dfrac{27}{10}$

**39.** $\begin{aligned} \dfrac{6 + |8 - 2| + 3^2}{18 - 3} &= \dfrac{6 + |6| + 3^2}{18 - 3} \\ &= \dfrac{6 + 6 + 3^2}{18 - 3} \\ &= \dfrac{6 + 6 + 9}{18 - 3} \\ &= \dfrac{21}{15} \\ &= \dfrac{3 \cdot 7}{3 \cdot 5} \\ &= \dfrac{7}{5} \end{aligned}$

**41.** $\begin{aligned} 2 + 3[10(4 \cdot 5 - 16) - 30] &= 2 + 3[10(20 - 16) - 30] \\ &= 2 + 3[10(4) - 30] \\ &= 2 + 3[40 - 30] \\ &= 2 + 3[10] \\ &= 2 + 30 \\ &= 32 \end{aligned}$

**43.** $\begin{aligned} \left(\dfrac{2}{3}\right)^3 + \dfrac{1}{9} + \dfrac{1}{3} \cdot \dfrac{4}{3} &= \dfrac{8}{27} + \dfrac{1}{9} + \dfrac{1}{3} \cdot \dfrac{4}{3} \\ &= \dfrac{8}{27} + \dfrac{1}{9} + \dfrac{4}{9} \\ &= \dfrac{8}{27} + \dfrac{3}{27} + \dfrac{12}{27} \\ &= \dfrac{23}{27} \end{aligned}$

**45.**
  **a.** $(6 + 2) \cdot (5 + 3) = 8 \cdot 8 = 64$

  **b.** $(6 + 2) \cdot 5 + 3 = 8 \cdot 5 + 3 = 40 + 3 = 43$

  **c.** $6 + 2 \cdot 5 + 3 = 6 + 10 + 3 = 19$

  **d.** $6 + 2 \cdot (5 + 3) = 6 + 2 \cdot 8 = 6 + 16 = 22$

**47.** Let $y = 3$.
$3y = 3(3) = 9$

**49.** Let $x = 1$ and $z = 5$.
$\dfrac{z}{5x} = \dfrac{5}{5(1)} = \dfrac{5}{5} = 1$

**51.** Let $x = 1$.
$3x - 2 = 3(1) - 2 = 3 - 2 = 1$

**53.** Let $x = 1$ and $y = 3$.
$|2x + 3y| = |2(1) + 3(3)| = |2 + 9| = |11| = 11$

**55.** Let $x = 1$, $y = 3$, and $z = 5$.
$xy + z = 1 \cdot 3 + 5 = 3 + 5 = 8$

**57.** Let $y = 3$.
$5y^2 = 5(3)^2 = 5(9) = 45$

**59.** Let $x = 12$, $y = 8$, and $z = 4$.
$\dfrac{x}{z} + 3y = \dfrac{12}{4} + 3(8) = 3 + 24 = 27$

**61.** Let $x = 12$ and $y = 8$.
$\begin{aligned} x^2 - 3y + x &= (12)^2 - 3(8) + 12 \\ &= 144 - 24 + 12 \\ &= 132 \end{aligned}$

**63.** Let $x = 12$, $y = 8$, and $z = 4$.

$$\frac{x^2 + z}{y^2 + 2z} = \frac{(12)^2 + 4}{(8)^2 + 2(4)} = \frac{144 + 4}{64 + 8} = \frac{148}{72} = \frac{37}{18}$$

**65.** Evaluate $16t^2$ for each value of $t$.

$t = 1$: $16(1)^2 = 16(1) = 16$

$t = 2$: $16(2)^2 = 16(4) = 64$

$t = 3$: $16(3)^2 = 16(9) = 144$

$t = 4$: $16(4)^2 = 16(16) = 256$

| Time $t$ (in seconds) | Distance $16t^2$ (in feet) |
| --- | --- |
| 1 | 16 |
| 2 | 64 |
| 3 | 144 |
| 4 | 256 |

**67.** Let $x = 5$.

$3x + 30 = 9x$

$3(5) + 30 \stackrel{?}{=} 9(5)$

$15 + 30 \stackrel{?}{=} 45$

$45 = 45$, true

5 is a solution of the equation.

**69.** Let $x = 0$.

$2x + 6 = 5x - 1$

$2(0) + 6 \stackrel{?}{=} 5(0) - 1$

$0 + 6 \stackrel{?}{=} 0 - 1$

$6 = -1$, false

0 is not a solution of the equation.

**71.** Let $x = 8$.

$2x - 5 = 5$

$2(8) - 5 \stackrel{?}{=} 5$

$16 - 5 \stackrel{?}{=} 5$

$9 = 5$, false

8 is not a solution of the equation.

**73.** Let $x = 2$.

$x + 6 = x + 6$

$2 + 6 \stackrel{?}{=} 2 + 6$

$8 = 8$, true

2 is a solution of the equation.

**75.** Let $x = 0$.

$x = 5x + 15$

$(0) \stackrel{?}{=} 5(0) + 15$

$0 \stackrel{?}{=} 0 + 15$

$0 = 15$, false

0 is not a solution of the equation.

**77.** Fifteen more than a number is written as $x + 15$.

**79.** Five subtracted from a number is written as $x - 5$.

**81.** The ratio of a number and 4 is written as $\frac{x}{4}$.

**83.** Three times a number, increased by 22 is written as $3x + 22$.

**85.** One increased by two equals the quotient of nine and three is written as $1 + 2 = 9 \div 3$.

**87.** Three is not equal to four divided by two is written as $3 \neq 4 \div 2$.

**89.** The sum of 5 and a number is 20 is written as $5 + x = 20$.

**91.** The product 7.6 and a number is 17 is written as $7.6x = 17$.

**93.** Thirteen minus three times a number is 13 is written as $13 - 3x = 13$.

**95.** To simplify the expression $1 + 3 \cdot 6$, first multiply.

**97.** To simplify the expression $(20 - 4) \cdot 2$, first subtract.

**99.** No; answers may vary.

| | Length, $l$ | Width, $w$ | Perimeter of Rectangle: $2l + 2w$ | Area of Rectangle: $lw$ |
|---|---|---|---|---|
| **101.** | 4 in. | 3 in. | $2l + 2w$ $= 2(4 \text{ in.}) + 2(3 \text{ in.})$ $= 8 \text{ in.} + 6 \text{ in.}$ $= 14 \text{ in.}$ | $lw$ $= (4 \text{ in.})(3 \text{ in.})$ $= 12 \text{ sq in.}$ |
| **103.** | 5.3 in. | 1.7 in. | $2l + 2w$ $= 2(5.3 \text{ in.}) + 2(1.7 \text{ in.})$ $= 10.6 \text{ in.} + 3.4 \text{ in.}$ $= 14 \text{ in.}$ | $lw$ $= (5.3 \text{ in.})(1.7 \text{ in.})$ $= 9.01 \text{ sq in.}$ |

**105.** Rectangles with the same perimeter can have different areas.

**107.** $(20 - 4) \cdot 4 \div 2 = 16 \cdot 4 \div 2 = 64 \div 2 = 32$

**109. a.** $5x + 6$ is an expression since it does not contain the equal symbol, "=."

    **b.** $2a = 7$ is an equation since it contains the equal symbol.

    **c.** $3a + 2 = 9$ is an equation since it contains the equal symbol.

    **d.** $4x + 3y - 8z$ is an expression since it does not contain the equal symbol.

    **e.** $5^2 - 2(6 - 2)$ is an expression since it does not contain the equal symbol.

**111.** answers may vary

**113.** answers may vary; for example, $-2(5) - 1$: $-2(5) - 1 = -10 - 1 = -11$

**115.** Let $l = 120$ and $w = 100$.
$lw = (120)(100) = 12,000 \text{ sq ft}$

**117.** Let $d = 432$ and $t = 8.5$.
$$\frac{d}{t} = \frac{432}{8.5} = 51$$
The rate is 51 mph.

**Section 1.5 Practice**

**1.**

$2 + 4 = 6$

**2.**

$-2 + (-3) = -5$

**3. a.** $-5 + (-8)$
Add the absolute values.
$5 + 8 = 13$
The common sign is negative, so
$-5 + (-8) = -13$.

**b.** $-31 + (-1)$
Add the absolute values.
$31 + 1 = 32$
The common sign is negative, so
$-31 + (-1) = -32$.

**4.**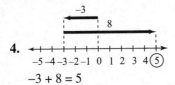
$-3 + 8 = 5$

**5. a.** $15 + (-18)$
Subtract the absolute values.
$18 - 15 = 3$
Use the sign of the number with the largest absolute value.
$15 + (-18) = -3$

**b.** $-19 + 20 = 20 - 19 = 1$

**c.** $-0.6 + 0.4 = -(0.6 - 0.4) = -(0.2) = -0.2$

**6. a.** $-\dfrac{3}{5} + \left(-\dfrac{2}{5}\right) = -\dfrac{5}{5} = -1$

**b.** $3 + (-9) = -6$

**c.** $2.2 + (-1.7) = 0.5$

**d.** $-\dfrac{2}{7} + \dfrac{3}{10} = -\dfrac{20}{70} + \dfrac{21}{70} = \dfrac{1}{70}$

**7. a.** $8 + (-5) + (-9) = 3 + (-9) = -6$

**b.** $[-8 + 5] + \left[-5 + |-2|\right] = [-3] + [-5 + 2]$
$\qquad\qquad\qquad\qquad\quad = -3 + [-3]$
$\qquad\qquad\qquad\qquad\quad = -6$

**8.** $-7 + 4 + 7 = -3 + 7 = 4$
The temperature at 8 A.M. was 4°F.

**9. a.** The opposite of $-\dfrac{5}{9}$ is $\dfrac{5}{9}$.

**b.** The opposite of 8 is −8.

**c.** The opposite of 6.2 is −6.2.

**d.** The opposite of −3 is 3.

**10. a.** Since $|-15| = 15,\ -|-15| = -15$.

**b.** $-\left(-\dfrac{3}{5}\right) = \dfrac{3}{5}$

**c.** $-(-5y) = 5y$

**d.** $-(-8) = 8$

**Vocabulary, Readiness & Video Check 1.5**

**1.** Two numbers that are the same distance from 0 but lie on opposite sides of 0 are called <u>opposites</u>.

**2.** If $n$ is a number, then $n + (-n) = \underline{0}$.

**3.** If $n$ is a number, then $-(-n) = \underline{n}$.

**4.** The sum of two negative numbers is always <u>a negative number</u>.

**5.** absolute values

**6.** Negative; when you add two numbers with different signs, the sign of the sum is the same as the sign of the number with the larger absolute value and −8.4 has a larger absolute value than 6.3.

**7.** Negative temperatures; the high temperature for the day was −6°F.

**8.** Example 13 is an example of the opposite of the *absolute value* of −a, not the opposite of −a. The absolute value of −a is positive, so its opposite is negative, therefore the answers to Examples 12 and 13 have different signs.

**Exercise Set 1.5**

**1.** $6 + 3 = 9$

**3.** $-6 + (-8) = -14$

**5.** $8 + (-7) = 1$

**7.** $-14 + 2 = -12$

**9.** $-2 + (-3) = -5$

**11.** $-9 + (-3) = -12$

**13.** $-7 + 3 = -4$

**15.** $10 + (-3) = 7$

**17.** $5 + (-7) = -2$

**19.** $-16 + 16 = 0$

**21.** $27 + (-46) = -19$

**23.** $-18 + 49 = 31$

**25.** $-33 + (-14) = -47$

**27.** $6.3 + (-8.4) = -2.1$

**29.** $|-8| + (-16) = 8 + (-16) = -8$

**31.** $117 + (-79) = 38$

**33.** $-9.6 + (-3.5) = -13.1$

**35.** $-\dfrac{3}{8} + \dfrac{5}{8} = \dfrac{2}{8} = \dfrac{1}{4}$

**37.** $-\dfrac{7}{16} + \dfrac{1}{4} = -\dfrac{7}{16} + \dfrac{1 \cdot 4}{4 \cdot 4} = -\dfrac{7}{16} + \dfrac{4}{16} = -\dfrac{3}{16}$

**39.** $-\dfrac{7}{10} + \left(-\dfrac{3}{5}\right) = -\dfrac{7}{10} + \left(-\dfrac{3 \cdot 2}{5 \cdot 2}\right)$

$\qquad\qquad\qquad = -\dfrac{7}{10} + \left(-\dfrac{6}{10}\right)$

$\qquad\qquad\qquad = -\dfrac{13}{10}$

**41.** $-15 + 9 + (-2) = -6 + (-2) = -8$

**43.** $-21 + (-16) + (-22) = -37 + (-22) = -59$

**45.** $-23 + 16 + (-2) = -7 + (-2) = -9$

**47.** $|5 + (-10)| = |-5| = 5$

**49.** $6 + (-4) + 9 = 2 + 9 = 11$

**51.** $[-17 + (-4)] + [-12 + 15] = [-21] + [3] = -18$

**53.** $|9 + (-12)| + |-16| = |-3| + 16 = 3 + 16 = 19$

**55.** $-1.3 + [0.5 + (-0.3) + 0.4] = -1.3 + [0.2 + 0.4]$

$\qquad\qquad\qquad\qquad\qquad\quad = -1.3 + [0.6]$

$\qquad\qquad\qquad\qquad\qquad\quad = -0.7$

**57.** $-15 + 9 = -6$
The high temperature in Anoka was $-6°$.

**59.** $-512 + 658 = 146$
Your elevation is 146 feet.

**61.** $2.5 + 9 + (-14.2) + (-4.2)$
$\quad = 11.5 + (-14.2) + (-4.2)$
$\quad = -2.7 + (-4.2)$
$\quad = -6.9$
The total net income for fiscal year 2009 was $-\$6.9$ million.

**63.** $-6 + (-5) + (-3) + (-2) = -16$
Her score was $-16$ or 16 under par.

**65.** The opposite of 6 is $-6$.

**67.** The opposite of $-2$ is 2.

**69.** The opposite of 0 is 0.

**71.** Since $|-6|$ is 6, the opposite of $|-6|$ is $-6$.

**73.** $-|-2| = -2$

**75.** $-|0| = -0 = 0$

**77.** $-\left|-\dfrac{2}{3}\right| = -\dfrac{2}{3}$

**79.** Let $x = -4$.
$\qquad x + 9 = 5$
$\qquad (-4) + 9 \overset{?}{=} 5$
$\qquad\qquad 5 = 5$, true
$-4$ is a solution of the equation.

**81.** Let $y = -1$.
$\qquad y + (-3) = -7$
$\qquad (-1) + (-3) \overset{?}{=} -7$
$\qquad\qquad -4 = -7$, false
$-1$ is not a solution of the equation.

**83.** Look for the tallest bar. The temperature is the highest in July.

**85.** Look for the bar whose length has a positive value closest to 0; October

**87.** $[(-9.1) + 14.4 + 8.8] \div 3 = [5.3 + 8.8] \div 3$
$\qquad\qquad\qquad\qquad\qquad = [14.1] \div 3$
$\qquad\qquad\qquad\qquad\qquad = 4.7$
The average was $4.7°$F.

**89.** $7 + (-10) = -3$

**91.** $-10 + (-12) = -22$

**93.** Since $a$ is a positive number, $-a$ is a <u>negative</u> number.

**95.** Since $a$ is a positive number, $a + a$ is a <u>positive</u> number.

**97.** True

**99.** False; for example, $4 + (-2) = 2 > 0$.

**101.** answers may vary

**103.** answers may vary

## Section 1.6 Practice

**1. a.** $-7 - 6 = -7 + (-6) = -13$

   **b.** $-8 - (-1) = -8 + 1 = -7$

   **c.** $9 - (-3) = 9 + 3 = 12$

   **d.** $5 - 7 = 5 + (-7) = -2$

**2. a.** $8.4 - (-2.5) = 8.4 + 2.5 = 10.9$

   **b.** $-\dfrac{5}{8} - \left(-\dfrac{1}{8}\right) = -\dfrac{5}{8} + \dfrac{1}{8} = -\dfrac{4}{8} = -\dfrac{1}{2}$

   **c.** $-\dfrac{3}{4} - \dfrac{1}{5} = -\dfrac{3}{4} + \left(-\dfrac{1}{5}\right)$
$$= -\dfrac{15}{20} + \left(-\dfrac{4}{20}\right)$$
$$= -\dfrac{19}{20}$$

**3.** $-2 - 5 = -2 + (-5) = -7$

**4. a.** $-15 - 2 - (-4) + 7 = -15 + (-2) + 4 + 7 = -6$

   **b.** $3.5 + (-4.1) - (-6.7) = 3.5 + (-4.1) + 6.7$
$$= 6.1$$

**5. a.** $-4 + [(-8 - 3) - 5] = -4 + [(-8 + (-3)) - 5]$
$$= -4 + [(-11) - 5]$$
$$= -4 + [-11 + (-5)]$$
$$= -4 + [-16]$$
$$= -20$$

   **b.** $|-13| - 3^2 + [2 - (-7)] = 13 - 9 + [2 + 7]$
$$= 13 - 9 + 9$$
$$= 13$$

**6. a.** $\dfrac{7 - x}{2y + x} = \dfrac{7 - (-3)}{2(4) + (-3)} = \dfrac{7 + 3}{8 + (-3)} = \dfrac{10}{5} = 2$

   **b.** $y^2 + x = (4)^2 + (-3) = 16 + (-3) = 13$

**7.** $282 - (-75) = 282 + 75 = 357$
The overall change was $357.

**8. a.** $x = 90° - 62° = 28°$

   **b.** $y = 180° - 43° = 137°$

## Vocabulary, Readiness & Video Check 1.6

**1.** 7 minus a number <u>$7 - x$</u>

**2.** 7 subtracted from a number <u>$x - 7$</u>.

**3.** A number decreased by 7 <u>$x - 7$</u>

**4.** 7 less a number <u>$7 - x$</u>

**5.** A number less than 7 <u>$7 - x$</u>

**6.** A number subtracted from 7 <u>$7 - x$</u>

**7.** To evaluate $x - y$ for $x = -10$ and $y = -14$, we replace $x$ with $-10$ and $y$ with $-14$ and evaluate <u>$-10 - (-14)$</u>.  d

**8.** The expression $-5 - 10$ equals <u>$-5 + (-10)$</u>.  c

**9.** To subtract two real numbers, change the operation to <u>addition</u> and take the <u>opposite</u> of the second number.

**10.** $-10 + (8) + (-4) + (-20)$; it's rewritten to change the subtraction operations to addition and turn the problem into an addition of real numbers problem.

**11.** There's a minus sign in the numerator and the replacement value is negative (notice parentheses are used around the replacement value), and it's always good to be careful when working with negative signs.

**12.** This means that the overall vertical altitude change of the jet is actually a decrease in altitude from when the Example started.

**13.** In Example 9, you have two supplementary angles and know the measure of one of them. From the definition, you know that the two supplementary angles must sum to 180°. Therefore you can subtract the known angle measure from 180° to get the measure of the other angle.

**Exercise Set 1.6**

**1.** $-6 - 4 = -6 + (-4) = -10$

**3.** $4 - 9 = 4 + (-9) = -5$

**5.** $16 - (-3) = 16 + 3 = 19$

**7.** $\dfrac{1}{2} - \dfrac{1}{3} = \dfrac{1}{2} + \left(-\dfrac{1}{3}\right)$

$\phantom{7.} = \dfrac{1 \cdot 3}{2 \cdot 3} + \left(-\dfrac{1 \cdot 2}{3 \cdot 2}\right)$

$\phantom{7.} = \dfrac{3}{6} + \left(-\dfrac{2}{6}\right)$

$\phantom{7.} = \dfrac{1}{6}$

**9.** $-16 - (-18) = -16 + 18 = 2$

**11.** $-6 - 5 = -6 + (-5) = -11$

**13.** $7 - (-4) = 7 + 4 = 11$

**15.** $-6 - (-11) = -6 + 11 = 5$

**17.** $16 - (-21) = 16 + 21 = 37$

**19.** $9.7 - 16.1 = 9.7 + (-16.1) = -6.4$

**21.** $-44 - 27 = -44 + (-27) = -71$

**23.** $-21 - (-21) = -21 + 21 = 0$

**25.** $-2.6 - (-6.7) = -2.6 + 6.7 = 4.1$

**27.** $-\dfrac{3}{11} - \left(-\dfrac{5}{11}\right) = -\dfrac{3}{11} + \dfrac{5}{11} = \dfrac{2}{11}$

**29.** $-\dfrac{1}{6} - \dfrac{3}{4} = -\dfrac{1}{6} + \left(-\dfrac{3}{4}\right)$

$\phantom{29.} = -\dfrac{1 \cdot 2}{6 \cdot 2} + \left(-\dfrac{3 \cdot 3}{4 \cdot 3}\right)$

$\phantom{29.} = -\dfrac{2}{12} + \left(-\dfrac{9}{12}\right)$

$\phantom{29.} = -\dfrac{11}{12}$

**31.** $8.3 - (-0.62) = 8.3 + 0.62 = 8.92$

**33.** $8 - (-5) = 8 + 5 = 13$

**35.** $-6 - (-1) = -6 + 1 = -5$

**37.** $7 - 8 = 7 + (-8) = -1$

**39.** $-8 - 15 = -8 + (-15) = -23$

**41.** $-10 - (-8) + (-4) - 20 = -10 + 8 + (-4) + (-20)$

$\phantom{41.} = -2 + (-4) + (-20)$

$\phantom{41.} = -6 + (-20)$

$\phantom{41.} = -26$

**43.** $5 - 9 + (-4) - 8 - 8 = 5 + (-9) + (-4) + (-8) + (-8)$

$\phantom{43.} = -4 + (-4) + (-8) + (-8)$

$\phantom{43.} = -8 + (-8) + (-8)$

$\phantom{43.} = -16 + (-8)$

$\phantom{43.} = -24$

**45.** $-6 - (2 - 11) = -6 - (-9) = -6 + 9 = 3$

**47.** $3^3 - 8 \cdot 9 = 27 - 8 \cdot 9 = 27 - 72 = 27 + (-72) = -45$

**49.** $2 - 3(8 - 6) = 2 - 3(2) = 2 - 6 = 2 + (-6) = -4$

**51.** $(3 - 6) + 4^2 = [3 + (-6)] + 4^2$

$\phantom{51.} = [-3] + 4^2$

$\phantom{51.} = [-3] + 16$

$\phantom{51.} = 13$

**53.** $-2 + [(8 - 11) - (-2 - 9)]$

$\phantom{53.} = -2 + [(8 + (-11)) - (-2 + (-9))]$

$\phantom{53.} = -2 + [(-3) - (-11)]$

$\phantom{53.} = -2 + [(-3) + 11]$

$\phantom{53.} = -2 + [8]$

$\phantom{53.} = 6$

**55.** $|-3| + 2^2 + [-4 - (-6)] = 3 + 2^2 + [-4 + 6]$
$$= 3 + 2^2 + [2]$$
$$= 3 + 4 + [2]$$
$$= 7 + [2]$$
$$= 9$$

**57.** Let $x = -5$ and $y = 4$.
$x - y = -5 - 4 = -5 + (-4) = -9$

**59.** Let $x = -5$, $y = 4$, and $t = 10$.
$|x| + 2t - 8y = |-5| + 2(10) - 8(4)$
$$= 5 + 2(10) - 8(4)$$
$$= 5 + 20 - 32$$
$$= 25 - 32$$
$$= 25 + (-32)$$
$$= -7$$

**61.** Let $x = -5$ and $y = 4$.
$$\frac{9 - x}{y + 6} = \frac{9 - (-5)}{4 + 6} = \frac{9 + 5}{4 + 6} = \frac{14}{10} = \frac{2 \cdot 7}{2 \cdot 5} = \frac{7}{5}$$

**63.** Let $x = -5$ and $y = 4$.
$y^2 - x = 4^2 - (-5) = 16 + 5 = 21$

**65.** Let $x = -5$ and $t = 10$.
$$\frac{|x - (-10)|}{2t} = \frac{|-5 - (-10)|}{2(10)}$$
$$= \frac{|-5 + 10|}{2(10)}$$
$$= \frac{|5|}{2(10)}$$
$$= \frac{5}{20}$$
$$= \frac{5}{4 \cdot 5}$$
$$= \frac{1}{4}$$

**67.** The change in temperature is the difference between the last temperature and the first temperature.
$-56 - 44 = -56 + (-44) = -100$
The temperature dropped 100°F.

**69.** $136 - (-129) = 136 + 129 = 265$
136°F is 265°F warmer than −129°F.

**71.** $13,796 - (-21,857) = 13,796 + 21,857 = 35,653$
The difference in elevation is 35,653 feet.

**73.** $-250 + 120 - 178 = -250 + 120 + (-178)$
$$= -130 + (-178)$$
$$= -308$$
The overall vertical change is −308 feet.

**75.** $19,340 - (-512) = 19,340 + 512 = 19,852$
Mt. Kilimanjaro is 19,852 feet higher.

**77.** $y = 180 - 50 = 180 + (-50) = 130$
The supplementary angle is 130°.

**79.** $x = 90 - 60 = 90 + (-60) = 30$
The complementary angle is 30°.

**81.** Let $x = -4$.
$x - 9 = 5$
$-4 - 9 \stackrel{?}{=} 5$
$-4 + (-9) \stackrel{?}{=} 5$
$-13 = 5$, false
−4 is not a solution of the equation.

**83.** Let $x = -2$.
$-x + 6 = -x - 1$
$-(-2) + 6 \stackrel{?}{=} -(-2) - 1$
$2 + 6 \stackrel{?}{=} 2 + (-1)$
$8 = 1$, false
−2 is not a solution of the equation.

**85.** Let $x = 2$.
$-x - 13 = -15$
$-2 - 13 \stackrel{?}{=} -15$
$-2 + (-13) \stackrel{?}{=} -15$
$-15 = -15$, true
2 is a solution of the equation.

**87.** The sum of −5 and a number is $-5 + x$.

**89.** Subtract a number from −20 is $-20 - x$.

**91.**

| Month | Monthly Increase or Decrease |
|---|---|
| February | $-23.7 - (-19.3) = -23.7 + 19.3 = -4.4°$ |
| March | $-21.1 - (-23.7) = -21.1 + 23.7 = 2.6°$ |
| April | $-9.1 - (-21.1) = -9.1 + 21.1 = 12°$ |
| May | $14.4 - (-9.1) = 14.4 + 9.1 = 23.5°$ |
| June | $29.7 - 14.4 = 29.7 + (-14.4) = 15.3°$ |

**93.** The largest positive number corresponds to May.

**95.** answers may vary

**97.** $9 - (-7) = 9 + 7 = 16$

**99.** $10 - 30 = 10 + (-30) = -20$

**101.** true; answers may vary

**103.** false; answers may vary

**105.** Since 56,875 is less than 87,262, the answer is negative.
$56,875 - 87,262 = -30,387$

**Integrated Review**

**1.** The opposite of a positive number is a <u>negative</u> number.

**2.** The sum of two negative numbers is a <u>negative</u> number.

**3.** The absolute value of a negative number is a <u>positive</u> number.

**4.** The absolute value of zero is <u>0</u>.

**5.** The reciprocal of a positive number is a <u>positive</u> number.

**6.** The sum of a number and its opposite is <u>0</u>.

**7.** The absolute value of a positive number is a <u>positive</u> number.

**8.** The opposite of a negative number is a <u>positive</u> number.

|  | Number | Opposite | Absolute Value |
|---|---|---|---|
| **9.** | $\frac{1}{7}$ | $-\frac{1}{7}$ | $\frac{1}{7}$ |
| **10.** | $-\frac{12}{5}$ | $\frac{12}{5}$ | $\frac{12}{5}$ |
| **11.** | $3$ | $-3$ | $3$ |
| **12.** | $-\frac{9}{11}$ | $\frac{9}{11}$ | $\frac{9}{11}$ |

**13.** $-19 + (-23) = -42$

**14.** $7 - (-3) = 7 + 3 = 10$

**15.** $-15 + 17 = 2$

**16.** $-8 - 10 = -8 + (-10) = -18$

**17.** $18 + (-25) = -7$

**18.** $-2 + (-37) = -39$

**19.** $-14 - (-12) = -14 + 12 = -2$

**20.** $5 - 14 = 5 + (-14) = -9$

**21.** $4.5 - 7.9 = 4.5 + (-7.9) = -3.4$

**22.** $-8.6 - 1.2 = -8.6 + (-1.2) = -9.8$

**23.** $-\dfrac{3}{4} - \dfrac{1}{7} = -\dfrac{21}{28} - \dfrac{4}{28} = -\dfrac{21}{28} + \left(-\dfrac{4}{28}\right) = -\dfrac{25}{28}$

**24.** $\dfrac{2}{3} - \dfrac{7}{8} = \dfrac{16}{24} - \dfrac{21}{24} = \dfrac{16}{24} + \left(-\dfrac{21}{24}\right) = -\dfrac{5}{24}$

**25.** $\begin{aligned}-9 - (-7) + 4 - 6 &= -9 + 7 + 4 - 6 \\ &= -9 + 7 + 4 + (-6) \\ &= -4\end{aligned}$

**26.** $\begin{aligned}11 - 20 + (-3) - 12 &= 11 + (-20) + (-3) + (-12) \\ &= -9 + (-3) + (-12) \\ &= -12 + (-12) \\ &= -24\end{aligned}$

**27.** $\begin{aligned}24 - 6(14 - 11) &= 24 - 6[14 + (-11)] \\ &= 24 - 6(3) \\ &= 24 - 18 \\ &= 24 + (-18) \\ &= 6\end{aligned}$

**28.** $\begin{aligned}30 - 5(10 - 8) &= 30 - 5[10 + (-8)] \\ &= 30 - 5(2) \\ &= 30 - 10 \\ &= 30 + (-10) \\ &= 20\end{aligned}$

**29.** $(7 - 17) + 4^2 = [7 + (-17)] + 4^2 = (-10) + 16 = 6$

**30.** $\begin{aligned}9^2 + (10 - 30) &= 9^2 + [10 + (-30)] \\ &= 81 + (-20) \\ &= 61\end{aligned}$

**31.** $\begin{aligned}|-9| + 3^2 + (-4 - 20) &= 9 + 9 + [-4 + (-20)] \\ &= 9 + 9 + (-24) \\ &= 18 + (-24) \\ &= -6\end{aligned}$

**32.** $\begin{aligned}|-4 - 5| + 5^2 + (-50) &= |-4 + (-5)| + 5^2 + (-50) \\ &= |-9| + 25 + (-50) \\ &= 9 + 25 + (-50) \\ &= 34 + (-50) \\ &= -16\end{aligned}$

**33.** $\begin{aligned}-7 + [(1 - 2) + (-2 - 9)] &= -7 + [(-1) + (-11)] \\ &= -7 + [-12] \\ &= -19\end{aligned}$

**34.** $\begin{aligned}-6 + [(-3 + 7) + (4 - 15)] &= -6 + [(4) + (-11)] \\ &= -6 + (-7) \\ &= -13\end{aligned}$

**35.** $1 - 5 = 1 + (-5) = -4$

**36.** $-3 - (-2) = -3 + 2 = -1$

**37.** $\dfrac{1}{4} - \left(-\dfrac{2}{5}\right) = \dfrac{1}{4} + \dfrac{2}{5} = \dfrac{5}{20} + \dfrac{8}{20} = \dfrac{13}{20}$

**38.** $-\dfrac{5}{8} - \left(\dfrac{1}{10}\right) = -\dfrac{25}{40} - \dfrac{4}{40} = -\dfrac{25}{40} + \left(-\dfrac{4}{40}\right) = -\dfrac{29}{40}$

**39.** $\begin{aligned}2(19 - 17)^3 &- 3(-7 + 9)^2 \\ &= 2[19 + (-17)]^3 - 3(-7 + 9)^2 \\ &= 2(2)^3 - 3(2)^2 \\ &= 2(8) - 3(4) \\ &= 16 - 12 \\ &= 16 + (-12) \\ &= 4\end{aligned}$

**40.** $\begin{aligned}3(10 - 9)^2 &+ 6(20 - 19)^3 \\ &= 3[10 + (-9)]^2 + 6[20 + (-19)]^3 \\ &= 3(1)^2 + 6(1)^3 \\ &= 3 + 6 \\ &= 9\end{aligned}$

**41.** $x - y = -2 - (-1) = -2 + 1 = -1$

**42.** $x + y = -2 + (-1) = -3$

**43.** $y + z = -1 + 9 = 8$

**44.** $z - y = 9 - (-1) = 9 + 1 = 10$

**45.** $\dfrac{|5z - x|}{y - x} = \dfrac{|5(9) - (-2)|}{-1 - (-2)} = \dfrac{|45 + 2|}{-1 + 2} = \dfrac{|47|}{1} = 47$

**46.** $\dfrac{|-x - y + z|}{2z} = \dfrac{|-(-2) - (-1) + 9|}{2(9)}$

$= \dfrac{|2 + 1 + 9|}{18}$

$= \dfrac{|12|}{18}$

$= \dfrac{12}{18}$

$= \dfrac{2}{3}$

**Section 1.7 Practice**

**1. a.** $8(-5) = -40$

  **b.** $(-3)(-4) = 12$

  **c.** $(-6)(9) = -54$

**2. a.** $(-1)(-5)(-6) = 5(-6) = -30$

  **b.** $(-3)(-2)(4) = 6(4) = 24$

  **c.** $(-4)(0)(5) = 0(5) = 0$

  **d.** $(-2)(-3) - (-4)(5) = 6 - (-20)$
$$= 6 + 20$$
$$= 26$$

**3. a.** $(0.23)(-0.2) = -[(0.23)(0.2)] = -0.046$

  **b.** $\left(-\dfrac{3}{5}\right) \cdot \left(\dfrac{4}{9}\right) = -\dfrac{3 \cdot 4}{5 \cdot 9} = -\dfrac{12}{45} = -\dfrac{4}{15}$

  **c.** $\left(-\dfrac{7}{12}\right)(-24) = \dfrac{7 \cdot 24}{12 \cdot 1} = 7 \cdot 2 = 14$

**4. a.** $(-6)^2 = (-6)(-6) = 36$

  **b.** $-6^2 = -(6 \cdot 6) = -(36) = -36$

  **c.** $(-4)^3 = (-4)(-4)(-4) = 16(-4) = -64$

  **d.** $-4^3 = -(4 \cdot 4 \cdot 4) = -[16(4)] = -64$

**5. a.** The reciprocal of $\dfrac{8}{3}$ is $\dfrac{3}{8}$ since $\dfrac{8}{3} \cdot \dfrac{3}{8} = 1$.

  **b.** The reciprocal of 15 is $\dfrac{1}{15}$ since $15 \cdot \dfrac{1}{15} = 1$.

  **c.** The reciprocal of $-\dfrac{2}{7}$ is $-\dfrac{7}{2}$ since
$$\left(-\dfrac{2}{7}\right)\left(-\dfrac{7}{2}\right) = 1.$$

  **d.** The reciprocal of $-5$ is $-\dfrac{1}{5}$ since
$$(-5)\left(-\dfrac{1}{5}\right) = 1.$$

**6. a.** $\dfrac{16}{-2} = 16\left(-\dfrac{1}{2}\right) = -8$

  **b.** $24 \div (-6) = 24\left(-\dfrac{1}{6}\right) = -4$

  **c.** $\dfrac{-35}{-7} = \dfrac{35}{7} = \dfrac{5 \cdot 7}{7} = 5$

**7. a.** $\dfrac{-18}{-6} = \dfrac{18}{6} = \dfrac{3 \cdot 6}{6} = 3$

  **b.** $\dfrac{-48}{3} = -\dfrac{48}{3} = -\dfrac{3 \cdot 16}{3} = -16$

  **c.** $\dfrac{3}{5} \div \left(-\dfrac{1}{2}\right) = \dfrac{3}{5} \cdot (-2) = -\dfrac{6}{5}$

  **d.** $-\dfrac{4}{9} \div 8 = -\dfrac{4}{9} \cdot \dfrac{1}{8} = -\dfrac{4}{9 \cdot 4 \cdot 2} = -\dfrac{1}{9 \cdot 2} = -\dfrac{1}{18}$

**8. a.** $\dfrac{0}{-2} = 0$

  **b.** $\dfrac{-4}{0}$ is undefined.

  **c.** $\dfrac{-5}{6(0)} = \dfrac{-5}{0}$ is undefined.

**9. a.** $\dfrac{(-8)(-11) - 4}{-9 - (-4)} = \dfrac{88 - 4}{-9 + 4} = \dfrac{84}{-5} = -\dfrac{84}{5}$

**b.**  $\dfrac{3(-2)^3 - 9}{-6 + 3} = \dfrac{3(-8) - 9}{-3}$

$\qquad\qquad = \dfrac{-24 - 9}{-3}$

$\qquad\qquad = \dfrac{-33}{-3}$

$\qquad\qquad = 11$

**10. a.**  $7y - x = 7(-2) - (-5) = -14 + 5 = -9$

**b.**  $x^2 - y^3 = (-5)^2 - (-2)^3$

$\qquad\quad = 25 - (-8)$

$\qquad\quad = 25 + 8$

$\qquad\quad = 33$

**c.**  $\dfrac{2x}{3y} = \dfrac{2(-5)}{3(-2)} = \dfrac{-10}{-6} = \dfrac{5}{3}$

**11.**  total score $= 4 \cdot (-13) = -52$
The card player's total score was $-52$.

## Calculator Explorations

**1.**  $-38(26 - 27) = 38$

**2.**  $-59(-8) + 1726 = 2198$

**3.**  $134 + 25(68 - 91) = -441$

**4.**  $45(32) - 8(218) = -304$

**5.**  $\dfrac{-50(294)}{175 - 265} = 163.\overline{3}$

**6.**  $\dfrac{-444 - 444.8}{-181 - 324} = 1.76$

**7.**  $9^5 - 4550 = 54,499$

**8.**  $5^8 - 6259 = 384,366$

**9.**  $(-125)^2 = 15,625$

**10.**  $-125^2 = -15,625$

## Vocabulary, Readiness & Video Check 1.7

**1.**  If $n$ is a real number, then $n \cdot 0 = \underline{0}$ and $0 \cdot n = \underline{0}$.

**2.**  If $n$ is a real number, but not 0, then $\dfrac{0}{n} = \underline{0}$ and we say $\dfrac{n}{0}$ is <u>undefined</u>.

**3.**  The product of two negative numbers is a <u>positive</u> number.

**4.**  The quotient of two negative numbers is a <u>positive</u> number.

**5.**  The quotient of a positive number and a negative number is a <u>negative</u> number.

**6.**  The product of a positive number and a negative number is a <u>negative</u> number.

**7.**  The reciprocal of a positive number is a <u>positive</u> number.

**8.**  The opposite of a positive number is a <u>negative</u> number.

**9.**  The parentheses, or lack of them, determine the base of the expression. In Example 6, $(-2)^4$, the base is $-2$ and all of $-2$ is raised to the 4th power. In Example 7, $-2^4$, the base is 2 and only 2 is raised to the 4th power.

**10.**  Remember, the product of a number and its reciprocal is 1, *not* $-1$. $\dfrac{2}{3} \cdot \dfrac{3}{2} = 1$, as needed.

**11.**  Yes; because division of real numbers is defined in terms of multiplication.

**12.**  The replacement values are negative and both will be squared. Therefore they must be placed in parentheses so the entire value, including the negative, is squared.

**13.**  The football team lost 4 years on each play and a loss of yardage is represented by a negative number.

## Exercise Set 1.7

**1.**  $-6(4) = -24$

**3.**  $2(-1) = -2$

**5.**  $-5(-10) = 50$

**7.**  $-3 \cdot 4 = -12$

**9.** $-7 \cdot 0 = 0$

**11.** $2(-9) = -18$

**13.** $-\dfrac{1}{2}\left(-\dfrac{3}{5}\right) = \dfrac{1 \cdot 3}{2 \cdot 5} = \dfrac{3}{10}$

**15.** $-\dfrac{3}{4}\left(-\dfrac{8}{9}\right) = \dfrac{3 \cdot 8}{4 \cdot 9} = \dfrac{24}{36} = \dfrac{2 \cdot 12}{3 \cdot 12} = \dfrac{2}{3}$

**17.** $5(-1.4) = -7$

**19.** $-0.2(-0.7) = 0.14$

**21.** $-10(80) = -800$

**23.** $4(-7) = -28$

**25.** $(-5)(-5) = 25$

**27.** $\dfrac{2}{3}\left(-\dfrac{4}{9}\right) = -\dfrac{2 \cdot 4}{3 \cdot 9} = -\dfrac{8}{27}$

**29.** $-11(11) = -121$

**31.** $-\dfrac{20}{25}\left(\dfrac{5}{16}\right) = -\dfrac{20 \cdot 5}{25 \cdot 16} = -\dfrac{100}{400} = -\dfrac{1}{4}$

**33.** $(-1)(2)(-3)(-5) = -2(-3)(-5) = 6(-5) = -30$

**35.** $(-2)(5) - (-11)(3) = -10 - (-33) = -10 + 33 = 23$

**37.** $(-6)(-1)(-2) - (-5) = -12 + 5 = -7$

**39.** True; example: $(-2)(-2)(-2) = -8$
False; example: $(-2)(-2)(-2)(-2) = 16$

**41.** False

**43.** $(-2)^4 = (-2)(-2)(-2)(-2)$
$\phantom{(-2)^4} = 4(-2)(-2)$
$\phantom{(-2)^4} = -8(-2)$
$\phantom{(-2)^4} = 16$

**45.** $-1^5 = -(1)(1)(1)(1)(1) = -1$

**47.** $(-5)^2 = (-5)(-5) = 25$

**49.** $-7^2 = -(7)(7) = -49$

**51.** Reciprocal of 9 is $\dfrac{1}{9}$ since $9 \cdot \dfrac{1}{9} = 1$.

**53.** Reciprocal of $\dfrac{2}{3}$ is $\dfrac{3}{2}$ since $\dfrac{2}{3} \cdot \dfrac{3}{2} = 1$.

**55.** Reciprocal of $-14$ is $-\dfrac{1}{14}$ since $-14 \cdot -\dfrac{1}{14} = 1$.

**57.** Reciprocal of $-\dfrac{3}{11}$ is $-\dfrac{11}{3}$ since $-\dfrac{3}{11} \cdot -\dfrac{11}{3} = 1$.

**59.** Reciprocal of $0.2$ is $\dfrac{1}{0.2}$ since $0.2 \cdot \dfrac{1}{0.2} = 1$.

**61.** Reciprocal of $\dfrac{1}{-6.3}$ is $-6.3$ since
$\dfrac{1}{-6.3} \cdot -6.3 = 1$.

**63.** $\dfrac{18}{-2} = 18 \cdot -\dfrac{1}{2} = -9$

**65.** $\dfrac{-16}{-4} = -16 \cdot -\dfrac{1}{4} = 4$

**67.** $\dfrac{-48}{12} = -48 \cdot \dfrac{1}{12} = -4$

**69.** $\dfrac{0}{-4} = 0 \cdot -\dfrac{1}{4} = 0$

**71.** $-\dfrac{15}{3} = -15 \cdot \dfrac{1}{3} = -5$

**73.** $\dfrac{5}{0}$ is undefined.

**75.** $\dfrac{-12}{-4} = -12 \cdot -\dfrac{1}{4} = 3$

**77.** $\dfrac{30}{-2} = 30 \cdot -\dfrac{1}{2} = -15$

**79.** $\dfrac{6}{7} \div -\dfrac{1}{3} = \dfrac{6}{7} \cdot \left(-\dfrac{3}{1}\right) = -\dfrac{6 \cdot 3}{7 \cdot 1} = -\dfrac{18}{7}$

**81.** $-\dfrac{5}{9} \div \left(-\dfrac{3}{4}\right) = -\dfrac{5}{9} \cdot \left(-\dfrac{4}{3}\right) = \dfrac{5 \cdot 4}{9 \cdot 3} = \dfrac{20}{27}$

**83.** $-\dfrac{4}{9} \div \dfrac{4}{9} = -\dfrac{4}{9} \cdot \dfrac{9}{4} = -1$

**85.** $\dfrac{-9(-3)}{-6} = \dfrac{27}{-6} = -\dfrac{9}{2}$

**87.** $\dfrac{12}{9-12} = \dfrac{12}{-3} = -4$

**89.** $\dfrac{-6^2+4}{-2} = \dfrac{-36+4}{-2} = \dfrac{-32}{-2} = 16$

**91.** $\dfrac{8+(-4)^2}{4-12} = \dfrac{8+16}{4-12} = \dfrac{24}{-8} = -3$

**93.** $\dfrac{22+(3)(-2)}{-5-2} = \dfrac{22+(-6)}{-5-2} = \dfrac{16}{-7} = -\dfrac{16}{7}$

**95.** $\dfrac{-3-5^2}{2(-7)} = \dfrac{-3-25}{2(-7)} = \dfrac{-3+(-25)}{-14} = \dfrac{-28}{-14} = 2$

**97.** $\dfrac{6-2(-3)}{4-3(-2)} = \dfrac{6-(-6)}{4-(-6)} = \dfrac{6+6}{4+6} = \dfrac{12}{10} = \dfrac{6}{5}$

**99.** $\dfrac{-3-2(-9)}{-15-3(-4)} = \dfrac{-3-(-18)}{-15-(-12)} = \dfrac{-3+18}{-15+12} = \dfrac{15}{-3} = -5$

**101.** $\dfrac{|5-9|+|10-15|}{|2(-3)|} = \dfrac{|-4|+|-5|}{|-6|} = \dfrac{4+5}{6} = \dfrac{9}{6} = \dfrac{3}{2}$

**103.** Let $x = -5$ and $y = -3$.
$3x + 2y = 3(-5) + 2(-3) = -15 + (-6) = -21$

**105.** Let $x = -5$ and $y = -3$.
$$2x^2 - y^2 = 2(-5)^2 - (-3)^2$$
$$= 2(25) - 9$$
$$= 50 + (-9)$$
$$= 41$$

**107.** Let $x = -5$ and $y = -3$.
$x^3 + 3y = (-5)^3 + 3(-3) = -125 + (-9) = -134$

**109.** Let $x = -5$ and $y = -3$.
$\dfrac{2x-5}{y-2} = \dfrac{2(-5)-5}{-3-2} = \dfrac{-10-5}{-3-2} = \dfrac{-15}{-5} = 3$

**111.** Let $x = -5$ and $y = -3$.
$\dfrac{-3-y}{x-4} = \dfrac{-3-(-3)}{-5-4} = \dfrac{-3+3}{-5-4} = \dfrac{0}{-9} = 0$

**113.** The product of $-71$ and a number is $-71 \cdot x$ or $-71x$.

**115.** Subtract a number from $-16$ is $-16 - x$.

**117.** $-29$ increased by a number is $-29 + x$.

**119.** Divide a number by $-33$ is $\dfrac{x}{-33}$ or $x \div (-33)$.

**121.** A loss of 4 yards is represented by $-4$.
$3 \cdot (-4) = -12$
The team had a total loss of 12 yards.

**123.** Each move of 20 feet down is represented by $-20$.
$5 \cdot (-20) = -100$
The diver is at a depth of 100 feet.

**125.** Let $x = 7$.
$-5x = -35$
$-5(7) \overset{?}{=} -35$
$-35 = -35$, true
7 is a solution of the equation.

**127.** Let $x = -20$.
$$\dfrac{x}{10} = 2$$
$$\dfrac{-20}{10} \overset{?}{=} 2$$
$$-2 = 2, \text{ false}$$
$-20$ is not a solution of the equation.

**129.** Let $x = 5$.
$-3x - 5 = -20$
$-3(5) - 5 \overset{?}{=} -20$
$-15 - 5 \overset{?}{=} -20$
$-20 = -20$, true
5 is a solution of the equation.

**131.** $2(-81) = -162$
The surface temperature of Jupiter is $-162°$F.

**133.** answers may vary

**135.** $-1$ and $1$ are their own reciprocals.

**137.** Since $q$ is negative, $r$ is negative, and $t$ is positive, then $\dfrac{q}{r \cdot t}$ is positive.

**139.** It is not possible to determine whether $q + t$ is positive or negative.

**141.** Since $q$ is negative, $r$ is negative, and $t$ is positive, then $t(q + r)$ is negative.

**143.**
$$-2 + \frac{-15}{3} = \frac{-2 \cdot 3}{1 \cdot 3} + \frac{-15}{3}$$
$$= \frac{-6 + (-15)}{3}$$
$$= \frac{-21}{3}$$
$$= -7$$

**145.** $2[-5 + (-3)] = 2(-8) = -16$

**Section 1.8 Practice**

**1. a.** $x \cdot 8 = \underline{8 \cdot x}$

   **b.** $x + 17 = \underline{17 + x}$

**2. a.** $(2 + 9) + 7 = \underline{2 + (9 + 7)}$

   **b.** $-4 \cdot (2 \cdot 7) = \underline{(-4 \cdot 2) \cdot 7}$

**3. a.** $(5 + x) + 9 = (x + 5) + 9 = x + (5 + 9) = x + 14$

   **b.** $5(-6x) = [5 \cdot (-6)]x = -30x$

**4. a.** $5(x - y) = 5(x) - 5(y) = 5x - 5y$

   **b.** $-6(4 + 2t) = -6(4) + (-6)(2t) = -24 - 12t$

   **c.** $2(3x - 4y - z) = 2(3x) + 2(-4y) + 2(-z)$
   $$= 6x - 8y - 2z$$

   **d.** $(3 - y) \cdot (-1) = 3(-1) + (-y)(-1) = -3 + y$

   **e.** $-(x - 7 + 2s) = (-1)(x - 7 + 2s)$
   $$= (-1)x + (-1)(-7) + (-1)(2s)$$
   $$= -x + 7 - 2s$$

   **f.** $\frac{1}{2}(2x + 4) + 9 = \frac{1}{2}(2x) + \frac{1}{2}(4) + 9$
   $$= x + 2 + 9$$
   $$= x + 11$$

**5. a.** $5 \cdot w + 5 \cdot 3 = 5(w + 3)$

   **b.** $9w + 9z = 9 \cdot w + 9 \cdot z = 9(w + z)$

**6. a.** $(7 \cdot 3x) \cdot 4 = (3x \cdot 7) \cdot 4$; commutative property of multiplication

   **b.** $6 + (3 + y) = (6 + 3) + y$; associative property of addition

   **c.** $8 + (t + 0) = 8 + t$; identity element for addition

   **d.** $-\frac{3}{4} \cdot \left(-\frac{4}{3}\right) = 1$; multiplicative inverse property

   **e.** $(2 + x) + 5 = 5 + (2 + x)$; commutative property of addition

   **f.** $3 + (-3) = 0$; additive inverse property

   **g.** $(-3b) \cdot 7 = (-3 \cdot 7) \cdot b$; commutative and associative properties of multiplication

**Vocabulary, Readiness & Video Check 1.8**

**1.** $x + 5 = 5 + x$ is a true statement by the <u>commutative property of addition</u>.

**2.** $x \cdot 5 = 5 \cdot x$ is a true statement by the <u>commutative property of multiplication</u>.

**3.** $3(y + 6) = 3 \cdot y + 3 \cdot 6$ is true by the <u>distributive property</u>.

**4.** $2 \cdot (x \cdot y) = (2 \cdot x) \cdot y$ is a true statement by the <u>associative property of multiplication</u>.

**5.** $x + (7 + y) = (x + 7) + y$ is a true statement by the <u>associative property of addition</u>.

**6.** The numbers $-\frac{2}{3}$ and $-\frac{3}{2}$ are called <u>reciprocals or multiplicative inverses</u>.

**7.** The numbers $-\frac{2}{3}$ and $\frac{2}{3}$ are called <u>opposites or additive inverses</u>.

**8.** order; grouping

**9.** 2 is outside the parentheses, so the point is made that you should only distribute the $-9$ to the terms within the parentheses and not also to the 2.

**10.** The identity element for addition is <u>0</u> because if we add <u>0</u> to any real number, the result is that real number.
The identity element for multiplication is <u>1</u> because any real number times <u>1</u> gives a result of that original real number.

**Exercise Set 1.8**

**1.** $x + 16 = 16 + x$

**3.** $-4 \cdot y = y \cdot (-4)$

**5.** $xy = yx$

**7.** $2x + 13 = 13 + 2x$

**9.** $(xy) \cdot z = x \cdot (yz)$

**11.** $2 + (a + b) = (2 + a) + b$

**13.** $4 \cdot (ab) = 4a \cdot (b)$

**15.** $(a + b) + c = a + (b + c)$

**17.** $8 + (9 + b) = (8 + 9) + b = 17 + b$

**19.** $4(6y) = (4 \cdot 6)y = 24y$

**21.** $\dfrac{1}{5}(5y) = \left(\dfrac{1}{5} \cdot 5\right)y = 1 \cdot y = y$

**23.** $(13 + a) + 13 = (a + 13) + 13$
$= a + (13 + 13)$
$= a + 26$

**25.** $-9(8x) = (-9 \cdot 8)x = -72x$

**27.** $\dfrac{3}{4}\left(\dfrac{4}{3}s\right) = \left(\dfrac{3}{4} \cdot \dfrac{4}{3}\right)s = 1s = s$

**29.** $\dfrac{2}{3} + \left(\dfrac{4}{3} + x\right) = \dfrac{2}{3} + \dfrac{4}{3} + x = \dfrac{6}{3} + x = 2 + x$

**31.** $4(x + y) = 4x + 4y$

**33.** $9(x - 6) = 9x - 9 \cdot 6 = 9x - 54$

**35.** $2(3x + 5) = 2(3x) + 2(5) = 6x + 10$

**37.** $7(4x - 3) = 7(4x) - 7(3) = 28x - 21$

**39.** $3(6 + x) = 3(6) + 3x = 18 + 3x$

**41.** $-2(y - z) = -2y - (-2)z = -2y + 2z$

**43.** $-7(3y + 5) = -7(3y) + (-7)(5) = -21y - 35$

**45.** $5(x + 4m + 2) = 5x + 5(4m) + 5(2)$
$= 5x + 20m + 10$

**47.** $-4(1 - 2m + n) = -4(1) - (-4)(2m) + (-4)n$
$= -4 + 8m - 4n$

**49.** $-(5x + 2) = -1(5x + 2)$
$= -1(5x) + (-1)(2)$
$= -5x - 2$

**51.** $-(r - 3 - 7p) = -1(r - 3 - 7p)$
$= -1r - (-1)(3) - (-1)(7p)$
$= -r + 3 + 7p$

**53.** $\dfrac{1}{2}(6x + 8) = \dfrac{1}{2}(6x) + \dfrac{1}{2}(8)$
$= \left(\dfrac{1}{2} \cdot 6\right)x + \left(\dfrac{1}{2} \cdot 8\right)$
$= 3x + 4$

**55.** $-\dfrac{1}{3}(3x - 9y) = -\dfrac{1}{3}(3x) - \left(-\dfrac{1}{3}\right)(9y)$
$= \left(-\dfrac{1}{3} \cdot 3\right)x - \left(-\dfrac{1}{3} \cdot 9\right)y$
$= -1 \cdot x + 3 \cdot y$
$= -x + 3y$

**57.** $3(2r + 5) - 7 = 3(2r) + 3(5) - 7$
$= 6r + 15 + (-7)$
$= 6r + 8$

**59.** $-9(4x + 8) + 2 = -9(4x) + (-9)(8) + 2$
$= -36x - 72 + 2$
$= -36x - 70$

**61.** $-4(4x + 5) - 5 = -4(4x) + (-4)(5) - 5$
$= -16x + (-20) + (-5)$
$= -16x - 25$

**63.** $4 \cdot 1 + 4 \cdot y = 4(1 + y)$

**65.** $11x + 11y = 11(x + y)$

**67.** $(-1) \cdot 5 + (-1) \cdot x = -1(5 + x) = -(5 + x)$

**69.** $30a + 30b = 30(a + b)$

**71.** $3 \cdot 5 = 5 \cdot 3$; commutative property of multiplication

**73.** $2 + (x + 5) = (2 + x) + 5$; associative property of addition

**75.** $9(3 + 7) = 9 \cdot 3 + 9 \cdot 7$; distributive property

**77.** $(4 \cdot y) \cdot 9 = 4 \cdot (y \cdot 9)$; associative property of multiplication

**79.** $0 + 6 = 6$; identity element of addition

**81.** $-4(y + 7) = -4 \cdot y + (-4) \cdot 7$; distributive property

**83.** $-4 \cdot (8 \cdot 3) = (8 \cdot -4) \cdot 3$; associative and commutative properties of multiplication

**85.**

| Expression | Opposite | Reciprocal |
|---|---|---|
| 8 | −8 | $\frac{1}{8}$ |

**87.**

| Expression | Opposite | Reciprocal |
|---|---|---|
| $x$ | $-x$ | $\frac{1}{x}$ |

**89.**

| Expression | Opposite | Reciprocal |
|---|---|---|
| $2x$ | $-2x$ | $\frac{1}{2x}$ |

**91.** False; the opposite of $-\frac{a}{2}$ is $\frac{a}{2}$. $-\frac{2}{a}$ is the reciprocal of $-\frac{a}{2}$.

**93.** "Taking a test" and "studying for the test" are not commutative, since the order in which they are performed affects the outcome.

**95.** "Putting on your left shoe" and "putting on your right shoe" are commutative, since the order in which they are performed does not affect the outcome.

**97.** "Mowing the lawn" and "trimming the hedges" are commutative, since the order in which they are performed does not affect the outcome.

**99.** "Dialing a number" and "turning on the cell phone" are not commutative, since the order in which they are performed affects the outcome.

**101. a.** The property illustrated is the commutative property of addition since the order in which they are added changed.

**b.** The property illustrated is the commutative property of addition since the order in which they are added changed.

**c.** The property illustrated is the associative property of addition since the grouping of addition changed.

**103.** answers may vary

**105.** answers may vary

**Chapter 1 Vocabulary Check**

**1.** The symbols ≠, <, and > are called inequality symbols.

**2.** A mathematical statement that two expressions are equal is called an equation.

**3.** The absolute value of a number is the distance between that number and 0 on the number line.

**4.** A symbol used to represent a number is called a variable.

**5.** Two numbers that are the same distance from 0 but lie on opposite sides of 0 are called opposites.

**6.** The number in a fraction above the fraction bar is called the numerator.

**7.** A solution of an equation is a value for the variable that makes the equation a true statement.

**8.** Two numbers whose product is 1 are called reciprocals.

**9.** In $2^3$, the 2 is called the base and the 3 is called the exponent.

**10.** The number in a fraction below the fraction bar is called the denominator.

**11.** Parentheses and brackets are examples of grouping symbols.

**12.** A set is a collection of objects.

**Chapter 1 Review**

**1.** $8 < 10$ since 8 is to the left of 10 on the number line.

**2.** $7 > 2$ since 7 is to the right of 2 on the number line.

**3.** $-4 > -5$ since $-4$ is to the right of $-5$ on the number line.

**4.** $\dfrac{12}{2} > -8$ since $6 > -8$.

**5.** $|-7| < |-8|$ since $7 < 8$.

**6.** $|-9| > -9$ since $9 > -9$.

**7.** $-|-1| = -1$ since $-1 = -1$.

**8.** $|-14| = -(-14)$ since $14 = 14$.

**9.** $1.2 > 1.02$ since $1.2$ is to the right of $1.02$ on the number line.

**10.** $-\dfrac{3}{2} < -\dfrac{3}{4}$ since $-\dfrac{3}{2}$ is to the left of $-\dfrac{3}{4}$ on the number line.

**11.** Four is greater than or equal to negative three is written as $4 \geq -3$.

**12.** Six is not equal to five is written as $6 \neq 5$.

**13.** $0.03$ is less than $0.3$ is written as $0.03 < 0.3$.

**14.** $400 > 155$ or $155 < 400$

**15. a.** The natural numbers are 1 and 3.

    **b.** The whole numbers are 0, 1, and 3.

    **c.** The integers are $-6$, 0, 1, and 3.

    **d.** The rational numbers are $-6$, 0, 1, $1\dfrac{1}{2}$, 3, and 9.62.

    **e.** The irrational number is $\pi$.

    **f.** The real numbers are all numbers in the given set.

**16. a.** The natural numbers are 2 and 5.

    **b.** The whole numbers are 2 and 5.

    **c.** The integers are $-3$, 2, and 5.

    **d.** The rational numbers are $-3$, $-1.6$, 2, 5, $\dfrac{11}{2}$, and 15.1.

    **e.** The irrational numbers are $\sqrt{5}$ and $2\pi$.

**f.** The real numbers are all numbers in the given set.

**17.** Look for the negative number with the greatest absolute value. The greatest loss was on Friday.

**18.** Look for the largest positive number. The greatest gain was on Wednesday.

**19.** $36 = 4 \cdot 9 = 2 \cdot 2 \cdot 3 \cdot 3$

**20.** $120 = 8 \cdot 15 = 2 \cdot 2 \cdot 2 \cdot 3 \cdot 5$

**21.** $\dfrac{8}{15} \cdot \dfrac{27}{30} = \dfrac{8 \cdot 27}{15 \cdot 30} = \dfrac{2 \cdot 4 \cdot 3 \cdot 3 \cdot 3}{3 \cdot 5 \cdot 2 \cdot 3 \cdot 5} = \dfrac{12}{25}$

**22.** $\dfrac{7}{8} \div \dfrac{21}{32} = \dfrac{7}{8} \cdot \dfrac{32}{21} = \dfrac{7 \cdot 32}{8 \cdot 21} = \dfrac{7 \cdot 8 \cdot 4}{8 \cdot 3 \cdot 7} = \dfrac{4}{3}$

**23.** 
$$
\begin{aligned}
\dfrac{7}{15} + \dfrac{5}{6} &= \dfrac{7 \cdot 2}{15 \cdot 2} + \dfrac{5 \cdot 5}{6 \cdot 5} \\
&= \dfrac{14}{30} + \dfrac{25}{30} \\
&= \dfrac{14 + 25}{30} \\
&= \dfrac{39}{30} \\
&= \dfrac{3 \cdot 13}{3 \cdot 10} \\
&= \dfrac{13}{10}
\end{aligned}
$$

**24.** 
$$
\begin{aligned}
\dfrac{3}{4} - \dfrac{3}{20} &= \dfrac{3 \cdot 5}{4 \cdot 5} - \dfrac{3}{20} \\
&= \dfrac{15}{20} - \dfrac{3}{20} \\
&= \dfrac{15 - 3}{20} \\
&= \dfrac{12}{20} \\
&= \dfrac{3 \cdot 4}{5 \cdot 4} \\
&= \dfrac{3}{5}
\end{aligned}
$$

**25.** $2\dfrac{3}{4}+6\dfrac{5}{8}=\dfrac{11}{4}+\dfrac{53}{8}$

$\qquad\qquad=\dfrac{11\cdot 2}{4\cdot 2}+\dfrac{53}{8}$

$\qquad\qquad=\dfrac{22}{8}+\dfrac{53}{8}$

$\qquad\qquad=\dfrac{22+53}{8}$

$\qquad\qquad=\dfrac{75}{8}$

$\qquad\qquad=9\dfrac{3}{8}$

**26.** $7\dfrac{1}{6}-2\dfrac{2}{3}=\dfrac{43}{6}-\dfrac{8}{3}$

$\qquad\qquad=\dfrac{43}{6}-\dfrac{8\cdot 2}{3\cdot 2}$

$\qquad\qquad=\dfrac{43}{6}-\dfrac{16}{6}$

$\qquad\qquad=\dfrac{43-16}{6}$

$\qquad\qquad=\dfrac{27}{6}$

$\qquad\qquad=\dfrac{9\cdot 3}{2\cdot 3}$

$\qquad\qquad=\dfrac{9}{2}$

$\qquad\qquad=4\dfrac{1}{2}$

**27.** $5\div\dfrac{1}{3}=5\cdot\dfrac{3}{1}=15$

**28.** $2\cdot 8\dfrac{3}{4}=2\cdot\dfrac{35}{4}=\dfrac{2\cdot 35}{2\cdot 2}=\dfrac{35}{2}=17\dfrac{1}{2}$

**29.** $1-\dfrac{1}{6}-\dfrac{1}{4}=\dfrac{12}{12}-\dfrac{1\cdot 2}{6\cdot 2}-\dfrac{1\cdot 3}{4\cdot 3}$

$\qquad\qquad=\dfrac{12}{12}-\dfrac{2}{12}-\dfrac{3}{12}$

$\qquad\qquad=\dfrac{12-2-3}{12}$

$\qquad\qquad=\dfrac{7}{12}$

The unknown part is $\dfrac{7}{12}$.

**30.** $1-\dfrac{1}{2}-\dfrac{1}{5}=\dfrac{10}{10}-\dfrac{5}{10}-\dfrac{2}{10}=\dfrac{10-5-2}{10}=\dfrac{3}{10}$

The unknown part is $\dfrac{3}{10}$.

**31.** $P=2l+2w$

$P=2\left(1\dfrac{1}{3}\right)+2\left(\dfrac{7}{8}\right)$

$\qquad=\dfrac{2}{1}\cdot\dfrac{4}{3}+\dfrac{2}{1}\cdot\dfrac{7}{8}$

$\qquad=\dfrac{8}{3}+\dfrac{14}{8}$

$\qquad=\dfrac{8\cdot 8}{3\cdot 8}+\dfrac{14\cdot 3}{8\cdot 3}$

$\qquad=\dfrac{64}{24}+\dfrac{42}{24}$

$\qquad=\dfrac{64+42}{24}$

$\qquad=\dfrac{106}{24}$

$\qquad=4\dfrac{10}{24}$

$\qquad=4\dfrac{5}{12}$ meters

$A=lw$

$A=1\dfrac{1}{3}\cdot\dfrac{7}{8}$

$\qquad=\dfrac{4}{3}\cdot\dfrac{7}{8}$

$\qquad=\dfrac{4\cdot 7}{3\cdot 2\cdot 4}$

$\qquad=\dfrac{7}{6}$

$\qquad=1\dfrac{1}{6}$ sq meters

**32.** $P=$ the sum of the lengths of the sides

$P=\dfrac{5}{11}+\dfrac{8}{11}+\dfrac{3}{11}+\dfrac{3}{11}+\dfrac{2}{11}+\dfrac{5}{11}=\dfrac{26}{11}=2\dfrac{4}{11}$ in.

$A=$ the sum of the two areas, each given by $lw$

$A=\dfrac{5}{11}\cdot\dfrac{5}{11}+\dfrac{3}{11}\cdot\dfrac{3}{11}=\dfrac{25}{121}+\dfrac{9}{121}=\dfrac{34}{121}$ sq in.

**33.** $2\dfrac{1}{2}+3\dfrac{1}{16}+1\dfrac{3}{4}+2\dfrac{9}{16}+1\dfrac{13}{16}+2\dfrac{7}{16}$

$=\dfrac{5}{2}+\dfrac{49}{16}+\dfrac{7}{4}+\dfrac{41}{16}+\dfrac{29}{16}+\dfrac{39}{16}$

$=\dfrac{40}{16}+\dfrac{49}{16}+\dfrac{28}{16}+\dfrac{41}{16}+\dfrac{29}{16}+\dfrac{39}{16}$

$=\dfrac{226}{16}$

$=14\dfrac{2}{16}$

$=14\dfrac{1}{8}$ lb

**34.** $2\dfrac{1}{8}+2\dfrac{3}{16}=\dfrac{17}{.8}+\dfrac{35}{16}=\dfrac{34}{16}+\dfrac{35}{16}=\dfrac{69}{16}=4\dfrac{5}{16}$ lb

**35.** Total weight = weight of boys + weight of girls

$=\dfrac{226}{16}+\dfrac{69}{16}$

$=\dfrac{295}{16}$

$=18\dfrac{7}{16}$ lb

**36.** Look for the largest number. Baby C weighed the most.

**37.** Look for the smallest number. Baby E weighed the least.

**38.** $3\dfrac{1}{16}-1\dfrac{3}{4}=\dfrac{49}{16}-\dfrac{7}{4}=\dfrac{49}{16}-\dfrac{28}{16}=\dfrac{21}{16}=1\dfrac{5}{16}$ lb

**39.** $6\cdot 3^2+2\cdot 8=6\cdot 9+2\cdot 8=54+16=70$
The answer is c.

**40.** $68-5\cdot 2^3=68-5\cdot 8=68-40=68+(-40)=28$
The answer is b.

**41.** $\left(\dfrac{2}{7}\right)^2=\dfrac{2}{7}\cdot\dfrac{2}{7}=\dfrac{4}{49}$

**42.** $\left(\dfrac{3}{4}\right)^3=\dfrac{3}{4}\cdot\dfrac{3}{4}\cdot\dfrac{3}{4}=\dfrac{27}{64}$

**43.** $3(1+2\cdot 5)+4=3(1+10)+4$
$=3(11)+4$
$=33+4$
$=37$

**44.** $8+3(2\cdot 6-1)=8+3(12-1)$
$=8+3(11)$
$=8+33$
$=41$

**45.** $\dfrac{4+|6-2|+8^2}{4+6\cdot 4}=\dfrac{4+|4|+64}{4+24}$

$=\dfrac{4+4+64}{4+24}$

$=\dfrac{72}{28}$

$=\dfrac{4\cdot 18}{4\cdot 7}$

$=\dfrac{18}{7}$

**46.** $5[3(2+5)-5]=5[3(7)-5]$
$=5[21-5]$
$=5[16]$
$=80$

**47.** The difference of twenty and twelve is equal to the product of two and four is written as $20-12=2\cdot 4$.

**48.** The quotient of nine and two is greater than negative five is written as $\dfrac{9}{2}>-5$.

**49.** Let $x=6$ and $y=2$.
$2x+3y=2(6)+3(2)=12+6=18$

**50.** Let $x=6$, $y=2$, and $z=8$.
$x(y+2z)=6[2+2(8)]=6[2+16]=6[18]=108$

**51.** Let $x=6$, $y=2$, and $z=8$.
$\dfrac{x}{y}+\dfrac{z}{2y}=\dfrac{6}{2}+\dfrac{8}{2(2)}=\dfrac{6}{2}+\dfrac{8}{4}=3+2=5$

**52.** Let $x=6$ and $y=2$.
$x^2-3y^2=(6)^2-3(2)^2$
$=36-3(4)$
$=36-12$
$=36+(-12)$
$=24$

**53.** Replace $a$ with 37 and $b$ with 80.
$$\begin{aligned}
180 - a - b &= 180 - 37 - 80 \\
&= 180 + (-37) + (-80) \\
&= 143 + (-80) \\
&= 63
\end{aligned}$$
The measure of the unknown angle is 63°.

**54.** Replace $a$ with 93, $b$ with 80, and $c$ with 82.
$$\begin{aligned}
360 - a - b - c &= 360 - 93 - 80 - 82 \\
&= 360 + (-93) + (-80) + (-82) \\
&= 267 + (-80) + (-82) \\
&= 187 + (-82) \\
&= 105
\end{aligned}$$
The measure of the unknown angle is 105°.

**55.** Let $x = 3$.
$$\begin{aligned}
7x - 3 &= 18 \\
7(3) - 3 &\overset{?}{=} 18 \\
21 - 3 &\overset{?}{=} 18 \\
18 &= 18, \text{ true}
\end{aligned}$$
3 is a solution to the equation.

**56.** Let $x = 1$.
$$\begin{aligned}
3x^2 + 4 &= x - 1 \\
3(1)^2 + 4 &\overset{?}{=} 1 - 1 \\
3 + 4 &\overset{?}{=} 0 \\
7 &= 0, \text{ false}
\end{aligned}$$
1 is not a solution to the equation.

**57.** The additive inverse of $-9$ is 9.

**58.** The additive inverse of $\dfrac{2}{3}$ is $-\dfrac{2}{3}$.

**59.** The additive inverse of $|-2|$ is $-2$ since $|-2| = 2$.

**60.** The additive inverse of $-|-7|$ is 7 since $-|-7| = -7$.

**61.** $-15 + 4 = -11$

**62.** $-6 + (-11) = -17$

**63.** $\begin{aligned}[t]
\frac{1}{16} + \left(-\frac{1}{4}\right) &= \frac{1}{16} + \left(-\frac{1 \cdot 4}{4 \cdot 4}\right) \\
&= \frac{1}{16} + \left(-\frac{4}{16}\right) \\
&= -\frac{3}{16}
\end{aligned}$

**64.** $-8 + |-3| = -8 + 3 = -5$

**65.** $-4.6 + (-9.3) = -13.9$

**66.** $-2.8 + 6.7 = 3.9$

**67.** $6 - 20 = 6 + (-20) = -14$

**68.** $-3.1 - 8.4 = -3.1 + (-8.4) = -11.5$

**69.** $-6 - (-11) = -6 + 11 = 5$

**70.** $4 - 15 = 4 + (-15) = -11$

**71.** $\begin{aligned}[t]
-21 - 16 + 3(8 - 2) &= -21 + (-16) + 3[8 + (-2)] \\
&= -21 + (-16) + 3[6] \\
&= -21 + (-16) + 18 \\
&= -37 + 18 \\
&= -19
\end{aligned}$

**72.** $\begin{aligned}[t]
\frac{11 - (-9) + 6(8 - 2)}{2 + 3 \cdot 4} &= \frac{11 + 9 + 6[8 + (-2)]}{2 + 3 \cdot 4} \\
&= \frac{11 + 9 + 6[6]}{2 + 3 \cdot 4} \\
&= \frac{11 + 9 + 36}{2 + 12} \\
&= \frac{56}{14} \\
&= 4
\end{aligned}$

**73.** Replace $x$ with 3, $y$ with $-6$, and $z$ with $-9$.
$$\begin{aligned}
2x^2 - y + z &= 2(3)^2 - (-6) + (-9) \\
&= 2(9) + 6 + (-9) \\
&= 18 + 6 + (-9) \\
&= 24 + (-9) \\
&= 15
\end{aligned}$$
The answer is a.

**74.** Replace $x$ with 3 and $y$ with $-6$.
$$\begin{aligned}
\frac{|y - 4x|}{2x} &= \frac{|-6 - 4(3)|}{2(3)} \\
&= \frac{|-6 - 12|}{6} \\
&= \frac{|-6 + (-12)|}{6} \\
&= \frac{|-18|}{6} \\
&= \frac{18}{6} \\
&= 3
\end{aligned}$$
The answer is a.

**75.** $50+1+(-2)+5+1+(-4)$
$= 51+(-2)+5+1+(-4)$
$= 49+5+1+(-4)$
$= 54+1+(-4)$
$= 55+(-4)$
$= 51$
The price at the end of the week is $51.

**76.** $50+1+(-2)+5 = 51+(-2)+5 = 49+5 = 54$
The price at the end of the day on Wednesday is $54.

**77.** The multiplicative inverse of $-6$ is $-\dfrac{1}{6}$ since
$$-6 \cdot -\frac{1}{6} = 1.$$

**78.** The multiplicative inverse of $\dfrac{3}{5}$ is $\dfrac{5}{3}$ since
$$\frac{3}{5} \cdot \frac{5}{3} = 1.$$

**79.** $6(-8) = -48$

**80.** $(-2)(-14) = 28$

**81.** $\dfrac{-18}{-6} = 3$

**82.** $\dfrac{42}{-3} = -14$

**83.** $\dfrac{4 \cdot (-3)+(-8)}{2+(-2)} = \dfrac{-12+(-8)}{2+(-2)} = \dfrac{-20}{0}$
The expression is undefined.

**84.** $\dfrac{3(-2)^2-5}{-14} = \dfrac{3(4)-5}{-14} = \dfrac{12-5}{-14} = \dfrac{7}{-14} = -\dfrac{1}{2}$

**85.** $\dfrac{-6}{0}$ is undefined.

**86.** $\dfrac{0}{-2} = 0$

**87.** $-4^2-(-3+5)\div(-1)\cdot 2 = -16-(2)\div(-1)\cdot 2$
$= -16+2\cdot 2$
$= -16+4$
$= -12$

**88.** $-5^2-(2-20)\div(-3)\cdot 3 = -25-(-18)\div(-3)\cdot 3$
$= -25-6\cdot 3$
$= -25-18$
$= -43$

**89.** Let $x = -5$ and $y = -2$.
$x^2-y^4 = (-5)^2-(-2)^4 = 25-16 = 9$

**90.** Let $x = -5$ and $y = -2$.
$x^2-y^3 = (-5)^2-(-2)^3 = 25-(-8) = 25+8 = 33$

**91.** $-7x$ or $-7 \cdot x$

**92.** $\dfrac{x}{13}$ or $x \div (-13)$

**93.** $-20-x$

**94.** $-1+x$

**95.** $-6+5 = 5+(-6)$; commutative property of addition

**96.** $6 \cdot 1 = 6$; multiplicative identity property

**97.** $3(8-5) = 3 \cdot 8 - 3 \cdot 5$; distributive property

**98.** $4+(-4) = 0$; additive inverse property

**99.** $2+(3+9) = (2+3)+9$; associative property of addition

**100.** $2 \cdot 8 = 8 \cdot 2$; commutative property of multiplication

**101.** $6(8+5) = 6 \cdot 8 + 6 \cdot 5$; distributive property

**102.** $(3 \cdot 8) \cdot 4 = 3 \cdot (8 \cdot 4)$; associative property of multiplication

**103.** $4 \cdot \dfrac{1}{4} = 1$; multiplicative inverse property

**104.** $8+0 = 8$; additive identity property

**105.** $5(y-2) = 5(y)+5(-2) = 5y-10$

**106.** $-3(z+y) = -3(z)+(-3)(y) = -3z-3y$

**107.** $-(7-x+4z) = (-1)(7)+(-1)(-x)+(-1)(4z)$
$= -7+x-4z$

**108.** $\frac{1}{2}(6z-10) = \frac{1}{2}(6z) + \frac{1}{2}(-10) = 3z - 5$

**109.** $-4(3x+5) - 7 = -4(3x) + (-4)(5) - 7$
$= -12x - 20 - 7$
$= -12x - 27$

**110.** $-8(2y+9) - 1 = -8(2y) + (-8)(9) - 1$
$= -16y - 72 - 1$
$= -16y - 73$

**111.** $-|-11| < |11.4|$ since $-|-11| = -11$ and $|11.4| = 11.4$.

**112.** $-1\frac{1}{2} > -2\frac{1}{2}$ since $-1\frac{1}{2}$ is to the right of $-2\frac{1}{2}$ on the number line.

**113.** $-7.2 + (-8.1) = -15.3$

**114.** $14 - 20 = 14 + (-20) = -6$

**115.** $4(-20) = -80$

**116.** $\frac{-20}{4} = -5$

**117.** $-\frac{4}{5}\left(\frac{5}{16}\right) = -\frac{4}{16} = -\frac{1}{4}$

**118.** $-0.5(-0.3) = 0.15$

**119.** $8 \div 2 \cdot 4 = 4 \cdot 4 = 16$

**120.** $(-2)^4 = (-2)(-2)(-2)(-2) = 16$

**121.** $\frac{-3-2(-9)}{-15-3(-4)} = \frac{-3+18}{-15+12} = \frac{15}{-3} = -5$

**122.** $5 + 2[(7-5)^2 + (1-3)] = 5 + 2[2^2 + (-2)]$
$= 5 + 2[4 + (-2)]$
$= 5 + 2[2]$
$= 5 + 4$
$= 9$

**123.** $-\frac{5}{8} \div \frac{3}{4} = -\frac{5}{8} \cdot \frac{4}{3} = -\frac{20}{24} = -\frac{5}{6}$

**124.** $\frac{-15 + (-4)^2 + |-9|}{10 - 2 \cdot 5} = \frac{-15 + 16 + 9}{10 - 10} = \frac{1+9}{0}$ is undefined.

**125.** $7\frac{1}{2} - 6\frac{1}{8} = \frac{15}{2} - \frac{49}{8}$
$= \frac{15 \cdot 4}{2 \cdot 4} - \frac{49}{8}$
$= \frac{60}{8} - \frac{49}{8}$
$= \frac{60 - 49}{8}$
$= \frac{11}{8}$
$= 1\frac{3}{8}$ ft

**Chapter 1 Test**

1. The absolute value of negative seven is greater than five is written as $|-7| > 5$.

2. The sum of nine and five is greater than or equal to four is written as $(9 + 5) \geq 4$.

3. $-13 + 8 = -5$

4. $-13 - (-2) = -13 + 2 = -11$

5. $12 \div 4 \cdot 3 - 6 \cdot 2 = 3 \cdot 3 - 6 \cdot 2 = 9 - 12 = -3$

6. $(13)(-3) = -39$

7. $(-6)(-2) = 12$

8. $\frac{|-16|}{-8} = \frac{16}{-8} = -2$

9. $\frac{-8}{0}$ is undefined.

10. $\frac{|-6| + 2}{5 - 6} = \frac{6 + 2}{5 + (-6)} = \frac{8}{-1} = -8$

11. $\frac{1}{2} - \frac{5}{6} = \frac{1 \cdot 3}{2 \cdot 3} - \frac{5}{6} = \frac{3}{6} - \frac{5}{6} = \frac{3-5}{6} = \frac{-2}{6} = -\frac{1}{3}$

**12.** $5\dfrac{3}{4} - 1\dfrac{1}{8} = \dfrac{23}{4} - \dfrac{9}{8}$

$\qquad\qquad = \dfrac{2 \cdot 23}{2 \cdot 4} - \dfrac{9}{8}$

$\qquad\qquad = \dfrac{46}{8} - \dfrac{9}{8}$

$\qquad\qquad = \dfrac{46 + (-9)}{8}$

$\qquad\qquad = \dfrac{37}{8}$

$\qquad\qquad = 4\dfrac{5}{8}$

**13.** $-0.6 + 1.875 = 1.275$

**14.** $3(-4)^2 - 80 = 3(16) - 80 = 48 + (-80) = -32$

**15.** $6[5 + 2(3 - 8) - 3] = 6\{5 + 2[3 + (-8)] + (-3)\}$

$\qquad\qquad\qquad\quad = 6\{5 + 2[-5] + (-3)\}$

$\qquad\qquad\qquad\quad = 6\{5 + (-10) + (-3)\}$

$\qquad\qquad\qquad\quad = 6\{-5 + (-3)\}$

$\qquad\qquad\qquad\quad = 6\{-8\}$

$\qquad\qquad\qquad\quad = -48$

**16.** $\dfrac{-12 + 3 \cdot 8}{4} = \dfrac{-12 + 24}{4} = \dfrac{12}{4} = 3$

**17.** $\dfrac{(-2)(0)(-3)}{-6} = \dfrac{0(-3)}{-6} = \dfrac{0}{-6} = 0$

**18.** $-3 > -7$ since $-3$ is to the right of $-7$ on the number line.

**19.** $4 > -8$ since $4$ is to the right of $-8$ on the number line.

**20.** $2 < |-3|$ since $2 < 3$.

**21.** $|-2| = -1 - (-3)$ since $|-2| = 2$ and $-1 - (-3) = -1 + 3 = 2$.

**22.** $2221 < 10{,}993$

**23. a.** The natural numbers are 1 and 7.

**b.** The whole numbers are 0, 1 and 7.

**c.** The integers are $-5, -1, 0, 1$, and 7.

**d.** The rational numbers are $-5, -1, \dfrac{1}{4}, 0, 1, 7$, and 11.6.

**e.** The irrational numbers are $\sqrt{7}$ and $3\pi$.

**f.** The real numbers are all numbers in the given set.

**24.** Let $x = 6$ and $y = -2$.
$x^2 + y^2 = (6)^2 + (-2)^2 = 36 + 4 = 40$

**25.** Let $x = 6$, $y = -2$ and $z = -3$.
$x + yz = 6 + (-2)(-3) = 6 + 6 = 12$

**26.** Let $x = 6$ and $y = -2$.
$2 + 3x - y = 2 + 3(6) - (-2)$
$\qquad\qquad = 2 + 18 + 2$
$\qquad\qquad = 20 + 2$
$\qquad\qquad = 22$

**27.** Let $x = 6$, $y = -2$ and $z = -3$.
$\dfrac{y + z - 1}{x} = \dfrac{-2 + (-3) - 1}{6} = \dfrac{-5 + (-1)}{6} = \dfrac{-6}{6} = -1$

**28.** $8 + (9 + 3) = (8 + 9) + 3$; associative property of addition

**29.** $6 \cdot 8 = 8 \cdot 6$; commutative property of multiplication

**30.** $-6(2 + 4) = -6 \cdot 2 + (-6) \cdot 4$; distributive property

**31.** $\dfrac{1}{6}(6) = 1$; multiplicative inverse property

**32.** The opposite of $-9$ is 9.

**33.** The reciprocal of $-\dfrac{1}{3}$ is $-3$.

**34.** Look for the negative number that has the greatest absolute value. The second down had the greatest loss of yardage.

**35.** Gains: 5, 29
Losses: $-10, -2$
Total gain or loss $= 5 + (-10) + (-2) + 29$
$\qquad\qquad\qquad\quad = (-5) + (-2) + 29$
$\qquad\qquad\qquad\quad = -7 + 29$
$\qquad\qquad\qquad\quad = 22$ yards gained
Yes, they scored a touchdown.

**36.** Since $-14 + 31 = 17$, the temperature at noon was 17°.

**37.** $356 + 460 + (-166) = 816 + (-166) = 650$
The net income was $650 million.

**38.** Change in value per share $= -1.50$
Change in total value $= 280(-1.50) = -420$
She had a total loss of $420.

# Chapter 2

## Section 2.1 Practice

**1. a.** The numerical coefficient of $t$ is 1, since $t$ is $1t$.

**b.** The numerical coefficient of $-7x$ is $-7$.

**c.** The numerical coefficient of $-\dfrac{w}{5}$ is $-\dfrac{1}{5}$,

since $-\dfrac{w}{5}$ means $-\dfrac{1}{5} \cdot w$.

**d.** The numerical coefficient of $43x^4$ is 43.

**e.** The numerical coefficient of $-b$ is $-1$, since $-b$ is $-1b$.

**2. a.** $-4xy$ and $5yx$ are like terms, since $xy = yx$ by the commutative property.

**b.** $5q$ and $-3q^2$ are unlike terms, since the exponents on $q$ are not the same.

**c.** $3ab^2$, $-2ab^2$, and $43ab^2$ are like terms, since each variable and its exponent match.

**d.** $y^5$ and $\dfrac{y^5}{2}$ are like terms, since the exponents on $y$ are the same.

**3. a.** $-3y + 11y = (-3 + 11)y = 8y$

**b.** $4x^2 + x^2 = 4x^2 + 1x^2 = (4+1)x^2 = 5x^2$

**c.** $5x - 3x^2 + 8x^2 = 5x + (-3+8)x^2 = 5x + 5x^2$

**d.** $20y^2 + 2y^2 - y^2 = 20y^2 + 2y^2 - 1y^2$
$= (20 + 2 - 1)y^2$
$= 21y^2$

**4. a.** $3y + 8y - 7 + 2 = (3+8)y + (-7+2) = 11y - 5$

**b.** $6x - 3 - x - 3 = 6x - 1x + (-3-3)$
$= (6-1)x + (-3-3)$
$= 5x - 6$

**c.** $\dfrac{3}{4}t - t = \dfrac{3}{4}t - 1t = \left(\dfrac{3}{4} - 1\right)t = -\dfrac{1}{4}t$

**d.** $9y + 3.2y + 10 + 3 = (9 + 3.2)y + (10 + 3)$
$= 12.2y + 13$

**e.** $5z - 3z^4$
These two terms cannot be combined because they are unlike terms.

**5. a.** $3(2x - 7) = 3(2x) + 3(-7) = 6x - 21$

**b.** $-5(x - 0.5z - 5)$
$= -5(x) + (-5)(-0.5z) + (-5)(-5)$
$= -5x + 2.5z + 25$

**c.** $-(2x - y + z - 2)$
$= -1(2x - y + z - 2)$
$= -1(2x) - 1(-y) - 1(z) - 1(-2)$
$= -2x + y - z + 2$

**6. a.** $4(9x + 1) + 6 = 36x + 4 + 6 = 36x + 10$

**b.** $-7(2x - 1) - (6 - 3x) = -14x + 7 - 6 + 3x$
$= -11x + 1$

**c.** $8 - 5(6x + 5) = 8 - 30x - 25 = -30x - 17$

**7.** "Subtract $7x - 1$ from $2x + 3$" translates to
$(2x + 3) - (7x - 1) = 2x + 3 - 7x + 1 = -5x + 4$

**8. a.**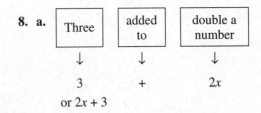

or $2x + 3$

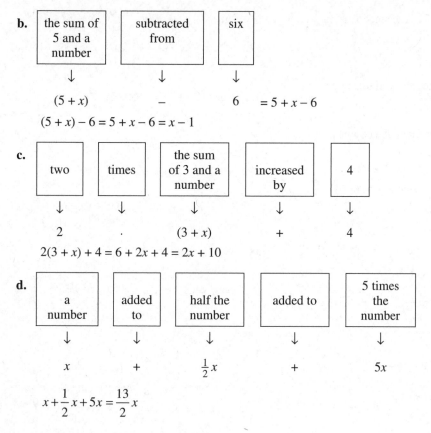

**b.**

| the sum of 5 and a number | subtracted from | six |
|---|---|---|
| ↓ | ↓ | ↓ |
| $(5 + x)$ | $-$ | $6$ |

$= 5 + x - 6$

$(5 + x) - 6 = 5 + x - 6 = x - 1$

**c.**

| two | times | the sum of 3 and a number | increased by | 4 |
|---|---|---|---|---|
| ↓ | ↓ | ↓ | ↓ | ↓ |
| 2 | · | $(3 + x)$ | $+$ | 4 |

$2(3 + x) + 4 = 6 + 2x + 4 = 2x + 10$

**d.**

| a number | added to | half the number | added to | 5 times the number |
|---|---|---|---|---|
| ↓ | ↓ | ↓ | ↓ | ↓ |
| $x$ | $+$ | $\frac{1}{2}x$ | $+$ | $5x$ |

$x + \dfrac{1}{2}x + 5x = \dfrac{13}{2}x$

## Vocabulary, Readiness & Video Check 2.1

1. $23y^2 + 10y - 6$ is called an <u>expression</u> while $23y^2$, $10y$, and $-6$ are each called a <u>term</u>.

2. To simplify $x + 4x$, we <u>combine like terms</u>.

3. The term $y$ has an understood <u>numerical coefficient</u> of 1.

4. The terms $7z$ and $7y$ are <u>unlike</u> terms and the terms $7z$ and $-z$ are <u>like</u> terms.

5. For the term $-\dfrac{1}{2}xy^2$, the number $-\dfrac{1}{2}$ is the <u>numerical coefficient</u>.

6. $5(3x - y)$ equals $15x - 5y$ by the <u>distributive</u> property.

7. Although these terms have exactly the same variables, the exponents on each are not exactly the same—the exponents on $x$ differ in each term.

8. distributive property

9. $-1$

10. The sum of 5 times a number and $-2$, plus 7 times the number; $5x + (-2) + 7x$; because there are like terms.

## Exercise Set 2.1

1. The numerical coefficient of $-7y$ is $-7$.

**3.** The numerical coefficient of $x$ is 1, since $x = 1x$.

**5.** The numerical coefficient of $17x^2y$ is 17.

**7.** $5y$ and $-y$ are like terms, since the variable and its exponent match.

**9.** $2z$ and $3z^2$ are unlike terms, since the exponents on $z$ are not the same.

**11.** $8wz$ and $\frac{1}{7}zw$ are like terms, since $wz = zw$ by the commutative property.

**13.** $7y + 8y = (7 + 8)y = 15y$

**15.** $8w - w + 6w = (8 - 1 + 6)w = 13w$

**17.** $3b - 5 - 10b - 4 = 3b - 10b - 5 - 4$
$$= (3 - 10)b - 9$$
$$= -7b - 9$$

**19.** $m - 4m + 2m - 6 = (1 - 4 + 2)m - 6 = -m - 6$

**21.** $5g - 3 - 5 - 5g = (5g - 5g) + (-3 - 5)$
$$= (5 - 5)g + (-8)$$
$$= 0g - 8$$
$$= -8$$

**23.** $6.2x - 4 + x - 1.2 = 6.2x + x - 4 - 1.2$
$$= (6.2 + 1)x - 5.2$$
$$= 7.2x - 5.2$$

**25.** $6x - 5x + x - 3 + 2x = 6x - 5x + x + 2x - 3$
$$= (6 - 5 + 1 + 2)x - 3$$
$$= 4x - 3$$

**27.** $7x^2 + 8x^2 - 10x^2 = (7 + 8 - 10)x^2 = 5x^2$

**29.** $6x + 0.5 - 4.3x - 0.4x + 3$
$$= 6x - 4.3x - 0.4x + 0.5 + 3$$
$$= (6 - 4.3 - 0.4)x + (0.5 + 3)$$
$$= 1.3x + 3.5$$

**31.** $5(y - 4) = 5(y) - 5(4) = 5y - 20$

**33.** $-2(x + 2) = -2(x) + (-2)(2) = -2x - 4$

**35.** $7(d - 3) + 10 = 7d - 21 + 10 = 7d - 11$

**37.** $-5(2x - 3y + 6) = -5(2x) - (-5)(3y) + (-5)(6)$
$$= -10x + 15y - 30$$

**39.** $-(3x - 2y + 1) = -3x + 2y - 1$

**41.** $5(x + 2) - (3x - 4) = 5x + 10 - 3x + 4$
$$= 2x + 14$$

**43.** $\quad 6x + 7 \quad$ added to $\quad 4x - 10$
$$\downarrow \qquad\qquad \downarrow \qquad\qquad \downarrow$$
$$(6x + 7) \quad + \quad (4x - 10) = 6x + 4x + 7 - 10$$
$$= 10x - 3$$

**45.** $\quad 3x - 8 \quad$ minus $\quad 7x + 1$
$$\downarrow \qquad\qquad \downarrow \qquad\qquad \downarrow$$
$$(3x - 8) \quad - \quad (7x + 1) = 3x - 8 - 7x - 1$$
$$= 3x - 7x - 8 - 1$$
$$= -4x - 9$$

**47.** $\quad m - 9 \quad$ minus $\quad 5m - 6$
$$\downarrow \qquad\qquad \downarrow \qquad\qquad \downarrow$$
$$(m - 9) \quad - \quad (5m - 6) = m - 9 - 5m + 6$$
$$= m - 5m - 9 + 6$$
$$= -4m - 3$$

**49.** $2k - k - 6 = (2 - 1)k - 6 = k - 6$

**51.** $-9x + 4x + 18 - 10x = -9x + 4x - 10x + 18$
$$= (-9 + 4 - 10)x + 18$$
$$= -15x + 18$$

**53.** $-4(3y - 4) + 12y = -4(3y) - (-4)(4) + 12y$
$$= -12y + 16 + 12y$$
$$= -12y + 12y + 16$$
$$= 16$$

**55.** $3(2x - 5) - 5(x - 4) = 6x - 15 - 5x + 20 = x + 5$

**57.** $-2(3x - 4) + 7x - 6 = -6x + 8 + 7x - 6 = x + 2$

**59.** $5k - (3k - 10) = 5k - 3k + 10 = 2k + 10$

**61.** $(3x + 4) - (6x - 1) = 3x + 4 - 6x + 1 = -3x + 5$

**63.** $3.4m - 4 - 3.4m - 7 = 3.4m - 3.4m - 4 - 7 = -11$

**65.** $\frac{1}{3}(7y-1)+\frac{1}{6}(4y+7)=\frac{7}{3}y-\frac{1}{3}+\frac{4}{6}y+\frac{7}{6}$

$$=\frac{7}{3}y+\frac{2}{3}y-\frac{1}{3}+\frac{7}{6}$$

$$=\frac{9}{3}y-\frac{2}{6}+\frac{7}{6}$$

$$=3y+\frac{5}{6}$$

**67.** $2+4(6x-6)=2+24x-24=-22+24x$

**69.** $0.5(m+2)+0.4m=0.5m+1+0.4m=0.9m+1$

**71.** $10-3(2x+3y)=10-6x-9y$

**73.** $6(3x-6)-2(x+1)-17x=18x-36-2x-2-17x$

$$=18x-2x-17x-36-2$$

$$=-x-38$$

**75.** $\frac{1}{2}(12x-4)-(x+5)=6x-2-x-5=5x-7$

**77.**

| twice a number | decreased by | 4 |
|:---:|:---:|:---:|
| ↓ | ↓ | ↓ |
| $2x$ | − | 4 |

**79.**

| seven | added to | double a number |
|:---:|:---:|:---:|
| ↓ | ↓ | ↓ |
| 7 | + | $2x$ |

**81.**

| three-fourths of a number | increased by | 12 |
|:---:|:---:|:---:|
| ↓ | ↓ | ↓ |
| $\frac{3}{4}x$ | + | 12 |

**83.**

| 5 times a number | added to | −2 | added to | 7 times the number |
|:---:|:---:|:---:|:---:|:---:|
| ↓ | ↓ | ↓ | ↓ | ↓ |
| $5x$ | + | −2 | + | $7x$ |

$5x+(-2)+7x=12x-2$

**85.**

| 8 | times | the sum of a number and 6 |
|:---:|:---:|:---:|
| ↓ | ↓ | ↓ |
| 8 | · | $(x+6)$ |

$8(x+6)=8x+48$

**87.**

| double a number | minus | the sum of the number and 10 |
|:---:|:---:|:---:|
| ↓ | ↓ | ↓ |
| $2x$ | − | $(x+10)$ |

$2x - (x + 10) = 2x - x - 10 = x - 10$

**89.**

| 2 | added to | 3 times a number | added to | −9 | added to | 4 times the number |
|:---:|:---:|:---:|:---:|:---:|:---:|:---:|
| ↓ | ↓ | ↓ | ↓ | ↓ | ↓ | ↓ |
| 2 | + | $3x$ | + | −9 | + | $4x$ |

$2 + 3x + (-9) + 4x = 7x - 7$

**91.** $y - x^2 = 3 - (-1)^2 = 3 - 1 = 2$

**93.** $a - b^2 = 2 - (-5)^2 = 2 - 25 = -23$

**95.** $yz - y^2 = (-5)(0) - (-5)^2 = 0 - 25 = -25$

**97.** $5x + (4x - 1) + 5x + (4x - 1) = 5x + 4x - 1 + 5x + 4x - 1$
$$= 18x - 2$$
The perimeter is $(18x - 2)$ feet.

**99.** 1 cone + 1 cylinder $\overset{?}{=}$ 3 cubes
   1 cube + 2 cubes $\overset{?}{=}$ 3 cubes
      3 cubes = 3 cubes: Balanced

**101.** 2 cylinders + 1 cube $\overset{?}{=}$ 3 cones + 2 cubes
   $2 \cdot 2$ cubes + 1 cube $\overset{?}{=}$ 3 cubes + 2 cubes
      4 cubes + 1 cube $\overset{?}{=}$ 3 cubes + 2 cubes
         5 cubes = 5 cubes: Balanced

**103.** answers may vary

**105.** $12(x + 2) + (3x - 1) = 12x + 24 + 3x - 1 = 15x + 23$
The total length is $(15x + 23)$ inches.

**107.** answers may vary

**109.** $5b^2c^3 + 8b^3c^2 - 7b^3c^2 = 5b^2c^3 + b^3c^2$

**111.** $3x - (2x^2 - 6x) + 7x^2 = 3x - 2x^2 + 6x + 7x^2$
$$= 5x^2 + 9x$$

**113.** $-(2x^2y + 3z) + 3z - 5x^2y = -2x^2y - 3z + 3z - 5x^2y$
$$= -7x^2y$$

**Section 2.2 Practice**

1.      $x+3=-5$
$$x+3-3=-5-3$$
$$x=-8$$
Check:   $x+3=-5$
$$-8+3\overset{?}{=}-5$$
$$-5=-5$$
The solution is $-8$.

2.      $y-0.3=-2.1$
$$y-0.3+0.3=-2.1+0.3$$
$$y=-1.8$$
Check:     $y-0.3=-2.1$
$$-1.8-0.3\overset{?}{=}-2.1$$
$$-2.1=-2.1$$
The solution is $-1.8$.

3.         $\dfrac{2}{5}=x+\dfrac{3}{10}$
$$\dfrac{2}{5}-\dfrac{3}{10}=x+\dfrac{3}{10}-\dfrac{3}{10}$$
$$\dfrac{2}{2}\cdot\dfrac{2}{5}-\dfrac{3}{10}=x$$
$$\dfrac{4}{10}-\dfrac{3}{10}=x$$
$$\dfrac{1}{10}=x$$
Check: $\dfrac{2}{5}=x+\dfrac{3}{10}$
$$\dfrac{2}{5}\overset{?}{=}\dfrac{1}{10}+\dfrac{3}{10}$$
$$\dfrac{2}{5}=\dfrac{2}{5}$$
The solution is $\dfrac{1}{10}$.

4.      $4t+7=5t-3$
$$4t+7-4t=5t-3-4t$$
$$7=t-3$$
$$7+3=t-3+3$$
$$10=t$$
Check:     $4t+7=5t-3$
$$4(10)+7\overset{?}{=}5(10)-3$$
$$40+7\overset{?}{=}50-3$$
$$47=47$$
The solution is 10.

5.   $8x-5x-3+9=x+x+3-7$
$$3x+6=2x-4$$
$$3x+6-2x=2x-4-2x$$
$$x+6=-4$$
$$x+6-6=-4-6$$
$$x=-10$$
Check:
$$8x-5x-3+9=x+x+3-7$$
$$8(-10)-5(-10)-3+9\overset{?}{=}-10+(-10)+3-7$$
$$-80+50-3+9\overset{?}{=}-10+(-10)+3-7$$
$$-24=-24$$
The solution is $-10$.

6.      $4(2a-3)-(7a+4)=2$
$$4(2a)+4(-3)-7a-4=2$$
$$8a-12-7a-4=2$$
$$a-16=2$$
$$a-16+16=2+16$$
$$a=18$$
Check by replacing $a$ with 18 in the original equation.

7.       $12-x=20$
$$12-x-12=20-12$$
$$-x=8$$
$$x=-8$$
Check:     $12-x=20$
$$12-(-8)\overset{?}{=}20$$
$$20=20$$
The solution is $-8$.

8. **a.**    The other number is $9-2=7$.

    **b.**    The other number is $9-x$.

    **c.**    The other piece has length $(9-x)$ feet.

9.   The speed of the French TGV is $(s-3.8)$ mph.

**Vocabulary, Readiness & Video Check 2.2**

1. The difference between an equation and an expression is that an <u>equation</u> contains an equal sign, whereas an <u>expression</u> does not.

2. <u>Equivalent</u> equations are equations that have the same solution.

3. A value of the variable that makes the equation a true statement is called a <u>solution</u> of the equation.

   

**4.** The process of finding the solution of an equation is called <u>solving</u> the equation for the variable.

**5.** By the <u>addition</u> property of equality, $x = -2$ and $x + 10 = -2 + 10$ are equivalent equations.

**6.** The equations $x = \dfrac{1}{2}$ and $\dfrac{1}{2} = x$ are equivalent equations. The statement is true.

**7.** The addition property of equality means that if we have an equation, we can add the same real number to <u>both sides</u> of the equation and have an equivalent equation.

**8.** To confirm our solution, we replace the variable with the solution in the original equation to make sure we have a true statement.

**9.** $\dfrac{1}{7}x$

**Exercise Set 2.2**

**1.**
$$x + 7 = 10$$
$$x + 7 - 7 = 10 - 7$$
$$x = 3$$
Check: $x + 7 = 10$
$$3 + 7 \stackrel{?}{=} 10$$
$$10 = 10$$
The solution is 3.

**3.**
$$x - 2 = -4$$
$$x - 2 + 2 = -4 + 2$$
$$x = -2$$
Check: $x - 2 = -4$
$$-2 - 2 \stackrel{?}{=} -4$$
$$-4 = -4$$
The solution is $-2$.

**5.**
$$-2 = t - 5$$
$$-2 + 5 = t - 5 + 5$$
$$3 = t$$
Check: $-2 = t - 5$
$$-2 \stackrel{?}{=} 3 - 5$$
$$-2 = -2$$
The solution is 3.

**7.**
$$r - 8.6 = -8.1$$
$$r - 8.6 + 8.6 = -8.1 + 8.6$$
$$r = 0.5$$
Check: $x - 8.6 = -8.1$
$$0.5 - 8.6 \stackrel{?}{=} -8.1$$
$$-8.1 = -8.1$$
The solution is 0.5.

**9.**
$$\frac{3}{4} = \frac{1}{3} + f$$
$$\frac{3}{4} - \frac{1}{3} = \frac{1}{3} - \frac{1}{3} + f$$
$$\frac{9}{12} - \frac{4}{12} = f$$
$$\frac{5}{12} = f$$
Check: $\dfrac{3}{4} = \dfrac{1}{3} + f$
$$\frac{3}{4} \stackrel{?}{=} \frac{1}{3} + \frac{5}{12}$$
$$\frac{3}{4} \stackrel{?}{=} \frac{4}{12} + \frac{5}{12}$$
$$\frac{3}{4} \stackrel{?}{=} \frac{9}{12}$$
$$\frac{3}{4} = \frac{3}{4}$$
The solution is $\dfrac{5}{12}$.

**11.**
$$5b - 0.7 = 6b$$
$$5b - 5b - 0.7 = 6b - 5b$$
$$-0.7 = b$$
Check: $\qquad 5b - 0.7 = 6b$
$$5(-0.7) - 0.7 \stackrel{?}{=} 6(-0.7)$$
$$-3.5 - 0.7 \stackrel{?}{=} -4.2$$
$$-4.2 = -4.2$$
The solution is $-0.7$.

**13.**
$$7x - 3 = 6x$$
$$7x - 6x - 3 = 6x - 6x$$
$$x - 3 = 0$$
$$x - 3 + 3 = 0 + 3$$
$$x = 3$$
Check: $7x - 3 = 6x$
$$7(3) - 3 \stackrel{?}{=} 6(3)$$
$$21 - 3 \stackrel{?}{=} 18$$
$$18 = 18$$
The solution is 3.

**15.**
$$y + \frac{11}{25} = -\frac{3}{25}$$
$$y + \frac{11}{25} - \frac{11}{25} = -\frac{3}{25} - \frac{11}{25}$$
$$y = -\frac{14}{25}$$

Check:     $y + \frac{11}{25} = -\frac{3}{25}$

$$-\frac{14}{25} + \frac{11}{25} \stackrel{?}{=} -\frac{3}{25}$$
$$-\frac{3}{25} = -\frac{3}{25}$$

The solution is $-\frac{14}{25}$.

**17.** $7x + 2x = 8x - 3$
$$9x = 8x - 3$$
$$9x - 8x = 8x - 8x - 3$$
$$x = -3$$
Check:          $7x + 2x = 8x - 3$
$$7(-3) + 2(-3) \stackrel{?}{=} 8(-3) - 3$$
$$-21 - 6 \stackrel{?}{=} -24 - 3$$
$$-27 = -27$$
The solution is $-3$.

**19.** $\frac{5}{6}x + \frac{1}{6}x = -9$
$$\frac{6}{6}x = -9$$
$$x = -9$$

Check:          $\frac{5}{6}x + \frac{1}{6}x = -9$
$$\frac{5}{6}(-9) + \frac{1}{6}(-9) \stackrel{?}{=} -9$$
$$-\frac{45}{6} - \frac{9}{6} \stackrel{?}{=} -9$$
$$-9 = -9$$
The solution is $-9$.

**21.**    $2y + 10 = 5y - 4y$
$$2y + 10 = y$$
$$2y - y + 10 = y - y$$
$$y + 10 = 0$$
$$y + 10 - 10 = 0 - 10$$
$$y = -10$$
Check:          $2y + 10 = 5y - 4y$
$$2(-10) + 10 \stackrel{?}{=} 5(-10) - 4(-10)$$
$$-20 + 10 \stackrel{?}{=} -50 + 40$$
$$-10 = -10$$
The solution is $-10$.

**23.**          $-5(n - 2) = 8 - 4n$
$$-5n + 10 = 8 - 4n$$
$$-5n + 5n + 10 = 8 - 4n + 5n$$
$$10 = 8 + n$$
$$10 - 8 = 8 - 8 + n$$
$$2 = n$$
Check:  $-5(n - 2) = 8 - 4n$
$$-5(2 - 2) \stackrel{?}{=} 8 - 4(2)$$
$$-5(0) \stackrel{?}{=} 8 - 8$$
$$0 = 0$$
The solution is 2.

**25.**          $\frac{3}{7}x + 2 = -\frac{4}{7}x - 5$
$$\frac{3}{7}x + 2 + \frac{4}{7}x = -\frac{4}{7}x - 5 + \frac{4}{7}x$$
$$\frac{7}{7}x + 2 = -5$$
$$x + 2 - 2 = -5 - 2$$
$$x = -7$$
Check:     $\frac{3}{7}x + 2 = -\frac{4}{7}x - 5$
$$\frac{3}{7}(-7) + 2 \stackrel{?}{=} -\frac{4}{7}(-7) - 5$$
$$-3 + 2 \stackrel{?}{=} 4 - 5$$
$$-1 = -1$$
The solution is $-7$.

**27.**          $5x - 6 = 6x - 5$
$$5x - 6 - 5x = 6x - 5 - 5x$$
$$-6 = x - 5$$
$$-6 + 5 = x - 5 + 5$$
$$-1 = x$$
Check:     $5x - 6 = 6x - 5$
$$5(-1) - 6 \stackrel{?}{=} 6(-1) - 5$$
$$-5 - 6 \stackrel{?}{=} -6 - 5$$
$$-11 = -11$$
The solution is $-1$.

**29.** $8y + 2 - 6y = 3 + y - 10$
$$2y + 2 = y - 7$$
$$2y - y + 2 = y - y - 7$$
$$y + 2 = -7$$
$$y + 2 - 2 = -7 - 2$$
$$y = -9$$
Check:          $8y + 2 - 6y = 3 + y - 10$
$$8(-9) + 2 - 6(-9) \stackrel{?}{=} 3 + (-9) - 10$$
$$-72 + 2 + 54 \stackrel{?}{=} -16$$
$$-16 = -16$$
The solution is $-9$.

**31.**
$$-3(x-4) = -4x$$
$$-3x+12 = -4x$$
$$-3x+12+3x = -4x+3x$$
$$12 = -x$$
$$x = -12$$
Check: $-3(x-4) = -4x$
$$-3(-12-4) \stackrel{?}{=} -4(-12)$$
$$-3(-16) \stackrel{?}{=} 48$$
$$48 = 48$$
The solution is $-12$.

**33.**
$$\frac{3}{8}x - \frac{1}{6} = -\frac{5}{8}x - \frac{2}{3}$$
$$\frac{3}{8}x + \frac{5}{8}x - \frac{1}{6} = -\frac{5}{8}x + \frac{5}{8}x - \frac{2}{3}$$
$$\frac{8}{8}x - \frac{1}{6} = -\frac{2}{3}$$
$$x - \frac{1}{6} + \frac{1}{6} = -\frac{2}{3} + \frac{1}{6}$$
$$x = -\frac{4}{6} + \frac{1}{6}$$
$$x = -\frac{3}{6}$$
$$x = -\frac{1}{2}$$

Check:
$$\frac{3}{8}x - \frac{1}{6} = -\frac{5}{8}x - \frac{2}{3}$$
$$\frac{3}{8}\left(-\frac{1}{2}\right) - \frac{1}{6} \stackrel{?}{=} -\frac{5}{8}\left(-\frac{1}{2}\right) - \frac{2}{3}$$
$$-\frac{3}{16} - \frac{1}{6} \stackrel{?}{=} \frac{5}{16} - \frac{2}{3}$$
$$-\frac{9}{48} - \frac{8}{48} \stackrel{?}{=} \frac{15}{48} - \frac{32}{48}$$
$$-\frac{17}{48} = -\frac{17}{48}$$

The solution is $-\frac{1}{2}$.

**35.**
$$2(x-4) = x+3$$
$$2x-8 = x+3$$
$$2x-x-8 = x-x+3$$
$$x-8 = 3$$
$$x-8+8 = 3+8$$
$$x = 11$$
Check: $2(x-4) = x+3$
$$2(11-4) \stackrel{?}{=} 11+3$$
$$2(7) \stackrel{?}{=} 14$$
$$14 = 14$$
The solution is 11.

**37.**
$$3(n-5) - (6-2n) = 4n$$
$$3n-15-6+2n = 4n$$
$$5n-21 = 4n$$
$$5n-4n-21 = 4n-4n$$
$$n-21 = 0$$
$$n-21+21 = 0+21$$
$$n = 21$$
Check: $3(n-5) - (6-2n) = 4n$
$$3(21-5) - (6-2(21)) \stackrel{?}{=} 4(21)$$
$$3(16) - (6-42) \stackrel{?}{=} 84$$
$$48 - (-36) \stackrel{?}{=} 84$$
$$84 = 84$$
The solution is 21.

**39.**
$$-2(x+6) + 3(2x-5) = 3(x-4) + 10$$
$$-2x-12+6x-15 = 3x-12+10$$
$$4x-27 = 3x-2$$
$$4x-3x-27 = 3x-3x-2$$
$$x-27 = -2$$
$$x-27+27 = -2+27$$
$$x = 25$$
Check: $-2(x+6) + 3(2x-5) = 3(x-4)+10$
$$-2(25+6) + 3(2(25)-5) \stackrel{?}{=} 3(25-4)+10$$
$$-2(31) + 3(50-5) \stackrel{?}{=} 3(21)+10$$
$$-62 + 3(45) \stackrel{?}{=} 63+10$$
$$-62 + 135 \stackrel{?}{=} 73$$
$$73 = 73$$
The solution is 25.

**41.**
$$-11 = 3+x$$
$$-11-3 = 3+x-3$$
$$-14 = x$$

**43.**
$$x - \frac{2}{5} = -\frac{3}{20}$$
$$x - \frac{2}{5} + \frac{2}{5} = -\frac{3}{20} + \frac{2}{5}$$
$$x = -\frac{3}{20} + \frac{8}{20}$$
$$x = \frac{5}{20}$$
$$x = \frac{1}{4}$$

**45.**
$$3x-6 = 2x+5$$
$$3x-2x-6 = 2x-2x+5$$
$$x-6 = 5$$
$$x-6+6 = 5+6$$
$$x = 11$$

**47.** $13x - 9 + 2x - 5 = 12x - 1 + 2x$
$15x - 14 = 14x - 1$
$15x - 14 - 14x = 14x - 1 - 14x$
$x - 14 = -1$
$x - 14 + 14 = -1 + 14$
$x = 13$

**49.** $7(6 + w) = 6(2 + w)$
$42 + 7w = 12 + 6w$
$42 + 7w - 6w = 12 + 6w - 6w$
$42 + w = 12$
$42 - 42 + w = 12 - 42$
$w = -30$

**51.** $n + 4 = 3.6$
$n + 4 - 4 = 3.6 - 4$
$n = -0.4$

**53.** $10 - (2x - 4) = 7 - 3x$
$10 - 2x + 4 = 7 - 3x$
$14 - 2x = 7 - 3x$
$14 - 2x + 3x = 7 - 3x + 3x$
$14 + x = 7$
$14 - 14 + x = 7 - 14$
$x = -7$

**55.** $\dfrac{1}{3} = x + \dfrac{2}{3}$
$\dfrac{1}{3} - \dfrac{2}{3} = x + \dfrac{2}{3} - \dfrac{2}{3}$
$-\dfrac{1}{3} = x$

**57.** $-6.5 - 4x - 1.6 - 3x = -6x + 9.8$
$-8.1 - 7x = -6x + 9.8$
$-8.1 - 7x + 7x = -6x + 7x + 9.8$
$-8.1 = x + 9.8$
$-8.1 - 9.8 = x + 9.8 - 9.8$
$-17.9 = x$

**59.** $-3\left(x - \dfrac{1}{4}\right) = -4x$
$-3x + \dfrac{3}{4} = -4x$
$-3x + 4x + \dfrac{3}{4} = -4x + 4x$
$x + \dfrac{3}{4} = 0$
$x + \dfrac{3}{4} - \dfrac{3}{4} = 0 - \dfrac{3}{4}$
$x = -\dfrac{3}{4}$

**61.** $7(m - 2) - 6(m + 1) = -20$
$7m - 14 - 6m - 6 = -20$
$m - 20 = -20$
$m - 20 + 20 = -20 + 20$
$m = 0$

**63.** $0.8t + 0.2(t - 0.4) = 1.75$
$0.8t + 0.2t - 0.08 = 1.75$
$1t - 0.08 = 1.75$
$t - 0.08 + 0.08 = 1.75 + 0.08$
$t = 1.83$

**65.** The other number is $20 - p$.

**67.** The length of the other piece is $(10 - x)$ feet.

**69.** The supplement of the angle $x°$ is $(180 - x)°$.

**71.** If the number of undergraduate students is $n$, and the number of graduate students is 28,000 fewer than $n$, then the number of graduate students is $n - 28{,}000$.

**73.** The area of the Sahara Desert is $7x$ square miles.

**75.** The reciprocal of $\dfrac{5}{8}$ is $\dfrac{8}{5}$ since $\dfrac{5}{8} \cdot \dfrac{8}{5} = 1$.

**77.** The reciprocal of 2 is $\dfrac{1}{2}$ since $2 \cdot \dfrac{1}{2} = 1$.

**79.** The reciprocal of $-\dfrac{1}{9}$ is $-9$ since $-\dfrac{1}{9} \cdot -9 = 1$.

**81.** $\dfrac{3x}{3} = x$

**83.** $-5\left(-\dfrac{1}{5}y\right) = y$

**85.** $\dfrac{3}{5}\left(\dfrac{5}{3}x\right) = x$

**87.** $180 - [x + (2x + 7)] = 180 - [x + 2x + 7]$
$$= 180 - [3x + 7]$$
$$= 180 - 3x - 7$$
$$= 173 - 3x$$
The third angle is $(173 - 3x)°$.

**89.** answers may vary

**91.** answers may vary

**93.** $200 + 150 + 400 + x = 1000$
$$750 + x = 1000$$
$$750 - 750 + x = 1000 - 750$$
$$x = 250$$
The fluid needed by the patient is 250 ml.

**95.** answers may vary

**97.** $x - 4 = -9$
$$x - 4 + (4) = -9 + (4)$$
$$x = -5$$
The answer is 4.

**99.** answers may vary

**101.** Check $x = 0.05$.
$$1.23x - 0.06 = 2.6x - 0.1285$$
$$1.23(0.05) - 0.06 \stackrel{?}{=} 2.6(0.05) - 0.1285$$
$$0.0615 - 0.06 \stackrel{?}{=} 0.13 - 0.1285$$
$$0.0015 = 0.0015$$
Solution

**103.** Check $a = 6.3$.
$$3(a + 4.6) = 5a + 2.5$$
$$3(6.3 + 4.6) \stackrel{?}{=} 5(6.3) + 2.5$$
$$3(10.9) \stackrel{?}{=} 31.5 + 2.5$$
$$32.7 = 34$$
Not a solution

## Section 2.3 Practice

**1.**    $\dfrac{4}{5}x = 16$
$$\dfrac{5}{4} \cdot \dfrac{4}{5}x = \dfrac{5}{4} \cdot 16$$
$$\left(\dfrac{5}{4} \cdot \dfrac{4}{5}\right)x = \dfrac{5}{4} \cdot 16$$
$$1x = 20$$
$$x = 20$$

Check:    $\dfrac{4}{5}x = 16$
$$\dfrac{4}{5} \cdot 20 \stackrel{?}{=} 16$$
$$16 = 16$$
The solution is 20.

**2.** $8x = -96$
$$\dfrac{8x}{8} = \dfrac{-96}{8}$$
$$x = -12$$
Check:     $8x = -96$
$$8(-12) \stackrel{?}{=} -96$$
$$-96 = -96$$
The solution is $-12$.

**3.**    $\dfrac{x}{5} = 13$
$$5 \cdot \dfrac{x}{5} = 5 \cdot 13$$
$$x = 65$$

Check:    $\dfrac{x}{5} = 13$
$$\dfrac{65}{5} \stackrel{?}{=} 13$$
$$13 = 13$$
The solution is 65.

**4.** $2.7x = 4.05$
$$\dfrac{2.7x}{2.7} = \dfrac{4.05}{2.7}$$
$$x = 1.5$$
The solution is 1.5.
Check by replacing $x$ with 1.5 in the original equation.

**5.**
$$-\frac{5}{3}x = \frac{4}{7}$$
$$-\frac{3}{5} \cdot -\frac{5}{3}x = -\frac{3}{5} \cdot \frac{4}{7}$$
$$x = -\frac{12}{35}$$

Check by replacing $x$ with $-\dfrac{12}{35}$ in the original equation. The solution is $-\dfrac{12}{35}$.

**6.**
$$-y + 3 = -8$$
$$-y + 3 - 3 = -8 - 3$$
$$-y = -11$$
$$\frac{-y}{-1} = \frac{-11}{-1}$$
$$y = 11$$

To check, replace $y$ with 11 in the original equation. The solution is 11.

**7.**
$$6b - 11b = 18 + 2b - 6 + 9$$
$$-5b = 21 + 2b$$
$$-5b - 2b = 21 + 2b - 2b$$
$$-7b = 21$$
$$\frac{-7b}{-7} = \frac{21}{-7}$$
$$b = -3$$

Check by replacing $b$ with $-3$ in the original equation. The solution is $-3$.

**8.**
$$10x - 4 = 7x + 14$$
$$10x - 4 - 7x = 7x + 14 - 7x$$
$$3x - 4 = 14$$
$$3x - 4 + 4 = 14 + 4$$
$$3x = 18$$
$$\frac{3x}{3} = \frac{18}{3}$$
$$x = 6$$

To check, replace $x$ with 6 in the original equation to see that a true statement results. The solution is 6.

**9.**
$$4(3x - 2) = -1 + 4$$
$$4(3x) - 4(2) = -1 + 4$$
$$12x - 8 = 3$$
$$12x - 8 + 8 = 3 + 8$$
$$12x = 11$$
$$\frac{12x}{12} = \frac{11}{12}$$
$$x = \frac{11}{12}$$

To check, replace $x$ with $\dfrac{11}{12}$ in the original equation to see that a true statement results. The solution is $\dfrac{11}{12}$.

**10.** Let $x$ = first integer.
$x + 2$ = second even integer.
$x + 4$ = third even integer.
$x + (x + 2) + (x + 4) = 3x + 6$

**Vocabulary, Readiness & Video Check 2.3**

**1.** By the <u>multiplication</u> property of equality,
$y = \dfrac{1}{2}$ and $5 \cdot y = 5 \cdot \dfrac{1}{2}$ are equivalent equations.

**2.** The equations $\dfrac{z}{4} = 10$ and $4 \cdot \dfrac{z}{4} = 10$ are not equivalent equations. The statement is false.

**3.** The equations $-7x = 30$ and $\dfrac{-7x}{-7} = \dfrac{30}{7}$ are not equivalent equations. The statement is false.

**4.** By the <u>multiplication</u> property of equality,
$9x = -63$ and $\dfrac{9x}{9} = \dfrac{-63}{9}$ are equivalent equations.

**5.** We can multiply both sides of an equation by the <u>same</u> nonzero number and have an equivalent equation.

**6.** addition property; multiplication property; answers may vary

**7.** $(x + 1) + (x + 3) = 2x + 4$

**Exercise Set 2.3**

**1.**   $-5x = -20$

$$\frac{-5x}{-5} = \frac{-20}{-5}$$

$$x = 4$$

Check:   $-5x = -20$

$$-5(4) \stackrel{?}{=} -20$$

$$-20 = -20$$

The solution is 4.

**3.**   $3x = 0$

$$\frac{3x}{3} = \frac{0}{3}$$

$$x = 0$$

Check:   $3x = 0$

$$3(0) \stackrel{?}{=} 0$$

$$0 = 0$$

The solution is 0.

**5.**   $-x = -12$

$$\frac{-x}{-1} = \frac{-12}{-1}$$

$$x = 12$$

Check:   $-x = -12$

$$-(12) \stackrel{?}{=} -12$$

$$-12 = -12$$

The solution is 12.

**7.**       $\frac{2}{3}x = -8$

$$\frac{3}{2}\left(\frac{2}{3}x\right) = \frac{3}{2}(-8)$$

$$x = -12$$

Check:       $\frac{2}{3}x = -8$

$$\frac{2}{3}(-12) \stackrel{?}{=} -8$$

$$-8 = -8$$

The solution is $-12$.

**9.**       $\frac{1}{2} = \frac{1}{6}d$

$$6\left(\frac{1}{2}\right) = 6\left(\frac{1}{6}d\right)$$

$$3 = d$$

Check:   $\frac{1}{2} = \frac{1}{6}d$

$$\frac{1}{2} \stackrel{?}{=} \frac{1}{6}(3)$$

$$\frac{1}{2} = \frac{1}{2}$$

The solution is 3.

**11.**       $\frac{a}{2} = 1$

$$2\left(\frac{a}{2}\right) = 2(1)$$

$$a = 2$$

Check:   $\frac{a}{2} = 1$

$$\frac{2}{2} \stackrel{?}{=} 1$$

$$1 = 1$$

The solution is 2.

**13.**       $\frac{k}{-7} = 0$

$$-7\left(\frac{k}{-7}\right) = -7(0)$$

$$k = 0$$

Check:   $\frac{k}{-7} = 0$

$$\frac{0}{-7} \stackrel{?}{=} 0$$

$$0 = 0$$

The solution is 0.

**15.**   $10.71 = 1.7x$

$$\frac{10.71}{1.7} = \frac{1.7x}{1.7}$$

$$6.3 = x$$

Check:   $10.71 = 1.7x$

$$10.71 \stackrel{?}{=} 1.7(6.3)$$

$$10.71 = 10.71$$

The solution is 6.3.

**17.**       $2x - 4 = 16$

$$2x - 4 + 4 = 16 + 4$$

$$2x = 20$$

$$\frac{2x}{2} = \frac{20}{2}$$

$$x = 10$$

Check:     $2x - 4 = 16$

$2(10) - 4 \overset{?}{=} 16$

$20 - 4 \overset{?}{=} 16$

$16 = 16$

The solution is 10.

**19.**     $-x + 2 = 22$

$-x + 2 - 2 = 22 - 2$

$-x = 20$

$x = -20$

Check:     $-x + 2 = 22$

$-(-20) + 2 \overset{?}{=} 22$

$20 + 2 \overset{?}{=} 22$

$22 = 22$

The solution is $-20$.

**21.**     $6a + 3 = 3$

$6a + 3 - 3 = 3 - 3$

$6a = 0$

$\dfrac{6a}{6} = \dfrac{0}{6}$

$a = 0$

Check:     $6a + 3 = 3$

$6(0) + 3 \overset{?}{=} 3$

$0 + 3 \overset{?}{=} 3$

$3 = 3$

The solution is 0.

**23.**     $\dfrac{x}{3} - 2 = -5$

$\dfrac{x}{3} - 2 + 2 = -5 + 2$

$\dfrac{x}{3} = -3$

$3 \cdot \dfrac{x}{3} = 3 \cdot -3$

$x = -9$

Check:     $\dfrac{x}{3} - 2 = -5$

$\dfrac{-9}{3} - 2 \overset{?}{=} -5$

$-3 - 2 \overset{?}{=} -5$

$-5 = -5$

The solution is $-9$.

**25.**     $6z - z = -2 + 2z - 1 - 6$

$5z = 2z - 9$

$5z - 2z = 2z - 9 - 2z$

$3z = -9$

$\dfrac{3z}{3} = \dfrac{-9}{3}$

$z = -3$

Check:     $6z - z = -2 + 2z - 1 - 6$

$6(-3) - (-3) \overset{?}{=} -2 + 2(-3) - 1 - 6$

$-18 + 3 \overset{?}{=} -2 - 6 - 1 - 6$

$-15 = -15$

The solution is $-3$.

**27.**     $1 = 0.4x - 0.6x - 5$

$1 = -0.2x - 5$

$1 + 5 = -0.2x - 5 + 5$

$6 = -0.2x$

$\dfrac{6}{-0.2} = \dfrac{-0.2x}{-0.2}$

$-30 = x$

Check: $1 = 0.4x - 0.6x - 5$

$1 \overset{?}{=} 0.4(-30) - 0.6(-30) - 5$

$1 \overset{?}{=} -12 + 18 - 5$

$1 = 1$

The solution is $-30$.

**29.**     $\dfrac{2}{3}y - 11 = -9$

$\dfrac{2}{3}y - 11 + 11 = -9 + 11$

$\dfrac{2}{3}y = 2$

$\dfrac{3}{2} \cdot \dfrac{2}{3}y = \dfrac{3}{2} \cdot 2$

$y = 3$

Check:     $\dfrac{2}{3}y - 11 = -9$

$\dfrac{2}{3} \cdot 3 - 11 \overset{?}{=} -9$

$2 - 11 \overset{?}{=} -9$

$-9 = -9$

The solution is 3.

**31.** $\dfrac{3}{4}t - \dfrac{1}{2} = \dfrac{1}{3}$

$\dfrac{3}{4}t - \dfrac{1}{2} + \dfrac{1}{2} = \dfrac{1}{3} + \dfrac{1}{2}$

$\dfrac{3}{4}t = \dfrac{5}{6}$

$\dfrac{4}{3} \cdot \dfrac{3}{4}t = \dfrac{4}{3} \cdot \dfrac{5}{6}$

$t = \dfrac{10}{9}$

Check: $\dfrac{3}{4}t - \dfrac{1}{2} = \dfrac{1}{3}$

$\dfrac{3}{4} \cdot \dfrac{10}{9} - \dfrac{1}{2} \overset{?}{=} \dfrac{1}{3}$

$\dfrac{5}{6} - \dfrac{1}{2} \overset{?}{=} \dfrac{1}{3}$

$\dfrac{2}{6} \overset{?}{=} \dfrac{1}{3}$

$\dfrac{1}{3} = \dfrac{1}{3}$

The solution is $\dfrac{10}{9}$.

**33.** $8x + 20 = 6x + 18$

$8x + 20 - 6x = 6x + 18 - 6x$

$2x + 20 = 18$

$2x + 20 - 20 = 18 - 20$

$2x = -2$

$\dfrac{2x}{2} = \dfrac{-2}{2}$

$x = -1$

**35.** $3(2x + 5) = -18 + 9$

$6x + 15 = -9$

$6x + 15 - 15 = -9 - 15$

$6x = -24$

$\dfrac{6x}{6} = \dfrac{-24}{6}$

$x = -4$

**37.** $2x - 5 = 20x + 4$

$2x - 5 - 2x = 20x + 4 - 2x$

$-5 = 18x + 4$

$-5 - 4 = 18x + 4 - 4$

$-9 = 18x$

$\dfrac{-9}{18} = \dfrac{18x}{18}$

$-\dfrac{1}{2} = x$

**39.** $2 + 14 = -4(3x - 4)$

$16 = -12x + 16$

$16 - 16 = -12x + 16 - 16$

$0 = -12x$

$\dfrac{0}{-12} = \dfrac{-12x}{-12}$

$0 = x$

**41.** $-6y - 3 = -5y - 7$

$-6y - 3 + 6y = -5y - 7 + 6y$

$-3 = y - 7$

$-3 + 7 = y - 7 + 7$

$4 = y$

**43.** $\dfrac{1}{2}(2x - 1) = -\dfrac{1}{7} - \dfrac{3}{7}$

$x - \dfrac{1}{2} = -\dfrac{4}{7}$

$x - \dfrac{1}{2} + \dfrac{1}{2} = -\dfrac{4}{7} + \dfrac{1}{2}$

$x = -\dfrac{8}{14} + \dfrac{7}{14}$

$x = -\dfrac{1}{14}$

**45.** $-10z - 0.5 = -20z + 1.6$

$-10z - 0.5 + 20z = -20z + 1.6 + 20z$

$10z - 0.5 = 1.6$

$10z - 0.5 + 0.5 = 1.6 + 0.5$

$10z = 2.1$

$\dfrac{10z}{10} = \dfrac{2.1}{10}$

$z = 0.21$

**47.** $-4x + 20 = 4x - 20$

$-4x + 20 + 4x = 4x - 20 + 4x$

$20 = 8x - 20$

$20 + 20 = 8x - 20 + 20$

$40 = 8x$

$\dfrac{40}{8} = \dfrac{8x}{8}$

$5 = x$

**49.** $42 = 7x$

$\dfrac{42}{7} = \dfrac{7x}{7}$

$6 = x$

**51.**
$$4.4 = -0.8x$$
$$\frac{4.4}{-0.8} = \frac{-0.8x}{-0.8}$$
$$-5.5 = x$$

**53.**
$$6x + 10 = -20$$
$$6x + 10 - 10 = -20 - 10$$
$$6x = -30$$
$$\frac{6x}{6} = \frac{-30}{6}$$
$$x = -5$$

**55.**
$$5 - 0.3k = 5$$
$$5 - 5 - 0.3k = 5 - 5$$
$$-0.3k = 0$$
$$\frac{-0.3k}{-0.3} = \frac{0}{-0.3}$$
$$k = 0$$

**57.**
$$13x - 5 = 11x - 11$$
$$13x - 5 - 11x = 11x - 11 - 11x$$
$$2x - 5 = -11$$
$$2x - 5 + 5 = -11 + 5$$
$$2x = -6$$
$$\frac{2x}{2} = \frac{-6}{2}$$
$$x = -3$$

**59.**
$$9(3x + 1) = 4x - 5x$$
$$27x + 9 = -x$$
$$27x + 9 - 27x = -x - 27x$$
$$9 = -28x$$
$$\frac{9}{-28} = \frac{-28x}{-28}$$
$$-\frac{9}{28} = x$$

**61.**
$$-\frac{3}{7}p = -2$$
$$-\frac{7}{3}\left(-\frac{3}{7}p\right) = -\frac{7}{3}(-2)$$
$$p = \frac{14}{3}$$

**63.**
$$-\frac{4}{3}x = 12$$
$$-\frac{3}{4}\left(-\frac{4}{3}x\right) = -\frac{3}{4}(12)$$
$$x = -9$$

**65.**
$$-2x + \frac{1}{2} = \frac{7}{2}$$
$$-2x + \frac{1}{2} - \frac{1}{2} = \frac{7}{2} - \frac{1}{2}$$
$$-2x = \frac{6}{2}$$
$$-2x = 3$$
$$\frac{-2x}{-2} = \frac{3}{-2}$$
$$x = -\frac{3}{2}$$

**67.**
$$10 = 2x - 1 \quad \frac{11}{2}.$$
$$10 + 1 = 2x - 1 + 1$$
$$11 = 2x$$
$$\frac{11}{2} = \frac{2x}{2}$$
$$\frac{11}{2} = x$$

**69.**
$$10 - 3x - 6 - 9x = 7$$
$$4 - 12x = 7$$
$$4 - 4 - 12x = 7 - 4$$
$$-12x = 3$$
$$\frac{-12x}{-12} = \frac{3}{-12}$$
$$x = -\frac{1}{4}$$

**71.**
$$z - 5z = 7z - 9 - z$$
$$-4z = 6z - 9$$
$$-4z - 6z = 6z - 6z - 9$$
$$-10z = -9$$
$$\frac{-10z}{-10} = \frac{-9}{-10}$$
$$z = \frac{9}{10}$$

**73.**
$$-x - \frac{4}{5} = x + \frac{1}{2} + \frac{2}{5}$$
$$-x - \frac{4}{5} = x + \frac{9}{10}$$
$$-x - \frac{4}{5} + x = x + \frac{9}{10} + x$$
$$-\frac{4}{5} = 2x + \frac{9}{10}$$
$$-\frac{8}{10} - \frac{9}{10} = 2x + \frac{9}{10} - \frac{9}{10}$$
$$-\frac{17}{10} = 2x$$
$$\frac{1}{2} \cdot \left(-\frac{17}{10}\right) = \frac{1}{2} \cdot 2x$$
$$-\frac{17}{20} = x$$

**75.**
$$-15 + 37 = -2(x + 5)$$
$$22 = -2x - 10$$
$$22 + 10 = -2x - 10 + 10$$
$$32 = -2x$$
$$\frac{32}{-2} = \frac{-2x}{-2}$$
$$-16 = x$$

**77.** Sum = first integer + second integer
Sum = $x + (x + 2) = x + x + 2 = 2x + 2$

**79.** Sum = first integer + third integer
Sum = $x + (x + 2) = x + x + 2 = 2x + 2$

**81.** Let $x$ = first room number.
$x + 2$ = second room number
$x + 4$ = third room number
$x + 6$ = fourth room number
$x + 8$ = fifth room number
$x + (x + 2) + (x + 4) + (x + 6) + (x + 8) = 5x + 20$

**83.** $5x + 2(x - 6) = 5x + 2 \cdot x + 2 \cdot (-6)$
$$= 5x + 2x - 12$$
$$= 7x - 12$$

**85.** $6(2z + 4) + 20 = 6 \cdot 2z + 6 \cdot 4 + 20$
$$= 12z + 24 + 20$$
$$= 12z + 44$$

**87.** $-(3a - 3) + 2a - 6 = -3a + 3 + 2a - 6$
$$= -3a + 2a + 3 - 6$$
$$= -a - 3$$

**89.** $(-3)^2 = (-3)(-3) = 9$
$-3^2 = -3 \cdot 3 = -9$
$(-3)^2 > -3^2$

**91.** $(-2)^3 = (-2)(-2)(-2) = -8$
$-2^3 = -2 \cdot 2 \cdot 2 = -8$
$(-2)^3 = -2^3$

**93.**   $6x = $ _____
$6(-8) = $ _____
$-48 = $ _____

**95.** answers may vary

**97.** answers may vary

**99.** $9x = 2100$
$$\frac{9x}{9} = \frac{2100}{9}$$
$$x = \frac{700}{3}$$

Each dose should be $\frac{700}{3}$ milligrams.

**101.** $-3.6x = 10.62$
$$\frac{-3.6x}{-3.6} = \frac{10.62}{-3.6}$$
$$x = -2.95$$

**103.**   $7x - 5.06 = -4.92$
$$7x - 5.06 + 5.06 = -4.92 + 5.06$$
$$7x = 0.14$$
$$\frac{7x}{7} = \frac{0.14}{7}$$
$$x = 0.02$$

**Section 2.4 Practice**

**1.** $2(4a - 9) + 3 = 5a - 6$
$$8a - 18 + 3 = 5a - 6$$
$$8a - 15 = 5a - 6$$
$$8a - 15 - 5a = 5a - 6 - 5a$$
$$3a - 15 = -6$$
$$3a - 15 + 15 = -6 + 15$$
$$3a = 9$$
$$\frac{3a}{3} = \frac{9}{3}$$
$$a = 3$$

Check: $2(4a - 9) + 3 = 5a - 6$

$2[4(3) - 9] + 3 \overset{?}{=} 5(3) - 6$

$2(12 - 9) + 3 \overset{?}{=} 15 - 6$

$2(3) + 3 \overset{?}{=} 9$

$6 + 3 \overset{?}{=} 9$

$9 = 1$

The solution is 3 or the solution set is {3}.

**2.**　　$7(x - 3) = -6x$

$7x - 21 = -6x$

$7x - 21 - 7x = -6x - 7x$

$-21 = -13x$

$\dfrac{-21}{-13} = \dfrac{-13x}{-13}$

$\dfrac{21}{13} = x$

Check:　　$7(x - 3) = -6x$

$7\left(\dfrac{21}{13} - 3\right) \overset{?}{=} -6\left(\dfrac{21}{13}\right)$

$7\left(\dfrac{21}{13} - \dfrac{39}{13}\right) \overset{?}{=} -\dfrac{126}{13}$

$7\left(-\dfrac{18}{13}\right) \overset{?}{=} -\dfrac{126}{13}$

$-\dfrac{126}{13} = -\dfrac{126}{13}$

The solution is $\dfrac{21}{13}$.

**3.**　　$\dfrac{3}{5}x - 2 = \dfrac{2}{3}x - 1$

$15\left(\dfrac{3}{5}x - 2\right) = 15\left(\dfrac{2}{3}x - 1\right)$

$15\left(\dfrac{3}{5}x\right) - 15(2) = 15\left(\dfrac{2}{3}x\right) - 15(1)$

$9x - 30 = 10x - 15$

$9x - 30 - 9x = 10x - 15 - 9x$

$-30 = x - 15$

$-30 + 15 = x - 15 + 15$

$-15 = x$

Check:　　$\dfrac{3}{5}x - 2 = \dfrac{2}{3}x - 1$

$\dfrac{3}{5} \cdot -15 - 2 \overset{?}{=} \dfrac{2}{3} \cdot -15 - 1$

$-9 - 2 \overset{?}{=} -10 - 1$

$-11 = -11$

The solution is $-15$.

**4.**　　$\dfrac{4(y + 3)}{3} = 5y - 7$

$3 \cdot \dfrac{4(y + 3)}{3} = 3 \cdot (5y - 7)$

$4(y + 3) = 3(5y - 7)$

$4y + 12 = 15y - 21$

$4y + 12 - 4y = 15y - 21 - 4y$

$12 = 11y - 21$

$12 + 21 = 11y - 21 + 21$

$33 = 11y$

$\dfrac{33}{11} = \dfrac{11y}{11}$

$3 = y$

To check, replace $y$ with 3 in the original equation. The solution is 3.

**5.**　　$0.35x + 0.09(x + 4) = 0.30(12)$

$100[0.35x + 0.09(x + 4)] = 100[0.03(12)]$

$35x + 9(x + 4) = 3(12)$

$35x + 9x + 36 = 36$

$44x + 36 = 36$

$44x + 36 - 36 = 36 - 36$

$44x = 0$

$\dfrac{44x}{44} = \dfrac{0}{44}$

$x = 0$

To check, replace $x$ with 0 in the original equation. The solution is 0.

**6.**　　$4(x + 4) - x = 2(x + 11) + x$

$4x + 16 - x = 2x + 22 + x$

$3x + 16 = 3x + 22$

$3x + 16 - 3x = 3x + 22 - 3x$

$16 = 22$

There is no solution.

**7.**　　$12x - 18 = 9(x - 2) + 3x$

$12x - 18 = 9x - 18 + 3x$

$12x - 18 = 12x - 18$

$12x - 18 + 18 = 12x - 18 + 18$

$12x = 12x$

$12x - 12x = 12x - 12x$

$0 = 0$

The solution is all real numbers.

**Calculator Explorations**

**1.** Solution $(-24 = -24)$

**2.** Solution $(-4 = -4)$

**3.** Not a solution $(19.4 \ne 10.4)$

**4.** Not a solution $(-11.9 \neq -60.1)$

**5.** Solution $(17{,}061 = 17{,}061)$

**6.** Solution $(-316 = -316)$

**Vocabulary, Readiness & Video Check 2.4**

**1.** $x = -7$ is an equation.

**2.** $x - 7$ is an expression.

**3.** $4y - 6 + 9y + 1$ is an expression.

**4.** $4y - 6 = 9y + 1$ is an equation.

**5.** $\dfrac{1}{x} - \dfrac{x-1}{8}$ is an expression.

**6.** $\dfrac{1}{x} - \dfrac{x-1}{8} = 6$ is an equation.

**7.** $0.1x + 9 = 0.2x$ is an equation.

**8.** $0.1x^2 + 9y - 0.2x^2$ is an expression.

**9.** 3; distributive property, addition property of equality, multiplication property of equality

**10.** Because both sides have more than one term, you need to apply the distributive property to make sure you multiply every single term in the equation by the LCD.

**11.** The number of decimal places in each number helps you determine what power of 10 you can multiply through by so you are no longer dealing with decimals.

**12.** When solving a linear equation and all variable terms, subtract out:

   **a.** If you have a true statement, then the equation has <u>all real numbers</u> as a solution.

   **b.** If you have a false statement, then the equation has <u>no</u> solution.

**Exercise Set 2.4**

**1.**
$$-4y + 10 = -2(3y + 1)$$
$$-4y + 10 = -6y - 2$$
$$-4y + 10 - 10 = -6y - 2 - 10$$
$$-4y = -6y - 12$$
$$-4y + 6y = -6y - 12 + 6y$$
$$2y = -12$$
$$\frac{2y}{2} = \frac{-12}{2}$$
$$y = -6$$

**3.**
$$15x - 8 = 10 + 9x$$
$$15x - 8 + 8 = 10 + 9x + 8$$
$$15x = 18 + 9x$$
$$15x - 9x = 18 + 9x - 9x$$
$$6x = 18$$
$$\frac{6x}{6} = \frac{18}{6}$$
$$x = 3$$

**5.**
$$-2(3x - 4) = 2x$$
$$-6x + 8 = 2x$$
$$-6x + 6x + 8 = 2x + 6x$$
$$8 = 8x$$
$$\frac{8}{8} = \frac{8x}{8}$$
$$1 = x$$

**7.**
$$5(2x - 1) - 2(3x) = 1$$
$$10x - 5 - 6x = 1$$
$$4x - 5 = 1$$
$$4x - 5 + 5 = 1 + 5$$
$$4x = 6$$
$$\frac{4x}{4} = \frac{6}{4}$$
$$x = \frac{3}{2}$$

**9.**
$$-6(x - 3) - 26 = -8$$
$$-6x + 18 - 26 = -8$$
$$-6x - 8 = -8$$
$$-6x - 8 + 8 = -8 + 8$$
$$-6x = 0$$
$$\frac{-6x}{-6} = \frac{0}{-6}$$
$$x = 0$$

**11.**
$$8 - 2(a+1) = 9 + a$$
$$8 - 2a - 2 = 9 + a$$
$$-2a + 6 = 9 + a$$
$$-2a + 6 - 6 = 9 + a - 6$$
$$-2a = 3 + a$$
$$-2a - a = 3 + a - a$$
$$-3a = 3$$
$$\frac{-3a}{-3} = \frac{3}{-3}$$
$$a = -1$$

**13.**
$$4x + 3 = -3 + 2x + 14$$
$$4x + 3 = 2x + 11$$
$$4x - 2x + 3 = 2x - 2x + 11$$
$$2x + 3 = 11$$
$$2x + 3 - 3 = 11 - 3$$
$$2x = 8$$
$$\frac{2x}{2} = \frac{8}{2}$$
$$x = 4$$

**15.**
$$-2y - 10 = 5y + 18$$
$$-2y - 5y - 10 = 5y - 5y + 18$$
$$-7y - 10 = 18$$
$$-7y - 10 + 10 = 18 + 10$$
$$-7y = 28$$
$$\frac{-7y}{-7} = \frac{28}{-7}$$
$$y = -4$$

**17.**
$$\frac{2}{3}x + \frac{4}{3} = -\frac{2}{3}$$
$$3\left(\frac{2}{3}x + \frac{4}{3}\right) = 3\left(-\frac{2}{3}\right)$$
$$2x + 4 = -2$$
$$2x + 4 - 4 = -2 - 4$$
$$2x = -6$$
$$\frac{2x}{2} = \frac{-6}{2}$$
$$x = -3$$

**19.**
$$\frac{3}{4}x - \frac{1}{2} = 1$$
$$4\left(\frac{3}{4}x - \frac{1}{2}\right) = 4(1)$$
$$3x - 2 = 4$$
$$3x - 2 + 2 = 4 + 2$$
$$3x = 6$$
$$\frac{3x}{3} = \frac{6}{3}$$
$$x = 2$$

**21.**
$$0.50x + 0.15(70) = 35.5$$
$$100[0.50x + 0.15(70)] = 100(35.5)$$
$$50x + 15(70) = 3550$$
$$50x + 1050 = 3550$$
$$50x + 1050 - 1050 = 3550 - 1050$$
$$50x = 2500$$
$$\frac{50x}{50} = \frac{2500}{50}$$
$$x = 50$$

**23.**
$$\frac{2(x+1)}{4} = 3x - 2$$
$$4\left[\frac{2(x+1)}{4}\right] = 4(3x - 2)$$
$$2(x+1) = 12x - 8$$
$$2x + 2 = 12x - 8$$
$$2x - 12x + 2 = 12x - 12x - 8$$
$$-10x + 2 = -8$$
$$-10x + 2 - 2 = -8 - 2$$
$$-10x = -10$$
$$\frac{-10x}{-10} = \frac{-10}{-10}$$
$$x = 1$$

**25.**
$$x + \frac{7}{6} = 2x - \frac{7}{6}$$
$$6\left(x + \frac{7}{6}\right) = 6\left(2x - \frac{7}{6}\right)$$
$$6x + 7 = 12x - 7$$
$$6x - 12x + 7 = 12x - 12x - 7$$
$$-6x + 7 = -7$$
$$-6x + 7 - 7 = -7 - 7$$
$$-6x = -14$$
$$\frac{-6x}{-6} = \frac{-14}{-6}$$
$$x = \frac{7}{3}$$

**27.**
$$0.12(y-6)+0.06y=0.08y-0.70$$
$$100[0.12(y-6)+0.06y]=100[0.08y-0.70]$$
$$12(y-6)+6y=8y-70$$
$$12y-72+6y=8y-70$$
$$18y-72=8y-70$$
$$18y-8y-72=8y-8y-70$$
$$10y-72=-70$$
$$10y-72+72=-70+72$$
$$10y=2$$
$$\frac{10y}{10}=\frac{2}{10}$$
$$y=\frac{1}{5}=0.2$$

**29.**
$$4(3x+2)=12x+8$$
$$12x+8=12x+8$$
$$12x+8-12x=12x+8-12x$$
$$8=8$$
All real numbers are solutions.

**31.**
$$\frac{x}{4}+1=\frac{x}{4}$$
$$4\left(\frac{x}{4}+1\right)=4\left(\frac{x}{4}\right)$$
$$x+4=x$$
$$x-x+4=x-x$$
$$4=0$$
There is no solution.

**33.**
$$3x-7=3(x+1)$$
$$3x-7=3x+3$$
$$3x-3x-7=3x-3x+3$$
$$-7=3$$
There is no solution.

**35.**
$$-2(6x-5)+4=-12x+14$$
$$-12x+10+4=-12x+14$$
$$-12x+14=-12x+14$$
$$-12x+14+12x=-12x+14+12x$$
$$14=14$$
All real numbers are solutions.

**37.**
$$\frac{6(3-z)}{5}=-z$$
$$5\left[\frac{6(3-z)}{5}\right]=5(-z)$$
$$6(3-z)=5(-z)$$
$$18-6z=-5z$$
$$18-6z+6z=-5z+6z$$
$$18=z$$

**39.**
$$-3(2t-5)+2t=5t-4$$
$$-6t+15+2t=5t-4$$
$$-4t+15=5t-4$$
$$-4t+15+4=5t-4+4$$
$$-4t+19=5t$$
$$-4t+19+4t=5t+4t$$
$$19=9t$$
$$\frac{19}{9}=\frac{9t}{9}$$
$$\frac{19}{9}=t$$

**41.**
$$5y+2(y-6)=4(y+1)-2$$
$$5y+2y-12=4y+4-2$$
$$7y-12=4y+2$$
$$7y-12+12=4y+2+12$$
$$7y=4y+14$$
$$7y-4y=4y+14-4y$$
$$3y=14$$
$$\frac{3y}{3}=\frac{14}{3}$$
$$y=\frac{14}{3}$$

**43.**
$$\frac{3(x-5)}{2}=\frac{2(x+5)}{3}$$
$$6\left[\frac{3(x-5)}{2}\right]=6\left[\frac{2(x+5)}{3}\right]$$
$$9(x-5)=4(x+5)$$
$$9x-45=4x+20$$
$$9x-4x-45=4x-4x+20$$
$$5x-45=20$$
$$5x-45+45=20+45$$
$$5x=65$$
$$\frac{5x}{5}=\frac{65}{5}$$
$$x=13$$

**45.**
$$0.7x-2.3=0.5$$
$$10(0.7x-2.3)=10(0.5)$$
$$7x-23=5$$
$$7x-23+23=5+23$$
$$7x=28$$
$$\frac{7x}{7}=\frac{28}{7}$$
$$x=4$$

**47.**
$$5x - 5 = 2(x + 1) + 3x - 7$$
$$5x - 5 = 2x + 2 + 3x - 7$$
$$5x - 5 = 5x - 5$$
$$5x - 5x - 5 = 5x - 5x - 5$$
$$-5 = -5$$
All real numbers are solutions.

**49.**
$$4(2n + 1) = 3(6n + 3) + 1$$
$$8n + 4 = 18n + 9 + 1$$
$$8n + 4 = 18n + 10$$
$$8n + 4 - 4 = 18n + 10 - 4$$
$$8n = 18n + 6$$
$$8n - 18n = 18n + 6 - 18n$$
$$-10n = 6$$
$$\frac{-10n}{-10} = \frac{6}{-10}$$
$$n = -\frac{3}{5}$$

**51.**
$$x + \frac{5}{4} = \frac{3}{4}x$$
$$4\left(x + \frac{5}{4}\right) = 4\left(\frac{3}{4}x\right)$$
$$4x + 5 = 3x$$
$$4x + 5 - 4x = 3x - 4x$$
$$5 = -x$$
$$\frac{5}{-1} = \frac{-x}{-1}$$
$$-5 = x$$

**53.**
$$\frac{x}{2} - 1 = \frac{x}{5} + 2$$
$$10\left(\frac{x}{2} - 1\right) = 10\left(\frac{x}{5} + 2\right)$$
$$5x - 10 = 2x + 20$$
$$5x - 10 + 10 = 2x + 20 + 10$$
$$5x = 2x + 30$$
$$5x - 2x = 2x + 30 - 2x$$
$$3x = 30$$
$$\frac{3x}{3} = \frac{30}{3}$$
$$x = 10$$

**55.**
$$2(x + 3) - 5 = 5x - 3(1 + x)$$
$$2x + 6 - 5 = 5x - 3 - 3x$$
$$2x + 1 = 2x - 3$$
$$2x - 2x + 1 = 2x - 2x - 3$$
$$1 = -3$$
There is no solution.

**57.**
$$0.06 - 0.01(x + 1) = -0.02(2 - x)$$
$$100[0.06 - 0.01(x + 1)] = 100[-0.02(2 - x)]$$
$$6 - (x + 1) = -2(2 - x)$$
$$6 - x - 1 = -4 + 2x$$
$$5 - x = -4 + 2x$$
$$5 - x - 2x = -4 + 2x - 2x$$
$$5 - 3x = -4$$
$$5 - 5 - 3x = -4 - 5$$
$$-3x = -9$$
$$\frac{-3x}{-3} = \frac{-9}{-3}$$
$$x = 3$$

**59.**
$$\frac{9}{2} + \frac{5}{2}y = 2y - 4$$
$$2\left(\frac{9}{2} + \frac{5}{2}y\right) = 2(2y - 4)$$
$$9 + 5y = 4y - 8$$
$$9 + 5y - 4y = 4y - 8 - 4y$$
$$9 + y = -8$$
$$9 + y - 9 = -8 - 9$$
$$y = -17$$

**61.**
$$-2y - 10 = 5y + 18$$
$$-2y - 10 - 18 = 5y + 18 - 18$$
$$-2y - 28 = 5y$$
$$-2y - 28 + 2y = 5y + 2y$$
$$-28 = 7y$$
$$\frac{-28}{7} = \frac{7y}{7}$$
$$-4 = y$$

**63.**
$$0.6x - 0.1 = 0.5x + 0.2$$
$$10(0.6x - 0.1) = 10(0.5x + 0.2)$$
$$6x - 1 = 5x + 2$$
$$6x - 5x - 1 = 5x - 5x + 2$$
$$x - 1 = 2$$
$$x - 1 + 1 = 2 + 1$$
$$x = 3$$

**65.**
$$0.02(6t - 3) = 0.12(t - 2) + 0.18$$
$$100[0.02(6t - 3)] = 100[0.12(t - 2) + 0.18]$$
$$2(6t - 3) = 12(t - 2) + 18$$
$$12t - 6 = 12t - 24 + 18$$
$$12t - 6 = 12t - 6$$
$$12t - 12t - 6 = 12t - 12t - 6$$
$$-6 = -6$$
All real numbers are solutions.

**67.** $\underset{\downarrow}{-8}$ $\underset{\downarrow}{\text{minus}}$ $\underset{\downarrow}{\text{a number}}$

$\quad -8 \quad\; - \qquad\; x$

**69.** $\underset{\downarrow}{-3}$ $\underset{\downarrow}{\text{plus}}$ $\underset{\downarrow}{\substack{\text{twice a} \\ \text{number}}}$

$\quad -3 \quad\; + \qquad 2x$

**71.** $\underset{\downarrow}{9}$ $\underset{\downarrow}{\text{times}}$ $\underset{\downarrow}{\text{a number}}$ $\underset{\downarrow}{\text{plus}}$ $\underset{\downarrow}{20}$

$\quad 9 \qquad \cdot \qquad\; (x \qquad + \quad 20) = 9(x+20)$

**73.** $x+(2x-3)+(3x-5) = x+2x-3+3x-5$
$$= 6x-8$$
The perimeter is $(6x - 8)$ meters.

**75. a.** $\quad x+3 = x+3$
$$x+3-x = x+3-x$$
$$3 = 3$$
$$3-3 = 3-3$$
$$0 = 0$$
All real numbers are solutions.

**b.** answers may vary

**c.** answers may vary

**77.** $\quad 5x+1 = 5x+1$
$$5x+1-5x = 5x+1-5x$$
$$1 = 1$$
All real numbers are solutions. The answer is a.

**79.** $\quad 2x-6x-10 = -4x+3-10$
$$-4x-10 = -4x-7$$
$$-4x-10+4x = -4x-7+4x$$
$$-10 = -7$$
There is no solution. The answer is b.

**81.** $\quad 9x-20 = 8x-20$
$$9x-20-8x = 8x-20-8x$$
$$x-20 = -20$$
$$x-20+20 = -20+20$$
$$x = 0$$
The answer is c.

**83.** answers may vary

**85. a.** Since the perimeter is the sum of the lengths of the sides, $x + x + x + 2x + 2x = 28$.

**b.**   $7x = 28$

$$\frac{7x}{7} = \frac{28}{7}$$

$$x = 4$$

**c.**   $2x = 2(4) = 8$

The lengths are $x = 4$ centimeters and $2x = 8$ centimeters.

**87.** answers may vary

**89.**
$$1000(7x - 10) = 50(412 + 100x)$$
$$7000x - 10,000 = 20,600 + 5000x$$
$$7000x - 5000x - 10,000 = 20,600 + 5000x - 5000x$$
$$2000x - 10,000 = 20,600$$
$$2000x - 10,000 + 10,000 = 20,600 + 10,000$$
$$2000x = 30,600$$
$$\frac{2000x}{2000} = \frac{30,600}{2000}$$
$$x = 15.3$$

**91.**
$$0.035x + 5.112 = 0.010x + 5.107$$
$$1000(0.035x + 5.112) = 1000(0.010x + 5.107)$$
$$35x + 5112 = 10x + 5107$$
$$35x - 10x + 5112 = 10x - 10x + 5107$$
$$25x + 5112 = 5107$$
$$25x + 5112 - 5112 = 5107 - 5112$$
$$25x = -5$$
$$\frac{25x}{25} = \frac{-5}{25}$$
$$x = -\frac{1}{5} = -0.2$$

**93.**
$$x(x - 3) = x^2 + 5x + 7$$
$$x^2 - 3x = x^2 + 5x + 7$$
$$x^2 - x^2 - 3x = x^2 - x^2 + 5x + 7$$
$$-3x = 5x + 7$$
$$-3x - 5x = 5x - 5x + 7$$
$$-8x = 7$$
$$\frac{-8x}{-8} = \frac{7}{-8}$$
$$x = -\frac{7}{8}$$

**95.**
$$2z(z+6) = 2z^2 + 12z - 8$$
$$2z^2 + 12z = 2z^2 + 12z - 8$$
$$2z^2 - 2z^2 + 12z = 2z^2 - 2z^2 + 12z - 8$$
$$12z = 12z - 8$$
$$12z - 12z = 12z - 12z - 8$$
$$0 = -8$$
There is no solution.

## Integrated Review

**1.**
$$x - 10 = -4$$
$$x - 10 + 10 = -4 + 10$$
$$x = 6$$

**2.**
$$y + 14 = -3$$
$$y + 14 - 14 = -3 - 14$$
$$y = -17$$

**3.**
$$9y = 108$$
$$\frac{9y}{9} = \frac{108}{9}$$
$$y = 12$$

**4.**
$$-3x = 78$$
$$\frac{-3x}{-3} = \frac{78}{-3}$$
$$x = -26$$

**5.**
$$-6x + 7 = 25$$
$$-6x + 7 - 7 = 25 - 7$$
$$-6x = 18$$
$$\frac{-6x}{-6} = \frac{18}{-6}$$
$$x = -3$$

**6.**
$$5y - 42 = -47$$
$$5y - 42 + 42 = -47 + 42$$
$$5y = -5$$
$$\frac{5y}{5} = \frac{-5}{5}$$
$$y = -1$$

**7.**
$$\frac{2}{3}x = 9$$
$$\frac{3}{2}\left(\frac{2}{3}x\right) = \frac{3}{2}(9)$$
$$x = \frac{27}{2}$$

**8.**
$$\frac{4}{5}z = 10$$
$$\frac{5}{4}\left(\frac{4}{5}z\right) = \frac{5}{4}(10)$$
$$z = \frac{25}{2}$$

**9.**
$$\frac{r}{-4} = -2$$
$$-4\left(\frac{r}{-4}\right) = -4(-2)$$
$$r = 8$$

**10.**
$$\frac{y}{-8} = 8$$
$$-8\left(\frac{y}{-8}\right) = -8(8)$$
$$y = -64$$

**11.**
$$6 - 2x + 8 = 10$$
$$-2x + 14 = 10$$
$$-2x + 14 - 14 = 10 - 14$$
$$-2x = -4$$
$$\frac{-2x}{-2} = \frac{-4}{-2}$$
$$x = 2$$

**12.**
$$-5 - 6y + 6 = 19$$
$$-6y + 1 = 19$$
$$-6y + 1 - 1 = 19 - 1$$
$$-6y = 18$$
$$\frac{-6y}{-6} = \frac{18}{-6}$$
$$y = -3$$

**13.**
$$2x - 7 = 2x - 27$$
$$2x - 2x - 7 = 2x - 2x - 27$$
$$-7 = -27$$
There is no solution.

**14.**
$$3 + 8y = 8y - 2$$
$$3 + 8y - 8y = 8y - 8y - 2$$
$$3 = -2$$
There is no solution.

**15.**
$$-3a + 6 + 5a = 7a - 8a$$
$$2a + 6 = -a$$
$$2a - 2a + 6 = -a - 2a$$
$$6 = -3a$$
$$\frac{6}{-3} = \frac{-3a}{-3}$$
$$-2 = a$$

**16.**
$$4b - 8 - b = 10b - 3b$$
$$3b - 8 = 7b$$
$$3b - 3b - 8 = 7b - 3b$$
$$-8 = 4b$$
$$\frac{-8}{4} = \frac{4b}{4}$$
$$-2 = b$$

**17.**
$$-\frac{2}{3}x = \frac{5}{9}$$
$$-\frac{3}{2}\left(-\frac{2}{3}x\right) = -\frac{3}{2}\left(\frac{5}{9}\right)$$
$$x = -\frac{5}{6}$$

**18.**
$$-\frac{3}{8}y = -\frac{1}{16}$$
$$-\frac{8}{3}\left(-\frac{3}{8}y\right) = -\frac{8}{3}\left(-\frac{1}{16}\right)$$
$$y = \frac{1}{6}$$

**19.**
$$10 = -6n + 16$$
$$10 - 16 = -6n + 16 - 16$$
$$-6 = -6n$$
$$\frac{-6}{-6} = \frac{-6n}{-6}$$
$$1 = n$$

**20.**
$$-5 = -2m + 7$$
$$-5 - 7 = -2m + 7 - 7$$
$$-12 = -2m$$
$$\frac{-12}{-2} = \frac{-2m}{-2}$$
$$6 = m$$

**21.**
$$3(5c - 1) - 2 = 13c + 3$$
$$15c - 3 - 2 = 13c + 3$$
$$15c - 5 = 13c + 3$$
$$15c - 13c - 5 = 13c - 13c + 3$$
$$2c - 5 = 3$$
$$2c - 5 + 5 = 3 + 5$$
$$2c = 8$$
$$\frac{2c}{2} = \frac{8}{2}$$
$$c = 4$$

**22.**
$$4(3t + 4) - 20 = 3 + 5t$$
$$12t + 16 - 20 = 3 + 5t$$
$$12t - 4 = 3 + 5t$$
$$12t - 5t - 4 = 3 + 5t - 5t$$
$$7t - 4 = 3$$
$$7t - 4 + 4 = 3 + 4$$
$$7t = 7$$
$$\frac{7t}{7} = \frac{7}{7}$$
$$t = 1$$

**23.**
$$\frac{2(z + 3)}{3} = 5 - z$$
$$3\left[\frac{2(z + 3)}{3}\right] = 3(5 - z)$$
$$2z + 6 = 15 - 3z$$
$$2z + 3z + 6 = 15 - 3z + 3z$$
$$5z + 6 = 15$$
$$5z + 6 - 6 = 15 - 6$$
$$5z = 9$$
$$\frac{5z}{5} = \frac{9}{5}$$
$$z = \frac{9}{5}$$

**24.**
$$\frac{3(w + 2)}{4} = 2w + 3$$
$$4\left[\frac{3(w + 2)}{4}\right] = 4(2w + 3)$$
$$3w + 6 = 8w + 12$$
$$3w - 8w + 6 = 8w - 8w + 12$$
$$-5w + 6 = 12$$
$$-5w + 6 - 6 = 12 - 6$$
$$-5w = 6$$
$$\frac{-5w}{-5} = \frac{6}{-5}$$
$$w = -\frac{6}{5}$$

    

**25.**  $-2(2x-5) = -3x+7-x+3$
$-4x+10 = -4x+10$
$-4x+4x+10 = -4x+4x+10$
$10 = 10$
All real numbers are solutions.

**26.**  $-4(5x-2) = -12x+4-8x+4$
$-20x+8 = -20x+8$
$-20x+20x+8 = -20x+20x+8$
$8 = 8$
All real numbers are solutions.

**27.**  $0.02(6t-3) = 0.04(t-2)+0.02$
$100[0.02(6t-3)] = 100[0.04(t-2)+0.02]$
$2(6t-3) = 4(t-2)+2$
$12t-6 = 4t-8+2$
$12t-6 = 4t-6$
$12t-4t-6 = 4t-4t-6$
$8t-6 = -6$
$8t-6+6 = -6+6$
$8t = 0$
$\dfrac{8t}{8} = \dfrac{0}{8}$
$t = 0$

**28.**  $0.03(m+7) = 0.02(5-m)+0.03$
$100[0.03(m+7)] = 100[0.02(5-m)+0.03]$
$3(m+7) = 2(5-m)+3$
$3m+21 = 10-2m+3$
$3m+21 = 13-2m$
$3m+2m+21 = 13-2m+2m$
$5m+21 = 13$
$5m+21-21 = 13-21$
$5m = -8$
$\dfrac{5m}{5} = \dfrac{-8}{5}$
$m = -\dfrac{8}{5} = -1.6$

**29.**  $-3y = \dfrac{4(y-1)}{5}$
$5(-3y) = 5\left[\dfrac{4(y-1)}{5}\right]$
$-15y = 4y-4$
$-15y-4y = 4y-4y-4$
$-19y = -4$
$\dfrac{-19y}{-19} = \dfrac{-4}{-19}$
$y = \dfrac{4}{19}$

**30.**  $-4x = \dfrac{5(1-x)}{6}$
$6(-4x) = 6\left[\dfrac{5(1-x)}{6}\right]$
$-24x = 5-5x$
$-24x+5x = 5-5x+5x$
$-19x = 5$
$\dfrac{-19x}{-19} = \dfrac{5}{-19}$
$x = -\dfrac{5}{19}$

**31.**  $\dfrac{5}{3}x - \dfrac{7}{3} = x$
$3\left(\dfrac{5}{3}x - \dfrac{7}{3}\right) = 3(x)$
$5x-7 = 3x$
$5x-5x-7 = 3x-5x$
$-7 = -2x$
$\dfrac{-7}{-2} = \dfrac{-2x}{-2}$
$\dfrac{7}{2} = x$

**32.**  $\dfrac{7}{5}n + \dfrac{3}{5} = -n$
$5\left(\dfrac{7}{5}n + \dfrac{3}{5}\right) = 5(-n)$
$7n+3 = -5n$
$7n-7n+3 = -5n-7n$
$3 = -12n$
$\dfrac{3}{-12} = \dfrac{-12n}{-12}$
$-\dfrac{1}{4} = n$

**33.**  $9(3x-1) = -4+49$
$27x-9 = 45$
$27x-9+9 = 45+9$
$27x = 54$
$\dfrac{27x}{27} = \dfrac{54}{27}$
$x = 2$

**34.**
$$12(2x+1) = -6+66$$
$$24x+12 = 60$$
$$24x+12-12 = 60-12$$
$$24x = 48$$
$$\frac{24x}{24} = \frac{48}{24}$$
$$x = 2$$

**35.**
$$\frac{1}{10}(3x-7) = \frac{3}{10}x+5$$
$$10\left[\frac{1}{10}(3x-7)\right] = 10\left(\frac{3}{10}x+5\right)$$
$$3x-7 = 3x+50$$
$$3x-7-3x = 3x+50-3x$$
$$-7 = 50$$
There is no solution.

**36.**
$$\frac{1}{7}(2x-5) = \frac{2}{7}x+1$$
$$7\left[\frac{1}{7}(2x-5)\right] = 7\left(\frac{2}{7}x+1\right)$$
$$2x-5 = 2x+7$$
$$2x-5-2x = 2x+7-2x$$
$$-5 = 7$$
There is no solution.

**37.**
$$5+2(3x-6) = -4(6x-7)$$
$$5+6x-12 = -24x+28$$
$$6x-7 = -24x+28$$
$$6x-7+24x = -24x+28+24x$$
$$30x-7 = 28$$
$$30x-7+7 = 28+7$$
$$30x = 35$$
$$\frac{30x}{30} = \frac{35}{30}$$
$$x = \frac{7}{6}$$

**38.**
$$3+5(2x-4) = -7(5x+2)$$
$$3+10x-20 = -35x-14$$
$$10x-17 = -35x-14$$
$$10x-17+35x = -35x-14+35x$$
$$45x-17 = -14$$
$$45x-17+17 = -14+17$$
$$45x = 3$$
$$\frac{45x}{45} = \frac{3}{45}$$
$$x = \frac{1}{15}$$

## Section 2.5 Practice

**1.** Let $x$ = the number.
$$3x-6 = 2x+3$$
$$3x-6-2x = 2x+3-2x$$
$$x-6 = 3$$
$$x-6+6 = 3+6$$
$$x = 9$$
The number is 9.

**2.** Let $x$ = the number.
$$3x-4 = 2(x-1)$$
$$3x-4 = 2x-2$$
$$3x-4-2x = 2x-2-2x$$
$$x-4 = -2$$
$$x-4+4 = -2+4$$
$$x = 2$$
The number is 2.

**3.** Let $x$ = the length of short piece,
then $4x$ = the length of long piece.
$$x+4x = 45$$
$$5x = 45$$
$$\frac{5x}{5} = \frac{45}{5}$$
$$x = 9$$
$$4x = 4(9) = 36$$
The short piece is 9 inches and the long piece is 36 inches.

**4.** Let $x$ = number of Republican governors, then
$x-9$ = number of Democratic governors.
$$x+x-9 = 49$$
$$2x-9 = 49$$
$$2x-9+9 = 49+9$$
$$2x = 58$$
$$\frac{2x}{2} = \frac{58}{2}$$
$$x = 29$$
$$x-9 = 20$$
There were 29 Republican and 20 Democratic governors.

**5.** $x =$ degree measure of first angle
$3x =$ degree measure of second angle
$x + 55 =$ degree measure of third angle
$$x + 3x + (x + 55) = 180$$
$$5x + 55 = 180$$
$$5x + 55 - 55 = 180 - 55$$
$$5x = 125$$
$$\frac{5x}{5} = \frac{125}{5}$$
$$x = 25$$
$3x = 3(25) = 75$
$x + 55 = 25 + 55 = 80$
The measures of the angles are 25°, 75°, and 80°.

**6.** Let $x =$ the first even integer, then
$x + 2 =$ the second even integer, and
$x + 4 =$ the third even integer.
$$x + (x + 2) + (x + 4) = 144$$
$$3x + 6 = 144$$
$$3x + 6 - 6 = 144 - 6$$
$$3x = 138$$
$$\frac{3x}{3} = \frac{138}{3}$$
$$x = 46$$
$x + 2 = 46 + 2 = 48$
$x + 4 = 46 + 4 = 50$
The integers are 46, 48, and 50.

**Vocabulary, Readiness & Video Check 2.5**

1. $2x$; $2x - 31$

2. $3x$; $3x + 17$

3. $x + 5$; $2(x + 5)$

4. $x - 11$; $7(x - 11)$

5. $20 - y$; $\dfrac{20 - y}{3}$ or $(20 - y) \div 3$

6. $-10 + y$; $\dfrac{-10 + y}{9}$ or $(-10 + y) \div 9$

7. in the statement of the application

8. The original application asks for the measure of two supplementary angles. The solution of $x = 43$ only gives us the measure of one of the angles.

**9.** That the 3 angle measures are consecutive even integers and that they sum to 180°.

**Exercise Set 2.5**

**1.** Let $x =$ the number.
$$6x + 1 = 5x$$
$$6x + 1 - 5x = 5x - 5x$$
$$x + 1 = 0$$
$$x + 1 - 1 = 0 - 1$$
$$x = -1$$
The number is $-1$.

**3.** Let $x =$ the number.
$$3x - 6 = 2x + 8$$
$$3x - 6 - 2x = 2x + 8 - 2x$$
$$x - 6 = 8$$
$$x - 6 + 6 = 8 + 6$$
$$x = 14$$
The number is 14.

**5.** Let $x =$ the number.
$$2(x - 8) = 3(x + 3)$$
$$2x - 16 = 3x + 9$$
$$2x - 2x - 16 = 3x - 2x + 9$$
$$-16 = x + 9$$
$$-16 - 9 = x + 9 - 9$$
$$-25 = x$$
The number is $-25$.

**7.** Let $x =$ the number.
$$2(-2 + x) = x - \frac{1}{2}$$
$$-4 + 2x = x - \frac{1}{2}$$
$$-4 + 2x - x = x - x - \frac{1}{2}$$
$$-4 + x = -\frac{1}{2}$$
$$-4 + 4 + x = -\frac{1}{2} + 4$$
$$x = -\frac{1}{2} + \frac{8}{2}$$
$$x = \frac{7}{2}$$
The number is $\dfrac{7}{2}$.

**9.** The sum of the three lengths is 25 inches.

$$x + 2x + 1 + 5x = 25$$
$$1 + 8x = 25$$
$$1 + 8x - 1 = 25 - 1$$
$$8x = 24$$
$$\frac{8x}{8} = \frac{24}{8}$$
$$x = 3$$

$2x = 2(3) = 6$

$1 + 5x = 1 + 5(3) = 1 + 15 = 16$

The lengths are 3 inches, 6 inches, and 16 inches.

**11.** Let $x$ be the length of the first piece. Then the second piece is $2x$ and the third piece is $5x$. The sum of the lengths is 40 inches.

$$x + 2x + 5x = 40$$
$$8x = 40$$
$$\frac{8x}{8} = \frac{40}{8}$$
$$x = 5$$

$2x = 2(5) = 10$

$5x = 5(5) = 25$

The 1st piece is 5 inches, 2nd piece is 10 inches, and 3rd piece is 25 inches.

**13.**

$$x + x + 23{,}873 = 39{,}547$$
$$2x + 23{,}873 = 39{,}547$$
$$2x + 23{,}873 - 23{,}873 = 39{,}547 - 23{,}873$$
$$2x = 15{,}674$$
$$\frac{2x}{2} = \frac{15{,}674}{2}$$
$$x = 7837$$

In 2010, 7837 screens were 3D.

**15.** Let $x$ be the measure of each of the two equal angles. Then $2x + 30$ is the measure of the third angle. Their sum is $180°$.

$$x + x + 2x + 30 = 180$$
$$4x + 30 = 180$$
$$4x + 30 - 30 = 180 - 30$$
$$4x = 150$$
$$\frac{4x}{4} = \frac{150}{4}$$
$$x = 37.5$$

$2x + 30 = 2(37.5) + 30 = 75 + 30 = 105$

The 1st angle measures $37.5°$, the 2nd angle measures $37.5°$, and the 3rd angle measures $105°$.

| | First Integer | Next Integers | | | Indicated Sum |
|---|---|---|---|---|---|
| **17.** Three consecutive integers: | Integer: $x$ | $x + 1$ | $x + 2$ | | Sum of the three consecutive integers simplified: $(x + 1) + (x + 2) = 2x + 3$ |
| **19.** Three consecutive even integers: | Even integer: $x$ | $x + 2$ | $x + 4$ | | Sum of the first and third even consecutive integers, simplified: $x + (x + 4) = 2x + 4$ |
| **21.** Four consecutive integers: | Integer: $x$ | $x + 1$ | $x + 2$ | $x + 3$ | Sum of the four consecutive integers, simplified: $x + (x + 1) + (x + 2) + (x + 3) = 4x + 6$ |
| **23.** Three consecutive odd integers: | Odd integer: $x$ | $x + 2$ | $x + 4$ | | Sum of the second and third consecutive odd integers, simplified: $(x + 2) + (x + 4) = 2x + 6$ |

**25.** Let $x$ = the number of the left page and
$x + 1$ = the number of the right page.
$$x + x + 1 = 469$$
$$2x + 1 = 469$$
$$2x + 1 - 1 = 469 - 1$$
$$2x = 468$$
$$\frac{2x}{2} = \frac{468}{2}$$
$$x = 234$$
$x + 1 = 234 + 1 = 235$
The page numbers are 234 and 235.

**27.** Let $x$ = the code for Belgium,
$x + 1$ = the code for France,
$x + 2$ = the code for Spain.
$$x + x + 1 + x + 2 = 99$$
$$3x + 3 = 99$$
$$3x + 3 - 3 = 99 - 3$$
$$3x = 96$$
$$\frac{3x}{3} = \frac{96}{3}$$
$$x = 32$$
$x + 1 = 32 + 1 = 33$
$x + 2 = 32 + 2 = 34$
The codes are Belgium: 32; France: 33; Spain: 34.

**29.** Let $x$ represent the area of the Gobi Desert, in square miles. Then $7x$ represents the area of the Sahara Desert.
$$x + 7x = 4,000,000$$
$$8x = 4,000,000$$
$$\frac{8x}{8} = \frac{4,000,000}{8}$$
$$x = 500,000$$
$7x = 7(500,000) = 3,500,000$
The Gobi Desert's area is 500,000 square miles and the Sahara Desert's area is 3,500,000 square miles.

**31.** Let $x$ be the length of the shorter piece. Then $2x + 2$ is the length of the longer piece. The measures sum to 17 feet.
$$x + 2x + 2 = 17$$
$$3x + 2 = 17$$
$$3x + 2 - 2 = 17 - 2$$
$$3x = 5$$
$$\frac{3x}{3} = \frac{15}{3}$$
$$x = 5$$
$2x + 2 = 2(5) + 2 = 10 + 2 = 12$
The pieces measure 5 feet and 12 feet.

**33.** Let $x$ = the number.
$$10 - 5x = 3x$$
$$10 - 5x + 5x = 3x + 5x$$
$$10 = 8x$$
$$\frac{10}{8} = \frac{8x}{8}$$
$$\frac{5}{4} = x$$

The number is $\dfrac{5}{4}$.

**35.** Let $x$ = carats in Angola, then $4x$ = carats in Botswana.
$$x + 4x = 40,000,000$$
$$5x = 40,000,000$$
$$\frac{5x}{5} = \frac{40,000,000}{5}$$
$$x = 8,000,000$$
$4x = 4(8,000,000) = 32,000,000$
Botswana produces 32,000,000 carats and Angola produces 8,000,000 carats.

**37.** Let $x$ = the measure of the smallest angle, $x + 2$ = the measure of the second, and $x + 4$ = the measure of the third.
$$x + x + 2 + x + 4 = 180$$
$$3x + 6 = 180$$
$$3x + 6 - 6 = 180 - 6$$
$$3x = 174$$
$$\frac{3x}{3} = \frac{174}{3}$$
$$x = 58$$
$x + 2 = 58 + 2 = 60$
$x + 4 = 58 + 4 = 62$
The angles are 58°, 60°, and 62°.

**39.** Let $x$ = first integer (South Korea),
$x + 1$ = second integer (Russia),
$x + 2$ = third integer (Austria).
$$x + (x + 1) + (x + 2) = 45$$
$$3x + 3 = 45$$
$$3 + 3 - 3 = 45 - 3$$
$$3x = 42$$
$$\frac{3x}{3} = \frac{42}{3}$$
$$x = 14$$
$x + 1 = 14 + 1 = 15$
$x + 2 = 14 + 2 = 16$
The number of medals for each country is South Korea: 14; Russia: 15; Austria: 16.

**41.** Let $x$ = the number.
$$3(x + 5) = 2x - 1$$
$$3x + 15 = 2x - 1$$
$$3x + 15 - 2x = 2x - 1 - 2x$$
$$x + 15 = -1$$
$$x + 15 - 15 = -1 - 15$$
$$x = -16$$
The number is $-16$.

**43.** Let $x$ = smaller angle, then $3x + 8$ = larger angle.
$$x + (3x + 8) = 180$$
$$4x + 8 = 180$$
$$4x + 8 - 8 = 180 - 8$$
$$4x = 172$$
$$\frac{4x}{4} = \frac{172}{4}$$
$$x = 43$$
$3x + 8 = 3(43) + 8 = 137$
The angles measure 43° and 137°.

**45.** Let $x$ = the number.
$$\frac{x}{4} + \frac{1}{2} = \frac{3}{4}$$
$$4\left(\frac{x}{4} + \frac{1}{2}\right) = 4\left(\frac{3}{4}\right)$$
$$x + 2 = 3$$
$$x + 2 - 2 = 3 - 2$$
The number is 1.

   

**47.**
$$\frac{2}{3} + 4x = 5x - \frac{5}{6}$$
$$6 \cdot \left(\frac{2}{3} + 4x\right) = 6 \cdot \left(5x - \frac{5}{6}\right)$$
$$4 + 24x = 30x - 5$$
$$4 + 24x - 24x = 30x - 5 - 24x$$
$$4 = 6x - 5$$
$$4 + 5 = 6x - 5 + 5$$
$$9 = 6x$$
$$\frac{9}{6} = \frac{6x}{6}$$
$$\frac{3}{2} = x$$

The number is $\frac{3}{2}$.

**49.** Let $x$ = speed of TGV, then
$x + 3.8$ = speed of Maglev.
$$x + x + 3.8 = 718.2$$
$$2x + 3.8 = 718.2$$
$$2x + 3.8 - 3.8 = 718.2 - 3.8$$
$$2x = 714.4$$
$$\frac{2x}{2} = \frac{714.4}{2}$$
$$x = 357.2$$
$x + 3.8 = 357.2 + 3.8 = 361$
The speed of the TGV is 357.2 mph and the speed of the Maglev is 361 mph.

**51.** Let $x$ = the number.
$$\frac{1}{3} \cdot x = \frac{5}{6}$$
$$3 \cdot \frac{1}{3} x = 3 \cdot \frac{5}{6}$$
$$x = \frac{5}{2}$$

The number is $\frac{5}{2}$.

**53.** Let $x$ = number of counties in Montana and
$x + 2$ = number in California.
$$x + x + 2 = 114$$
$$2x + 2 = 114$$
$$2x + 2 - 2 = 114 - 2$$
$$2x = 112$$
$$\frac{2x}{2} = \frac{112}{2}$$
$$x = 56$$
$x + 2 = 56 + 2 = 58$
There are 56 counties in Montana and 58 counties in California.

**55.** Let $x$ = smaller angles, then
$x + 76.5$ = third angle.
$$x + x + (x + 76.5) = 180$$
$$3x + 76.5 = 180$$
$$3x + 76.5 - 76.5 = 180 - 76.5$$
$$3x = 103.5$$
$$\frac{3x}{3} = \frac{103.5}{3}$$
$$x = 34.5$$
$x + 76.5 = 34.5 + 76.5 = 111$
The third angle measures 111°.

**57.** Let $x$ = length of first piece,
then $4x$ = length of second piece,
and $5x$ = length of third piece.
$$x + 4x + 5x = 30$$
$$10x = 30$$
$$\frac{10x}{10} = \frac{30}{10}$$
$$x = 3$$
$4x = 4(3) = 12$
$5x = 5(3) = 15$
The first piece is 3 feet, the second piece is 12 feet, and the third piece is 15 feet.

**59.** The longest bar represents the album Eagles: *Their Greatest Hits*, so Eagles: *Their Greatest Hits* is the best selling album of all time.

**61.** Let $x$ represent sales of *The Wall*. Then $x + 4$ is the sales of *Thriller*.
$$x + x + 4 = 50$$
$$2x + 4 = 50$$
$$2x + 4 - 4 = 50 - 4$$
$$2x = 46$$
$$\frac{2x}{2} = \frac{46}{2}$$
$$x = 23$$
$x + 4 = 23 + 4 = 27$
*Thriller* brought in $27 million and *The Wall* brought in $23 million.

**63.** answers may vary

**65.** Replace $W$ by 7 and $L$ by 10.
$2W + 2L = 2(7) + 2(10) = 14 + 20 = 34$

**67.** Replace $r$ by 15.
$\pi r^2 = \pi(15)^2 = \pi(225) = 225\pi$

**69.** Let $x$ represent the width. Then $1.6x$ represents the length. The perimeter is $2 \cdot$ length $+ 2 \cdot$ width.

$$2(1.6x) + 2x = 78$$
$$3.2x + 2x = 78$$
$$5.2x = 78$$
$$\frac{5.2x}{5.2} = \frac{78}{5.2}$$
$$x = 15$$

$1.6x = 1.6(15) = 24$

The dimensions of the garden are 15 feet by 24 feet.

**71.** 90 chirps every minute is $\dfrac{90 \text{ chirps}}{1 \text{ min}}$. There are 60 minutes in one hour.

$$\frac{90 \text{ chirps}}{1 \text{ min}} \cdot 60 \text{ min} = 5400 \text{ chirps}$$

At this rate, there are 5400 chirps each hour.

$24 \cdot 5400 = 129{,}600$

There are 129,600 chirps in one 24-hour day.

$365 \cdot 129{,}600 = 47{,}304{,}000$

There are 47,304,000 chirps in one year.

**73.** answers may vary

**75.** answers may vary

**77.** Measurements may vary. Rectangle (c) best approximates the shape of a golden rectangle.

## Section 2.6 Practice

**1.** Let $d = 580$ and $r = 5$.
$$d = r \cdot t$$
$$580 = 5t$$
$$\frac{580}{5} = \frac{5t}{5}$$
$$116 = t$$

It takes 116 seconds or 1 minute 56 seconds.

**2.** Let $l = 40$ and $P = 98$.
$$P = 2l + 2w$$
$$98 = 2 \cdot 40 + 2w$$
$$98 = 80 + 2w$$
$$98 - 80 = 80 + 2w - 80$$
$$18 = 2w$$
$$\frac{18}{2} = \frac{2w}{2}$$
$$9 = w$$

The dog run is 9 feet wide.

**3.** Let $C = 8$.
$$F = \frac{9}{5}C + 32$$
$$F = \frac{9}{5} \cdot 8 + 32$$
$$F = \frac{72}{5} + \frac{160}{5}$$
$$F = \frac{232}{5} = 46.4$$

The equivalent temperature is $46.4°$F.

**4.** Let $w =$ width of sign, then $5w + 3 =$ length of sign.
$$P = 2l + 2w$$
$$66 = 2(5w + 3) + 2w$$
$$66 = 10w + 6 + 2w$$
$$66 = 12w + 6$$
$$66 - 6 = 12w + 6 - 6$$
$$60 = 12w$$
$$\frac{60}{12} = \frac{12w}{12}$$
$$5 = w$$

$5w + 3 = 5(5) + 3 = 28$

The sign has length 28 inches and width 5 inches.

**5.**
$$I = Prt$$
$$\frac{I}{Pt} = \frac{Prt}{Pt}$$
$$\frac{I}{Pt} = r \text{ or } r = \frac{I}{Pt}$$

**6.**
$$H = 5as + 10a$$
$$H - 10a = 5as + 10a - 10a$$
$$H - 10a = 5as$$
$$\frac{H - 10a}{5a} = \frac{5as}{5a}$$
$$\frac{H - 10a}{5a} = s \text{ or } s = \frac{H - 10a}{5a}$$

**7.**
$$N = F + d(n - 1)$$
$$N - F = F + d(n - 1) - F$$
$$N - F = d(n - 1)$$
$$\frac{N - F}{n - 1} = \frac{d(n - 1)}{n - 1}$$
$$\frac{N - F}{n - 1} = d \text{ or } d = \frac{N - F}{n - 1}$$

**8.**
$$A = \frac{1}{2}a(b+B)$$
$$2 \cdot A = 2 \cdot \frac{1}{2}a(b+B)$$
$$2A = a(b+B)$$
$$2A = ab + aB$$
$$2A - ab = ab + aB - ab$$
$$2A - ab = aB$$
$$\frac{2A - ab}{a} = \frac{aB}{a}$$
$$\frac{2A - ab}{a} = B \text{ or } B = \frac{2A - ab}{a}$$

**Vocabulary, Readiness & Video Check 2.6**

1. A formula is an equation that describes known <u>relationships</u> among quantities.

2. This is a distance, rate, and time problem. The distance is given in miles and the time is given in hours, so the rate that we are finding must be in miles per hour (mph).

3. To show that the process of solving this equation for *x*—dividing both sides by 5, the coefficient of *x*—is the same process used to solve a formula for a specific variable. Treat whatever is multiplied by that specific variable as the coefficient—the coefficient is all the factors except that specific variable.

**Exercise Set 2.6**

**1.** Let $A = 45$ and $b = 15$.
$$A = bh$$
$$45 = 15h$$
$$\frac{45}{15} = \frac{15h}{15}$$
$$3 = h$$

**3.** Let $S = 102$, $l = 7$, and $w = 3$.
$$S = 4lw + 2wh$$
$$102 = 4(7)(3) + 2(3)h$$
$$102 = 84 + 6h$$
$$102 - 84 = 84 - 84 + 6h$$
$$18 = 6h$$
$$\frac{18}{6} = \frac{6h}{6}$$
$$3 = h$$

**5.** Let $A = 180$, $B = 11$, and $b = 7$.
$$A = \frac{1}{2}h(B+b)$$
$$180 = \frac{1}{2}h(11+7)$$
$$2(180) = 2\left[\frac{1}{2}h(18)\right]$$
$$360 = 18h$$
$$\frac{360}{18} = \frac{18h}{18}$$
$$20 = h$$

**7.** Let $P = 30$, $a = 8$, and $b = 10$.
$$P = a + b + c$$
$$30 = 8 + 10 + c$$
$$30 = 18 + c$$
$$30 - 18 = 18 - 18 + c$$
$$12 = c$$

**9.** Let $C = 15.7$, and $\pi \approx 3.14$.
$$C = 2\pi r$$
$$15.7 \approx 2(3.14)r$$
$$15.7 \approx 6.28r$$
$$\frac{15.7}{6.28} \approx \frac{6.28r}{6.28}$$
$$2.5 \approx r$$

**11.** Let $I = 3750$, $P = 25{,}000$, and $R = 0.05$.
$$I = PRT$$
$$3750 = 25{,}000(0.05)T$$
$$3750 = 1250T$$
$$\frac{3750}{1250} = \frac{1250T}{1250}$$
$$3 = T$$

**13.** Let $V = 565.2$, $r = 6$, and $\pi \approx 3.14$.
$$V = \frac{1}{3}\pi r^2 h$$
$$565.2 \approx \frac{1}{3}(3.14)(6)^2 h$$
$$565.2 \approx 37.68h$$
$$\frac{565.2}{37.68} \approx \frac{37.68h}{37.68}$$
$$15 \approx h$$

**15.**
$$f = 5gh$$
$$\frac{f}{5g} = \frac{5gh}{5g}$$
$$\frac{f}{5g} = h$$

**17.**  $V = lwh$

$$\frac{V}{lh} = \frac{lwh}{lh}$$

$$\frac{V}{lh} = w$$

**19.**      $3x + y = 7$

$$3x - 3x + y = 7 - 3x$$

$$y = 7 - 3x$$

**21.**      $A = P + PRT$

$$A - P = P - P + PRT$$

$$A - P = PRT$$

$$\frac{A - P}{PT} = \frac{PRT}{PT}$$

$$\frac{A - P}{PT} = R$$

**23.**  $V = \frac{1}{3}Ah$

$$3V = 3\left(\frac{1}{3}Ah\right)$$

$$3V = Ah$$

$$\frac{3V}{h} = \frac{Ah}{h}$$

$$\frac{3V}{h} = A$$

**25.**          $P = a + b + c$

$$P - (b + c) = a + b + c - (b + c)$$

$$P - b - c = a + b + c - b - c$$

$$P - b - c = a$$

**27.**            $S = 2\pi rh + 2\pi r^2$

$$S - 2\pi r^2 = 2\pi rh + 2\pi r^2 - 2\pi r^2$$

$$S - 2\pi r^2 = 2\pi rh$$

$$\frac{S - 2\pi r^2}{2\pi r} = \frac{2\pi rh}{2\pi r}$$

$$\frac{S - 2\pi r^2}{2\pi r} = h$$

**29.**  Use $A = lw$ when $A = 10{,}080$ and $w = 84$.

$$A = lw$$

$$10{,}080 = l(84)$$

$$\frac{10{,}080}{84} = \frac{84l}{84}$$

$$120 = l$$

The length (height) of the sign is 120 feet.

**31. a.**      $A = lw$          $P = 2l + 2w$
               $A = 11.5(9)$     $P = 2(11.5) + 2(9)$
               $A = 103.5$       $P = 23 + 18$
                                 $P = 41$

  The area is 103.5 square feet and the perimeter is 41 feet.

**b.**  Baseboards have to do with perimeter because they are installed around the edges. Carpet has to do with area because it is installed in the middle of the room.

**33. a.**  $A = \frac{1}{2}h(b_1 + b_2)$      $P = l_1 + l_2 + l_3 + l_4$

$P = 24 + 20 + 56 + 20$

  $A = \frac{1}{2}(12)(56 + 24)$   $P = 120$

$A = 6(80)$

$A = 480$

  The area is 480 square inches and the perimeter is 120 inches.

**b.**  The frame has to do with perimeter because it surrounds the edge of the picture. The glass has to do with area because it covers the entire picture.

**35.**  Let $F = 14$.

$$14 = \frac{9}{5}C + 32$$

$$5(14) = 5\left(\frac{9}{5}\right)C + 5(32)$$

$$70 = 9C + 160$$

$$70 - 160 = 9C + 160 - 160$$

$$-90 = 9C$$

$$\frac{-90}{9} = \frac{9C}{9}$$

$$-10 = C$$

The equivalent temperature is $-10°\text{C}$.

**37.**  Let $P = 260$ and $w = \frac{2}{3}l$.

$$P = 2l + 2w$$

$$260 = 2l + 2\left(\frac{2}{3}l\right)$$

$$260 = \frac{10}{3}l$$

$$3(260) = 3\left(\frac{10}{3}l\right)$$

$$780 = 10l$$

$$\frac{780}{10} = \frac{10l}{10}$$

$$78 = l$$

$w = \frac{2}{3}l = \frac{2}{3}(78) = 52$

The width is 52 feet and the length is 78 feet.

**39.** Let $P = 102$, $a$ = the length of the shortest side, $b = 2a$, and $c = a + 30$.

$$P = a + b + c$$
$$102 = a + 2a + a + 30$$
$$102 = 4a + 30$$
$$102 - 30 = 4a + 30 - 30$$
$$72 = 4a$$
$$\frac{72}{4} = \frac{4a}{4}$$
$$18 = a$$

$b = 2a = 2(18) = 36$
$c = a + 30 = 18 + 30 = 48$

The lengths are 18 feet, 36 feet, and 48 feet.

**41.** Let $d = 138$ and $t = 2.5$.

$$d = rt$$
$$138 = r \cdot 2.5$$
$$\frac{138}{2.5} = \frac{r \cdot 2.5}{2.5}$$
$$55.2 = r$$

The speed is 55.2 mph.

**43.** Let $l = 8$, $w = 6$, and $h = 3$.

$V = lwh$
$V = 8(6)(3) = 144$

Let $x$ = number of piranha and volume per fish = 1.5.

$$144 = 1.5x$$
$$\frac{144}{1.5} = \frac{1.5x}{1.5}$$
$$96 = x$$

96 piranhas can be placed in the tank.

**45.** Use $N = 86$.

$$T = 50 + \frac{N - 40}{4}$$
$$T = 50 + \frac{86 - 40}{4}$$
$$T = 50 + \frac{46}{4}$$
$$T = 50 + 11.5$$
$$T = 61.5$$

The temperature is 61.5° Fahrenheit.

**47.** Use $T = 55$.

$$T = 50 + \frac{N - 40}{4}$$
$$55 = 50 + \frac{N - 40}{4}$$
$$55 - 50 = 50 + \frac{N - 40}{4} - 50$$
$$5 = \frac{N - 40}{4}$$
$$4 \cdot 5 = 4 \cdot \frac{N - 40}{4}$$
$$20 = N - 40$$
$$20 + 40 = N - 40 + 40$$
$$60 = N$$

There are 60 chirps per minute.

**49.** As the number of cricket chirps per minute increases, the air temperature of their environment <u>increases</u>.

**51.** Let $h = 60$, $B = 130$, and $b = 70$.

$$A = \frac{1}{2}(B + b)h$$
$$A = \frac{1}{2}(130 + 70)60 = \frac{1}{2}(200)(60) = 6000$$

Let $x$ = number of bags of fertilizer and the area per bag = 4000.

$$4000x = 6000$$
$$\frac{4000x}{4000} = \frac{6000}{4000}$$
$$x = 1.5$$

Two bags must be purchased.

**53.** Let $d = 16$, so $r = 8$.

$A = \pi r^2 = \pi(8)^2 = 64\pi$

Let $d = 10$, so $r = 5$.

$A = 2\pi r^2 = 2\pi(5)^2 = 50\pi$

One 16-inch pizza has more area and therefore gives more pizza for the price.

**55.**
$$x + x + x + 2.5x + 2.5x = 48$$
$$8x = 48$$
$$\frac{8x}{8} = \frac{48}{8}$$
$$x = 6$$

$2.5x = 2.5(6) = 15$

Three sides measure 6 meters and two sides measure 15 meters.

**57.** $r = 0.5$ and $d = 11$.
$$d = rt$$
$$11 = 0.5t$$
$$\frac{11}{0.5} = \frac{0.5t}{0.5}$$
$$22 = t$$
It will take 22 hours.

**59.** Let $x$ = the length of a side of the square and $x + 5$ = the length of a side of the triangle.
$$P(\text{triangle}) = P(\text{square}) + 7$$
$$3(x+5) = 4x + 7$$
$$3x + 15 = 4x + 7$$
$$3x - 3x + 15 = 4x - 3x + 7$$
$$15 = x + 7$$
$$15 - 7 = x + 7 - 7$$
$$8 = x$$
$$x + 5 = 8 + 5 = 13$$
The side of the triangle is 13 inches.

**61.** Let $d = 135$ and $r = 60$.
$$d = rt$$
$$135 = 60t$$
$$\frac{135}{60} = \frac{60t}{60}$$
$$2.25 = t$$
It would take 2.25 hours.

**63.** Let $A = 1,813,500$ and $w = 150$.
$$A = lw$$
$$1,813,500 = l(150)$$
$$\frac{1,813,500}{150} = \frac{150l}{150}$$
$$12,090 = l$$
The length is 12,090 feet.

**65.** Let $F = 122$.
$$122 = \frac{9}{5}C + 32$$
$$5(122) = 5\left(\frac{9}{5}\right)C + 5(32)$$
$$610 = 9C + 160$$
$$610 - 160 = 9C + 160 - 160$$
$$450 = 9C$$
$$\frac{450}{9} = \frac{9C}{9}$$
$$50 = C$$
The equivalent temperature is 50°C.

**67.** Let $C = 167$.
$$F = \frac{9}{5}C + 32$$
$$= \frac{9}{5}(167) + 32$$
$$= 300.6 + 32$$
$$= 332.6$$
The equivalent temperature is 332.6°F.

**69.** Use $V = \frac{4}{3}\pi r^3$ when $r = \frac{9.5}{2} = 4.75$ and $\pi = 3.14$.
$$V = \frac{4}{3}\pi r^3 = \frac{4}{3}(3.14)(4.75)^3 \approx 449$$
The volume of the sphere is 449 cubic inches.

**71.** $32\% = 0.32$

**73.** $200\% = 2.00$ or $2$

**75.** $0.17 = 0.17(100\%) = 17\%$

**77.** $7.2 = 7.2(100\%) = 720\%$

**79.** Use $V = lwh$. If the length is doubled, the new length is $2l$. If the width and height are doubled, the new width and height are $2w$ and $2h$, respectively.
$$V = (2l)(2w)(2h) = 2 \cdot 2 \cdot 2lwh = 8lwh$$
The volume of the box is multiplied by 8.

**81.** Replace $T$ with $N$ and solve for $N$.
$$T = 50 + \frac{N-40}{4}$$
$$N = 50 + \frac{N-40}{4}$$
$$N - 50 = 50 + \frac{N-40}{4} - 50$$
$$N - 50 = \frac{N-40}{4}$$
$$4(N-50) = 4 \cdot \frac{N-40}{4}$$
$$4N - 200 = N - 40$$
$$4N - 200 - N = N - 40 - N$$
$$3N - 200 = -40$$
$$3N - 200 + 200 = -40 + 200$$
$$3N = 160$$
$$\frac{3N}{3} = \frac{160}{3}$$
$$N = 53\frac{1}{3}$$
They are the same when the number of cricket chirps per minute is $53\frac{1}{3}$.

**83.**

$$N = R + \frac{V}{G}$$

$$N - R = R + \frac{V}{G} - R$$

$$N - R = \frac{V}{G}$$

$$G(N - R) = G \cdot \frac{V}{G}$$

$$G(N - R) = V$$

**85.** $\square - \bigcirc \cdot \square = \triangle$

$$-\bigcirc \cdot \square = \triangle - \square$$

$$\frac{-\bigcirc \square}{-\square} = \frac{\triangle - \square}{-\square}$$

$$\bigcirc = \frac{\square - \triangle}{\square}$$

**87.** Let $d = 93{,}000{,}000$ and $r = 186{,}000$.

$$d = rt$$

$$93{,}000{,}000 = 186{,}000t$$

$$\frac{93{,}000{,}000}{186{,}000} = \frac{186{,}000t}{186{,}000}$$

$$500 = t$$

It will take 500 seconds or $8\frac{1}{3}$ minutes.

**89.** Let $t = 365$ and $r = 20$.

$$d = rt = 20(365) = 7300 \text{ inches}$$

$$\frac{7300 \text{ inches}}{1} \cdot \frac{1 \text{ foot}}{12 \text{ inch}} \approx 608.33 \text{ feet}$$

It moves about 608.33 feet.

**91.** Use $d = rt$, when $r = 581$ and $d = 42.8$.

$$d = rt$$

$$42.8 = 581t$$

$$\frac{42.8}{581} = \frac{581t}{581}$$

$$0.0737 \approx t$$

$$0.0737(60) \approx 4.42$$

It would last 4.42 minutes.

**93.** Use $d = rt$ when $d = 303$ and $t = 8\frac{1}{2}$.

$$d = rt$$

$$303 = r \cdot 8\frac{1}{2}$$

$$303 = \frac{17}{2}r$$

$$\frac{2}{17} \cdot 303 = \frac{2}{17} \cdot \frac{17}{2}r$$

$$\frac{606}{17} = r$$

$$35\frac{11}{17} = r$$

The average rate during the flight was

$35\frac{11}{17}$ mph.

**Section 2.7 Practice**

**1.** Let $x =$ the unknown percent.

$$35 = x \cdot 56$$

$$\frac{35}{56} = \frac{56x}{56}$$

$$0.625 = x$$

The number 35 is 62.5% of 56.

**2.** Let $x =$ the unknown number.

$$198 = 55\% \cdot x$$

$$198 = 0.55x$$

$$\frac{198}{0.55} = \frac{0.55x}{0.55}$$

$$360 = x$$

The number 198 is 55% of 360.

**3. a.** From the circle graph, we see that 40% of pets owned are freshwater fish and 2% are saltwater fish; thus 40% + 2% = 42% of pets owned are freshwater fish or saltwater fresh.

   **b.** The circle graph percents have a sum of 100%; thus the percent of pets that are not equines is 100% − 2% = 98%.

   **c.** To find the number of dogs owned, we find 21% of 377.41
= (0.21)(377.41)
= 9.2561
≈ 79.3
Thus, about 79.3 million dogs are owned in the United States.

**4.** Let $x$ = discount.

$x = 85\% \cdot 480$

$x = 0.85 \cdot 480$

$x = 408$

The discount is \$408.

New price = \$480 − \$408 = \$72

**5.** Increase = 2710 − 1900 = 810

Let $x$ = percent increase.

$810 = x \cdot 1900$

$\dfrac{810}{1900} = \dfrac{1900x}{1900}$

$0.426 \approx x$

The percent increase is 42.6%.

**6.** Let $x$ = number of digital 3D screens last year.

$x + 1.38x = 8459$

$2.38x = 8459$

$\dfrac{2.38x}{2.38} = \dfrac{8459}{2.38}$

$x \approx 3554$

There were 3554 screens last year.

**7.** Let $x$ = number of liters of 2% solution.

| Eyewash | No. of gallons | $\cdot$ | Acid Strength | $=$ | Amt. of Acid |
|---------|----------------|---------|---------------|-----|--------------|
| 2%      | $x$            |         | 2%            |     | $0.02x$      |
| 5%      | $6 - x$        |         | 5%            |     | $0.05(6 - x)$ |
| Mix: 3% | 6              |         | 3%            |     | $0.03(6)$    |

$0.02x + 0.05(6 - x) = 0.03(6)$

$0.02x + 0.3 - 0.05x = 0.18$

$-0.03x + 0.3 = 0.18$

$-0.03x + 0.3 - 0.3 = 0.18 - 0.3$

$-0.03x = -0.12$

$\dfrac{-0.03x}{-0.03} = \dfrac{-0.12}{-0.03}$

$x = 4$

$6 - x = 6 - 4 = 2$

She should mix 4 liters of 2% eyewash with 2 liters of 5% eyewash.

## Vocabulary, Readiness & Video Check 2.7

**1.** No, 25% + 25% + 40% = 90% ≠ 100%.

**2.** No, 30% + 30% + 30% = 90% ≠ 100%.

**3.** Yes, 25% + 25% + 25% + 25% = 100%.

**4.** Yes, 40% + 50% + 10% = 100%.

5. a.   equals; =

   b.   multiplication; ·

   c.   Drop the percent symbol and move the decimal point two places to the left.

6. a.   You also find a discount amount by multiplying the (discount) percent by the original price.

   b.   For discount, the new price is the original price minus the discount amount, so you *subtract* from the original price rather than *add* as with mark-up.

7.   You must first find the actual amount of increase in price by subtracting the original price from the new price.

8.
| Alloy | Ounces | Copper Strength | Amount of Copper |
|-------|--------|-----------------|-------------------|
| 10%   | $x$    | 0.10            | $0.10x$           |
| 30%   | 400    | 0.30            | $0.30(400)$       |
| 20%   | $x + 400$ | 0.20         | $0.20(x + 400)$   |

**Exercise Set 2.7**

1.   Let $x$ be the unknown number.
$x = 16\% \cdot 70$
$x = 0.16 \cdot 70$
$x = 11.2$
11.2 is 16% of 70.

3.   Let $x$ be the unknown percent.
$28.6 = x \cdot 52$
$\dfrac{28.6}{52} = \dfrac{52x}{52}$
$0.55 = x$
$55\% = x$
The number 28.6 is 55% of 52.

5.   Let $x$ be the unknown number.
$45 = 25\% \cdot x$
$45 = 0.25 \cdot x$
$\dfrac{45}{0.25} = \dfrac{0.25x}{0.25}$
$180 = x$
45 is 25% of 180.

7.   From the graph, 4% of adults spend more than 121 minutes on the phone each day.

9.   37% of adults talk 16–60 minutes on the phone each day.
$37\%$ of $27,000 = 37\% \cdot 27,000$
$\qquad\qquad\qquad\quad = 0.37 \cdot 27,000$
$\qquad\qquad\qquad\quad = 9990$
You would expect 9990 of the adults in Florence to talk 16–60 minutes each day.

11.  Let $x$ = amount of discount.
$x = 8\% \cdot 18,500$
$x = 0.08 \cdot 18,500$
$x = 1480$
New price $= 18,500 - 1480 = 17,020$
The discount was \$1480 and the new price is \$17,020.

13.  Let $x$ = tip.
$x = 15\% \cdot 40.50$
$x = 0.15 \cdot 40.5$
$x = 6.075 \approx 6.08$
Total $= 40.50 + 6.08 = 46.58$
The total cost is \$46.58.

15.  Decrease $= 337 - 304 = 33$
Let $x$ = percent.
$33 = x \cdot 337$
$\dfrac{33}{337} = \dfrac{337x}{337}$
$0.098 \approx x$
The percent decrease is 9.8%.

17.  Decrease $= 40 - 28 = 12$
Let $x$ = percent.
$12 = x \cdot 40$
$\dfrac{12}{40} = \dfrac{40x}{40}$
$0.3 = x$
The percent decrease is 30%.

19.  Let $x$ = the original price and $0.25x$ = the discount.
$x - 0.25x = 78$
$\quad 0.75x = 78$
$\dfrac{0.75x}{0.75} = \dfrac{78}{0.75}$
$\qquad x = 104$
The original price was \$104.

**21.** Let $x$ = last year's salary, and $0.04x$ = pay raise.

$$x + 0.04x = 44,200$$
$$1.04x = 44,200$$
$$\frac{1.04x}{1.04} = \frac{44,200}{1.04}$$
$$x = 42,500$$

Last year's salary was $42,500.

**23.** Let $x$ = the amount of pure acid.

| | No. of gallons | · Strength | = | Amt. of Acid |
|---|---|---|---|---|
| 100% | $x$ | 1.00 | | $x$ |
| 40% | 2 | 0.4 | | 2(0.4) |
| 70% | $x + 2$ | 0.7 | | 0.7(x + 2) |

$$x + 2(0.4) = 0.7(x + 2)$$
$$x + 0.8 = 0.7x + 1.4$$
$$x - 0.7x + 0.8 = 0.7x - 0.7x + 1.4$$
$$0.3x + 0.8 = 1.4$$
$$0.3x + 0.8 - 0.8 = 1.4 - 0.8$$
$$0.3x = 0.6$$
$$\frac{0.3x}{0.3} = \frac{0.6}{0.3}$$
$$x = 2$$

Mix 2 gallons of pure acid.

**25.** Let $x$ = the number of pounds at $7/lb.

| | No. of lb | · Cost/lb | = | Value |
|---|---|---|---|---|
| $7/lb | $x$ | 7 | | $7x$ |
| $4/lb | 14 | 4 | | 4(14) |
| $5/lb | $x + 14$ | 5 | | 5(x + 14) |

$$7x + 4(14) = 5(x + 14)$$
$$7x + 56 = 5x + 70$$
$$7x - 5x + 56 = 5x - 5x + 70$$
$$2x + 56 = 70$$
$$2x + 56 - 56 = 70 - 56$$
$$2x = 14$$
$$\frac{2x}{2} = \frac{14}{2}$$
$$x = 7$$

Add 7 pounds of $7/pound coffee.

**27.** Let $x$ = the number.

$$x = 23\% \cdot 20$$
$$x = 0.23 \cdot 20$$
$$x = 4.6$$

23% of 20 is 4.6.

**29.** Let $x$ = the number.

$$40 = 80\% \cdot x$$
$$40 = 0.8x$$
$$\frac{40}{0.8} = \frac{0.8x}{0.8}$$
$$50 = x$$

40 is 80% of 50.

**31.** Let $x$ = the percent.

$$144 = x \cdot 480$$
$$\frac{144}{480} = \frac{480x}{480}$$
$$0.3 = x$$

144 is 30% of 480.

**33.** From the graph, the height of the bar is about 23. Therefore, 23% of online purchases were in the category of electronic equipment.

**35.** 41% of $220,500 = 0.41 \cdot 220,500 \approx 90,405$

We predict 90,405 people will purchase books online.

**37.**

| Top Cranberry-Producing States Forecast in 2010 (in millions of pounds) | | |
|---|---|---|
| | Millions of Pounds | Percent of Total (rounded to nearest percent) |
| Wisconsin | 435 | $\frac{435}{736} \approx 0.5910 \approx 59\%$ |
| Oregon | 39 | $\frac{39}{736} \approx 0.0529 \approx 5\%$ |
| Massachusetts | 195 | $\frac{195}{736} \approx 0.2649 \approx 26\%$ |
| Washington | 14 | $\frac{14}{736} \approx 0.0190 \approx 2\%$ |
| New Jersey | 53 | $\frac{53}{736} \approx 7\%$ |
| | 736 | 99% due to rounding |

**39.** Let $x$ = the decrease in price.
$x = 0.25(256) = 64$
The decrease in price is $64.
The sale price is $256 - 64 = \$192$.

**41.** percent decrease $= \dfrac{\text{amount of decrease}}{\text{original amount}}$

$= \dfrac{23.5 - 17.1}{23.5}$

$= \dfrac{6.4}{23.5}$

$\approx 0.272$

Head lettuce consumption decreased by 27.2%.

**43.** Let $x$ = the number of vehicles in 2002.
$x + 3\% \cdot x = 246$
$x + 0.03x = 246$
$1.03x = 246$
$\dfrac{1.03x}{1.03} = \dfrac{246}{1.03}$
$x \approx 239$

There were about 239 registered vehicles in the United States in 2002.

**45.** percent increase $= \dfrac{\text{amount of increase}}{\text{original amount}}$

$= \dfrac{144 - 36}{36}$

$= \dfrac{108}{36}$

$= 3$

The area increased by 300%.

**47.** Let $x$ be the ounces of alloy that is 20% copper.

|  | ounces | concentration | amount |
|---|---|---|---|
| 20% copper | $x$ | 20% | $0.2x$ |
| 50% copper | 200 | 50% | $0.5(200)$ |
| 30% copper | $200 + x$ | 30% | $0.3(200 + x)$ |

The amount of copper being combined must be the same as that in the mixture.

$$0.2x + 0.5(200) = 0.3(200 + x)$$
$$0.2x + 100 = 60 + 0.3x$$
$$0.2x + 100 - 0.2x = 60 + 0.3x - 0.2x$$
$$100 = 60 + 0.1x$$
$$100 - 60 = 60 + 0.1x - 60$$
$$40 = 0.1x$$
$$\frac{40}{0.1} = \frac{0.1x}{0.1}$$
$$400 = x$$

Thus 400 ounces should be used.

**49.** $\text{percent decrease} = \dfrac{\text{amount of decrease}}{\text{original amount}}$

$$= \frac{6.3 - 2.1}{6.3}$$
$$= \frac{4.2}{6.3}$$
$$\approx 0.667$$

The percent decrease in the number of farms is 66.7%.

**51.** Let $x$ be the prior number of employees.

$$x - 0.35x = 78$$
$$0.65x = 78$$
$$\frac{0.65x}{0.65} = \frac{78}{0.65}$$
$$x = 120$$

There were 120 employees prior to the layoffs.

**53.** $42\% \cdot 860 = 0.42 \cdot 860 = 361.2$

You would expect 361 students to rank flexible hours as their top priority.

**55.** Let $x$ be the ounces of self-tanning lotion.

|  | ounces | cost ($) | value |
|---|---|---|---|
| self-tanning | $x$ | 3 | $3x$ |
| everyday | 800 | 0.30 | $0.3(800)$ |
| experimental | $800 + x$ | 1.20 | $1.2(800 + x)$ |

The value of the lotions being combined must be the same as the value of the mixture.

$$3x + 0.3(800) = 1.2(800 + x)$$
$$3x + 240 = 960 + 1.2x$$
$$3x + 240 - 1.2x = 960 + 1.2x - 1.2x$$
$$1.8x + 240 = 960$$
$$1.8x + 240 - 240 = 960 - 240$$
$$1.8x = 720$$
$$\frac{1.8x}{1.8} = \frac{720}{1.8}$$
$$x = 400$$

Therefore, 400 ounces of the self-tanning lotion should be used.

**57.** increase $= 48\% \cdot 577 = 0.48 \cdot 577 = 276.96$
$577 + 276.96 = 853.96$
The Naga Jolokia pepper measures 854 thousand Scoville units.

**59.** $-5 > -7$

**61.** $|-5| = -(-5)$

**63.** $(-3)^2 = 9; -3^2 = -9$

$(-3)^2 > -3^2$

**65.** no; answers may vary

**67.** no; answers may vary

**69.** 230 mg is what percent of 2400 mg?
Let $x$ represent the unknown percent.
$$x \cdot 2400 = 230$$
$$\frac{2400x}{2400} = \frac{230}{2400}$$
$$x = 0.0958\overline{3}$$
This food contains 9.6% of the daily value of sodium in one serving.

**71.** 35 is what percent of 130? Let $x$ be the unknown percent.
$$35 = x \cdot 130$$
$$\frac{35}{130} = \frac{130x}{130}$$
$$0.269 \approx x$$
The percent calories from fat is 26.9%. Yes, this food satisfies the recommendation since $26.9\% \leq 30\%$.

**73.** 12 g $\cdot$ 4 calories/gram = 48 calories
48 of the 280 calories come from protein.
$$\frac{48}{280} \approx 0.171$$
17.1% of the calories in this food come from protein.

**Section 2.8 Practice**

**1.** Let $x$ = time down, then $x + 1$ = time up.

| | Rate | $\cdot$ Time | = Distance |
|---|---|---|---|
| Up | 1.5 | $x + 1$ | $1.5(x + 1)$ |
| Down | 4 | $x$ | $4x$ |

$$d = d$$
$$1.5(x + 1) = 4x$$
$$1.5x + 1.5 = 4x$$
$$1.5 = 2.5x$$
$$\frac{1.5}{2.5} = \frac{2.5x}{2.5}$$
$$0.6 = x$$

Total Time $= x + 1 + x = 0.6 + 1 + 0.6 = 2.2$
The entire hike took 2.2 hours.

**2.** Let $x$ = speed of eastbound train, then
$x - 10$ = speed of westbound train.

| | $r$ | $\cdot$ $t$ | = $d$ |
|---|---|---|---|
| East | $x$ | 1.5 | $1.5x$ |
| West | $x - 10$ | 1.5 | $1.5(x - 10)$ |

$$1.5x + 1.5(x - 10) = 171$$
$$1.5x + 1.5x - 15 = 171$$
$$3x - 15 = 171$$
$$3x = 186$$
$$\frac{3x}{3} = \frac{186}{3}$$
$$x = 62$$
$x - 10 = 62 - 10 = 52$
The eastbound train is traveling at 62 mph and the westbound train is traveling at 52 mph.

**3.** Let $x$ = the number of \$20 bills, then
$x + 47$ = number of \$5 bills.

| Denomination | Number | Value |
|---|---|---|
| \$5 bills | $x + 47$ | $5(x + 47)$ |
| \$20 bills | $x$ | $20x$ |

$$5(x+47)+20x=1710$$
$$5x+235+20x=1710$$
$$235+25x=1710$$
$$25x=1475$$
$$x=59$$
$$x+47=59+47=106$$

There are 106 \$5 bills and 59 \$20 bills.

**4.** Let $x$ = amount invested at 11.5%, then
$30,000 - x$ = amount invested at 6%.

| | Principal · | Rate · | Time = | Interest |
|---|---|---|---|---|
| 11.5% | $x$ | 0.115 | 1 | $x(0.115)(1)$ |
| 6% | $30,000 - x$ | 0.06 | 1 | $0.06(30,000 - x)(1)$ |
| Total | 30,000 | | | 2790 |

$$0.115x+0.06(30,000-x)=2790$$
$$0.115x+1800-0.06x=2790$$
$$1800+0.055x=2790$$
$$0.055x=990$$
$$\frac{0.055x}{0.055}=\frac{990}{0.055}$$
$$x=18,000$$

$30,000 - x = 30,000 - 18,000 = 12,000$

She invested \$18,000 at 11.5% and \$12,000 at 6%.

**Vocabulary, Readiness & Video Check 2.8**

**1.**

| | $r$ · | $t$ = | $d$ |
|---|---|---|---|
| bus | 55 | $x$ | $55x$ |
| car | 50 | $x+3$ | $50(x+3)$ |

$$55x = 50(x+3)$$

**2.** The important thing is to remember the difference between the *number* of bills you have and the *value* of the bills.

**3.**

| $P$ · | $R$ · | $T$ = | $I$ |
|---|---|---|---|
| $x$ | 0.06 | 1 | $0.06x$ |
| $36,000 - x$ | 0.04 | 1 | $0.04(36,000 - x)$ |

$$0.06x = 0.04(36,000 - x)$$

**Exercise Set 2.8**

**1.** Let $x$ = the time traveled by the jet plane.

| | Rate | · Time | = Distance |
|------|------|--------|-----------|
| Jet | 500 | $x$ | $500x$ |
| Prop | 200 | $x + 2$ | $200(x + 2)$ |

$$d = d$$
$$500x = 200(x + 2)$$
$$500x = 200x + 400$$
$$300x = 400$$
$$\frac{300x}{300} = \frac{400}{300}$$
$$x = \frac{4}{3}$$

The jet traveled for $\frac{4}{3}$ hours.

$$d = rt$$
$$d = 500\left(\frac{4}{3}\right) = 666\frac{2}{3}$$

The planes are $666\frac{2}{3}$ miles from the starting

point.

**3.** Let $x$ = the average speed on the winding road
and $x + 20$ on the level.

| | Rate | · Time | = Distance |
|---------|--------|------|-----------|
| Winding | $x$ | 4 | $4x$ |
| Level | $x + 20$ | 3 | $3(x + 20)$ |
| Total | | | 305 |

$$4x + 3(x + 20) = 305$$
$$4x + 3x + 60 = 305$$
$$7x + 60 = 305$$
$$7x = 245$$
$$\frac{7x}{7} = \frac{245}{7}$$
$$x = 35$$
$$x + 20 = 35 + 20 = 55$$

The average speed on level road was 55 mph.

**5.** The value of $y$ dimes is $0.10y$.

**7.** The value of $x + 7$ nickels is $0.05(x + 7)$.

**9.** The value of $4y$ $20 bills is $20(4y)$ or $80y$.

**11.** The value of $35 - x$ $50 bills is $50(35 - x)$.

**13.** Let $x$ = number of $10 bills, then
$20 + x$ number of $5 bills.

| | Number of Bills | Value of Bills |
|-----------|-----------------|----------------|
| $5 bills | $20 + x$ | $5(20 + x)$ |
| $10 bills | $x$ | $10x$ |
| Total | | 280 |

$$5(20 + x) + 10x = 280$$
$$100 + 5x + 10x = 280$$
$$100 + 15x = 280$$
$$15x = 180$$
$$x = 12$$
$$20 + x = 32$$

There are 12 $10 bills and 32 $5 bills.

**15.** Let $x$ = the amount invested at 9% for one year.

| | Principal | · Rate | = Interest |
|-------|-----------|--------|-----------|
| 9% | $x$ | 0.09 | $0.09x$ |
| 8% | $25,000 - x$ | 0.08 | $0.08(25,000 - x)$ |
| Total | 25,000 | | 2135 |

$$0.09x + 0.08(25,000 - x) = 2135$$
$$0.09x + 2000 - 0.08x = 2135$$
$$0.01x + 2000 = 2135$$
$$0.01x = 135$$
$$\frac{0.01x}{0.01} = \frac{135}{0.01}$$
$$x = 13,500$$
$$25,000 - x = 25,000 - 13,500 = 11,500$$

She invested $11,500 at 8% and $13,500 at 9%.

**17.** Let $x$ = the amount invested at 11% for one year.

| | Principal | · Rate | = Interest |
|-------|-----------|--------|-----------|
| 11% | $x$ | 0.11 | $0.11x$ |
| 4% | $10,000 - x$ | $-0.04$ | $-0.04(10,000 - x)$ |
| Total | 10,000 | | 650 |

$$0.11x - 0.04(10,000 - x) = 650$$
$$0.11x - 400 + 0.04x = 650$$
$$0.15x - 400 = 650$$
$$0.15x = 1050$$
$$\frac{0.15x}{0.15} = \frac{1050}{0.15}$$
$$x = 7000$$
$$10,000 - x = 10,000 - 7000 = 3000$$

He invested $7000 at 11% and $3000 at 4%.

**19.** Let $x$ = the number of adult tickets, then $500 - x$ = the number of child tickets.

| | Number · | Rate = | Cost |
|---|---|---|---|
| Adult | $x$ | 43 | $43x$ |
| Child | $500 - x$ | 28 | $28(500 - x)$ |
| Total | 500 | | 16,805 |

$$43x + 28(500 - x) = 16,805$$
$$43x + 14,000 - 28x = 16,805$$
$$14,000 + 15x = 16,805$$
$$15x = 2805$$
$$x = 187$$
$$500 - x = 500 - 187 = 313$$

Sales included 187 adult tickets and 313 child tickets.

**21.** Let $x$ = the time traveled.

| | Rate · | Time = | Distance |
|---|---|---|---|
| Car A | 56 | $x$ | $56x$ |
| Car B | 47 | $x$ | $47x$ |

The total distance is 206 miles.
$$56x + 47x = 206$$
$$103x = 206$$
$$\frac{103x}{103} = \frac{206}{103}$$
$$x = 2$$

The two cars will be 206 miles apart in 2 hours.

**23.** Let $x$ = the amount invested at 10% for one year.

| | Principal · | Rate = | Interest |
|---|---|---|---|
| 10% | $x$ | 0.10 | $0.10x$ |
| 8% | $54,000 - x$ | 0.08 | $0.08(54,000 - x)$ |

$$0.10x = 0.08(54,000 - x)$$
$$0.10x = 4320 - 0.08x$$
$$0.18x = 4320$$
$$\frac{0.18x}{0.18} = \frac{4320}{0.18}$$
$$x = 24,000$$
$$54,000 - x = 54,000 - 24,000 = 30,000$$

Invest $30,000 at 8% and $24,000 at 10%.

**25.** Let $x$ = the time they are able to talk.

| | Rate · | Time = | Distance |
|---|---|---|---|
| Alan | 55 | $x$ | $55x$ |
| Dave | 65 | $x - 1$ | $65(x - 1)$ |
| Total | | | 250 |

$$55x + 65(x - 1) = 250$$
$$55x + 65x - 65 = 250$$
$$120x - 65 = 250$$
$$120x = 315$$
$$\frac{120x}{120} = \frac{315}{120}$$
$$x = 2\frac{5}{8}$$

They can talk for $2\frac{5}{8}$ hours or $2$ hours $37\frac{1}{2}$ minutes.

**27.** Let $x$ = the speed of the slower train.

| | Rate · | Time = | Distance |
|---|---|---|---|
| Train A | $x$ | 2.5 | $2.5x$ |
| Train B | $x + 10$ | 2.5 | $2.5(x + 10)$ |

The total distance is 205 miles.
$$2.5x + 2.5(x + 10) = 205$$
$$2.5x + 2.5x + 25 = 205$$
$$5x + 25 = 205$$
$$5x = 180$$
$$\frac{5x}{5} = \frac{180}{5}$$
$$x = 36$$
$$x + 10 = 46$$

The speeds of the trains are 36 mph and 46 mph.

**29.** Let $x$ = number of nickels, then
$3x$ = number of dimes.

|        | Number | Value     |
|--------|--------|-----------|
| Nickels | $x$    | $0.05x$   |
| Dimes   | $3x$   | $0.10(3x)$ |
| Total   |        | 56.35     |

$$0.05x + 0.10(3x) = 56.35$$
$$0.05x + 0.3x = 56.35$$
$$0.35x = 56.35$$
$$x = 161$$
$3x = 3(161) = 483$
They collected 161 nickels and 483 dimes.

**31.** Let $x$ = time traveled.

| | Rate | · Time | = Distance |
|-------|------|------|------|
| Truck | 52   | $x$  | $52x$ |
| Van   | 63   | $x$  | $63x$ |

The total distance is 460 miles.
$$52x + 63x = 460$$
$$115x = 460$$
$$\frac{115x}{115} = \frac{460}{115}$$
$$x = 4$$
The truck and the van will be 460 miles apart in 4 hours.

**33.** Let $x$ = time traveled.

| | Rate | · Time | = Distance |
|-------|------|------|------|
| Car A | 70   | $x$  | $70x$ |
| Car B | 58   | $x$  | $58x$ |

They are traveling in the same direction, so find the difference of their distances.
$$70x - 58x = 30$$
$$12x = 30$$
$$\frac{12x}{12} = \frac{30}{12}$$
$$x = 2.5$$
They will be 30 miles apart in 2.5 hours.

**35.** Let $x$ = the amount invested at 9% for one year.

| | Principal · | Rate = | Interest |
|-------|------|------|------|
| 9%    | $x$  | 0.09 | $0.09x$ |
| 6%    | 3000 | 0.06 | $0.06(3000)$ |
| Total |      |      | 585 |

$$0.09x + 0.06(3000) = 585$$
$$0.09x + 180 = 585$$
$$0.09x = 405$$
$$\frac{0.09x}{0.09} = \frac{405}{0.09}$$
$$x = 4500$$
Should invest \$4500 at 9%.

**37.** Let $x$ = the rate of hiker 1.

| | Rate | · Time | = Distance |
|---------|--------|------|------------|
| Hiker 1 | $x$    | 2    | $2x$       |
| Hiker 2 | $x + 1.1$ | 2 | $2(x + 1.1)$ |
| Total   |        |      | 11         |

$$2x + 2(x + 1.1) = 11$$
$$2x + 2x + 2.2 = 11$$
$$4x + 2.2 = 11$$
$$4x = 8.8$$
$$\frac{4x}{4} = \frac{8.8}{4}$$
$$x = 2.2$$
$x + 1.1 = 2.2 + 1.1 = 3.3$
Hiker 1: 2.2 mph; Hiker 2: 3.3 mph

**39.** Let $x$ = the time spent rowing upstream.

| | Rate · | Time = | Distance |
|------------|------|---------|------------|
| Upstream   | 5    | $x$     | $5x$       |
| Downstream | 11   | $4 - x$ | $11(4 - x)$ |

$$5x = 11(4 - x)$$
$$5x = 44 - 11x$$
$$16x = 44$$
$$\frac{16x}{16} = \frac{44}{16}$$
$$x = 2.75$$
He rowed upstream for 2.75 hours.
$d = rt$
$d = 5(2.75) = 13.75$
He rowed 13.75 miles each way for a total of 27.5 miles.

**41.** $3 + (-7) = -4$

**43.** $\dfrac{3}{4} - \dfrac{3}{16} = \dfrac{4}{4} \cdot \dfrac{3}{4} - \dfrac{3}{16} = \dfrac{12}{16} - \dfrac{3}{16} = \dfrac{12-3}{16} = \dfrac{9}{16}$

**45.** $-5 - (-1) = -5 + 1 = -4$

**47.** Let $x$ = number of \$100 bills, then
$x + 46$ = number of \$50 bills, and
$7x$ = number of \$20 bills.

| | Number | Value |
|---|---|---|
| \$100 bills | $x$ | $100x$ |
| \$50 bills | $x + 46$ | $50(x + 46)$ |
| \$20 bills | $7x$ | $20(7x)$ |
| Total | | $9550$ |

$100x + 50(x + 46) + 20(7x) = 9550$
$100x + 50x + 2300 + 140x = 9550$
$290x + 2300 = 9550$
$290x = 7250$
$x = 25$

$x + 46 = 71$
$7x = 7(25) = 175$
There were 25 \$100 bills, 71 \$50 bills, and 175 \$20 bills.

**49.** $\quad R = C$
$24x = 100 + 20x$
$4x = 100$
$\dfrac{4x}{x} = \dfrac{100}{4}$
$x = 25$
Should sell 25 skateboards to break even.

**51.** $\quad R = C$
$7.50x = 4.50x + 2400$
$3x = 2400$
$\dfrac{3x}{3} = \dfrac{2400}{3}$
$x = 800$
Should sell 800 books to break even.

**53.** Answers may vary

**Section 2.9 Practice**

**1.** $x < 5$
Place a parenthesis at 5 since the inequality symbol is $<$. Shade to the left of 5. The solution set is $(-\infty, 5)$.

**2.** $\quad x + 11 \ge 6$
$x + 11 - 11 \ge 6 - 11$
$x \ge -5$
The solution set is $[-5, \infty)$.

**3.** $\quad -5x \ge -15$
$\dfrac{-5x}{-5} \le \dfrac{-15}{-5}$
$x \le 3$
The solution set is $(-\infty, 3]$.

**4.** $\quad 3x > -9$
$\dfrac{3x}{3} > \dfrac{-9}{3}$
$x > -3$
The solution set is $(-3, \infty)$.

**5.** $\quad 45 - 7x \le -4$
$45 - 7x - 45 \le -4 - 45$
$-7x \le -49$
$\dfrac{-7x}{-7} \ge \dfrac{-49}{-7}$
$x \ge 7$
The solution set is $[7, \infty)$.

**6.** $\quad 3x + 20 \le 2x + 13$
$3x + 20 - 2x \le 2x + 13 - 2x$
$x + 20 \le 13$
$x + 20 - 20 \le 13 - 20$
$x \le -7$
The solution set is $(-\infty, -7]$.

**7.**
$$6 - 5x > 3(x - 4)$$
$$6 - 5x > 3x - 12$$
$$6 - 5x - 3x > 3x - 12 - 3x$$
$$6 - 8x > -12$$
$$6 - 8x - 6 > -12 - 6$$
$$-8x > -18$$
$$\frac{-8x}{-8} < \frac{-18}{-8}$$
$$x < \frac{9}{4}$$

The solution set is $\left(-\infty, \frac{9}{4}\right)$.

$\frac{9}{4}$

**8.**
$$3(x - 4) - 5 \le 5(x - 1) - 12$$
$$3x - 12 - 5 \le 5x - 5 - 12$$
$$3x - 17 \le 5x - 17$$
$$3x - 17 - 5x \le 5x - 17 - 5x$$
$$-2x - 17 \le -17$$
$$-2x - 17 + 17 \le -17 + 17$$
$$-2x \le 0$$
$$\frac{-2x}{-2} \ge \frac{0}{-2}$$
$$x \ge 0$$

The solution set is $[0, \infty)$.

$0$

**9.** $-3 \le x < 1$

Graph all numbers greater than or equal to $-3$ and less than 1. Place a bracket at $-3$ and a parenthesis at 1.

The solution set is $[-3, 1)$.

$-3$     $1$

**10.**
$$-4 < 3x + 2 \le 8$$
$$-4 - 2 < 3x + 2 - 2 \le 8 - 2$$
$$-6 < 3x \le 6$$
$$\frac{-6}{3} < \frac{3x}{3} \le \frac{6}{3}$$
$$-2 < x \le 2$$

The solution set is $(-2, 2]$.

$-2$     $2$

**11.**
$$1 < \frac{3}{4}x + 5 < 6$$
$$4(1) < 4\left(\frac{3}{4}x + 5\right) < 4(6)$$
$$4 < 3x + 20 < 24$$
$$4 - 20 < 3x + 20 - 20 < 24 - 20$$
$$-16 < 3x < 4$$
$$\frac{-16}{3} < \frac{3x}{3} < \frac{4}{3}$$
$$-\frac{16}{3} < x < \frac{4}{3}$$

The solution set is $\left(-\frac{16}{3}, \frac{4}{3}\right)$.

$-\frac{16}{3}$     $\frac{4}{3}$

$-6$     $0$   $2$

**12.** Let $x$ = the number.
$$35 - 2x > 15$$
$$35 - 2x - 35 > 15 - 35$$
$$-2x > -20$$
$$\frac{-2x}{-2} < \frac{-20}{-2}$$
$$x < 10$$

All numbers less than 10.

**13.** Let $x$ = number of classes.
$$300 + 375x \le 1500$$
$$300 + 375x - 300 \le 1500 - 300$$
$$375x \le 1200$$
$$\frac{375x}{375} \le \frac{1200}{375}$$
$$x \le 3.2$$

Kasonga can afford at most 3 community college classes this semester.

**Vocabulary, Readiness & Video Check 2.9**

**1.** $6x - 7(x + 9)$ is an expression.

**2.** $6x = 7(x + 9)$ is an equation.

**3.** $6x < 7(x + 9)$ is an inequality.

**4.** $5y - 2 \ge -38$ is an inequality.

**5.** $-5$ is not a solution to $x \ge -3$.

**6.** $|-6| = 6$ is not a solution to $x < 6$.

**7.** The graph of Example 1 is shaded from $-\infty$ to and including $-1$, as indicated by a bracket. To write interval notation, you write down what is shaded for the inequality from left to right. A parenthesis is always used with $-\infty$, so from the graph, the interval notation is $(-\infty, -1]$.

**8.** Step 5 is where you apply the multiplication property of inequality. If a negative number is multiplied or divided when applying this property, you need to make sure you remember to reverse the direction of the inequality symbol.

**9.** You would divide the left, middle, and right by $-3$ instead of 3, which would reverse the directions of both inequality symbols.

**10.** no; greater than; $\leq$

**Exercise Set 2.9**

**1.** $[2, \infty), x \geq 2$

**3.** $(-\infty, -5), x < -5$

**5.** $x \leq -1, (-\infty, -1]$

**7.** $x < \dfrac{1}{2}, \left(-\infty, \dfrac{1}{2}\right)$

**9.** $y \geq 5, [5, \infty)$

**11.** $2x < -6$
$x < -3, (-\infty, -3)$

**13.** $x - 2 \geq -7$
$x \geq -5, [-5, \infty)$

**15.** $-8x \leq 16$
$\dfrac{-8x}{-8} \geq \dfrac{16}{-8}$
$x \geq -2, [-2, \infty)$

**17.** $3x - 5 > 2x - 8$
$x - 5 > -8$
$x > -3, (-3, \infty)$

**19.** $4x - 1 \leq 5x - 2x$
$4x - 1 \leq 3x$
$x - 1 \leq 0$
$x \leq 1, (-\infty, 1]$

**21.** $x - 7 < 3(x + 1)$
$x - 7 < 3x + 3$
$-2x - 7 < 3$
$-2x < 10$
$\dfrac{-2x}{-2} > \dfrac{10}{-2}$
$x > -5, (-5, \infty)$

**23.** $-6x + 2 \geq 2(5 - x)$
$-6x + 2 \geq 10 - 2x$
$-4x + 2 \geq 10$
$-4x \geq 8$
$\dfrac{-4x}{-4} \leq \dfrac{8}{-4}$
$x \leq -2, (\infty, -2]$

**25.** $4(3x - 1) \leq 5(2x - 4)$
$12x - 4 \leq 10x - 20$
$2x - 4 \leq -20$
$2x \leq -16$
$x \leq -8, (-\infty, -8]$

**27.** $3(x+2)-6 > -2(x-3)+14$
$3x+6-6 > -2x+6+14$
$3x > -2x+20$
$5x > 20$
$x > 4, (4, \infty)$

**29.** $-2x \le -40$
$\dfrac{-2x}{-2} \ge \dfrac{-40}{-2}$
$x \ge 20, [20, \infty)$

**31.** $-9+x > 7$
$x > 16, (16, \infty)$

**33.** $3x-7 < 6x+2$
$-3x-7 < 2$
$-3x < 9$
$\dfrac{-3x}{-3} > \dfrac{9}{-3}$
$x > -3, (-3, \infty)$

**35.** $5x-7x \ge x+2$
$-2x \ge x+2$
$-3x \ge 2$
$\dfrac{-3x}{-3} \le \dfrac{2}{-3}$
$x \le -\dfrac{2}{3}, \left(-\infty, -\dfrac{2}{3}\right]$

**37.** $\dfrac{3}{4}x > 2$
$x > \dfrac{8}{3}, \left(\dfrac{8}{3}, \infty\right)$

**39.** $3(x-5) < 2(2x-1)$
$3x-15 < 4x-2$
$-x-15 < -2$
$-x < 13$
$\dfrac{-x}{-1} > \dfrac{13}{-1}$
$x > -13, (-13, \infty)$

**41.** $4(2x+1) < 4$
$8x+4 < 4$
$8x < 0$
$x < 0, (-\infty, 0)$

**43.** $-5x+4 \ge -4(x-1)$
$-5x+4 \ge -4x+4$
$-x+4 \ge 4$
$-x \ge 0$
$\dfrac{-x}{-1} \le \dfrac{0}{-1}$
$x \le 0, (-\infty, 0]$

**45.** $-2(x-4)-3x < -(4x+1)+2x$
$-2x+8-3x < -4x-1+2x$
$-5x+8 < -2x-1$
$-3x+8 < -1$
$-3x < -9$
$\dfrac{-3x}{-3} > \dfrac{-9}{-3}$
$x > 3, (3, \infty)$

**47.**
$$\frac{1}{4}(x+4) < \frac{1}{5}(2x+3)$$
$$20 \cdot \frac{1}{4}(x+4) < 20 \cdot \frac{1}{5}(2x+3)$$
$$5(x+4) < 4(2x+3)$$
$$5x+20 < 8x+12$$
$$5x+20-5x < 8x+12-5x$$
$$20 < 3x+12$$
$$20-12 < 3x+12-12$$
$$8 < 3x$$
$$\frac{8}{3} < \frac{3x}{3}$$
$$\frac{8}{3} < x, \left(\frac{8}{3}, \infty\right)$$

**49.** $-1 < x < 3, (-1, 3)$

**51.** $0 \le y < 2, [0, 2)$

**53.** $-3 < 3x < 6$
$-1 < x < 2, (-1, 2)$

**55.** $2 \le 3x-10 \le 5$
$12 \le 3x \le 15$
$4 \le x \le 5, [4, 5]$

**57.** $-4 < 2(x-3) \le 4$
$-4 < 2x-6 \le 4$
$2 < 2x \le 10$
$1 < x \le 5, (1, 5]$

**59.** $-2 < 3x-5 < 7$
$3 < 3x < 12$
$1 < x < 4, (1, 4)$

**61.** $-6 < 3(x-2) \le 8$
$-6 < 3x-6 \le 8$
$0 < 3x \le 14$
$$0 < x \le \frac{14}{3}, \left(0, \frac{14}{3}\right]$$

**63.** Let $x$ be the number.
$$2x+6 > -14$$
$$2x+6-6 > -14-6$$
$$2x > -20$$
$$\frac{2x}{2} > \frac{-20}{2}$$
$$x > -10$$
All numbers greater than $-10$ make this statement true.

**65.** Use $P = 2l + 2w$ when $w = 15$ and $P \le 100$.
$$2l+2(15) \le 100$$
$$2l+30 \le 100$$
$$2l+30-30 \le 100-30$$
$$2l \le 70$$
$$\frac{2l}{2} \le \frac{70}{2}$$
$$l \le 35$$
The maximum length of the rectangle is 35 cm.

**67.** Let $x$ be the score in his third game.
$$\frac{146+201+x}{3} \ge 180$$
$$\frac{347+x}{3} \ge 180$$
$$3 \cdot \frac{347+x}{3} \ge 3 \cdot 180$$
$$347+x \ge 540$$
$$347+x-347 \ge 540-347$$
$$x \ge 193$$
He must bowl at least 193 in the third game.

**69.** Let $x$ represent the number of people. Then the cost is $50 + 34x$.
$$50+34x \le 3000$$
$$50+34x-50 \le 3000-50$$
$$34x \le 2950$$
$$\frac{34x}{34} \le \frac{2950}{34}$$
$$x \le \frac{2950}{34} \approx 86.76$$
They can invite at most 86 people.

**71.** Let $x$ represent the number of minutes.

$$5.8x \geq 200$$

$$\frac{5.8x}{5.8} \geq \frac{200}{5.8}$$

$$x \geq \frac{200}{5.8} \approx 35$$

The person must walk at least 35 minutes.

**73.** Let $x =$ the unknown number.

$$-5 < 2x + 1 < 7$$
$$-6 < 2x < 6$$
$$-3 < x < 3$$

All numbers between $-3$ and $3$

**75.** $(2)^3 = (2)(2)(2) = 8$

**77.** $(1)^{12} = (1)(1)(1)(1)(1)(1)(1)(1)(1)(1)(1)(1) = 1$

**79.** $\left(\frac{4}{7}\right)^2 = \left(\frac{4}{7}\right)\left(\frac{4}{7}\right) = \frac{16}{49}$

**81.** Since $3 < 5$, $3(-4) > 5(-4)$.

**83.** If $m \leq n$, then $-2m \geq -2n$.

**85.** Reverse the direction of the inequality symbol when multiplying or dividing by a negative number.

**87.** Let $x$ be the score on his final exam. Since the final counts as two tests, his final course average is $\dfrac{75 + 83 + 85 + 2x}{5}$.

$$\frac{75 + 83 + 85 + 2x}{5} \geq 80$$

$$\frac{243 + 2x}{5} \geq 80$$

$$5\left(\frac{243 + 2x}{5}\right) \geq 5(80)$$

$$243 + 2x \geq 400$$
$$243 + 2x - 243 \geq 400 - 243$$
$$2x \geq 157$$
$$\frac{2x}{2} \geq \frac{157}{2}$$
$$x \geq 78.5$$

His final exam score must be at least 78.5 for him to get a B.

**89.** answers may vary

**91.** answers may vary

**93.** $C = 3.14d$
$$2.9 \leq 3.14d \leq 3.1$$
$$0.924 \leq d \leq 0.987$$
The diameter must be between 0.924 cm and 0.987 cm.

**95.** $x(x+4) > x^2 - 2x + 6$
$$x^2 + 4x > x^2 - 2x + 6$$
$$4x > -2x + 6$$
$$6x > 6$$
$$x > 1, \ (1, \infty)$$

**97.** $x^2 + 6x - 10 < x(x - 10)$
$$x^2 + 6x - 10 < x^2 - 10x$$
$$6x - 10 < -10x$$
$$16x - 10 < 0$$
$$16x < 10$$
$$x < \frac{10}{6}$$
$$x < \frac{5}{8}, \ \left(-\infty, \frac{5}{8}\right)$$

## Chapter 2 Vocabulary Check

1. Terms with the same variables raised to exactly the same powers are called like terms.

2. If terms are not like terms, they are unlike terms.

3. A linear equation in one variable can be written in the form $ax + b = c$.

4. A linear inequality in one variable can be written in the form $ax + b < c$, (or $>$, $\leq$, $\geq$).

5. Inequalities containing two inequality symbols are called compound inequalities.

6. An equation that describes a known relationship among quantities is called a formula.

7. The numerical coefficient of a term is its numerical factor.

8. Equations that have the same solution are called equivalent equations.

9. The solutions to the equation $x + 5 = x + 5$ are all real numbers.

**10.** The solution to the equation $x + 5 = x + 4$ is <u>no solution</u>.

**11.** If both sides of an inequality are multiplied or divided by the same positive number, the direction of the inequality symbol is <u>the same</u>.

**12.** If both sides of an inequality are multiplied by the same negative number, the direction of the inequality symbol is <u>reversed</u>.

**Chapter 2 Review**

**1.** $5x - x + 2x = 6x$

**2.** $0.2z - 4.6x - 7.4z = -4.6x - 7.2z$

**3.** $\dfrac{1}{2}x + 3 + \dfrac{7}{2}x - 5 = \dfrac{8}{2}x - 2 = 4x - 2$

**4.** $\dfrac{4}{5}y + 1 + \dfrac{6}{5}y + 2 = \dfrac{10}{5}y + 3 = 2y + 3$

**5.** $2(n - 4) + n - 10 = 2n - 8 + n - 10 = 3n - 18$

**6.** $3(w + 2) - (12 - w) = 3w + 6 - 12 + w = 4w - 6$

**7.** $(x + 5) - (7x - 2) = x + 5 - 7x + 2 = -6x + 7$

**8.** $(y - 0.7) - (1.4y - 3) = y - 0.7 - 1.4y + 3$
$= -0.4y + 2.3$

**9.** Three times a number decreased by 7 is $3x - 7$.

**10.** Twice the sum of a number and 2.8 added to 3 times the number is $2(x + 2.8) + 3x$.

**11.** $\begin{aligned} 8x + 4 &= 9x \\ 8x + 4 - 8x &= 9x - 8x \\ 4 &= x \end{aligned}$

**12.** $\begin{aligned} 5y - 3 &= 6y \\ 5y - 3 - 5y &= 6y - 5y \\ -3 &= y \end{aligned}$

**13.** $\begin{aligned} \dfrac{2}{7}x + \dfrac{5}{7}x &= 6 \\ \dfrac{7}{7}x &= 6 \\ x &= 6 \end{aligned}$

**14.** $\begin{aligned} 3x - 5 &= 4x + 1 \\ -5 &= x + 1 \\ -6 &= x \end{aligned}$

**15.** $\begin{aligned} 2x - 6 &= x - 6 \\ x - 6 &= -6 \\ x &= 0 \end{aligned}$

**16.** $\begin{aligned} 4(x + 3) &= 3(1 + x) \\ 4x + 12 &= 3 + 3x \\ x + 12 &= 3 \\ x &= -9 \end{aligned}$

**17.** $\begin{aligned} 6(3 + n) &= 5(n - 1) \\ 18 + 6n &= 5n - 5 \\ 18 + n &= -5 \\ n &= -23 \end{aligned}$

**18.** $\begin{aligned} 5(2 + x) - 3(3x + 2) &= -5(x - 6) + 2 \\ 10 + 5x - 9x - 6 &= -5x + 30 + 2 \\ -4x + 4 &= -5x + 32 \\ x + 4 &= 32 \\ x &= 28 \end{aligned}$

**19.** $\begin{aligned} x - 5 &= 3 \\ x - 5 + \underline{5} &= 3 + \underline{5} \\ x &= 8 \end{aligned}$

**20.** $\begin{aligned} x + 9 &= -2 \\ x + 9 - \underline{9} &= -2 - \underline{9} \\ x &= -11 \end{aligned}$

**21.** $10 - x$; choice b.

**22.** $x - 5$; choice a.

**23.** Complementary angles sum to $90°$.
$(90 - x)°$; choice b.

**24.** Supplementary angles sum to $180°$.
$180 - (x + 5) = 180 - x - 5 = 175 - x$
$(175 - x)°$; choice c.

**25.** $\begin{aligned} \dfrac{3}{4}x &= -9 \\ \dfrac{4}{3}\left(\dfrac{3}{4}x\right) &= \dfrac{4}{3}(-9) \\ x &= -12 \end{aligned}$

**26.** $\begin{aligned} \dfrac{x}{6} &= \dfrac{2}{3} \\ 6 \cdot \dfrac{x}{6} &= 6 \cdot \dfrac{2}{3} \\ x &= 4 \end{aligned}$

**27.** $-5x = 0$

$$\frac{-5x}{-5} = \frac{0}{-5}$$

$$x = 0$$

**28.** $-y = 7$

$$\frac{-y}{-1} = \frac{7}{-1}$$

$$y = -7$$

**29.** $0.2x = 0.15$

$$\frac{0.2x}{0.2} = \frac{0.15}{0.2}$$

$$x = 0.75$$

**30.** $\dfrac{-x}{3} = 1$

$$-3 \cdot \frac{-x}{3} = -3 \cdot 1$$

$$x = -3$$

**31.** $-3x + 1 = 19$

$$-3x = 18$$

$$\frac{-3x}{-3} = \frac{18}{-3}$$

$$x = -6$$

**32.** $5x + 25 = 20$

$$5x = -5$$

$$\frac{5x}{5} = \frac{-5}{5}$$

$$x = -1$$

**33.** $7(x - 1) + 9 = 5x$

$$7x - 7 + 9 = 5x$$

$$7x + 2 = 5x$$

$$2 = -2x$$

$$\frac{2}{-2} = \frac{-2x}{-2}$$

$$-1 = x$$

**34.** $7x - 6 = 5x - 3$

$$2x - 6 = -3$$

$$2x = 3$$

$$\frac{2x}{2} = \frac{3}{2}$$

$$x = \frac{3}{2}$$

**35.** $-5x + \dfrac{3}{7} = \dfrac{10}{7}$

$$7\left(-5x + \frac{3}{7}\right) = 7 \cdot \frac{10}{7}$$

$$-35x + 3 = 10$$

$$-35x = 7$$

$$x = -\frac{7}{35}$$

$$x = -\frac{1}{5}$$

**36.** $5x + x = 9 + 4x - 1 + 6$

$$6x = 4x + 14$$

$$2x = 14$$

$$x = 7$$

**37.** Let $x$ = the first integer, then
$x + 1$ = the second integer, and
$x + 2$ = the third integer.
sum = $x + (x + 1) + (x + 2) = 3x + 3$

**38.** Let $x$ = the first integer, then
$x + 2$ = the second integer
$x + 4$ = the third integer
$x + 6$ = the fourth integer.
sum = $x + (x + 6) = 2x + 6$

**39.** $\dfrac{5}{3}x + 4 = \dfrac{2}{3}x$

$$3\left(\frac{5}{3}x + 4\right) = 3\left(\frac{2}{3}x\right)$$

$$5x + 12 = 2x$$

$$12 = -3x$$

$$-4 = x$$

**40.** $\dfrac{7}{8}x + 1 = \dfrac{5}{8}x$

$$8\left(\frac{7}{8}x + 1\right) = 8\left(\frac{5}{8}x\right)$$

$$7x + 8 = 5x$$

$$8 = -2x$$

$$-4 = x$$

**41.** $-(5x + 1) = -7x + 3$

$$-5x - 1 = -7x + 3$$

$$2x - 1 = 3$$

$$2x = 4$$

$$x = 2$$

**42.**  $-4(2x+1) = -5x+5$
$-8x-4 = -5x+5$
$-3x-4 = 5$
$-3x = 9$
$x = -3$

**43.**  $-6(2x-5) = -3(9+4x)$
$-12x+30 = -27-12x$
$30 = -27$
There is no solution.

**44.**  $3(8y-1) = 6(5+4y)$
$24y-3 = 30+24y$
$-3 = 30$
There is no solution.

**45.**  $\dfrac{3(2-z)}{5} = z$
$3(2-z) = 5z$
$6-3z = 5z$
$6 = 8z$
$\dfrac{6}{8} = z$
$\dfrac{3}{4} = z$

**46.**  $\dfrac{4(n+2)}{5} = -n$
$4(n+2) = -5n$
$4n+8 = -5n$
$8 = -9n$
$-\dfrac{8}{9} = n$

**47.**  $0.5(2n-3)-0.1 = 0.4(6+2n)$
$10[0.5(2n-3)-0.1] = 10[0.4(6+2n)]$
$5(2n-3)-1 = 4(6+2n)$
$10n-15-1 = 24+8n$
$10n-16 = 24+8n$
$2n-16 = 24$
$2n = 40$
$n = 20$

**48.**  $-9-5a = 3(6a-1)$
$-9-5a = 18a-3$
$-9 = 23a-3$
$-6 = 23a$
$-\dfrac{6}{23} = a$

**49.**  $\dfrac{5(c+1)}{6} = 2c-3$
$5(c+1) = 6(2c-3)$
$5c+5 = 12c-18$
$-7c+5 = -18$
$-7c = -23$
$c = \dfrac{23}{7}$

**50.**  $\dfrac{2(8-a)}{3} = 4-4a$
$2(8-a) = 3(4-4a)$
$16-2a = 12-12a$
$10a+16 = 12$
$10a = -4$
$a = \dfrac{-4}{10}$
$a = -\dfrac{2}{5}$

**51.**  $200(70x-3560) = -179(150x-19,300)$
$14,000x-712,000 = -26,850x+3,454,700$
$40,850x-712,000 = 3,454,700$
$40,850x = 4,166,700$
$x = 102$

**52.**  $1.72y-0.04y = 0.42$
$1.68y = 0.42$
$y = 0.25$

**53.**  Let $x$ = length of a side of the square, then
$50.5 + 10x$ = the height.
$x+(50.5+10x) = 7327$
$11x+50.5 = 7327$
$11x = 7276.5$
$x = 661.5$
$50.5 + 10x = 50.5 + 10(661.5) = 6665.5$
The height is 6665.5 inches.

**54.**  Let $x$ = the length of the shorter piece and
$2x$ = the length of the other.
$x+2x = 12$
$3x = 12$
$x = 4$
$2x = 2(4) = 8$
The lengths are 4 feet and 8 feet.

**55.** Let $x$ = number of Keebler plants, then
$2x - 1$ = number of Kellogg plants.
$$x + (2x - 1) = 53$$
$$3x - 1 = 53$$
$$3x = 54$$
$$x = 18$$
$2x - 1 = 2(18) - 1 = 35$
There were 18 Keebler plants and 35 Kellogg plants.

**56.** Let $x$ = first integer, then
$x + 1$ = second integer, and
$x + 2$ = third integer.
$$x + (x + 1) + (x + 2) = -114$$
$$3x + 3 = -114$$
$$3x = -117$$
$$x = -39$$
$x + 1 = -39 + 1 = -38$
$x + 2 = -39 + 2 = -37$
The integers are $-39, -38, -37$.

**57.** Let $x$ = the unknown number.
$$\frac{x}{3} = x - 2$$
$$3 \cdot \frac{x}{3} = 3(x - 2)$$
$$x = 3x - 6$$
$$-2x = -6$$
$$x = 3$$
The number is 3.

**58.** Let $x$ = the unknown number.
$$2(x + 6) = -x$$
$$2x + 12 = -x$$
$$12 = -3x$$
$$-4 = x$$
The number is $-4$.

**59.** Let $P = 46$ and $l = 14$.
$$P = 2l + 2w$$
$$46 = 2(14) + 2w$$
$$46 = 28 + 2w$$
$$18 = 2w$$
$$9 = w$$

**60.** Let $V = 192$, $l = 8$, and $w = 6$.
$$V = lwh$$
$$192 = 8(6)h$$
$$192 = 48h$$
$$4 = h$$

**61.**
$$y = mx + b$$
$$y - b = mx$$
$$\frac{y - b}{x} = m$$

**62.**
$$r = vst - 5$$
$$r + 5 = vst$$
$$\frac{r + 5}{vt} = s$$

**63.** $2y - 5x = 7$
$$-5x = -2y + 7$$
$$x = \frac{-2y + 7}{-5}$$
$$x = \frac{2y - 7}{5}$$

**64.** $3x - 6y = -2$
$$-6y = -3x - 2$$
$$y = \frac{-3x - 2}{-6}$$
$$y = \frac{3x + 2}{6}$$

**65.** $C = \pi D$
$$\frac{C}{D} = \pi$$

**66.** $C = 2\pi r$
$$\frac{C}{2r} = \pi$$

**67.** Let $V = 900$, $l = 20$, and $h = 3$.
$$V = lwh$$
$$900 = 20w(3)$$
$$900 = 60w$$
$$15 = w$$
The width is 15 meters.

**68.** Let $x$ = width, then $x + 6$ = length.
$$60 = 2x + 2(x + 6)$$
$$60 = 2x + 2x + 12$$
$$60 = 4x + 12$$
$$48 = 4x$$
$$12 = x$$
$x + 6 = 12 + 6 = 18$
The dimensions are 18 feet by 12 feet.

**69.** Let $d = 10{,}000$ and $r = 125$.

$$d = rt$$
$$10{,}000 = 125t$$
$$80 = t$$

It will take 80 minutes or 1 hour and 20 minutes.

**70.** Let $F = 104$.

$$C = \frac{5}{9}(F - 32)$$
$$= \frac{5}{9}(104 - 32)$$
$$= \frac{5}{9}(72)$$
$$= 40$$

The temperature was 40°C.

**71.** Let $x =$ the percent.

$$9 = x \cdot 45$$
$$\frac{9}{45} = \frac{45x}{45}$$
$$0.2 = x$$

9 is 20% of 45.

**72.** Let $x =$ the percent.

$$59.5 = x \cdot 85$$
$$\frac{59.5}{85} = \frac{85x}{85}$$
$$0.7 = x$$

59.5 is 70% of 85.

**73.** Let $x =$ the number.

$$137.5 = 125\% \cdot x$$
$$137.5 = 1.25x$$
$$\frac{137.5}{1.25} = \frac{1.25x}{1.25}$$
$$110 = x$$

137.5 is 125% of 110.

**74.** Let $x =$ the number.

$$768 = 60\% \cdot x$$
$$768 = 0.6x$$
$$\frac{768}{0.6} = \frac{0.6x}{0.6}$$
$$1280 = x$$

768 is 60% of 1280.

**75.** Let $x =$ mark-up.

$$x = 11\% \cdot 1900$$
$$x = 0.11 \cdot 1900$$
$$x = 209$$

New price $= 1900 + 209 = 2109$

The mark-up is $209 and the new price is $2109.

    

**76.** Find 79% of 76,000.
$0.79 \cdot 76{,}000 = 60{,}040$
We would expect 60,040 people in that city to use the Internet.

**77.** Let $x$ = gallons of 40% solution.

| Strength | gallons | Concentration | |
|---|---|---|---|
| 40% | $x$ | 0.4 | $0.4x$ |
| 10% | $30 - x$ | 0.1 | $0.1(30 - x)$ |
| 20% | 30 | 0.2 | $0.2(30)$ |

$$0.4x + 0.1(30 - x) = 0.2(30)$$
$$0.4x + 3 - 0.1x = 6$$
$$0.3x + 3 = 6$$
$$0.3x = 3$$
$$x = 10$$
$30 - x = 30 - 10 = 20$
Mix 10 gallons of 40% acid solution with 20 gallons of 10% acid solution.

**78.** Increase $= 21.0 - 20.7 = 0.3$
Let $x$ = percent.
$$0.3 = x \cdot 20.7$$
$$\frac{0.3}{20.7} = \frac{20.7x}{20.7}$$
$$0.0145 \approx x$$
The percent increase is 1.45%.

**79.** From the graph, the height of 'Almost hit a car' is 18%.

**80.** Choose the tallest bar. The most common effect is swerving into another lane.

**81.** Find 21% of 4600.
$0.21 \cdot 4600 = 966$
We would expect 966 customers to have cut someone off.

**82.** Find 41% of 4600.
$0.41 \cdot 4600 = 1886$
We would expect 1886 customers to have sped up.

**83.**  $\text{percent decrease} = \dfrac{\text{amount of decrease}}{\text{original amount}}$
$$= \frac{250 - 170}{250}$$
$$= \frac{80}{250}$$
$$= 0.32$$
The percent decrease is 32%.

**84.** Let $x$ = original price.

$$x - 0.20x = 19.20$$
$$0.80x = 19.20$$
$$\frac{0.80x}{0.80} = \frac{19.20}{0.80}$$
$$x = 24$$

The original price was \$24.

**85.** Let $x$ = time up, then $3 - x$ = time down.

Rate · Time = Distance

|      | Rate | Time | Distance |
|------|------|------|----------|
| Up   | 10   | $x$  | $10x$    |
| Down | 50   | $3-x$ | $50(3-x)$ |

$$d = d$$
$$10x = 50(3 - x)$$
$$10x = 150 - 50x$$
$$60x = 150$$
$$x = 2.5$$

$$\begin{aligned}
\text{Total distance} &= 10x + 50(3 - x) \\
&= 10(2.5) + 50(3 - 2.5) \\
&= 25 + 50(0.5) \\
&= 25 + 25 \\
&= 50
\end{aligned}$$

The distance traveled was 50 km.

**86.** Let $x$ = the amount invested at 10.5% for one year.

| | Principal · | Rate = | Interest |
|---|---|---|---|
| 10.5% | $x$ | 0.105 | 0.105 |
| 8.5% | $50,000 - x$ | 0.085 | $0.085(50,000 - x)$ |
| Total | 50,000 | | 4550 |

$$0.105x + 0.085(50,000 - x) = 4550$$
$$0.105x + 4250 - 0.085x = 4550$$
$$0.02x + 4250 = 4550$$
$$0.02x = 300$$
$$x = 15,000$$

$50,000 - x = 50,000 - 15,000 = 35,000$
Invest \$35,000 at 8.5% and \$15,000 at 10.5%.

**87.** Let $x$ = the number of dimes,
$2x$ = the number of quarters, and
$500 - x - 2x$ the number of nickels.

| | No. of Coins | · Value = | Amt. of Money |
|---|---|---|---|
| Dimes | $x$ | 0.1 | $0.1x$ |
| Quarters | $2x$ | 0.25 | $0.25(2x)$ |
| Nickels | $500 - 3x$ | 0.05 | $0.05(500 - 3x)$ |
| Total | 500 | | 88 |

$$0.1x + 0.25(2x) + 0.05(500 - 3x) = 88$$
$$0.1x + 0.5x + 25 - 0.15x = 88$$
$$0.45x + 25 = 88$$
$$0.45x = 63$$
$$x = 140$$
$$500 - 3x = 500 - 3(140) = 500 - 420 = 80$$
There were 80 nickels in the pay phone.

**88.** Let $x$ = the time traveled by the Amtrak train.

| | Rate · | Time = | Distance |
|---|---|---|---|
| Amtrak | 60 | $x$ | $60x$ |
| Freight | 45 | $x + 1.5$ | $45(x + 1.5)$ |

$$d = d$$
$$60x = 45(x + 1.5)$$
$$60x = 45x + 67.5$$
$$15x = 67.5$$
$$x = 4.5$$
It will take 4.5 hours.

**89.** $x > 0$, $(0, \infty)$

**90.** $x \le -2$, $(-\infty, -2]$

**91.** $0.5 \le y < 1.5$, $[0.5, 1.5)$

**92.** $-1 < x < 1$, $(-1, 1)$

**93.** $-3x > 12$

$$\frac{-3x}{-3} < \frac{12}{-3}$$

$$x < -4, \ (-\infty, \ -4)$$

$-4$

**94.** $-2x \geq -20$

$$\frac{-2x}{-2} \leq \frac{-20}{-2}$$

$$x \leq 10, \ (-\infty, \ 10]$$

$10$

**95.** $x + 4 \geq 6x - 16$

$$-5x + 4 \geq -16$$

$$-5x \geq -20$$

$$\frac{-5x}{-5} \leq \frac{-20}{-5}$$

$$x \leq 4, \ (-\infty, \ 4]$$

$4$

**96.** $5x - 7 > 8x + 5$

$$-3x - 7 > 5$$

$$-3x > 12$$

$$\frac{-3x}{-3} < \frac{12}{-3}$$

$$x < -4, \ (-\infty, \ -4)$$

$-4$

**97.** $-3 < 4x - 1 < 2$

$$-2 < 4x < 3$$

$$-\frac{1}{2} < x < \frac{3}{4}, \ \left(-\frac{1}{2}, \ \frac{3}{4}\right)$$

$-\dfrac{1}{2}$      $\dfrac{3}{4}$

**98.** $2 \leq 3x - 4 < 6$

$$6 \leq 3x < 10$$

$$2 \leq x < \frac{10}{3}, \ \left[2, \ \frac{10}{3}\right)$$

$2$      $\dfrac{10}{3}$

**99.** $4(2x - 5) \leq 5x - 1$

$$8x - 20 \leq 5x - 1$$

$$3x - 20 \leq -1$$

$$3x \leq 19$$

$$x \leq \frac{19}{3}, \ \left(-\infty, \ \frac{19}{3}\right]$$

$\dfrac{19}{3}$

**100.** $-2(x - 5) > 2(3x - 2)$

$$-2x + 10 > 6x - 4$$

$$-8x + 10 > -4$$

$$-8x > -14$$

$$\frac{-8x}{-8} < \frac{-14}{-8}$$

$$x < \frac{7}{4}, \ \left(-\infty, \ \frac{7}{4}\right)$$

$\dfrac{7}{4}$

**101.** Let $x =$ the amount of sales then
$0.05x =$ her commission.

$$175 + 0.05x \geq 300$$

$$0.05x \geq 125$$

$$x \geq 2500$$

Sales must be at least \$2500.

**102.** Let $x =$ her score on the fourth round.

$$\frac{76 + 82 + 79 + x}{4} < 80$$

$$237 + x < 320$$

$$x < 83$$

Her score must be less than 83.

**103.** $6x + 2x - 1 = 5x + 11$

$$8x - 1 = 5x + 11$$

$$3x - 1 = 11$$

$$3x = 12$$

$$x = 4$$

**104.** $2(3y - 4) = 6 + 7y$

$$6y - 8 = 6 + 7y$$

$$-8 = 6 + y$$

$$-14 = y$$

    

**105.**
$$4(3-a)-(6a+9)=-12a$$
$$12-4a-6a-9=-12a$$
$$3-10a=-12a$$
$$3=-2a$$
$$-\frac{3}{2}=a$$

**106.**
$$\frac{x}{3}-2=5$$
$$\frac{x}{3}=7$$
$$3\cdot\frac{x}{3}=3\cdot7$$
$$x=21$$

**107.**
$$2(y+5)=2y+10$$
$$2y+10=2y+10$$
$$10=10$$
All real numbers are solutions.

**108.**
$$7x-3x+2=2(2x-1)$$
$$4x+2=4x-2$$
$$2=-2$$
There is no solution.

**109.** Let $x$ = the number.
$$6+2x=x-7$$
$$6+x=-7$$
$$x=-13$$
The number is $-13$.

**110.** Let $x$ = length of shorter piece, then
$4x + 3$ = length of longer piece.
$$x+(4x+3)=23$$
$$5x+3=23$$
$$5x=20$$
$$x=4$$
$4x + 3 = 4(4) + 3 = 19$
The shorter piece is 4 inches and the longer piece is 19 inches.

**111.**
$$V=\frac{1}{3}Ah$$
$$3\cdot V=3\cdot\frac{1}{3}Ah$$
$$3V=Ah$$
$$\frac{3V}{A}=\frac{Ah}{A}$$
$$\frac{3V}{A}=h$$

**112.** Let $x$ = the number.
$$x=26\%\cdot85$$
$$x=0.26\cdot85$$
$$x=22.1$$
22.1 is 26% of 85.

**113.** Let $x$ = the number.
$$72=45\%\cdot x$$
$$72=0.45x$$
$$\frac{72}{0.45}=\frac{0.45x}{0.45}$$
$$160=x$$
72 is 45% of 160.

**114.** Increase $= 282 - 235 = 47$
Let $x$ = percent.
$$47=x\cdot235$$
$$\frac{47}{235}=\frac{235x}{235}$$
$$0.2=x$$
The percent increase is 20%.

**115.**
$$4x-7>3x+2$$
$$x-7>2$$
$$x>9,\ (9,\infty)$$

**116.**
$$-5x<20$$
$$\frac{-5x}{-5}>\frac{20}{-5}$$
$$x>-4,\ (-4,\infty)$$

**117.**
$$-3(1+2x)+x\geq-(3-x)$$
$$-3-6x+x\geq-3+x$$
$$-3-5x\geq-3+x$$
$$-5x\geq x$$
$$-6x\geq0$$
$$\frac{-6x}{-6}\leq\frac{0}{-6}$$
$$x\leq0,\ (-\infty,0]$$

**Chapter 2 Test**

**1.** $2y-6-y-4=y-10$

**2.** $2.7x+6.1+3.2x-4.9=5.9x+1.2$

**3.** $4(x-2)-3(2x-6)=4x-8-6x+18$
$$=-2x+10$$

**4.** $7 + 2(5y - 3) = 7 + 10y - 6 = 10y + 1$

**5.**
$$-\frac{4}{5}x = 4$$
$$-\frac{5}{4} \cdot \left(-\frac{4}{5}x\right) = -\frac{5}{4} \cdot 4$$
$$x = -5$$

**6.** $4(n - 5) = -(4 - 2n)$
$$4n - 20 = -4 + 2n$$
$$2n - 20 = -4$$
$$2n = 16$$
$$n = 8$$

**7.** $5y - 7 + y = -(y + 3y)$
$$6y - 7 = -4y$$
$$-7 = -10y$$
$$\frac{7}{10} = y$$

**8.** $4z + 1 - z = 1 + z$
$$3z + 1 = 1 + z$$
$$2z + 1 = 1$$
$$2z = 0$$
$$z = 0$$

**9.** $\dfrac{2(x + 6)}{3} = x - 5$
$$2(x + 6) = 3(x - 5)$$
$$2x + 12 = 3x - 15$$
$$12 = x - 15$$
$$27 = x$$

**10.** $\dfrac{1}{2} - x + \dfrac{3}{2} = x - 4$
$$2\left(\frac{1}{2} - x + \frac{3}{2}\right) = 2(x - 4)$$
$$1 - 2x + 3 = 2x - 8$$
$$-2x + 4 = 2x - 8$$
$$-4x + 4 = -8$$
$$-4x = -12$$
$$x = 3$$

**11.** $-0.3(x - 4) + x = 0.5(3 - x)$
$$10[-0.3(x - 4) + x] = 10[0.5(3 - x)]$$
$$-3(x - 4) + 10x = 5(3 - x)$$
$$-3x + 12 + 10x = 15 - 5x$$
$$7x + 12 = 15 - 5x$$
$$12x + 12 = 15$$
$$12x = 3$$
$$x = \frac{3}{12} = \frac{1}{4} = 0.25$$

**12.** $-4(a + 1) - 3a = -7(2a - 3)$
$$-4a - 4 - 3a = -14a + 21$$
$$-7a - 4 = -14a + 21$$
$$7a - 4 = 21$$
$$7a = 25$$
$$a = \frac{25}{7}$$

**13.** $-2(x - 3) = x + 5 - 3x$
$$-2x + 6 = -2x + 5$$
$$6 = 5$$
There is no solution.

**14.** Let $y = -14$, $m = -2$, and $b = -2$.
$$y = mx + b$$
$$-14 = -2x - 2$$
$$-12 = -2x$$
$$6 = x$$

**15.** $V = \pi r^2 h$
$$\frac{V}{\pi r^2} = \frac{\pi r^2 h}{\pi r^2}$$
$$\frac{V}{\pi r^2} = h$$

**16.** $3x - 4y = 10$
$$-4y = -3x + 10$$
$$y = \frac{-3x + 10}{-4}$$
$$y = \frac{3x - 10}{4}$$

**17.** $3x - 5 \geq 7x + 3$
$$-4x - 5 \geq 3$$
$$-4x \geq 8$$
$$\frac{-4x}{-4} \leq \frac{8}{-4}$$
$$x \leq -2, \ (-\infty, -2]$$

**18.**
$$x + 6 > 4x - 6$$
$$-3x + 6 > -6$$
$$-3x > -12$$
$$\frac{-3x}{-3} < \frac{-12}{-3}$$
$$x < 4, (-\infty, 4)$$

**19.**
$$-2 < 3x + 1 < 8$$
$$-3 < 3x < 7$$
$$-1 < x < \frac{7}{3}, \left(-1, \frac{7}{3}\right)$$

**20.**
$$\frac{2(5x + 1)}{3} > 2$$
$$2(5x + 1) > 6$$
$$10x + 2 > 6$$
$$10x > 4$$
$$x > \frac{4}{10} = \frac{2}{5}, \left(\frac{2}{5}, \infty\right)$$

**21.** Let $x =$ the number.
$$x + \frac{2}{3}x = 35$$
$$3\left(x + \frac{2}{3}x\right) = 3(35)$$
$$3x + 2x = 105$$
$$5x = 105$$
$$x = 21$$
The number is 21.

**22.** Let $x =$ width, then $x + 2 =$ length.
$$P = 2w + 2l$$
$$252 = 2x + 2(x + 2)$$
$$252 = 2x + 2x + 4$$
$$252 = 4x + 4$$
$$252 - 4 = 4x + 4 - 4$$
$$248 = 4x$$
$$\frac{248}{4} = \frac{4x}{4}$$
$$62 = x$$
$$64 = x + 2$$
The dimensions of the deck are 62 feet by 64 feet.

**23.** Let $x =$ one area code, then
$2x =$ other area code.
$$x + 2x = 1203$$
$$3x = 1203$$
$$\frac{3x}{3} = \frac{1203}{3}$$
$$x = 401$$
$$2x = 2(401) = 802$$
The area codes are 401 and 802.

**24.** Let $x =$ the amount invested at 10% for one year.

| Principal | · Rate | = Interest | |
|---|---|---|---|
| 10% | $x$ | 0.10 | $0.1x$ |
| 12% | $2x$ | 0.12 | $0.12(2x)$ |
| Total | | | 2890 |

$$0.1x + 0.12(2x) = 2890$$
$$0.1x + 0.24x = 2890$$
$$0.34x = 2890$$
$$x = 8500$$
$$2x = 2(8500) = 17,000$$
He invested \$8500 at 10% and \$17,000 at 12%.

**25.** Let $x =$ the time they travel.

| Rate | · Time | = Distance | |
|---|---|---|---|
| Train 1 | 50 | $x$ | $50x$ |
| Train 2 | 64 | $x$ | $64x$ |
| Total | | | 285 |

$$50x + 64x = 285$$
$$114x = 285$$
$$x = 2\frac{1}{2}$$

They must travel for $2\frac{1}{2}$ hours.

**26.** From the graph, 69% are classified as weak.
Find 69% of 800.
$$69\% \cdot 800 = 0.69 \cdot 800 = 552$$
You would expect 552 of the 800 to be classified as weak.

**27.** Let $x$ be the unknown percent.
$$72 = x \cdot 180$$
$$\frac{72}{180} = \frac{180x}{180}$$
$$0.4 = x$$
72 is 40% of 180.

**28.** $\text{percent decrease} = \dfrac{\text{amount of decrease}}{\text{original amount}}$

$$= \dfrac{225 - 189}{225}$$

$$= \dfrac{36}{225}$$

$$= 0.16$$

The percent decrease is 16%.

## Chapter 2 Cumulative Review

**1. a.** the natural numbers are 11 and 112.

   **b.** The whole numbers are 0, 11, and 112.

   **c.** The integers are −3, −2, 0, 11, and 112.

   **d.** The rational numbers are −3, −2, −1.5, 0, $\dfrac{1}{4}$, 11, and 112.

   **e.** The irrational number is $\sqrt{2}$.

   **f.** All the numbers in the given set are real numbers.

**2. a.** The natural numbers are 2, 7, and 8.

   **b.** The whole numbers are 0, 2, 7, and 8.

   **c.** The integers are −185, 0, 2, 7, and 8.

   **d.** The rational numbers are −185, $-\dfrac{1}{5}$, 0, 2, 7, and 8.

   **e.** The irrational number is $\sqrt{3}$.

   **f.** All the numbers in the given set are real numbers.

**3. a.** $|4| = 4$

   **b.** $|-5| = 5$

   **c.** $|0| = 0$

   **d.** $\left|-\dfrac{1}{2}\right| = \dfrac{1}{2}$

   **e.** $|5.6| = 5.6$

**4. a.** $|5| = 5$

   **b.** $|-8| = 8$

   **c.** $\left|-\dfrac{2}{3}\right| = \dfrac{2}{3}$

**5. a.** $40 = 2 \cdot 2 \cdot 2 \cdot 5$

   **b.** $63 = 3 \cdot 3 \cdot 7$

**6. a.** $44 = 2 \cdot 2 \cdot 11$

   **b.** $90 = 2 \cdot 3 \cdot 3 \cdot 5$

**7.** $\dfrac{2}{5} = \dfrac{2}{5} \cdot \dfrac{4}{4} = \dfrac{8}{20}$

**8.** $\dfrac{2}{3} = \dfrac{2}{3} \cdot \dfrac{8}{8} = \dfrac{16}{24}$

**9.** $3[4 + 2(10 - 1)] = 3[4 + 2(9)]$
$$= 3[4 + 18]$$
$$= 3[22]$$
$$= 66$$

**10.** $5[16 - 4(2 + 1)] = 5[16 - 4(3)]$
$$= 5[16 - 12]$$
$$= 5[4]$$
$$= 20$$

**11.** Let $x = 2$.
$$3x + 10 = 8x$$
$$3(2) + 10 \overset{?}{=} 8(2)$$
$$6 + 10 \overset{?}{=} 16$$
$$16 = 16$$
2 is a solution of the equation.

**12.** Let $x = 3$.
$$5x - 2 = 4x$$
$$5(3) - 2 \overset{?}{=} 4(3)$$
$$15 - 2 \overset{?}{=} 12$$
$$13 \neq 12$$
3 is not a solution of the equation.

**13.** $-1 + (-2) = -3$

**14.** $(-2) + (-8) = -10$

**15.** $-4 + 6 = 2$

**16.** $-3 + 10 = 7$

**17. a.** $-(-10) = 10$

   **b.** $-\left(-\dfrac{1}{2}\right) = \dfrac{1}{2}$

**c.** $-(-2x) = 2x$

**d.** $-|-6| = -(6) = -6$

**18. a.** $-(-5) = 5$

**b.** $-\left(-\dfrac{2}{3}\right) = \dfrac{2}{3}$

**c.** $-(-a) = a$

**d.** $-|-3| = -(3) = -3$

**19. a.** $5.3 - (-4.6) = 5.3 + 4.6 = 9.9$

**b.** $-\dfrac{3}{10} - \dfrac{5}{10} = -\dfrac{3}{10} + \left(-\dfrac{5}{10}\right)$
$$= \dfrac{-3-5}{10}$$
$$= -\dfrac{8}{10}$$
$$= -\dfrac{4}{5}$$

**c.** $-\dfrac{2}{3} - \left(-\dfrac{4}{5}\right) = -\dfrac{2}{3} \cdot \dfrac{5}{5} + \dfrac{4}{5} \cdot \dfrac{3}{3}$
$$= -\dfrac{10}{15} + \dfrac{12}{15}$$
$$= \dfrac{2}{15}$$

**20. a.** $-2.7 - 8.4 = -2.7 + (-8.4) = -11.1$

**b.** $-\dfrac{4}{5} - \left(-\dfrac{3}{5}\right) = -\dfrac{4}{5} + \dfrac{3}{5} = \dfrac{-4+3}{5} = -\dfrac{1}{5}$

**c.** $\dfrac{1}{4} - \left(-\dfrac{1}{2}\right) = \dfrac{1}{4} + \dfrac{1}{2} \cdot \dfrac{2}{2} = \dfrac{1}{4} + \dfrac{2}{4} = \dfrac{3}{4}$

**21. a.** $x = 90 - 38 = 90 + (-38) = 52$
The complementary angle is 52°.

**b.** $y = 180 - 62 = 180 + (-62) = 118$
The supplementary angle is 118°.

**22. a.** $x = 90 - 72 = 90 + (-72) = 18$
The complementary angle is 18°.

**b.** $y = 180 - 47 = 180 + (-47) = 133$
The supplementary angle is 133°.

**23. a.** $(-1.2)(0.05) = -0.06$

**b.** $\dfrac{2}{3} \cdot \left(-\dfrac{7}{10}\right) = -\dfrac{2 \cdot 7}{3 \cdot 10} = -\dfrac{14}{30} = -\dfrac{7}{15}$

**c.** $\left(-\dfrac{4}{5}\right)(-20) = \dfrac{4 \cdot 20}{5} = \dfrac{80}{5} = 16$

**24. a.** $(4.5)(-0.08) = -0.36$

**b.** $-\dfrac{3}{4} \cdot \left(-\dfrac{8}{17}\right) = \dfrac{3 \cdot 8}{4 \cdot 17} = \dfrac{24}{68} = \dfrac{6}{17}$

**25. a.** $\dfrac{-24}{-4} = 6$

**b.** $\dfrac{-36}{3} = -12$

**c.** $\dfrac{2}{3} \div \left(-\dfrac{5}{4}\right) = \dfrac{2}{3}\left(-\dfrac{4}{5}\right) = -\dfrac{8}{15}$

**d.** $-\dfrac{3}{2} \div 9 = -\dfrac{3}{2} \div \dfrac{9}{1} = -\dfrac{3}{2} \cdot \dfrac{1}{9} = -\dfrac{3}{18} = -\dfrac{1}{6}$

**26. a.** $\dfrac{-32}{8} = -4$

**b.** $\dfrac{-108}{-12} = 9$

**c.** $-\dfrac{5}{7} \div \left(\dfrac{-9}{2}\right) = -\dfrac{5}{7}\left(-\dfrac{2}{9}\right) = \dfrac{10}{63}$

**27. a.** $x + 5 = 5 + x$

**b.** $3 \cdot x = x \cdot 3$

**28. a.** $y + 1 = 1 + y$

**b.** $y \cdot 4 = 4 \cdot y$

**29. a.** $8 \cdot 2 + 8 \cdot x = 8(2 + x)$

**b.** $7s + 7t = 7(s + t)$

**30. a.** $4 \cdot y + 4 \cdot \dfrac{1}{3} = 4\left(y + \dfrac{1}{3}\right)$

**b.** $0.10x + 0.10y = 0.10(x + y)$

**31.** $(2x - 3) - (4x - 2) = 2x - 3 - 4x + 2 = -2x - 1$

**32.** $(-5x+1)-(10x+3) = -5x+1-10x-3$
$$= -15x-2$$

**33.** $\dfrac{1}{2} = x - \dfrac{3}{4}$
$$4\left(\dfrac{1}{2}\right) = 4(x) - 4\left(\dfrac{3}{4}\right)$$
$$2 = 4x - 3$$
$$5 = 4x$$
$$\dfrac{5}{4} = x$$

**34.** $\dfrac{5}{6} + x = \dfrac{2}{3}$
$$6\left(\dfrac{5}{6}\right) + 6(x) = 6\left(\dfrac{2}{3}\right)$$
$$5 + 6x = 4$$
$$6x = -1$$
$$x = -\dfrac{1}{6}$$

**35.** $6(2a-1)-(11a+6) = 7$
$$12a-6-11a-6 = 7$$
$$a-12 = 7$$
$$a = 19$$

**36.** $-3x+1-(-4x-6) = 10$
$$-3x+1+4x+6 = 10$$
$$x+7 = 10$$
$$x = 3$$

**37.** $\dfrac{y}{7} = 20$
$$y = 140$$

**38.** $\dfrac{x}{4} = 18$
$$x = 72$$

**39.** $4(2x-3)+7 = 3x+5$
$$8x-12+7 = 3x+5$$
$$8x-5 = 3x+5$$
$$5x-5 = 5$$
$$5x = 10$$
$$x = 2$$

**40.** $6x+5 = 4(x+4)-1$
$$6x+5 = 4x+16-1$$
$$6x+5 = 4x+15$$
$$2x+5 = 15$$
$$2x = 10$$
$$x = 5$$

**41.** Let $x = $ a number.
$$2(x+4) = 4x-12$$
$$2x+8 = 4x-12$$
$$8 = 2x-12$$
$$20 = 2x$$
$$10 = x$$
The number is 10.

**42.** Let $x = $ a number.
$$x+4 = 3x-8$$
$$4 = 2x-8$$
$$12 = 2x$$
$$6 = x$$
The number is 6.

**43.** $V = lwh$
$$\dfrac{V}{wh} = \dfrac{lwh}{wh}$$
$$\dfrac{V}{wh} = l$$

**44.** $C = 2\pi r$
$$\dfrac{C}{2\pi} = \dfrac{2\pi r}{2\pi}$$
$$\dfrac{C}{2\pi} = r$$

**45.** $x+4 \le -6$
$$x \le -10, \ (-\infty, -10]$$

**46.** $x-3 > 2$
$$x > 5, \ (5, \infty)$$

    

# Chapter 3

## Section 3.1 Practice

1. **a.** We look for the shortest bar, which is the bar representing the Africa/Middle East region. We move from the right edge of this bar vertically downward to the Internet user axis. This region has approximately 145 million Internet users.

   **b.** The Asia/Oceania/Australia region has approximately 785 million Internet users. The Africa/Middle East region has approximately 145 million Internet users. We subtract 785 − 145 = 640 or 640 million. The Asia/Oceania/Australia region has 640 million more Internet users than the Africa/Middle East region.

2. **a.** We locate the number 40 along the time axis and move vertically upward until the line is reached. From this point on the line, we move horizontally to the left until the pulse rate axis is reached. Reading the number of beats per minute, we find that the pulse rate is 70 beats per minute 40 minutes after a cigarette is lit.

   **b.** The number 0 on the time axis corresponds to the time when the cigarette is being lit. We move vertically upward to the point on the line and then horizontally to the left to the pulse rate axis. The pulse rate is 60 beats per minute when the cigarette is being lit.

   **c.** We find the highest point of the line graph, which represents the highest pulse rate. From this point, we move vertically downward to the time axis. We find the pulse rate is the highest at 5 minutes, which means 5 minutes after lighting a cigarette.

3. **a.** Point (4, −3) lies in quadrant IV.

   **b.** Point (−3, 5) lies in quadrant II.

   **c.** Point (0, 4) lies on an axis, so it is not in any quadrant.

   **d.** Point (−6, 1) lies in quadrant II.

   **e.** Point (−2, 0) lies on an axis, so it is not in any quadrant.

   **f.** Point (5, 5) lies in quadrant I.

   **g.** Point $\left(3\frac{1}{2}, 1\frac{1}{2}\right)$ lies in quadrant I.

   **h.** Point (−4, −5) lies in quadrant III.

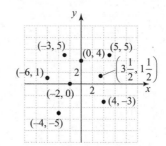

4. **a.** The ordered pairs are (2004, 65), (2005, 67), (2006, 96), (2007, 86), (2008, 79), (2009, 79), and (2010, 72).

   **b.** We plot the ordered pairs. We label the horizontal axis "Year" and the vertical axis "Wildfires (in thousands)."

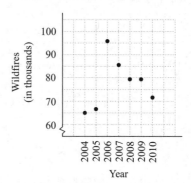

5. **a.** Let $x = 3$ and $y = 1$.
   $$x + 3y = 6$$
   $$3 + 3(1) = 6$$
   $$3 + 3 = 6$$
   $$6 = 6 \quad \text{true}$$
   Yes, (3, 1) is a solution.

   **b.** Let $x = 6$ and $y = 0$.
   $$x + 3y = 6$$
   $$6 + 3(0) = 6$$
   $$6 + 0 = 6$$
   $$6 = 6 \quad \text{true}$$
   Yes, (6, 0) is a solution.

**c.** Let $x = -2$ and $y = \dfrac{2}{3}$.

$$x + 3y = 6$$
$$-2 + 3\left(\dfrac{2}{3}\right) = 6$$
$$-2 + 2 = 6$$
$$0 = 6 \quad \text{false}$$

No, $\left(-2, \dfrac{2}{3}\right)$ is not a solution.

**6. a.** Let $x = 0$ and solve for $y$.
$$2x - y = 8$$
$$2(0) - y = 8$$
$$0 - y = 8$$
$$-y = 8$$
$$y = -8$$
The ordered pair is $(0, -8)$.

**b.** Let $y = 4$ and solve for $x$.
$$2x - y = 8$$
$$2x - 4 = 8$$
$$2x = 12$$
$$x = 6$$
The ordered pair is $(6, 4)$.

**c.** Let $x = -3$ and solve for $y$.
$$2x - y = 8$$
$$2(-3) - y = 8$$
$$-6 - y = 8$$
$$-y = 14$$
$$y = -14$$
The ordered pair is $(-3, -14)$.

**7. a.** Replace $x$ with $-2$ in the equation and solve for $y$.
$$y = -4x$$
$$y = -4(-2)$$
$$y = 8$$
The ordered pair is $(-2, 8)$.

**b.** Replace $y$ with $-12$ in the equation and solve for $x$.
$$y = -4x$$
$$-12 = -4x$$
$$3 = x$$
The ordered pair is $(3, -12)$.

**c.** Replace $x$ with $0$ in the equation and solve for $y$.
$$y = -4x$$
$$y = -4(0)$$
$$y = 0$$
The ordered pair is $(0, 0)$.

The completed table is shown below.

| $x$ | $y$ |
|-----|-----|
| $-2$ | $8$ |
| $3$ | $-12$ |
| $0$ | $0$ |

**8. a.** Let $x = -10$.
$$y = \dfrac{1}{5}x - 2$$
$$y = \dfrac{1}{5}(-10) - 2$$
$$y = -2 - 2$$
$$y = -4$$
Ordered pair: $(-10, -4)$

**b.** Let $x = 0$.
$$y = \dfrac{1}{5}x - 2$$
$$y = \dfrac{1}{5}(0) - 2$$
$$y = 0 - 2$$
$$y = -2$$
Ordered pair: $(0, -2)$

**c.** Let $y = 0$.
$$y = \dfrac{1}{5}x - 2$$
$$0 = \dfrac{1}{5}x - 2$$
$$2 = \dfrac{1}{5}x$$
$$10 = x$$
Ordered pair: $(10, 0)$

The completed table is shown below.

| $x$ | $y$ |
|-----|-----|
| $-10$ | $-4$ |
| $0$ | $-2$ |
| $10$ | $0$ |

9. When $x = 0$,
$y = -1800x + 12{,}000$
$y = -1800 \cdot 0 + 12{,}000$
$y = 0 + 12{,}000$
$y = 12{,}000$

When $x = 1$,
$y = -1800x + 12{,}000$
$y = -1800 \cdot 1 + 12{,}000$
$y = -1800 + 12{,}000$
$y = 10{,}200$

When $x = 2$,
$y = -1800x + 12{,}000$
$y = -1800 \cdot 2 + 12{,}000$
$y = -3600 + 12{,}000$
$y = 8400$

When $x = 3$,
$y = -1800x + 12{,}000$
$y = -1800 \cdot 3 + 12{,}000$
$y = -5400 + 12{,}000$
$y = 6600$

When $x = 4$,
$y = -1800x + 12{,}000$
$y = -1800 \cdot 4 + 12{,}000$
$y = -7200 + 12{,}000$
$y = 4800$

The completed table is shown below.

| $x$ | 0 | 1 | 2 | 3 | 4 |
|---|---|---|---|---|---|
| $y$ | 12,000 | 10,200 | 8400 | 6600 | 4800 |

## Vocabulary, Readiness & Video Check 3.1

1. The horizontal axis is called the x-axis and the vertical axis is called the y-axis.

2. The intersection of the horizontal axis and the vertical axis is a point called the origin.

3. The axes divide the plane into regions called quadrants. There are four of these regions.

4. In the ordered pair of numbers $(-2, 5)$, the number $-2$ is called the x-coordinate and the number 5 is called the y-coordinate.

5. Each ordered pair of numbers corresponds to one point in the plane.

6. An ordered pair is a solution of an equation in two variables if replacing the variables by the coordinates of the ordered pair results in a true statement.

7. horizontal: top tourist countries; vertical: number of arrivals (in millions) to these countries

8. Origin; left or right; up or down

9. Data occurring in pairs of numbers can be written as ordered pairs, called paired data, and then graphed on a coordinate system.

10. $(7, 0)$ and $(0, 7)$; since one of these points is a solution and one is not, it shows that it is very important to remember that the first number is the *x*-value and the second number is the *y*-value and not to mix them up.

11. a linear equation in one variable

## Exercise Set 3.1

1. The tallest bar corresponds to France, so France is the most popular tourist destination.

3. Find the bars that have heights greater than 40. France, U.S., Spain, Italy, and China have more than 40 million tourists per year.

5. The height of the bar is near 43, so approximately 43 million tourists go to Italy each year.

7. The line is at about 71,000 for year 2009. Thus, the attendance was 71,000 in 2009.

9. The highest point corresponds to 2011, and is at a height of about 103,000.

11. From 2000 on the year axis, we move vertically up to the point on the line graph. When we move horizontally to the vertical axis. The number of students per teacher was approximately 15.9 in 2002.

13. The number of students per teacher shows the greatest decrease between 1998 and 2000. Notice that the line graph is steepest between 1998 and 2000.

15. The points on the line graph for 1998 through 2012 lie above the horizontal line at 15 on the vertical axis. The point for 2014 is the first that lies below this horizontal line. The first year shown that the number of students per teacher fell below 15 was 2014.

**17. a.** Point (1, 5) lies in quadrant I.

  **b.** Point (−5, −2) lies in quadrant III.

  **c.** Point (−3, 0) lies on the *x*-axis, so it is not in any quadrant.

  **d.** Point (0, −1) lies on the *y*-axis, so it is not in any quadrant.

  **e.** Point (2, −4) lies in quadrant IV.

  **f.** Point $\left(-1, 4\frac{1}{2}\right)$ lies in quadrant II.

  **g.** (3.7, 2.2) lies in quadrant I.

  **h.** Point $\left(\frac{1}{2}, -3\right)$ lies in quadrant IV.

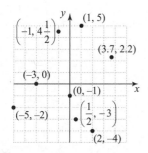

**19.** Point *A* lies at the origin. Its coordinates are given by the ordered pair (0, 0).

**21.** Point *C* lies three units to the right and two units above the origin. Its coordinates are given by the ordered pair (3, 2).

**23.** Point *E* lies two units to the left and two units below the origin. Its coordinates are given by the ordered pair (−2, −2).

**25.** Point *G* lies two units to the right and one unit below the origin. Its coordinates are given by the ordered pair (2, −1).

**27.** Point *B* lies on the *y*-axis three units below the origin. Its coordinates are given by the ordered pair (0, −3).

**29.** Point *D* lies one unit to the right and three units above the origin. Its coordinates are given by the ordered pair (1, 3).

**31.** Point *F* lies three units to the left and one unit below the origin. Its coordinates are given by the ordered pair (−3, −1).

**33. a.** The ordered pairs are (2006, 25.5), (2007, 26.3), (2008, 27.7), (2009, 29.4) and (2010, 31.8).

  **b.** The ordered pair (2010, 31.8) indicates that the worldwide box office in 2010 was $31.8 billion.

  **c.**

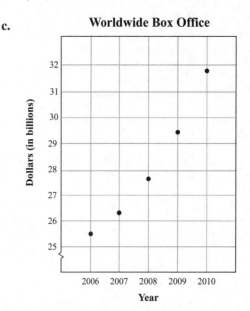

  **d.** The worldwide box office increased every year.

**35. a.** The ordered pairs are (0.50, 10), (0.75, 12), (1.00, 15), (1.25, 16), (1.50, 18), (1.50, 19), (1.75, 19), and (2.00, 20).

  **b.** The ordered pair (1.25, 16) indicates that when Minh studied 1.25 hours, her quiz score was 16.

  **c.**

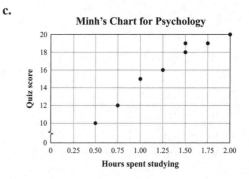

  **d.** answers may vary

**37. a.** The ordered pairs are (2313, 2), (2085, 1), (2711, 21), (2869, 39), (2920, 42), (4038, 99), (1783, 0), and (2493, 9).

**b.** We plot the ordered pairs. We label the horizontal axis "Distance from Equator (in miles)" and the vertical axis "Average Annual Snowfall (in inches)."

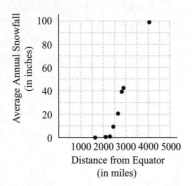

**c.** The farther from the equator, the more snowfall.

**39.** For (3, 1), let $x = 3$ and $y = 1$.
$$2x + y = 7$$
$$2(3) + 1 = 7$$
$$6 + 1 = 7$$
$$7 = 7 \quad \text{true}$$
Yes, (3, 1) is a solution.

For (7, 0), let $x = 7$ and $y = 0$.
$$2x + y = 7$$
$$2(7) + 0 = 7$$
$$14 + 0 = 7$$
$$14 = 7 \quad \text{false}$$
No, (7, 0) is not a solution.

For (0, 7), let $x = 0$ and $y = 7$.
$$2x + y = 7$$
$$2(0) + 7 = 7$$
$$0 + 7 = 7$$
$$7 = 7 \quad \text{true}$$
Yes, (0, 7) is a solution.

**41.** For (0, 0), let $x = 0$ and $y = 0$.
$$x = -\frac{1}{3}y$$
$$0 = -\frac{1}{3}(0)$$
$$0 = 0 \quad \text{true}$$
Yes, (0, 0) is a solution.

For (3, −9), let $x = 3$ and $y = -9$.

$$x = -\frac{1}{3}y$$
$$3 = -\frac{1}{3}(-9)$$
$$3 = 3 \quad \text{true}$$
Yes, (3, −9) is a solution.

**43.** For (4, 5), let $x = 4$ and $y = 5$.
$$x = 5$$
$$4 = 5 \quad \text{false}$$
No, (4, 5) is not a solution.

For (5, 4), let $x = 5$ and $y = 4$.
$$x = 5$$
$$5 = 5 \quad \text{true}$$
Yes, (5, 4) is a solution.

For (5, 0), let $x = 5$ and $y = 0$.
$$x = 5$$
$$5 = 5 \quad \text{true}$$
Yes, (5, 0) is a solution.

**45.** Replace $y$ with −2 and solve for $x$.
$$x - 4y = 4$$
$$x - 4(-2) = 4$$
$$x + 8 = 4$$
$$x = -4$$
The ordered pair is (−4, −2).

Replace $x$ with 4 and solve for $y$.
$$x - 4y = 4$$
$$4 - 4y = 4$$
$$-4y = 0$$
$$y = 0$$
The ordered pair is (4, 0).

**47.** Replace $x$ with −8 and solve for $y$.
$$y = \frac{1}{4}x - 3$$
$$y = \frac{1}{4}(-8) - 3$$
$$y = -2 - 3$$
$$y = -5$$
The ordered pair is (−8, −5).

Replace $y$ with 1 and solve for $x$.

$$y = \frac{1}{4}x - 3$$
$$1 = \frac{1}{4}x - 3$$
$$4 = \frac{1}{4}x$$
$$16 = x$$

The ordered pair is (16, 1).

**49.** Replace $x$ with 0 and solve for $y$.
$$y = -7x$$
$$y = -7(0)$$
$$y = 0$$
The ordered pair is (0, 0).

Replace $x$ with $-1$ and solve for $y$.
$$y = -7x$$
$$y = -7(-1)$$
$$y = 7$$
The ordered pair is $(-1, 7)$.

Replace $y$ with 2 and solve for $x$.
$$y = -7x$$
$$2 = -7x$$
$$-\frac{2}{7} = x$$

The ordered pair is $\left(-\frac{2}{7}, 2\right)$.
The completed table is shown below.

| $x$ | $y$ |
|-----|-----|
| 0 | 0 |
| $-1$ | 7 |
| $-\frac{2}{7}$ | 2 |

**51.** Replace $x$ with 0 and solve for $y$.
$$y = -x + 2$$
$$y = -0 + 2$$
$$y = 2$$
The ordered pair is (0, 2).

Replace $y$ with 0 and solve for $x$.
$$y = -x + 2$$
$$0 = -x + 2$$
$$x = 2$$
The ordered pair is (2, 0).

Replace $x$ with $-3$ and solve for $y$.

$$y = -x + 2$$
$$y = -(-3) + 2$$
$$y = 3 + 2$$
$$y = 5$$
The ordered pair is $(-3, 5)$.
The completed table is shown below.

| $x$ | $y$ |
|-----|-----|
| 0 | 2 |
| 2 | 0 |
| $-3$ | 5 |

**53.** Replace $x$ with 0 and solve for $y$.
$$y = \frac{1}{2}x$$
$$y = \frac{1}{2}(0)$$
$$y = 0$$
The ordered pair is (0, 0).

Replace $x$ with $-6$ and solve for $y$.
$$y = \frac{1}{2}x$$
$$y = \frac{1}{2}(-6)$$
$$y = -3$$
The ordered pair is $(-6, -3)$.

Replace $y$ with 1 and solve for $x$.
$$y = \frac{1}{2}x$$
$$1 = \frac{1}{2}x$$
$$2 = x$$
The ordered pair is (2, 1).
The completed table is shown below.

| $x$ | $y$ |
|-----|-----|
| 0 | 0 |
| $-6$ | $-3$ |
| 2 | 1 |

**55.** Replace $x$ with 0 and solve for $y$.
$$x + 3y = 6$$
$$0 + 3y = 6$$
$$3y = 6$$
$$y = 2$$
The ordered pair is (0, 2).

Replace $y$ with 0 and solve for $x$.

$$x + 3y = 6$$
$$x + 3(0) = 6$$
$$x + 0 = 6$$
$$x = 6$$

The ordered pair is (6, 0).

Replace $y$ with 1 and solve for $x$.
$$x + 3y = 6$$
$$x + 3(1) = 6$$
$$x + 3 = 6$$
$$x = 3$$

The ordered pair is (3, 1).
The completed table is shown below.

| $x$ | $y$ |
|-----|-----|
| 0   | 2   |
| 6   | 0   |
| 3   | 1   |

**57.** Replace $x$ with 0 and solve for $y$.
$$y = 2x - 12$$
$$y = 2(0) - 12$$
$$y = 0 - 12$$
$$y = -12$$

The ordered pair is (0, −12).

Replace $y$ with −2 and solve for $x$.
$$y = 2x - 12$$
$$-2 = 2x - 12$$
$$10 = 2x$$
$$5 = x$$

The ordered pair is (5, −2).

Replace $x$ with 3 and solve for $y$.
$$y = 2x - 12$$
$$y = 2(3) - 12$$
$$y = 6 - 12$$
$$y = -6$$

The ordered pair is (3, −6).
The completed table is shown below.

| $x$ | $y$ |
|-----|-----|
| 0   | −12 |
| 5   | −2  |
| 3   | −6  |

**59.** Replace $x$ with 0 and solve for $y$.
$$2x + 7y = 5$$
$$2(0) + 7y = 5$$
$$7y = 5$$
$$y = \frac{5}{7}$$

The ordered pair is $\left(0, \frac{5}{7}\right)$.

Replace $y$ with 0 and solve for $x$.
$$2x + 7y = 5$$
$$2x + 7(0) = 5$$
$$2x = 5$$
$$x = \frac{5}{2}$$

The ordered pair is $\left(\frac{5}{2}, 0\right)$.

Replace $y$ with 1 and solve for $x$.
$$2x + 7y = 5$$
$$2x + 7(1) = 5$$
$$2x + 7 = 5$$
$$2x = -2$$
$$x = -1$$

The ordered pair is (−1, 1).
The completed table is shown below.

| $x$           | $y$           |
|---------------|---------------|
| 0             | $\frac{5}{7}$ |
| $\frac{5}{2}$ | 0             |
| −1            | 1             |

**61.** Replace $y$ with 0 and solve for $x$.
$$x = -5y$$
$$x = -5(0)$$
$$x = 0$$

The ordered pair is (0, 0).

Replace $y$ with 1 and solve for $x$.
$$x = -5y$$
$$x = -5(1)$$
$$x = -5$$

The ordered pair is (−5, 1).

Replace $x$ with 10 and solve for $y$.
$$x = -5y$$
$$10 = -5y$$
$$-2 = y$$

The ordered pair is (10, −2).
The completed table is shown below.

| x | y |
|---|---|
| 0 | 0 |
| −5 | 1 |
| 10 | −2 |

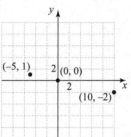

The completed table is shown below.

| x | y |
|---|---|
| 0 | 2 |
| −3 | 1 |
| −6 | 0 |

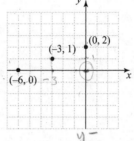

**63.** Replace *x* with 0 and solve for *y*.

$$y = \frac{1}{3}x + 2$$

$$y = \frac{1}{3}(0) + 2$$

$$y = 2$$

The ordered pair is (0, 2).

Replace *x* with −3 and solve for *y*.

$$y = \frac{1}{3}x + 2$$

$$y = \frac{1}{3}(-3) + 2$$

$$y = -1 + 2$$

$$y = 1$$

The ordered pair is (−3, 1).

Replace *y* with 0 and solve for *x*.

$$y = \frac{1}{3}x + 2$$

$$0 = \frac{1}{3}x + 2$$

$$-\frac{1}{3}x = 2$$

$$x = -6$$

The ordered pair is (−6, 0).

**65. a.** When *x* = 100,
$$y = 80x + 5000$$
$$y = 80(100) + 5000$$
$$y = 8000 + 5000$$
$$y = 13,000$$

When *x* = 200,
$$y = 80x + 5000$$
$$y = 80(200) + 5000$$
$$y = 16,000 + 5000$$
$$y = 21,000$$

When *x* = 300,
$$y = 80x + 5000$$
$$y = 80(300) + 5000$$
$$y = 24,000 + 5000$$
$$y = 29,000$$

The completed table is shown below.

| x | 100 | 200 | 300 |
|---|-----|-----|-----|
| y | 13,000 | 21,000 | 29,000 |

**b.** Replace *y* with 8600 and solve for *x*.
$$y = 80x + 5000$$
$$8600 = 80x + 5000$$
$$3600 = 80x$$
$$45 = x$$

Thus, 45 computer desks can be produced for $8600.

**67. a.**

| $x$ | 1 | 3 | 5 |
|---|---|---|---|
| $y = 0.24x + 5.28$ | $0.24(1) + 5.28$ $= 0.24 + 5.28$ $= 5.52$ | $0.24(3) + 5.28$ $= 0.72 + 5.28$ $= 6.00$ | $0.24(5) + 5.28$ $= 1.2 + 5.28$ $= 6.48$ |

**b.** Find $x$ when $y = 7.50$.
$$750 = 0.24x + 5.28$$
$$2.22 = 0.24x$$
$$9 \approx x$$
$$2000 + 9 = 2009$$
The average cinema admission price was $7.50 in year 9 or 2009.

**c.** Find $x$ when $y = 9.00$.
$$9.00 = 0.24x + 5.28$$
$$3.72 = 0.24x$$
$$15.5 = x$$
$$16 \approx x$$
$$2000 + 16 = 2016$$
The average cinema admission price is predicted to be $9.00 in year 16 or 2016.

**d.** $(5, 6.48)$ means that in 2005, the average cinema price was $6.48.

**69.** Ten years after 2000, or in 2010, there were 3755 Walmart stores.

**71.** In year 7, there appear to be approximately 3450 Walmart stores. In year 8 there appear to be approximately 3550 Walmart stores. The increase is approximately $3550 - 3450 = 100$ stores.

In year 8, there appear to be approximately 3550 Walmart stores. In year 9, there appear to be approximately 3655 Walmart stores. The increase is approximately $3655 - 3550 = 105$ stores.

In year 9, there appear to be approximately 3655 Walmart stores. In year 10, there appear to be approximately 3755 Walmart stores. The increase is approximately $3755 - 3655 = 100$ stores.

**73.** The coordinates of points whose graphs lie on the $x$-axis all have $y$-values of 0.

**75.** Subtract $x$ from each side.
$$x + y = 5$$
$$y = 5 - x$$

**77.** Subtract $2x$ from each side. Then divide each side by 4.
$$2x + 4y = 5$$
$$4y = -2x + 5$$
$$y = -\frac{1}{2}x + \frac{5}{4}$$

**79.** Divide each side by $-5$.
$$10x = -5y$$
$$-2x = y$$
$$y = -2x$$

**81.** Subtract $x$ from each side. Then divide each side by $-3$.

$$x - 3y = 6$$
$$-3y = -x + 6$$
$$y = \frac{1}{3}x - 2$$

**83.** False; the point $(-1, 5)$ lies in quadrant II.

**85.** True

**87.** In quadrant III, both coordinates are negative: (negative, negative).

**89.** In quadrant IV, the $x$-coordinate is positive and the $y$-coordinate is negative: (positive, negative).

**91.** At the origin, both coordinates are zero: $(0, 0)$.

**93.** A point of the form (0, number) is located on the $y$-axis.

**95.** No; answers may vary.

**97.** Answers may vary

**99.** The point four units to the right of the $y$-axis and seven units below the $x$-axis has ordered pair $(4, -7)$.

**101.** The length of the rectangle is $3 - (-1) = 4$ and the width of the rectangle is $5 - (-4) = 9$.

Perimeter $= 2(\text{length}) + 2(\text{width})$
$$= 2(4) + 2(9)$$
$$= 8 + 18$$
$$= 26$$

The perimeter is 26 units.

**103.**

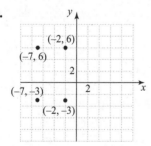

**a.** The fourth vertex is $(-2, 6)$. The rectangle is 9 units by 5 units.

**b.** The perimeter is $9 + 5 + 9 + 5 = 28$ units.

**c.** The area is $9 \times 5 = 45$ square units.

**Section 3.2 Practice**

**1. a.** $3x + 2.7y = -5.3$ is a linear equation in two variables because it is written in the form $Ax + By = C$ with $A = 3$, $B = 2.7$, and $C = -5.3$.

**b.** $x^2 + y = 8$ is not a linear equation in two variables because $x$ is squared.

**c.** $y = 12$ is a linear equation in two variables because it can be written in the form $Ax + By = C$: $0x + y = 12$.

**d.** $5x = -3y$ is a linear equation in two variables because it can be written in the form $Ax + By = C$: $5x + 3y = 0$.

**2.** Find three ordered pair solutions.

Let $x = 0$.
$$x + 3y = 9$$
$$0 + 3y = 9$$
$$3y = 9$$
$$y = 3$$

Let $x = 3$.
$$x + 3y = 9$$
$$3 + 3y = 9$$
$$3y = 6$$
$$y = 2$$

Let $y = 1$.
$$x + 3y = 9$$
$$x + 3(1) = 9$$
$$x + 3 = 9$$
$$x = 6$$

The ordered pairs are $(0, 3)$, $(3, 2)$, and $(6, 1)$.

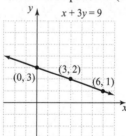

**3.** Find three ordered pair solutions.

Let $x = 0$.
$$3x - 4y = 12$$
$$3(0) - 4y = 12$$
$$-4y = 12$$
$$y = -3$$

Let $y = 0$.
$$3x - 4y = 12$$
$$3x - 4(0) = 12$$
$$3x = 12$$
$$x = 4$$

Let $x = 2$.
$$3x - 4y = 12$$
$$3(2) - 4y = 12$$
$$6 - 4y = 12$$
$$-4y = 6$$
$$y = -\frac{6}{4} = -\frac{3}{2}$$

The ordered pairs are $(0, -3)$, $(4, 0)$, and $\left(2, -\dfrac{3}{2}\right)$.

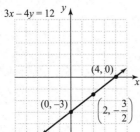

4. Find three ordered pair solutions.
   If $x = 1$, $y = -2(1) = -2$.
   If $x = 0$, $y = -2(0) = 0$.
   If $x = -1$, $y = -2(-1) = 2$.

| $x$ | $y$ |
|-----|-----|
| 1   | -2  |
| 0   | 0   |
| -1  | 2   |

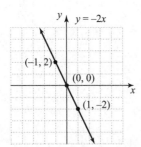

5. Find three ordered pair solutions.
   If $x = 2$, $y = \dfrac{1}{2}(2) + 3 = 1 + 3 = 4$.

   If $x = 0$, $y = \dfrac{1}{2}(0) + 3 = 0 + 3 = 3$.

   If $x = -4$, $y = \dfrac{1}{2}(-4) + 3 = -2 + 3 = 1$.

| $x$ | $y$ |
|-----|-----|
| 2   | 4   |
| 0   | 3   |
| -4  | 1   |

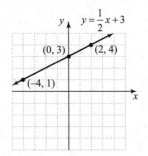

6. The equation $x = -2$ can be written in standard form as $x + 0y = -2$. No matter what value replaces $y$, $x$ is always $-2$. It is a vertical line. Plot points $(-2, 2)$, $(-2, 0)$, and $(-2, -4)$, for example.

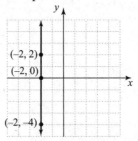

7. Find three ordered pair solutions.
   If $x = 1$, $y = -2(1) + 3 = -2 + 3 = 1$.
   If $x = 0$, $y = -2(0) + 3 = 0 + 3 = 3$.
   If $x = 3$, $y = -2(3) + 3 = -6 + 3 = -3$.

| $x$ | $y$ |
|-----|-----|
| 1   | 1   |
| 0   | 0   |
| 3   | -3  |

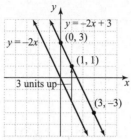

The graph of $y = -2x + 3$ is the same as the graph of $y = -2x$ except that the graph of $y = -2x + 3$ is moved three units upward.

**8. a.** Find three ordered pair solutions.
If $x = 0$, $y = 17.5(0) + 515 = 515$.
If $x = 6$,
$y = 17.5(6) + 515 = 620$.
If $x = 10$,
$y = 17.5(10) + 515 = 690$.

| $x$ | $y$ |
|-----|-----|
| 0 | 515 |
| 6 | 620 |
| 10 | 690 |

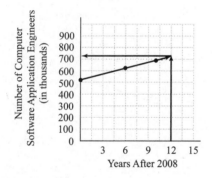

**b.** $2020 - 2008 = 12$
The graph shows that we predict approximately 725 thousand computer software application engineers in the year 2020.

**Calculator Explorations**

**1.** $y = -3x + 7$

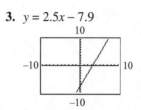

**2.** $y = -x + 5$

**3.** $y = 2.5x - 7.9$

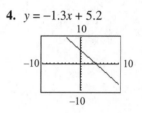

**4.** $y = -1.3x + 5.2$

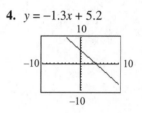

**5.** $y = -\dfrac{3}{10}x + \dfrac{32}{5}$

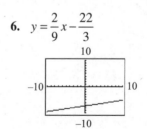

**6.** $y = \dfrac{2}{9}x - \dfrac{22}{3}$

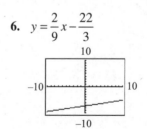

**Vocabulary, Readiness & Video Check 3.2**

**1.** In the definition, $x$ and $y$ both have an understood power of 1. Example 3 shows an equation where $y$ has a power of 2, so it is not a linear equation in two variables.

**2.** Find 3 points in order to check your work. Make sure the points lie along one straight line—if not, an algebraic mistake was probably made.

**3.** An infinite number of points make up the line and each point corresponds to an ordered pair that is a solution of the linear equation in two variables.

**Exercise Set 3.2**

1. Yes; it can be written in the form $Ax + By = C$.

3. Yes; it can be written in the form $Ax + By = C$.

5. No; $x$ is squared.

7. Yes; it can be written in the form $Ax + By = C$.

9. Let $y = 0$.
$$x - y = 6$$
$$x - 0 = 6$$
$$x = 6$$

Let $x = 4$.          Let $y = -1$
$$x - y = 6$$          $$x - y = 6$$
$$4 - y = 6$$          $$x - (-1) = 6$$
$$-y = 2$$          $$x + 1 = 6$$
$$y = -2$$          $$x = 5$$

| $x$ | $y$ |
|-----|-----|
| 6 | 0 |
| 4 | -2 |
| 5 | -1 |

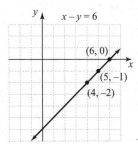

11. $y = -4x$
If $x = 1$, $y = -4(1) = -4$.
If $x = 0$, $y = -4(0) = 0$.
If $x = -1$, $y = -4(-1) = 4$.

| $x$ | $y$ |
|-----|-----|
| 1 | -4 |
| 0 | 0 |
| -1 | 4 |

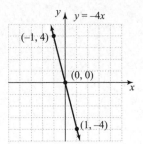

13. $y = \dfrac{1}{3}x$

If $x = 0$, $y = \dfrac{1}{3}(0) = 0$.

If $x = 6$, $y = \dfrac{1}{3}(6) = 2$.

If $x = -3$, $y = \dfrac{1}{3}(-3) = -1$.

| $x$ | $y$ |
|-----|-----|
| 0 | 0 |
| 6 | 2 |
| -3 | -1 |

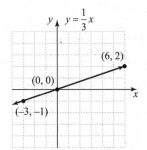

15. $y = -4x + 3$
If $x = 0$, $y = -4(0) + 3 = 0 + 3 = 3$.
If $x = 1$, $y = -4(1) + 3 = -4 + 3 = -1$.
If $x = 2$, $y = -4(2) + 3 = -8 + 3 = -5$.

| $x$ | $y$ |
|-----|-----|
| 0 | 3 |
| 1 | -1 |
| 2 | -5 |

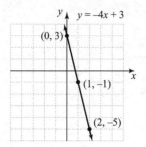

**17.** $x + y = 1$

| x | y |
|---|---|
| 0 | 1 |
| 1 | 0 |
| 2 | -1 |

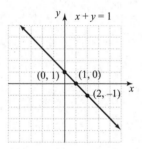

**19.** $x - y = -2$

| x | y |
|---|---|
| -2 | 0 |
| 0 | 2 |
| 2 | 4 |

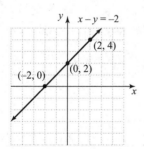

**21.** $x - 2y = 6$

| x | y |
|---|---|
| -4 | -5 |
| 0 | -3 |
| 4 | -1 |

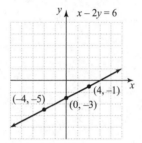

**23.** $y = 6x + 3$

| x | y |
|---|---|
| -1 | -3 |
| 0 | 3 |
| 1 | 9 |

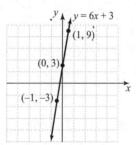

**25.** $x = -4$

| x | y |
|---|---|
| -4 | -1 |
| -4 | 0 |
| -4 | 2 |

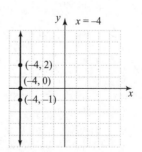

**27.** $y = 3$

| x | y |
|---|---|
| −1 | 3 |
| 0 | 3 |
| 2 | 3 |

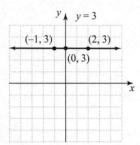

**29.** $y = x$

| x | y |
|---|---|
| −1 | −1 |
| 0 | 0 |
| 2 | 2 |

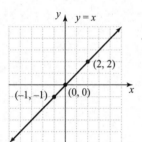

**31.** $x = -3y$

| x | y |
|---|---|
| −6 | 2 |
| 0 | 0 |
| 6 | −2 |

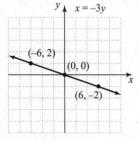

**33.** $x + 3y = 9$

| x | y |
|---|---|
| −9 | 6 |
| 0 | 3 |
| 3 | 2 |

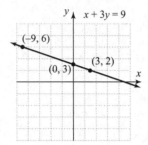

**35.** $y = \dfrac{1}{2}x + 2$

| x | y |
|---|---|
| −4 | 0 |
| 0 | 2 |
| 4 | 4 |

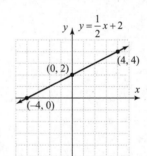

**37.** $3x - 2y = 12$

| x | y |
|---|---|
| 0 | −6 |
| 2 | −3 |
| 4 | 0 |

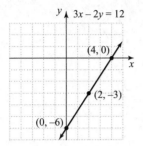
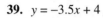

**39.** $y = -3.5x + 4$

| $x$ | $y$ |
| --- | --- |
| 0 | 4 |
| 1 | 0.5 |
| 2 | -3 |

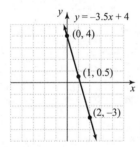

**41.**

| $y = 5x$ | |
| --- | --- |
| $x$ | $y$ |
| -1 | -5 |
| 0 | 0 |
| 1 | 5 |

| $y = 5x + 4$ | |
| --- | --- |
| $x$ | $y$ |
| -1 | -1 |
| 0 | 4 |
| 1 | 9 |

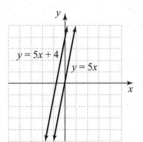

Answers may vary; possible answer: The graph of $y = 5x + 4$ is the same as the graph of $y = 5x$ except it is moved 4 units upward.

**43.**

| $y = -2x$ | |
| --- | --- |
| $x$ | $y$ |
| -2 | 4 |
| 0 | 0 |
| 2 | -4 |

| $y = -2x - 3$ | |
| --- | --- |
| $x$ | $y$ |
| -2 | 1 |
| 0 | -3 |
| 2 | -7 |

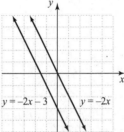

Answers may vary; possible answer: The graph of $y = -2x - 3$ is the same as the graph of $y = -2x$ except it is moved 3 units downward.

**45.**

| $y = \frac{1}{2}x$ | |
| --- | --- |
| $x$ | $y$ |
| -4 | -2 |
| 0 | 0 |
| 4 | 2 |

| $y = \frac{1}{2}x + 2$ | |
| --- | --- |
| $x$ | $y$ |
| -4 | 0 |
| 0 | 2 |
| 4 | 4 |

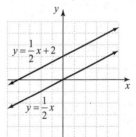

Answers may vary; possible answer: The graph of $y = \frac{1}{2}x + 2$ is the same as the graph of $y = \frac{1}{2}x$ except it is moved 2 units upward.

**47.** Comparing $y = 5x + 5$ to $y = mx + b$, we see that $b = 5$. We see that graph c crosses the $y$-axis at $(0, 5)$.

**49.** Comparing $y = 5x - 1$ to $y = mx + b$, we see that $b = -1$. We see that graph d crosses the $y$-axis at $(0, -1)$.

**51. a.** Using the equation, let $x = 8$.
$y = x + 23$
$y = 8 + 23 = 31$
The ordered pair is (8, 31).

**b.** Eight years after 2000, in 2008, there were 31 million joggers.

**c.** The year 2017 is 17 years after 2000, so let $x = 17$.
$y = x + 23$
$y = 17 + 23 = 40$
If the trend continues, there will be 40 million joggers in 2017.

**53. a.** $y = 2.2x + 190$
Let $x = 8$.
$y = 2.2(8) + 190 = 17.6 + 190 = 207.6$
The ordered pair is (8, 207.6).

**b.** In 2008, there were 207.6 million people with driver's licenses.

**c.** Let $x = 2016 - 2000 = 16$.
$y = 2.2(16) + 190 = 35.2 + 190 = 225.2$
In 2016, it is predicted that there will be 225.2 million people with driver's licenses.

**55.**

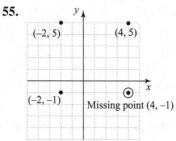

The fourth vertex is at (4, −1).

**57.** $x - y = -3$
$x = 0$: $0 - y = -3$
$\qquad\quad y = 3$
$y = 0$: $x - 0 = -3$
$\qquad\quad x = -3$

| $x$ | $y$ |
|---|---|
| 0 | 3 |
| −3 | 0 |

**59.** $y = 2x$
$x = 0$: $y = 2(0) = 0$
$y = 0$: $0 = 2x$
$\qquad\quad 0 = x$

| $x$ | $y$ |
|---|---|
| 0 | 0 |
| 0 | 0 |

**61.** The equation is $y = x + 5$.

| $x$ | $y$ |
|---|---|
| −2 | 3 |
| 0 | 5 |
| 2 | 7 |

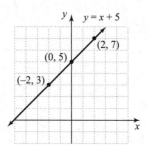

**63.** The equation is $2x + 3y = 6$.

| $x$ | $y$ |
|---|---|
| 3 | 0 |
| 0 | 2 |
| −3 | 4 |

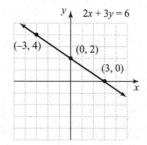

**65.** $x + y + 5 + 5 = 22$
$\quad\; x + y + 10 = 22$
$\qquad\quad x + y = 12$
Let $x = 3$.
$\quad 3 + y = 12$
$\qquad\quad y = 9$ centimeters

**67.** answers may vary

**69.** $y = x^2$

| $x$ | $y$ |
|-----|-----|
| 0   | 0   |
| 1   | 1   |
| −1  | 1   |
| 2   | 4   |
| −2  | 4   |

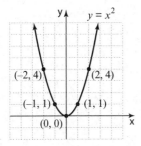

### Section 3.3 Practice

**1.** The graph crosses the $x$-axis at the point (−4, 0).
The $x$-intercept is (−4, 0).
The graph crosses the $y$-axis at the point (0, −6).
The $y$-intercept is (0, −6).

**2.** The graph crosses the $x$-axis at the point (−2, 0)
and at the point (2, 0). The $x$-intercepts are
(−2, 0) and (2, 0).
The graph crosses the $y$-axis at the point (0, −3).
The $y$-intercept is (0, −3).

**3.** The graph crosses both the $x$-axis and the $y$-axis
at the point (0, 0). The $x$-intercept is (0, 0), and
the $y$-intercept is (0, 0).

**4.** The graph does not cross the $x$-axis. There is no
$x$-intercept. The graph crosses the $y$-axis at the
point (0, 3). The $y$-intercept is (0, 3).

**5.** The graph crosses the $x$-axis at the point (−1, 0)
and at the point (5, 0). The $x$-intercepts are
(−1, 0) and (5, 0).
The graph crosses the $y$-axis at the point (0, −2)
and at the point (0, 2). The $y$-intercepts are
(0, −2) and (0, 2).

**6.** Let $y = 0$.                Let $x = 0$.
$\quad x + 2y = -4$              $\quad x + 2y = -4$
$\quad x + 2(0) = -4$            $\quad 0 + 2y = -4$
$\quad x + 0 = -4$               $\quad 2y = -4$
$\quad x = -4$                   $\quad y = -2$

The $x$-intercept is (−4, 0), and the $y$-intercept is
(0, −2).
Let $x = 2$.
$\quad x + 2y = -4$
$\quad 2 + 2y = -4$
$\quad 2y = -6$
$\quad y = -3$

| $x$ | $y$ |
|-----|-----|
| −4  | 0   |
| 0   | −2  |
| 2   | −3  |

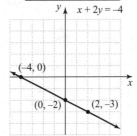

**7.** Let $y = 0$.                Let $x = 0$.
$\quad x = 3y$                   $\quad x = 3y$
$\quad x = 3(0)$                 $\quad 0 = 3y$
$\quad x = 0$                    $\quad 0 = y$

Both the $x$-intercept and the $y$-intercept are
(0, 0).
Let $y = -1$                     Let $y = 1$.
$\quad x = 3(-1)$                $\quad x = 3(1)$
$\quad x = -3$                   $\quad x = 3$

| $x$ | $y$ |
|-----|-----|
| 0   | 0   |
| 3   | 1   |
| −3  | −1  |

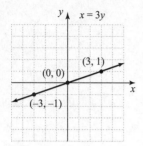

**8.** Let $y = 0$.       Let $x = 0$.
$3x = 2y + 4$      $3x = 2y + 4$
$3x = 2(0) + 4$      $3(0) = 2y + 4$
$3x = 4$          $-4 = 2y$
$x = \dfrac{4}{3}$        $-2 = y$

Let $x = 2$.
$3x = 2y + 4$
$3(2) = 2y + 4$
$6 = 2y + 4$
$2 = 2y$
$1 = y$

| $x$ | $y$ |
|-----|-----|
| 0 | $-2$ |
| $\dfrac{4}{3}$ | 0 |
| 2 | 1 |

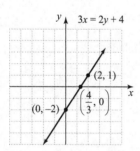

**9.** For any $y$-value chosen, notice that $x$ is $-2$.

| $x$ | $y$ |
|-----|-----|
| $-2$ | $-4$ |
| $-2$ | 0 |
| $-2$ | 4 |

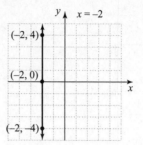

**10.** For any $x$-value chosen, notice that $y$ is 2.

| $x$ | $y$ |
|-----|-----|
| $-5$ | 2 |
| 0 | 2 |
| 5 | 2 |

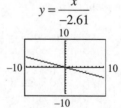

**Calculator Explorations**

**1.** $x = 3.78y$

$y = \dfrac{x}{3.78}$

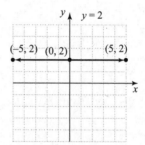

**2.** $-2.61y = x$

$y = \dfrac{x}{-2.61}$

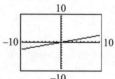

**3.** $3x + 7y = 21$

$7y = -3x + 21$

$y = -\dfrac{3}{7}x + 3$

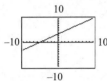

**4.** $-4x + 6y = 21$

$6y = 4x + 21$

$y = \dfrac{2}{3}x + \dfrac{7}{2}$

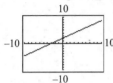

**5.** $-2.2x + 6.8y = 15.5$

$6.8y = 2.2x + 15.5$

$y = \dfrac{2.2}{6.8}x + \dfrac{15.5}{6.8}$

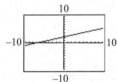

**6.** $5.9x - 0.8y = -10.4$

$-0.8y = -5.9x - 10.4$

$y = \dfrac{5.9}{0.8}x + \dfrac{10.4}{0.8}$

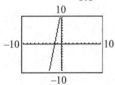

**Vocabulary, Readiness & Video Check 3.3**

**1.** An equation that can be written in the form $Ax + By = C$ is called a <u>linear</u> equation in two variables.

**2.** The form $Ax + By = C$ is called <u>standard</u> form.

**3.** The graph of the equation $y = -1$ is a <u>horizontal</u> line.

**4.** The graph of the equation $x = 5$ is a <u>vertical</u> line.

**5.** A point where a graph crosses the *y*-axis is called a <u>*y*-intercept</u>.

**6.** A point where a graph crosses the *x*-axis is called an <u>*x*-intercept</u>.

**7.** Given an equation of a line, to find the *x*-intercept (if there is one), let <u>*y*</u> = 0 and solve for <u>*x*</u>.

**8.** Given an equation of a line, to find the *y*-intercept (if there is one), let <u>*x*</u> = 0 and solve for <u>*y*</u>.

**9.** Because *x*-intercepts lie on the *x*-axis; because *y*-intercepts lie on the *y*-axis.

**10.** Using a third point as a check that your points lie along a straight line is always good practice.

**11.** For a horizontal line, the coefficient of *x* will be 0 and the coefficient of *y* will be 1; for a vertical line, the coefficient of *y* will be 0 and the coefficient of *x* will be 1.

**Exercise Set 3.3**

**1.** *x*-intercept: (−1, 0); *y*-intercept: (0, 1)

**3.** *x*-intercept: (−2, 0), (2, 0)

**5.** *x*-intercepts: (−2, 0), (1, 0), (3, 0)
*y*-intercept: (0, 3)

**7.** *x*-intercepts: (−1, 0), (1, 0)
*y*-intercepts: (0, 1), (0, −2)

**9.** Infinite; because the line could be vertical (*x* = 0) or horizontal (*y* = 0).

**11.** 0; because the circle could completely reside within one quadrant.

**13.** $x - y = 3$
$y = 0: x - 0 = 3, x = 3$
$x = 0: 0 - y = 3, y = -3$
*x*-intercept: (3, 0); *y*-intercept: (0, −3)

| x | y |
|---|---|
| 3 | 0 |
| 0 | −3 |

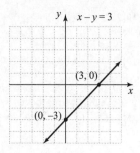

**15.** $x = 5y$
$y = 0$: $x = 5(0) = 0$
$x = 0$: $0 = 5y$, $y = 0$
*x*-intercept: $(0, 0)$; *y*-intercept: $(0, 0)$
$y = 1$: $x = 5(1) = 5$

| x | y |
|---|---|
| 0 | 0 |
| 5 | 1 |

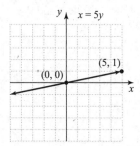

**17.** $-x + 2y = 6$
$y = 0$: $-x + 2(0) = 6$, $x = -6$
$x = 0$: $-0 + 2y = 6$, $y = 3$
*x*-intercept: $(-6, 0)$; *y*-intercept: $(0, 3)$

| x | y |
|---|---|
| −6 | 0 |
| 0 | 3 |

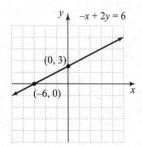

**19.** $2x - 4y = 8$
$y = 0$: $2x - 4(0) = 8$, $x = 4$
$x = 0$: $2(0) - 4y = 8$, $y = -2$
*x*-intercept: $(4, 0)$; *y*-intercept: $(0, -2)$

| x | y |
|---|---|
| 4 | 0 |
| 0 | −2 |

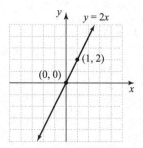

**21.** $y = 2x$
$y = 0$: $0 = 2x$, $0 = x$
$x = 0$: $y = 2(0)$, $y = 0$
*x*-intercept: $(0, 0)$; *y*-intercept: $(0, 0)$
$x = 1$: $y = 2(1)$, $y = 2$

| x | y |
|---|---|
| 0 | 0 |
| 1 | 2 |

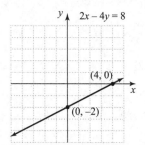

**23.** $y = 3x + 6$
$y = 0$: $0 = 3x + 6$, $-6 = 3x$, $-2 = x$
$x = 0$: $y = 3(0) + 6$, $y = 6$
*x*-intercept: $(-2, 0)$; *y*-intercept: $(0, 6)$

| x | y |
|---|---|
| −2 | 0 |
| 0 | 6 |

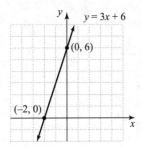

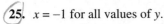

**25.** $x = -1$ for all values of $y$.

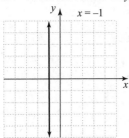

**27.** $y = 0$ for all values of $x$.

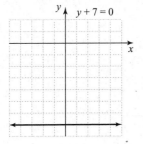

**29.** $y + 7 = 0$

$y = -7$ for all values of $x$.

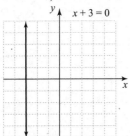

**31.** $x + 3 = 0$; $x = -3$ for all values of $y$.

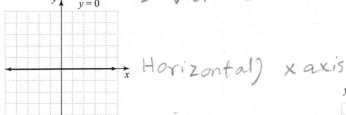

y axis → vertical

Horizontal) x axis

**33.** $x = y$

$x$-intercept: $(0, 0)$; $y$-intercept: $(0, 0)$

Second point: $(4, 4)$

| $x$ | $y$ |
|-----|-----|
| 4 | 4 |
| 0 | 0 |

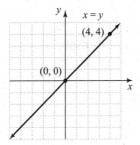

**35.** $x + 8y = 8$

$x$-intercept: $(8, 0)$; $y$-intercept: $(0, 1)$

| $x$ | $y$ |
|-----|-----|
| 8 | 0 |
| 0 | 1 |

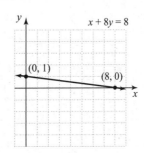

**37.** $5 = 6x - y$

$x$-intercept: $\left(\dfrac{5}{6}, 0\right)$; $y$-intercept: $(0, -5)$

| $x$ | $y$ |
|-----|-----|
| $\dfrac{5}{6}$ | 0 |
| 0 | $-5$ |

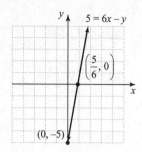

**39.** $-x + 10y = 11$

$x$-intercept: $(-11, 0)$; $y$-intercept: $\left(0, \dfrac{11}{10}\right)$

| $x$ | $y$ |
|-----|-----|
| $-11$ | $0$ |
| $0$ | $\dfrac{11}{10}$ |

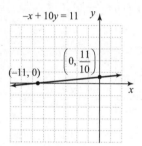

**41.** $x = -4\dfrac{1}{2}$ for all values of $y$.

| $x$ | $y$ |
|-----|-----|
| $-4\dfrac{1}{2}$ | $0$ |
| $-4\dfrac{1}{2}$ | $3$ |

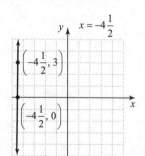

**43.** $y = 3\dfrac{1}{4}$ for all values of $x$.

| $x$ | $y$ |
|-----|-----|
| $0$ | $3\dfrac{1}{4}$ |
| $2$ | $3\dfrac{1}{4}$ |

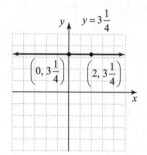

**45.** $y = -\dfrac{2}{3}x + 1$

$x$-intercept: $\left(\dfrac{3}{2}, 0\right)$; $y$-intercept: $(0, 1)$

| $x$ | $y$ |
|-----|-----|
| $\dfrac{3}{2}$ | $0$ |
| $0$ | $1$ |

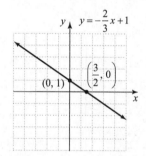

**47.** $4x - 6y + 2 = 0$

$x$-intercept: $\left(-\dfrac{1}{2}, 0\right)$; $y$-intercept: $\left(0, \dfrac{1}{3}\right)$

| $x$ | $y$ |
|-----|-----|
| $-\dfrac{1}{2}$ | $0$ |
| $0$ | $\dfrac{1}{3}$ |

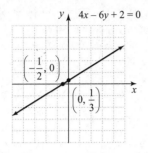

**49.** $y = 3$

The graph is a horizontal line with $y$-intercept $(0, 3)$.

C

**51.** $x = -1$

The graph is a vertical line with $x$-intercept $(-1, 0)$.

E

**53.** $y = 2x + 3$

The $y$-intercept is $(0, 3)$ and the $x$-intercept is $\left(-\dfrac{3}{2}, 0\right)$.

B

**55.** $\dfrac{-6-3}{2-8} = \dfrac{-9}{-6} = \dfrac{3}{2}$

**57.** $\dfrac{-8-(-2)}{-3-(-2)} = \dfrac{-6}{-1} = 6$

**59.** $\dfrac{0-6}{5-0} = \dfrac{-6}{5} = -\dfrac{6}{5}$

**61.** False; for example, the horizontal line $y = 2$ does not have an $x$-intercept.

**63.** True

**65.** $3x + 6y = 1200$

$x = 0:\ 3(0) + 6y = 1200$
$\qquad\qquad\quad 6y = 1200$
$\qquad\qquad\quad\ y = 200$

The ordered pair $(0, 200)$ corresponds to manufacturing 0 chairs and 200 desks.

**67.** Manufacturing 50 desks corresponds to $y = 50$.

$3x + 6y = 1200$

$y = 50:\ 3x + 6(50) = 1200$
$\qquad\qquad\quad 3x + 300 = 1200$
$\qquad\qquad\qquad\ \ 3x = 900$
$\qquad\qquad\qquad\ \ \ x = 300$

When 50 desks are manufactured, 300 chairs can be manufactured.

**69.** $y = -0.025x + 1.55$

**a.** $y = 0:\qquad 0 = -0.025x + 1.55$
$\qquad\qquad\quad 0.025x = 1.55$
$\qquad\qquad\qquad\ \ x = 62$

The $x$-intercept is $(62, 0)$.

**b.** 62 years after 2002 (2064); 0 people will attend movies at the theatre.

**c.** answers may vary

**71.** Parallel to $x = 5$ is vertical.

$x$-intercept is $(1, 0)$, so $x = 1$ for all values of $y$. The equation is $x = 1$.

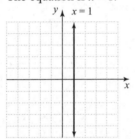

**73.** answers may vary

**75.** answers may vary

## Section 3.4 Practice

**1.** If we let $(x_1, y_1)$ be $(-4, 11)$, then $x_1 = -4$ and $y_1 = 11$. Also, let $(x_2, y_2)$ be $(2, 5)$ so that $x_2 = 2$ and $y_2 = 5$.

$m = \dfrac{y_2 - y_1}{x_2 - x_1} = \dfrac{5 - 11}{2 - (-4)} = \dfrac{-6}{6} = -1$

The slope of the line is $-1$.

**2.** Let $(x_1, y_1)$ be $(3, 1)$ and $(x_2, y_2)$ be $(-3, -1)$.

$m = \dfrac{y_2 - y_1}{x_2 - x_1} = \dfrac{-1 - 1}{-3 - 3} = \dfrac{-2}{-6} = \dfrac{1}{3}$

3. $y = \dfrac{2}{3}x - 2$

   The equation is in slope-intercept form,

   $y = mx + b$. The coefficient of $x$, $\dfrac{2}{3}$, is the slope.

   The constant term, $-2$, is the $y$-value of the
   $y$-intercept, $(0, -2)$.

4. Write the equation in slope-intercept form by
   solving the equation for $y$.
   $$-y = -6x + 5$$
   $$\dfrac{-y}{-1} = \dfrac{-6x}{-1} + \dfrac{5}{-1}$$
   $$y = 6x - 5$$

   The coefficient of $x$, 6, is the slope. The constant
   term, $-5$, is the $y$-value of the $y$-intercept,
   $(0, -5)$.

5. Write the equation in slope-intercept form by
   solving the equation for $y$.
   $$5x + 2y = 8$$
   $$2y = -5x + 8$$
   $$\dfrac{2y}{2} = \dfrac{-5x}{2} + \dfrac{8}{2}$$
   $$y = -\dfrac{5}{2}x + 4$$

   The coefficient of $x$, $-\dfrac{5}{2}$, is the slope, and the

   $y$-intercept is $(0, 4)$.

6. Recall that $y = 3$ is a horizontal line. Two
   ordered pair solutions of $y = 3$ and $(1, 3)$ and
   $(3, 3)$.
   $$m = \dfrac{y_2 - y_1}{x_2 - x_1} = \dfrac{3 - 3}{3 - 1} = \dfrac{0}{2} = 0$$
   The slope of the line $y = 3$ is 0.

7. Recall that the graph of $x = -4$ is a vertical line.
   Two ordered pair solutions of $x = -4$ and $(-4, 1)$
   and $(-4, 3)$.
   $$m = \dfrac{y_2 - y_1}{x_2 - x_1} = \dfrac{3 - 1}{-4 - (-4)} = \dfrac{2}{0}$$
   The slope of the vertical line $x = -4$ is undefined.

8. a. The slope of the line $y = -5x + 1$ is $-5$. We
      solve the second equation for $y$.
      $$x - 5y = 10$$
      $$-5y = -x + 10$$
      $$\dfrac{-5y}{-5} = \dfrac{-x}{-5} + \dfrac{10}{-5}$$
      $$y = \dfrac{1}{5}x - 2$$

      The slope of the second line is $\dfrac{1}{5}$. Since the

      product of the slopes is $\dfrac{1}{5}(-5) = -1$, the

      lines are perpendicular.

   b. Solve each equation for $y$.
      $$\begin{array}{ll} x + y = 11 & 2x + y = 11 \\ y = -x + 11 & y = -2x + 11 \end{array}$$
      The slopes are $-1$ and $-2$. The slopes are not
      the same, and their product is not $-1$. Thus,
      the lines are neither parallel nor
      perpendicular.

   c. Solve each equation for $y$.
      $$\begin{array}{ll} 2x + 3y = 21 & 6y = -4x - 2 \\ 3y = -2x + 21 & \dfrac{6y}{6} = \dfrac{-4x}{6} - \dfrac{2}{6} \\ \dfrac{3y}{3} = \dfrac{-2x}{3} + \dfrac{21}{3} & y = -\dfrac{2}{3}x - \dfrac{1}{3} \\ y = -\dfrac{2}{3}x + 7 & \end{array}$$

      The slopes are $-\dfrac{2}{3}$ and $-\dfrac{2}{3}$. Since the lines

      have the same slope and different
      $y$-intercepts, they are parallel.

9. $\text{grade} = \dfrac{\text{rise}}{\text{run}} = \dfrac{1794}{7176} = 0.25 = 25\%$

   The grade is 25%.

10. Use $(2, 2)$ and $(6, 5)$ to calculate slope.
    $$m = \dfrac{5 - 2}{6 - 2} = \dfrac{3}{4} = \dfrac{0.75 \text{ dollar}}{1 \text{ pound}}$$
    The Wash-n-Fold charges $0.75 per pound of
    laundry.

**Calculator Explorations**

1. $y_1 = 3.8x$
   $y_2 = 3.8x - 3$
   $y_3 = 3.8x + 9$

   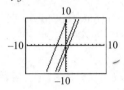

2. $y_1 = -4.9x$
   $y_2 = -4.9x + 1$
   $y_3 = -4.9x + 8$

   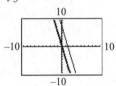

3. $y_1 = \dfrac{1}{4}x$

   $y_2 = \dfrac{1}{4}x + 5$

   $y_3 = \dfrac{1}{4}x - 8$

   (graph from $-10$ to $10$)

4. $y_1 = -\dfrac{3}{4}x$

   $y_2 = -\dfrac{3}{4}x - 5$

   $y_3 = -\dfrac{3}{4}x + 6$

   (graph from $-10$ to $10$)

**Vocabulary, Readiness & Video Check 3.4**

1. The measure of the steepness or tilt of a line is called <u>slope</u>.

2. If an equation is written in the form $y = mx + b$, the value of the letter <u>$m$</u> is the value of the slope of the graph.

3. The slope of a horizontal line is <u>0</u>.

4. The slope of a vertical line is <u>undefined</u>.

5. If the graph of a line moves upward from left to right, the line has <u>positive</u> slope.

6. If the graph of a line moves downward from left to right, the line has <u>negative</u> slope.

7. Given two points of a line, slope $= \dfrac{\text{change in } \underline{y}}{\text{change in } \underline{x}}$.

8. Whatever $y$-value you decide to start with in the numerator, you *must* start with the corresponding $x$-value in the denominator.

9. Solve the equation for $y$; the slope is the coefficient of $x$.

10. Zero slope indicates $m = 0$ and a horizontal line; undefined slope indicates $m$ is undefined and a vertical line; no slope refers to an undefined slope.

11. Slope-intercept form; this form makes the slope easy to see, and you need to compare slopes to determine if two lines are parallel or perpendicular.

12. Step 4: INTERPRET the results.

**Exercise Set 3.4**

1. $(x_1, y_1) = (-1, 5)$ and $(x_2, y_2) = (6, -2)$

   $m = \dfrac{y_2 - y_1}{x_2 - x_1} = \dfrac{-2 - 5}{6 - (-1)} = \dfrac{-7}{7} = -1$

3. $(x_1, y_1) = (-4, 3)$ and $(x_2, y_2) = (-4, 5)$

   $m = \dfrac{y_2 - y_1}{x_2 - x_1} = \dfrac{5 - 3}{-4 - (-4)} = \dfrac{2}{0}$

   The slope is undefined.

5. $(x_1, y_1) = (-2, 8)$ and $(x_2, y_2) = (1, 6)$

   $m = \dfrac{y_2 - y_1}{x_2 - x_1} = \dfrac{6 - 8}{1 - (-2)} = \dfrac{-2}{3} = -\dfrac{2}{3}$

**7.** $(x_1, y_1) = (5, 1)$ and $(x_2, y_2) = (-2, 1)$

$$m = \frac{y_2 - y_1}{x_2 - x_1} = \frac{1-1}{-2-5} = \frac{0}{-7} = 0$$

**9.** $(x_1, y_1) = (-1, 2)$ and $(x_2, y_2) = (2, -2)$

$$m = \frac{y_2 - y_1}{x_2 - x_1} = \frac{-2-2}{2-(-1)} = \frac{-4}{3} = -\frac{4}{3}$$

**11.** $(x_1, y_1) = (2, 3)$ and $(x_2, y_2) = (2, -1)$

$$m = \frac{y_2 - y_1}{x_2 - x_1} = \frac{-1-3}{2-2} = \frac{-4}{0}$$

The slope is undefined.

**13.** $(x_1, y_1) = (-3, -2)$ and $(x_2, y_2) = (-1, 3)$

$$m = \frac{y_2 - y_1}{x_2 - x_1} = \frac{3-(-2)}{-1-(-3)} = \frac{5}{2}$$

**15.** The line goes down. The slope is negative.

**17.** The line is vertical. The slope is undefined.

**19.** The slope is positive. The line is "upward."

**21.** The slope is 0. The line is horizontal.

**23.** The slope of line 1 is positive, and the slope of line 2 is negative. Thus, line 1 has the greater slope.

**25.** Both line 1 and line 2 have positive slopes, but line 2 is steeper than line 1. Thus, line 2 has the greater slope.

**27.** $(0, 0)$ and $(2, 2)$

$$m = \frac{y_2 - y_1}{x_2 - x_1} = \frac{2-0}{2-0} = \frac{2}{2} = 1$$

**D**

**29.** A vertical line has undefined slope.

**B**

**31.** $(2, 0)$ and $(4, -1)$

$$m = \frac{y_2 - y_1}{x_2 - x_1} = \frac{-1-0}{4-2} = -\frac{1}{2}$$

**E**

**33.** $x = 6$ is a vertical line, so it has an undefined slope.

**35.** $y = -4$ is a horizontal line, so it has a slope $m = 0$.

**37.** $x = -3$ is a vertical line, so it has an undefined slope.

**39.** $y = 0$ is a horizontal line, so it has a slope $m = 0$.

**41.** $y = 5x - 2$
The equation is in slope-intercept form. The coefficient of $x$, 5, is the slope.

**43.** $y = -0.3x + 2.5$
The equation is in slope-intercept form. The coefficient of $x$, $-0.3$, is the slope.

**45.** Solve for $y$.
$$2x + y = 7$$
$$y = -2x + 7$$
The coefficient of $x$, $-2$, is the slope.

**47.** Solve for $y$.
$$2x - 3y = 10$$
$$-3y = -2x + 10$$
$$\frac{-3y}{-3} = \frac{-2x}{-3} + \frac{10}{-3}$$
$$y = \frac{2}{3}x - \frac{10}{3}$$

The coefficient of $x$, $\frac{2}{3}$, is the slope.

**49.** The graph of $x = 1$ is a vertical line. The slope is undefined.

**51.** Solve for $y$.
$$x = 2y$$
$$\frac{1}{2}x = y \text{ or } y = \frac{1}{2}x$$

The coefficient of $x$, $\frac{1}{2}$, is the slope.

**53.** The graph of $y = -3$ is a horizontal line. The slope is 0.

**55.** Solve for $y$.
$$-3x - 4y = 6$$
$$-4y = 3x + 6$$
$$\frac{-4y}{-4} = \frac{3x}{-4} + \frac{6}{-4}$$
$$y = -\frac{3}{4}x - \frac{3}{2}$$

The coefficient of $x$, $-\frac{3}{4}$, is the slope.

**57.** Solve for $y$.

$$20x - 5y = 1.2$$
$$-5y = -20x + 1.2$$
$$\frac{-5y}{-5} = \frac{-20x}{-5} + \frac{1.2}{-5}$$
$$y = 4x - 0.24$$

The coefficient of $x$, 4, is the slope.

**59.** $(-3, -3)$ and $(0, 0)$

$$m = \frac{y_2 - y_1}{x_2 - x_1} = \frac{0 - (-3)}{0 - (-3)} = \frac{3}{3} = 1$$

**a.** $m = 1$

**b.** $m = -1$

**61.** $(-8, -4)$ and $(3, 5)$

$$m = \frac{y_2 - y_1}{x_2 - x_1} = \frac{5 - (-4)}{3 - (-8)} = \frac{9}{11}$$

**a.** $m = \dfrac{9}{11}$

**b.** $m = -\dfrac{11}{9}$

**63.** $y = \dfrac{2}{9}x + 3, \; y = -\dfrac{2}{9}x$

The slopes are $\dfrac{2}{9}$ and $-\dfrac{2}{9}$. The slopes are not the same, and their product is not $-1$. The lines are neither parallel nor perpendicular.

**65.** The slope of $y = 3x - 9$ is 3. Solve the other equation for $y$.

$$x - 3y = -6$$
$$-3y = -x - 6$$
$$\frac{-3y}{-3} = -\frac{x}{-3} - \frac{6}{-3}$$
$$y = \frac{1}{3}x + 2$$

The slope is $\dfrac{1}{3}$. The slopes are not the same, and their product is not $-1$. The lines are neither parallel nor perpendicular.

**67.** Solve the equations for $y$.

$$6x = 5y + 1 \qquad\qquad -12x + 10y = 1$$
$$6x - 1 = 5y \qquad\qquad 10y = 12x + 1$$
$$\frac{6x}{5} - \frac{1}{5} = \frac{5y}{5} \qquad \frac{10y}{10} = \frac{12x}{10} + \frac{1}{10}$$
$$y = \frac{6}{5}x - \frac{1}{5} \qquad\qquad y = \frac{6}{5}x + \frac{1}{10}$$

The lines have the same slope, $\dfrac{6}{5}$, but different $y$-intercepts. The lines are parallel.

**69.** Solve the equations for $y$.

$$6 + 4x = 3y \qquad\qquad 3x + 4y = 8$$
$$\frac{6}{3} + \frac{4x}{3} = \frac{3y}{3} \qquad\qquad 4y = -3x + 8$$
$$y = \frac{4}{3}x + 2 \qquad\qquad \frac{4y}{4} = -\frac{3x}{4} + \frac{8}{4}$$
$$y = -\frac{3}{4}x + 2$$

The slopes are $\dfrac{4}{3}$ and $-\dfrac{3}{4}$. Their product is $-1$, so the lines are perpendicular.

**71.** $\text{pitch} = \dfrac{6}{10} = \dfrac{3}{5}$

**73.** $\text{grade} = \dfrac{\text{rise}}{\text{run}} = \dfrac{2}{16} = 0.125 = 12.5\%$

**75.** $\text{grade} = \dfrac{\text{rise}}{\text{run}} = \dfrac{2580}{6450} = 0.40 = 40\%$

**77.** Canton Avenue:

$$\text{grade} = \frac{\text{rise}}{\text{run}} = \frac{11 \text{ meters}}{30 \text{ meters}} \approx 0.37 = 37\%$$

The grade of Canton Avenue is 37%.

Baldwin Street:

$$\text{grade} = \frac{\text{rise}}{\text{run}} = \frac{1 \text{ meter}}{2.86 \text{ meters}} \approx 0.35 = 35\%$$

The grade of Baldwin Street is 35%.

**79.** $m = \dfrac{y_2 - y_1}{x_2 - x_1} = \dfrac{115 - 110}{2010 - 2005} = \dfrac{5}{5} = \dfrac{1}{1}$ or $1$

Every 1 year there are 1 million more U.S. households with televisions.

**81.** Use $(5000, 2350)$ and $(20{,}000, 9400)$ to calculate slope.

$$m = \frac{9400 - 2350}{20{,}000 - 5000} = \frac{7050}{15{,}000} = \frac{0.47 \text{ dollar}}{1 \text{ mile}}$$

It costs $0.47 per 1 mile to own and operate a compact car.

**83.** $y - (-6) = 2(x - 4)$
$\phantom{y} y + 6 = 2x - 8$
$\phantom{y + 6 ==} y = 2x - 14$

**85.** $y - 1 = -6(x - (-2))$
$\phantom{y} y - 1 = -6(x + 2)$
$\phantom{y} y - 1 = -6x - 12$
$\phantom{y -- 1} y = -6x - 11$

**87.** (2, 1) and (0, 0): $m = \dfrac{0-1}{0-2} = \dfrac{-1}{-2} = \dfrac{1}{2}$

(2, 1) and (-2, -1): $m = \dfrac{-1-1}{-2-2} = \dfrac{-2}{-4} = \dfrac{1}{2}$

(2, 1) and (-4, -2): $m = \dfrac{-2-1}{-4-2} = \dfrac{-3}{-6} = \dfrac{1}{2}$

(0, 0) and (-2, -1): $m = \dfrac{-1-0}{-2-0} = \dfrac{-1}{-2} = \dfrac{1}{2}$

(0, 0) and (-4, -2): $m = \dfrac{-2-0}{-4-0} = \dfrac{-2}{-4} = \dfrac{1}{2}$

(-2, -1) and (-4, -2): $m = \dfrac{-2-(-1)}{-4-(-2)} = \dfrac{-1}{-2} = \dfrac{1}{2}$

Since the slope of the line between each pair of points is the same, the points lie on the same line.

**89.** answers may vary

**91.** The line slopes down between 2005 and 2006, so there was a decrease in average fuel economy between 2005 and 2006.

**93.** The lowest point on the graph corresponds to 2000. The average fuel economy for that year was 28.5 miles per gallon.

**95.** Of the line segments listed, the segment from 2008 to 2009 is the steepest and therefore has the greatest slope.

**97.** $\text{pitch} = \dfrac{\text{rise}}{\text{run}}$

$\dfrac{1}{3} = \dfrac{x}{18}$
$3x = 18$
$\phantom{3}x = 6$

**99. a.** (2007, 2207) and (2010, 2333)

**b.** $m = \dfrac{y_2 - y_1}{x_2 - x_1}$

$= \dfrac{2333 - 2207}{2010 - 2007}$

$= \dfrac{126}{3}$

$= 42$

The slope is 42.

**c.** For the years 2007 through 2010, the number of heart transplants increased at a rate of 42 per year.

**101.** (1, 1), (-4, 4) and (-3, 0)

$m_1 = \dfrac{0-1}{-3-1} = \dfrac{1}{4}, \ m_2 = \dfrac{0-4}{-3-(-4)} = -4$

$m_1 m_2 = -1$, so the sides are perpendicular.

**103.** (2.1, 6.7) and (-8.3, 9.3)

$m = \dfrac{y_2 - y_1}{x_2 - x_1} = \dfrac{9.3 - 6.7}{-8.3 - 2.1} = \dfrac{2.6}{-10.4} = -0.25$

**105.** (2.3, 0.2) and (7.9, 5.1)

$m = \dfrac{y_2 - y_1}{x_2 - x_1} = \dfrac{5.1 - 0.2}{7.9 - 2.3} = \dfrac{4.9}{5.6} = 0.875$

**107.** $y = -\dfrac{1}{3}x + 2$
$y = -2x + 2$
$y = -4x + 2$

The line becomes steeper.

**Integrated Review**

**1.** (0, 0) and (2, 4)

$m = \dfrac{y_2 - y_1}{x_2 - x_1} = \dfrac{4-0}{2-0} = \dfrac{4}{2} = 2$

**2.** Horizontal line, $m = 0$

**3.** $(0, 1)$ and $(3, -1)$

$$m = \frac{y_2 - y_1}{x_2 - x_1} = \frac{-1-1}{3-0} = -\frac{2}{3}$$

**4.** Vertical line, slope is undefined.

**5.** $y = -2x$
   $m = -2, b = 0$

| $x$ | $y$ |
|-----|-----|
| 0   | 0   |
| 1   | $-2$ |
| $-1$ | 2  |

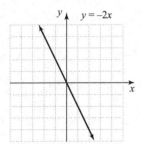

**6.** $x + y = 3$
   $\quad y = -x + 3$
   $m = -1, b = 3$

| $x$ | $y$ |
|-----|-----|
| 0   | 3   |
| 3   | 0   |
| 1   | 2   |

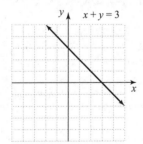

**7.** $x = -1$ for all values of $y$.
   Vertical line; slope is undefined.

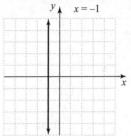

**8.** $y = 4$ for all values of $x$.
   Horizontal line; $m = 0$

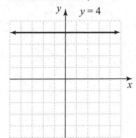

**9.** $x - 2y = 6$
   $\quad -2y = -x + 6$
   $\quad y = \dfrac{1}{2}x - 3$
   $m = \dfrac{1}{2}, b = -3$

| $x$ | $y$ |
|-----|-----|
| 0   | $-3$ |
| 2   | $-2$ |
| 4   | $-1$ |

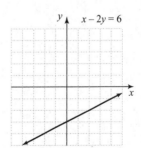

**10.** $y = 3x + 2$
$m = 3, b = 2$

| x | y |
|---|---|
| 0 | 2 |
| −1 | −1 |
| −2 | −4 |

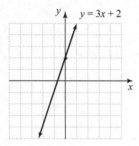

**11.** $5x + 3y = 15$

| x | y |
|---|---|
| 0 | 5 |
| 3 | 0 |

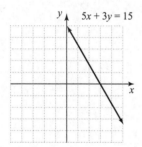

**12.** $2x − 4y = 8$

| x | y |
|---|---|
| 0 | −2 |
| 4 | 0 |

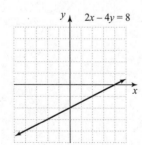

**13.** The slope of the first line is $-\dfrac{1}{5}$. Solve the second equation for *y*.

$$3x = -15y$$
$$\frac{3x}{-15} = \frac{-15y}{-15}$$
$$y = -\frac{1}{5}x$$

The slope of the second line is also $-\dfrac{1}{5}$. Since the lines have the same slope but different *y*-intercepts, the lines are parallel.

**14.** Solve the equations for *y*.

$$x - y = \frac{1}{2} \qquad\qquad 3x - y = \frac{1}{2}$$
$$-y = -x + \frac{1}{2} \qquad\qquad -y = -3x + \frac{1}{2}$$
$$y = x - \frac{1}{2} \qquad\qquad y = 3x - \frac{1}{2}$$

The slopes are 1 and 3. Since the slopes are not equal and their product is not −1, the lines are neither parallel nor perpendicular.

**15.** **a.** Let $x = 0$.
$y = 1.7(0) + 587 = 587$
The *y*-intercept is (0, 587).

**b.** In 2000, there were 587 thousand public bridges in the United States.

**c.** The equation is in slope-intercept form. The coefficient of *x*, 1.7, is the slope.

**d.** For the years 2000 through 2009, the number of public bridges increased at a rate of 1.7 thousand per year.

**16.** **a.** Let $x = 9$.
$y = 3.3(9) - 3.1 = 29.7 - 3.1 = 26.6$
The ordered pair is (9, 26.6).

**b.** In 2009, the revenue for online advertising was $26.6 billion.

**Section 3.5 Practice**

**1.** $y = \dfrac{2}{3}x - 5$

The slope is $\dfrac{2}{3}$, and the $y$-intercept is $(0, -5)$.

We plot $(0, -5)$. From this point, we move up 2 units and then right 3 units. We stop at the point $(3, -3)$.

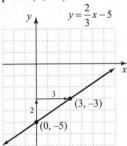

**2.** Solve the equation for $y$.

$3x - y = 2$

$-y = -3x + 2$

$y = 3x - 2$

The slope is 3, and the $y$-intercept is $(0, -2)$. We plot $(0, -2)$. From this point, we move up 3 units and then right 1 unit. We stop at the point $(1, 1)$.

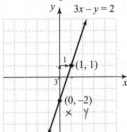

**3.** $y$-intercept: $(0, 7)$; slope: $\dfrac{1}{2}$

Let $m = \dfrac{1}{2}$ and $b = 7$.

$y = mx + b$

$y = \dfrac{1}{2}x + 7$

**4.** Line passing through $(2, 3)$ with slope 4

$y - y_1 = m(x - x_1)$

$y - 3 = 4(x - 2)$

$y - 3 = 4x - 8$

$-4x + y = -5$

$4x - y = 5$

**5.** Line through $(-1, 6)$ and $(3, 1)$

$m = \dfrac{1 - 6}{3 - (-1)} = \dfrac{-5}{4} = -\dfrac{5}{4}$

Use the slope $-\dfrac{5}{4}$ and the point $(3, 1)$.

$y - y_1 = m(x - x_1)$

$y - 1 = -\dfrac{5}{4}(x - 3)$

$4(y - 1) = 4\left(-\dfrac{5}{4}\right)(x - 3)$

$4y - 4 = -5(x - 3)$

$4y - 4 = -5x + 15$

$5x + 4y = 19$

**6.** The equation of a vertical line can be written in the form $x = c$, so an equation for a vertical line passing through $(3, -2)$ is $x = 3$.

**7.** Since the graph of $y = -2$ is a horizontal line, any line parallel to it is also vertical. The equation of a horizontal line can be written in the form $y = c$. An equation for the horizontal line passing through $(4, 3)$ is $y = 3$.

**8. a.** Write two ordered pairs, $(30, 150{,}000)$ and $(50, 120{,}000)$.

$m = \dfrac{120{,}000 - 150{,}000}{50 - 30}$

$= \dfrac{-30{,}000}{20}$

$= -1500$

Use the slope $-1500$ and the point $(30, 150{,}000)$.

$y - y_1 = m(x - x_1)$

$y - 150{,}000 = -1500(x - 30)$

$y - 150{,}000 = -1500x + 45{,}000$

$y = -1500x + 195{,}000$

**b.** Find $y$ when $x = 60$.

$y = -1500x + 195{,}000$

$y = -1500(60) + 195{,}000$

$y = -90{,}000 + 195{,}000$

$y = 105{,}000$

To sell 60 condos per month, the price should be $105,000.

**Calculator Explorations**

1. $y_1 = x$, $y_2 = 6x$, $y_3 = -6x$

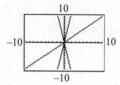

2. $y_1 = -x$, $y_2 = -5x$, $y_3 = -10x$

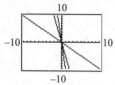

3. $y_1 = \frac{1}{2}x + 2$, $y_2 = \frac{3}{4}x + 2$, $y_3 = x + 2$

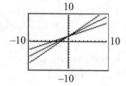

4. $y_1 = x + 1$, $y_2 = \frac{5}{4}x + 1$, $y_3 = \frac{5}{2}x + 1$

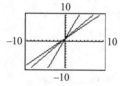

5. $y_1 = -7x + 5$, $y_2 = 7x + 5$

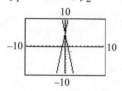

6. $y_1 = 3x - 1$, $y_2 = -3x - 1$

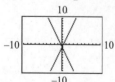

**Vocabulary, Readiness & Video Check 3.5**

1. The form $y = mx + b$ is called <u>slope-intercept</u> form. When a linear equation in two variables is written in this form, <u>m</u> is the slope of its graph and $(0, \underline{b})$ is its $y$-intercept.

2. The form $y - y_1 = m(x - x_1)$ is called <u>point-slope</u> form. When a linear equation in two variables is written in this form, <u>m</u> is the slope of its graph and $(x_1, y_1)$ is a point on the graph.

3. Start by graphing the <u>$y$-intercept</u>. Find another point by applying the slope to this point—rewrite the slope as a <u>fraction</u> if necessary.

4. $\left(0, -\frac{1}{6}\right)$

5. Write the equation with $x$- and $y$-terms on one side of the equal sign and a constant on the other side.

6. Yes, if one of the points given is the $y$-intercept. You will need to use the slope formula to find the slope, but then you'll have the slope and $y$-intercept for the slope-intercept form.

7. Example 6: $y = -3$; Example 7: $x = -2$

8. You need to know what your variables stand for in order to solve part b of the Example, and that depends on how you set up your ordered pairs in part a.

**Exercise Set 3.5**

1. $y = 2x + 1$

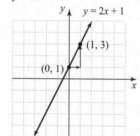

Copyright © 2013 Pearson Education, Inc.

**3.** $y = \dfrac{2}{3}x + 5$

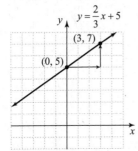

**5.** $y = -5x$

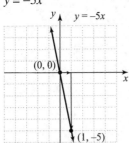

**7.** $4x + y = 6$

$y = -4x + 6$

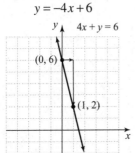

**9.** $4x - 7y = -14$

$-7y = -4x - 14$

$y = \dfrac{4}{7}x + 2$

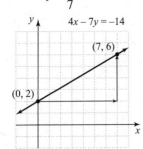

**11.** $x = \dfrac{5}{4}y$

$\dfrac{4}{5}x = y$

$y = \dfrac{4}{5}x$

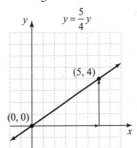

**13.** $m = 5,\ b = 3$

$y = mx + b$

$y = 5x + 3$

**15.** $m = -4,\ b = -\dfrac{1}{6}$

$y = mx + b$

$y = -4x + \left(-\dfrac{1}{6}\right)$

$y = -4x - \dfrac{1}{6}$

**17.** $m = \dfrac{2}{3},\ b = 0$

$y = mx + b$

$y = \dfrac{2}{3}x + 0$

$y = \dfrac{2}{3}x$

**19.** $m = 0,\ b = -8$

$y = mx + b$

$y = 0x + (-8)$

$y = -8$

**21.** $m = -\dfrac{1}{5},\ b = \dfrac{1}{9}$

$y = mx + b$

$y = -\dfrac{1}{5}x + \dfrac{1}{9}$

**23.** $m = 6$; $(2, 2)$

$$y - y_1 = m(x - x_1)$$
$$y - 2 = 6(x - 2)$$
$$y - 2 = 6x - 12$$
$$-6x + y = -10 \text{ or } 6x - y = 10$$

**25.** $m = -8$; $(-1, -5)$

$$y - y_1 = m(x - x_1)$$
$$y - (-5) = -8(x - (-1))$$
$$y + 5 = -8x - 8$$
$$8x + y = -13$$

**27.** $m = \dfrac{3}{2}$; $(5, -6)$

$$y - y_1 = m(x - x_1)$$
$$y - (-6) = \frac{3}{2}(x - 5)$$
$$2(y + 6) = 3(x - 5)$$
$$2y + 12 = 3x - 15$$
$$-3x + 2y = -27$$
$$3x - 2y = 27$$

**29.** $m = -\dfrac{1}{2}$; $(-3, 0)$

$$y - y_1 = m(x - x_1)$$
$$y - 0 = -\frac{1}{2}(x - (-3))$$
$$y = -\frac{1}{2}(x + 3)$$
$$-2y = x + 3$$
$$-x - 2y = 3$$
$$x + 2y = -3$$

**31.** $(3, 2)$ and $(5, 6)$

$$m = \frac{y_2 - y_1}{x_2 - x_1} = \frac{6 - 2}{5 - 3} = \frac{4}{2} = 2$$
$$m = 2; (3, 2)$$
$$y - y_1 = m(x - x_1)$$
$$y - 2 = 2(x - 3)$$
$$y - 2 = 2x - 6$$
$$-2x + y = -4$$
$$2x - y = 4$$

**33.** $(-1, 3)$ and $(-2, -5)$

$$m = \frac{y_2 - y_1}{x_2 - x_1} = \frac{-5 - 3}{-2 - (-1)} = \frac{-8}{-1} = 8$$
$$m = 8; (-1, 3)$$
$$y - y_1 = m(x - x_1)$$
$$y - 3 = 8(x - (-1))$$
$$y - 3 = 8x + 8$$
$$-8x + y = 11$$
$$8x - y = -11$$

**35.** $(2, 3)$ and $(-1, -1)$

$$m = \frac{y_2 - y_1}{x_2 - x_1} = \frac{-1 - 3}{-1 - 2} = \frac{-4}{-3} = \frac{4}{3}$$
$$m = \frac{4}{3}; (2, 3)$$
$$y - y_1 = m(x - x_1)$$
$$y - 3 = \frac{4}{3}(x - 2)$$
$$3(y - 3) = 4(x - 2)$$
$$3y - 9 = 4x - 8$$
$$-4x + 3y = 1$$
$$4x - 3y = -1$$

**37.** $(0, 0)$ and $\left(-\dfrac{1}{8}, \dfrac{1}{13}\right)$

$$m = \frac{\frac{1}{13} - 0}{-\frac{1}{8} - 0} = \frac{\frac{1}{13}}{-\frac{1}{8}} = \frac{1}{13}\left(-\frac{8}{1}\right) = -\frac{8}{13}$$
$$m = -\frac{8}{13}; (0, 0)$$
$$y - y_1 = m(x - x_1)$$
$$y - 0 = -\frac{8}{13}(x - 0)$$
$$y = -\frac{8}{13}x$$
$$13y = -8x$$
$$8x + 13y = 0$$

**39.** Vertical line, point $(0, 2)$

$$x = c$$
$$x = 0$$

**41.** Horizontal line, point $(-1, 3)$

$$y = c$$
$$y = 3$$

**43.** Vertical line, point $\left(-\dfrac{7}{3}, -\dfrac{2}{5}\right)$

$x = c$

$x = -\dfrac{7}{3}$

**45.** $y = 5$ is horizontal.
Parallel to $y = 5$ is horizontal; $y = c$.
Point $(1, 2)$
$y = 2$

**47.** $x = -3$ is vertical.
Perpendicular to $x = -3$ is horizontal; $y = c$.
Point $(-2, 5)$
$y = 5$

**49.** $x = 0$ is vertical.
Parallel to $x = 0$ is vertical; $x = c$.
Point $(6, -8)$
$x = 6$

**51.** $m = -\dfrac{1}{2}; \left(0, \dfrac{5}{3}\right)$

$y = mx + b$

$y = -\dfrac{1}{2}x + \dfrac{5}{3}$

**53.** $(10, 7)$ and $(7, 10)$

$m = \dfrac{y_2 - y_1}{x_2 - x_1} = \dfrac{10 - 7}{7 - 10} = \dfrac{3}{-3} = -1$

$m = -1; (10, 7)$

$y - y_1 = m(x - x_1)$

$y - 7 = -1(x - 10)$

$y - 7 = -x + 10$

$y = -x + 17$

**55.** Undefined slope, through $\left(-\dfrac{3}{4}, 1\right)$

A line with undefined slope is vertical. A vertical line has an equation of the form $x = c$.

$x = -\dfrac{3}{4}$

**57.** $m = 1; (-7, 9)$

$y - y_1 = m(x - x_1)$

$y - 9 = 1[x - (-7)]$

$y - 9 = x + 7$

$y = x + 16$

**59.** $m = -5, b = 7$

$y = mx + b$

$y = -5x + 7$

**61.** $x$-axis is horizontal.
Parallel to $x$-axis is horizontal; $y = c$.
Point $(6, 7)$
$y = 7$

**63.** $(2, 3)$ and $(0, 0)$

$m = \dfrac{y_2 - y_1}{x_2 - x_1} = \dfrac{3 - 0}{2 - 0} = \dfrac{3}{2}; \ b = 0$

$y = mx + b$

$y = \dfrac{3}{2}x + 0$

$y = \dfrac{3}{2}x$

**65.** $y$-axis is vertical.
Perpendicular to $y$-axis is horizontal; $y = c$.
Point $(-2, -3)$
$y = -3$

**67.** $m = -\dfrac{4}{7}; \ (-1, -2)$

$y - y_1 = m(x - x_1)$

$y - (-2) = -\dfrac{4}{7}[x - (-1)]$

$y + 2 = -\dfrac{4}{7}x - \dfrac{4}{7}$

$y = -\dfrac{4}{7}x - \dfrac{4}{7} - 2$

$y = -\dfrac{4}{7}x - \dfrac{18}{7}$

**69. a.** $(1, 32)$ and $(3, 96)$

$m = \dfrac{y_2 - y_1}{x_2 - x_1} = \dfrac{96 - 32}{3 - 1} = \dfrac{64}{2} = 32$

$m = 32; (1, 32)$

$s - s_1 = m(t - t_1)$

$s - 32 = 32(t - 1)$

$s - 32 = 32t - 32$

$s = 32t$

**b.** If $t = 4$, then $s = 32(4) = 128$ ft/sec.

**71. a.** Use $(0, 356)$ and $(2, 290)$.

$m = \dfrac{290 - 356}{2 - 0} = \dfrac{-66}{2} = -33$

$b = 356$

$y = mx + b$

$y = -33x + 356$

**b.** Let $x = 2015 - 2007 = 8$
$$y = -33(8) + 356$$
$$= -264 + 356$$
$$= 92$$
We predict there will be 92 thousand or 92,000 hybrids sold in 2015.

**73. a.** Use (0, 85) and (5, 88).
$$m = \frac{88 - 85}{5 - 0} = \frac{3}{5} = 0.6$$
$$b = 85$$
$$y = mx + b$$
$$y = 0.6x + 85$$

**b.** Let $x = 2020 - 2006 = 14$.
$$y = 0.6(14) + 85 = 8.4 + 85 = 93.4$$
We predict there will be 93.4 persons per square mile in 2020.

**75. a.** Use (0, 60) and (8, 46).
$$m = \frac{46 - 60}{8 - 0} = \frac{-14}{8} = -1.75$$
$$b = 60$$
$$y = mx + b$$
$$y = -1.75x + 60$$

**b.** Let $x = 2017 - 2001 = 16$
$$y = -1.75(16) + 60 = -28 + 60 = 32$$
We predict that the newspaper circulation in 2017 will be 32 million.

**77. a.** The ordered pairs are (3, 10,000) and (5, 8000).
$$m = \frac{S_2 - S_1}{p_2 - p_1}$$
$$= \frac{8000 - 10,000}{5 - 3}$$
$$= \frac{-2000}{2}$$
$$= -1000$$
$$S - S_1 = m(p - p_1)$$
$$S - 10,000 = -1000(p - 3)$$
$$S - 10,000 = -1000p + 3000$$
$$S = -1000p + 13,000$$

**b.** $p = 3.50$: $S = -1000(3.50) + 13,000$
$$S = -3500 + 13,000$$
$$S = 9500$$
9500 Fun Noodles will be sold when the price is $3.50 each.

**79.** If $x = 2$, then
$$x^2 - 3x + 1 = (2)^2 - 3(2) + 1 = 4 - 6 + 1 = -1$$

**81.** If $x = -1$, then
$$x^2 - 3x + 1 = (-1)^2 - 3(-1) + 1 = 1 + 3 + 1 = 5$$

**83.** No

**85.** Yes

**87.** $y - 7 = 4(x + 3)$ is in <u>point-slope</u> form.

**89.** $y = \frac{3}{4}x - \frac{1}{3}$ is in <u>slope-intercept</u> form.

**91.** $y = \frac{1}{2}$ is a <u>horizontal</u> line.

**93.** answers may vary

**95.** $y = 3x - 1$, $m_1 = 3$

**a.** Parallel: $m_2 = m_1 = 3$; (−1, 2)
$$y - y_1 = m_2(x - x_1)$$
$$y - 2 = 3(x - (-1))$$
$$y - 2 = 3x + 3$$
$$-3x + y = 5$$
$$3x - y = -5$$

**b.** Perpendicular: $m_2 = -\frac{1}{m_1} = -\frac{1}{3}$; (−1, 2)
$$y - y_1 = m_2(x - x_1)$$
$$y - 2 = -\frac{1}{3}(x - (-1))$$
$$3(y - 2) = -1(x + 1)$$
$$3y - 6 = -x - 1$$
$$x + 3y = 5$$

**97.** $3x + 2y = 7$, $y = -\frac{3}{2}x + \frac{7}{2}$, $m_1 = -\frac{3}{2}$

**a.** Parallel: $m_2 = m_1 = -\frac{3}{2}$; (3, −5)
$$y - y_1 = m_2(x - x_1)$$
$$y - (-5) = -\frac{3}{2}(x - 3)$$
$$2(y + 5) = -3(x - 3)$$
$$2y + 10 = -3x + 9$$
$$3x + 2y = -1$$

**b.** Perpendicular: $m_2 = -\dfrac{1}{m_1} = \dfrac{2}{3};\ (3, -5)$

$$y - y_1 = m_2(x - x_1)$$
$$y - (-5) = \frac{2}{3}(x - 3)$$
$$3(y + 5) = 2(x - 3)$$
$$3y + 15 = 2x - 6$$
$$2x - 3y = 21$$

**Section 3.6 Practice**

1. The domain is the set of all $x$-values $\{0, 1, 5\}$.
   The range is the set of all $y$-values: $\{-2, 0, 3, 4\}$.

2. **a.** $\{(4, 1), (3, -2), (8, 5), (-5, 3)\}$
   Each $x$-value is assigned to only one $y$-value, so this set of ordered pairs is a function.

   **b.** $\{(1, 2), (-4, 3), (0, 8), (1, 4)\}$
   The $x$-value 1 is assigned to two $y$-values, 2 and 4, so this set of ordered pairs is not a function.

3. **a.** This is the graph of the relation $\{(-2, 1), (3, -3), (3, 2)\}$. The $x$-coordinate 3 is paired with two $y$-coordinates, $-3$ and 2, so this is not the graph of a function.

   **b.** This is the graph of the relation $\{(-2, 1), (0, 1), (1, -3), (3, 2)\}$. Each $x$-coordinate has exactly one $y$-coordinate, so this is the graph of a function.

4. **a.** This is the graph of a function since no vertical line will intersect this graph more than once.

   **b.** This is the graph of a function since no vertical line will intersect this graph more than once.

   **c.** This is the graph of a function since no vertical line will intersect this graph more than once.

   **d.** This is not the graph of a function. Vertical lines can be drawn that intersect the graph in two points. An example of one is shown.

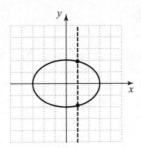

5. **a.** $y = 2x$ is a function because its graph is a nonvertical line.

   **b.** $y = -3x - 1$ is a function because its graph is a nonvertical line.

   **c.** $y = 8$ is a function because its graph is a nonvertical line.

   **d.** $x = 2$ is not a function because its graph is a vertical line.

6. **a.** Since June is the sixth month, we look for 6 on the horizontal axis. From this point, we move vertically upward until the graph is reached. From the point on the graph, we move horizontally to the left to the vertical axis. The vertical axis there reads about 69°F.

   **b.** We find 26°F on the temperature axis and move horizontally to the right. We eventually reach the point corresponding to 2, or February.

   **c.** Yes, this is the graph of a function. It passes the vertical line test.

7. $h(x) = x^2 + 5$

   **a.** $h(2) = 2^2 + 5 = 4 + 5 = 9$
   $(2, 9)$

   **b.** $h(-5) = (-5)^2 + 5 = 25 + 5 = 30$
   $(-5, 30)$

   **c.** $h(0) = 0^2 + 5 = 0 + 5 = 5$
   $(0, 5)$

8. **a.** $h(x) = 6x + 3$
   In this function, $x$ can be any real number. The domain of $h(x)$ is the set of all real numbers, or $(-\infty, \infty)$ in interval notation.

b. $f(x) = \dfrac{1}{x^2}$

Recall that we cannot divide by 0 so that the domain of $f(x)$ is the set of all real numbers except 0. In interval notation, we write $(-\infty, 0) \cup (0, \infty)$.

9. a.

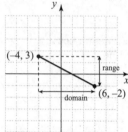

The domain is $[-4, 6]$.
The range is $[-2, 3]$.

b.

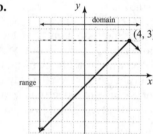

The domain is $(-\infty, \infty)$.
The range is $(-\infty, 3]$.

**Vocabulary, Readiness & Video Check 3.6**

1. A set of ordered pairs is called a <u>relation</u>.

2. A set of ordered pairs that assigns to each $x$-value exactly one $y$-value is called a <u>function</u>.

3. The set of all $y$-coordinates of a relation is called the <u>range</u>.

4. The set of all $x$-coordinates of a relation is called the <u>domain</u>.

5. All linear equations are functions except those whose graphs are <u>vertical</u> lines.

6. All linear equations are functions except those whose equations are of the form <u>$x = c$</u>.

7. A relation is a set of ordered pairs and an equation in two variables defines a set of ordered pairs. therefore, an equation in two variables can also define a relation.

8. Yes, this is a function. the definition restricts $x$-values to be assigned to exactly one $y$-value, but it makes no such restriction on the $y$-values.

9. A vertical line represents one $x$-value paired with many $y$-values. A function only allows an $x$-value paired with exactly one $y$-value, so if a vertical line intersects a graph more than once, there's an $x$-value paired with more than one $y$-value, and we don't have a function.

10. $f(-2) = 6$ corresponds to $(-2, 6)$, and $f(3) = 11$ corresponds to $(3, 11)$.

**Exercise Set 3.6**

1. $\{(2, 4), (0, 0), (-7, 10), (10, -7)\}$
   Domain: $\{-7, 0, 2, 10\}$
   Range: $\{-7, 0, 4, 10\}$

3. $\{(0, -2), (1, -2), (5, -2)\}$
   Domain: $\{0, 1, 5\}$
   Range: $\{-2\}$

5. Every point has a unique $x$-value: it is a function.

7. Two or more points have the same $x$-value: it is not a function.

9. No; two points have $x$-coordinate 1.

11. Yes; no two points have the same $x$-coordinate.

13. Yes; no vertical line can be drawn that intersects the graph more than once.

15. No; there are many vertical lines that intersect the graph twice, $x = 1$, for example.

17. Yes; $y = x + 1$ is a non-vertical line.

19. Yes; $y - x = 7$ is a non-vertical line.

21. Yes; $y = 6$ is a non-vertical line.

23. No; $x = -2$ is a vertical line.

25. No; does not pass the vertical line test.

27. The point on the graph above June corresponds to approximately 9:30 p.m. on the time axis.

29. The sunset is at approximately 3 p.m. twice, on January 1 and on December 1.

31. Yes; it passes the vertical line test.

33. $4.25 per hour; the segment representing dates before October 1996 corresponds to 4.25 on the vertical axis.

35. 2009; the first line segment above 7.00 on the vertical axis represents dates beginning July 24, 2009.

37. yes; answers may vary

39. According to the graph, the postage would be $1.50.

41. From the graph, it would cost $1 to mail a large envelope that weighs more than 1 ounce and less than or equal to 2 ounces.

43. yes; answers may vary

45. $f(x) = 2x - 5$
$f(-2) = 2(-2) - 5 = -4 - 5 = -9$
$f(0) = 2(0) - 5 = -5$
$f(3) = 2(3) - 5 = 6 - 5 = 1$

47. $f(x) = x^2 + 2$
$f(-2) = (-2)^2 + 2 = 4 + 2 = 6$
$f(0) = (0)^2 + 2 = 2$
$f(3) = (3)^2 + 2 = 9 + 2 = 11$

49. $f(x) = 3x$
$f(-2) = 3(-2) = -6$
$f(0) = 3(0) = 0$
$f(3) = 3(3) = 9$

51. $f(x) = |x|$
$f(-2) = |-2| = 2$
$f(0) = |0| = 0$
$f(3) = |3| = 3$

53. $h(x) = -5x$
$h(-1) = -5(-1) = 5$
$h(0) = -5(0) = 0$
$h(4) = -5(4) = -20$

55. $h(x) = 2x^2 + 3$
$h(-1) = 2(-1)^2 + 3 = 2 + 3 = 5$
$h(0) = 2(0)^2 + 3 = 3$
$h(4) = 2(4)^2 + 3 = 2 \cdot 16 + 3 = 32 + 3 = 35$

57. $f(3) = 6$ corresponds to the ordered pair $(3, 6)$.

59. $g(0) = -\dfrac{1}{2}$ corresponds to the ordered pair $\left(0, -\dfrac{1}{2}\right)$.

61. $h(-2) = 9$ corresponds to the ordered pair $(-2, 9)$.

63. $(-\infty, \infty)$

65. $x + 5 \neq 0 \Rightarrow x \neq -5$, therefore $(-\infty, -5) \cup (-5, \infty)$ or all real numbers except $-5$.

67. $(-\infty, \infty)$

69. D: $(-\infty, \infty)$, R: $x \geq -4$, $[-4, \infty)$

71. D: $(-\infty, \infty)$, R: $(-\infty, \infty)$

73. D: $(-\infty, \infty)$, R: $\{2\}$

75. When $x = 0$, $y = -1$, so the ordered-pair solution is $(0, -1)$.

77. When $x = 0$, $y = -1$, so $f(0) = -1$.

79. When $y = 0$, $x = -1$ and $x = 5$.

81. $(-2, 1)$

83. $(-3, -1)$

85. $f(-5) = 12$

87. $(3, -4)$

89. $f(5) = 0$

91. $H(x) = 2.59x + 47.24$

   **a.** $H(46) = 2.59(46) + 47.24 = 166.38$ cm

   **b.** $H(39) = 2.59(39) + 47.24 = 148.25$ cm

93. Answers may vary

95. $y = x + 7$
$f(x) = x + 7$

97. $f(x) = 2x + 7$

   **a.** $f(2) = 2(2) + 7 = 11$

   **b.** $f(a) = 2(a) + 7 = 2a + 7$

**99.** $h(x) = x^2 + 7$

    **a.** $h(3) = (3)^2 + 7 = 16$

    **b.** $h(a) = (a)^2 + 7 = a^2 + 7$

## Chapter 3 Vocabulary Check

1. An ordered pair is a <u>solution</u> of an equation in two variables if replacing the variables by the coordinates of the ordered pair results in a true statement.

2. The vertical number line in the rectangle coordinate system is called the <u>y-axis</u>.

3. A <u>linear</u> equation can be written in the form $Ax + By = C$.

4. An <u>x-intercept</u> is a point of the graph where the graph crosses the $x$-axis.

5. The form $Ax + By = C$ is called <u>standard</u> form.

6. A <u>y-intercept</u> is a point of the graph where the graph crosses the $y$-axis.

7. The equation $y = 7x - 5$ is written in <u>slope-intercept</u> form.

8. The equation $y + 1 = 7(x - 2)$ is written in <u>point-slope</u> form.

9. To find an $x$-intercept of a graph, let <u>$y$</u> $= 0$.

10. The horizontal number line in the rectangular coordinate system is called the <u>x-axis</u>.

11. To find a $y$-intercept of a graph, let <u>$x$</u> $= 0$.

12. The <u>slope</u> of a line measures the steepness or tilt of a line.

13. A set of ordered pairs that assigns to each $x$-value exactly one $y$-value is called a <u>function</u>.

14. The set of all $x$-coordinates of a relation is called the <u>domain</u> of the relation.

15. The set of all $y$-coordinates of a relation is called the <u>range</u> of the relation.

16. A set of ordered pairs is called a <u>relation</u>.

## Chapter 3 Review

**1–6.**

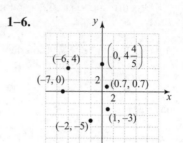

**7. a.** (5.00, 50), (8.50, 100), (20.00, 250), (27.00, 500)

    **b.**

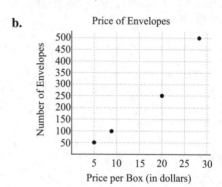

**8. a.** (2005, 13.8), (2006, 13.6), (2007, 14.1), (2008, 13.9), (2009, 14.6), (2010, 14.6)

    **b.**

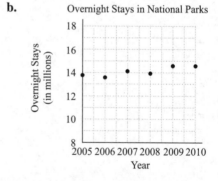

**9.** $7x - 8y = 56$
(0, 56)
$7(0) - 8(56) \overset{?}{=} 56$
    $-448 \neq 0$   No

(8, 0)
$7(8) - 8(0) \overset{?}{=} 56$
    $56 = 56$   Yes

**10.** $-2x + 5y = 10$
$(-5, 0)$
$-2(-5) + 5(0) \stackrel{?}{=} 10$
$\qquad\qquad 10 = 10$   Yes

$(1, 1)$
$-2(1) + 5(1) \stackrel{?}{=} 10$
$\qquad\qquad 3 \neq 10$   No

**11.** $x = 13$
$(13, 5)$
$(13) \stackrel{?}{=} 13$
$\quad 13 = 13$   Yes

$(13, 13)$
$(13) \stackrel{?}{=} 13$
$\quad 13 = 13$   Yes

**12.** $y = 2$
$(7, 2)$
$(2) \stackrel{?}{=} 2$
$\quad 2 = 2$   Yes

$(2, 7)$
$(7) \stackrel{?}{=} 2$
$\quad 7 \neq 2$   No

**13.** $-2 + y = 6x,\ x = 7$
$-2 + y = 6(7)$
$-2 + y = 42$
$\qquad y = 44$
$(7, 44)$

**14.** $y = 3x + 5,\ y = -8$
$-8 = 3x + 5$
$-13 = 3x$
$-\dfrac{13}{3} = x$
$\left(-\dfrac{13}{3}, -8\right)$

**15.** $9 = -3x + 4y$
$y = 0:\ 9 = -3x + 4(0),\ 9 = -3x,\ -3 = x$
$y = 3:\ 9 = -3x + 4(3),\ 9 = -3x + 12,\ -3 = -3x,$
$\qquad 1 = x$
$x = 9:\ 9 = -3(9) + 4y,\ 9 = -27 + 4y,\ 36 = 4y,$
$\qquad 9 = y$

| $x$ | $y$ |
|-----|-----|
| $-3$ | $0$ |
| $1$ | $3$ |
| $9$ | $9$ |

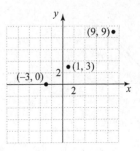

**16.** $x = 2y$
$y = 0:\ x = 2(0) = 0$
$y = 5:\ x = 2(5) = 10$
$y = -5:\ x = 2(-5) = -10$

| $x$ | $y$ |
|-----|-----|
| $0$ | $0$ |
| $10$ | $5$ |
| $-10$ | $-5$ |

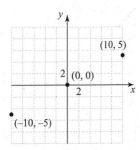

**17.** $y = 5x + 2000$
$x = 1:\ y = 5(1) + 2000 = 2005$
$x = 100:\ y = 5(100) + 2000 = 2500$
$x = 1000:\ y = 5(1000) + 2000 = 7000$

| $x$ | $1$ | $100$ | $1000$ |
|-----|-----|-------|--------|
| $y$ | $2005$ | $2500$ | $7000$ |

**18.** $y = 5x + 2000$
Let $y = 6430$.
$6430 = 5x + 2000$
$4430 = 5x$
$\ 886 = x$
886 compact disc holders can be produced.

**19.** $x - y = 1$

| $x$ | $y$ |
|-----|-----|
| $1$ | $0$ |
| $0$ | $-1$ |

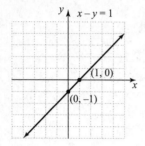

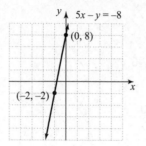

**20.** $x + y = 6$

| x | y |
|---|---|
| 6 | 0 |
| 0 | 6 |

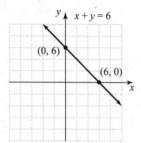

**23.** $x = 3y$

| x | y |
|---|---|
| 0 | 0 |
| 6 | 2 |

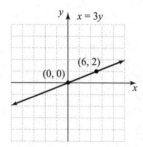

**21.** $x - 3y = 12$

| x | y |
|---|---|
| 12 | 0 |
| 0 | -4 |

**24.** $y = -2x$

| x | y |
|---|---|
| 0 | 0 |
| 4 | -8 |

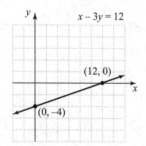

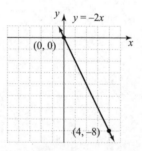

**22.** $5x - y = -8$

| x | y |
|---|---|
| -2 | -2 |
| 0 | 8 |

**25.** $2x - 3y = 6$

| x | y |
|---|---|
| 0 | -2 |
| 3 | 0 |

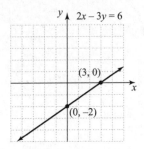

**26.** $4x - 3y = 12$

| $x$ | $y$ |
|-----|-----|
| 0   | −4  |
| 3   | 0   |

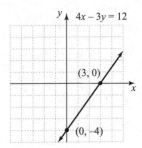

**27.** $x$-intercept: $(4, 0)$
   $y$-intercept: $(0, -2)$

**28.** $y$-intercept: $(0, -3)$

**29.** $x$-intercepts: $(-2, 0), (2, 0)$
   $y$-intercepts: $(0, 2), (0, -2)$

**30.** $x$-intercepts: $(-1, 0), (2, 0), (3, 0)$
   $y$-intercept: $(0, -2)$

**31.** $x - 3y = 12$

| $x$ | $y$ |
|-----|-----|
| 0   | −4  |
| 12  | 0   |

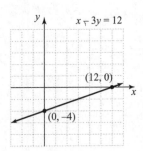

**32.** $-4x + y = 8$

| $x$ | $y$ |
|-----|-----|
| 0   | 8   |
| −2  | 0   |

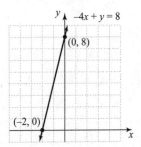

**33.** $y = -3$ for all $x$

| $x$ | $y$ |
|-----|-----|
| 0   | −3  |

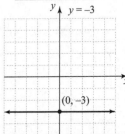

**34.** $x = 5$ for all $y$

| $x$ | $y$ |
|-----|-----|
| 5   | 0   |

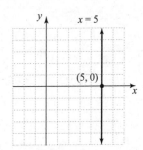

**35.** $y = -3x$

Find a second point.

| $x$ | $y$ |
|-----|-----|
| 0   | 0   |
| 3   | -9  |

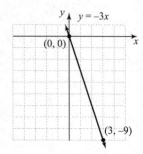

**36.** $x = 5y$

Find a second point.

| $x$ | $y$ |
|-----|-----|
| 0   | 0   |
| 5   | 1   |

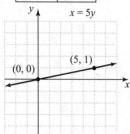

**37.** $x - 2 = 0$

$x = 2$ for all $y$

| $x$ | $y$ |
|-----|-----|
| 2   | 0   |

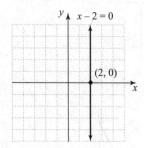

**38.** $y + 6 = 0$

$y = -6$ for all $x$

| $x$ | $y$ |
|-----|-----|
| 0   | -6  |

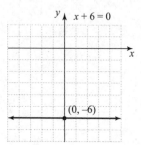

**39.** $(-1, 2)$, and $(3, -1)$

$$m = \frac{y_2 - y_1}{x_2 - x_1} = \frac{-1 - 2}{3 - (-1)} = -\frac{3}{4}$$

**40.** $(-2, -2)$ and $(3, -1)$

$$m = \frac{y_2 - y_1}{x_2 - x_1} = \frac{-1 - (-2)}{3 - (-2)} = \frac{1}{5}$$

**41.** $m = 0$

**d**

**42.** $m = -1$

**b**

**43.** Slope is undefined.

**c**

**44.** $m = 3$

**a**

**45.** $(2, 5)$ and $(6, 8)$

$$m = \frac{y_2 - y_1}{x_2 - x_1} = \frac{8 - 5}{6 - 2} = \frac{3}{4}$$

**46.** $(4, 7)$ and $(1, 2)$

$$m = \frac{y_2 - y_1}{x_2 - x_1} = \frac{2 - 7}{1 - 4} = \frac{-5}{-3} = \frac{5}{3}$$

**47.** $(1, 3)$ and $(-2, -9)$

$$m = \frac{y_2 - y_1}{x_2 - x_1} = \frac{-9 - 3}{-2 - 1} = \frac{-12}{-3} = 4$$

**48.** $(-4, 1)$, and $(3, -6)$

$$m = \frac{y_2 - y_1}{x_2 - x_1} = \frac{-6 - 1}{3 - (-4)} = \frac{-7}{7} = -1$$

**49.** $y = 3x + 7$

The equation is in slope-intercept form. The slope is the coefficient of $x$, or 3.

**50.** Solve for $y$.

$x - 2y = 4$

$-2y = -x + 4$

$y = \frac{1}{2}x - 2$

The slope is $\frac{1}{2}$.

**51.** $y = -2$

This is the equation of a horizontal line. The slope is 0.

**52.** $x = 0$

This is the equation of a vertical line. The slope is undefined.

**53.** Solve the equations for $y$.

$\begin{array}{ll} x - y = 6 & x + y = 3 \\ -y = -x + 6 & y = -x + 3 \\ y = x - 6 & \end{array}$

The slopes are 1 and $-1$. Since their product is $-1$, the lines are perpendicular.

**54.** Solve the equations for $y$.

$\begin{array}{ll} 3x + y = 7 & -3x - y = 10 \\ y = -3x + 7 & -y = 3x + 10 \\ & y = -3x - 10 \end{array}$

The slopes are both $-3$. Since the lines have the same slope but different $y$-intercepts, they are parallel.

**55.** The first line, $y = 4x + \frac{1}{2}$, has slope 4. Solve the

second equation for $y$.

$4x + 2y = 1$

$2y = -4x + 1$

$y = -2x + \frac{1}{2}$

The second line has slope $-2$. Since the slopes are not the same and their product is not $-1$, the lines are neither parallel nor perpendicular.

**56.** $x = 4$, $y = -2$

The first equation's graph is a vertical line, and the second equation's graph is a horizontal line. These lines are perpendicular.

**57.** $m = \frac{y_2 - y_1}{x_2 - x_1} = \frac{785 - 565}{2009 - 2004} = \frac{220}{5} = 44$

Every 1 year, 44 thousand (44,000) more students graduate with an associate's degree.

**58.** $m = \frac{y_2 - y_1}{x_2 - x_1} = \frac{16,900 - 14,800}{2010 - 2004} = \frac{2100}{6} = 350$

Every 1 year, 350 more people get kidney transplants.

**59.** $3x + y = 7$

$y = -3x + 7$

$y = mx + b$

$m = -3$, $y$-intercept $= (0, 7)$

**60.** $x - 6y = -1$

$-6y = -x - 1$

$y = \frac{1}{6}x + \frac{1}{6}$

$y = mx + b$

$m = \frac{1}{6}$, $y$-intercept $= \left(0, \frac{1}{6}\right)$

**61.** $y = 2$

$y = mx + b$

$m = 0$, $y$-intercept $= (0, 2)$

**62.** $x = -5$

$y = mx + b$

$m$ is undefined.

There is no $y$-intercept.

**63.** $y = 3x - 1$

$y = mx + b$

$m = 3$, $b = -1$

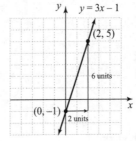

**64.** $y = -3x$
$y = mx + b$
$m = -3, b = 0$

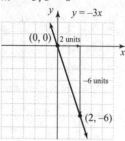

**65.** $5x - 3y = 15$
$-3y = -5x + 15$
$y = \frac{5}{3}x - 5$
$y = mx + b$
$m = \frac{5}{3}, b = -5$

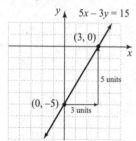

**66.** $-x + 2y = 8$
$2y = x + 8$
$y = \frac{1}{2}x + 4$
$y = mx + b$
$m = \frac{1}{2}, b = 4$

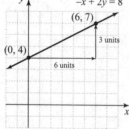

**67.** $m = -5, b = \frac{1}{2}$
$y = mx + b$
$y = -5x + \frac{1}{2}$

**68.** $m = \frac{2}{3}, b = 6$
$y = mx + b$
$y = \frac{2}{3}x + 6$

**69.** $y = -4x$
$m = -4, b = 0$
**c**

**70.** $y = 2x$
$m = 2, b = 0$
**a**

**71.** $y = 2x - 1$
$m = 2, b = -1$
**b**

**72.** $y = -2x + 1$
$m = -2, b = 1$
**d**

**73.** $y = 56x + 1859$
$y = mx + b$
$y$-intercept $= (0, 1859)$

**74.** Since $x = 0$ represents 1995, then in 1995, the average cost for tuition and fees at a public two-year college was $1859.

**75.** $m = -3; (0, -5)$
$y = mx + b$
$y = -3x - 5$
$3x + y = -5$

**76.** $m = \frac{1}{2}; \left(0, -\frac{7}{2}\right)$
$y = mx + b$
$y = \frac{1}{2}x - \frac{7}{2}$
$2y = x - 7$
$x - 2y = 7$

**77.** Horizontal line, point $(-2, -3)$
$y = c$
$y = -3$

**78.** Horizontal line, point $(0, 0)$
$y = c$
$y = 0$

**79.** $m = -6$; $(2, -1)$
$$y - y_1 = m(x - x_1)$$
$$y - (-1) = -6(x - 2)$$
$$y + 1 = -6x + 12$$
$$6x + y = 11$$

**80.** $m = 12$; $\left(\dfrac{1}{2}, 5\right)$
$$y - y_1 = m(x - x_1)$$
$$y - 5 = 12\left(x - \dfrac{1}{2}\right)$$
$$y - 5 = 12x - 6$$
$$12x - y = 1$$

**81.** $(0, 6)$ and $(6, 0)$
$$m = \dfrac{y_2 - y_1}{x_2 - x_1} = \dfrac{0 - 6}{6 - 0} = \dfrac{-6}{6} = -1$$
$$m = -1; \ (0, 6)$$
$$y - y_1 = m(x - x_1)$$
$$y - 6 = -1(x - 0)$$
$$y - 6 = -x$$
$$x + y = 6$$

**82.** $(0, -4)$ and $(-8, 0)$
$$m = \dfrac{y_2 - y_1}{x_2 - x_1} = \dfrac{0 - (-4)}{-8 - 0} = \dfrac{4}{-8} = -\dfrac{1}{2}$$
$$m = -\dfrac{1}{2}; \ (0, -4)$$
$$y - y_1 = m(x - x_1)$$
$$y - (-4) = -\dfrac{1}{2}(x - 0)$$
$$y + 4 = -\dfrac{1}{2}x$$
$$2y + 8 = -x$$
$$x + 2y = -8$$

**83.** Vertical line, point $(5, 7)$
$$x = c$$
$$x = 5$$

**84.** Horizontal line, point $(-6, 8)$
$$y = c$$
$$y = 8$$

**85.** $y = 8$ is horizontal.
Perpendicular to $y = 8$ is vertical; $x = c$.
Point $(6, 0)$
$$x = 6$$

**86.** $x = -2$ is vertical.
Perpendicular to $x = -2$ is horizontal; $y = c$,
point $(10, 12)$
$$y = 12$$

**87.** Two points have the same $x$-value: it is not a function.

**88.** Every point has a unique $x$-value: it is a function.

**89.** Yes; $7x - 6y = 1$ is a non-vertical line.

**90.** Yes; $y = 7$ is a non-vertical line.

**91.** No; $x = 2$ is a vertical line.

**92.** Yes; for each value of $x$ there is only one value of $y$.

**93.** No; the graph does not pass the vertical line test.

**94.** Yes; the graph passes the vertical line test.

**95.** $f(x) = -2x + 6$

   **a.** $f(0) = -2(0) + 6 = 6$

   **b.** $f(-2) = -2(-2) + 6 = 4 + 6 = 10$

   **c.** $f\left(\dfrac{1}{2}\right) = -2\left(\dfrac{1}{2}\right) + 6 = -1 + 6 = 5$

**96.** $h(x) = -5 - 3x$

   **a.** $h(2) = -5 - 3(2) = -11$

   **b.** $h(-3) = -5 - 3(-3) = 4$

   **c.** $h(0) = -5 - 3(0) = -5$

**97.** $g(x) = x^2 + 12x$

   **a.** $g(3) = (3)^2 + 12(3) = 45$

   **b.** $g(-5) = (-5)^2 + 12(-5) = -35$

   **c.** $g(0) = (0)^2 + 12(0) = 0$

**98.** $h(x) = 6 - |x|$

   **a.** $h(-1) = 6 - |-1| = 6 - 1 = 5$

   **b.** $h(1) = 6 - |1| = 6 - 1 = 5$

**c.** $h(-4) = 6 - |-4| = 6 - 4 = 2$

**99.** $(-\infty, \infty)$

**100.** $x - 2 \neq 0 \Rightarrow x \neq 2$, therefore $(-\infty, 2) \cup (2, \infty)$ or all real numbers except 2.

**101.** D: $[-3, 5]$, R: $[-4, 2]$

**102.** D: $(-\infty, \infty)$, R: $x \geq 0$, $[0, \infty)$

**103.** D: $\{3\}$, R: $(-\infty, \infty)$

**104.** D: $(-\infty, \infty)$, R: $x \leq 2$, $(-\infty, 2]$

**105.** $2x - 5y = 9$

Let $y = 1$.      Let $x = 2$.
$2x - 5(1) = 9$      $2(2) - 5y = 9$
$2x - 5 = 9$         $4 - 5y = 9$
$2x = 14$           $-5y = 5$
$x = 7$             $y = -1$

Let $y = -3$.
$2x - 5(-3) = 9$
$2x + 15 = 9$
$2x = -6$
$x = -3$

| $x$ | $y$ |
|-----|-----|
| 7 | 1 |
| 2 | -1 |
| -3 | -3 |

**106.** $x = -3y$

Let $x = 0$.      Let $y = 1$.
$0 = -3y$      $x = -3(1)$
$0 = y$         $x = -3$

Let $x = 6$.
$6 = -3y$
$-2 = y$

| $x$ | $y$ |
|-----|-----|
| 0 | 0 |
| -3 | 1 |
| 6 | -2 |

**107.** $2x - 3y = 6$

Let $y = 0$.      Let $x = 0$.
$2x - 3(0) = 6$      $2(0) - 3y = 6$
$2x = 6$          $-3y = 6$
$x = 3$           $y = -2$

$x$-intercept: $(3, 0)$
$y$-intercept: $(0, -2)$

**108.** $-5x + y = 10$

Let $y = 0$.      Let $x = 0$.
$-5x + 0 = 10$      $-5(0) + y = 10$
$-5x = 10$        $y = 10$
$x = -2$

$x$-intercept: $(-2, 0)$
$y$-intercept: $(0, 10)$

**109.** $x - 5y = 10$

| $x$ | $y$ |
|-----|-----|
| 10 | 0 |
| 0 | -2 |

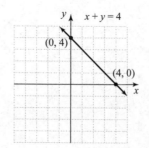

**110.** $x + y = 4$

| $x$ | $y$ |
|-----|-----|
| 4 | 0 |
| 0 | 4 |

**111.** $y = -4x$

| $x$ | $y$ |
|-----|-----|
| 0   | 0   |
| 1   | $-4$ |

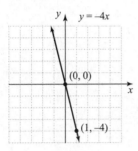

**112.** $2x + 3y = -6$

| $x$ | $y$ |
|-----|-----|
| $-3$ | 0   |
| 0   | $-2$ |

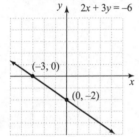

**113.** $x = 3$

This is the equation of a vertical line with $x$-intercept $(3, 0)$.

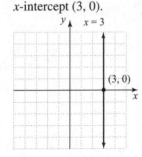

**114.** $y = -2$

This is the equation of a horizontal line with $y$-intercept $(0, -2)$.

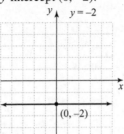

**115.** $(3, -5)$ and $(-4, 2)$

$$m = \frac{y_2 - y_1}{x_2 - x_1} = \frac{2 - (-5)}{-4 - 3} = \frac{7}{-7} = -1$$

**116.** $(1, 3)$ and $(-6, -8)$

$$m = \frac{y_2 - y_1}{x_2 - x_1} = \frac{-8 - 3}{-6 - 1} = \frac{-11}{-7} = \frac{11}{7}$$

**117.** $(0, -4)$ and $(2, 0)$

$$m = \frac{y_2 - y_1}{x_2 - x_1} = \frac{0 - (-4)}{2 - 0} = \frac{4}{2} = 2$$

**118.** $(0, 2)$ and $(6, 0)$

$$m = \frac{y_2 - y_1}{x_2 - x_1} = \frac{0 - 2}{6 - 0} = \frac{-2}{6} = -\frac{1}{3}$$

**119.** Solve for $y$.

$$-2x + 3y = -15$$
$$3y = 2x - 15$$
$$y = \frac{2}{3}x - 5$$

The slope is $\frac{2}{3}$. The $y$-intercept is $(0, -5)$.

**120.** Solve for $y$.

$$6x + y - 2 = 0$$
$$y = -6x + 2$$

The slope is $-6$. The $y$-intercept is $(0, 2)$.

**121.** $m = -5; (3, -7)$

$$y - y_1 = m(x - x_1)$$
$$y - (-7) = -5(x - 3)$$
$$y + 7 = -5x + 15$$
$$5x + y = 8$$

**122.** $m = 3; (0, 6)$
$$y = mx + b$$
$$y = 3x + 6$$
$$3x - y = -6$$

**123.** $(-3, 9)$ and $(-2, 5)$
$$m = \frac{y_2 - y_1}{x_2 - x_1} = \frac{5 - 9}{-2 - (-3)} = \frac{-4}{1} = -4$$
$m = -4; (-2, 5)$
$$y - y_1 = m(x - x_1)$$
$$y - 5 = -4(x - (-2))$$
$$y - 5 = -4(x + 2)$$
$$y - 5 = -4x - 8$$
$$4x + y = -3$$

**124.** $(3, 1)$ and $(5, -9)$
$$m = \frac{y_2 - y_1}{x_2 - x_1} = \frac{-9 - 1}{5 - 3} = \frac{-10}{2} = -5$$
$m = -5; (3, 1)$
$$y - y_1 = m(x - x_1)$$
$$y - 1 = -5(x - 3)$$
$$y - 1 = -5x + 15$$
$$5x + y = 16$$

**125.** Use $(0, 2134)$ and $(7, 3800)$.
$$m = \frac{y_2 - y_1}{x_2 - x_1} = \frac{3800 - 2134}{7 - 0} = \frac{1666}{7} = 238$$
$$b = 2134$$
$$y = mx + b$$
$$y = 238x + 2134$$

**126.** Let $x = 2014 - 2002 = 12$
$y = 238(12) + 2134 = 2856 + 2134 = 4990$
In 2014, we predict the yogurt production to be 4990 million pounds.

**Chapter 3 Test**

**1.** $y = \dfrac{1}{2} x$

$m = \dfrac{1}{2}; \ b = 0$

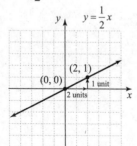

**2.** $2x + y = 8$

| $x$ | $y$ |
|-----|-----|
| 4   | 0   |
| 0   | 8   |

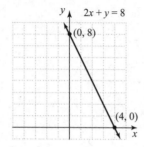

**3.** $5x - 7y = 10$

| $x$  | $y$  |
|------|------|
| 2    | 0    |
| -5   | -5   |

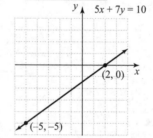

**4.** $y = -1$ for all values of $x$.

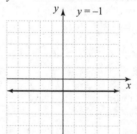

**5.** $x - 3 = 0$

$x = 3$ for all values of $y$.

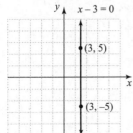

**6.** $(-1, -1)$ and $(4, 1)$

$$m = \frac{y_2 - y_1}{x_2 - x_1} = \frac{1 - (-1)}{4 - (-1)} = \frac{2}{5}$$

**7.** Horizontal line: $m = 0$

**8.** $(6, -5)$ and $(-1, 2)$

$$m = \frac{y_2 - y_1}{x_2 - x_1} = \frac{2 - (-5)}{-1 - 6} = \frac{7}{-7} = -1$$

**9.** $-3x + y = 5$

$\quad\quad y = 3x + 5$

$y = mx + b$

$m = 3$

**10.** $x = 6$ is a vertical line. The slope is undefined.

**11.** $7x - 3y = 2$

$\quad\quad -3y = -7x + 2$

$\quad\quad\quad y = \frac{7}{3}x - \frac{2}{3}$

$y = mx + b$

$m = \frac{7}{3}, b = -\frac{2}{3}, \left(0, -\frac{2}{3}\right)$

**12.** $y = 2x - 6, \ m_1 = 2$

$-4x = 2y, \ -2x = y$

$y = -2x, \ m_2 = -2$

$m_1 \neq m_2$ and $m_1 m_2 \neq -1$, neither

**13.** $m = -\frac{1}{4}; \ (2, 2)$

$\quad y - y_1 = m(x - x_1)$

$\quad\quad y - 2 = -\frac{1}{4}(x - 2)$

$\quad 4(y - 2) = -(x - 2)$

$\quad\quad 4y - 8 = -x + 2$

$\quad\quad x + 4y = 10$

**14.** $(0, 0)$ and $(6, -7)$

$$m = \frac{y_2 - y_1}{x_2 - x_1} = \frac{-7 - 0}{6 - 0} = -\frac{7}{6}$$

$m = -\frac{7}{6}; \ (0, 0)$

$\quad y - y_1 = m(x - x_1)$

$\quad\quad y - 0 = -\frac{7}{6}(x - 0)$

$\quad\quad\quad 6y = -7x$

$\quad 7x + 6y = 0$

**15.** $(2, -5)$ and $(1, 3)$

$$m = \frac{y_2 - y_1}{x_2 - x_1} = \frac{3 - (-5)}{1 - 2} = \frac{8}{-1} = -8$$

$m = -8; \ (1, 3)$

$\quad y - y_1 = m(x - x_1)$

$\quad\quad y - 3 = -8(x - 1)$

$\quad\quad y - 3 = -8x + 8$

$\quad 8x + y = 11$

**16.** $x = 7$ is vertical.

Parallel to $x = 7$ is vertical;

$x = c$, point $(-5, -1)$

$x = -5$

**17.** $m = \frac{1}{8}, \ b = 12$

$\quad\quad y = mx + b$

$\quad\quad y = \frac{1}{8}x + 12$

$\quad 8y = x + 96$

$\quad x - 8y = -96$

**18.** Yes; it passes the vertical line test.

**19.** No; it does not pass the vertical line test.

**20.** $h(x) = x^3 - x$

**a.** $h(-1) = (-1)^3 - (-1) = -1 + 1 = 0$

**b.** $h(0) = (0)^3 - (0) = 0$

**c.** $h(4) = (4)^3 - (4) = 64 - 4 = 60$

**21.** $x + 1 \neq 0 \Rightarrow x \neq -1$, therefore $(-\infty, -1) \cup (-1, \infty)$ or all real numbers except $-1$.

**22. a.** The graph crosses the *x*-axis at 0 and 4.
   *x*-intercepts: (0, 0), (4, 0)
   The graph crosses the *y*-axis at 0. The *y*-intercept is (0, 0).

   **b.** Domain: $(-\infty, \infty)$; range: $x \le 4$, $(-\infty, 4]$

**23. a.** The graph crosses the *x*-axis at 2, so the *x*-intercept is (2, 0). The graph crosses the *y*-axis at −2, so the
   *y*-intercept is (0, −2).

   **b.** Domain: $(-\infty, \infty)$; range: $(-\infty, \infty)$

**24.** $f(7) = 20$ corresponds to the ordered pair (7, 20).

**25.** The bar for Denmark extends to about 210 on the horizontal axis. The average water use per person per day in
   Denmark is approximately 210 liters.

**26.** The bar for Australia extends to about 490 on the horizontal axis. The average water use per person per day in
   Australia is approximately 490 liters.

**27.** The highest point on the graph corresponds to 7 on the horizontal axis, denoting July. The average high
   temperature is the greatest in July.

**28.** April corresponds to 4 on the horizontal axis. Moving horizontally to the left from the point on the graph above 4,
   we reach approximately 63 on the vertical axis. The average high temperature for April is approximately 63°F.

**29.** The points for months 1, 2, 3, 11, and 12 lie below 60 on the vertical axis. Thus, the average high temperature is
   below 60°F in January, February, March, November, and December.

**30. a.** The ordered pairs are (2003, 66.0), (2004, 65.4), (2005, 65.4), (2006, 65.6), (2007, 64.9), (2008, 63.7),
   (2009, 62.1).

   **b.**

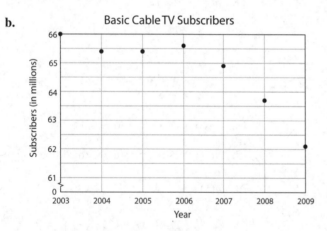

**31. a.** $m = \dfrac{\text{rise}}{\text{run}} = \dfrac{1340 - 1380}{2010 - 2005} = \dfrac{-40}{5} = \dfrac{-8}{1}$ or $-8$
   For every 1 year, 8 million fewer tickets are sold.

   **b.** (0, 1380), (5, 1340)

c. $m = \dfrac{y_2 - y_1}{x_2 - x_1} = \dfrac{1340 - 1380}{5 - 0} = \dfrac{-40}{5} = -8$

$b = 1380$

$y = mx + b$

$y = -8x + 1380$

d. Let $x = 2015 - 2005 = 10$

$y = -8(10) + 1380 = -80 + 1380 = 1300$

In 2015, we predict that 1300 million movie tickets will be sold in the United States and Canada.

## Chapter 3 Cumulative Review

1. a. $2 < 3$

   b. $7 > 4$

   c. $72 > 27$

2. $\dfrac{56}{64} = \dfrac{7 \cdot 8}{8 \cdot 8} = \dfrac{7}{8}$

3. $\dfrac{2}{15} \cdot \dfrac{5}{13} = \dfrac{2 \cdot 5}{3 \cdot 5 \cdot 13} = \dfrac{2}{39}$

4. $\dfrac{10}{3} + \dfrac{5}{21} = \dfrac{10 \cdot 7}{3 \cdot 7} + \dfrac{5}{21}$

   $= \dfrac{70 + 5}{21}$

   $= \dfrac{75}{21}$

   $= \dfrac{3 \cdot 25}{3 \cdot 7}$

   $= \dfrac{25}{7}$

   $= 3\dfrac{4}{7}$

5. $\dfrac{3 + |4 - 3| + 2^2}{6 - 3} = \dfrac{3 + |1| + 2^2}{6 - 3} = \dfrac{3 + 1 + 4}{6 - 3} = \dfrac{8}{3}$

6. $16 - 3 \cdot 3 + 2^4 = 16 - 3 \cdot 3 + 16$

   $= 16 - 9 + 16$

   $= 23$

7. a. $-8 + (-11) = -19$

   b. $-5 + 35 = 30$

   c. $0.6 + (-1.1) = -0.5$

d. $-\dfrac{7}{10} + \left(-\dfrac{1}{10}\right) = -\dfrac{8}{10} = -\dfrac{4}{5}$

e. $11.4 + (-4.7) = 6.7$

f. $-\dfrac{3}{8} + \dfrac{2}{5} = -\dfrac{3 \cdot 5}{8 \cdot 5} + \dfrac{2 \cdot 8}{5 \cdot 8} = \dfrac{-15 + 16}{40} = \dfrac{1}{40}$

8. $|9 + (-20)| + |-10| = |-11| + |-10| = 11 + 10 = 21$

9. a. $-14 - 8 + 10 - (-6) = -14 + (-8) + 10 + 6$

   $= -6$

   b. $1.6 - (-10.3) + (-5.6) = 1.6 + 10.3 + (-5.6)$

   $= 6.3$

10. $-9 - (3 - 8) = -9 - (-5) = -9 + 5 = -4$

11. Let $x = -2$ and $y = -4$.

   a. $5x - y = 5(-2) - (-4) = -10 + 4 = -6$

   b. $x^4 - y^2 = (-2)^4 - (-4)^2 = 16 - 16 = 0$

   c. $\dfrac{3x}{2y} = \dfrac{3(-2)}{2(-4)} = \dfrac{-6}{-8} = \dfrac{3}{4}$

12. $\dfrac{x}{-10} = 2$

   Let $x = -20$.

   $\dfrac{-20}{-10} \overset{?}{=} 2$

   $2 = 2$   True

   $-20$ is a solution to the equation.

13. a. $10 + (x + 12) = 10 + x + 12 = x + 22$

   b. $-3(7x) = -21x$

14. $(12 + x) - (4x - 7) = 12 + x - 4x + 7 = 19 - 3x$

15. a. $-3y: -3$

   b. $22z^4: 22$

   c. $y = 1y: 1$

   d. $-x = -1x: -1$

   e. $\dfrac{x}{7} = \dfrac{1}{7}x: \dfrac{1}{7}$

**16.** $-5(x-7) = -5x - (-5)(7) = -5x + 35$

**17.** $y + 0.6 = -1.0$
$y = -1.6$

**18.** $5(3+z) - (8z+9) = -4$
$15 + 5z - 8z - 9 = -4$
$-3z + 6 = -4$
$-3z = -10$
$z = \dfrac{10}{3}$

**19.** $-\dfrac{2}{3}x = -\dfrac{5}{2}$
$6\left(-\dfrac{2}{3}x\right) = 6\left(-\dfrac{5}{2}\right)$
$-4x = -15$
$x = \dfrac{15}{4}$

**20.** $\dfrac{x}{4} - 1 = -7$
$4\left(\dfrac{x}{4}\right) - 4(1) = 4(-7)$
$x - 4 = -28$
$x = -24$

**21.** Sum
= first integer + second integer + third integer
Sum $= x + (x+1) + (x+2)$
$= x + x + 1 + x + 2$
$= 3x + 3$

**22.** $\dfrac{x}{3} - 2 = \dfrac{x}{3}$
$3\left(\dfrac{x}{3}\right) - 3(2) = 3\left(\dfrac{x}{3}\right)$
$x - 6 = x$
$-6 = 0$
This is false. There is no solution.

**23.** $\dfrac{2(a+3)}{3} = 6a + 2$
$2(a+3) = 3(6a+2)$
$2a + 6 = 18a + 6$
$-16a + 6 = 6$
$-16a = 0$
$a = 0$

**24.** $x + 2y = 6$
$x - x + 2y = 6 - x$
$2y = 6 - x$
$\dfrac{2y}{2} = \dfrac{6-x}{2}$
$y = \dfrac{6-x}{2}$

**25.** Let $x$ = the number of Republican representatives, then $x - 49$ = the number of Democratic representatives.
$x + x - 49 = 435$
$2x - 49 = 435$
$2x = 484$
$x = 242$
$x - 49 = 193$
There were 242 Republican representatives and 193 Democratic representatives.

**26.** $5(x+4) \ge 4(2x+3)$
$5x + 20 \ge 8x + 12$
$-3x + 20 \ge 12$
$-3x \ge -8$
$\dfrac{-3x}{-3} \le \dfrac{-8}{-3}$
$x \le \dfrac{8}{3}, \left(-\infty, \dfrac{8}{3}\right]$

**27.** The perimeter of a rectangle is given by the formula $P = 2l + 2w$. Let $l$ = the length of the garden.
$P = 2l + 2w$
$140 = 2l + 2w$
$140 = 2l + 2(30)$
$140 = 2l + 60$
$80 = 2l$
$40 = l$
The length of the garden is 40 feet.

**28.** $-3 < 4x - 1 \le 2$
$-2 < 4x \le 3$
$-\dfrac{1}{2} < x \le \dfrac{3}{4}, \left(-\dfrac{1}{2}, \dfrac{3}{4}\right]$

**29.** $y = mx + b$
$y - b = mx + b - b$
$y - b = mx$
$\dfrac{y-b}{m} = \dfrac{mx}{m}$
$\dfrac{y-b}{m} = x$

**30.** $y = -5x$

| $x$ | $y$ |
|---|---|
| 0 | 0 |
| $-1$ | 5 |
| 2 | $-10$ |

**31.** Let $x$ = the amount of 70% acid.

No. of liters · Strength = Amt of Acid

| 70% | $x$ | 0.7 | $0.7x$ |
|---|---|---|---|
| 40% | $12 - x$ | 0.4 | $0.4(12 - x)$ |
| 50% | 12 | 0.5 | $0.5(12)$ |

$$0.7x + 0.4(12 - x) = 0.5(12)$$
$$0.7x + 4.8 - 0.4x = 6$$
$$0.3x + 4.8 = 6$$
$$0.3x = 1.2$$
$$x = 4$$

$12 - x = 12 - 4 = 8$

Mix 4 liters of 70% acid with 8 liters of 40% acid.

**32.** $y = -3x + 5$

| $x$ | $y$ |
|---|---|
| $-1$ | 8 |
| 0 | 5 |
| 1 | 2 |

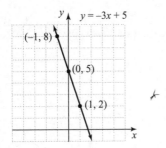

**33.** $x \geq -1$, $[-1, \infty)$

**34.** $2x + 4y = -8$

$x$-intercept, $y = 0$

$2x + 4(0) = -8 \Rightarrow x = -4$: $(-4, 0)$

$y$-intercept, $x = 0$

$2(0) + 4y = -8 \Rightarrow y = -2$: $(0, -2)$

**35.** $-1 \leq 2x - 3 < 5$

$2 \leq 2x < 8$

$1 \leq x < 4$, $[1, 4)$

**36.** $x = 2$

$x = 2$ for all values of $y$.

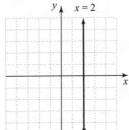

**37.** **a.** $x - 2y = 6$

$(6, 0)$

$(6) - 2(0) \overset{?}{=} 6$

$6 = 6$ Yes

**b.** $x - 2y = 6$

$(0, 3)$

$(0) - 2(3) \overset{?}{=} 6$

$-6 \neq 6$ No

**c.** $x - 2y = 6$

$\left(1, -\dfrac{5}{2}\right)$

$(1) - 2\left(-\dfrac{5}{2}\right) \overset{?}{=} 6$

$1 + 5 \overset{?}{=} 6$

$6 = 6$ Yes

**38.** $(0, 5)$ and $(-5, 4)$

$$m = \frac{y_2 - y_1}{x_2 - x_1} = \frac{4 - 5}{-5 - 0} = \frac{-1}{-5} = \frac{1}{5}$$

**39.** **a.** linear; because it can be written in the form $Ax + By = C$.

**b.** linear; because it can be written in the form $Ax + By = C$.

**c.** not linear; because $y$ is squared.

**d.** linear; because it can be written in the form $Ax + By = C$.

**40.** $x = -10$ is a vertical line. The slope is undefined.

**41.** $y = -1$ is horizontal, slope is 0.

**42.** $2x - 5y = 10$

$\quad\quad -5y = -2x + 10$

$\quad\quad\quad y = \dfrac{2}{5}x - 2$

$\quad y = mx + b$

$\quad m = \dfrac{2}{5}, \ b = -2$

The slope is $\dfrac{2}{5}$.

The $y$-intercept is $(0, -2)$.

**43.** $m = \dfrac{1}{4}; \ b = -3$

$\quad y = mx + b$

$\quad y = \dfrac{1}{4}x + (-3)$

$\quad y = \dfrac{1}{4}x - 3$

**44.** $(2, 3)$ and $(0, 0)$

$m = \dfrac{y_2 - y_1}{x_2 - x_1} = \dfrac{0 - 3}{0 - 2} = \dfrac{-3}{-2} = \dfrac{3}{2}$

Point: $(0, 0)$

$\quad y - y_1 = m(x - x_1)$

$\quad\quad y - 0 = \dfrac{3}{2}(x - 0)$

$\quad\quad\quad 2y = 3x$

$\quad 3x - 2y = 0$

# Chapter 4

## Section 4.1 Practice

**1.** $\begin{cases} 4x - y = 2 \\ y = 3x \end{cases}$

(4, 12)

$4(4) - 12 \overset{?}{=} 2$

$16 - 12 \overset{?}{=} 2$

$4 = 2$  False

(4, 12) is not a solution of the system.

**2.** $\begin{cases} x - 3y = -7 \\ 2x + 9y = 1 \end{cases}$

(−4, 1)

$-4 - 3(1) \overset{?}{=} -7$ $\qquad$ $2(-4) + 9(1) \overset{?}{=} 1$

$-4 - 3 \overset{?}{=} -7$ $\qquad\qquad$ $-8 + 9 \overset{?}{=} 1$

$\qquad -7 = -7$  True $\qquad\qquad$ $1 = 1$  True

(−4, 1) is a solution of the system.

**3.** $\begin{cases} x - y = 3 \\ x + 2y = 18 \end{cases}$

$x - y = 3$

| $x$ | $y$ |
|----|----|
| −4 | −7 |
| 0 | −3 |
| 4 | 1 |

$x + 2y = 18$

| $x$ | $y$ |
|----|----|
| −4 | 11 |
| 0 | 9 |
| 4 | 7 |

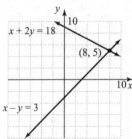

The two lines appear to intersect at (8, 5).

$x - y = 3$ $\qquad\qquad$ $x + 2y = 18$

$8 - 5 \overset{?}{=} 3$ $\qquad\qquad$ $8 + 2(5) \overset{?}{=} 18$

$\quad 3 = 3$  True $\qquad\qquad$ $8 + 10 \overset{?}{=} 18$

$\qquad\qquad\qquad\qquad\qquad 18 = 18$  True

(8, 5) is the solution of the system.

**4.** $\begin{cases} -4x + 3y = -3 \\ \qquad\quad y = -5 \end{cases}$

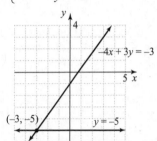

The two lines appear to intersect at (−3, −5). Check.

$-4x + 3y = -3$ $\qquad\qquad$ $y = -5$

$-4(-3) + 3(-5) \overset{?}{=} -3$ $\qquad$ $-5 = -5$  True

$12 - 15 \overset{?}{=} -3$

$\qquad -3 = -3$  True

(−3, −5) is the solution of the system.

**5.** $\begin{cases} 3y = 9x \\ 6x - 2y = 12 \end{cases}$

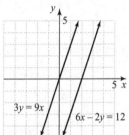

The lines appear to be parallel. To confirm this, write both equations in slope-intercept form.

$3y = 9x$ $\qquad\qquad$ $6x - 2y = 12$

$\quad y = 3x$ $\qquad\qquad\quad$ $-2y = -6x + 12$

$\qquad\qquad\qquad\qquad\quad$ $y = 3x - 6$

The slopes are the same, so the lines are parallel. Thus, there is no solution of the system and the system is inconsistent.

**6.** $\begin{cases} \qquad x - y = 4 \\ -2x + 2y = -8 \end{cases}$

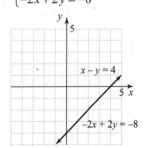

The graphs appear to be identical. To confirm this, write both equations in slope-intercept form.

$$x - y = 4 \qquad\qquad -2x + 2y = -8$$
$$-y = -x + 4 \qquad\quad -x + y = -4$$
$$y = x - 4 \qquad\qquad y = x - 4$$

The equations are identical. Thus, there is an infinite number of solutions of the system; the system is consistent; the equations are dependent.

**7.** $\begin{cases} 5x + 4y = 6 \\ x - y = 3 \end{cases}$

Write each equation in slope-intercept form.

$$5x + 4y = 6 \qquad\qquad x - y = 3$$
$$4y = -5x + 6 \qquad\quad -y = -x + 3$$
$$y = -\frac{5}{4}x + \frac{3}{2} \qquad\qquad y = x - 3$$

The slopes are not equal, so the two lines are neither parallel nor identical and must intersect. Therefore, this system has one solution and is consistent.

**8.** $\begin{cases} -\dfrac{2}{3}x + y = 6 \\ 3y = 2x + 5 \end{cases}$

Write each equation in slope-intercept form.

$$-\frac{2}{3}x + y = 6 \qquad\qquad 3y = 2x + 5$$
$$y = \frac{2}{3}x + 6 \qquad\qquad y = \frac{2}{3}x + \frac{5}{3}$$

The slope of each line is $\dfrac{2}{3}$, but they have different $y$-intercepts. Therefore, the lines are parallel. The system has no solution and is inconsistent.

## Calculator Explorations

**1.** $\begin{cases} y = -2.68x + 1.21 \\ y = 5.22x - 1.68 \end{cases}$

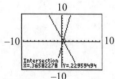

The approximate point of intersection is $(0.37, 0.23)$.

**2.** $\begin{cases} y = 4.25x + 3.89 \\ y = -1.88x + 3.21 \end{cases}$

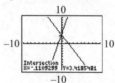

The approximate point of intersection is $(-0.11, 3.42)$.

**3.** $\begin{cases} 4.3x - 2.9y = 5.6 \\ 8.1x + 7.6y = -14.1 \end{cases}$

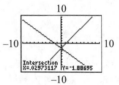

The approximate point of intersection is $(0.03, -1.89)$.

**4.** $\begin{cases} -3.6x - 8.6y = 10 \\ -4.5x + 9.6y = -7.7 \end{cases}$

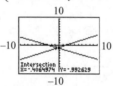

The approximate point of intersection is $(-0.41, -0.99)$.

## Vocabulary, Readiness & Video Check 4.1

1. In a system of linear equations in two variables, if the graphs of the equations are the same, the equations are <u>dependent</u> equations.

2. Two or more linear equations are called a <u>system of linear equations</u>.

3. A system of equations that has at least one solution is called a <u>consistent</u> system.

4. A <u>solution</u> of a system of two equations in two variables is an ordered pair of numbers that is a solution of both equations in the system.

5. A system of equations that has no solution is called an <u>inconsistent</u> system.

6. In a system of linear equations in two variables, if the graphs of the equations are different, the equations are <u>independent</u> equations.

7. The ordered pair must satisfy all equations of the system in order to be a solution of the system, so we must check that the ordered pair is a solution of both equations.

8. Graphing is not the most accurate method, especially if your graph is off just slightly, or the point of intersection does not have integer coordinates.

9. Writing the equations of a system in slope-intercept form lets you see their slope and $y$-intercept. Different slopes mean one solution; same slope with different $y$-intercepts means no solution; same slope with same $y$-intercept means infinite number of solutions.

**Exercise Set 4.1**

1. One solution, $(-1, 3)$

3. Infinite number of solutions

5. **a.** Let $x = 2$ and $y = 4$.
$$
\begin{array}{ll}
x + y = 8 & 3x + 2y = 21 \\
2 + 4 \stackrel{?}{=} 8 & 3(2) + 2(4) \stackrel{?}{=} 21 \\
6 = 8 \text{ False} & 6 + 8 \stackrel{?}{=} 21 \\
& 14 = 21 \text{ False}
\end{array}
$$
(2, 4) is not a solution of the system.

    **b.** Let $x = 5$ and $y = 3$.
$$
\begin{array}{ll}
x + y = 8 & 3x + 2y = 21 \\
5 + 3 \stackrel{?}{=} 8 & 3(5) + 2(3) \stackrel{?}{=} 21 \\
8 = 8 \text{ True} & 15 + 6 \stackrel{?}{=} 21 \\
& 21 = 21 \text{ True}
\end{array}
$$
(5, 3) is a solution of the system.

7. **a.** Let $x = 3$ and $y = 4$.
$$
\begin{array}{ll}
3x - y = 5 & x + 2y = 11 \\
3(3) - 4 \stackrel{?}{=} 5 & 3 + 2(4) \stackrel{?}{=} 11 \\
9 - 4 \stackrel{?}{=} 5 & 3 + 8 \stackrel{?}{=} 11 \\
5 = 5 \text{ True} & 11 = 11 \text{ True}
\end{array}
$$
(3, 4) is a solution of the system.

    **b.** Let $x = 0$ and $y = -5$.
$$
\begin{array}{ll}
3x - y = 5 & x + 2y = 11 \\
3(0) - (-5) \stackrel{?}{=} 5 & 0 + 2(-5) \stackrel{?}{=} 11 \\
0 + 5 \stackrel{?}{=} 5 & 0 - 10 \stackrel{?}{=} 11 \\
5 = 5 \text{ True} & -10 = 11 \text{ False}
\end{array}
$$
(0, −5) is not a solution of the system.

9. **a.** Let $x = -3$ and $y = -3$.
$$
\begin{array}{ll}
2y = 4x + 6 & 2x - y = -3 \\
2(-3) \stackrel{?}{=} 4(-3) + 6 & 2(-3) - (-3) \stackrel{?}{=} -3 \\
-6 \stackrel{?}{=} -12 + 6 & -6 + 3 \stackrel{?}{=} -3 \\
-6 = -6 \text{ True} & -3 = -3 \text{ True}
\end{array}
$$
(−3, −3) is a solution of the system.

    **b.** Let $x = 0$ and $y = 3$.
$$
\begin{array}{ll}
2y = 4x + 6 & 2x - y = -3 \\
2(3) \stackrel{?}{=} 4(0) + 6 & 2(0) - 3 \stackrel{?}{=} -3 \\
6 \stackrel{?}{=} 0 + 6 & 0 - 3 \stackrel{?}{=} -3 \\
6 = 6 \text{ True} & -3 = -3 \text{ True}
\end{array}
$$
(0, 3) is a solution of the system.

11. **a.** Let $x = -2$ and $y = 0$.
$$
\begin{array}{ll}
-2 = x - 7y & 6x - y = 13 \\
-2 \stackrel{?}{=} -2 - 7(0) & 6(-2) - 0 \stackrel{?}{=} 13 \\
-2 = -2 \text{ True} & -12 = 13 \text{ False}
\end{array}
$$
(−2, 0) is not a solution of the system.

    **b.** Let $x = \dfrac{1}{2}$ and $y = \dfrac{5}{14}$.
$$
\begin{array}{ll}
-2 = x - 7y & 6x - y = 13 \\
-2 \stackrel{?}{=} \dfrac{1}{2} - 7\left(\dfrac{5}{14}\right) & 6\left(\dfrac{1}{2}\right) - \left(\dfrac{5}{14}\right) \stackrel{?}{=} 13 \\
2 \stackrel{?}{=} \dfrac{1}{2} - \dfrac{5}{2} = -\dfrac{4}{2} & 3 - \dfrac{5}{14} \stackrel{?}{=} 13 \\
-2 = -2 \text{ True} & \dfrac{37}{14} = 13 \text{ False}
\end{array}
$$
$\left(\dfrac{1}{2}, \dfrac{5}{14}\right)$ is not a solution of the system.

13. $\begin{cases} x + y = 4 \\ x - y = 2 \end{cases}$

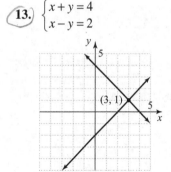

The solution of the system is (3, 1), consistent and independent.

    **163**

**15.** $\begin{cases} x+y=6 \\ -x+y=-6 \end{cases}$

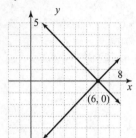

The solution of the system is (6, 0), consistent and independent.

**17.** $\begin{cases} y=2x \\ 3x-y=-2 \end{cases}$

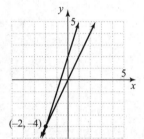

The solution of the system is (−2, −4), consistent and independent.

**19.** $\begin{cases} y=x+1 \\ y=2x-1 \end{cases}$

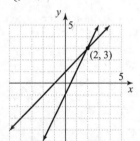

The solution of the system is (2, 3), consistent and independent.

**21.** $\begin{cases} 2x+y=0 \\ 3x+y=1 \end{cases}$

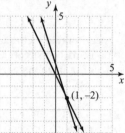

The solution of the system is (1, −2), consistent and independent.

**23.** $\begin{cases} y=-x-1 \\ y=2x+5 \end{cases}$

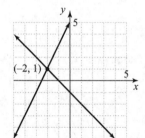

The solution of the system is (−2, 1), consistent and independent.

**25.** $\begin{cases} x+y=5 \\ x+y=6 \end{cases}$

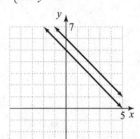

There is no solution, inconsistent and independent.

**27.** $\begin{cases} 2x - y = 6 \\ \qquad y = 2 \end{cases}$

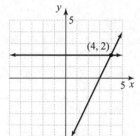

The solution of the system is (4, 2), consistent and independent.

**29.** $\begin{cases} \ x - 2y = 2 \\ 3x + 2y = -2 \end{cases}$

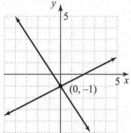

The solution of the system is (0, −1), consistent and independent.

**31.** $\begin{cases} 2x + y = 4 \\ 6x = -3y + 6 \end{cases}$

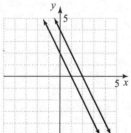

There is no solution, inconsistent and independent.

**33.** $\begin{cases} \ y - 3x = -2 \\ 6x - 2y = 4 \end{cases}$

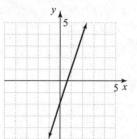

There is an infinite number of solutions, consistent and dependent.

**35.** $\begin{cases} x = 3 \\ y = -1 \end{cases}$

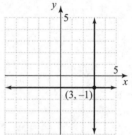

The solution of the system is (3, −1), consistent and independent.

**37.** $\begin{cases} y = x - 2 \\ y = 2x + 3 \end{cases}$

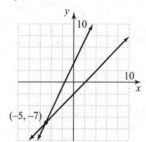

The solution of the system is (−5, −7), consistent and independent.

    

**39.** $\begin{cases} 2x - 3y = -2 \\ -3x + 5y = 5 \end{cases}$

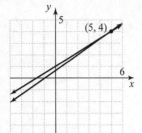

The solution of the system is (5, 4), consistent and independent.

**41.** $\begin{cases} 6x - y = 4 \\ \dfrac{1}{2}y = -2 + 3x \end{cases}$

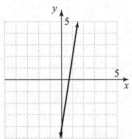

There is an infinite number of solutions, consistent and dependent.

**43.** $\begin{cases} 4x + y = 24 \\ x + 2y = 2 \end{cases} \rightarrow \begin{cases} y = -4x + 24 \\ y = -\dfrac{1}{2}x + 1 \end{cases}$

The lines are intersecting; there is one solution.

**45.** $\begin{cases} 2x + y = 0 \\ 2y = 6 - 4x \end{cases} \rightarrow \begin{cases} y = -2x \\ y = -2x + 3 \end{cases}$

The lines are parallel; there is no solution.

**47.** $\begin{cases} 6x - y = 4 \\ \dfrac{1}{2}y = -2 + 3x \end{cases} \rightarrow \begin{cases} y = 6x - 4 \\ y = 6x - 4 \end{cases}$

The lines are identical; there is an infinite number of solutions.

**49.** $\begin{cases} x = 5 \\ y = -2 \end{cases}$

The lines are intersecting; there is one solution.

**51.** $\begin{cases} 3y - 2x = 3 \\ x + 2y = 9 \end{cases} \rightarrow \begin{cases} y = \dfrac{2}{3}x + 1 \\ y = -\dfrac{1}{2}x + \dfrac{9}{2} \end{cases}$

The lines are intersecting; there is one solution.

**53.** $\begin{cases} 6y + 4x = 6 \\ 3y - 3 = -2x \end{cases} \rightarrow \begin{cases} y = -\dfrac{2}{3}x + 1 \\ y = -\dfrac{2}{3}x + 1 \end{cases}$

The lines are identical; there is an infinite number of solutions.

**55.** $\begin{cases} x + y = 4 \\ x + y = 3 \end{cases} \rightarrow \begin{cases} y = -x + 4 \\ y = -x + 3 \end{cases}$

The lines are parallel; there is no solution.

**57.** $5(x - 3) + 3x = 1$
$5x - 15 + 3x = 1$
$8x - 15 = 1$
$8x = 16$
$x = 2$
The solution is 2.

**59.** $4\left(\dfrac{y + 1}{2}\right) + 3y = 0$
$2(y + 1) + 3y = 0$
$2y + 2 + 3y = 0$
$5y + 2 = 0$
$5y = -2$
$y = -\dfrac{2}{5}$

The solution is $-\dfrac{2}{5}$.

**61.** $8a - 2(3a - 1) = 6$
$8a - 6a + 2 = 6$
$2a + 2 = 6$
$2a = 4$
$a = 2$
The solution is 2.

**63.** answers may vary

**65.** answers may vary

**67.** answers may vary

**69.** The lines cross at a point between 1988 and 1989, and again between 2001 and 2002. The number of pounds of imported fishery products was equal to the domestic catch between 1988 and 1989, and also between 2001 and 2002.

**71.** The average attendance per game for the Texas Rangers was greater than the average attendance per game for the Minnesota Twins in 2003, 2004, 2005, 2006, and 2007.

**73.** answers may vary

**75. a.** (4, 9) appears in both tables, so it is a solution of the system.

**b.**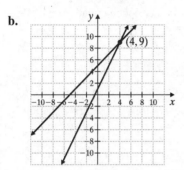

**c.** Yes; the two lines intersect at (4, 9).

**77.** answers may vary

## Section 4.2 Practice

**1.** $\begin{cases} 2x - y = 9 \\ x = y + 1 \end{cases}$

Substitute $y + 1$ for $x$ in the first equation.
$$2x - y = 9$$
$$2(y+1) - y = 9$$
$$2y + 2 - y = 9$$
$$y + 2 = 9$$
$$y = 7$$

Let $y = 7$ in the second equation.
$$x = y + 1 = 7 + 1 = 8$$
The solution of the system is (8, 7).
Check.

| $2x - y = 9$ | $x = y + 1$ |
|---|---|
| $2(8) - 7 \stackrel{?}{=} 9$ | $8 \stackrel{?}{=} 7 + 1$ |
| $16 - 7 \stackrel{?}{=} 9$ | $8 = 8$   True |
| $9 = 9$   True | |

The solution of the system is (8, 7).

**2.** $\begin{cases} 7x - y = -15 \\ y = 2x \end{cases}$

Substitute $2x$ for $y$ in the first equation.
$$7x - y = -15$$
$$7x - 2x = -15$$
$$5x = -15$$
$$x = -3$$

Let $x = -3$ in the second equation.
$$y = 2x = 2(-3) = -6$$
The solution of the system is $(-3, -6)$.

**3.** $\begin{cases} x + 3y = 6 \\ 2x + 3y = 10 \end{cases}$

Solve the first equation for $x$.
$$x + 3y = 6$$
$$x = -3y + 6$$

Substitute $-3y + 6$ for $x$ in the second equation.
$$2x + 3y = 10$$
$$2(-3y + 6) + 3y = 10$$
$$-6y + 12 + 3y = 10$$
$$-3y + 12 = 10$$
$$-3y = -2$$
$$y = \frac{2}{3}$$

Let $y = \frac{2}{3}$ in the equation for $x$.
$$x = 3y + 6 = -3\left(\frac{2}{3}\right) + 6 = -2 + 6 = 4$$

The solution of the system is $\left(4, \dfrac{2}{3}\right)$.

**4.** $\begin{cases} 5x + 3y = -9 \\ -2x + y = 8 \end{cases}$

Solve the second equation for $y$.
$$-2x + y = 8$$
$$y = 2x + 8$$

Substitute $2x + 8$ for $y$ in the first equation.
$$5x + 3y = -9$$
$$5x + 3(2x + 8) = -9$$
$$5x + 6x + 24 = -9$$
$$11x + 24 = -9$$
$$11x = -33$$
$$x = -3$$

Let $x = -3$ in the equation for $y$.
$$y = 2x + 8 = 2(-3) + 8 = -6 + 8 = 2$$
The solution of the system is $(-3, 2)$.

**5.** $\begin{cases} \dfrac{1}{4}x - y = 2 \\ x = 4y + 8 \end{cases}$

Substitute $4y + 8$ for $x$ in the first equation.

$$\frac{1}{4}x - y = 2$$

$$\frac{1}{4}(4y + 8) - y = 2$$

$$y + 2 - y = 2$$

$$2 = 2$$

The two linear equations are equivalent. Thus, the system has an infinite number of solutions.

6. $\begin{cases} 4x - 3y = 12 \\ -8x + 6y = -30 \end{cases}$

Solve the first equation for $x$.

$$4x - 3y = 12$$

$$4x = 3y + 12$$

$$x = \frac{3}{4}y + 3$$

Substitute $\frac{3}{4}y + 3$ for $x$ in the second equation.

$$-8x + 6y = -30$$

$$-8\left(\frac{3}{4}y + 3\right) + 6y = -30$$

$$-6y - 24 + 6y = -30$$

$$-24 = -30$$

The false statement $-24 = -30$ indicates that the system has no solution and is inconsistent.

**Vocabulary, Readiness & Video Check 4.2**

1. Since $x = 1$, $y = 4x = 4(1) = 4$ and the solution is $(1, 4)$.

2. There is no solution, since $0 = 34$ is a false statement.

3. There is an infinite number of solutions, since the statement $0 = 0$ is true for all values of the variables.

4. Since $y = 0$, $x = y + 5 = 0 + 5 = 5$ and the solution is $(5, 0)$.

5. Since $x = 0$ and $x + y = 0$, $y = -x = -0 = 0$ and the solution is $(0, 0)$.

6. There is an infinite number of solutions, since the statement $0 = 0$ is true for all values of the variables.

7. You solved one equation for a variable. Now be sure to substitute this expression for the variable into the *other* equation.

**Exercise Set 4.2**

1. $\begin{cases} x + y = 3 \\ x = 2y \end{cases}$

Substitute $2y$ for $x$ in the first equation.

$$2y + y = 3$$

$$3y = 3$$

$$y = 1$$

Let $y = 1$ in the second equation.

$$x = 2(1) = 2$$

The solution is $(2, 1)$.

3. $\begin{cases} x + y = 6 \\ y = -3x \end{cases}$

Substitute $-3x$ for $y$ in the first equation.

$$x + (-3x) = 6$$

$$-2x = 6$$

$$x = -3$$

Let $x = -3$ in the second equation.

$$y = -3(-3) = 9$$

The solution is $(-3, 9)$.

5. $\begin{cases} y = 3x + 1 \\ 4y - 8x = 12 \end{cases}$

Substitute $3x + 1$ for $y$ in the second equation.

$$4(3x + 1) - 8x = 12$$

$$12x + 4 - 8x = 12$$

$$4x + 4 = 12$$

$$4x = 8$$

$$x = 2$$

Let $x = 2$ in the first equation.

$$y = 3(2) + 1 = 7$$

The solution is $(2, 7)$.

7. $\begin{cases} y = 2x + 9 \\ y = 7x + 10 \end{cases}$

Substitute $2x + 9$ for $y$ in the second equation.

$$2x + 9 = 7x + 10$$

$$-5x + 9 = 10$$

$$-5x = 1$$

$$x = -\frac{1}{5}$$

Let $x = -\frac{1}{5}$ in the first equation.

$$y = 2\left(-\frac{1}{5}\right) + 9 = -\frac{2}{5} + \frac{45}{5} = \frac{43}{5}$$

The solution is $\left(-\frac{1}{5}, \frac{43}{5}\right)$.

**9.** $\begin{cases} 3x - 4y = 10 \\ y = x - 3 \end{cases}$

Substitute $x - 3$ for $y$ in the first equation.

$3x - 4(x - 3) = 10$
$3x - 4x + 12 = 10$
$-x = -2$
$x = 2$

Let $x = 2$ in the second equation
$y = 2 - 3 = -1$
The solution is $(2, -1)$.

**11.** $\begin{cases} x + 2y = 6 \\ 2x + 3y = 8 \end{cases}$

Solve the first equation for $x$.

$x = 6 - 2y$

Substitute $6 - 2y$ for $x$ in the second equation.

$2(6 - 2y) + 3y = 8$
$12 - 4y + 3y = 8$
$-y = -4$
$y = 4$

Let $y = 4$ in $x = 6 - 2y$.
$x = 6 - 2(4) = -2$
The solution is $(-2, 4)$.

**13.** $\begin{cases} 3x + 2y = 16 \\ x = 3y - 2 \end{cases}$

Substitute $3y - 2$ for $x$ in the first equation.

$3(3y - 2) + 2y = 16$
$9y - 6 + 2y = 16$
$11y = 22$
$y = 2$

Let $y = 2$ in the second equation.
$x = 3(2) - 2 = 4$
The solution is $(4, 2)$.

**15.** $\begin{cases} 2x - 5y = 1 \\ 3x + y = -7 \end{cases}$

Solve the second equation for $y$.

$y = -7 - 3x$

Substitute $-7 - 3x$ for $y$ in the first equation.

$2x - 5(-7 - 3x) = 1$
$2x + 35 + 15x = 1$
$17x = -34$
$x = -2$

Let $x = -2$ in $y = -7 - 3x$.
$y = -7 - 3(-2) = -1$
The solution is $(-2, -1)$.

**17.** $\begin{cases} 4x + 2y = 5 \\ -2x = y + 4 \end{cases}$

Solve the second equation for $y$.

$y = -2x - 4$

Substitute $-2x - 4$ for $y$ in the first equation.

$4x + 2(-2x - 4) = 5$
$4x - 4x - 8 = 5$
$-8 = 5$   False

The system has no solution.

**19.** $\begin{cases} 4x + y = 11 \\ 2x + 5y = 1 \end{cases}$

Solve the first equation for $y$.

$y = 11 - 4x$

Substitute $11 - 4x$ for $y$ in the second equation.

$2x + 5(11 - 4x) = 1$
$2x + 55 - 20x = 1$
$-18x = -54$
$x = 3$

Let $x = 3$ in $y = 11 - 4x$.
$y = 11 - 4(3) = -1$
The solution is $(3, -1)$.

**21.** $\begin{cases} x + 2y + 5 = -4 + 5y - x \\ \quad 2x + 9 = 3y \\ \\ \quad 2x + x = y + 4 \\ \quad -4 + 3x = y \end{cases}$

Substitute $3x - 4$ for $y$ in $2x + 9 = 3y$.

$2x + 9 = 3(3x - 4)$
$2x + 9 = 9x - 12$
$21 = 7x$
$3 = x$

Let $x = 3$ in $3x - 4 = y$.
$y = 3(3) - 4 = 5$
The solution of the system is $(3, 5)$.

**23.** $\begin{cases} 6x - 3y = 5 \\ x + 2y = 0 \end{cases}$

Solve the second equation for $x$.

$x = -2y$

Substitute $-2y$ for $x$ in the first equation.

$6(-2y) - 3y = 5$
$-12y - 3y = 5$
$-15y = 5$
$y = -\dfrac{1}{3}$

Let $y = -\dfrac{1}{3}$ in $x = -2y$.

$x = -2\left(-\dfrac{1}{3}\right) = \dfrac{2}{3}$

The solution is $\left(\dfrac{2}{3}, -\dfrac{1}{3}\right)$.

**25.** $\begin{cases} 3x - y = 1 \\ 2x - 3y = 10 \end{cases}$

Solve the first equation for $y$.

$y = 3x - 1$

Substitute $3x - 1$ for $y$ in the second equation.

$2x - 3(3x - 1) = 10$

$\quad 2x - 9x + 3 = 10$

$\qquad\qquad -7x = 7$

$\qquad\qquad\quad x = -1$

Let $x = -1$ in $y = 3x - 1$.

$y = 3(-1) - 1 = -4$

The solution is $(-1, -4)$.

**27.** $\begin{cases} -x + 2y = 10 \\ -2x + 3y = 18 \end{cases}$

Solve the first equation for $x$.

$x = 2y - 10$

Substitute $2y - 10$ for $x$ in the second equation.

$-2(2y - 10) + 3y = 18$

$\quad -4y + 20 + 3y = 18$

$\qquad\qquad\quad -y = -2$

$\qquad\qquad\quad\ y = 2$

Let $y = 2$ in $x = 2y - 10$.

$x = 2(2) - 10 = -6$

The solution is $(-6, 2)$.

**29.** $\begin{cases} 5x + 10y = 20 \\ 2x + 6y = 10 \end{cases}$

Solve the first equation for $x$.

$x + 2y = 4$

$\quad x = 4 - 2y$

Substitute $4 - 2y$ for $x$ in the second equation.

$2(4 - 2y) + 6y = 10$

$\quad 8 - 4y + 6y = 10$

$\qquad\qquad 2y = 2$

$\qquad\qquad\ y = 1$

Let $y = 1$ in $x = 4 - 2y$.

$x = 4 - 2(1) = 2$

The solution is $(2, 1)$.

**31.** $\begin{cases} 3x + 6y = 9 \\ 4x + 8y = 16 \end{cases}$

Solve the first equation for $x$.

$x + 2y = 3$

$\quad x = 3 - 2y$

Substitute $3 - 2y$ for $x$ in the second equation.

$4(3 - 2y) + 8y = 16$

$\quad 12 - 8y + 8y = 16$

$\qquad\qquad\quad 12 = 16 \quad \text{False}$

The system has no solution.

**33.** $\begin{cases} \dfrac{1}{3}x - y = 2 \\ x - 3y = 6 \end{cases}$

Solve the second equation for $x$.

$x = 6 + 3y$

Substitute $6 + 3y$ for $x$ in the first equation.

$\dfrac{1}{3}(6 + 3y) - y = 2$

$\qquad 2 + y - y = 2$

$\qquad\qquad\quad 2 = 2$

The equations in the original system are equivalent and there are an infinite number of solutions.

**35.** $\begin{cases} x = \dfrac{3}{4}y - 1 \\ 8x - 5y = -6 \end{cases}$

Substitute $\dfrac{3}{4}y - 1$ for $x$ in the second equation.

$8\left(\dfrac{3}{4}y - 1\right) - 5y = -6$

$\qquad 6y - 8 - 5y = -6$

$\qquad\qquad\qquad y = 2$

Let $y = 2$ in the first equation.

$x = \dfrac{3}{4}(2) - 1 = \dfrac{1}{2}$

The solution of the system is $\left(\dfrac{1}{2}, 2\right)$.

**37.** $\begin{cases} -5y + 6y = 3x + 2(x - 5) - 3x + 5 \\ \qquad\quad y = 3x + 2x - 10 - 3x + 5 \\ \qquad\quad y = 2x - 5 \\ \\ 4(x + y) - x + y = -12 \\ 4x + 4y - x + y = -12 \\ \qquad\quad 3x + 5y = -12 \end{cases}$

Substitute $2x - 5$ for $y$ in the second equation.

$$3x + 5(2x - 5) = -12$$
$$3x + 10x - 25 = -12$$
$$13x = 13$$
$$x = 1$$
Let $x = 1$ in $y = 2x - 5$.
$$y = 2(1) - 5 = -3$$
The solution is $(1, -3)$.

**39.**
$$3x + 2y = 6$$
$$-2(3x + 2y) = -2(6)$$
$$-6x - 4y = -12$$

**41.**
$$-4x + y = 3$$
$$3(-4x + y) = 3(3)$$
$$-12x + 3y = 9$$

**43.** $3n + 6m$
$$\underline{2n - 6m}$$
$$5n$$

**45.** $-5a - 7b$
$$\underline{5a - 8b}$$
$$-15b$$

**47.** answers may vary

**49.** No; answers may vary.

**51.** **c**; answers may vary.

**53.** **a.** $\begin{cases} y = 3.9x + 443 \\ y = 14.2x + 314 \end{cases}$

Substitute $14.2x + 314$ for $y$ in the first equation.
$$14.2x + 314 = 3.9x + 443$$
$$10.3x = 129$$
$$x \approx 12.52$$
Let $x = 12.52$ in $y = 3.9x + 443$.
$$y \approx 3.9(12.52) + 443 \approx 491.828$$
The solution is $(13, 492)$.

**b.** In $1970 + 13 = 1983$, the number of men and the number of women receiving bachelor's degrees was the same.

**c.**

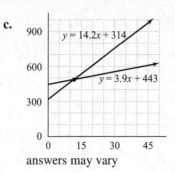

answers may vary

**55.** $\begin{cases} y = 5.1x + 14.56 \\ y = -2x - 3.9 \end{cases}$

Substitute $-2x - 3.9$ for $y$ in the first equation.
$$-2x - 3.9 = 5.1x + 14.56$$
$$-7.1x = 18.46$$
$$x = -2.6$$
Let $x = -2.6$ in $y = -2x - 3.9$.
$$y = -2(-2.6) - 3.9 = 1.3$$
The solution is $(-2.6, 1.3)$.

**57.** $\begin{cases} 3x + 2y = 14.05 \\ 5x + y = 18.5 \end{cases}$

Solve the second equation for $y$.
$$y = -5x + 18.5$$
Substitute $-5x + 18.5$ for $y$ in the first equation.
$$3x + 2(-5x + 18.5) = 14.05$$
$$3x - 10x + 37 = 14.05$$
$$-7x = -22.95$$
$$x \approx 3.279$$
Let $x = 3.279$ in $y = -5x + 18.5$.
$$y \approx -5(3.279) + 18.5 \approx 2.105$$
The solution is approximately $(3.28, 2.11)$.

**Section 4.3 Practice**

**1.** $\begin{cases} x - y = 2 \\ x + y = 8 \end{cases}$

Add the left sides of the equations together and the right sides of the equations together.
$$x - y = 2$$
$$\underline{x + y = 8}$$
$$2x = 10$$
$$x = 5$$
Let $x = 5$ in the first equation.
$$x - y = 2$$
$$5 - y = 2$$
$$3 = y$$
The solution is $(5, 3)$.
Check.

       

$x - y = 2$
$5 - 3 \overset{?}{=} 2$
$\qquad 2 = 2$ True

$x + y = 8$
$5 + 3 \overset{?}{=} 8$
$\qquad 8 = 8$ True

The solution of the system is (5, 3).

**2.** $\begin{cases} x - 2y = 11 \\ 3x - y = 13 \end{cases}$

Multiply both sides of the first equation by −3 and add to the second equation.

$-3x + 6y = -33$
$\underline{3x - y = 13}$
$\qquad 5y = -20$
$\qquad\quad y = -4$

Let $y = -4$ in the first equation.
$x - 2y = 11$
$x - 2(-4) = 11$
$x + 8 = 11$
$\qquad x = 3$

The solution of the system is (3, −4).

**3.** $\begin{cases} x - 3y = 5 \\ 2x - 6y = -3 \end{cases}$

Multiply both sides of the first equation by −2 and add to the second equation.

$-2x + 6y = -10$
$\underline{2x - 6y = -3}$
$\qquad 0 = -13$ False

The system has no solution.

**4.** $\begin{cases} 4x - 3y = 5 \\ -8x + 6y = -10 \end{cases}$

Multiply the first equation by 2 and add to the second equation.

$8x - 6y = 10$
$\underline{-8x + 6y = -10}$
$\qquad 0 = 0$ True

The equations are equivalent, so the system has an infinite number of solutions.

**5.** $\begin{cases} 4x + 3y = 14 \\ 3x - 2y = 2 \end{cases}$

Multiply the first equation by 2 and the second equation by 3 and add.

$8x + 6y = 28$
$\underline{9x - 6y = 6}$
$17x \qquad = 34$
$\qquad x = 2$

Let $x = 2$ in the second equation.

$3x - 2y = 2$
$3(2) - 2y = 2$
$6 - 2y = 2$
$-2y = -4$
$\quad y = 2$

The solution of the system is (2, 2).

**6.** $\begin{cases} -2x + \dfrac{3y}{2} = 5 \\ -\dfrac{x}{2} - \dfrac{y}{4} = \dfrac{1}{2} \end{cases}$

Clear fractions by multiplying the first equation by 2 and the second by 4.

$\begin{cases} -4x + 3y = 10 \\ -2x - y = 2 \end{cases}$

Multiply the second simplified equation by 3 and add.

$-4x + 3y = 10$
$\underline{-6x - 3y = 6}$
$-10x \qquad = 16$
$\qquad x = -\dfrac{16}{10} = -\dfrac{8}{5}$

Now multiply the second simplified equation by −2 and add.

$-4x + 3y = 10$
$\underline{4x + 2y = -4}$
$\qquad 5y = 6$
$\qquad\; y = \dfrac{6}{5}$

The solution of the system is $\left( -\dfrac{8}{5}, \dfrac{6}{5} \right)$.

**Vocabulary, Readiness & Video Check 4.3**

**1.** $\begin{cases} 3x - 2y = -9 \\ x + 5y = 14 \end{cases}$

Multiply the second equation by −3, then add the resulting equations.

$3x - 2y = -9$
$\underline{-3x - 15y = -42}$
$\qquad -17y = -51$

The $y$'s are not eliminated; the statement is false.

**2.** $\begin{cases} 3x - 2y = -9 \\ x + 5y = 14 \end{cases}$

Multiply the second equation by −3, then add the resulting equations.

$$3x - 2y = -9$$
$$\underline{-3x - 15y = -42}$$
$$-17y = -51$$

The statement is true.

**3.** $\begin{cases} 3x - 2y = -9 \\ x + 5y = 14 \end{cases}$

Multiply the first equation by 5 and the second equation by 2, then add the two new equations.
$$15x - 10y = -45$$
$$\underline{2x + 10y = \phantom{0}28}$$
$$17x \phantom{+10y} = -17$$

The statement is true.

**4.** $\begin{cases} 3x - 2y = -9 \\ x + 5y = 14 \end{cases}$

Multiply the first equation by 5 and the second equation by −2, then add the two new equations.
$$15x - 10y = -45$$
$$\underline{-2x - 10y = -28}$$
$$13x - 20y = -73$$

The $y$'s are not eliminated; the statement is false.

**5.** The multiplication property of equality; be sure to multiply *both* sides of the equation by the number chosen.

## Exercise Set 4.3

**1.** $\begin{cases} 3x + y = 5 \\ 6x - y = 4 \end{cases}$

$$3x + y = 5$$
$$\underline{6x - y = 4}$$
$$9x \phantom{+ y} = 9$$
$$x = 1$$

Let $x = 1$ in the first equation.
$$3(1) + y = 5$$
$$3 + y = 5$$
$$y = 2$$

The solution of the system is (1, 2).

**3.** $\begin{cases} x - 2y = 8 \\ -x + 5y = -17 \end{cases}$

$$x - 2y = 8$$
$$\underline{-x + 5y = -17}$$
$$3y = -9$$
$$y = -3$$

Let $y = -3$ in the first equation.

$$x - 2(-3) = 8$$
$$x + 6 = 8$$
$$x = 2$$

The solution of the system is (2, −3).

**5.** $\begin{cases} 3x + y = -11 \\ 6x - 2y = -2 \end{cases}$

Multiply the first equation by 2.
$$6x + 2y = -22$$
$$\underline{6x - 2y = \phantom{0}-2}$$
$$12x \phantom{+ 2y} = -24$$
$$x = -2$$

Let $x = -2$ in the first equation.
$$3(-2) + y = -11$$
$$-6 + y = -11$$
$$y = -5$$

The solution of the system is (−2, −5).

**7.** $\begin{cases} 3x + 2y = 11 \\ 5x - 2y = 29 \end{cases}$

$$3x + 2y = 11$$
$$\underline{5x - 2y = 29}$$
$$8x \phantom{+ 2y} = 40$$
$$x = 5$$

Let $x = 5$ in the first equation.
$$3(5) + 2y = 11$$
$$15 + 2y = 11$$
$$2y = -4$$
$$y = -2$$

The solution of the system is (5, −2).

**9.** $\begin{cases} x + 5y = 18 \\ 3x + 2y = -11 \end{cases}$

Multiply the first equation by −3.
$$-3x - 15y = -54$$
$$\underline{3x + \phantom{0}2y = -11}$$
$$-13y = -65$$
$$y = 5$$

Let $y = 5$ in the first equation.
$$x + 5(5) = 18$$
$$x + 25 = 18$$
$$x = -7$$

The solution of the system is (−7, 5).

**11.** $\begin{cases} x+y=6 \\ x-y=6 \end{cases}$

$x+y=6$
$\underline{x-y=6}$
$2x \phantom{+y}=12$
$\phantom{2}x=6$

Let $x = 6$ in the first equation.
$6+y=6$
$\phantom{6+}y=0$
The solution of the system is (6, 0).

**13.** $\begin{cases} 2x+3y=0 \\ 4x+6y=3 \end{cases}$

Multiply the first equation by $-2$.
$-4x-6y=0$
$\underline{\phantom{-}4x+6y=3}$
$\phantom{-4x-6y}0=3 \quad$ False
The system has no solution.

**15.** $\begin{cases} -x+5y=-1 \\ 3x-15y=3 \end{cases}$

Multiply the first equation by 3.
$-3x+15y=-3$
$\underline{\phantom{-}3x-15y=3}$
$\phantom{-3x+15y}0=0$
There are an infinite number of solutions.

**17.** $\begin{cases} 3x-2y=7 \\ 5x+4y=8 \end{cases}$

Multiply the first equation by 2.
$6x-4y=14$
$\underline{5x+4y=8}$
$11x \phantom{-4y}=22$
$\phantom{11}x=2$

Let $x = 2$ in the first equation.
$3(2)-2y=7$
$\phantom{3}6-2y=7$
$\phantom{3(2)}-2y=1$
$\phantom{3(2)}y=-\dfrac{1}{2}$

The solution of the system is $\left(2, -\dfrac{1}{2}\right)$.

**19.** $\begin{cases} 8x=-11y-16 \\ 2x+3y=-4 \end{cases}$

Add $11y$ to both sides of the first equation and multiply the second equation by $-4$, then add.

$8x+11y=-16$
$\underline{-8x-12y=16}$
$\phantom{8x+11}-y=0$
$\phantom{8x+11}y=0$

Let $y = 0$ in the first equation.
$8x=-11(0)-16$
$8x=-16$
$\phantom{8}x=-2$
The solution of the system is (−2, 0).

**21.** $\begin{cases} 4x-3y=7 \\ 7x+5y=2 \end{cases}$

Multiply the first equation by 5 and the second equation by 3.
$20x-15y=35$
$\underline{21x+15y=6}$
$41x \phantom{+15y}=41$
$\phantom{41}x=1$

Let $x = 1$ in the first equation.
$4x-3y=7$
$4(1)-3y=7$
$\phantom{4(1)}4-3y=7$
$\phantom{4(1)4}-3y=3$
$\phantom{4(1)4}y=-1$
The solution of the system is (1, −1).

**23.** $\begin{cases} 4x-6y=8 \\ 6x-9y=12 \end{cases}$

Multiply the first equation by 3 and the second equation by $-2$.
$12x-18y=24$
$\underline{-12x+18y=-24}$
$\phantom{12x-18y}0=0$

The equations in the original system are equivalent and there is an infinite number of solutions.

**25.** $\begin{cases} 2x-5y=4 \\ 3x-2y=4 \end{cases}$

Multiply the first equation by $-3$ and the second equation by 2.
$-6x+15y=-12$
$\underline{\phantom{-}6x-4y=8}$
$\phantom{-6x+1}11y=-4$
$\phantom{-6x+11y}y=-\dfrac{4}{11}$

Multiply the first equation by $-2$ and the second equation by 5.

$$-4x+10y=-8$$
$$\underline{15x-10y=20}$$
$$11x\phantom{+10y}=12$$
$$x=\frac{12}{11}$$

The solution of the system is $\left(\dfrac{12}{11},\,-\dfrac{4}{11}\right)$.

**27.** $\begin{cases}\dfrac{x}{3}+\dfrac{y}{6}=1\\[2mm]\dfrac{x}{2}-\dfrac{y}{4}=0\end{cases}$

Multiply the first equation by 6 and the second equation by 4.

$$2x+y=6$$
$$\underline{2x-y=0}$$
$$4x\phantom{+y}=6$$
$$x=\frac{3}{2}$$

Multiply the second equation of the simplified system by $-1$.

$$2x+y=6$$
$$\underline{-2x+y=0}$$
$$2y=6$$
$$y=3$$

The solution of the system is $\left(\dfrac{3}{2},\,3\right)$.

**29.** $\begin{cases}\dfrac{10}{3}x+4y=-4\\[2mm]5x+6y=-6\end{cases}$

Multiply the first equation by 3 and the second equation by $-2$.

$$10x+12y=-12$$
$$\underline{-10x-12y=12}$$
$$0=0$$

The system has an infinite number of solutions.

**31.** $\begin{cases}x-\dfrac{y}{3}=-1\\[2mm]-\dfrac{x}{2}+\dfrac{y}{8}=\dfrac{1}{4}\end{cases}$

Multiply the first equation by 3 and the second equation by 8.

$$3x-y=-3$$
$$\underline{-4x+y=2}$$
$$-x\phantom{+y}=-1$$
$$x=1$$

Multiply the first equation of the simplified

---

system by 4 and the second equation by 3.

$$12x-4y=-12$$
$$\underline{-12x+3y=\phantom{0}6}$$
$$-y=-6$$
$$y=6$$

The solution of the system is $(1, 6)$.

**33.** $\begin{cases}-4(x+2)=3y\\2x-2y=3\end{cases}\rightarrow\begin{cases}-4x-8=3y\\2x-2y=3\end{cases}$

$$\rightarrow\begin{cases}-4x-3y=8\\2x-2y=3\end{cases}$$

Multiply the second equation by 2.

$$-4x-3y=8$$
$$\underline{4x-4y=6}$$
$$-7y=14$$
$$y=-2$$

Let $y=-2$ in the second equation.

$$2x-2(-2)=3$$
$$2x+4=3$$
$$2x=-1$$
$$x=-\frac{1}{2}$$

The solution of the system is $\left(-\dfrac{1}{2},\,-2\right)$.

**35.** $\begin{cases}\dfrac{x}{3}-y=2\\[2mm]-\dfrac{x}{2}+\dfrac{3y}{2}=-3\end{cases}$

Multiply the first equation by 3 and the second equation by 2.

$$x-3y=6$$
$$\underline{-2x+3y=-6}$$
$$0=0$$

The equations of the original system are equivalent and there is an infinite number of solutions.

**37.** $\begin{cases}\dfrac{3}{5}x-y=-\dfrac{4}{5}\\[2mm]3x+\dfrac{y}{2}=-\dfrac{9}{5}\end{cases}$

Multiply the first equation by 5 and the second equation by 10.

$$3x-5y=-4$$
$$\underline{30x+5y=-18}$$
$$33x\phantom{+5y}=-22$$
$$x=-\frac{2}{3}$$

Let $x = -\dfrac{2}{3}$ in $30x + 5y = -18$.

$$30\left(-\dfrac{2}{3}\right) + 5y = -18$$
$$-20 + 5y = -18$$
$$5y = 2$$
$$y = \dfrac{2}{5}$$

The solution of the system is $\left(-\dfrac{2}{3}, \dfrac{2}{5}\right)$.

**39.** $\begin{cases} 3.5x + 2.5y = 17 \\ -1.5x - 7.5y = -33 \end{cases}$

Multiply the first equation by 6 and the second equation by 2.

$$\begin{array}{r} 21x + 15y = 102 \\ -3x - 15y = -66 \\ \hline 18x \quad\quad = 36 \\ x = 2 \end{array}$$

Let $x = 2$ in $-3x - 15y = -66$.
$$-3(2) - 15y = -66$$
$$-6 - 15y = -66$$
$$-15y = -60$$
$$y = 4$$

The solution of the system is (2, 4).

**41.** $\begin{cases} 0.02x + 0.04y = 0.09 \\ -0.1x + 0.3y = 0.8 \end{cases}$

Multiply the first equation by 100 and the second equation by 20.

$$\begin{array}{r} 2x + 4y = 9 \\ -2x + 6y = 16 \\ \hline 10y = 25 \\ y = \dfrac{5}{2} = 2.5 \end{array}$$

Let $y = 2.5$ in $2x + 4y = 9$.
$$2x + 4(2.5) = 9$$
$$2x + 10 = 9$$
$$2x = -1$$
$$x = -\dfrac{1}{2} = -0.5$$

The solution of the system is (−0.5, 2.5).

**43.** $\begin{cases} 2x - 3y = -11 \\ y = 4x - 3 \end{cases}$

Substitute $4x - 3$ for $y$ in the first equation.
$$2x - 3(4x - 3) = -11$$
$$2x - 12x + 9 = -11$$
$$-10x = -20$$
$$x = 2$$

Let $x = 2$ in the second equation.
$$y = 4(2) - 3 = 5$$
The solution of the system is (2, 5).

**45.** $\begin{cases} x + 2y = 1 \\ 3x + 4y = -1 \end{cases}$

Multiply the first equation by −2.
$$\begin{array}{r} -2x - 4y = -2 \\ 3x + 4y = -1 \\ \hline x \quad\quad = -3 \end{array}$$

Let $x = -3$ in the first equation.
$$-3 + 2y = 1$$
$$2y = 4$$
$$y = 2$$

The solution is (−3, 2).

**47.** $\begin{cases} 2y = x + 6 \\ 3x - 2y = -6 \end{cases}$

Subtract $x$ from both sides of the first equation.
$$\begin{array}{r} -x + 2y = 6 \\ 3x - 2y = -6 \\ \hline 2x \quad\quad = 0 \\ x = 0 \end{array}$$

Let $x = 0$ in the first equation.
$$2y = 0 + 6$$
$$2y = 6$$
$$y = 3$$

The solution of the system is (0, 3).

**49.** $\begin{cases} y = 2x - 3 \\ y = 5x - 18 \end{cases}$

Substitute $5x - 18$ for $y$ in the first equation.
$$5x - 18 = 2x - 3$$
$$3x = 15$$
$$x = 5$$

Let $x = 5$ in the second equation.
$$y = 5(5) - 18 = 7$$
The solution of the system is (5, 7).

**51.** $\begin{cases} x+\dfrac{1}{6}y=\dfrac{1}{2} \\ 3x+2y=3 \end{cases}$

Multiply the first equation by $-12$.

$\begin{array}{l} -12x-2y=-6 \\ \underline{\phantom{-}3x+2y=3} \\ \phantom{-}{-9x}\phantom{+2y}=-3 \end{array}$

$$x=\dfrac{1}{3}$$

Substitute $\dfrac{1}{3}$ for $x$ in the second equation.

$3\left(\dfrac{1}{3}\right)+2y=3$

$\phantom{3\left(\dfrac{1}{3}\right)}1+2y=3$

$\phantom{3\left(\dfrac{1}{3}\right)11}2y=2$

$\phantom{3\left(\dfrac{1}{3}\right)111}y=1$

The solution of the system is $\left(\dfrac{1}{3},1\right)$.

**53.** $\begin{cases} \dfrac{x+2}{2}=\dfrac{y+11}{3} \\ \dfrac{x}{2}=\dfrac{2y+16}{6} \end{cases}$

Multiply the first equation by 6 and the second equation by $-6$.

$\begin{cases} 3(x+2)=2(y+11) \\ 3x+6=2y+22 \\ 3x-2y=16 \\ \\ -3x=-2y-16 \\ -3x+2y=-16 \end{cases}$

Add the two equations.

$\begin{array}{l} 3x-2y=16 \\ \underline{-3x+2y=-16} \\ \phantom{3x-2y}0=0 \end{array}$

There is an infinite number of solutions.

**55.** $\begin{cases} 2x+3y=14 \\ 3x-4y=-69.1 \end{cases}$

Multiply the first equation by 3 and the second equation by $-2$.

$\begin{array}{l} 6x+9y=42 \\ \underline{-6x+8y=138.2} \\ \phantom{6x+}17y=180.2 \\ \phantom{6x+11}y=10.6 \end{array}$

Let $y = 10.6$ in the first equation.

$2x+3(10.6)=14$

$\phantom{2x+}2x+31.8=14$

$\phantom{2x+31.81}2x=-17.8$

$\phantom{2x+31.8}x=-8.9$

The solution of the system is $(-8.9, 10.6)$.

**57.** Let $x =$ a number.
$2x + 6 = x - 3$

**59.** Let $x =$ a number.
$20 - 3x = 2$

**61.** Let $n =$ a number.
$4(n + 6) = 2n$

**63.** $\begin{cases} 4x+2y=-7 \\ 3x-y=-12 \end{cases}$

To eliminate $y$, multiply the second equation by 2.

$6x - 2y = -24$

**65.** $\begin{cases} 3x+8y=-5 \\ 2x-4y=3 \end{cases} = \begin{cases} 3x+8y=-5 \\ 4x-8y=6 \end{cases}$

The correct answer is **b**; answers may vary

**67.** answers may vary

**69.** $\begin{cases} x+y=5 \\ 3x+3y=b \end{cases}$

Multiply the first equation by $-3$.

$\begin{array}{l} -3x-3y=-15 \\ \underline{\phantom{-}3x+3y=b} \\ \phantom{-3x-3y}0=b-15 \end{array}$

**a.** The system has an infinite number of solutions if this statement is true.
$b = 15$

**b.** The system has no solution if this statement is false. $b =$ any real number except 15.

**71.** $\begin{cases} 1.2x+3.4y=27.6 \\ 7.2x-1.7y=-46.56 \end{cases}$

Multiply the second equation by 2.

$\begin{array}{l} 1.2x+3.4y=27.6 \\ \underline{14.4x-3.4y=-93.12} \\ 15.6x\phantom{-3.4y}=-65.52 \end{array}$

$\phantom{15.6}x=-4.2$

Let $x = -4.2$ in the first equation.

$$1.2(-4.2)+3.4y=27.6$$
$$-5.04+3.4y=27.6$$
$$3.4y=32.64$$
$$y=9.6$$

The solution of the system is $(-4.2, 9.6)$.

**73. a.** $\begin{cases} 0.05x-y=-21.6 \\ 0.58x-y=-18.1 \end{cases}$

Multiply the first equation by $-1$.

$$-0.05x+y=21.6$$
$$\underline{0.58x-y=-18.1}$$
$$0.53x\qquad=3.5$$
$$x\approx 7$$

Let $x=7$ in the first equation.
$$0.05(7)-y=-21.6$$
$$0.35-y=-21.6$$
$$-y=-21.95$$
$$y\approx 22$$

The solution of the system is approximately $(7, 22)$.

**b.** In 2015 (2008 + 7), the percent of workers age 25–34 and the percent of workers age 55 and older will be the same.

**c.** There will be approximately 22% of workers for each of these age groups.

**Integrated Review**

**1.** $\begin{cases} 2x-3y=-11 \\ y=4x-3 \end{cases}$

Substitute $4x-3$ for $y$ in the first equation.
$$2x-3(4x-3)=-11$$
$$2x-12x+9=-11$$
$$-10x=-20$$
$$x=2$$
Let $x=2$ in the second equation.
$$y=4(2)-3=5$$
The solution of the system is $(2, 5)$.

**2.** $\begin{cases} 4x-5y=6 \\ y=3x-10 \end{cases}$

Substitute $3x-10$ for $y$ in the first equation.
$$4x-5(3x-10)=6$$
$$4x-15x+50=6$$
$$-11x=-44$$
$$x=4$$
Let $x=4$ in the second equation.
$$y=3(4)-10=2$$
The solution of the system is $(4, 2)$.

**3.** $\begin{cases} x+y=3 \\ x-y=7 \end{cases}$

$$x+y=3$$
$$\underline{x-y=7}$$
$$2x\qquad=10$$
$$x=5$$

Let $x=5$ in the first equation.
$$5+y=3$$
$$y=-2$$
The solution of the system is $(5, -2)$.

**4.** $\begin{cases} x-y=20 \\ x+y=-8 \end{cases}$

$$x-y=20$$
$$\underline{x+y=-8}$$
$$2x\qquad=12$$
$$x=6$$

Let $x=6$ in the second equation.
$$6+y=-8$$
$$y=-14$$
The solution of the system is $(6, -14)$.

**5.** $\begin{cases} x+2y=1 \\ 3x+4y=-1 \end{cases}$

Solve the first equation for $x$.
$$x=1-2y$$
Substitute $1-2y$ for $x$ in the second equation.
$$3(1-2y)+4y=-1$$
$$3-6y+4y=-1$$
$$-2y=-4$$
$$y=2$$
Let $y=2$ in $x=1-2y$.
$$x=1-2(2)=-3$$
The solution is $(-3, 2)$.

**6.** $\begin{cases} x+3y=5 \\ 5x+6y=-2 \end{cases}$

Solve the first equation for $x$.
$$x=5-3y$$
Substitute $5-3y$ for $x$ in the second equation.
$$5(5-3y)+6y=-2$$
$$25-15y+6y=-2$$
$$-9y=-27$$
$$y=3$$
Let $y=3$ in $x=5-3y$.
$$x=5-3(3)=-4$$
The solution is $(-4, 3)$.

**7.** $\begin{cases} y = x + 3 \\ 3x - 2y = -6 \end{cases}$

Substitute $x + 3$ for $y$ in the second equation.
$3x - 2(x + 3) = -6$
$\quad 3x - 2x - 6 = -6$
$\qquad\qquad\quad x = 0$
Let $x = 0$ in the first equation.
$y = 0 + 3 = 3$
The solution is (0, 3).

**8.** $\begin{cases} y = -2x \\ 2x - 3y = -16 \end{cases}$

Substitute $-2x$ for $y$ in the second equation.
$2x - 3(-2x) = -16$
$\quad\quad 2x + 6x = -16$
$\qquad\qquad 8x = -16$
$\qquad\qquad\; x = -2$
Let $x = -2$ in the first equation.
$y = -2(-2) = 4$
The solution is (−2, 4).

**9.** $\begin{cases} y = 2x - 3 \\ y = 5x - 18 \end{cases}$

Substitute $5x - 18$ for $y$ in the first equation.
$5x - 18 = 2x - 3$
$\qquad 3x = 15$
$\qquad\; x = 5$
Let $x = 5$ in the second equation.
$y = 5(5) - 18 = 7$
The solution is (5, 7).

**10.** $\begin{cases} y = 6x - 5 \\ y = 4x - 11 \end{cases}$

Substitute $6x - 5$ for $y$ in the second equation.
$6x - 5 = 4x - 11$
$\quad 2x = -6$
$\quad\; x = -3$
Let $x = -3$ in the first equation.
$y = 6(-3) - 5 = -23$
The solution is (−3, −23).

**11.** $\begin{cases} x + \dfrac{1}{6}y = \dfrac{1}{2} \\ 3x + 2y = 3 \end{cases}$

Multiply the first equation by 6.
$\begin{cases} 6x + y = 3 \\ 3x + 2y = 3 \end{cases}$

Multiply the first equation of the simplified system by −2.

$\begin{array}{r} -12x - 2y = -6 \\ 3x + 2y = 3 \\ \hline -9x \quad\quad = -3 \end{array}$

$\qquad x = \dfrac{1}{3}$

Multiply the second equation of the simplified system by −2.

$\begin{cases} 6x + y = 3 \\ -6x - 4y = -6 \end{cases}$

$\begin{array}{r} -3y = -3 \\ y = 1 \end{array}$

The solution of the system is $\left( \dfrac{1}{3}, 1 \right)$.

**12.** $\begin{cases} x + \dfrac{1}{3}y = \dfrac{5}{12} \\ 8x + 3y = 4 \end{cases}$

Multiply the first equation by 12.
$\begin{cases} 12x + 4y = 5 \\ 8x + 3y = 4 \end{cases}$

Multiply the first equation of the simplified system by 2 and the second equation by −3.

$\begin{array}{r} 24x + 8y = 10 \\ -24x - 9y = -12 \\ \hline -y = -2 \\ y = 2 \end{array}$

Multiply the first equation of the simplified system by 3 and the second equation by −4.

$\begin{array}{r} 36x + 12y = 15 \\ -32x - 12y = -16 \\ \hline 4x \quad\quad = -1 \end{array}$

$\qquad x = -\dfrac{1}{4}$

The solution of the system is $\left( -\dfrac{1}{4}, 2 \right)$.

**13.** $\begin{cases} x - 5y = 1 \\ -2x + 10y = 3 \end{cases}$

Multiply the first equation by 2.

$\begin{array}{r} 2x - 10y = 2 \\ -2x + 10y = 3 \\ \hline 0 = 5 \quad \text{False} \end{array}$

The system has no solution.

    

**14.** $\begin{cases} -x + 2y = 3 \\ 3x - 6y = -9 \end{cases}$

Multiply the first equation by 3.

$$-3x + 6y = 9$$
$$\underline{3x - 6y = -9}$$
$$0 = 0$$

The equations in the original system are equivalent and there is an infinite number of solutions.

**15.** $\begin{cases} 0.2x - 0.3y = -0.95 \\ 0.4x + 0.1y = 0.55 \end{cases}$

Multiply both equations by 10.

$\begin{cases} 2x - 3y = -9.5 \\ 4x + y = 5.5 \end{cases}$

Multiply the first equation of the simplified system by −2.

$$-4x + 6y = 19$$
$$\underline{4x + y = 5.5}$$
$$7y = 24.5$$
$$y = 3.5$$

Multiply the second equation of the simplified system by 3.

$$2x - 3y = -9.5$$
$$\underline{12x + 3y = 16.5}$$
$$14x \quad\quad = 7$$
$$x = 0.5$$

The solution of the system is (0.5, 3.5).

**16.** $\begin{cases} 0.08x - 0.04y = -0.11 \\ 0.02x - 0.06y = -0.09 \end{cases}$

Multiply both equations by 100.

$\begin{cases} 8x - 4y = -11 \\ 2x - 6y = -9 \end{cases}$

Multiply the second equation of the simplified system by −4.

$$8x - 4y = -11$$
$$\underline{-8x + 24y = 36}$$
$$20y = 25$$
$$y = 1.25$$

Multiply the first equation of the simplified system by −3 and the second equation by 2.

$$-24x + 12y = 33$$
$$\underline{4x - 12y = -18}$$
$$-20x \quad\quad = 15$$
$$x = -0.75$$

The solution of the system is (−0.75, 1.25).

**17.** $\begin{cases} x = 3y - 7 \\ 2x - 6y = -14 \end{cases}$

Substitute $3y - 7$ for $x$ in the second equation.

$$2(3y - 7) - 6y = -14$$
$$6y - 14 - 6y = -14$$
$$-14 = -14$$

The equations in the original system are equivalent and there is an infinite number of solutions.

**18.** $\begin{cases} y = \dfrac{x}{2} - 3 \\ 2x - 4y = 0 \end{cases}$

Substitute $\dfrac{x}{2} - 3$ for $y$ in the second equation.

$$2x - 4\left(\frac{x}{2} - 3\right) = 0$$
$$2x - 2x + 12 = 0$$
$$12 = 0 \quad \text{False}$$

There is no solution.

**19.** $\begin{cases} 2x + 5y = -1 \\ 3x - 4y = 33 \end{cases}$

Multiply the first equation by 4 and the second equation by 5.

$$8x + 20y = -4$$
$$\underline{15x - 20y = 165}$$
$$23x \quad\quad = 161$$
$$x = 7$$

Let $x = 7$ in the first equation.

$$2(7) + 5y = -1$$
$$14 + 5y = -1$$
$$5y = -15$$
$$y = -3$$

The solution of the system is (7, −3).

**20.** $\begin{cases} 7x - 3y = 2 \\ 6x + 5y = -21 \end{cases}$

Multiply the first equation by 5 and the second equation by 3.

$$35x - 15y = 10$$
$$\underline{18x + 15y = -63}$$
$$53x \quad\quad = -53$$
$$x = -1$$

Let $x = -1$ in the first equation.

$$7(-1)-3y=2$$
$$-7-3y=2$$
$$-3y=9$$
$$y=-3$$

The solution of the system is $(-1, -3)$.

**21.** answers may vary

**22.** answers may vary

**Section 4.4 Practice**

**1.** Let $x$ be one number and $y$ be the other.

The system is $\begin{cases} x+y=30 \\ x-y=6 \end{cases}$.

Add.
$$x+y=30$$
$$\underline{x-y=6}$$
$$2x\phantom{+y}=36$$
$$x=18$$

Let $x = 18$ in the first equation.
$$x+y=30$$
$$18+y=30$$
$$y=12$$

The numbers are 18 and 12.

**2.** Let $x$ = price for adult admission and $y$ = price per child admission.
$$\begin{cases} 3x+3y=75 \\ 2x+4y=62 \end{cases}$$

Multiply the first equation by 2 and the second equation by $-3$.
$$6x+6y=150$$
$$\underline{-6x-12y=-186}$$
$$-6y=-36$$
$$y=6$$

Let $y = 6$ in the second equation.
$$2x+4y=62$$
$$2x+4(6)=62$$
$$2x+24=62$$
$$2x=38$$
$$x=19$$

**a.** $x = 19$, so the adult price is $19.

**b.** $y = 6$, so the child price is $6.

**c.** $5(19) + 15(6) = 95 + 90 = 185 < 200$
No, the regular rates are less than the group rate.

**3.** Let $x$ and $y$ be the speed of the hikers. Let the slower hiker be $y$; then $y = x - 2$.

| $r$ | $\cdot$ | $t$ | $=$ | $d$ |
|---|---|---|---|---|
| $x$ | | 4 | | $4x$ |
| $y$ | | 4 | | $4y$ |

The total distance is 22 miles, so the system is:
$$\begin{cases} 4x+4y=22 \\ y=x-2 \end{cases} \rightarrow \begin{cases} 2x+2y=11 \\ -x+y=-2 \end{cases}$$

Multiply $-x + y = -2$ by 2.
$$2x+2y=11$$
$$\underline{-2x+2y=-4}$$
$$4y=7$$
$$y=\frac{7}{4}=1.75$$

Let $y = 1.75$ in $-x + y = -2$.
$$-x+1.75=-2$$
$$-x=-3.75$$
$$x=3.75$$

The speeds are 1.75 miles per hour and 3.75 miles per hour.

**4.** Let $x$ = pounds of Kona and $y$ = pounds of Blue. Then, $x + y = 20$ pounds of mix. $20x$ is the total price of the Kona. $28y$ is the total price of the Blue. $22(20)$ is the total price of the blend. Then, $20x + 28y = 440$.

The system is $\begin{cases} x+y=20 \\ 20x+28y=440 \end{cases}$.

Multiply the first equation by $-20$.
$$-20x-20y=-400$$
$$\underline{20x+28y=440}$$
$$8y=40$$
$$y=5$$

Let $y = 5$ in the first equation.
$$x+y=20$$
$$x+5=20$$
$$x=15$$

Jemima should use 15 pounds of Kona and 5 pounds of Blue Mountain.

**Vocabulary, Readiness & Video Check 4.4**

**1.** Up to now we've been working with one variable/unknown and one equation. Because systems involve two equations with two unknowns, for these applications, you need to choose two variables to represent two unknowns and translate the problem into two equations.

**Exercise Set 4.4**

**1. a.** $l - w = 8 - 5 = 3$
$$P = 2l + 2w$$
$$= 2(8) + 2(5)$$
$$= 13 + 10$$
$$= 23 \neq 30$$

**b.** $l - w = 8 - 7 = 1 \neq 3$

**c.** $l - w = 9 - 6 = 3$
$$P = 2l + 2w = 2(9) + 2(6) = 18 + 12 = 30$$

Choice **c** is correct.

**3. a.** $2d + 3n = 2(3) + 3(4) = 6 + 12 = 18 \neq 17$

**b.** $2d + 3n = 2(4) + 3(3) = 8 + 9 = 17$
$5d + 4n = 5(4) + 4(3) = 20 + 12 = 32$

**c.** $2d + 3n = 2(2) + 3(5) = 4 + 15 = 19 \neq 17$

Choice **b** is correct.

**5. a.** $80 + 20 = 100$
$$80d + 20q = 80(0.10) + 20(0.25)$$
$$= 8 + 5$$
$$= 13$$

**b.** $20 + 44 = 64 \neq 100$

**c.** $60 + 40 = 100$
$$60d + 40q = 60(0.10) + 40(0.25)$$
$$= 6 + 10$$
$$= 16 \neq 13$$

Choice **a** is correct.

**7.** Let $x =$ the larger number and
$y =$ the smaller number.
$$\begin{cases} x + y = 15 \\ x - y = 7 \end{cases}$$

**9.** Let $x =$ the amount invested in the larger account and $y =$ the amount invested in the smaller account.
$$\begin{cases} x + y = 6500 \\ x = y + 800 \end{cases}$$

**11.** Let $x =$ the first number and
$y =$ the second number.
$$\begin{cases} x + y = 83 \\ x - y = 17 \end{cases}$$

$$\begin{array}{r} x + y = 83 \\ x - y = 17 \\ \hline 2x \quad\quad = 100 \\ x = 50 \end{array}$$

Let $x = 50$ in the first equation.
$$50 + y = 83$$
$$y = 33$$

The numbers are 50 and 33.

**13.** Let $x =$ the first number and
$y =$ the second number.
$$\begin{cases} x + 2y = 8 \\ 2x + y = 25 \end{cases}$$

Multiply the first equation by $-2$.
$$\begin{array}{r} -2x - 4y = -16 \\ 2x \;\; + y = 25 \\ \hline -3y = 9 \\ y = -3 \end{array}$$

Let $y = -3$ in the first equation.
$$x + 2(-3) = 8$$
$$x - 6 = 8$$
$$x = 14$$

The numbers are 14 and $-3$.

**15.** Let $x$ be the number of runs that Miguel Cabrera batted in and $y$ be the number that Alex Rodriguez batted in.
$$\begin{cases} y = x - 1 \\ x + y = 251 \end{cases}$$

Substitute $x - 1$ for $y$ in the second equation and solve for $x$.
$$x + y = 251$$
$$x + x - 1 = 251$$
$$2x = 252$$
$$x = 126$$

Now solve for $y$.
$$y = x - 1 = 126 - 1 = 125$$
Miguel Cabrera batted in 126 runs and Alex Rodriguez batted in 125 runs.

**17.** Let $x =$ the price of an adult's ticket and $y =$ the price of a child's ticket.
$$\begin{cases} 3x + 4y = 159 \\ 2x + 3y = 112 \end{cases}$$

Multiply the first equation by $-2$ and the second equation by 3.
$$\begin{array}{r} -6x - 8y = -318 \\ 6x + 9y = 336 \\ \hline y = 18 \end{array}$$

Let $y = 18$ in the first equation.

$$3x + 4(18) = 159$$
$$3x + 72 = 159$$
$$3x = 87$$
$$x = 29$$

An adult's ticket is \$29 and a child's ticket is \$18.

19. Let $x$ = the number of quarters and $y$ = the number of nickels.

$$\begin{cases} x + y = 80 \\ 0.25x + 0.05y = 14.6 \end{cases}$$

Solve the first equation for $y$.
$$y = 80 - x$$
Substitute $80 - x$ for $y$ in the second equation.
$$0.25x + 0.05(80 - x) = 14.6$$
$$0.25x + 4 - 0.05x = 14.6$$
$$0.20x = 10.6$$
$$x = 53$$
Let $x = 53$ in $y = 80 - x$.
$$y = 80 - 53$$
$$y = 27$$
There are 53 quarters and 27 nickels.

21. Let $x$ be the value of one McDonald's share and let $y$ be the value of one Ohio Art Company share.

$$\begin{cases} 35x + 69y = 2814 \\ x = y + 70 \end{cases}$$

Substitute $y + 70$ for $x$ in the first equation and solve for $y$.
$$35x + 69y = 2814$$
$$35(y + 70) + 69y = 2814$$
$$35y + 2450 + 69y = 2814$$
$$2450 + 104y = 2814$$
$$104y = 364$$
$$y = 3.5$$
Now solve for $x$.
$$x = y + 70 = 3.5 + 70 = 73.5$$
On that day, the closing price of the McDonald's stock was \$73.50 per share and the closing price of The Ohio Art Company stock was \$3.50 per share.

23. Let $x$ be the daily fee and $y$ be the mileage charge.

$$\begin{cases} 4x + 450y = 240.50 \\ 3x + 200y = 146.00 \end{cases}$$

Multiply the first equation by 3 and the second by $-4$.

$$\begin{cases} 3(4x + 450y) = 3(240.50) \\ -4(3x + 200y) = -4(146.00) \end{cases}$$

$$\rightarrow \begin{cases} 12x + 1350y = 721.5 \\ -12x - 800y = -584 \end{cases}$$

Add the equations to eliminate $x$ and solve for $y$.
$$12x + 1350y = 721.5$$
$$\underline{-12x - 800y = -584}$$
$$550y = 137.5$$
$$y = 0.25$$
Now solve for $x$.
$$3x + 200y = 146$$
$$3x + 200(0.25) = 146$$
$$3x + 50 = 146$$
$$3x = 96$$
$$x = 32$$
There is a \$32 daily fee and a \$0.25 per mile mileage charge.

25.

| | $d$ | $=$ | $r$ | $\cdot$ | $t$ |
|---|---|---|---|---|---|
| Downstream | 18 | | $x + y$ | | 2 |
| Upstream | 18 | | $x - y$ | | $4\frac{1}{2}$ |

$$\begin{cases} 2(x + y) = 18 \\ \dfrac{9}{2}(x - y) = 18 \end{cases}$$

Multiply the first equation by $\dfrac{1}{2}$ and the second equation by $\dfrac{2}{9}$.

$$x + y = 9$$
$$\underline{x - y = 4}$$
$$2x \quad\;\; = 13$$
$$x = 6.5$$
Let $x = 6.5$ in $x + y = 9$.
$$6.5 + y = 9$$
$$y = 2.5$$
Pratap can row 6.5 miles per hour in still water. The rate of the current is 2.5 miles per hour.

27.

| | $d$ | $=$ | $r$ | $\cdot$ | $t$ |
|---|---|---|---|---|---|
| With the wind | 780 | | $x + y$ | | $1\frac{1}{2}$ |
| Into the wind | 780 | | $x - y$ | | 2 |

$$\begin{cases} \dfrac{3}{2}(x + y) = 780 \\ 2(x - y) = 780 \end{cases}$$

Multiply the first equation by $\dfrac{2}{3}$ and the second equation by $\frac{1}{2}$.

$$x + y = 520$$
$$\underline{x - y = 390}$$
$$2x \quad\ = 910$$
$$x = 455$$

Let $x = 455$ in $x + y = 520$.
$$455 + y = 520$$
$$y = 65$$

The plane can fly 455 miles per hour in still air. The speed of the wind is 65 miles per hour.

29. Let $x$ = the time spent walking and $y$ = the time spent on the bicycle.

| | $r$ | $\cdot$ | $t$ | $=$ | $d$ |
|---|---|---|---|---|---|
| Walking | 4 | | $x$ | | $4x$ |
| Biking | 24 | | $y$ | | $24y$ |

$$\begin{cases} x + y = 6 \\ 4x + 24y = 114 \end{cases}$$

Multiply the first equation by $-4$.
$$-4x - 4y = -24$$
$$\underline{4x + 24y = 114}$$
$$20y = 90$$
$$y = 4.5$$

He spent $4\frac{1}{2}$ hours on the bicycle.

31. Let $x$ = liters of 4% solution and $y$ = liters of 12% solution.

| Concentration Rate | Liters of Solution | Liters of Pure Acid |
|---|---|---|
| 0.04 | $x$ | $0.04x$ |
| 0.12 | $y$ | $0.12y$ |
| 0.09 | 12 | $0.09(12)$ |

$$\begin{cases} x + y = 12 \\ 0.04x + 0.12y = 0.09(12) \end{cases}$$

Multiply the first equation by $-4$ and the second equation by 100.

$$-4x - 4y = -48$$
$$\underline{4x + 12y = 108}$$
$$8y = 60$$
$$y = 7.5$$

Let $y = 7.5$ in the first equation.
$$x + 7.5 = 12$$
$$x = 4.5$$

$4\frac{1}{2}$ liters of 4% solution and $7\frac{1}{2}$ liters of 12% solution should be mixed.

33. Let $x$ = pounds of \$4.95 per pound beans and $y$ = pounds of \$2.65 per pound beans.

| | Cost Rate | Pounds of Beans | Dollars Cost |
|---|---|---|---|
| High Quality | 4.95 | $x$ | $4.95x$ |
| Low Quality | 2.65 | $y$ | $2.65y$ |
| Mixture | 3.95 | 200 | $3.95(200)$ |

$$\begin{cases} x + y = 200 \\ 4.95x + 2.65y = 3.95(200) \end{cases}$$

Solve the first equation for $y$.
$$y = 200 - x$$
Substitute $200 - x$ for $y$ in the second equation.
$$4.95x + 2.65(200 - x) = 3.95(200)$$
$$4.95x + 530 - 2.65x = 790$$
$$2.30x = 260$$
$$x \approx 113.04$$
Let $x = 113.04$ in the first equation.
$$113.04 + y = 200$$
$$y \approx 86.96$$
He needs 113 pounds of \$4.95 per pound beans and 87 pounds of \$2.65 per pound beans.

35. Let $x$ = the first angle and $y$ = the second angle.
$$\begin{cases} x + y = 90 \\ x = 2y \end{cases}$$
Substitute $2y$ for $x$ in the first equation.
$$2y + y = 90$$
$$3y = 90$$
$$y = 30$$
Let $y = 30$ in the second equation.
$$x = 2(30) = 60$$
The angles are 60° and 30°.

**37.** Let $x$ = the first angle and $y$ = the second angle.
$$\begin{cases} x+y=90 \\ x=3y+10 \end{cases}$$
Substitute $3y + 10$ for $x$ in the first equation.
$$3y+10+y=90$$
$$4y=80$$
$$y=20$$
Let $y = 20$ in the second equation.
$x = 3(20) + 10 = 70$
The angles are 70° and 20°.

**39.** Let $x$ = the number sold at \$9.50 and $y$ = the number sold at \$7.50.
$$\begin{cases} x+y=90 \\ 9.5x+7.5y=721 \end{cases}$$
Solve the first equation for $y$.
$$y=90-x$$
Substitute $90 - x$ for $y$ in the second equation.
$$9.5x+7.5(90-x)=721$$
$$9.5x+675-7.5x=721$$
$$2x=46$$
$$x=23$$
Let $x = 23$ in $y = 90 - x$.
$$y=90-23=67$$
They sold 23 at \$9.50 and 67 at \$7.50.

**41.** Let $x$ = the rate of the faster group and $y$ = the rate of the slower group.

| | $r$ | $\cdot$ $t$ | $=$ $d$ |
|---|---|---|---|
| Faster group | $x$ | 240 | $240x$ |
| Slower group | $y$ | 240 | $240y$ |

$$\begin{cases} x=y+\dfrac{1}{2} \\ 240x+240y=1200 \end{cases}$$
Substitute $y+\dfrac{1}{2}$ for $x$ in the second equation.
$$240\left(y+\dfrac{1}{2}\right)+240y=1200$$
$$240y+120+240y=1200$$
$$480y=1080$$
$$y=\dfrac{1080}{480}=2\dfrac{1}{4}$$
Let $y=2\dfrac{1}{4}$ in the first equation.

$$x=2\dfrac{1}{4}+\dfrac{1}{2}=2\dfrac{3}{4}$$
The rate of the faster group is $2\dfrac{3}{4}$ miles per hour. The rate of the slower group is $2\dfrac{1}{4}$ miles per hour.

**43.** Let $x$ = gallons of 30% solution and $y$ = gallons of 60% solution.

| Concentration Rate | Gallons of Solution | Gallons of Pure Fertilizer |
|---|---|---|
| 0.30 | $x$ | $0.30x$ |
| 0.60 | $y$ | $0.60y$ |
| 0.50 | 150 | $0.50(150)$ |

$$\begin{cases} x+y=150 \\ 0.30x+0.60y=0.50(150) \end{cases}$$
Multiply the first equation by $-3$ and the second equation by 10.
$$-3x-3y=-450$$
$$\underline{3x+6y=750}$$
$$3y=300$$
$$y=100$$
Let $y = 100$ in the first equation.
$$x+100=150$$
$$x=50$$
50 gallons of 30% solution and 100 gallons of 60% solution.

**45.** Let $x$ = the width and $y$ = the length.
$$\begin{cases} 2x+2y=144 \\ y=x+12 \end{cases}$$
Substitute $x + 12$ for $y$ in the first equation.
$$2x+2(x+12)=144$$
$$2x+2x+24=144$$
$$4x=120$$
$$x=30$$
Let $x = 30$ in the second equation.
$$y=30+12=42$$
The width is 30 inches and the length is 42 inches.

**47.** $-3x<-9$
$$\dfrac{-3x}{-3}>\dfrac{-9}{-3}$$
$$x>3,\ (3,\ \infty)$$

**49.** $4(2x-1) \geq 0$

$8x - 4 \geq 0$

$8x \geq 4$

$x \geq \dfrac{1}{2}, \left[\dfrac{1}{2}, \infty\right)$

**51.** The minimum price is $0.49.
The maximum price is $0.65.
$0.72 > 0.65$   Impossible
$0.29 < 0.49$   Impossible
$0.49 < 0.58 < 0.65$   Possible
The answer is **a**.

**53.** Let $x$ = the width and $y$ = the length.
$$\begin{cases} 2x + y = 33 \\ y = 2x - 3 \end{cases}$$
Substitute $2x - 3$ for $y$ in the first equation.
$2x + 2x - 3 = 33$

$4x = 36$

$x = 9$

Let $x = 9$ in the second equation.
$y = 2(9) - 3 = 15$
The width is 9 feet and the length is 15 feet.

**55. a.** $\begin{cases} y = 0.82x + 17.2 \\ y = 0.33x + 30.5 \end{cases}$

Substitute $0.82x + 17.2$ for $y$ in the second equation.
$0.82x + 17.2 = 0.33x + 30.5$

$0.49x = 13.3$

$x \approx 27.14$

Let $x = 27.14$ in $y = 0.82x + 17.2$.
$y = 0.82(27.14) + 17.2$
$y \approx 39.4548$
The solution is approximately $(27.1, 39.5)$.

**b.** For viewers who are 27.1 years over 18 (or 45.1 years of age) the percent who watch cable news and network news is the same, 39.5%.

**c.** Answers may vary

## Section 4.5 Practice

**1.** $x - 4y > 8$

**a.** $(-3, 2)$: $-3 - 4(2) > 8$

$-3 - 8 > 8$

$-11 > 8$   False

$(-3, 2)$ is not a solution of the inequality.

**b.** $(9, 0)$: $9 - 4(0) > 8$

$9 - 0 > 8$

$9 > 8$   True

$(9, 0)$ is a solution of the inequality.

**2.** Graph the boundary line, $x + y = 5$, with a dashed line.
Test $(0, 0)$: $x + y > 5$

$0 + 0 > 5$

$0 > 5$ False

Shade the half-plane not containing $(0, 0)$.

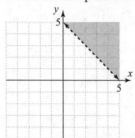

**3.** Graph the boundary line, $3x - y = 4$, with a solid line.
Test $(0, 0)$:   $3x - y \geq 4$

$3(0) - 0 \geq 4$

$0 \geq 4$  False

Shade the half-plane not containing $(0, 0)$.

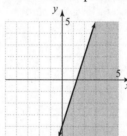

**4.** Graph the boundary line, $x = 3y$, with a dashed line.
Test $(2, 0)$: $x > 3y$

$2 > 3(0)$

$2 > 0$  True

Shade the half-plane containing $(2, 0)$.

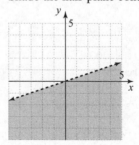

**5.** Graph the boundary line, $3x + 4y = 12$, with a solid line.

Test (0, 0):     $3x + 4y \geq 12$
$3(0) + 4(0) \geq 12$
$0 \geq 12$  False

Shade the half-plane not containing (0, 0).

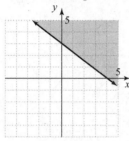

**6.** Graph the boundary line, $x = 3$, with a dashed line.

Test (0, 0): $x > 3$
$0 > 3$  False

Shade the half-plane not containing (0, 0).

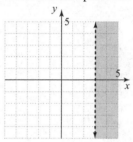

**7.** Graph the boundary line, $y = \dfrac{3}{4}x - 1$, with a dashed line.

Test (0, 0): $y < \dfrac{3}{4}x - 1$
$0 < \dfrac{3}{4}(0) - 1$
$0 < -1$  False

Shade the half-plane not containing (0, 0).

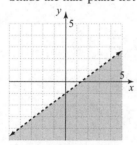

**Vocabulary, Readiness & Video Check 4.5**

**1.** The statement $5x - 6y < 7$ is an example of a <u>linear inequality in two variables</u>.

**2.** A boundary line divides a plane into two regions called <u>half-planes</u>.

**3.** The graph of $5x - 6y < 7$ does not include its corresponding boundary line. The statement is false.

**4.** When graphing a linear inequality to determine which side of the boundary line to shade, choose a point *not* on the boundary line. The statement is true.

**5.** The boundary line for the inequality $5x - 6y < 7$ is the graph of $5x - 6y = 7$. The statement is true.

**6.** The graph of <u>$y < 3$</u> is

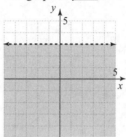

**7.** An ordered pair is a solution of an inequality if replacing the variables with the coordinates of the ordered pair results in a true statement.

**8.** You find the boundary line equation by replacing the inequality symbol with =. The points on this line are solutions (line is solid) if the inequality is $\geq$ or $\leq$; they are not solutions (line is dashed) if the inequality is $>$ or $<$.

**Exercise Set 4.5**

**1.** $x - y > 3$
$(2, -1), \ 2 - (-1) \overset{?}{>} 3$
$2 + 1 \overset{?}{>} 3$
$3 \overset{?}{>} 3,$  False
$(2, -1)$ is not a solution.

$(5, 1), \ 5 - 1 \overset{?}{>} 3$
$4 \overset{?}{>} 3,$   True
$(5, 1)$ is a solution.

**3.** $3x - 5y \leq -4$

$$(-1, -1), \quad 3(-1) - 5(-1) \overset{?}{\leq} -4$$
$$-3 + 5 \overset{?}{\leq} -4$$
$$2 \overset{?}{\leq} -4, \quad \text{False}$$

$(-1, -1)$ is not a solution.

$$(4, 0), \quad 3(4) - 5(0) \overset{?}{\leq} -4$$
$$12 - 0 \overset{?}{\leq} -4$$
$$12 \overset{?}{\leq} -4, \text{ False}$$

$(4, 0)$ is not a solution.

**5.** $x < -y$

$$(0, 2), \quad 0 \overset{?}{<} -2, \text{ False}$$
$(0, 2)$ is not a solution.
$$(-5, 1), \quad -5 \overset{?}{<} -1, \text{ True}$$
$(-5, 1)$ is a solution.

**7.** $x + y \leq 1$
Test $(0, 0)$
$$0 + 0 \overset{?}{\leq} 1, \text{ True}$$
Shade below.

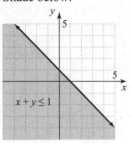

**9.** $2x + y > -4$
Test $(0, 0)$
$$2(0) + 0 \overset{?}{>} -4$$
True
Shade above.

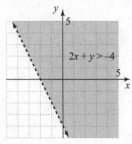

**11.** $x + 6y \leq -6$
Test $(0, 0)$
$$0 + 6(0) \overset{?}{\leq} -6$$
False
Shade below.

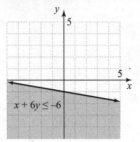

**13.** $2x + 5y > -10$
Test $(0, 0)$
$$2(0) + 5(0) \overset{?}{>} -10$$
True
Shade above.

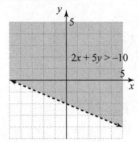

**15.** $x + 2y \leq 3$
Test $(0, 0)$
$$0 + 2(0) \overset{?}{\leq} 3$$
True
Shade below.

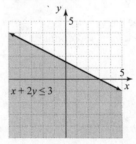

**17.** $2x + 7y > 5$
Test $(0, 0)$
$$2(0) + 7(0) \overset{?}{>} 5$$
False
Shade above.

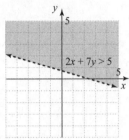

**19.** $x - 2y \geq 3$
Test $(0, 0)$
$$(0) - 2(0) \overset{?}{\geq} 3$$
False
Shade below.

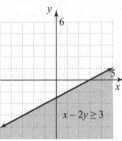

**21.** $5x + y < 3$
Test $(0, 0)$
$$5(0) + 0 \overset{?}{<} 3$$
True
Shade below.

**23.** $4x + y < 8$
Test $(0, 0)$
$$4(09) + 0 \overset{?}{<} 8$$
True
Shade below.

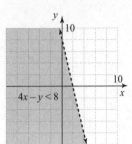

**25.** $y \geq 2x$
Test $(1, 0)$
$$0 \overset{?}{\geq} 2(1)$$
False
Shade above.

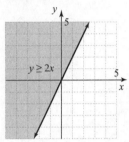

**27.** $x \geq 0$
Shade right.

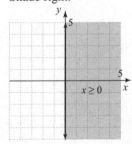

**29.** $y \leq -3$
Shade below.

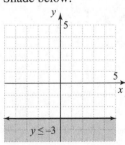

**31.** $2x - 7y > 0$
Test $(1, 0)$
$$2(1) - 7(0) \overset{?}{>} 0$$
True

Shade below.

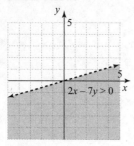

**33.** $3x - 7y \geq 0$
Test $(1, 0)$

$$3(1) - 7(0) \overset{?}{\geq} 0$$

True
Shade below.

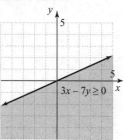

**35.** $x > y$
Test $(0, 1)$

$$0 \overset{?}{>} 1$$

False
Shade below.

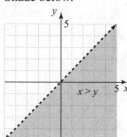

**37.** $x - y \leq 6$
Test $(0, 0)$

$$0 - 0 \overset{?}{\leq} 6$$

True
Shade above.

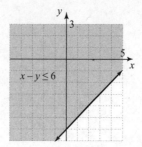

**39.** $-\dfrac{1}{4}y + \dfrac{1}{3}x > 1$
Test $(0, 0)$

$$-\dfrac{1}{4}(0) + \dfrac{1}{3}(0) \overset{?}{>} 1$$

False
Shade below.

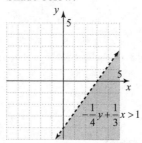

**41.** $-x < 0.4y$
Test $(1, 0)$

$$-(1) \overset{?}{<} 0$$

True
Shade above.

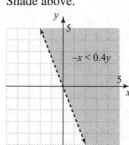

**43. e**

**45. c**

**47. f**

**49.** $2^3 = 2 \cdot 2 \cdot 2 = 8$

**51.** $(-2)^5 = (-2)(-2)(-2)(-2)(-2) = -32$

**53.** $3 \cdot 4^2 = 3 \cdot 4 \cdot 4 = 48$

**55.** Let $x = -5$.
$x^2 = (-5)(-5) = 25$

**57.** Let $x = -1$.
$2x^3 = 2(-1)(-1)(-1) = -2$

**59.** Yes, since the inequality is $\geq$, the graph includes the boundary line.

**61.** Yes, since the inequality is $\geq$, the graph includes the boundary line.

**63.**　$3x + 4y < 8$; $(1,1)$
$3(1) + 4(1) < 8$
$3 + 4 < 8$
$7 < 8$　True
$(1, 1)$ is included in the graph.

**65.**　$y \geq -\dfrac{1}{2}x$; $(1, 1)$

$y \geq -\dfrac{1}{2}(1)$

$y \geq -\dfrac{1}{2}$　True

$(1, 1)$ is included in the graph.

**67.** The inequality is $x + y \geq 13$.

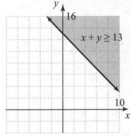

**69.** answers may vary

**71.**

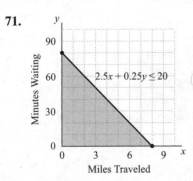

**73.** answers may vary

**75. a.**　$x + y \leq 5000$

**b.**

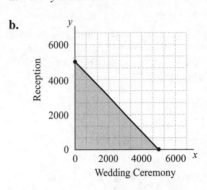

**c.**　answers may vary

**Section 4.6 Practice**

**1.** $\begin{cases} 4x \leq y \\ x + 3y \geq 9 \end{cases}$

Graph $4x \leq y$ with a solid line.
Test $(1, 0)$
$$4(1) \overset{?}{\leq} 0$$
False
Shade above.
Graph $x + 3y \geq 9$ with a solid line.
Test $(0, 0)$
$$0 + 3(0) \overset{?}{\geq} 9$$
False
Shade above.
The solution of the system is the darker shaded region and includes parts of both boundary lines.

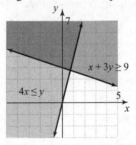

**2.** $\begin{cases} x - y > 4 \\ x + 3y < -4 \end{cases}$

Graph both inequalities using dashed lines. The solution of the system is the darker shaded region which does not include any of the boundary lines.

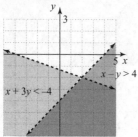

**3.** $\begin{cases} y \le 6 \\ -2x + 5y > 10 \end{cases}$

Graph both inequalities. The solution of the system is the darker shaded region.

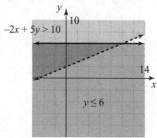

**Vocabulary, Readiness & Video Check 4.6**

**1.** No; we can choose any test point except a point on the second inequality's own boundary line.

**Exercise Set 4.6**

**1.** $\begin{cases} y \ge x + 1 \\ y \ge 3 - x \end{cases}$

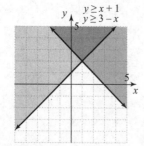

**3.** $\begin{cases} y < 3x - 4 \\ y \le x + 2 \end{cases}$

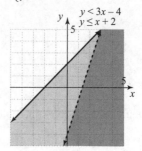

**5.** $\begin{cases} y \le -2x - 2 \\ y \ge x + 4 \end{cases}$

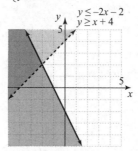

**7.** $\begin{cases} y \ge -x + 2 \\ y \le 2x + 5 \end{cases}$

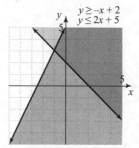

**9.** $\begin{cases} x \ge 3y \\ x + 3y \le 6 \end{cases}$

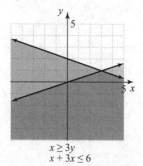

**11.** $\begin{cases} y + 2x \ge 0 \\ 5x - 3y \le 12 \end{cases}$

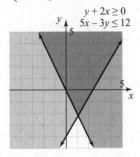

**13.** $\begin{cases} 3x - 4y \ge -6 \\ 2x + y \le 7 \end{cases}$

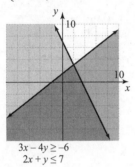

**15.** $\begin{cases} x \le 2 \\ y \ge -3 \end{cases}$

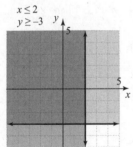

**17.** $\begin{cases} y \ge 1 \\ x < -3 \end{cases}$

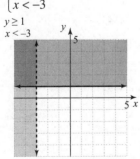

**19.** $\begin{cases} 2x + 3y < -8 \\ x \ge -4 \end{cases}$

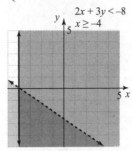

**21.** $\begin{cases} 2x - 5y \le 9 \\ y \le -3 \end{cases}$

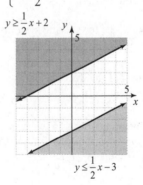

**23.** $\begin{cases} y \ge \dfrac{1}{2}x + 2 \\ y \le \dfrac{1}{2}x - 3 \end{cases}$

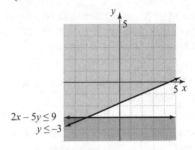

**25.** $4^2 = (4)(4) = 16$

**27.** $(6x)^2 = (6x)(6x) = 36x^2$

**29.** $(10y^3)^2 = (10y^3)(10y^3) = 100y^6$

**31.** C

**33.** D

**35.** answers may vary

**37.** $\begin{cases} 2x - y \le 6 \\ \quad x \ge 3 \\ \quad y > 2 \end{cases}$

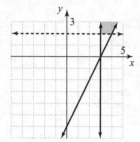

**Chapter 4 Vocabulary Check**

1. In a system of linear equations in two variables, if the graphs of the equations are the same, the equations are <u>dependent</u> equations.

2. Two or more linear equations are called a <u>system of linear equations</u>.

3. A system of equations that has at least one solution is called a <u>consistent</u> system.

4. A <u>solution</u> of a system of two equations in two variables is an ordered pair of numbers that is a solution of both equations in the system.

5. Two algebraic methods for solving systems of equations are <u>addition</u> and <u>substitution</u>.

6. A system of equations that has no solution is called an <u>inconsistent</u> system.

7. In a system of linear equations in two variables, if the graphs of the equations are different, the equations are <u>independent</u> equations.

8. Two or more linear inequalities are called a <u>system of linear equalities</u>.

**Chapter 4 Review**

1. **a.** Let $x = 12$ and $y = 4$.
$$2x - 3y = 12$$
$$2(12) - 3(4) \stackrel{?}{=} 12$$
$$24 - 12 \stackrel{?}{=} 12$$
$$12 = 12 \quad \text{True}$$

$$3x + 4y = 1$$
$$3(12) + 4(4) \stackrel{?}{=} 1$$
$$36 + 16 \stackrel{?}{=} 1$$
$$52 = 1 \quad \text{False}$$
$(12, 4)$ is not a solution of the system.

**b.** Let $x = 3$ and $y = -2$.
$$2x - 3y = 12$$
$$2(3) - 3(-2) \stackrel{?}{=} 12$$
$$6 + 6 \stackrel{?}{=} 12$$
$$2 = 12 \quad \text{True}$$

$$3x + 4y = 1$$
$$3(3) + 4(-2) \stackrel{?}{=} 1$$
$$9 - 8 \stackrel{?}{=} 1$$
$$1 = 1 \quad \text{True}$$
$(3, -2)$ is a solution of the system.

**c.** Let $x = -3$ and $y = 6$.
$$2x - 3y = 12$$
$$2(-3) - 3(6) \stackrel{?}{=} 12$$
$$-6 - 18 \stackrel{?}{=} 12$$
$$-24 = 12 \quad \text{False}$$

$$3x + 4y = 1$$
$$3(-3) + 4(6) \stackrel{?}{=} 1$$
$$-9 + 24 \stackrel{?}{=} 1$$
$$15 = 1 \quad \text{False}$$
$(-3, 6)$ is not a solution of the system.

2. **a.** Let $x = \dfrac{3}{4}$ and $y = -3$.
$$4x + y = 0$$
$$4\left(\frac{3}{4}\right) - 3 \stackrel{?}{=} 0$$
$$3 - 3 \stackrel{?}{=} 0$$
$$0 = 0 \quad \text{True}$$

$$-8x - 5y = 9$$
$$-8\left(\frac{3}{4}\right) - 5(-3) \stackrel{?}{=} 9$$
$$-6 + 15 \stackrel{?}{=} 9$$
$$9 = 9 \quad \text{True}$$
$\left(\dfrac{3}{4}, -3\right)$ is a solution of the system.

**b.** Let $x = -2$ and $y = 8$.
$$4x + y = 0$$
$$4(-2) + 8 \stackrel{?}{=} 0$$
$$-8 + 8 \stackrel{?}{=} 0$$
$$0 = 0 \quad \text{True}$$

$$-8x - 5y = 9$$
$$-8(-2) - 5(8) \stackrel{?}{=} 9$$
$$16 - 40 \stackrel{?}{=} 9$$
$$-24 = 9 \quad \text{False}$$

$(-2, 8)$ is not a solution of the system.

**c.** Let $x = \dfrac{1}{2}$ and $y = -2$.

$$4x + y = 0$$
$$4\left(\dfrac{1}{2}\right) - 2 \stackrel{?}{=} 0$$
$$2 - 2 \stackrel{?}{=} 0$$
$$0 = 0 \quad \text{True}$$

$$-8x - 5y = 9$$
$$-8\left(\dfrac{1}{2}\right) - 5(-2) \stackrel{?}{=} 9$$
$$-4 + 10 \stackrel{?}{=} 9$$
$$6 = 9 \quad \text{False}$$

$\left(\dfrac{1}{2}, -2\right)$ is not a solution of the system.

**3. a.** Let $x = -6$ and $y = -8$.

$$5x - 6y = 18$$
$$5(-6) - 6(-8) \stackrel{?}{=} 18$$
$$-30 + 48 \stackrel{?}{=} 18$$
$$18 = 18 \quad \text{True}$$

$$2y - x = -4$$
$$2(-8) - (-6) \stackrel{?}{=} -4$$
$$-16 + 6 \stackrel{?}{=} -4$$
$$-10 = -4 \quad \text{False}$$

$(-6, -8)$ is not a solution of the system.

**b.** Let $x = 3$ and $y = \dfrac{5}{2}$.

$$5x - 6y = 18$$
$$5(3) - 6\left(\dfrac{5}{2}\right) \stackrel{?}{=} 18$$
$$15 - 15 \stackrel{?}{=} 18$$
$$0 = 18 \quad \text{False}$$

$$2y - x = -4$$
$$2\left(\dfrac{5}{2}\right) - 3 \stackrel{?}{=} -4$$
$$5 - 3 \stackrel{?}{=} -4$$
$$2 = -4 \quad \text{False}$$

$\left(3, \dfrac{5}{2}\right)$ is not a solution of the system.

**c.** Let $x = 3$ and $y = -\dfrac{1}{2}$.

$$5x - 6y = 18$$
$$5(3) - 6\left(-\dfrac{1}{2}\right) \stackrel{?}{=} 18$$
$$15 + 3 \stackrel{?}{=} 18$$
$$18 = 18 \quad \text{True}$$

$$2y - x = -4$$
$$2\left(-\dfrac{1}{2}\right) - 3 \stackrel{?}{=} -4$$
$$-1 - 3 \stackrel{?}{=} -4$$
$$-4 = -4 \quad \text{True}$$

$\left(3, -\dfrac{1}{2}\right)$ is a solution of the system.

**4. a.** Let $x = 2$ and $y = 2$.

$$2x + 3y = 1 \qquad\qquad 3y - x = 4$$
$$2(2) + 3(2) \stackrel{?}{=} 1 \qquad 3(2) - 2 \stackrel{?}{=} 4$$
$$4 + 6 \stackrel{?}{=} 1 \qquad\qquad 6 - 2 \stackrel{?}{=} 4$$
$$10 = 1 \quad \text{False} \qquad 4 = 4 \quad \text{True}$$

$(2, 2)$ is not a solution of the system.

**b.** Let $x = -1$ and $y = 1$.

$$2x + 3y = 1 \qquad\qquad 3y - x = 4$$
$$2(-1) + 3(1) \stackrel{?}{=} 1 \qquad 3(1) - (-1) \stackrel{?}{=} 4$$
$$-2 + 3 \stackrel{?}{=} 1 \qquad\qquad 3 + 1 \stackrel{?}{=} 4$$
$$1 = 1 \quad \text{True} \qquad 4 = 4 \quad \text{True}$$

$(-1, 1)$ is a solution of the system.

**c.** Let $x = 2$ and $y = -1$.

$$2x + 3y = 1$$
$$2(2) + 3(-1) \stackrel{?}{=} 1$$
$$4 - 3 \stackrel{?}{=} 1$$
$$1 = 1 \quad \text{True}$$

$$3y - x = 4$$
$$3(-1) - 2 \stackrel{?}{=} 4$$
$$-3 - 2 \stackrel{?}{=} 4$$
$$-5 = 4 \quad \text{False}$$

$(2, -1)$ is not a solution of the system.

5. $\begin{cases} x+y=5 \\ x-1=y \end{cases}$

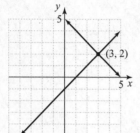

The solution of the system is (3, 2).

6. $\begin{cases} x+y=3 \\ x-y=-1 \end{cases}$

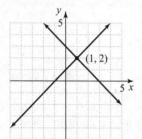

The solution of the system is (1, 2).

7. $\begin{cases} x=5 \\ y=-1 \end{cases}$

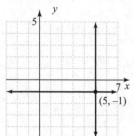

The solution of the system is (5, −1).

8. $\begin{cases} x=-3 \\ y=2 \end{cases}$

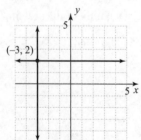

The solution of the system is (−3, 2).

9. $\begin{cases} 2x+y=5 \\ x=-3y \end{cases}$

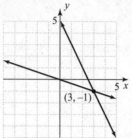

The solution of the system is (3, −1).

10. $\begin{cases} 3x+y=-2 \\ y=-5x \end{cases}$

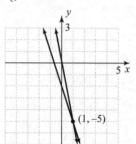

The solution of the system is (1, −5).

11. $\begin{cases} y=3x \\ -6x+2y=6 \end{cases}$

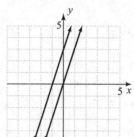

There is no solution.

12. $\begin{cases} x-2y=2 \\ -2x+4y=-4 \end{cases}$

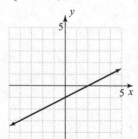

There is an infinite number of solutions.

**13.** $\begin{cases} y = 2x + 6 \\ 3x - 2y = -11 \end{cases}$

Substitute $2x + 6$ for $y$ in the second equation.
$3x - 2(2x + 6) = -11$
$\quad 3x - 4x - 12 = -11$
$\qquad\qquad -x = 1$
$\qquad\qquad\quad x = -1$
Let $x = -1$ in the first equation.
$y = 2(-1) + 6 = 4$
The solution is $(-1, 4)$.

**14.** $\begin{cases} y = 3x - 7 \\ 2x - 3y = 7 \end{cases}$

Substitute $3x - 7$ for $y$ in the second equation.
$2x - 3(3x - 7) = 7$
$\quad 2x - 9x + 21 = 7$
$\qquad\qquad -7x = -14$
$\qquad\qquad\quad x = 2$
Let $x = 2$ in the first equation.
$y = 3(2) - 7 = -1$
The solution is $(2, -1)$.

**15.** $\begin{cases} x + 3y = -3 \\ 2x + y = 4 \end{cases}$

Solve the first equation for $x$.
$x = -3y - 3$
Substitute $-3y - 3$ for $x$ in the second equation.
$2(-3y - 3) + y = 4$
$\quad -6y - 6 + y = 4$
$\qquad\qquad -5y = 10$
$\qquad\qquad\quad y = -2$
Let $y = -2$ in $x = -3y - 3$.
$x = -3(-2) - 3 = 3$
The solution is $(3, -2)$.

**16.** $\begin{cases} 3x + y = 11 \\ x + 2y = 12 \end{cases}$

Solve the first equation for $y$.
$y = 11 - 3x$
Substitute $11 - 3x$ for $y$ in the second equation.
$x + 2(11 - 3x) = 12$
$\quad x + 22 - 6x = 12$
$\qquad\qquad -5x = -10$
$\qquad\qquad\quad x = 2$
Let $x = 2$ in $y = 11 - 3x$.
$y = 11 - 3(2) = 5$
The solution is $(2, 5)$.

**17.** $\begin{cases} 4y = 2x + 6 \\ x - 2y = -3 \end{cases}$

Solve the second equation for $x$.
$x = 2y - 3$
Substitute $2y - 3$ for $x$ in the first equation.
$4y = 2(2y - 3) + 6$
$4y = 4y - 6 + 6$
$\ 0 = 0$
The system has an infinite number of solutions.

**18.** $\begin{cases} 9x = 6y + 3 \\ 6x - 4y = 2 \end{cases}$

Solve the first equation for $y$.
$\qquad 9x = 6y + 3$
$\quad 9x - 3 = 6y$
$\dfrac{3}{2}x - \dfrac{1}{2} = y$
Substitute $\dfrac{3}{2}x - \dfrac{1}{2}$ for $y$ in the second equation.
$6x - 4\left(\dfrac{3}{2}x - \dfrac{1}{2}\right) = 2$
$\qquad\quad 6x - 6x + 2 = 2$
$\qquad\qquad\qquad\quad 2 = 2$
The system has an infinite number of solutions.

**19.** $\begin{cases} x + y = 6 \\ y = -x - 4 \end{cases}$

Substitute $-x - 4$ for $y$ in the first equation.
$x + (-x - 4) = 6$
$\quad x - x - 4 = 6$
$\qquad\qquad -4 = 6 \ \ \text{False}$
There is no solution.

**20.** $\begin{cases} -3x + y = 6 \\ y = 3x + 2 \end{cases}$

Substitute $3x + 2$ for $y$ in the first equation.
$-3x + (3x + 2) = 6$
$\quad -3x + 3x + 2 = 6$
$\qquad\qquad\qquad 2 = 6 \ \ \text{False}$
There is no solution.

**21.** $\begin{cases} 2x + 3y = -6 \\ x - 3y = -12 \end{cases}$

$\begin{array}{r} 2x + 3y = -6 \\ x - 3y = -12 \\ \hline 3x \quad\ \ = -18 \\ x = -6 \end{array}$

Let $x = -6$ in the first equation.

$2(-6)+3y=-6$
$-12+3y=-6$
$3y=6$
$y=2$

The solution of the system is $(-6, 2)$.

**22.** $\begin{cases} 4x+y=15 \\ -4x+3y=-19 \end{cases}$

$4x \;\;\;+y=15$
$-4x+3y=-19$
$\overline{\phantom{-4x+}4y=-4}$
$y=-1$

Let $y=-1$ in the first equation.
$4x+(-1)=15$
$4x-1=15$
$4x=16$
$x=4$

The solution of the system is $(4, -1)$.

**23.** $\begin{cases} 2x-3y=-15 \\ x+4y=31 \end{cases}$

Multiply the second equation by $-2$.
$2x-3y=-15$
$-2x-8y=-62$
$\overline{\phantom{2x-}-11y=-77}$
$y=7$

Let $y=7$ in the second equation.
$x+4(7)=31$
$x+28=31$
$x=3$

The solution of the system is $(3, 7)$.

**24.** $\begin{cases} x-5y=-22 \\ 4x+3y=4 \end{cases}$

Multiply the first equation by $-4$.
$-4x+20y=88$
$4x \;\;\;+3y=4$
$\overline{\phantom{-4x+}23y=92}$
$y=4$

Let $y=4$ in the first equation.
$x-5(4)=-22$
$x-20=-22$
$x=-2$

The solution of the system is $(-2, 4)$.

**25.** $\begin{cases} 2x-6y=-1 \\ -x+3y=\dfrac{1}{2} \end{cases}$

Multiply the second equation by 2.

$2x-6y=-1$
$-2x+6y=1$
$\overline{\phantom{2x-6y}0=0}$

There is an infinite number of solutions.

**26.** $\begin{cases} 0.6x-0.3y=-1.5 \\ 0.04x-0.02y=-0.1 \end{cases}$

Multiply the first equation by 20 and the second equation by $-300$.
$12x-6y=-30$
$-12x+6y=30$
$\overline{\phantom{12x-6y}0=0}$

There are an infinite number of solutions.

**27.** $\begin{cases} \dfrac{3}{4}x+\dfrac{2}{3}y=2 \\ x+\dfrac{y}{3}=6 \end{cases}$

Multiply the first equation by 12 and the second equation by 3.
$\begin{cases} 9x+8y=24 \\ 3x+y=18 \end{cases}$

Multiply the second equation in the simplified system by $-3$.
$9x+8y=24$
$-9x-3y=-54$
$\overline{\phantom{9x+}5y=-30}$
$y=-6$

Let $y=-6$ in $3x+y=18$.
$3x+(-6)=18$
$3x=24$
$x=8$

The solution of the system is $(8, -6)$.

**28.** $\begin{cases} 10x+2y=0 \\ 3x+5y=33 \end{cases}$

Multiply the first equation by $-5$ and the second equation by 2.
$-50x-10y=0$
$6x+10y=66$
$\overline{-44x\phantom{+10y}=66}$
$x=-\dfrac{3}{2}$

Let $x=-\dfrac{3}{2}$ in the first equation.

$$10\left(-\frac{3}{2}\right)+2y=0$$

$$-15+2y=0$$

$$2y=15$$

$$y=\frac{15}{2}$$

The solution is $\left(-\frac{3}{2},\frac{15}{2}\right)$.

29. Let $x$ = the larger number and $y$ = the smaller number.

$$\begin{cases} x+y=16 \\ 3x-y=72 \end{cases}$$

$$\begin{array}{r} x+y=16 \\ 3x-y=72 \\ \hline 4x\quad=88 \end{array}$$

$$x=22$$

Let $x = 22$ in the first equation.

$$22+y=16$$

$$y=-6$$

The numbers are $-6$ and $22$.

30. Let $x$ = the number of orchestra seats and $y$ = the number of balcony seats.

$$\begin{cases} x+y=360 \\ 45x+35y=15,150 \end{cases}$$

Solve the first equation for $x$.

$$x=360-y$$

Substitute $360-y$ for $x$ in the second equation.

$$45(360-y)+35y=15,150$$

$$16,200-45y+35y=15,150$$

$$-10y=-1050$$

$$y=105$$

Let $y=105$ in $x=360-y$.

$$x=360-105=255$$

There are 255 orchestra seats and 105 balcony seats.

31. Let $x$ = the riverboat's speed in still water and $y$ = the rate of the current.

|          | $d$ | = | $r$   | · | $t$ |
|----------|-----|---|-------|---|-----|
| Downriver | 340 |   | $x+y$ |   | 14  |
| Upriver  | 340 |   | $x-y$ |   | 19  |

$$\begin{cases} 14(x+y)=340 \\ 19(x-y)=340 \end{cases}$$

Multiply the first equation by $\frac{1}{14}$ and the second equation by $\frac{1}{19}$.

$$\begin{array}{r} x+y=\dfrac{340}{14}\approx 24.29 \\ x-y=\dfrac{340}{19}\approx 17.89 \\ \hline 2x\qquad\approx 42.18 \end{array}$$

$$x\approx 21.09$$

Multiply the second equation of the simplified system by $-1$.

$$\begin{array}{r} x+y\approx 24.29 \\ -x+y\approx -17.89 \\ \hline 2y\approx 6.4 \end{array}$$

$$y\approx 3.2$$

The riverboat's speed in still water is 21.1 miles per hour. The rate of the current is 3.2 miles per hour.

32. Let $x$ = amount of 6% solution and $y$ = amount of 14% solution.

| Concentration Rate | Amount of Solution | Amount of Pure Acid |
|--------------------|--------------------|---------------------|
| 0.06               | $x$                | $0.06x$             |
| 0.14               | $y$                | $0.14y$             |
| 0.12               | 50                 | $0.12(50)$          |

$$\begin{cases} x+y=50 \\ 0.06x+0.14y=0.12(50) \end{cases}$$

Multiply the first equation by $-6$ and the second equation by 100.

$$\begin{array}{r} -6x-6y=-300 \\ 6x+14y=600 \\ \hline 8y=300 \end{array}$$

$$y=37.5$$

Let $y=37.5$ in the first equation.

$$x+37.5=50$$

$$x=12.5$$

$12\dfrac{1}{2}$ cc of 6% solution and $37\dfrac{1}{2}$ cc of 14% solution.

**33.** Let $x =$ the cost of an egg and
$y =$ the cost of a strip of bacon.
$$\begin{cases} 3x + 4y = 3.80 \\ 2x + 3y = 2.75 \end{cases}$$
Multiply the first equation by $-2$ and the second equation by 3.
$$-6x - 8y = -7.60$$
$$\underline{6x + 9y = 8.25}$$
$$y = 0.65$$
Let $y = 0.65$ in the first equation.
$$3x + 4(0.65) = 3.80$$
$$3x + 2.60 = 3.80$$
$$3x = 1.20$$
$$x = 0.40$$
An egg costs 40¢ and a strip of bacon costs 65¢.

**34.** Let $x =$ the time spent walking and
$y =$ the time spent jogging.

| | $r$ | $\cdot$ $t$ | $= d$ |
|---|---|---|---|
| Walking | 4 | $x$ | $4x$ |
| Jogging | 7.5 | $y$ | $7.5y$ |

$$\begin{cases} x + y = 3 \\ 4x + 7.5y = 15 \end{cases}$$
Multiply the first equation by $-4$.
$$-4x - 4y = -12$$
$$\underline{4x + 7.5y = 15}$$
$$3.5y = 3$$
$$y \approx 0.857$$
Let $y = 0.857$ in the first equation.
$$x + 0.857 = 3$$
$$x \approx 2.143$$
He spent 2.14 hours walking and 0.86 hours jogging.

**35.** $5x + 4y < 20$
Test $(0, 0)$
$$0 + 0 \overset{?}{<} 20$$
True
Shade below.

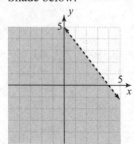

**36.** $x + 3y > 4$
Test $(0, 0)$
$$0 + 0 \overset{?}{>} 4$$
False
Shade above.

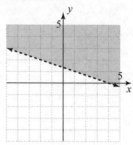

**37.** $y \geq -7$
Shade above.

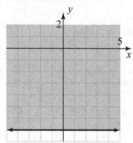

**38.** $y \leq -4$
Shade below.

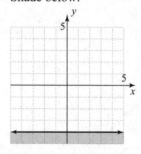

**39.** $-x \leq y$
Test $(1, 0)$
$$-1 \overset{?}{\leq} 0$$
True
Shade above.

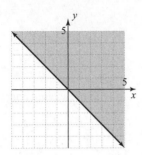

**40.** $x \geq -y$
Test $(1, 0)$
$\overset{?}{1 \geq 0}$
True
Shade above.

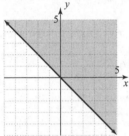

**41.** $\begin{cases} y \geq 2x - 3 \\ y \leq -2x + 1 \end{cases}$
$y \geq 2x - 3$
$y \leq -2x + 1$

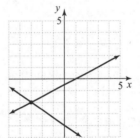

**42.** $\begin{cases} y \leq -3x - 3 \\ y \leq 2x + 7 \end{cases}$

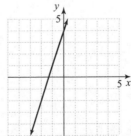
$y \leq -3x - 3$
$y \leq 2x + 7$

**43.** $\begin{cases} -3x + 2y > -1 \\ y < -2 \end{cases}$

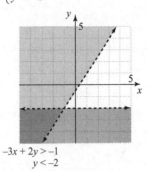
$-3x + 2y > -1$
$y < -2$

**44.** $\begin{cases} -2x + 3y > -7 \\ x \geq -2 \end{cases}$

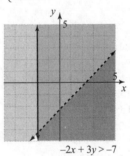
$-2x + 3y > -7$
$x \geq -2$

**45.** $\begin{cases} x - 2y = 1 \\ 2x + 3y = -12 \end{cases}$

The solution is $(-3, -2)$.

**46.** $\begin{cases} 3x - y = -4 \\ 6x - 2y = -8 \end{cases}$

There is an infinite number of solutions.

**47.** $\begin{cases} x + 4y = 11 \\ 5x - 9y = -3 \end{cases}$

Solve the first equation for $x$.

$x = 11 - 4y$

Substitute $11 - 4y$ for $x$ in the second equation.

$5(11 - 4y) - 9y = -3$

$55 - 20y - 9y = -3$

$-29y = -58$

$y = 2$

Let $y = 2$ in the first equation.

$x + 4(2) = 11$

$x + 8 = 11$

$x = 3$

The solution is (3, 2).

**48.** $\begin{cases} x + 9y = 16 \\ 3x - 8y = 13 \end{cases}$

Solve the first equation for $x$.

$x = 16 - 9y$

Substitute $16 - 9y$ for $x$ in the second equation.

$3(16 - 9y) - 8y = 13$

$48 - 27y - 8y = 13$

$-35y = -35$

$y = 1$

Let $y = 1$ in the first equation.

$x + 9(1) = 16$

$x + 9 = 16$

$x = 7$

The solution is (7, 1).

**49.** $\begin{cases} y = -2x \\ 4x + 7y = -15 \end{cases}$

Substitute $-2x$ for $y$ in the second equation.

$4x + 7(-2x) = -15$

$4x - 14x = -15$

$-10x = -15$

$x = \dfrac{3}{2} = 1\dfrac{1}{2}$

Let $x = \dfrac{3}{2}$ in the first equation.

$y = -2\left(\dfrac{3}{2}\right) = -3$

The solution is $\left(1\dfrac{1}{2}, -3\right)$.

**50.** $\begin{cases} 3y = 2x + 15 \\ -2x + 3y = 21 \end{cases}$

Solve the first equation for $x$.

$3y = 2x + 15$

$3y - 15 = 2x$

$\dfrac{3}{2}y - \dfrac{15}{2} = x$

Substitute $\dfrac{3}{2}y - \dfrac{15}{2}$ for $x$ in the second equation.

$-2\left(\dfrac{3}{2}y - \dfrac{15}{2}\right) + 3y = 21$

$-3y + 15 + 3y = 21$

$15 = 21$    False

The system has no solution.

**51.** $\begin{cases} 3x - y = 4 \\ 4y = 12x - 16 \end{cases}$

Solve the first equation for $y$.

$3x - 4 = y$

Substitute $3x - 4$ for $y$ in the second equation.

$4(3x - 4) = 12x - 16$

$12x - 16 = 12x - 16$

$0 = 0$

There is an infinite number of solutions.

**52.** $\begin{cases} x + y = 19 \\ x - y = -3 \end{cases}$

$\begin{array}{r} x + y = 19 \\ \underline{x - y = -3} \\ 2x \quad\quad = 16 \\ x = 8 \end{array}$

Let $x = 8$ in the first equation.

$8 + y = 19$

$y = 11$

The solution is (8, 11).

**53.** $\begin{cases} x - 3y = -11 \\ 4x + 5y = -10 \end{cases}$

Solve the first equation for $x$.

$x = 3y - 11$

Substitute $3y - 11$ for $x$ in the second equation.

$4(3y - 11) + 5y = -10$

$12y - 44 + 5y = -10$

$17y = 34$

$y = 2$

Let $y = 2$ in the first equation.

$x - 3(2) = -11$

$x - 6 = -11$

$x = -5$

The solution is (−5, 2).

**54.** $\begin{cases} -x - 15y = 44 \\ 2x + 3y = 20 \end{cases}$

Solve the first equation for $x$.
$-x - 15y = 44$
$\quad -x = 15y + 44$
$\quad\quad x = -15y - 44$

Substitute $-15y - 44$ for $x$ in the second equation.
$2(-15y - 44) + 3y = 20$
$\quad -30y - 88 + 3y = 20$
$\quad\quad\quad\quad -27y = 108$
$\quad\quad\quad\quad\quad\quad y = -4$

Let $y = -4$ in $x = -15y - 44$.
$x = -15(-4) - 44 = 60 - 44 = 16$
The solution is $(16, -4)$.

**55.** $\begin{cases} 2x + y = 3 \\ 6x + 3y = 9 \end{cases}$

Solve the first equation for $y$.
$y = -2x + 3$
Substitute $-2x + 3$ for $y$ in the second equation.
$6x + 3(-2x + 3) = 9$
$\quad 6x - 6x + 9 = 9$
$\quad\quad\quad\quad 9 = 9$
There is an infinite number of solutions.

**56.** $\begin{cases} -3x + y = 5 \\ -3x + y = -2 \end{cases}$

Multiply the first equation by $-1$.
$\quad 3x - y = -5$
$\underline{-3x + y = -2}$
$\quad\quad 0 = -7$   False
There is no solution.

**57.** Let $x$ = the larger number and $y$ = the smaller number.
$\begin{cases} x + y = 12 \\ x + 3y = 20 \end{cases}$

Multiply the first equation by $-1$.
$\quad -x - y = -12$
$\underline{\quad x + 3y = 20}$
$\quad\quad 2y = 8$
$\quad\quad\; y = 4$

Let $y = 4$ in the first equation.
$x + 4 = 12$
$\quad\; x = 8$
The numbers are 4 and 8.

**58.** Let $x$ = the smaller number and $y$ = the larger number.
$\begin{cases} x - y = -18 \\ 2x - y = -23 \end{cases}$

Multiply the first equation by $-1$.
$\quad -x + y = 18$
$\underline{\;\; 2x - y = -23}$
$\quad\; x \quad\quad = -5$

Let $x = -5$ in the first equation.
$-5 - y = -18$
$\quad\; -y = -13$
$\quad\quad y = 13$
The numbers are $-5$ and 13.

**59.** Let $x$ = the number of nickels and $y$ = the number of dimes.
$\begin{cases} x + y = 65 \\ 0.05x + 0.10y = 5.30 \end{cases}$

Multiply the first equation by $-5$ and the second equation by 100.
$\quad -5x - 5y = -325$
$\underline{\quad\; 5x + 10y = 530}$
$\quad\quad\quad 5y = 205$
$\quad\quad\quad\; y = 41$

Let $y = 41$ in the first equation.
$x + 41 = 65$
$\quad\quad x = 24$
There are 24 nickels and 41 dimes.

**60.** Let $x$ = the number of 13¢ stamps and $y$ = the number of 22¢ stamps.
$\begin{cases} x + y = 26 \\ 0.13x + 0.22y = 4.19 \end{cases}$

Multiply the first equation by $-13$ and the second equation by 100.
$\quad -13x - 13y = -338$
$\underline{\quad\; 13x + 22y = 419}$
$\quad\quad\quad 9y = 81$
$\quad\quad\quad\; y = 9$

Let $y = 9$ in the first equation.
$x + 9 = 26$
$\quad\; x = 17$
They purchased 17 13¢ stamps and 9 22¢ stamps.

**61.** $x + 6y < 6$
Test $(0, 0)$
$$0 + 6(0) \overset{?}{<} 6$$
True
Shade below.

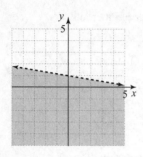

**62.** $x + y > -2$

Test $(0, 0)$

$$0 + 0 \overset{?}{>} -2$$

True

Shade above.

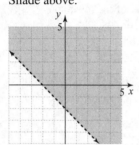

**Chapter 4 Test**

1. False; one solution, infinitely many solutions, or no solutions are the only possibilities.

2. False; a solution of a system of equations must be a solution of each equation in the system.

3. True

4. False; $x = 0$ is part of the solution.

5. Let $x = 1$ and $y = -1$.

| | |
|---|---|
| $2x - 3y = 5$ | $6x + y = 1$ |
| $2(1) - 3(-1) \overset{?}{=} 5$ | $6(1) + (-1) \overset{?}{=} 1$ |
| $2 + 3 \overset{?}{=} 5$ | $6 - 1 \overset{?}{=} 1$ |
| $5 = 5$   True | $5 = 1$   False |

$(1, -1)$ is not a solution of the system.

6. Let $x = 3$ and $y = -4$.

$$4x - 3y = 24$$
$$4(3) - 3(-4) \overset{?}{=} 24$$
$$12 + 12 \overset{?}{=} 24$$
$$24 = 24 \quad \text{True}$$

$$4x + 5y = -8$$
$$4(3) + 5(-4) \overset{?}{=} -8$$
$$12 - 20 \overset{?}{=} -8$$
$$-8 = -8 \quad \text{True}$$

$(3, -4)$ is a solution of the system.

7. $\begin{cases} x - y = 2 \\ 3x - y = -2 \end{cases}$

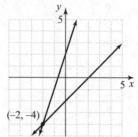

The solution of the system is $(-2, -4)$.

8. $\begin{cases} y = -3x \\ 3x + y = 6 \end{cases}$

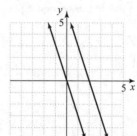

There is no solution.

9. $\begin{cases} 3x - 2y = -14 \\ y = x + 5 \end{cases}$

Substitute $x + 5$ for $y$ in the first equation.
$$3x - 2(x + 5) = -14$$
$$3x - 2x - 10 = -14$$
$$x = -4$$
Let $x = -4$ in $y = x + 5$.
$$y = -4 + 5 = 1$$
The solution is $(-4, 1)$.

10. $\begin{cases} \dfrac{1}{2}x + 2y = -\dfrac{15}{4} \\ 4x = -y \end{cases}$

Solve the second equation for $y$.
$$y = -4x$$
Substitute $-4x$ for $y$ in the first equation.

$$\frac{1}{2}x + 2(-4x) = -\frac{15}{4}$$

$$\frac{1}{2}x - 8x = -\frac{15}{4}$$

$$-\frac{15}{2}x = -\frac{15}{4}$$

$$x = \frac{1}{2}$$

Let $x = \frac{1}{2}$ in the equation $y = -4x$.

$$y = -4\left(\frac{1}{2}\right) = -2$$

The solution is $\left(\frac{1}{2}, -2\right)$.

**11.** $\begin{cases} x + y = 28 \\ x - y = 12 \end{cases}$

$$\begin{array}{r} x + y = 28 \\ x - y = 12 \\ \hline 2x \quad\quad = 40 \\ x = 20 \end{array}$$

Let $x = 20$ in the first equation.

$$20 + y = 28$$
$$y = 8$$

The solution is (20, 8).

**12.** $\begin{cases} 4x - 6y = 7 \\ -2x + 3y = 0 \end{cases}$

Multiply the second equation by 2.

$$\begin{array}{r} 4x - 6y = 7 \\ -4x + 6y = 0 \\ \hline 0 = 7 \end{array}$$

The system is inconsistent. There is no solution.

**13.** $\begin{cases} 3x + y = 7 \\ 4x + 3y = 1 \end{cases}$

Solve the first equation for $y$.

$$y = 7 - 3x$$

Substitute $7 - 3x$ for $y$ in the second equation.

$$4x + 3(7 - 3x) = 1$$
$$4x + 21 - 9x = 1$$
$$-5x = -20$$
$$x = 4$$

Let $x = 4$ in $y = 7 - 3x$.

$$y = 7 - 3(4) = -5$$

The solution is (4, −5).

**14.** $\begin{cases} 3(2x + y) = 4x + 20 \\ \quad 6x + 3y = 4x + 20 \\ \quad 2x + 3y = 20 \\ \\ \quad x - 2y = 3 \end{cases}$

Multiply the second equation by −2.

$$\begin{array}{r} 2x + 3y = 20 \\ -2x + 4y = -6 \\ \hline 7y = 14 \\ y = 2 \end{array}$$

Let $y = 2$ in the second equation.

$$x - 2(2) = 3$$
$$x - 4 = 3$$
$$x = 7$$

The solution of the system is (7, 2).

**15.** $\begin{cases} \dfrac{x-3}{2} = \dfrac{2-y}{4} \\ \dfrac{7-2x}{3} = \dfrac{y}{2} \end{cases}$

Multiply the first equation by 4 and the second equation by 6.

$$\begin{cases} 2(x - 3) = 2 - y \\ \quad 2x - 6 = 2 - y \\ \quad 2x + y = 8 \\ \\ 2(7 - 2x) = 3y \\ \quad 14 - 4x = 3y \\ \quad 4x + 3y = 14 \end{cases}$$

Multiply the first equation by −3.

$$\begin{array}{r} -6x - 3y = -24 \\ 4x + 3y = 14 \\ \hline -2x \quad\quad = -10 \\ x = 5 \end{array}$$

Let $x = 5$ in the first equation.

$$2(5) + y = 8$$
$$10 + y = 8$$
$$y = -2$$

The solution of the system is (5, −2).

**16.** $\begin{cases} 8x - 4y = 12 \\ y = 2x - 3 \end{cases}$

Substitute $2x - 3$ for $y$ in the first equation.

$$8x - 4(2x - 3) = 12$$
$$8x - 8x + 12 = 12$$
$$12 = 12$$

There is an infinite number of solutions.

**17.** $\begin{cases} 0.01x - 0.06y = -0.23 \\ 0.2x + 0.4y = 0.2 \end{cases}$

Multiply the first equation by 100 and the second equation by –5.

$x - 6y = -23$
$\underline{-x - 2y = -1}$
$\phantom{x} -8y = -24$
$\phantom{xxx} y = 3$

Let $y = 3$ in $x - 6y = -23$.
$x - 6(3) = -23$
$x - 18 = -23$
$\phantom{xx} x = -5$

The solution is (–5, 3).

**18.** $\begin{cases} x - \dfrac{2}{3}y = 3 \\ -2x + 3y = 10 \end{cases}$

Multiply the first equation by 9 and the second equation by 2.

$9x - 6y = 27$
$\underline{-4x + 6y = 20}$
$5x \phantom{xxxx} = 47$
$\phantom{xxx} x = \dfrac{47}{5}$

Let $x = \dfrac{47}{5}$ in the first equation.

$\dfrac{47}{5} - \dfrac{2}{3}y = 3$
$141 - 10y = 45$
$\phantom{xx} -10y = -96$
$\phantom{xxxx} y = \dfrac{48}{5}$

The solution is $\left( \dfrac{47}{5}, \dfrac{48}{5} \right)$.

**19.** Let $x$ = the larger number and $y$ = the smaller number.
$\begin{cases} x + y = 124 \\ x - y = 32 \end{cases}$

$x + y = 124$
$\underline{x - y = 32}$
$2x \phantom{xxx} = 156$
$\phantom{xx} x = 78$

Let $x = 78$ in the first equation.
$78 + y = 124$
$\phantom{xxx} y = 46$

The numbers are 78 and 46.

**20.** Let $x$ = cc's of 12% solution and $y$ = cc's of 16% solution.

| Concentration Rate | cc's of Solution | cc's of salt |
|---|---|---|
| 12% | $x$ | $0.12x$ |
| 22% | 80 | $0.22(80)$ |
| 16% | $y$ | $0.16y$ |

$\begin{cases} x + 80 = y \\ 0.12x + 0.22(80) = 0.16y \end{cases}$

Multiply the first equation by –16 and the second equation by 100.

$-16x - 1280 = -16y$
$\underline{12x + 1760 = 16y}$
$-4x + 480 = 0$
$\phantom{xx} -4x = -480$
$\phantom{xxx} x = 120$

Should add 120 cc's of 12% solution

**21.** Let $x$ = the number of thousands of farms in Texas and $y$ = the number of thousands of farms in Missouri.
$\begin{cases} x + y = 356 \\ x - y = 140 \end{cases}$

$x + y = 356$
$\underline{x - y = 140}$
$2x \phantom{xxx} = 496$
$\phantom{xx} x = 248$

Let $x = 248$ in the first equation.
$248 + y = 356$
$\phantom{xxx} y = 108$

There are 248 thousand farms in Texas and 108 thousand farms in Missouri.

**22.** Let $x$ = the speed of the faster hiker and $y$ = the speed of the slower hiker.

| | $r$ | $\cdot$ | $t$ | $=$ | $d$ |
|---|---|---|---|---|---|
| Faster | $x$ | | 4 | | $4x$ |
| Slower | $y$ | | 4 | | $4y$ |

$\begin{cases} 4x + 4y = 36 \\ x = 2y \end{cases}$

Substitute $2y$ for $x$ in the first equation.

$$4(2y) + 4y = 36$$
$$8y + 4y = 36$$
$$12y = 36$$
$$y = 3$$

Let $y = 3$ in the second equation.

$$x = 2(3) = 6$$

The speeds are 3 miles per hour and 6 miles per hour.

23. $x - y \geq -2$
Test $(0, 0)$
$0 - 0 \geq -2$
True
Shade below.

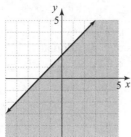

24. $y \geq -4x$
Test $(1, 0)$
$$0 \overset{?}{\geq} -4(1)$$
True
Shade above.

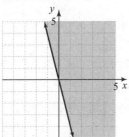

25. $2x - 3y > -6$
Test $(0, 0)$
$$2(0) - 3(0) \overset{?}{>} -6$$
True
Shade below.

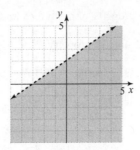

26. $\begin{cases} y + 2x \leq 4 \\ y \geq 2 \end{cases}$

$y + 2x \leq 4$
$y \geq 2$

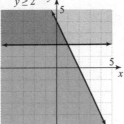

27. $\begin{cases} 2y - x \geq 1 \\ x + y \geq -4 \end{cases}$

$2y - x \geq 1$
$x + y \geq -4$

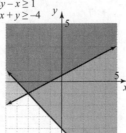

**Chapter 4 Cumulative Review**

1. **a.** $-1 < 0$

   **b.** $7 = \dfrac{14}{2}$

   **c.** $-5 > -6$

2. **a.** $5^2 = 5 \cdot 5 = 25$

   **b.** $2^5 = 2 \cdot 2 \cdot 2 \cdot 2 \cdot 2 = 32$

3. **a.** commutative property of multiplication

   **b.** associative property of addition

   **c.** identity element for addition

    

**d.**   commutative property of multiplication

**e.**   multiplicative inverse property

**f.**   additive inverse property

**g.**   commutative and associative properties of multiplication

**4.**  Let $x = 8$, $y = 5$.
$$y^2 - 3x = 5^2 - 3(8) = 25 - 24 = 1$$

**5.**  $(2x - 3) - (4x - 2) = 2x - 3 - 4x + 2 = -2x - 1$

**6.**  $7 - 12 + (-5) - 2 + (-2)$
$$= 7 + (-12) + (-5) + (-2) + (-2)$$
$$= 7 + (-21)$$
$$= -14$$

**7.**  $5t - 5 = 6t + 2$
$$-t - 5 = 2$$
$$-t = 7$$
$$t = -7$$

**8.**  Let $x = -7$, $y = -3$.
$$2y^2 - x^2 = 2(-3)^2 - (-7)^2$$
$$= 2(9) - 49$$
$$= 18 - 49$$
$$= -31$$

**9.**   $\dfrac{5}{2}x = 15$
$$\dfrac{2}{5} \cdot \dfrac{5}{2}x = \dfrac{2}{5} \cdot 15$$
$$x = 6$$

**10.**  $0.4y - 6.7 + y - 0.3 - 2.6y$
$$= 0.4y + y + (-2.6y) + (-6.7) + (-0.3)$$
$$= -1.2y - 7$$

**11.**   $\dfrac{x}{2} - 1 = \dfrac{2}{3}x - 3$
$$6\left(\dfrac{x}{2} - 1\right) = 6\left(\dfrac{2}{3}x - 3\right)$$
$$3x - 6 = 4x - 18$$
$$-x - 6 = -18$$
$$-x = -12$$
$$x = 12$$

**12.**  $7(x - 2) - 6(x + 1) = 20$
$$7x - 14 - 6x - 6 = 20$$
$$x - 20 = 20$$
$$x = 40$$

**13.**  Let $x$ = the number.
$$2(x + 4) = 4x - 12$$
$$2x + 8 = 4x - 12$$
$$-2x + 8 = -12$$
$$-2x = -20$$
$$x = 10$$
The number is 10.

**14.**  $5(y - 5) = 5y + 10$
$$5y - 25 = 5y + 10$$
$$-25 = 10$$
False statement; there is no solution.

**15.**      $y = mx + b$
$$y - b = mx + b - b$$
$$y - b = mx$$
$$\dfrac{y - b}{m} = \dfrac{mx}{m}$$
$$\dfrac{y - b}{m} = x$$

**16.**  Let $x$ = the number.
$$5(x - 1) = 6x$$
$$5x - 5 = 6x$$
$$-x - 5 = 0$$
$$-x = 5$$
$$x = -5$$
The number is $-5$.

**17.**  $-2x \le -4$
$$\dfrac{-2x}{-2} \ge \dfrac{-4}{-2}$$
$$x \ge 2, [2, \infty)$$

**18.**      $P = a + b + c$
$$P - a - c = a + b + c - a - c$$
$$P - a - c = b$$

**19.**  $x = -2y$

| $x$ | $y$ |
| --- | --- |
| 0 | 0 |
| $-4$ | 2 |

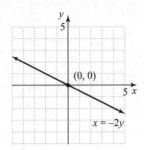

**20.** $3x + 7 \geq x - 9$

$\quad 2x + 7 \geq -9$

$\qquad 2x \geq -16$

$\qquad x \geq -8, [-8, \infty)$

**21.** $(-1, 5)$ and $(2, -3)$

$\quad m = \dfrac{y_2 - y_1}{x_2 - x_1} = \dfrac{-3 - 5}{2 - (-1)} = \dfrac{-8}{3} = -\dfrac{8}{3}$

**22.** $x - 3y = 3$

| $x$ | $y$ |
|-----|-----|
| 0 | −1 |
| 3 | 0 |
| 9 | 2 |

**23.** $y = \dfrac{3}{4}x + 6$

$\quad y = mx + b$

$\quad m = \dfrac{3}{4}$

**24.** $(-1, 3)$ and $(2, -8)$

$\quad m = \dfrac{y_2 - y_1}{x_2 - x_1} = \dfrac{-8 - 3}{2 - (-1)} = -\dfrac{11}{3}$

A parallel line has the same slope.

Slope is $-\dfrac{11}{3}$.

**25.** $3x - 4y = 4$

$\qquad -4y = -3x + 4$

$\qquad\quad y = \dfrac{-3x}{-4} + \dfrac{4}{-4}$

$\qquad\quad y = \dfrac{3}{4}x - 1$

$\quad y = mx + b$

$\quad m = \dfrac{3}{4}, \ b = -1$

Slope is $\dfrac{3}{4}$, $y$-intercept is $(0, -1)$.

**26.** $y = 7x + 0$

$\quad y = mx + b$

$\quad m = 7, b = 0$

Slope is 7, $y$-intercept is $(0, 0)$.

**27.** $m = -2$, with point $(-1, 5)$

$\quad y - y_1 = m(x - x_1)$

$\quad\ y - 5 = -2[x - (-1)]$

$\quad\ y - 5 = -2x - 2$

$\quad 2x + y = 3$

**28.** Line: $y = 4x - 5 \Rightarrow m_1 = 4$

Line 2: $-4x + y = 7 \Rightarrow y = 4x + 7 \Rightarrow m_2 = 4$

$\quad m_2 = m_1$

The lines are parallel.

**29.** A vertical line has an equation $x = c$.

Point, $(-1, 5)$

$x = -1$

**30.** $m = -5$, with point $(-2, 3)$

$\quad y - y_1 = m(x - x_1)$

$\quad\ y - 3 = -5[x - (-2)]$

$\quad\ y - 3 = -5x - 10$

$\qquad y = -5x - 7$

**31.** Domain is $\{-1, 0, 3\}$

Range is $\{-2, 0, 2, 3\}$

**32.** $f(x) = 5x^2 - 6$

$\quad f(0) = 5(0)^2 - 6 = -6$

$\quad f(-2) = 5(-2)^2 - 6 = 5(4) - 6 = 14$

**33. a.** function

**b.** not a function

**34. a.** not a function

**b.** function

**c.** not a function

**35.** $\begin{cases} 3x - y = 4 \\ y = 3x - 4, \ m = 3 \\[1em] x + 2y = 8 \\ y = -\dfrac{1}{2}x + 4, \ m = -\dfrac{1}{2} \end{cases}$

Because they have different slopes, there is only one solution.

**36. a.** Let $x = 1$ and $y = -4$.
$$2x - y = 6$$
$$2(1) - (-4) \stackrel{?}{=} 6$$
$$2 + 4 \stackrel{?}{=} 6$$
$$6 = 6 \quad \text{True}$$

$$3x + 2y = -5$$
$$3(1) + 2(-4) \stackrel{?}{=} -5$$
$$3 - 8 \stackrel{?}{=} -5$$
$$-5 = -5 \quad \text{True}$$
$(1, -4)$ is a solution of the system.

**b.** Let $x = 0$ and $y = 6$.
$$2x - y = 6 \qquad 3x + 2y = -5$$
$$2(0) - (6) \stackrel{?}{=} 6 \qquad \text{Test not needed}$$
$$0 - 6 \stackrel{?}{=} 6$$
$$-6 = 6 \quad \text{False}$$
$(0, 6)$ is not a solution of the system.

**c.** Let $x = 3$ and $y = 0$.
$$2x - y = 6$$
$$2(3) - (0) \stackrel{?}{=} 6$$
$$6 - 0 \stackrel{?}{=} 6$$
$$6 = 6 \quad \text{True}$$

$$3x + 2y = -5$$
$$3(3) + 2(0) \stackrel{?}{=} -5$$
$$9 + 0 \stackrel{?}{=} -5$$
$$9 = -5 \quad \text{False}$$
$(3, 0)$ is not a solution of the system.

**37.** $\begin{cases} x + 2y = 7 \\ 2x + 2y = 13 \end{cases}$

Solve the first equation for $x$.
$$x = 7 - 2y$$
Substitute $7 - 2y$ for $x$ in the second equation.
$$2(7 - 2y) + 2y = 13$$
$$14 - 4y + 2y = 13$$
$$-2y = -1$$
$$y = \frac{1}{2}$$

Let $y = \frac{1}{2}$ in $x = 7 - 2y$.

$$x = 7 - 2\left(\frac{1}{2}\right) = 6$$

The solution is $\left(6, \dfrac{1}{2}\right)$.

**38.** $\begin{cases} 3x - 4y = 10 \\ y = 2x \end{cases}$

Substitute $2x$ for $y$ in the first equation.

$$3x - 4(2x) = 10$$
$$3x - 8x = 10$$
$$-5x = 10$$
$$x = -2$$
Let $x = -2$ in the second equation.
$$y = 2(-2) = -4$$
The solution is $(-2, -4)$.

**39.** $\begin{cases} x + y = 7 \\ x - y = 5 \end{cases}$

$$\begin{array}{r} x + y = 7 \\ x - y = 5 \\ \hline 2x \quad\;\; = 12 \\ x = 6 \end{array}$$

Let $x = 6$ in the first equation.
$$6 + y = 7$$
$$y = 1$$
The solution to the system is $(6, 1)$.

**40.** $\begin{cases} x = 5y - 3 \\ x = 8y + 4 \end{cases}$

Substitute $8y + 4$ for $x$ in the first equation.
$$8y + 4 = 5y - 3$$
$$3y + 4 = -3$$
$$3y = -7$$
$$y = -\frac{7}{3}$$

Let $y = -\dfrac{7}{3}$ in the second equation.

$$x = 8\left(-\frac{7}{3}\right) + 4$$
$$x = -\frac{56}{3} + \frac{12}{3}$$
$$x = -\frac{44}{3}$$

The solution is $\left(-\dfrac{44}{3}, -\dfrac{7}{3}\right)$.

**41.** Let $x =$ the first number and $y =$ the second number.
$$\begin{cases} x + y = 37 \\ x - y = 21 \end{cases}$$

$$\begin{array}{r} x + y = 37 \\ x - y = 21 \\ \hline 2x \quad\;\; = 58 \\ x = 29 \end{array}$$

Let $x = 29$ in the first equation.

$29 + y = 37$

$y = 8$

The numbers are 29 and 8.

**42.** $x > 1$

Shade right.

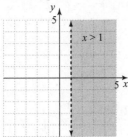

**43.** $2x - y \geq 3$

Test $(0, 0)$

$2(0) - 0 \overset{?}{\geq} 3$

False

Shade below.

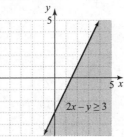

**44.** $\begin{cases} 2x + 3y < 6 \\ y < 2 \end{cases}$

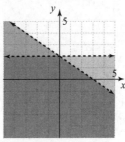

$2x + 3x < 6$

$y < 2$

# Chapter 5

**1. a.** $3^3 = 3 \cdot 3 \cdot 3 = 27$

**b.** Use 4 as a factor once, $4^1 = 4$

**c.** $(-8)^2 = (-8)(-8) = 64$

**d.** $-8^2 = -(8 \cdot 8) = -64$

**e.** $\left(\dfrac{3}{4}\right)^3 = \dfrac{3}{4} \cdot \dfrac{3}{4} \cdot \dfrac{3}{4} = \dfrac{27}{64}$

**f.** $(0.3)^4 = (0.3)(0.3)(0.3)(0.3) = 0.0081$

**g.** $3 \cdot 5^2 = 3 \cdot 25 = 75$

**2. a.** If $x$ is 3, $3x^4 = 3 \cdot (3)^4$
$$= 3 \cdot (3 \cdot 3 \cdot 3 \cdot 3)$$
$$= 3 \cdot 81$$
$$= 243$$

**b.** If $x$ is $-4$, $\dfrac{6}{x^2} = \dfrac{6}{(-4)^2} = \dfrac{6}{(-4)(-4)} = \dfrac{6}{16} = \dfrac{3}{8}$

**3. a.** $3^4 \cdot 3^6 = 3^{4+6} = 3^{10}$

**b.** $y^3 \cdot y^2 = y^{3+2} = y^5$

**c.** $z \cdot z^4 = z^1 \cdot z^4 = z^{1+4} = z^5$

**d.** $x^3 \cdot x^2 \cdot x^6 = x^{3+2+6} = x^{11}$

**e.** $(-2)^5 \cdot (-2)^3 = (-2)^{5+3} = (-2)^8$

**f.** $b^3 \cdot t^5$, cannot be simplified because $b$ and $t$ are different bases.

**4.** $(-5y^3)(-3y^4) = -5 \cdot y^3 \cdot -3 \cdot y^4$
$$= -5 \cdot -3 \cdot y^3 \cdot y^4$$
$$= 15y^7$$

**5. a.** $(y^7 z^3)(y^5 z) = (y^7 \cdot y^5) \cdot (z^3 \cdot z^1)$
$$= y^{12} \cdot z^4 \text{ or } y^{12} z^4$$

**b.** $(-m^4 n^4)(7mn^{10})$
$$= (-1 \cdot 7) \cdot (m^4 \cdot m^1) \cdot (n^4 \cdot n^{10})$$
$$= (-7) \cdot (m^5) \cdot (n^{14}) \text{ or } -7m^5 n^{14}$$

**6. a.** $(z^3)^7 = z^{3 \cdot 7} = z^{21}$

**b.** $(4^9)^2 = 4^{9 \cdot 2} = 4^{18}$

**c.** $[(-2)^3]^5 = (-2)^{3 \cdot 5} = (-2)^{15}$

**7. a.** $(pr)^5 = p^5 \cdot r^5 = p^5 r^5$

**b.** $(6b)^2 = 6^2 \cdot b^2 = 36b^2$

**c.** $\left(\dfrac{1}{4} x^2 y\right)^3 = \left(\dfrac{1}{4}\right)^3 \cdot (x^2)^3 \cdot y^3$
$$= \dfrac{1}{64} \cdot x^6 \cdot y^3$$
$$= \dfrac{1}{64} x^6 y^3$$

**d.** $(-3a^3 b^4 c)^4 = (-3)^4 \cdot (a^3)^4 \cdot (b^4)^4 \cdot c^4$
$$= 81a^{12} b^{16} c^4$$

**8. a.** $\left(\dfrac{x}{y^2}\right)^5 = \dfrac{x^5}{(y^2)^5} = \dfrac{x^5}{y^{10}}, \; y \neq 0$

**b.** $\left(\dfrac{2a^4}{b^3}\right)^5 = \dfrac{2^5 \cdot (a^4)^5}{(b^3)^5} = \dfrac{32a^{20}}{b^{15}}, \; b \neq 0$

**9. a.** $\dfrac{z^8}{z^4} = z^{8-4} = z^4$

**b.** $\dfrac{(-5)^5}{(-5)^3} = (-5)^{5-3} = (-5)^2 = 25$

**c.** $\dfrac{8^8}{8^6} = 8^{8-6} = 8^2 = 64$

**d.** $\dfrac{q^5}{t^2}$ cannot be simplified because $q$ and $t$ are different bases.

**e.** Begin by grouping common bases.

$$\frac{6x^3y^7}{xy^5} = 6 \cdot \frac{x^3}{x} \cdot \frac{y^7}{y^5} = 6 \cdot x^{3-1} \cdot y^{7-5} = 6x^2y^2$$

**10. a.** $-3^0 = -1 \cdot 3^0 = -1 \cdot 1 = -1$

**b.** $(-3)^0 = 1$

**c.** $8^0 = 1$

**d.** $(0.2)^0 = 1$

**e.** Assume that neither *a* nor *y* is zero.
$(7a^2y^4)^0 = 1$

**f.** $7y^0 = 7 \cdot y^0 = 7 \cdot 1 = 7$

**11. a.** $\left(\dfrac{z}{12}\right)^2 = \dfrac{z^2}{12^2} = \dfrac{z^2}{144}$

**b.** $(4x^6)^3 = 4^3 \cdot (x^6)^3 = 64x^{18}$

**c.** $y^{10} \cdot y^3 = y^{10+3} = y^{13}$

**12. a.** $8^2 - 8^0 = 64 - 1 = 63$

**b.** $(z^0)^6 + (4^0)^5 = 1^6 + 1^5 = 1 + 1 = 2$

**c.** $\left(\dfrac{5x^3}{15y^4}\right)^2 = \dfrac{5^2(x^3)^2}{15^2(y^4)^2} = \dfrac{25x^6}{225y^8} = \dfrac{x^6}{9y^8}$

**d.**
$$\begin{aligned}
\frac{(2z^8x^5)^4}{-16z^2x^{20}} &= \frac{2^4(z^8)^4(x^5)^4}{-16z^2x^{20}} \\
&= \frac{16z^{32}x^{20}}{-16z^2x^{20}} \\
&= -1 \cdot (z^{32-2}) \cdot (x^{20-20}) \\
&= -1 \cdot z^{30} \cdot x^0 \\
&= -1 \cdot z^{30} \cdot 1 \\
&= -z^{30}
\end{aligned}$$

**Vocabulary, Readiness & Video Check 5.1**

**1.** Repeated multiplication of the same factor can be written using an <u>exponent</u>.

**2.** In $5^2$, the 2 is called the <u>exponent</u> and the 5 is called the <u>base</u>.

**3.** To simplify $x^2 \cdot x^7$, keep the base and <u>add</u> the exponents.

**4.** To simplify $(x^3)^6$, keep the base and <u>multiply</u> the exponents.

**5.** The understood exponent on the term *y* is <u>1</u>.

**6.** If $x^{\square} = 1$, the exponent is <u>0</u>.

**7.** Example 4 can be written as $-4^2 = -1 \cdot 4^2$, which is similar to Example 7, $4 \cdot 3^2$, and shows why the negative sign should not be considered part of the base when there are no parentheses.

**8.** The properties allow us to reorder and regroup factors and put those with common bases together, making it easier to apply the product rule; yes, in Example 13.

**9.** Be careful not to confuse the power rule with the product rule. The power rule involves a power raised to a power (exponents are multiplied), and the product rule involves a product (exponents are added).

**10.** Remember to raise the −2 (or any number) to the power along with the variables.

**11.** the quotient rule

**12.** No, Example 30 is a fraction and does not use the quotient rule.

**Exercise Set 5.1**

**1.** In $3^2$, the base is 3 and the exponent is 2.

**3.** In $-4^2$, the base is 4 and the exponent is 2.

**5.** In $5x^2$, the base 5 has exponent 1 and the base *x* has exponent 2.

**7.** $7^2 = 7 \cdot 7 = 49$

**9.** $(-5)^1 = -5$

**11.** $-2^4 = -2 \cdot 2 \cdot 2 \cdot 2 = -16$

**13.** $(-2)^4 = (-2)(-2)(-2)(-2) = 16$

**15.** $(0.1)^5 = (0.1)(0.1)(0.1)(0.1)(0.1)$
$= 0.00001$

**17.** $\left(\frac{1}{3}\right)^4 = \left(\frac{1}{3}\right)\left(\frac{1}{3}\right)\left(\frac{1}{3}\right)\left(\frac{1}{3}\right) = \frac{1}{81}$

**19.** $7 \cdot 2^5 = 7 \cdot 2 \cdot 2 \cdot 2 \cdot 2 \cdot 2 = 224$

**21.** $-2 \cdot 5^3 = -2 \cdot 5 \cdot 5 \cdot 5 = -250$

**23.** $x^2 = (-2)^2 = (-2)(-2) = 4$

**25.** $5x^3 = 5(3)^3 = 5 \cdot 3 \cdot 3 \cdot 3 = 135$

**27.** $2xy^2 = 2(3)(5)^2 = 2(3)(5)(5) = 150$

**29.** $\frac{2z^4}{5} = \frac{2(-2)^4}{5} = \frac{2(-2)(-2)(-2)(-2)}{5} = \frac{32}{5}$

**31.** $x^2 \cdot x^5 = x^{2+5} = x^7$

**33.** $(-3)^3 \cdot (-3)^9 = (-3)^{3+9} = (-3)^{12}$

**35.** $(5y^4)(3y) = 5(3)y^{4+1} = 15y^5$

**37.** $(x^9 y)(x^{10} y^5) = x^{9+10} y^{1+5} = x^{19} y^6$

**39.** $(-8mn^6)(9m^2 n^2) = --8(9)m^{1+2} n^{6+2}$
$= -72m^3 n^8$

**41.** $(4z^{10})(-6z^7)(z^3) = 4(-6)z^{10+7+3} = -24z^{20}$

**43.** $A = (4x^2) \cdot (5x^3)$
$= (4 \cdot 5) \cdot (x^2 \cdot x^3)$
$= 20x^{2+3}$
$= 20x^5$
The area is $20x^5$ square feet.

**45.** $(x^9)^4 = x^{9 \cdot 4} = x^{36}$

**47.** $(pq)^8 = p^8 q^8$

**49.** $(2a^5)^3 = 2^3 \cdot (a^5)^3 = 8 \cdot a^{5 \cdot 3} = 8a^{15}$

**51.** $(x^2 y^3)^5 = (x^2)^5 \cdot (y^3)^5 = x^{2 \cdot 5} \cdot y^{3 \cdot 5} = x^{10} y^{15}$

**53.** $(-7a^2 b^5 c)^2 = (-7)^2 \cdot (a^2)^2 \cdot (b^5)^2 \cdot c^2$
$= 49a^{2 \cdot 2} b^{5 \cdot 2} c^2$
$= 49a^4 b^{10} c^2$

**55.** $\left(\frac{r}{s}\right)^9 = \frac{r^9}{s^9}$

**57.** $\left(\frac{mp}{n}\right)^5 = \frac{(mp)^5}{n^5} = \frac{m^5 \cdot p^5}{n^5} = \frac{m^5 p^5}{n^5}$

**59.** $\left(\frac{-2xz}{y^5}\right)^2 = \frac{(-2)^2 x^2 z^2}{y^{5 \cdot 2}} = \frac{4x^2 z^2}{y^{10}}$

**61.** $A = (8z^5)^2 = 8^2 \cdot (z^5)^2 = 64 \cdot z^{5 \cdot 2} = 64z^{10}$
The area is $64z^{10}$ square decimeters.

**63.** $V = (3y^4)^3 = 3^3 y^{4 \cdot 3} = 27y^{12}$
The volume is $27y^{12}$ cubic feet.

**65.** $\frac{x^3}{x} = \frac{x^3}{x^1} = x^{3-1} = x^2$

**67.** $\frac{(-4)^6}{(-4)^3} = (-4)^{6-3} = (-4)^3 = -64$

**69.** $\frac{p^7 q^{20}}{pq^{15}} = p^{7-1} q^{20-15} = p^6 q^5$

**71.** $\frac{7x^2 y^6}{14x^2 y^3} = \frac{7}{14} x^{2-2} y^{6-3} = \frac{1}{2} x^0 y^3 = \frac{y^3}{2}$

**73.** $7^0 = 1$

**75.** $(2x)^0 = 1$

**77.** $-7x^0 = -7(1) = -7$

**79.** $5^0 + y^0 = 1 + 1 = 2$

**81.** $-9^2 = -9 \cdot 9 = -81$

**83.** $\left(\dfrac{1}{4}\right)^3 = \dfrac{1}{4}\cdot\dfrac{1}{4}\cdot\dfrac{1}{4} = \dfrac{1}{64}$

**85.** $b^4 b^2 = b^{4+2} = b^6$

**87.** $a^2 a^3 a^4 = a^{2+3+4} = a^9$

**89.** $\begin{aligned}(2x^3)(-8x^4) &= (2\cdot -8)(x^3\cdot x^4)\\ &= -16x^{3+4}\\ &= -16x^7\end{aligned}$

**91.** $\begin{aligned}(a^7 b^{12})(a^4 b^8) &= a^7 a^4\cdot b^{12} b^8\\ &= a^{7+4} b^{12+8}\\ &= a^{11} b^{20}\end{aligned}$

**93.** $\begin{aligned}(-2mn^6)(-13m^8 n) &= (-2)(-13)(m\cdot m^8)(n^6\cdot n)\\ &= 26m^{1+8} n^{6+1}\\ &= 26m^9 n^7\end{aligned}$

**95.** $(z^4)^{10} = z^{4\cdot10} = z^{40}$

**97.** $(4ab)^3 = (4)^3 a^3 b^3 = 64a^3 b^3$

**99.** $\begin{aligned}(-6xyz^3)^2 &= (-6)^2 x^2 y^2 (z^3)^2\\ &= 36x^2 y^2 z^{3\cdot2}\\ &= 36x^2 y^2 z^6\end{aligned}$

**101.** $\dfrac{3x^5}{x^4} = 3\cdot\dfrac{x^5}{x^4} = 3x^{5-4} = 3x^1 = 3x$

**103.** $(9xy)^2 = 9^2\cdot x^2 y^2 = 81x^2 y^2$

**105.** $2^3 + 2^0 = (2\cdot2\cdot2) + 1 = 8 + 1 = 9$

**107.** $\left(\dfrac{3y^5}{6x^4}\right)^3 = \left(\dfrac{y^5}{2x^4}\right)^3 = \dfrac{(y^5)^3}{2^3(x^4)^3} = \dfrac{y^{5\cdot3}}{8x^{4\cdot3}} = \dfrac{y^{15}}{8x^{12}}$

**109.** $\begin{aligned}\dfrac{2x^3 y^2 z}{xyz} &= 2\cdot\dfrac{x^3}{x}\cdot\dfrac{y^2}{y}\cdot\dfrac{z}{z}\\ &= 2x^{3-1} y^{2-1} z^{1-1}\\ &= 2x^2 y^1 z^0\\ &= 2x^2 y\end{aligned}$

**111.** $(5^0)^3 + (y^0)^7 = 1^3 + 1^7 = 1 + 1 = 2$

**113.** $\left(\dfrac{5x^9}{10y^{11}}\right)^2 = \dfrac{5^2(x^9)^2}{10^2(y^{11})^2} = \dfrac{25x^{18}}{100y^{22}} = \dfrac{x^{18}}{4y^{22}}$

**115.** $\begin{aligned}\dfrac{(2a^5 b^3)^4}{-16a^{20} b^7} &= \dfrac{2^4(a^5)^4(b^3)^4}{-16a^{20} b^7}\\ &= \dfrac{16a^{20} b^{12}}{-16a^{20} b^7}\\ &= -1\cdot a^{20-20}\cdot b^{12-7}\\ &= -1\cdot a^0\cdot b^5\\ &= -1\cdot1\cdot b^5\\ &= -b^5\end{aligned}$

**117.** $y - 10 + y = y + y - 10 = 2y - 10$

**119.** $7x + 2 - 8x - 6 = 7x - 8x + 2 - 6 = -x - 4$

**121.** $2(x - 5) + 3(5 - x) = 2x - 10 + 15 - 3x = -x + 5$

**123.** $(x^{14})^{23} = x^{14\cdot23} = x^{322}$
Multiply the exponents; choice c.

**125.** $x^{14} + x^{23}$ cannot be simplified further; choice e.

**127.** answers may vary

**129.** answers may vary

**131.** $V = x^3 = 7^3 = 7\cdot7\cdot7 = 343$
The volume is 343 cubic meters.

**133.** Volume; volume measures capacity.

**135.** answers may vary

**137.** answers may vary

**139.** $x^{5a} x^{4a} = x^{5a+4a} = x^{9a}$

**141.** $(a^b)^5 = a^{b\cdot5} = a^{5b}$

**143.** $\dfrac{x^{9a}}{x^{4a}} = x^{9a-4a} = x^{5a}$

**145.** $A = P\left(1+\dfrac{r}{12}\right)^6$

$A = 1000\left(1+\dfrac{0.09}{12}\right)^6$

$= 1000(1.0075)^6$

$\approx 1045.85$

$1045.85 is needed to pay off the loan.

**Section 5.2 Practice**

**1. a.** The exponent on $y$ is 3, so the degree of $5y^3$ is 3.

**b.** $-3a^2b^5c$ can be written as $-3a^2b^5c^1$. The degree of the term is the sum of the exponents, so the degree is $2 + 5 + 1$ or 8.

**c.** The constant, 8, can be written as $8x^0$ (since $x^0 = 1$). The degree of 8 or $8x^0$ is 0.

**2. a.** The degree of the trinomial $5b^2 - 3b + 7$ is 2, the greatest degree of any of its terms.

**b.** Rewrite the binomial as $7t^1 + 3$, the degree is 1.

**c.** The degree of the polynomial $5x^2 + 3x - 6x^3 + 4$ is 3.

**3.**

| Term | numerical coefficient | degree of term |
|---|---|---|
| $-3x^3y^2$ | $-3$ | 5 |
| $4xy^2$ | 4 | 3 |
| $-y^2$ | $-1$ | 2 |
| $3x$ | 3 | 1 |
| $-2$ | $-2$ | 0 |

**4. a.** $-10x + 1 = -10(-3) + 1 = 30 + 1 = 31$

**b.** $2x^2 - 5x + 3 = 2(-3)^2 - 5(-3) + 3$
$= 2(9) + 15 + 3$
$= 18 + 15 + 3$
$= 36$

**5.** To find each height, we evaluate the polynomial when $t = 1$ and when $t = 2$.
$-16t^2 + 130 = -16(1)^2 + 130$
$= -16 + 130$
$= 114$
The height of the camera at 1 second is 114 feet.
$-16t^2 + 130 = -16(2)^2 + 130$
$= -16(4) + 130$
$= -64 + 130$
$= 66$
The height of the camera at 2 seconds is 66 feet.

**6. a.** $-4y + 2y = (-4 + 2)y = -2y$

**b.** These terms cannot be combined because $z$ and $5z^3$ are not like terms.

**c.** $15x^3 - x^3 = 15x^3 - 1x^3 = 14x^3$

**d.** $7a^2 - 5 - 3a^2 - 7 = 7a^2 - 3a^2 - 5 - 7$
$= 4a^2 - 12$

**e.** $\dfrac{3}{8}x^3 - x^2 + \dfrac{5}{6}x^4 + \dfrac{1}{12}x^3 - \dfrac{1}{2}x^4$
$= \left(\dfrac{5}{6} - \dfrac{1}{2}\right)x^4 + \left(\dfrac{3}{8} + \dfrac{1}{12}\right)x^3 - x^2$
$= \left(\dfrac{5}{6} - \dfrac{3}{6}\right)x^4 + \left(\dfrac{9}{24} + \dfrac{2}{24}\right)x^3 - x^2$
$= \dfrac{2}{6}x^4 + \dfrac{11}{24}x^3 - x^2$
$= \dfrac{1}{3}x^4 + \dfrac{11}{24}x^3 - x^2$

**7.** $9xy - 3x^2 - 4yx + 5y^2 = -3x^2 + (9-4)xy + 5y^2$
$= -3x^2 + 5xy + 5y^2$

**8.** $x \cdot x + 2 \cdot x + 2 \cdot 2 + 5 \cdot x + x \cdot 3x$
$= x^2 + 2x + 4 + 5x + 3x^2$
$= 4x^2 + 7x + 4$

**9.** $(-3x^2 - 4x + 9) + (2x^2 - 2x)$
$= -3x^2 - 4x + 9 + 2x^2 - 2x$
$= (-3x^2 + 2x^2) + (-4x - 2x) + 9$
$= -x^2 - 6x + 9$

**10.** $(-3x^3 + 7x^2 + 3x - 4) + (3x^2 - 9x)$

$= -3x^3 + 7x^2 + 3x - 4 + 3x^2 - 9x$

$= -3x^3 + (7x^2 + 3x^2) + (3x - 9x) - 4$

$= -3x^3 + 10x^2 - 6x - 4$

**11.** $5z^3 + 3z^2 + 4z$

$\dfrac{5z^2 + 4z}{5z^3 + 8z^2 + 8z}$

**12.** $(8x - 7) - (3x - 6) = (8x - 7) + [-(3x - 6)]$

$= (8x - 7) + (-3x + 6)$

$= (8x - 3x) + (-7 + 6)$

$= 5x - 1$

**13.** First, change the sign of each term of the second polynomial and then add.

$(3x^3 - 5x^2 + 4x) - (x^3 - x^2 + 6)$

$= (3x^3 - 5x^2 + 4x) + (-x^3 + x^2 - 6)$

$= 3x^3 - x^3 - 5x^2 + x^2 + 4x - 6$

$= 2x^3 - 4x^2 + 4x - 6$

**14.** Arrange the polynomials in vertical format, lining up like terms.

$$
\begin{array}{r}
-2z^2 - 8z + 5 \\
- (6z^2 + 3z - 6) \\
\hline
\end{array}
\qquad
\begin{array}{r}
-2z^2 \;\; -8z \;\; +5 \\
-6z^2 \;\; -3z \;\; +6 \\
\hline
-8z^2 - 11z + 11
\end{array}
$$

**15.** $[(8x - 11) + (2x + 5)] - (3x + 5)$

$= 8x - 11 + 2x + 5 - 3x - 5$

$= 8x + 2x - 3x - 11 + 5 - 5$

$= 7x - 11$

**16. a.** $(3a^2 - 4ab + 7b^2) + (-8a^2 + 3ab - b^2)$

$= 3a^2 - 4ab + 7b^2 - 8a^2 + 3ab - b^2$

$= -5a^2 - ab + 6b^2$

**b.** $(5x^2y^2 - 6xy - 4xy^2)$

$\qquad\qquad - (2x^2y^2 + 4xy - 5 + 6y^2)$

$= 5x^2y^2 - 6xy - 4xy^2 - 2x^2y^2$

$\qquad\quad - 4xy + 5 - 6y^2$

$= 3x^2y^2 - 10xy - 4xy^2 - 6y^2 + 5$

## Vocabulary, Readiness & Video Check 5.2

**1.** A <u>binomial</u> is a polynomial with exactly 2 terms.

**2.** A <u>monomial</u> is a polynomial with exactly one term.

**3.** A <u>trinomial</u> is a polynomial with exactly three terms.

**4.** The numerical factor of a term is called the <u>coefficient</u>.

**5.** A number term is also called a <u>constant</u>.

**6.** The degree of a polynomial is the <u>greatest</u> degree of any term of the polynomial.

**7.** The degree of the polynomial is the greatest degree of any of its terms, so we need to find the degree of each term first.

**8.** the replacement value for the variables

**9.** simplifying it

**10.** Addition; no, we subtract in Examples 9–11. To subtract, we first change the signs of the polynomial being subtracted and then add.

## Exercise Set 5.2

**1.** $x + 2$ is a binomial because it has two terms. The degree is 1 since $x$ is $x^1$.

**3.** $9m^3 - 5m^2 + 4m - 8$ is neither a monomial, a binomial, nor a trinomial because it has more than three terms. The degree is 3, the greatest degree of any of its terms.

**5.** $12x^4y - x^2y^2 - 12x^2y^4$ is a trinomial because it has three terms. The degree is 6, the greatest degree of any of its terms.

**7.** $3 - 5x^8$ is a binomial because it has two terms. The degree is 8, the greatest degree of any of its terms.

| | *Polynomial* | *Degree* |
|---|---|---|
| **9.** | $3xy^2 - 4$ | 3 |
| **11.** | $5a^2 - 2a + 1$ | 2 |

**13. a.** $5x - 6 = 5(0) - 6 = 0 - 6 = -6$

**b.** $5x - 6 = 5(-1) - 6 = -5 - 6 = -11$

**15. a.** $x^2 - 5x - 2 = (0)^2 - 5(0) - 2 = 0 - 0 - 2 = -2$

   **b.** $x^2 - 5x - 2 = (-1)^2 - 5(-1) - 2 = 1 + 5 - 2 = 4$

**17. a.** $-x^3 + 4x^2 - 15x + 1$
$$= -(0)^3 + 4(0)^2 - 15(0) + 1$$
$$= 0 + 0 - 0 + 1$$
$$= 1$$

   **b.** $-x^3 + 4x^2 - 15x + 1$
$$= -(-1)^3 + 4(-1)^2 - 15(-1) + 1$$
$$= -(-1) + 4(1) - 15(-1) + 1$$
$$= 1 + 4 + 15 + 1$$
$$= 21$$

**19.** $-16t^2 + 1150$

$t = 1;\ -16(1)^2 + 1150 = -16 + 1150 = 1134$
After 1 second, the height is 1134 feet.

**21.** $-16t^2 + 1150$

$t = 3;\ -16(3)^2 + 1150 = -144 + 1150 = 1006$
After 3 seconds, the height is 1006 feet.

**23.** $-7.5x^2 + 93x - 100 = -7.5(8)^2 + 93(8) - 100$
$$= -7.5(64) + 93(8) - 100$$
$$= -480 + 744 - 100$$
$$= 164$$
There were 164 thousand visitors in 2008.

**25.** $9x - 20x = (9 - 20)x = -11x$

**27.** $14x^2 + 9x^2 = (14 + 9)x^2 = 23x^2$

**29.** $15x^2 - 3x^2 - y = (15 - 3)x^2 - y = 12x^2 - y$

**31.** $8s - 5s + 4s = (8 - 5 + 4)s = 7s$

**33.** $0.1y^2 - 1.2y^2 + 6.7 - 1.9$
$$= (0.1 - 1.2)y^2 + (6.7 - 1.9)$$
$$= -1.1y^2 + 4.8$$

**35.** $\frac{2}{3}x^4 + 12x^3 + \frac{1}{6}x^4 - 19x^3 - 19$
$$= \frac{2}{3}x^4 + \frac{1}{6}x^4 + 12x^3 - 19x^3 - 19$$
$$= \left(\frac{4}{6} + \frac{1}{6}\right)x^4 + (12 - 19)x^3 - 19$$
$$= \frac{5}{6}x^4 - 7x^3 - 19$$

**37.** $\frac{2}{5}x^2 - \frac{1}{3}x^3 + x^2 - \frac{1}{4}x^3 + 6$
$$= \left(-\frac{1}{3} - \frac{1}{4}\right)x^3 + \left(\frac{2}{5} + 1\right)x^2 + 6$$
$$= \left(-\frac{4}{12} - \frac{3}{12}\right)x^3 + \left(\frac{2}{5} + \frac{5}{5}\right)x^2 + 6$$
$$= -\frac{7}{12}x^3 + \frac{7}{5}x^2 + 6$$

**39.** $6a^2 - 4ab + 7b^2 - a^2 - 5ab + 9b^2$
$$= (6 - 1)a^2 + (-4 - 5)ab + (7 + 9)b^2$$
$$= 5a^2 - 9ab + 16b^2$$

**41.** $(2x + 5) - (3x - 9) = (2x + 5) + (-3x + 9)$
$$= 2x + 5 - 3x + 9$$
$$= (2x - 3x) + (5 + 9)$$
$$= -x + 14$$

**43.** $(2x^2 + 5) - (3x^2 - 9) = (2x^2 + 5) + (-3x^2 + 9)$
$$= 2x^2 + 5 - 3x^2 + 9$$
$$= (2x^2 - 3x^2) + (5 + 9)$$
$$= -x^2 + 14$$

**45.** $(-7x + 5) + (-3x^2 + 7x + 5)$
$$= -7x + 5 - 3x^2 + 7x + 5$$
$$= -3x^2 + (-7x + 7x) + (5 + 5)$$
$$= -3x^2 + 10$$

**47.** $3x - (5x - 9) = 3x + (-5x + 9)$
$$= 3x - 5x + 9$$
$$= -2x + 9$$

**49.** $(2x^2 + 3x - 9) - (-4x + 7)$
$$= (2x^2 + 3x - 9) + (4x - 7)$$
$$= 2x^2 + 3x - 9 + 4x - 7$$
$$= 2x^2 + 3x + 4x - 9 - 7$$
$$= 2x^2 + 7x - 16$$

**51.**
$$3t^2 + 4$$
$$\underline{+\ 5t^2 - 8}$$
$$8t^2 - 4$$

**53.**
$$4z^2 - 8z + 3 \qquad\qquad 4z^2\ -8z + 3$$
$$\underline{-\ (6z^2 + 8z - 3)} \qquad \underline{+\ (-6z^2 - 8z + 3)}$$
$$\qquad\qquad\qquad\qquad\qquad -2z^2 - 16z + 6$$

**55.**
$$5x^3 - 4x^2 + 6x - 2 \qquad 5x^3 - 4x^2 + 6x - 2$$
$$\underline{-\ (3x^3 - 2x^2\ - x - 4)} \quad \underline{+\ (-3x^3 + 2x^2\ + x + 4)}$$
$$\qquad\qquad\qquad\qquad\qquad 2x^3 - 2x^2 + 7x + 2$$

**57.** $(81x^2 + 10) - (19x^2 + 5) = 81x^2 + 10 - 19x^2 - 5$
$$= 62x^2 + 5$$

**59.** $[(8x + 1) + (6x + 3)] - (2x + 2)$
$$= 8x + 1 + 6x + 3 - 2x - 2$$
$$= 8x + 6x - 2x + 1 + 3 - 2$$
$$= 12x + 2$$

**61.** $(2y + 20) + (5y - 30) = 2y + 20 + 5y - 30$
$$= 2y + 5y + 20 - 30$$
$$= 7y - 10$$

**63.** $(x^2 + 2x + 1) - (3x^2 - 6x + 2)$
$$= (x^2 + 2x + 1) + (-3x^2 + 6x - 2)$$
$$= x^2 + 2x + 1 - 3x^2 + 6x - 2$$
$$= -2x^2 + 8x - 1$$

**65.** $(3x^2 + 5x - 8) + (5x^2 + 9x + 12) - (8x^2 - 14)$
$$= (3x^2 + 5x - 8) + (5x^2 + 9x + 12) + (-8x^2 + 14)$$
$$= 3x^2 + 5x - 8 + 5x^2 + 9x + 12 - 8x^2 + 14$$
$$= 14x + 18$$

**67.** $(-a^2 + 1) - (a^2 - 3) + (5a^2 - 6a + 7)$
$$= (-a^2 + 1) + (-a^2 + 3) + (5a^2 - 6a + 7)$$
$$= -a^2 + 1 - a^2 + 3 + 5a^2 - 6a + 7$$
$$= 3a^2 - 6a + 11$$

**69.** $(7x - 3) - 4x = 7x - 3 - 4x = 3x - 3$

**71.** $(4x^2 - 6x + 1) + (3x^2 + 2x + 1)$
$$= 4x^2 - 6x + 1 + 3x^2 + 2x + 1$$
$$= 7x^2 - 4x + 2$$

**73.** $(7x^2 + 3x + 9) - (5x + 7)$
$$= (7x^2 + 3x + 9) + (-5x - 7)$$
$$= 7x^2 + 3x + 9 - 5x - 7$$
$$= 7x^2 - 2x + 2$$

**75.** $[(8y^2 + 7) + (6y + 9)] - (4y^2 - 6y - 3)$
$$= (8y^2 + 7) + (6y + 9) + (-4y^2 + 6y + 3)$$
$$= 8y^2 + 7 + 6y + 9 - 4y^2 + 6y + 3$$
$$= 4y^2 + 12y + 19$$

**77.** $2x \cdot 2x + x \cdot 7 + x \cdot x + x \cdot 5 = 4x^2 + 7x + x^2 + 5x$
$$= 5x^2 + 12x$$

**79.** $(9a + 6b - 5) + (-11a - 7b + 6)$
$$= 9a + 6b - 5 - 11a - 7b + 6$$
$$= -2a - b + 1$$

**81.** $(4x^2 + y^2 + 3) - (x^2 + y^2 - 2)$
$$= 4x^2 + y^2 + 3 - x^2 - y^2 + 2$$
$$= 3x^2 + 5$$

**83.** $(x^2 + 2xy - y^2) + (5x^2 - 4xy + 20y^2)$
$$= x^2 + 2xy - y^2 + 5x^2 - 4xy + 20y^2$$
$$= 6x^2 - 2xy + 19y^2$$

**85.** $(11r^2s + 16rs - 3 - 2r^2s^2) - (3sr^2 + 5 - 9r^2s^2)$
$$= 11r^2s + 16rs - 3 - 2r^2s^2 - 3sr^2 - 5 + 9r^2s^2$$
$$= 8r^2s + 16rs - 8 + 7r^2s^2$$

**87.** $7.75x + 9.16x^2 - 1.27 - 14.58x^2 - 18.34$
$$= (9.16 - 14.58)x^2 + 7.75x + (-1.27 - 18.34)$$
$$= -5.42x^2 + 7.75x - 19.61$$

**89.** $[(7.9y^4 - 6.8y^3 + 3.3y) + (6.1y^3 - 5)]$
$$\qquad - (4.2y^4 + 1.1y - 1)$$
$$= 7.9y^4 - 6.8y^3 + 3.3y + 6.1y^3 - 5 - 4.2y^4$$
$$\qquad - 1.1y + 1$$
$$= 3.7y^4 - 0.7y^3 + 2.2y - 4$$

**91.** $3x(2x) = 3 \cdot 2 \cdot x \cdot x = 6x^2$

**93.** $(12x^3)(-x^5) = (12x^3)(-1x^5)$
$$= (12)(-1)(x^3)(x^5)$$
$$= -12x^8$$

**95.** $10x^2(20xy^2) = 10 \cdot 20 x^2 \cdot x \cdot y^2 = 200x^3 y^2$

**97.** $9x + 10 + 3x + 12 + 4x + 15 + 2x + 7$
$= (9x + 3x + 4x + 2x) + (10 + 12 + 15 + 7)$
$= 18x + 44$

**99.** $(-x^2 + 3x) + (2x^2 + 5) + (4x - 1)$
$= -x^2 + 3x + 2x^2 + 5 + 4x - 1$
$= x^2 + 7x + 4$
The perimeter is $(x^2 + 7x + 4)$ feet.

**101.** $(4y^2 + 4y + 1) - (y^2 - 10)$
$= 4y^2 + 4y + 1 - y^2 + 10$
$= 3y^2 + 4y + 11$
The length of the remaining piece is
$(3y^2 + 4y + 11)$ meters.

**103.** $x = 2009 - 2004 = 5$
$0.16x^2 + 0.29x + 1.88$
$= 0.16(5)^2 + 0.29(5) + 1.88$
$= 4 + 1.45 + 1.88$
$= 7.33$
The number of cell phones recycled in 2009 is estimated at 7.33 million.

**105.** answers may vary

**107.** answers may vary

**109.** $10y - 6y^2 - y = (10 - 1)y - 6y^2 = 9y - 6y^2$
choice b

**111.** $(5x - 3) + (5x - 3) = (5x + 5x) + (-3 - 3)$
$= (5 + 5)x - 6$
$= 10x - 6$
choice e

**113. a.** $z + 3z = 1z + 3z = 4z$

**b.** $z \cdot 3z = z^1 \cdot 3z^1 = 3z^{1+1} = 3z^2$

**c.** $-z - 3z = -1z - 3z = -4z$

**d.** $(-z)(-3z) = (-z^1)(-3z^1) = 3z^{1+1} = 3z^2$

answers may vary

**115. a.** $m \cdot m \cdot m = m^1 \cdot m^1 \cdot m^1 = m^{1+1+1} = m^3$

**b.** $m + m + m = 1m + 1m + 1m = (1 + 1 + 1)m = 3m$

**c.** $(-m)(-m)(-m) = (-1 \cdot m^1)(-1 \cdot m^1)(-1 \cdot m^1)$
$= (-1)(-1)(-1)(m \cdot m \cdot m)$
$= -1m^3$
$= -m^3$

answers may vary

**117.** Since $3 + 4 = 7$, $3x^2 + 4x^2 = 7x^2$ is a true statement.

**119.** Since $2 + 4 = 6$ and $3 - 5 = -2$,
$2x^4 + 3x^3 - 5x^3 + 4x^4 = 6x^4 - 2x^3$ is a true statement.

**121.** $x^2 + x^2 + xy + xy + xy + xy = 2x^2 + 4xy$

**Section 5.3 Practice**

**1.** $5y \cdot 2y = (5 \cdot 2)(y \cdot y) = 10y^2$

**2.** $(5z^3) \cdot (-0.4z^5) = (5 \cdot -0.4)(z^3 \cdot z^5) = -2z^8$

**3.** $\left(-\frac{1}{9}b^6\right)\left(-\frac{7}{8}b^3\right) = \left(-\frac{1}{9} \cdot -\frac{7}{8}\right)(b^6 \cdot b^3) = \frac{7}{72}b^9$

**4. a.** $3x(9x^5 + 11) = 3x(9x^5) + 3x(11)$
$= 27x^6 + 33x$

**b.** $-6x^3(2x^2 - 9x + 2)$
$= -6x^3(2x^2) + (-6x^3)(-9x) + (-6x^2)(2)$
$= -12x^5 + 54x^4 - 12x^3$

**5.** Multiply each term of the first binomial by each term of the second.
$(5x - 2)(2x + 3)$
$= 5x(2x) + 5x(3) + (-2)(2x) + (-2)(3)$
$= 10x^2 + 15x - 4x - 6$
$= 10x^2 + 11x - 6$

6. Recall that $a^2 = a \cdot a$, so

$(5x - 3y)^2 = (5x - 3y)(5x - 3y)$. Multiply each term of the first binomial by each term of the second.

$(5x - 3y)(5x - 3y)$
$= 5x(5x) + 5x(-3y) + (-3y)(5x) + (-3y)(-3y)$
$= 25x^2 - 15xy - 15xy + 9y^2$
$= 25x^2 - 30xy + 9y^2$

7. Multiply each term of the first polynomial by each term of the second.

$(y + 4)(2y^2 - 3y + 5)$
$= y(2y^2) + y(-3y) + y(5) + 4(2y^2)$
$\qquad + 4(-3y) + 4(5)$
$= 2y^3 - 3y^2 + 5y + 8y^2 - 12y + 20$
$= 2y^3 + 5y^2 - 7y + 20$

8. Write $(s + 2t)^3$ as $(s + 2t)(s + 2t)(s + 2t)$.

$(s + 2t)(s + 2t)(s + 2t)$
$= (s^2 + 2st + 2st + 4t^2)(s + 2t)$
$= (s^2 + 4st + 4t^2)(s + 2t)$
$= (s^2 + 4st + 4t^2)s + (s^2 + 4st + 4t^2)(2t)$
$= s^3 + 4s^2t + 4st^2 + 2s^2t + 8st^2 + 8t^3$
$= s^3 + 6s^2t + 12st^2 + 8t^3$

9.
$$\begin{array}{r} 5x^2 - 3x + 5 \\ \times \qquad\qquad x - 4 \\ \hline -20x^2 + 12x - 20 \\ 5x^3 - 3x^2 + 5x \qquad\quad \\ \hline 5x^3 - 23x^2 + 17x - 20 \end{array}$$

10.
$$\begin{array}{r} x^3 - 2x^2 + 1 \\ \times \qquad\qquad x^2 + 2 \\ \hline 2x^3 - 4x^2 \qquad + 2 \\ x^5 - 2x^4 \qquad + x^2 \qquad\quad \\ \hline x^5 - 2x^4 + 2x^3 - 3x^2 + 2 \end{array}$$

11.
$$\begin{array}{r} 5x^2 + 2x - 2 \\ x^2 - x + 3 \\ \hline 15x^2 + 6x - 6 \\ -5x^3 - 2x^2 + 2x \qquad\quad \\ 5x^4 + 2x^3 - 2x^2 \qquad\qquad\quad \\ \hline 5x^4 - 3x^3 + 11x^2 + 8x - 6 \end{array}$$

## Vocabulary, Readiness & Video Check 5.3

1. The expression $5x(3x + 2)$ equals $5x \cdot 3x + 5x \cdot 2$ by the <u>distributive</u> property.

2. The expression $(x + 4)(7x - 1)$ equals $x(7x - 1) + 4(7x - 1)$ by the <u>distributive</u> property.

3. The expression $(5y - 1)^2$ equals $\underline{(5y - 1)(5y - 1)}$.

4. The expression $9x \cdot 3x$ equals $\underline{27x^2}$.

5. No; the monomials are unlike terms.

6. distributive property, product rule

7. Yes; the parentheses have been removed for the vertical format, but every term in the first polynomial is still distributed to every term in the second polynomial.

## Exercise Set 5.3

1. $-4n^3 \cdot 7n^7 = (-4 \cdot 7)(n^3 \cdot n^7) = -28n^{10}$

3. $(-3.1x^3)(4x^9) = (-3.1 \cdot 4)(x^3 \cdot x^9) = -12.4x^{12}$

5. $\left(-\dfrac{1}{3}y^2\right)\left(\dfrac{2}{5}y\right) = \left(-\dfrac{1}{3} \cdot \dfrac{2}{5}\right)(y^2 \cdot y) = -\dfrac{2}{15}y^3$

7. $(2x)(-3x^2)(4x^5) = (2 \cdot -3 \cdot 4)(x \cdot x^2 \cdot x^5) = -24x^8$

9. $3x(2x + 5) = 3x(2x) + 3x(5) = 6x^2 + 15x$

11. $-2a(a + 4) = -2a(a) + (-2a)(4) = -2a^2 - 8a$

13. $3x(2x^2 - 3x + 4) = 3x(2x^2) + 3x(-3x) + 3x(4)$
$\qquad\qquad\qquad = 6x^3 - 9x^2 + 12x$

15. $-2a^2(3a^2 - 2a + 3)$
$= -2a^2(3a^2) + (-2a^2)(-2a) + (-2a^2)(3)$
$= -6a^4 + 4a^3 - 6a^2$

17. $-y(4x^3 - 7x^2y + xy^2 + 3y^3)$
$= -y(4x^3) + (-y)(-7x^2y) + (-y)(xy^2)$
$\qquad\qquad\qquad + (-y)(3y^3)$
$= -4x^3y + 7x^2y^2 - xy^3 - 3y^4$

    

**19.** $\dfrac{1}{2}x^2(8x^2 - 6x + 1)$

$= \dfrac{1}{2}x^2(8x^2) + \dfrac{1}{2}x^2(-6x) + \dfrac{1}{2}x^2(1)$

$= 4x^4 - 3x^3 + \dfrac{1}{2}x^2$

**21.** $(x+4)(x+3) = x(x) + x(3) + 4(x) + 4(3)$

$= x^2 + 3x + 4x + 12$

$= x^2 + 7x + 12$

**23.** $(a+7)(a-2) = a(a) + a(-2) + 7(a) + 7(-2)$

$= a^2 - 2a + 7a - 14$

$= a^2 + 5a - 14$

**25.** $\left(x + \dfrac{2}{3}\right)\left(x - \dfrac{1}{3}\right)$

$= x(x) + x\left(-\dfrac{1}{3}\right) + \dfrac{2}{3}(x) + \dfrac{2}{3}\left(-\dfrac{1}{3}\right)$

$= x^2 - \dfrac{1}{3}x + \dfrac{2}{3}x - \dfrac{2}{9}$

$= x^2 + \dfrac{1}{3}x - \dfrac{2}{9}$

**27.** $(3x^2 + 1)(4x^2 + 7)$

$= 3x^2(4x^2) + 3x^2(7) + 1(4x^2) + 1(7)$

$= 12x^4 + 21x^2 + 4x^2 + 7$

$= 12x^4 + 25x^2 + 7$

**29.** $(2y - 4)^2$

$= (2y - 4)(2y - 4)$

$= 2y(2y) + 2y(-4) + (-4)(2y) + (-4)(-4)$

$= 4y^2 - 8y - 8y + 16$

$= 4y^2 - 16y + 16$

**31.** $(4x - 3)(3x - 5)$

$= 4x(3x) + 4x(-5) + (-3)(3x) + (-3)(-5)$

$= 12x^2 - 20x - 9x + 15$

$= 12x^2 - 29x + 15$

**33.** $(3x^2 + 1)^2 = (3x^2 + 1)(3x^2 + 1)$

$= 3x^2(3x^2) + 3x^2(1) + 1(3x^2) + 1(1)$

$= 9x^4 + 3x^2 + 3x^2 + 1$

$= 9x^4 + 6x^2 + 1$

**35. a.** $4y^2(-y^2) = 4(-1)y^{2+2} = -4y^4$

    **b.** $4y^2 - y^2 = 4y^2 - 1y^2 = (4-1)y^2 = 3y^2$

    **c.** answers may vary

**37.** $(x-2)(x^2 - 3x + 7)$

$= x(x^2) + x(-3x) + x(7) + (-2)(x^2)$

$\qquad\quad + (-2)(-3x) + (-2)(7)$

$= x^3 - 3x^2 + 7x - 2x^2 + 6x - 14$

$= x^3 - 5x^2 + 13x - 14$

**39.** $(x+5)(x^3 - 3x + 4)$

$= x(x^3) + x(-3x) + x(4) + 5(x^3) + 5(-3x) + 5(4)$

$= x^4 - 3x^2 + 4x + 5x^3 - 15x + 20$

$= x^4 + 5x^3 - 3x^2 - 11x + 20$

**41.** $(2a - 3)(5a^2 - 6a + 4)$

$= 2a(5a^2) + 2a(-6a) + 2a(4) + (-3)(5a^2)$

$\qquad\quad + (-3)(-6a) + (-3)(4)$

$= 10a^3 - 12a^2 + 8a - 15a^2 + 18a - 12$

$= 10a^3 - 27a^2 + 26a - 12$

**43.** $(x+2)^3 = (x+2)(x+2)(x+2)$

$= (x^2 + 2x + 2x + 4)(x + 2)$

$= (x^2 + 4x + 4)(x + 2)$

$= (x^2 + 4x + 4)x + (x^2 + 4x + 4)2$

$= x^3 + 4x^2 + 4x + 2x^2 + 8x + 8$

$= x^3 + 6x^2 + 12x + 8$

**45.** $(2y - 3)^3$

$= (2y - 3)(2y - 3)(2y - 3)$

$= (4y^2 - 6y - 6y + 9)(2y - 3)$

$= (4y^2 - 12y + 9)(2y - 3)$

$= (4y^2 - 12y + 9)2y + (4y^2 - 12y + 9)(-3)$

$= 8y^3 - 24y^2 + 18y - 12y^2 + 36y - 27$

$= 8y^3 - 36y^2 + 54y - 27$

**47.**

$$
\begin{array}{r}
2x - 11 \\
\times \quad\quad 6x + 1 \\
\hline
2x - 11 \\
12x^2 - 66x \quad\quad\;\; \\
\hline
12x^2 - 64x - 11
\end{array}
$$

**49.**
$$
\begin{array}{r}
2x^2 + 4x - 1 \\
\times\quad\quad\quad 5x + 1 \\
\hline
2x^2 + 4x - 1 \\
10x^3 + 20x^2 - 5x\quad\quad \\
\hline
10x^3 + 22x^2\ -x - 1
\end{array}
$$

**51.**
$$
\begin{array}{r}
2x^2 - 7x - 9 \\
\times\quad\quad\quad x^2 + 5x - 7 \\
\hline
-14x^2 + 49x + 63 \\
10x^3 - 35x^2 - 45x\quad\quad\quad \\
2x^4 - 7x^3\ -9x^2\quad\quad\quad\quad\quad \\
\hline
2x^4 + 3x^3 - 58x^2\ +4x + 63
\end{array}
$$

**53.** $-1.2y(-7y^6) = -1.2(-7)(y \cdot y^6) = 8.4y^7$

**55.** $-3x(x^2 + 2x - 8)$
$= -3x(x^2) + (-3x)(2x) + (-3x)(-8)$
$= -3x^3 - 6x^2 + 24x$

**57.** $(x + 19)(2x + 1) = x(2x) + x(1) + 19(2x) + 19(1)$
$\qquad\qquad\qquad = 2x^2 + x + 38x + 19$
$\qquad\qquad\qquad = 2x^2 + 39x + 19$

**59.** $\left(x + \dfrac{1}{7}\right)\left(x - \dfrac{3}{7}\right)$

$= x(x) + x\left(-\dfrac{3}{7}\right) + \dfrac{1}{7}(x) + \dfrac{1}{7}\left(-\dfrac{3}{7}\right)$

$= x^2 - \dfrac{3}{7}x + \dfrac{1}{7}(x) - \dfrac{3}{49}$

$= x^2 - \dfrac{2}{7}x - \dfrac{3}{49}$

**61.** $(3y + 5)^2 = (3y + 5)(3y + 5)$
$\qquad\qquad = 3y(3y) + 3y(5) + 5(3y) + 5(5)$
$\qquad\qquad = 9y^2 + 15y + 15y + 25$
$\qquad\qquad = 9y^2 + 30y + 25$

**63.** $(a + 4)(a^2 - 6a + 6)$
$= a(a^2) + a(-6a) + a(6) + 4(a^2) + 4(-6a) + 4(6)$
$= a^3 - 6a^2 + 6a + 4a^2 - 24a + 24$
$= a^3 - 2a^2 - 18a + 24$

**65.** $(2x - 5)^3$
$= (2x - 5)(2x - 5)(2x - 5)$
$= (4x^2 - 10x - 10x + 25)(2x - 5)$
$= (4x^2 - 20x + 25)(2x - 5)$
$= (4x^2 - 20x + 25)2x + (4x^2 - 20x + 25)(-5)$
$= 8x^3 - 40x^2 + 50x - 20x^2 + 100x - 125$
$= 8x^3 - 60x^2 + 150x - 125$

**67.** $(4x + 5)(8x^2 + 2x - 4)$
$= 4x(8x^2) + 4x(2x) + 4x(-4) + 5(8x^2)$
$\qquad\qquad + 5(2x) + 5(-4)$
$= 32x^3 + 8x^2 - 16x + 40x^2 + 10x - 20$
$= 32x^3 + 48x^2 - 6x - 20$

**69.**
$$
\begin{array}{r}
3x^2 + 2x - 4 \\
\times\quad\quad\quad 2x^2 - 4x + 3 \\
\hline
9x^2\ +6x - 12 \\
-12x^3 - 8x^2 + 16x\quad\quad\quad \\
6x^4 + 4x^3\ -8x^2\quad\quad\quad\quad\quad \\
\hline
6x^4 - 8x^3 - 7x^2\ +22x - 12
\end{array}
$$

**71.** $(2x - 5)(2x + 5)$
$= 2x(2x) + 2x(5) + (-5)(2x) + (-5)(5)$
$= 4x^2 + 10x - 10x - 25$
$= 4x^2 - 25$
The area is $(4x^2 - 25)$ square yards.

**73.** $\dfrac{1}{2}(3x - 2)(4x) = 2x(3x - 2)$
$\qquad\qquad\qquad = 2x(3x) + 2x(-2)$
$\qquad\qquad\qquad = 6x^2 - 4x$
The area is $(6x^2 - 4x)$ square inches.

**75.** Add: $5a + 15a = (5 + 15)a = 20a$
Subtract: $5a - 15a = (5 - 15)a = -10a$
Multiply: $5a \cdot 15a = 5 \cdot 15 \cdot a^1 \cdot a^1 = 75a^{1+1} = 75a^2$
Divide: $\dfrac{5a}{15a} = \dfrac{1}{3}a^{1-1} = \dfrac{1}{3}a^0 = \dfrac{1}{3} \cdot 1 = \dfrac{1}{3}$

**77.** Add: $-3y^5 + 9y^4$ cannot be simplified.

Subtract: $-3y^5 - 9y^4$ cannot be simplified.

Multiply: $-3y^5 \cdot 9y^4 = -3 \cdot 9 \cdot y^5 \cdot y^4$
$$= -27 y^{5+4}$$
$$= -27 y^9$$

Divide: $\dfrac{-3y^5}{9y^4} = -\dfrac{3}{9} \cdot y^{5-4} = -\dfrac{1}{3} y^1 = -\dfrac{y}{3}$

**79. a.** $(3x+5)+(3x+7) = 3x+5+3x+7$
$$= 6x+12$$

**b.** $(3x+5)(3x+7)$
$$= 3x(3x+7)+5(3x+7)$$
$$= 3x(3x)+3x(7)+5(3x)+5(7)$$
$$= 9x^2 + 21x + 15x + 35$$
$$= 9x^2 + 36x + 35$$

answers may vary

**81.** $(3x-1)+(10x-6) = 3x-1+10x-6 = 13x-7$

**83.** $(3x-1)(10x-6)$
$$= 3x(10x-6)+(-1)(10x-6)$$
$$= 3x(10x)+3x(-6)+(-1)(10x)+(-1)(-6)$$
$$= 30x^2 - 18x - 10x + 6$$
$$= 30x^2 - 28x + 6$$

**85.** $(3x-1)-(10x-6) = (3x-1)+(-10x+6)$
$$= 3x-1-10x+6$$
$$= -7x+5$$

**87.** left rectangle: $x \cdot x = x^2$
right rectangle: $x \cdot 3 = 3x$
left rectangle + right rectangle: $x^2 + 3x$

**89.** top left rectangle: $x \cdot x = x^2$
top right rectangle: $x \cdot 3 = 3x$
bottom left rectangle: $2 \cdot x = 2x$
bottom right rectangle: $2 \cdot 3 = 6$
entire figure: $x^2 + 3x + 2x + 6 = x^2 + 5x + 6$

**91.** $5a + 6a = (5+6)a = 11a$

**93.** $(5x)^2 + (2y)^2 = (5x)(5x)+(2y)(2y)$
$$= 25x^2 + 4y^2$$

**95. a.** $(a+b)(a-b) = a(a)+a(-b)+b(a)+b(-b)$
$$= a^2 - ab + ab - b^2$$
$$= a^2 - b^2$$

**b.** $(2x+3y)(2x-3y)$
$$= 2x(2x)+2x(-3y)+3y(2x)+3y(-3y)$$
$$= 4x^2 - 6xy + 6xy - 9y^2$$
$$= 4x^2 - 9y^2$$

**c.** $(4x+7)(4x-7)$
$$= 4x(4x)+4x(-7)+7(4x)+7(-7)$$
$$= 16x^2 - 28x + 28x - 49$$
$$= 16x^2 - 49$$

**d.** answers may vary

**97.** larger square: $(x+3)^2 = (x+3)(x+3)$
$$= x(x)+x(3)+3(x)+3(3)$$
$$= x^2 + 3x + 3x + 9$$
$$= x^2 + 6x + 9$$

smaller square: $2^2 = 2 \cdot 2 = 4$
shaded region: $x^2 + 6x + 9 - 4 = x^2 + 6x + 5$
The area of the shaded region is
$(x^2 + 6x + 5)$ square units.

**Section 5.4 Practice**

**1.** $(x+2)(x-5)$
$$= (x)(x)+(x)(-5)+(2)(x)+(2)(-5)$$
$$= x^2 - 5x + 2x - 10$$
$$= x^2 - 3x - 10$$

**2.** $(4x-9)(x-1)$
$$= 4x(x)+4x(-1)+(-9)(x)+(-9)(-1)$$
$$= 4x^2 - 4x - 9x + 9$$
$$= 4x^2 - 13x + 9$$

**3.** $3(x+5)(3x-1) = 3(3x^2 - x + 15x - 5)$
$$= 3(3x^2 + 14x - 5)$$
$$= 9x^2 + 42x - 15$$

**4.** $(4x-1)^2$
$$= (4x-1)(4x-1)$$
$$= (4x)(4x)+(4x)(-1)+(-1)(4x)+(-1)(-1)$$
$$= 16x^2 - 4x - 4x + 1$$
$$= 16x^2 - 8x + 1$$

**5. a.** $(b+3)^2 = b^2 + 2(b)(3) + 3^2 = b^2 + 6b + 9$

  **b.** $(x-y)^2 = x^2 - 2(x)(y) + y^2 = x^2 - 2xy + y^2$

  **c.** $(3y+2)^2 = (3y)^2 + 2(3y)(2) + 2^2$
$$= 9y^2 + 12y + 4$$

  **d.** $(a^2 - 5b)^2 = (a^2)^2 - 2(a^2)(5b) + (5b)^2$
$$= a^4 - 10a^2 b + 25b^2$$

**6. a.** $3(x+5)(x-5) = 3(x^2 - 5^2)$
$$= 3x(x^2 - 25)$$
$$= 3x^2 - 75$$

  **b.** $(4b-3)(4b+3) = (4b)^2 - 3^2 = 16b^2 - 9$

  **c.** $\left(x + \dfrac{2}{3}\right)\left(x - \dfrac{2}{3}\right) = x^2 - \left(\dfrac{2}{3}\right)^2 = x^2 - \dfrac{4}{9}$

  **d.** $(5s+t)(5s-t) = (5s)^2 - t^2 = 25s^2 - t^2$

  **e.** $(2y - 3z^2)(2y + 3z^2) = (2y)^2 - (3z^2)^2$
$$= 4y^2 - 9z^4$$

**7. a.** $(4x+3)(x-6) = 4x^2 - 24x + 3x - 18$
$$= 4x^2 - 21x - 18$$

  **b.** $(7b-2)^2 = (7b)^2 - 2(7b)(2) + 2^2$
$$= 49b^2 - 28b + 4$$

  **c.** $(x+0.4)(x-0.4) = x^2 - (0.4)^2 = x^2 - 0.16$

  **d.** $\left(x^2 - \dfrac{3}{7}\right)\left(3x^4 + \dfrac{2}{7}\right)$
$$= 3x^6 + \dfrac{2}{7}x^2 - \dfrac{9}{7}x^4 - \dfrac{6}{49}$$
$$= 3x^6 - \dfrac{9}{7}x^4 + \dfrac{2}{7}x^2 - \dfrac{6}{49}$$

  **e.** $(x+1)(x^2 + 5x - 2)$
$$= x(x^2 + 5x - 2) + 1(x^2 + 5x - 2)$$
$$= x^3 + 5x^2 - 2x + x^2 + 5x - 2$$
$$= x^3 + 6x^2 + 3x - 2$$

## Vocabulary, Readiness & Video Check 5.4

**1.** $(x+4)^2 = x^2 + 2(x)(4) + 4^2$
$$= x^2 + 8x + 16 \neq x^2 + 16$$
The statement is false.

**2.** $(x+6)(2x-1) = 2x^2 - x + 12x - 6$
$$= 2x^2 + 11x - 6$$
The statement is true.

**3.** $(x+4)(x-4) = x^2 - 4^2 = x^2 - 16 \neq x^2 + 16$
The statement is false.

**4.** $(x-1)(x^3 + 3x - 1)$
$$= x(x^3 + 3x - 1) - 1(x^3 + 3x - 1)$$
$$= x^4 + 3x^2 - x - x^3 - 3x + 1$$
$$= x^4 - x^3 + 3x^2 - 4x + 1$$
This is a polynomial of degree 4; the statement is false.

**5.** a binomial times a binomial

**6.** FOIL order for multiplication, distributive property

**7.** Multiplying gives you four terms, and the two like terms will always subtract out.

**8.** No; the FOIL method is used for multiplying a binomial and a binomial.

## Exercise Set 5.4

**1.** $(x+3)(x+4) = x^2 + 4x + 3x + 12 = x^2 + 7x + 12$

**3.** $(x-5)(x+10) = x^2 + 10x - 5x - 50$
$$= x^2 + 5x - 50$$

**5.** $(5x-6)(x+2) = 5x^2 + 10x - 6x - 12$
$$= 5x^2 + 4x - 12$$

**7.** $5(y-6)(4y-1) = 5(4y^2 - 1y - 24y + 6)$
$$= 5(4y^2 - 25y + 6)$$
$$= 20y^2 - 125y + 30$$

**9.** $(2x+5)(3x-1) = 6x^2 - 2x + 15x - 5$
$$= 6x^2 + 13x - 5$$

**11.** $\left(x - \dfrac{1}{3}\right)\left(x + \dfrac{2}{3}\right) = x^2 + \dfrac{2}{3}x - \dfrac{1}{3}x - \dfrac{2}{9}$

$\quad\quad\quad\quad\quad\quad = x^2 + \dfrac{1}{3}x - \dfrac{2}{9}$

**13.** $(x+2)^2 = x^2 + 2(x)(2) + 2^2 = x^2 + 4x + 4$

**15.** $(2x-1)^2 = (2x)^2 - 2(2x)(1) + (1)^2$

$\quad\quad\quad\quad = 4x^2 - 4x + 1$

**17.** $(3a-5)^2 = (3a)^2 - 2(3a)(5) + 5^2$

$\quad\quad\quad\quad = 9a^2 - 30a + 25$

**19.** $(5x+9)^2 = (5x)^2 + 2(5x)(9) + 9^2$

$\quad\quad\quad\quad = 25x^2 + 90x + 81$

**21.** $(a-7)(a+7) = a^2 - 7^2 = a^2 - 49$

**23.** $(3x-1)(3x+1) = (3x)^2 - 1^2 = 9x^2 - 1$

**25.** $\left(3x - \dfrac{1}{2}\right)\left(3x + \dfrac{1}{2}\right) = (3x)^2 - \left(\dfrac{1}{2}\right)^2 = 9x^2 - \dfrac{1}{4}$

**27.** $(9x+y)(9x-y) = (9x)^2 - y^2 = 81x^2 - y^2$

**29.** $(2x+0.1)(2x-0.1) = (2x)^2 - (0.1)^2 = 4x^2 - 0.01$

**31.** $(a+5)(a+4) = a^2 + 4a + 5a + 20 = a^2 + 9a + 20$

**33.** $(a+7)^2 = a^2 + 2(a)(7) + 7^2 = a^2 + 14a + 49$

**35.** $(4a+1)(3a-1) = 12a^2 - 4a + 3a - 1$

$\quad\quad\quad\quad\quad = 12a^2 - a - 1$

**37.** $(x+2)(x-2) = x^2 - 2^2 = x^2 - 4$

**39.** $(3a+1)^2 = (3a)^2 + 2(3a)(1) + 1^2 = 9a^2 + 6a + 1$

**41.** $(x^2+y)(4x-y^4) = 4x^3 - x^2y^4 + 4xy - y^5$

**43.** $(x+3)(x^2 - 6x + 1)$

$\quad = x(x^2 - 6x + 1) + 3(x^2 - 6x + 1)$

$\quad = x^3 - 6x^2 + x + 3x^2 - 18x + 3$

$\quad = x^3 - 3x^2 - 17x + 3$

**45.** $(2a-3)^2 = (2a)^2 - 2(2a)(3) + (3)^2$

$\quad\quad\quad\quad = 4a^2 - 12a + 9$

**47.** $(5x-6z)(5x+6z) = (5x)^2 - (6z)^2 = 25x^2 - 36z^2$

**49.** $(x^5-3)(x^5-5) = x^{10} - 5x^5 - 3x^5 + 15$

$\quad\quad\quad\quad\quad = x^{10} - 8x^5 + 15$

**51.** $(x+0.8)(x-0.8) = x^2 - 0.8^2 = x^2 - 0.64$

**53.** $(a^3+11)(a^4-3) = a^7 - 3a^3 + 11a^4 - 33$

**55.** $3(x-2)^2 = 3[x^2 - 2(x)(2) + 2^2]$

$\quad\quad\quad\quad = 3(x^2 - 4x + 4)$

$\quad\quad\quad\quad = 3x^2 - 12x + 12$

**57.** $(3b+7)(2b-5) = 6b^2 - 15b + 14b - 35$

$\quad\quad\quad\quad\quad = 6b^2 - b - 35$

**59.** $(7p-8)(7p+8) = (7p)^2 - (8)^2 = 49p^2 - 64$

**61.** $\left(\dfrac{1}{3}a^2 - 7\right)\left(\dfrac{1}{3}a^2 + 7\right) = \left(\dfrac{1}{3}a^2\right)^2 - (7)^2$

$\quad\quad\quad\quad\quad\quad = \dfrac{1}{9}a^4 - 49$

**63.** $5x^2(3x^2 - x + 2) = 5x^2(3x^2) + 5x^2(-x) + 5x^2(2)$

$\quad\quad\quad\quad\quad = 15x^4 - 5x^3 + 10x^2$

**65.** $(2r-3s)(2r+3s) = (2r)^2 - (3s)^2 = 4r^2 - 9s^2$

**67.** $(3x-7y)^2 = (3x)^2 - 2(3x)(7y) + (7y)^2$

$\quad\quad\quad\quad = 9x^2 - 42xy + 49y^2$

**69.** $(4x+5)(4x-5) = (4x)^2 - 5^2 = 16x^2 - 25$

**71.** $(8x+4)^2 = (8x)^2 + 2(8x)(4) + (4)^2$

$\quad\quad\quad\quad = 64x^2 + 64x + 16$

**73.** $\left(a - \dfrac{1}{2}y\right)\left(a + \dfrac{1}{2}y\right) = a^2 - \left(\dfrac{1}{2}y\right)^2 = a^2 - \dfrac{1}{4}y^2$

**75.** $\left(\dfrac{1}{5}x - y\right)\left(\dfrac{1}{5}x + y\right) = \left(\dfrac{1}{5}x\right)^2 - y^2 = \dfrac{1}{25}x^2 - y^2$

**77.** $(a+1)(3a^2 - a + 1)$
$= a(3a^2 - a + 1) + 1(3a^2 - a + 1)$
$= 3a^3 - a^2 + a + 3a^2 - a + 1$
$= 3a^3 + 2a^2 + 1$

**79.** $(2x+1)^2 = (2x)^2 + 2(2x)(1) + 1^2 = 4x^2 + 4x + 1$
The area is $(4x^2 + 4x + 1)$ square feet.

**81.** $\dfrac{50b^{10}}{70b^5} = \dfrac{50}{70}b^{10-5} = \dfrac{5b^5}{7}$

**83.** $\dfrac{8a^{17}b^{15}}{-4a^7b^{10}} = \dfrac{8}{-4}a^{17-7}b^{15-10} = -2a^{10}b^5$

**85.** $\dfrac{2x^4y^{12}}{3x^4y^4} = \dfrac{2}{3}x^{4-4}y^{12-4} = \dfrac{2y^8}{3}$

**87.** $(-1, 1)$ and $(2, 2)$
$m = \dfrac{y_2 - y_1}{x_2 - x_1} = \dfrac{2-1}{2-(-1)} = \dfrac{1}{3}$

**89.** $(-1, -2)$ and $(1, 0)$
$m = \dfrac{y_2 - y_1}{x_2 - x_1} = \dfrac{0-(-2)}{1-(-1)} = \dfrac{2}{2} = 1$

**91.** $(a-b)^2 = a^2 - 2ab + b^2$
Choice c.

**93.** $(a+b)^2 = a^2 + 2ab + b^2$
Choice d.

**95.** From FOIL, the first term in the result is
$(x^{\square})^2 = x^{2\square}$. Thus, $2\square = 4$ so $\square = 2$.

**97.** $(x^2 - 1)^2 - x^2 = ((x^2)^2 - 2(x^2)(1) + 1^2) - x^2$
$= (x^4 - 2x^2 + 1) - x^2$
$= x^4 - 2x^2 + 1 - x^2$
$= x^4 - 3x^2 + 1$
The area is $(x^4 - 3x^2 + 1)$ square meters.

**99.** $(5x-3)^2 - (x+1)^2$
$= (25x^2 - 30x + 9) - (x^2 + 2x + 1)$
$= 25x^2 - 30x + 9 - x^2 - 2x - 1$
$= (24x^2 - 32x + 8)$
The shaded area is
$(24x^2 - 32x + 8)$ square meters.

**101.** $(x+5)(x+5) = (x+5)^2$
$= x^2 + 2(x)(5) + 5^2$
$= x^2 + 10x + 25$
The area is $(x^2 + 10x + 25)$ square units.

**103.** answers may vary

**105.** answers may vary

**107.** answers may vary

**109.** $[(x+y)-3][(x+y)+3] = (x+y)^2 - 3^2$
$= x^2 + 2xy + y^2 - 9$

**111.** $[(a-3)+b][(a-3)-b] = (a-3)^2 - b^2$
$= a^2 - 6a + 9 - b^2$

**Integrated Review**

**1.** $(5x^2)(7x^3) = (5\cdot 7)(x^2 \cdot x^3) = 35x^5$

**2.** $(4y^2)(8y^7) = (4\cdot 8)(y^2 \cdot y^7) = 32y^9$

**3.** $-4^2 = -(4\cdot 4) = -16$

**4.** $(-4)^2 = (-4)(-4) = 16$

**5.** $(x-5)(2x+1) = 2x^2 + x - 10x - 5$
$= 2x^2 - 9x - 5$

**6.** $(3x-2)(x+5) = 3x^2 + 15x - 2x - 10$
$= 3x^2 + 13x - 10$

**7.** $(x-5) + (2x+1) = x - 5 + 2x + 1 = 3x - 4$

**8.** $(3x-2) + (x+5) = 3x - 2 + x + 5 = 4x + 3$

**9.** $\dfrac{7x^9y^{12}}{x^3y^{10}} = 7x^{9-3}y^{12-10} = 7x^6y^2$

**10.** $\dfrac{20a^2b^8}{14a^2b^2} = \dfrac{20}{14}a^{2-2}b^{8-2} = \dfrac{10b^6}{7}$

**11.** $(12m^7n^6)^2 = 12^2 m^{7\cdot2} n^{6\cdot2} = 144m^{14}n^{12}$

**12.** $(4y^9z^{10})^3 = 4^3 y^{9\cdot3} z^{10\cdot3} = 64y^{27}z^{30}$

**13.** $3(4y-3)(4y+3) = 3[(4y)^2 - 3^2]$
$\qquad\qquad\qquad = 3(16y^2 - 9)$
$\qquad\qquad\qquad = 48y^2 - 27$

**14.** $2(7x-1)(7x+1) = 2[(7x)^2 - 1^2]$
$\qquad\qquad\qquad = 2(49x^2 - 1)$
$\qquad\qquad\qquad = 98x^2 - 2$

**15.** $(x^7y^5)^9 = x^{7\cdot9} y^{5\cdot9} = x^{63}y^{45}$

**16.** $(3^1 x^9)^3 = 3^{1\cdot3} x^{9\cdot3} = 3^3 x^{27} = 27x^{27}$

**17.** $(7x^2 - 2x + 3) - (5x^2 + 9)$
$\qquad = 7x^2 - 2x + 3 - 5x^2 - 9$
$\qquad = 2x^2 - 2x - 6$

**18.** $(10x^2 + 7x - 9) - (4x^2 - 6x + 2)$
$\qquad = 10x^2 + 7x - 9 - 4x^2 + 6x - 2$
$\qquad = 6x^2 + 13x - 11$

**19.** $0.7y^2 - 1.2 + 1.8y^2 - 6y + 1 = 2.5y^2 - 6y - 0.2$

**20.** $7.8x^2 - 6.8x + 3.3 + 0.6x^2 - 9$
$\qquad = 8.4x^2 - 6.8x - 5.7$

**21.** $(x+4y)^2 = x^2 + 2(x)(4y) + (4y)^2$
$\qquad\qquad\quad = x^2 + 8xy + 16y^2$

**22.** $(y-9z)^2 = y^2 - 2(y)(9z) + (9z)^2$
$\qquad\qquad\quad = y^2 - 18yz + 81z^2$

**23.** $(x+4y) + (x+4y) = x + 4y + x + 4y = 2x + 8y$

**24.** $(y-9z) + (y-9z) = y - 9z + y - 9z = 2y - 18z$

**25.** $7x^2 - 6xy + 4(y^2 - xy) = 7x^2 - 6xy + 4y^2 - 4xy$
$\qquad\qquad\qquad\qquad = 7x^2 - 10xy + 4y^2$

**26.** $5a^2 - 3ab + 6(b^2 - a^2) = 5a^2 - 3ab + 6b^2 - 6a^2$
$\qquad\qquad\qquad\qquad\quad = -a^2 - 3ab + 6b^2$

**27.** $(x-3)(x^2 + 5x - 1)$
$\qquad = x(x^2 + 5x - 1) - 3(x^2 + 5x - 1)$
$\qquad = x^3 + 5x^2 - x - 3x^2 - 15x + 3$
$\qquad = x^3 + 2x^2 - 16x + 3$

**28.** $(x+1)(x^2 - 3x - 2)$
$\qquad = x(x^2 - 3x - 2) + 1(x^2 - 3x - 2)$
$\qquad = x^3 - 3x^2 - 2x + x^2 - 3x - 2$
$\qquad = x^3 - 2x^2 - 5x - 2$

**29.** $(2x^3 - 7)(3x^2 + 10)$
$\qquad = 2x^3(3x^2) + 2x^3(10) - 7(3x^2) - 7(10)$
$\qquad = 6x^5 + 20x^3 - 21x^2 - 70$

**30.** $(5x^3 - 1)(4x^4 + 5)$
$\qquad = 5x^3(4x^4) + 5x^3(5) - 1(4x^4) - 1(5)$
$\qquad = 20x^7 + 25x^3 - 4x^4 - 5$

**31.** $(2x-7)(x^2 - 6x + 1)$
$\qquad = 2x(x^2 - 6x + 1) - 7(x^2 - 6x + 1)$
$\qquad = 2x^3 - 12x^2 + 2x - 7x^2 + 42x - 7$
$\qquad = 2x^3 - 19x^2 + 44x - 7$

**32.** $(5x-1)(x^2 + 2x - 3)$
$\qquad = 5x(x^2 + 2x - 3) - 1(x^2 + 2x - 3)$
$\qquad = 5x^3 + 10x^2 - 15x - x^2 - 2x + 3$
$\qquad = 5x^3 + 9^2 - 17x + 3$

**33.** $5x^3 + 5y^3$ cannot be simplified.

**34.** $(5x^3)(5y^3) = 5 \cdot 5x^3 y^3 = 25x^3y^3$

**35.** $(5x^3)^3 = 5^3 x^{3\cdot3} = 125x^9$

**36.** $\dfrac{5x^3}{5y^3} = \dfrac{x^3}{y^3}$

**37.** $x + x = 2x$

**38.** $x \cdot x = x^2$

**Section 5.5 Practice**

1. **a.** $5^{-3} = \dfrac{1}{5^3} = \dfrac{1}{125}$

   **b.** $3y^{-4} = 3 \cdot \dfrac{1}{y^4} = \dfrac{3}{y^4}$

   **c.** $3^{-1} + 2^{-1} = \dfrac{1}{3} + \dfrac{1}{2} = \dfrac{2}{6} + \dfrac{3}{6} = \dfrac{5}{6}$

   **d.** $(-5)^{-2} = \dfrac{1}{(-5)^2} = \dfrac{1}{(-5)(-5)} = \dfrac{1}{25}$

   **e.** $x^{-5} = \dfrac{1}{x^5}$

2. **a.** $\dfrac{1}{s^{-5}} = \dfrac{s^5}{1} = s^5$

   **b.** $\dfrac{1}{2^{-3}} = \dfrac{2^3}{1} = 8$

   **c.** $\dfrac{x^{-7}}{y^{-5}} = \dfrac{y^5}{x^7}$

   **d.** $\dfrac{4^{-3}}{3^{-2}} = \dfrac{3^2}{4^3} = \dfrac{9}{64}$

3. **a.** $\dfrac{x^{-3}}{x^2} = x^{-3-2} = x^{-5} = \dfrac{1}{x^5}$

   **b.** $\dfrac{5}{y^{-7}} = 5 \cdot \dfrac{1}{y^{-7}} = 5 \cdot y^7 = 5y^7$

   **c.** $\dfrac{z}{z^{-4}} = \dfrac{z^1}{z^{-4}} = z^{1-(-4)} = z^5$

   **d.** $\left(\dfrac{5}{9}\right)^{-2} = \dfrac{5^{-2}}{9^{-2}} = \dfrac{9^2}{5^2} = \dfrac{81}{25}$

4. **a.** $(a^4 b^{-3})^{-5} = a^{-20} b^{15} = \dfrac{b^{15}}{a^{20}}$

   **b.** $\dfrac{x^2 (x^5)^3}{x^7} = \dfrac{x^2 \cdot x^{15}}{x^7}$
   $= \dfrac{x^{2+15}}{x^7}$
   $= \dfrac{x^{17}}{x^7}$
   $= x^{17-7}$
   $= x^{10}$

   **c.** $\left(\dfrac{5p^8}{q}\right)^{-2} = \dfrac{5^{-2}(p^8)^{-2}}{q^{-2}}$
   $= \dfrac{5^{-2}p^{-16}}{q^{-2}}$
   $= \dfrac{q^2}{5^2 p^{16}}$
   $= \dfrac{q^2}{25 p^{16}}$

   **d.** $\dfrac{6^{-2} x^{-4} y^{-7}}{6^{-3} x^3 y^{-9}} = 6^{-2-(-3)} x^{-4-3} y^{-7-(-9)}$
   $= 6^1 x^{-7} y^2$
   $= \dfrac{6y^2}{x^7}$

   **e.** $\left(\dfrac{-3x^4 y}{x^2 y^{-2}}\right)^3 = \dfrac{(-3)^3 x^{12} y^3}{x^6 y^{-6}}$
   $= \dfrac{-27 x^{12} y^3}{x^6 y^{-6}}$
   $= -27 x^{12-6} y^{3-(-6)}$
   $= -27 x^6 y^9$

5. **a.** $0.000007 = 7 \times 10^{-6}$
   The decimal point is moved 6 places, and the original number is less than 1, so the count is −6.

   **b.** $20,700,000 = 2.07 \times 10^7$
   The decimal point is moved 7 places, and the original number is 10 or greater, so the count is 7.

**c.**　$0.0043 = 4.3 \times 10^{-3}$
The decimal point is moved 3 places, and the original number is less than 1, so the count is –3.

**d.**　$812,000,000 = 8.12 \times 10^{8}$
The decimal point is moved 8 places, and the original number is 10 or greater, so the count is 8.

**6. a.**　Move the decimal point 4 places to the left.
$3.67 \times 10^{-4} = 0.000367$

**b.**　Move the decimal point 6 places to the right.
$8.954 \times 10^{6} = 8,954,000$

**c.**　Move the decimal point 5 places to the left.
$2.009 \times 10^{-5} = 0.00002009$

**d.**　Move the decimal point 3 places to the right.
$4.054 \times 10^{3} = 4054$

**7. a.**　$(5 \times 10^{-4})(8 \times 10^{6}) = (5 \cdot 8) \times (10^{-4} \cdot 10^{6})$
$= 40 \times 10^{2}$
$= 4000$

**b.**　$\dfrac{64 \times 10^{3}}{32 \times 10^{-7}} = \dfrac{64}{32} \times 10^{3-(-7)}$
$= 2 \times 10^{10}$
$= 20,000,000,000$

**Calculator Explorations**

1.　$5.31 \times 10^{3} = 5.31 \text{ EE } 3$

2.　$-4.8 \times 10^{14} = -4.8 \text{ EE } 14$

3.　$6.6 \times 10^{-9} = 6.6 \text{ EE } -9$

4.　$-9.9811 \times 10^{-2} = -9.9811 \text{ EE } -2$

5.　$3,000,000 \times 5,000,000 = 1.5 \times 10^{13}$

6.　$230,000 \times 1000 = 2.3 \times 10^{8}$

7.　$(3.26 \times 10^{6})(2.5 \times 10^{13}) = 8.15 \times 10^{19}$

8.　$(8.76 \times 10^{-4})(1.237 \times 10^{9}) = 1.083612 \times 10^{6}$

**Vocabulary, Readiness & Video Check 5.5**

1.　The expression $x^{-3}$ equals $\dfrac{1}{x^{3}}$.

2.　The expression $5^{-4}$ equals $\dfrac{1}{625}$.

3.　The number $3.021 \times 10^{-3}$ is written in <u>scientific notation</u>.

4.　The number 0.0261 is written in <u>standard form</u>.

5.　A negative exponent has nothing to do with the sign of the simplified result.

6.　power of a product rule, power rule for exponents, negative exponent definition, quotient rule for exponents

7.　When you move the decimal point to the left, the sign of the exponent will be positive; when you move the decimal point to the right, the sign of the exponent will be negative.

8.　the exponent on 10

9.　the quotient rule

**Exercise Set 5.5**

1.　$4^{-3} = \dfrac{1}{4^{3}} = \dfrac{1}{64}$

3.　$(-3)^{-4} = \dfrac{1}{(-3)^{4}} = \dfrac{1}{81}$

5.　$7x^{-3} = 7 \cdot \dfrac{1}{x^{3}} = \dfrac{7}{x^{3}}$

7.　$\left(\dfrac{1}{2}\right)^{-5} = \dfrac{1^{-5}}{2^{-5}} = \dfrac{2^{5}}{1^{5}} = 32$

9.　$\left(-\dfrac{1}{4}\right)^{-3} = \dfrac{(-1)^{-3}}{(4)^{-3}} = \dfrac{4^{3}}{(-1)^{3}} = \dfrac{64}{-1} = -64$

11.　$3^{-1} + 5^{-1} = \dfrac{1}{3} + \dfrac{1}{5} = \dfrac{5}{15} + \dfrac{3}{15} = \dfrac{8}{15}$

**13.** $\dfrac{1}{p^{-3}} = p^3$

**15.** $\dfrac{p^{-5}}{q^{-4}} = \dfrac{q^4}{p^5}$

**17.** $\dfrac{x^{-2}}{x} = x^{-2-1} = x^{-3} = \dfrac{1}{x^3}$

**19.** $\dfrac{z^{-4}}{z^{-7}} = z^{-4-(-7)} = z^3$

**21.** $3^{-2} + 3^{-1} = \dfrac{1}{3^2} + \dfrac{1}{3} = \dfrac{1}{9} + \dfrac{1}{3} = \dfrac{1}{9} + \dfrac{3}{9} = \dfrac{4}{9}$

**23.** $\dfrac{-1}{p^{-4}} = -1(p^4) = -p^4$

**25.** $-2^0 - 3^0 = -1(1) - 1 = -1 - 1 = -2$

**27.** $\dfrac{x^2 x^5}{x^3} = x^{2+5-3} = x^4$

**29.** $\dfrac{p^2 p}{p^{-1}} = p^{2+1-(-1)} = p^4$

**31.** $\dfrac{(m^5)^4 m}{m^{10}} = \dfrac{m^{20}m}{m^{10}} = m^{20+1-10} = m^{11}$

**33.** $\dfrac{r}{r^{-3}r^{-2}} = r^{1-(-3)-(-2)} = r^6$

**35.** $(x^5 y^3)^{-3} = x^{5(-3)} y^{3(-3)} = x^{-15} y^{-9} = \dfrac{1}{x^{15} y^9}$

**37.** $\dfrac{(x^2)^3}{x^{10}} = \dfrac{x^6}{x^{10}} = x^{6-10} = x^{-4} = \dfrac{1}{x^4}$

**39.** $\dfrac{(a^5)^2}{(a^3)^4} = \dfrac{a^{10}}{a^{12}} = a^{10-12} = a^{-2} = \dfrac{1}{a^2}$

**41.** $\dfrac{8k^4}{2k} = \dfrac{8}{2} \cdot k^{4-1} = 4k^3$

**43.** $\dfrac{-6m^4}{-2m^3} = \dfrac{-6}{-2} \cdot m^{4-3} = 3m$

**45.** $\dfrac{-24a^6 b}{6ab^2} = \dfrac{-24}{6} \cdot a^{6-1} b^{1-2} = -4a^5 b^{-1} = -\dfrac{4a^5}{b}$

**47.** $(-2x^3 y^{-4})(3x^{-1} y) = -2(3)x^{3+(-1)} y^{-4+1}$
$$= -6x^2 y^{-3}$$
$$= -\dfrac{6x^2}{y^3}$$

**49.** $(a^{-5} b^2)^{-6} = a^{-5(-6)} b^{2(-6)} = a^{30} b^{-12} = \dfrac{a^{30}}{b^{12}}$

**51.** $\left(\dfrac{x^{-2} y^4}{x^3 y^7}\right)^2 = \dfrac{x^{-2(2)} y^{4(2)}}{x^{3(2)} y^{7(2)}}$
$$= \dfrac{x^{-4} y^8}{x^6 y^{14}}$$
$$= x^{-4-6} y^{8-14}$$
$$= x^{-10} y^{-6}$$
$$= \dfrac{1}{x^{10} y^6}$$

**53.** $\dfrac{4^2 z^{-3}}{4^3 z^{-5}} = 4^{2-3} z^{-3-(-5)} = 4^{-1} z^2 = \dfrac{z^2}{4}$

**55.** $\dfrac{2^{-3} x^{-4}}{2^2 x} = 2^{-3-2} x^{-4-1}$
$$= 2^{-5} x^{-5}$$
$$= \dfrac{1}{2^5 x^5}$$
$$= \dfrac{1}{32x^5}$$

**57.** $\dfrac{7ab^{-4}}{7^{-1} a^{-3} b^2} = 7^{1-(-1)} a^{1-(-3)} b^{-4-2}$
$$= 7^2 a^4 b^{-6}$$
$$= \dfrac{49a^4}{b^6}$$

    

**59.** $\left(\dfrac{a^{-5}b}{ab^3}\right)^{-4} = \dfrac{a^{-5(-4)}b^{-4}}{a^{-4}b^{3(-4)}}$

$= \dfrac{a^{20}b^{-4}}{a^{-4}b^{-12}}$

$= a^{20-(-4)}b^{-4-(-12)}$

$= a^{24}b^8$

**61.** $\dfrac{(xy^3)^5}{(xy)^{-4}} = \dfrac{x^5y^{3(5)}}{x^{-4}y^{-4}}$

$= \dfrac{x^5y^{15}}{x^{-4}y^{-4}}$

$= x^{5-(-4)}y^{15-(-4)}$

$= x^9y^{19}$

**63.** $\dfrac{(-2xy^{-3})^{-3}}{(xy^{-1})^{-1}} = \dfrac{(-2)^{-3}x^{-3}y^9}{x^{-1}y^1}$

$= (-2)^{-3}x^{-3-(-1)}y^{9-1}$

$= (-2)^{-3}x^{-2}y^8$

$= \dfrac{y^8}{(-2)^3x^2}$

$= -\dfrac{y^8}{8x^2}$

**65.** $\dfrac{6x^2y^3}{-7xy^5} = -\dfrac{6}{7}x^{2-1}y^{3-5} = -\dfrac{6}{7}x^1y^{-2} = -\dfrac{6x}{7y^2}$

**67.** $\dfrac{(a^4b^{-7})^{-5}}{(5a^2b^{-1})^{-2}} = \dfrac{(a^4)^{-5}(b^{-7})^{-5}}{5^{-2}(a^2)^{-2}(b^{-1})^{-2}}$

$= \dfrac{a^{-20}b^{35}}{5^{-2}a^{-4}b^2}$

$= 5^2a^{-20-(-4)}b^{35-2}$

$= 5^2a^{-16}b^{33}$

$= \dfrac{25b^{33}}{a^{16}}$

**69.** $78,000 = 7.8 \times 10^4$

**71.** $0.00000167 = 1.67 \times 10^{-6}$

**73.** $0.00635 = 6.35 \times 10^{-3}$

**75.** $1,160,000 = 1.16 \times 10^6$

**77.** $2,000,000,000 = 2 \times 10^9$

**79.** $2400 = 2.4 \times 10^3$

**81.** $8.673 \times 10^{-10} = 0.0000000008673$

**83.** $3.3 \times 10^{-2} = 0.033$

**85.** $2.032 \times 10^4 = 20,320$

**87.** $7.0 \times 10^8 = 700,000,000$

**89.** $9.460 \times 10^{12} = 9,460,000,000,000$

**91.** $184,000,000,000 = 1.84 \times 10^{11}$

**93.** $1.55 \times 10^{11} = 155,000,000,000$

**95.** $3.5 \times 10^4 = 35,000$

**97.** $(1.2 \times 10^{-3})(3 \times 10^{-2}) = (1.2 \cdot 3) \times (10^{-3} \cdot 10^{-2})$

$= 3.6 \times 10^{-5}$

$= 0.000036$

**99.** $(4 \times 10^{-10})(7 \times 10^{-9}) = (4 \cdot 7) \times (10^{-10} \cdot 10^{-9})$

$= 28 \times 10^{-19}$

$= 2.8 \times 10^{-18}$

$= 0.0000000000000000028$

**101.** $\dfrac{8 \times 10^{-1}}{16 \times 10^5} = \dfrac{8}{16} \times 10^{-1-5}$

$= 0.5 \times 10^{-6}$

$= 5 \times 10^{-7}$

$= 0.0000005$

**103.** $\dfrac{1.4 \times 10^{-2}}{7 \times 10^{-8}} = \dfrac{1.4}{7} \times 10^{-2-(-8)}$

$= 0.2 \times 10^6$

$= 2.0 \times 10^5$

$= 200,000$

**105.** $\dfrac{5x^7}{3x^4} = \dfrac{5}{3} \cdot x^{7-4} = \dfrac{5x^3}{3}$

**107.** $\dfrac{15z^4y^3}{21zy} = \dfrac{15}{21}z^{4-1}y^{3-1} = \dfrac{5z^3y^2}{7}$

**109.** $\dfrac{1}{y}(5y^2 - 6y + 5) = \dfrac{1}{y}(5y^2) + \dfrac{1}{y}(-6y) + \dfrac{1}{y}(5)$

$$= 5y - 6 + \dfrac{5}{y}$$

**111.** $\left(\dfrac{3x^{-2}}{z}\right)^3 = \dfrac{3^3 x^{-6}}{z^3} = \dfrac{27}{x^6 z^3}$

The volume is $\dfrac{27}{x^6 z^3}$ cubic inches.

**113.** $(2a^3)^3 a^4 + a^5 a^8 = 2^3(a^3)^3 a^4 + a^{5+8}$

$$= 8a^9 a^4 + a^{13}$$
$$= 8a^{13} + a^{13}$$
$$= 9a^{13}$$

**115.** $x^{-5} = \dfrac{1}{x^5}$

**117.** answers may vary

**119. a.** $9.7 \times 10^{-2} = 0.097$

$1.3 \times 10^1 = 130$

$1.3 \times 10^1$ is larger.

**b.** $8.6 \times 10^5 = 860,000$

$4.4 \times 10^7 = 44,000,000$

$4.4 \times 10^7$ is larger.

**c.** $6.1 \times 10^{-2} = 0.061$

$5.6 \times 10^{-4} = 0.00056$

$6.1 \times 10^{-2}$ is larger.

**121.** answers may vary

**123.** $a^{-4m} \cdot a^{5m} = a^{-4m+5m} = a^m$

**125.** $(3y^{2z})^3 = 3^3 y^{2z \cdot 3} = 27y^{6z}$

**127.** $(2.63 \times 10^{12})(-1.5 \times 10^{-10}) = -394.5$

**129.** $t = \dfrac{d}{r}$

$t = \dfrac{238,857}{1.86 \times 10^5} = \dfrac{238,857}{186,000} \approx 1.3$

It takes 1.3 seconds for the reflected light of the moon to reach Earth.

**Section 5.6 Practice**

**1.** $\dfrac{8t^3 + 4t^2}{4t^2} = \dfrac{8t^3}{4t^2} + \dfrac{4t^2}{4t^2} = 2t + 1$

**Check:** $4t^2(2t + 1) = 4t^2(2t) + 4t^2(1)$
$$= 8t^3 + 4t^2$$

**2.** $\dfrac{16x^6 + 20x^3 - 12x}{4x^2} = \dfrac{16x^6}{4x^2} + \dfrac{20x^3}{4x^2} - \dfrac{12x}{4x^2}$

$$= 4x^4 + 5x - \dfrac{3}{x}$$

**Check:** $4x^2\left(4x^4 + 5x - \dfrac{3}{x}\right)$

$$= 4x^2(4x^4) + 4x^2(5x) - 4x^2\left(\dfrac{3}{x}\right)$$
$$= 16x^6 + 20x^3 - 12x$$

**3.** $\dfrac{15x^4y^4 - 10xy + y}{5xy} = \dfrac{15x^4y^4}{5xy} - \dfrac{10xy}{5xy} + \dfrac{y}{5xy}$

$$= 3x^3y^3 - 2 + \dfrac{1}{5x}$$

**Check:** $5xy\left(3x^3y^3 - 2 + \dfrac{1}{5x}\right)$

$$= 5xy(3x^3y^3) - 5xy(2) + 5xy\left(\dfrac{1}{5x}\right)$$
$$= 15x^4y^4 - 10xy + y$$

**4.**

$$\begin{array}{r} x + 3 \phantom{000000} \\ x+2 \overline{) x^2 + 5x + 6} \\ \underline{x^2 + 2x \phantom{0000}} \\ 3x + 6 \\ \underline{3x + 6} \\ 0 \end{array}$$

**Check:** $(x + 2) \cdot (x + 3) + 0 = x^2 + 5x + 6$

The quotient checks.

    

**5.**
$$
\begin{array}{r}
2x+3 \\
2x+1\overline{\smash{\big)}\,4x^2+8x-7} \\
\underline{4x^2+2x} \\
6x-7 \\
\underline{6x+3} \\
-10
\end{array}
$$

$$\frac{4x^2+8x-7}{2x+1}=2x+3+\frac{-10}{2x+1} \text{ or } 2x+3-\frac{10}{2x+1}$$

**Check:**

$$(2x+1)(2x+3)+(-10)=(4x^2+8x+3)-10$$
$$=4x^2+8x-7$$

The quotient checks.

**6.** Rewrite $11x-3+9x^3$ as $9x^3+0x^2+11x-3$.

$$
\begin{array}{r}
3x^2-2x+5 \\
3x+2\overline{\smash{\big)}\,9x^3+0x^2+11x\ \ -3} \\
\underline{9x^3+6x^2} \\
-6x^2+11x \\
\underline{-6x^2\ -4x} \\
15x\ \ -3 \\
\underline{15x+10} \\
-13
\end{array}
$$

$$\frac{11x-3+9x^3}{3x+2}=3x^2-2x+5+\frac{-13}{3x+2} \text{ or}$$

$$3x^2-2x+5-\frac{13}{3x+2}$$

**7.** Rewrite $x^2+2$ as $x^2+0x+2$.

$$
\begin{array}{r}
3x^2-2x-9 \\
x^2+0x+2\overline{\smash{\big)}\,3x^4-2x^3-3x^2\ \ +x+4} \\
\underline{3x^4+0x^3+6x^2} \\
-2x^3-9x^2\ \ +x \\
\underline{-2x^3+0x^2-4x} \\
-9x^2+5x\ \ +4 \\
\underline{-9x^2+0x-18} \\
5x+22
\end{array}
$$

$$\frac{3x^4-2x^3-3x^2+x+4}{x^2+2}=3x^2-2x-9+\frac{5x+22}{x^2+2}$$

**8.** Rewrite $x^3+27$ as $x^3+0x^2+0x+27$.

$$
\begin{array}{r}
x^2-3x+9 \\
x+3\overline{\smash{\big)}\,x^3+0x^2+0x+27} \\
\underline{x^3+3x^2} \\
-3x^2+0x \\
\underline{-3x^2-9x} \\
9x+27 \\
\underline{9x+27} \\
0
\end{array}
$$

$$\frac{x^3+27}{x+3}=x^2-3x+9$$

**Vocabulary, Readiness & Video Check 5.6**

**1.** In $6\overline{\smash{\big)}\,18}$ (with quotient $3$), the 18 is the <u>dividend</u>, the 3 is the <u>quotient</u> and the 6 is the <u>divisor</u>.

**2.** In $x+1\overline{\smash{\big)}\,x^2+3x+2}$ (with quotient $x+2$), the $x+1$ is the <u>divisor</u>, the $x^2+3x+2$ is the <u>dividend</u> and the $x+2$ is the <u>quotient</u>.

**3.** $\dfrac{a^6}{a^4}=a^{6-4}=2$

**4.** $\dfrac{p^8}{p^3}=p^{8-3}=p^5$

**5.** $\dfrac{y^2}{y}=\dfrac{y^2}{y^1}=y^{2-1}=y$

**6.** $\dfrac{a^3}{a}=\dfrac{a^3}{a^1}=a^{3-1}=a^2$

**7.** the common denominator

**8.** Filling in missing powers helps you keep like terms lined up and your work clear and neat.

**Exercise Set 5.6**

**1.** $\dfrac{12x^4+3x^2}{x}=\dfrac{12x^4}{x}+\dfrac{3x^2}{x}=12x^3+3x$

**3.** $\dfrac{20x^3 - 30x^2 + 5x + 5}{5} = \dfrac{20x^3}{5} - \dfrac{30x^2}{5} + \dfrac{5x}{5} + \dfrac{5}{5}$
$\qquad\qquad\qquad\qquad\quad = 4x^3 - 6x^2 + x + 1$

**5.** $\dfrac{15p^3 + 18p^2}{3p} = \dfrac{15p^3}{3p} + \dfrac{18p^2}{3p} = 5p^2 + 6p$

**7.** $\dfrac{-9x^4 + 18x^5}{6x^5} = \dfrac{-9x^4}{6x^5} + \dfrac{18x^5}{6x^5} = -\dfrac{3}{2x} + 3$

**9.** $\dfrac{-9x^5 + 3x^4 - 12}{3x^3} = \dfrac{-9x^5}{3x^3} + \dfrac{3x^4}{3x^3} - \dfrac{12}{3x^3}$
$\qquad\qquad\qquad\qquad\quad = -3x^2 + x - \dfrac{4}{x^3}$

**11.** $\dfrac{4x^4 - 6x^3 + 7}{-4x^4} = \dfrac{4x^4}{-4x^4} - \dfrac{6x^3}{-4x^4} + \dfrac{7}{-4x^4}$
$\qquad\qquad\qquad\qquad\quad = -1 + \dfrac{3}{2x} - \dfrac{7}{4x^4}$

**13.**
$$
\begin{array}{r}
x+1 \\
x+3\,\overline{)\,x^2 + 4x + 3} \\
\underline{x^2 + 3x}\phantom{ + 3} \\
x + 3 \\
\underline{x + 3} \\
0
\end{array}
$$

$\dfrac{x^2 + 4x + 3}{x + 3} = x + 1$

**15.**
$$
\begin{array}{r}
2x+3 \\
x+5\,\overline{)\,2x^2 + 13x + 15} \\
\underline{2x^2 + 10x}\phantom{ + 15} \\
3x + 15 \\
\underline{3x + 15} \\
0
\end{array}
$$

$\dfrac{2x^2 + 13x + 15}{x + 5} = 2x + 3$

**17.**
$$
\begin{array}{r}
2x+1 \\
x-4\,\overline{)\,2x^2 - 7x + 3} \\
\underline{2x^2 - 8x}\phantom{ + 3} \\
x + 3 \\
\underline{x - 4} \\
7
\end{array}
$$

$\dfrac{2x^2 - 7x + 3}{x - 4} = 2x + 1 + \dfrac{7}{x - 4}$

**19.**
$$
\begin{array}{r}
3a^2 - 3a + 1 \\
3a+2\,\overline{)\,9a^3 - 3a^2 - 3a + 4} \\
\underline{9a^3 + 6a^2}\phantom{ - 3a + 4} \\
-9a^2 - 3a \\
\underline{-9a^2 - 6a}\phantom{} \\
3a + 4 \\
\underline{3a + 2} \\
2
\end{array}
$$

$\dfrac{9a^3 - 3a^2 - 3a + 4}{3a + 2} = 3a^2 - 3a + 1 + \dfrac{2}{3a + 2}$

**21.**
$$
\begin{array}{r}
4x+3 \\
2x+1\,\overline{)\,8x^2 + 10x + 1} \\
\underline{8x^2 + 4x}\phantom{ + 1} \\
6x + 1 \\
\underline{6x + 3} \\
-2
\end{array}
$$

$\dfrac{8x^2 + 10x + 1}{2x + 1} = 4x + 3 - \dfrac{2}{2x + 1}$

**23.**
$$
\begin{array}{r}
2x^2 + 6x - 5 \\
x-2\,\overline{)\,2x^3 + 2x^2 - 17x + 8} \\
\underline{2x^3 - 4x^2}\phantom{ - 17x + 8} \\
6x^2 - 17x \\
\underline{6x^2 - 12x}\phantom{} \\
-5x + 8 \\
\underline{-5x + 10} \\
-2
\end{array}
$$

$\dfrac{2x^3 + 2x^2 - 17x + 8}{x - 2} = 2x^2 + 6x - 5 - \dfrac{2}{x - 2}$

**25.** Rewrite $x^2 - 36$ as $x^2 + 0x - 36$.

$$
\begin{array}{r}
x + 6 \\
x - 6 \overline{\smash{\big)}\ x^2 + 0x - 36} \\
\underline{x^2 - 6x} \\
6x - 36 \\
\underline{6x - 36} \\
0
\end{array}
$$

$$\frac{x^2 - 36}{x - 6} = x + 6$$

**27.** Rewrite $x^3 - 27$ as $x^3 + 0x^2 + 0x - 27$.

$$
\begin{array}{r}
x^2 + 3x + 9 \\
x - 3 \overline{\smash{\big)}\ x^3 + 0x^2 + 0x - 27} \\
\underline{x^3 - 3x^2} \\
3x^2 + 0x \\
\underline{3x^2 - 9x} \\
9x - 27 \\
\underline{9x - 27} \\
0
\end{array}
$$

$$\frac{x^3 - 27}{x - 3} = x^2 + 3x + 9$$

**29.** Rewrite $1 - 3x^2$ as $-3x^2 + 0x + 1$.

$$
\begin{array}{r}
-3x + 6 \\
x + 2 \overline{\smash{\big)}\ -3x^2 + 0x + 1} \\
\underline{-3x^2 - 6x} \\
6x + 1 \\
\underline{6x + 12} \\
-11
\end{array}
$$

$$\frac{1 - 3x^2}{x + 2} = -3x + 6 - \frac{11}{x + 2}$$

**31.** Rewrite $-4b + 4b^2 - 5$ as $4b^2 - 4b - 5$.

$$
\begin{array}{r}
2b - 1 \\
2b - 1 \overline{\smash{\big)}\ 4b^2 - 4b - 5} \\
\underline{4b^2 - 2b} \\
-2b - 5 \\
\underline{-2b + 1} \\
-6
\end{array}
$$

$$\frac{-4b + 4b^2 - 5}{2b - 1} = 2b - 1 - \frac{6}{2b - 1}$$

**33.** $\dfrac{a^2 b^2 - ab^3}{ab} = \dfrac{a^2 b^2}{ab} - \dfrac{ab^3}{ab} = ab - b^2$

**35.**
$$
\begin{array}{r}
4x + 9 \\
2x - 3 \overline{\smash{\big)}\ 8x^2 + 6x - 27} \\
\underline{8x^2 - 12x} \\
18x - 27 \\
\underline{18x - 27} \\
0
\end{array}
$$

$$\frac{8x^2 + 6x - 27}{2x - 3} = 4x + 9$$

**37.** $\dfrac{2x^2 y + 8x^2 y^2 - xy^2}{2xy} = \dfrac{2x^2 y}{2xy} + \dfrac{8x^2 y^2}{2xy} - \dfrac{xy^2}{2xy}$

$$= x + 4xy - \frac{y}{2}$$

**39.**
$$
\begin{array}{r}
2b^2 + b + 2 \\
b + 4 \overline{\smash{\big)}\ 2b^3 + 9b^2 + 6b - 4} \\
\underline{2b^3 + 8b^2} \\
b^2 + 6b \\
\underline{b^2 + 4b} \\
2b - 4 \\
\underline{2b + 8} \\
-12
\end{array}
$$

$$\frac{2b^3 + 9b^2 + 6b - 4}{b + 4} = 2b^2 + b + 2 - \frac{12}{b + 4}$$

**41.**
$$
\begin{array}{r}
5x - 2 \\
x + 6 \overline{\smash{\big)}\ 5x^2 + 28x - 10} \\
\underline{5x^2 + 30x} \\
-2x - 10 \\
\underline{-2x - 12} \\
2
\end{array}
$$

$$\frac{5x^2 + 28x - 10}{x + 6} = 5x - 2 + \frac{2}{x + 6}$$

**43.** $\dfrac{10x^3 - 24x^2 - 10x}{10x} = \dfrac{10x^3}{10x} - \dfrac{24x^2}{10x} - \dfrac{10x}{10x}$

$$= x^2 - \frac{12x}{5} - 1$$

**45.**
$$x+3 \overline{\smash{\big)}\, 6x^2 + 17x - 4} \quad \underset{6x-1}{}$$
$$\underline{6x^2 + 18x}$$
$$-x - 4$$
$$\underline{-x - 3}$$
$$-1$$

$$\frac{6x^2 + 17x - 4}{x + 3} = 6x - 1 - \frac{1}{x+3}$$

**47.**
$$5x - 2 \overline{\smash{\big)}\, 30x^2 - 17x + 2} \quad \underset{6x-1}{}$$
$$\underline{30x^2 - 12x}$$
$$-5x + 2$$
$$\underline{-5x + 2}$$
$$0$$

$$\frac{30x^2 - 17x + 2}{5x - 2} = 6x - 1$$

**49.** $\dfrac{3x^4 - 9x^3 + 12}{-3x} = \dfrac{3x^4}{-3x} - \dfrac{9x^3}{-3x} + \dfrac{12}{-3x}$
$$= -x^3 + 3x^2 - \frac{4}{x}$$

**51.**
$$x + 3 \overline{\smash{\big)}\, x^3 + 6x^2 + 18x + 27} \quad \underset{x^2+3x+9}{}$$
$$\underline{x^3 + 3x^2}$$
$$3x^2 + 18x$$
$$\underline{3x^2 + 9x}$$
$$9x + 27$$
$$\underline{9x + 27}$$
$$0$$

$$\frac{x^3 + 6x^2 + 18x + 27}{x + 3} = x^2 + 3x + 9$$

**53.** Rewrite $y^3 + 3y^2 + 4$ as $y^3 + 3y^2 + 0y + 4$.
$$y - 2 \overline{\smash{\big)}\, y^3 + 3y^2 + 0y + 4} \quad \underset{y^2+5y+10}{}$$
$$\underline{y^3 - 2y^2}$$
$$5y^2 + 0y$$
$$\underline{5y^2 - 10y}$$
$$10y + 4$$
$$\underline{10y - 20}$$
$$24$$

$$\frac{y^3 + 3y^2 + 4}{y - 2} = y^2 + 5y + 10 + \frac{24}{y - 2}$$

**55.** Rewrite $5 - 6x^2$ as $-6x^2 + 0x + 5$.
$$x - 2 \overline{\smash{\big)}\, -6x^2 + 0x + 5} \quad \underset{-6x-12}{}$$
$$\underline{-6x^2 + 12x}$$
$$-12x + 5$$
$$\underline{-12x + 24}$$
$$-19$$

$$\frac{5 - 6x^2}{x - 2} = -6x - 12 - \frac{19}{x - 2}$$

**57.** Rewrite $x^5 + x^2$ as $x^5 + 0x^4 + 0x^3 + x^2$.
$$x^2 + x \overline{\smash{\big)}\, x^5 + 0x^4 + 0x^3 + x^2} \quad \underset{x^3-x^2+x}{}$$
$$\underline{x^5 + x^4}$$
$$-x^4 + 0x^3$$
$$\underline{-x^4 - x^3}$$
$$x^3 + x^2$$
$$\underline{x^3 + x^2}$$
$$0$$

$$\frac{x^5 + x^2}{x^2 + x} = x^3 - x^2 + x$$

**59.** $2a(a^2 + 1) = 2a(a^2) + 2a(1) = 2a^3 + 2a$

**61.** $2x(x^2 + 7x - 5) = 2x(x^2) + 2x(7x) + 2x(-5)$
$$= 2x^3 + 14x^2 - 10x$$

**63.** $-3xy(xy^2 + 7x^2y + 8)$
$$= -3xy(xy^2) - 3xy(7x^2y) - 3xy(8)$$
$$= -3x^2y^3 - 21x^3y^2 - 24xy$$

    

**65.** $9ab(ab^2c + 4bc - 8)$

$= 9ab(ab^2c) + 9ab(4bc) + 9ab(-8)$

$= 9a^2b^3c + 36ab^2c - 72ab$

**67.** $P = 4s, \ s = \dfrac{P}{4}$

$\dfrac{12x^3 + 4x - 16}{4} = \dfrac{12x^3}{4} + \dfrac{4x}{4} - \dfrac{16}{4}$

$= 3x^3 + x - 4$

Each side is $(3x^3 + x - 4)$ feet.

**69.** $\dfrac{a+7}{7} = \dfrac{a}{7} + \dfrac{7}{7} = \dfrac{a}{7} + 1;$

choice c

**71.** answers may vary

**73.** $A = bh, \ h = \dfrac{A}{b}$

$h = \dfrac{10x^2 + 31x + 15}{5x + 3}$

$$\begin{array}{r} 2x + 5 \\ 5x+3 \overline{)10x^2 + 31x + 15} \\ \underline{10x^2 + 6x} \\ 25x + 15 \\ \underline{25x + 15} \\ 0 \end{array}$$

The height is $(2x + 5)$ meters.

**75.** $\dfrac{18x^{10a} - 12x^{8a} + 14x^{5a} - 2x^{3a}}{2x^{3a}}$

$= \dfrac{18x^{10a}}{2x^{3a}} - \dfrac{12x^{8a}}{2x^{3a}} + \dfrac{14x^{5a}}{2x^{3a}} - \dfrac{2x^{3a}}{2x^{3a}}$

$= 9x^{7a} - 6x^{5a} + 7x^{2a} - 1$

**Chapter 5 Vocabulary Check**

1. A <u>term</u> is a number or the product of numbers and variables raised to powers.

2. The <u>FOIL</u> method may be used when multiplying two binomials.

3. A polynomial with exactly 3 terms is called a <u>trinomial</u>.

4. The <u>degree of polynomial</u> is the greatest degree of any term of the polynomial.

5. A polynomial with exactly 2 terms is called a <u>binomial</u>.

6. The <u>coefficient</u> of a term is its numerical factor.

7. The <u>degree of a term</u> is the sum of the exponents on the variables in the term.

8. A polynomial with exactly 1 term is called a <u>monomial</u>.

9. Monomials, binomials, and trinomials are all examples of <u>polynomials</u>.

10. The <u>distributive</u> property is used to multiply $2x(x - 4)$.

**Chapter 5 Review**

1. In $7^9$, the base is 7 and the exponent is 9.

2. In $(-5)^4$, the base is $-5$ and the exponent is 4.

3. In $-5^4$, the base is 5 and the exponent is 4.

4. In $x^6$, the base is $x$ and the exponent is 6.

5. $8^3 = 8 \cdot 8 \cdot 8 = 512$

6. $(-6)^2 = (-6)(-6) = 36$

7. $-6^2 = -6 \cdot 6 = -36$

8. $-4^3 - 4^0 = -64 - 1 = -65$

9. $(3b)^0 = 1$

10. $\dfrac{8b}{8b} = 1$

11. $y^2 \cdot y^7 = y^{2+7} = y^9$

12. $x^9 \cdot x^5 = x^{9+5} = x^{14}$

13. $(2x^5)(-3x^6) = (2 \cdot -3)(x^5 \cdot x^6) = -6x^{11}$

14. $(-5y^3)(4y^4) = (-5 \cdot 4)(y^3 \cdot y^4) = -20y^7$

15. $(x^4)^2 = x^{4 \cdot 2} = x^8$

**16.** $(y^3)^5 = y^{3 \cdot 5} = y^{15}$

**17.** $(3y^6)^4 = 3^4(y^6)^4 = 81y^{24}$

**18.** $2^3(x^3)^3 = 8x^9$

**19.** $\dfrac{x^9}{x^4} = x^{9-4} = x^5$

**20.** $\dfrac{z^{12}}{z^5} = z^{12-5} = z^7$

**21.** $\dfrac{a^5 b^4}{ab} = a^{5-1}b^{4-1} = a^4 b^3$

**22.** $\dfrac{x^4 y^6}{xy} = x^{4-1}y^{6-1} = x^3 y^5$

**23.** $\dfrac{12xy^6}{3x^4 y^{10}} = \dfrac{12}{3}x^{1-3}y^{6-10} = 4x^{-2}y^{-4} = \dfrac{4}{x^2 y^4}$

**24.** $\dfrac{2x^7 y^8}{8xy^2} = \dfrac{2}{8}x^{7-1}y^{8-2} = \dfrac{x^6 y^6}{4}$

**25.** $5a^7(2a^4)^3 = 5a^7(2^3)(a^4)^3$
$\qquad\qquad\quad = (5 \cdot 8)(a^7 \cdot a^{12})$
$\qquad\qquad\quad = 40a^{19}$

**26.** $(2x)^2(9x) = (2^2 \cdot x^2)(9x)$
$\qquad\qquad\quad = (4 \cdot 9)(x^2 \cdot x)$
$\qquad\qquad\quad = 36x^3$

**27.** $(-5a)^0 + 7^0 + 8^0 = 1 + 1 + 1 = 3$

**28.** $8x^0 + 9^0 = 8(1) + 1 = 9$

**29.** $\left(\dfrac{3x^4}{4y}\right)^3 = \dfrac{3^3 x^{4 \cdot 3}}{4^3 y^3} = \dfrac{27x^{12}}{64y^3}$, choice b.

**30.** $\left(\dfrac{5a^6}{b^3}\right)^2 = \dfrac{5^2 a^{6 \cdot 2}}{b^{3 \cdot 2}} = \dfrac{25a^{12}}{b^6}$, choice c.

**31.** The degree of $-5x^4 y^3$ is $4 + 3 = 7$.

     **239**

**32.** The degree of $10x^3y^2z$ is $3 + 2 + 1 = 6$.

**33.** The degree of $35a^5bc^2$ is $5 + 1 + 2 = 8$.

**34.** The degree of $95xyz$ is $1 + 1 + 1 = 3$.

**35.** The degree is 5 because $y^5$ is the term with the highest degree.

**36.** The degree is 2 because $9y^2$ is the term with the highest degree.

**37.** The degree is 5 because $-28x^2y^3$ is the term with the highest degree.

**38.** The degree is 6 because $6x^2y^2z^2$ is the term with the highest degree.

**39.** 
$$-16t^2 + 4000 = -16(0)^2 + 4000$$
$$= 0 + 4000$$
$$= 4000$$

$$-16t^2 + 4000 = -16(1)^2 + 4000$$
$$= -16 + 4000$$
$$= 3984$$

$$-16t^2 + 4000 = -16(3)^2 + 4000$$
$$= -144 + 4000$$
$$= 3856$$

$$-16t^2 + 4000 = -16(5)^2 + 4000$$
$$= -400 + 4000$$
$$= 3600$$

| $t$ | 0 seconds | 1 second | 3 seconds | 5 seconds |
|---|---|---|---|---|
| $-16t^2 + 4000$ | 4000 feet | 3984 feet | 3856 feet | 3600 feet |

**40.** $2x^2 + 20x$:

$x = 1$: $2(1)^2 + 20(1) = 22$

$x = 3$: $2(3)^2 + 20(3) = 78$

$x = 5.1$: $2(5.1)^2 + 20(5.1) = 154.02$

$x = 10$: $2(10)^2 + 20(10) = 400$

**41.** $7a^2 - 4a^2 - a^2 = (7 - 4 - 1)a^2 = 2a^2$

**42.** $9y + y - 14y = (9 + 1 - 14)y = -4y$

**43.** $6a^2 + 4a + 9a^2 = (6 + 9)a^2 + 4a = 15a^2 + 4a$

**44.** $21x^2 + 3x + x^2 + 6 = (21 + 1)x^2 + 3x + 6$
$$= 22x^2 + 3x + 6$$

      

**45.** $4a^2b - 3b^2 - 8q^2 - 10a^2b + 7q^2$

$= (4a^2b - 10a^2b) - 3b^2 + (-8q^2 + 7q^2)$

$= -6a^2b - 3b^2 - q^2$

**46.** $2s^{14} + 3s^{13} + 12s^{12} - s^{10}$ cannot be combined.

**47.** $(3x^2 + 2x + 6) + (5x^2 + x)$

$= 3x^2 + 2x + 6 + 5x^2 + x$

$= 8x^2 + 3x + 6$

**48.** $(2x^5 + 3x^4 + 4x^3 + 5x^2) + (4x^2 + 7x + 6)$

$= 2x^5 + 3x^4 + 4x^3 + 5x^2 + 4x^2 + 7x + 6$

$= 2x^5 + 3x^4 + 4x^3 + 9x^2 + 7x + 6$

**49.** $(-5y^2 + 3) - (2y^2 + 4) = -5y^2 + 3 - 2y^2 - 4$

$= -7y^2 - 1$

**50.** $(3x^2 - 7xy + 7y^2) - (4x^2 - xy + 9y^2)$

$= 3x^2 - 7xy + 7y^2 - 4x^2 + xy - 9y^2$

$= -x^2 - 6xy - 2y^2$

**51.** $(-9x^2 + 6x + 2) + (4x^2 - x - 1)$

$= -9x^2 + 6x + 2 + 4x^2 - x - 1$

$= -9x^2 + 4x^2 + 6x - x + 2 - 1$

$= -5x^2 + 5x + 1$

**52.** $(8x^6 - 5xy - 10y^2) - (7x^6 - 9xy - 12y^2)$

$= 8x^6 - 5xy - 10y^2 - 7x^6 + 9xy + 12y^2$

$= 8x^6 - 7x^6 - 5xy + 9xy - 10y^2 + 12y^2$

$= x^6 + 4xy + 2y^2$

**53.** $(7x - 14y) - (3x - y) = 7x - 14y - 3x + y$

$= 4x - 13y$

**54.** $[(x^2 + 7x + 9) + (x^2 + 4)] - (4x^2 + 8x - 7)$

$= x^2 + 7x + 9 + x^2 + 4 - 4x^2 - 8x + 7$

$= -2x^2 - x + 20$

**55.** $4(2a + 7) = 4(2a) + 4(7) = 8a + 28$

**56.** $9(6a - 3) = 9(6a) - 9(3) = 54a - 27$

**57.** $-7x(x^2 + 5) = -7(x^2) - 7x(5) = -7x^3 - 35x$

**58.** $-8y(4y^2 - 6) = -8y(4y^2) - 8y(-6)$

$= -32y^3 + 48y$

**59.** $(3a^3 - 4a + 1)(-2a)$

$= 3a^3(-2a) - 4a(-2a) + 1(-2a)$

$= -6a^4 + 8a^2 - 2a$

**60.** $(6b^3 - 4b + 2)(7b) = 6b^3(7b) - 4b(7b) + 2(7b)$

$= 42b^4 - 28b^2 + 14b$

**61.** $(2x + 2)(x - 7) = 2x^2 - 14x + 2x - 14$

$= 2x^2 - 12x - 14$

**62.** $(2x - 5)(3x + 2) = 6x^2 + 4x - 15x - 10$

$= 6x^2 - 11x - 10$

**63.** $(x - 9)^2 = (x - 9)(x - 9)$

$= x^2 - 9x - 9x + 81$

$= x^2 - 18x + 81$

**64.** $(x - 12)^2 = (x - 12)(x - 12)$

$= x^2 - 12x - 12x + 144$

$= x^2 - 24x + 144$

**65.** $(4a - 1)(a + 7) = 4a^2 + 28a - a - 7$

$= 4a^2 + 27a - 7$

**66.** $(6a - 1)(7a + 3) = 42a^2 + 18a - 7a - 3$

$= 42a^2 + 11a - 3$

**67.** $(5x + 2)^2 = (5x + 2)(5x + 2)$

$= 25x^2 + 10x + 10x + 4$

$= 25x^2 + 20x + 4$

**68.** $(3x + 5)^2 = (3x + 5)(3x + 5)$

$= 9x^2 + 15x + 15x + 25$

$= 9x^2 + 30x + 25$

**69.** $(x + 7)(x^3 + 4x - 5)$

$= x(x^3 + 4x - 5) + 7(x^3 + 4x - 5)$

$= x^4 + 4x^2 - 5x + 7x^3 + 28x - 35$

$= x^4 + 7x^3 + 4x^2 + 23x - 35$

**70.** $(x+2)(x^5+x+1) = x(x^5+x+1) + 2(x^5+x+1)$
$\qquad = x^6 + x^2 + x + 2x^5 + 2x + 2$
$\qquad = x^6 + 2x^5 + x^2 + 3x + 2$

**71.** $(x^2+2x+4)(x^2+2x-4)$
$\quad = x^2(x^2+2x-4) + 2x(x^2+2x-4)$
$\qquad\qquad + 4(x^2+2x-4)$
$\quad = x^4 + 2x^3 - 4x^2 + 2x^3 + 4x^2 - 8x$
$\qquad\qquad + 4x^2 + 8x - 16$
$\quad = x^4 + 4x^3 + 4x^2 - 16$

**72.** $(x^3+4x+4)(x^3+4x-4)$
$\quad = x^3(x^3+4x-4) + 4x(x^3+4x-4)$
$\qquad\qquad + 4(x^3+4x-4)$
$\quad = x^6 + 4x^4 - 4x^3 + 4x^4 + 16x^2 - 16x + 4x^3$
$\qquad\qquad + 16x - 16$
$\quad = x^6 + 8x^4 + 16x^2 - 16$

**73.** $(x+7)^3 = (x+7)(x+7)(x+7)$
$\qquad = (x^2 + 7x + 7x + 49)(x+7)$
$\qquad = (x^2 + 14x + 49)(x+7)$
$\qquad = (x^2+14x+49)x + (x^2+14x+49)7$
$\qquad = x^3 + 14x^2 + 49x + 7x^2 + 98x + 343$
$\qquad = x^3 + 21x^2 + 147x + 343$

**74.** $(2x-5)^3$
$\quad = (2x-5)(2x-5)(2x-5)$
$\quad = (4x^2 - 10x - 10x + 25)(2x-5)$
$\quad = (4x^2 - 20x + 25)(2x-5)$
$\quad = (4x^2-20x+25)(2x) + (4x^2-20x+25)(-5)$
$\quad = 8x^3 - 40x^2 + 50x - 20x^2 + 100x - 125$
$\quad = 8x^3 - 60x^2 + 150x - 125$

**75.** $(x+7)^2 = x^2 + 2(x)(7) + 7^2 = x^2 + 14x + 49$

**76.** $(x-5)^2 = x^2 - 2(x)(5) + 5^2 = x^2 - 10x + 25$

**77.** $(3x-7)^2 = (3x)^2 - 2(3x)(7) + 7^2$
$\qquad = 9x^2 - 42x + 49$

**78.** $(4x+2)^2 = (4x)^2 + 2(4x)(2) + 2^2$
$\qquad = 16x^2 + 16x + 4$

**79.** $(5x-9)^2 = (5x)^2 - 2(5x)(9) + 9^2$
$\qquad = 25x^2 - 90x + 81$

**80.** $(5x+1)(5x-1) = (5x)^2 - 1^2 = 25x^2 - 1$

**81.** $(7x+4)(7x-4) = (7x)^2 - 4^2 = 49x^2 - 16$

**82.** $(a+2b)(a-2b) = a^2 - (2b)^2 = a^2 - 4b^2$

**83.** $(2x-6)(2x+6) = (2x)^2 - 6^2 = 4x^2 - 36$

**84.** $(4a^2-2b)(4a^2+2b) = (4a^2)^2 - (2b)^2$
$\qquad = 16a^4 - 4b^2$

**85.** $(3x-1)^2 = (3x)^2 - 2(3x)(1) + 1^2$
$\qquad = 9x^2 - 6x + 1$
The area is $(9x^2 - 6x + 1)$ square meters.

**86.** $(5x+2)(x-1) = 5x^2 - 5x + 2x - 2$
$\qquad = 5x^2 - 3x - 2$
The area is $(5x^2 - 3x - 2)$ square miles.

**87.** $7^{-2} = \dfrac{1}{7^2} = \dfrac{1}{49}$

**88.** $-7^{-2} = -\dfrac{1}{7^2} = -\dfrac{1}{49}$

**89.** $2x^{-4} = \dfrac{2}{x^4}$

**90.** $(2x)^{-4} = \dfrac{1}{(2x)^4} = \dfrac{1}{16x^4}$

**91.** $\left(\dfrac{1}{5}\right)^{-3} = \dfrac{1^{-3}}{5^{-3}} = \dfrac{5^3}{1^3} = 125$

**92.** $\left(\dfrac{-2}{3}\right)^{-2} = \dfrac{(-2)^{-2}}{3^{-2}} = \dfrac{3^2}{(-2)^2} = \dfrac{9}{4}$

**93.** $2^0 + 2^{-4} = 1 + \dfrac{1}{2^4} = \dfrac{16}{16} + \dfrac{1}{16} = \dfrac{17}{16}$

**94.** $6^{-1} - 7^{-1} = \dfrac{1}{6} - \dfrac{1}{7} = \dfrac{7}{42} - \dfrac{6}{42} = \dfrac{1}{42}$

**95.** $\dfrac{x^5}{x^{-3}} = x^{5-(-3)} = x^8$

**96.** $\dfrac{z^4}{z^{-4}} = z^{4-(-4)} = z^8$

**97.** $\dfrac{r^{-3}}{r^{-4}} = r^{-3-(-4)} = r$

**98.** $\dfrac{y^{-2}}{y^{-5}} = y^{-2-(-5)} = y^3$

**99.** $\left(\dfrac{bc^{-2}}{bc^{-3}}\right)^4 = \dfrac{b^4 c^{-8}}{b^4 c^{-12}} = b^{4-4} c^{-8-(-12)} = c^4$

**100.** $\left(\dfrac{x^{-3} y^{-4}}{x^{-2} y^{-5}}\right)^{-3} = \dfrac{x^9 y^{12}}{x^6 y^{15}}$

$\qquad\qquad\qquad = x^{9-6} y^{12-15}$

$\qquad\qquad\qquad = x^3 y^{-3}$

$\qquad\qquad\qquad = \dfrac{x^3}{y^3}$

**101.** $\dfrac{x^{-4} y^{-6}}{x^2 y^7} = x^{-4-2} y^{-6-7}$

$\qquad\qquad = x^{-6} y^{-13}$

$\qquad\qquad = \dfrac{1}{x^6 y^{13}}$

**102.** $\dfrac{a^5 b^{-5}}{a^{-5} b^5} = a^{5-(-5)} b^{-5-5} = a^{10} b^{-10} = \dfrac{a^{10}}{b^{10}}$

**103.** $a^{6m} a^{5m} = a^{6m+5m} = a^{11m}$

**104.** $\dfrac{(x^{5+h})^3}{x^5} = \dfrac{x^{3(5+h)}}{x^5}$

$\qquad\qquad = \dfrac{x^{15+3h}}{x^5}$

$\qquad\qquad = x^{15+3h-5}$

$\qquad\qquad = x^{10+3h}$

**105.** $(3xy^{2z})^3 = 3^3 x^3 y^{2z(3)} = 27 x^3 y^{6z}$

**106.** $a^{m+2} a^{m+3} = a^{(m+2)+(m+3)} = a^{2m+5}$

**107.** $0.00027 = 2.7 \times 10^{-4}$

**108.** $0.8868 = 8.868 \times 10^{-1}$

**109.** $80,800,000 = 8.08 \times 10^7$

**110.** $868,000 = 8.68 \times 10^5$

**111.** $91,000,000 = 9.1 \times 10^7$

**112.** $150,000 = 1.5 \times 10^5$

**113.** $8.67 \times 10^5 = 867,000$

**114.** $3.86 \times 10^{-3} = 0.00386$

**115.** $8.6 \times 10^{-4} = 0.00086$

**116.** $8.936 \times 10^5 = 893,600$

**117.** $1.43128 \times 10^{15} = 1,431,280,000,000,000$

**118.** $1 \times 10^{-10} = 0.0000000001$

**119.** $(8 \times 10^4)(2 \times 10^{-7}) = (8 \cdot 2) \times (10^4 \cdot 10^{-7})$

$\qquad\qquad\qquad\qquad\quad = 16 \times 10^{-3}$

$\qquad\qquad\qquad\qquad\quad = 0.016$

**120.** $\dfrac{8 \times 10^4}{2 \times 10^{-7}} = \dfrac{8}{2} \times 10^{4-(-7)}$

$\qquad\qquad\qquad = 4 \times 10^{11}$

$\qquad\qquad\qquad = 400,000,000,000$

**121.** $\dfrac{x^2 + 21x + 49}{7x^2} = \dfrac{x^2}{7x^2} + \dfrac{21x}{7x^2} + \dfrac{49}{7x^2}$

$\qquad\qquad\qquad = \dfrac{1}{7} + \dfrac{3}{x} + \dfrac{7}{x^2}$

**122.** $\dfrac{5a^3 b - 15ab^2 + 20ab}{-5ab} = \dfrac{5a^3 b}{-5ab} - \dfrac{15ab^2}{-5ab} + \dfrac{20ab}{-5ab}$

$\qquad\qquad\qquad\qquad\quad = -a^2 + 3b - 4$

    

**123.**

$$\begin{array}{r} a+1 \\ a-2\overline{\smash{\big)}\,a^2 \phantom{} -a+4} \\ \underline{a^2 -2a} \\ a+4 \\ \underline{a-2} \\ 6 \end{array}$$

$$(a^2 - a + 4) \div (a - 2) = a + 1 + \frac{6}{a-2}$$

**124.**

$$\begin{array}{r} 4x \\ x+5\overline{\smash{\big)}\,4x^2 +20x+7} \\ \underline{4x^2 +20x} \\ 7 \end{array}$$

$$(4x^2 + 20x + 7) \div (x + 5) = 4x + \frac{7}{x+5}$$

**125.**

$$\begin{array}{r} a^2 +3a+8 \\ a-2\overline{\smash{\big)}\,a^3 \phantom{} +a^2 \phantom{} +2a+6} \\ \underline{a^3 -2a^2} \\ 3a^2 +2a \\ \underline{3a^2 -6a} \\ 8a \phantom{} +6 \\ \underline{8a -16} \\ 22 \end{array}$$

$$\frac{a^3 + a^2 + 2a + 6}{a-2} = a^2 + 3a + 8 + \frac{22}{a-2}$$

**126.**

$$\begin{array}{r} 3b^2 -4b \\ 3b-2\overline{\smash{\big)}\,9b^3 -18b^2 +8b-1} \\ \underline{9b^3 \phantom{} -6b^2} \\ -12b^2 +8b \\ \underline{-12b^2 +8b} \\ -1 \end{array}$$

$$\frac{9b^3 - 18b^2 + 8b - 1}{3b-2} = 3b^2 - 4b - \frac{1}{3b-2}$$

**127.**

$$\begin{array}{r} 2x^3 -x^2 +2 \\ 2x-1\overline{\smash{\big)}\,4x^4 -4x^3 +x^2 +4x-3} \\ \underline{4x^4 -2x^3} \\ -2x^3 +x^2 \\ \underline{-2x^3 +x^2} \\ 4x-3 \\ \underline{4x-2} \\ -1 \end{array}$$

$$\frac{4x^4 - 4x^3 + x^2 + 4x - 3}{2x-1} = 2x^3 - x^2 + 2 - \frac{1}{2x-1}$$

**128.** Rewrite $-10x^2 - x^3 - 21x + 18$ as $-x^3 - 10x^2 - 21x + 18$.

$$\begin{array}{r} -x^2 -16x -117 \\ x-6\overline{\smash{\big)}\,-x^3 -10x^2 \phantom{} -21x+18} \\ \underline{-x^3 \phantom{} +6x^2} \\ -16x^2 -21x \\ \underline{-16x^2 +96x} \\ -117x \phantom{} +18 \\ \underline{-117x +702} \\ -684 \end{array}$$

$$\frac{-10x^2 - x^3 - 21x + 18}{x-6} = -x^2 - 16x - 117 - \frac{684}{x-6}$$

**129.**

$$\frac{15x^3 - 3x^2 + 60}{3x^2} = \frac{15x^3}{3x^2} - \frac{3x^2}{3x^2} + \frac{60}{3x^2}$$

$$= 5x - 1 + \frac{20}{x^2}$$

The width is $\left(5x - 1 + \dfrac{20}{x^2}\right)$ feet.

**130.**

$$\frac{21a^3b^6 + 3a - 3}{3} = \frac{21a^3b^6}{3} + \frac{3a}{3} - \frac{3}{3}$$

$$= 7a^3b^6 + a - 1$$

The length of a side is $(7a^3b^6 + a - 1)$ units.

**131.** $\left(-\dfrac{1}{2}\right)^3 = \left(-\dfrac{1}{2}\right)\left(-\dfrac{1}{2}\right)\left(-\dfrac{1}{2}\right) = -\dfrac{1}{8}$

**132.** $(4xy^2)(x^3y^5) = 4(x \cdot x^3)(y^2 \cdot y^5)$

$$= 4x^{1+3}y^{2+5}$$

$$= 4x^4y^7$$

**133.** $\dfrac{18x^9}{27x^3} = \dfrac{18}{27}x^{9-3} = \dfrac{2x^6}{3}$

**134.** $\left(\dfrac{3a^4}{b^2}\right)^3 = \dfrac{3^3(a^4)^3}{(b^2)^3} = \dfrac{27a^{12}}{b^6}$

**135.** $(2x^{-4}y^3)^{-4} = 2^{-4}(x^{-4})^{-4}(y^3)^{-4}$
$\qquad = \dfrac{1}{2^4}x^{16}y^{-12}$
$\qquad = \dfrac{x^{16}}{16y^{12}}$

**136.** $\dfrac{a^{-3}b^6}{9^{-1}a^{-5}b^{-2}} = 9a^{-3-(-5)}b^{6-(-2)} = 9a^2b^8$

**137.** $(6x+2)+(5x-7) = 6x+2+5x-7 = 11x-5$

**138.** $(-y^2-4)+(3y^2-6) = -y^2-4+3y^2-6$
$\qquad = 2y^2-10$

**139.** $(8y^2-3y+1)-(3y^2+2) = 8y^2-3y^2-3y+1-2$
$\qquad = 5y^2-3y-1$

**140.** $(5x^2+2x-6)-(-x-4) = 5x^2+2x-6+x+4$
$\qquad = 5x^2+3x-2$

**141.** $4x(7x^2+3) = 4x(7x^2)+4x(3)$
$\qquad = 28x^3+12x$

**142.** $(2x+5)(3x-2) = 6x^2-4x+15x-10$
$\qquad = 6x^2+11x-10$

**143.** $(x-3)(x^2+4x-6)$
$\qquad = x(x^2+4x-6)-3(x^2+4x-6)$
$\qquad = x^3+4x^2-6x-3x^2-12x+18$
$\qquad = x^3+x^2-18x+18$

**144.** $(7x-2)(4x-9) = 28x^2-63x-8x+18$
$\qquad = 28x^2-71x+18$

**145.** $(5x+4)^2 = (5x)^2+2(5x)(4)+4^2$
$\qquad = 25x^2+40x+16$

**146.** $(6x+3)(6x-3) = (6x)^2-(3)^2 = 36x^2-9$

**147.** $\dfrac{8a^4-2a^3+4a-5}{2a^3} = \dfrac{8a^4}{2a^3}-\dfrac{2a^3}{2a^3}+\dfrac{4a}{2a^3}-\dfrac{5}{2a^3}$
$\qquad = 4a-1+\dfrac{2}{a^2}-\dfrac{5}{2a^3}$

**148.**
$$
\begin{array}{r}
x-3 \\
x+5 \overline{\smash{)}\, x^2+2x+10} \\
\underline{x^2+5x} \\
-3x+10 \\
\underline{-3x-15} \\
25
\end{array}
$$

$\dfrac{x^2+2x+10}{x+5} = x-3+\dfrac{25}{x+5}$

**149.**
$$
\begin{array}{r}
2x^2+7x+5 \\
2x-3 \overline{\smash{)}\, 4x^3+8x^2\ -11x+4} \\
\underline{4x^3-6x^2} \\
14x^2-11x \\
\underline{14x^2-21x} \\
10x\ +4 \\
\underline{10x-15} \\
19
\end{array}
$$

$\dfrac{4x^3+8x^2-11x+4}{2x-3} = 2x^2+7x+5+\dfrac{19}{2x-3}$

## Chapter 5 Test

**1.** $2^5 = 2\cdot2\cdot2\cdot2\cdot2 = 32$

**2.** $(-3)^4 = (-3)(-3)(-3)(-3) = 81$

**3.** $-3^4 = -3\cdot3\cdot3\cdot3 = -81$

**4.** $4^{-3} = \dfrac{1}{4^3} = \dfrac{1}{64}$

**5.** $(3x^2)(-5x^9) = (3)(-5)(x^2\cdot x^9) = -15x^{11}$

**6.** $\dfrac{y^7}{y^2} = y^{7-2} = y^5$

**7.** $\dfrac{r^{-8}}{r^{-3}} = r^{-8-(-3)} = r^{-5} = \dfrac{1}{r^5}$

**8.** $\left(\dfrac{x^2 y^3}{x^3 y^{-4}}\right)^2 = \dfrac{x^4 y^6}{x^6 y^{-8}}$

$\qquad = x^{4-6} y^{6-(-8)}$

$\qquad = x^{-2} y^{14}$

$\qquad = \dfrac{y^{14}}{x^2}$

**9.** $\left(\dfrac{6^2 x^{-4} y^{-1}}{6^3 x^{-3} y^7}\right) = 6^{2-3} x^{-4-(-3)} y^{-1-7}$

$\qquad = 6^{-1} x^{-1} y^{-8}$

$\qquad = \dfrac{1}{6 x y^8}$

**10.** $563,000 = 5.63 \times 10^5$

**11.** $0.0000863 = 8.63 \times 10^{-5}$

**12.** $1.5 \times 10^{-3} = 0.0015$

**13.** $6.23 \times 10^4 = 62,300$

**14.** $(1.2 \times 10^5)(3 \times 10^{-7}) = (1.2)(3) \times 10^{5-7}$

$\qquad = 3.6 \times 10^{-2}$

$\qquad = 0.036$

**15. a.**

| Term | Numerical Coefficient | Degree of Term |
|------|-----------------------|----------------|
| $4xy^2$ | 4 | 3 |
| $7xyz$ | 7 | 3 |
| $x^3 y$ | 1 | 4 |
| $-2$ | $-2$ | 0 |

**b.** The degree is 4.

**16.** $5x^2 + 4xy - 7x^2 + 11 + 8xy$

$= (5x^2 - 7x^2) + (4xy + 8xy) + 11$

$= -2x^2 + 12xy + 11$

**17.** $(8x^3 + 7x^2 + 4x - 7) + (8x^3 - 7x - 6)$

$= 8x^3 + 7x^2 + 4x - 7 + 8x^3 - 7x - 6$

$= 16x^3 + 7x^2 - 3x - 13$

**18.** $\begin{array}{r} 5x^3 + x^2 + 5x - 2 \\ -(8x^3 - 4x^2 + x - 7) \\ \hline \end{array}$

$\begin{array}{r} 5x^3 + x^2 + 5x - 2 \\ -8x^3 + 4x^2 - x + 7 \\ \hline -3x^3 + 5x^2 + 4x + 5 \end{array}$

**19.** $[(8x^2 + 7x + 5) + (x^3 - 8)] - (4x + 2)$

$= 8x^2 + 7x + 5 + x^3 - 8 - 4x - 2$

$= x^3 + 8x^2 + 3x - 5$

**20.** $(3x + 7)(x^2 + 5x + 2)$

$= 3x(x^2 + 5x + 2) + 7(x^2 + 5x + 2)$

$= 3x^3 + 15x^2 + 6x + 7x^2 + 35x + 14$

$= 3x^3 + 22x^2 + 41x + 14$

**21.** $3x^2(2x^2 - 3x + 7)$

$= 3x^2(2x^2) + 3x^2(-3x) + 3x^2(7)$

$= 6x^4 - 9x^3 + 21x^2$

**22.** $(x + 7)(3x - 5) = 3x^2 - 5x + 21x - 35$

$\qquad = 3x^2 + 16x - 35$

**23.** $\left(3x - \dfrac{1}{5}\right)\left(3x + \dfrac{1}{5}\right) = (3x)^2 - \left(\dfrac{1}{5}\right)^2 = 9x^2 - \dfrac{1}{25}$

**24.** $(4x - 2)^2 = (4x)^2 - 2(4x)(2) + 2^2$

$\qquad = 16x^2 - 16x + 4$

**25.** $(8x + 3)^2 = (8x)^2 + 2(8x)(3) + (3)^2$

$\qquad = 64x^2 + 48x + 9$

**26.** $(x^2 - 9b)(x^2 + 9b) = (x^2)^2 - (9b)^2 = x^4 - 81b^2$

**27.** $-16t^2 + 1001$

$t = 0: \ -16(0)^2 + 1001 = 1001$ ft

$t = 1: \ -16(1)^2 + 1001 = 985$ ft

$t = 3: \ -16(3)^2 + 1001 = 857$ ft

$t = 5: \ -16(5)^2 + 1001 = 601$ ft

**28.** $(2x + 3)(2x - 3) = (2x)^2 - (3)^2 = 4x^2 - 9$

The area is $(4x^2 - 9)$ square inches.

**29.** $\dfrac{4x^2 + 24xy - 7x}{8xy} = \dfrac{4x^2}{8xy} + \dfrac{24xy}{8xy} - \dfrac{7x}{8xy}$

$\qquad\qquad\qquad = \dfrac{x}{2y} + 3 - \dfrac{7}{8y}$

**30.** $\begin{array}{r} x+2 \\ x+5\overline{\smash{\big)}\,x^2+7x+10} \\ \underline{x^2+5x}\phantom{+10} \\ 2x+10 \\ \underline{2x+10} \\ 0 \end{array}$

$\dfrac{x^2+7x+10}{x+5} = x+2$

**31.** Rewrite $27x^3 - 8$ as $27x^3 + 0x^2 + 0x - 8$.

$\begin{array}{r} 9x^2 - 6x + 4 \\ 3x+2\overline{\smash{\big)}\,27x^3 + 0x^2 + 0x - 8} \\ \underline{27x^3 + 18x^2}\phantom{+0x-8} \\ -18x^2 + 0x\phantom{-8} \\ \underline{-18x^2 - 12x}\phantom{-8} \\ 12x - 8 \\ \underline{12x + 8} \\ -16 \end{array}$

$\dfrac{27x^3 - 8}{3x+2} = 9x^2 - 6x + 4 - \dfrac{16}{3x+2}$

**Chapter 5 Cumulative Review**

**1. a.** $8 \ge 8$ is true since $8 = 8$.

  **b.** $8 \le 8$ is true since $8 = 8$.

  **c.** $23 \le 0$ is false.

  **d.** $23 \ge 0$ is true

**2. a.** $|-7.2| = 7.2$

  **b.** $|0| = 0$

  **c.** $\left|-\dfrac{1}{2}\right| = \dfrac{1}{2}$

**3. a.** $\dfrac{4}{5} \div \dfrac{5}{16} = \dfrac{4}{5} \cdot \dfrac{16}{5} = \dfrac{64}{25}$

  **b.** $\dfrac{7}{10} \div 14 = \dfrac{7}{10} \div \dfrac{14}{1} = \dfrac{7}{10} \cdot \dfrac{1}{14} = \dfrac{7}{10 \cdot 7 \cdot 2} = \dfrac{1}{20}$

  **c.** $\dfrac{3}{8} \div \dfrac{3}{10} = \dfrac{3}{8} \cdot \dfrac{10}{3} = \dfrac{3 \cdot 2 \cdot 5}{2 \cdot 4 \cdot 3} = \dfrac{5}{4}$

**4. a.** $\dfrac{3}{4} \cdot \dfrac{7}{21} = \dfrac{3 \cdot 7}{4 \cdot 3 \cdot 7} = \dfrac{1}{4}$

  **b.** $\dfrac{1}{2} \cdot 4\dfrac{5}{6} = \dfrac{1}{2} \cdot \dfrac{29}{6} = \dfrac{29}{12} = 2\dfrac{5}{12}$

**5. a.** $3^2 = 3 \cdot 3 = 9$

  **b.** $5^3 = 5 \cdot 5 \cdot 5 = 125$

  **c.** $2^4 = 2 \cdot 2 \cdot 2 \cdot 2 = 16$

  **d.** $7^1 = 7$

  **e.** $\left(\dfrac{3}{7}\right)^2 = \left(\dfrac{3}{7}\right)\left(\dfrac{3}{7}\right) = \dfrac{9}{49}$

**6.** Let $x = 5$ and $y = 1$.

$\dfrac{2x - 7y}{x^2} = \dfrac{2(5) - 7(1)}{5^2}$

$\qquad\quad = \dfrac{10 - 7}{25}$

$\qquad\quad = \dfrac{3}{25}$

**7. a.** $-3 + (-7) = -10$

  **b.** $-1 + (-20) = -21$

  **c.** $-2 + (-10) = -12$

**8.** $8 + 3(2 \cdot 6 - 1) = 8 + 3(12 - 1)$

$\qquad\qquad\qquad\; = 8 + 3(11)$

$\qquad\qquad\qquad\; = 8 + 33$

$\qquad\qquad\qquad\; = 41$

**9.** $-4 - 8 = -4 + (-8) = -12$

**10.** $x = 1$

$\qquad 5x^2 + 2 = x - 8$

$\qquad 5(1)^2 + 2 \overset{?}{=} 1 - 8$

$\qquad\quad 5 + 2 \overset{?}{=} -7$

$\qquad\qquad 7 \overset{?}{=} -7 \quad$ False

$x = 1$ is not a solution.

**11. a.** The reciprocal of 22 is $\dfrac{1}{22}$.

    **b.** The reciprocal of $\dfrac{3}{16}$ is $\dfrac{16}{3}$.

    **c.** The reciprocal of $-10$ is $-\dfrac{1}{10}$.

    **d.** The reciprocal of $-\dfrac{9}{13}$ is $-\dfrac{13}{9}$.

**12. a.** $7 - 40 = 7 + (-40) = -33$

    **b.** $-5 - (-10) = -5 + 10 = 5$

**13. a.** $5 + (4 + 6) = (5 + 4) + 6$

    **b.** $(-1 \cdot 2) \cdot 5 = -1 \cdot (2 \cdot 5)$

**14.** $\dfrac{4(-3) + (-8)}{5 + (-5)} = \dfrac{-12 + (-8)}{0}$ is undefined.

**15. a.** $\begin{aligned} 10 + (x + 12) &= 10 + (12 + x) \\ &= (10 + 12) + x \\ &= 22 + x \end{aligned}$

    **b.** $-3(7x) = (-3 \cdot 7)x = -21x$

**16.** $\begin{aligned} -2(x + 3y - z) &= -2(x) + (-2)(3y) - (-2)(z) \\ &= -2x - 6y + 2z \end{aligned}$

**17. a.** $5(3x + 2) = 5(3x) + 5(2) = 15x + 10$

    **b.** $\begin{aligned} &-2(y + 0.3z - 1) \\ &= -2(y) + (-2)(0.3z) - (-2)(1) \\ &= -2y - 0.6z + 2y \end{aligned}$

    **c.** $\begin{aligned} &-(9x + y - 2z + 6) \\ &= -1(9x + y - 2z + 6) \\ &= -1(9x) + (-1)(y) - (-1)(2z) + (-1)(6) \\ &= -9x - y + 2z - 6 \end{aligned}$

**18.** $\begin{aligned} 2(6x - 1) - (x - 7) &= 12x - 2 - x + 7 \\ &= 11x + 5 \end{aligned}$

**19.** $\begin{aligned} x - 7 &= 10 \\ x - 7 + 7 &= 10 + 7 \\ x &= 17 \end{aligned}$

**20.** Let $x = $ a number.

$(x + 7) - 2x$

**21.** $\begin{aligned} \dfrac{5}{2}x &= 15 \\ \dfrac{2}{5} \cdot \dfrac{5}{2}x &= \dfrac{2}{5} \cdot 15 \\ x &= 6 \end{aligned}$

**22.** $\begin{aligned} 2x + \dfrac{1}{8} &= x - \dfrac{3}{8} \\ x + \dfrac{1}{8} &= -\dfrac{3}{8} \\ x &= -\dfrac{4}{8} \\ x &= -\dfrac{1}{2} \end{aligned}$

**23.** $\begin{aligned} 4(2x - 3) + 7 &= 3x + 5 \\ 8x - 12 + 7 &= 3x + 5 \\ 8x - 5 &= 3x + 5 \\ 8x - 3x &= 5 + 5 \\ 5x &= 10 \\ \dfrac{5x}{5} &= \dfrac{10}{5} \\ x &= 2 \end{aligned}$

**24.** $\begin{aligned} 10 &= 5j - 2 \\ 12 &= 5j \\ \dfrac{12}{5} &= j \end{aligned}$

**25.** Let $x = $ the number.
$\begin{aligned} 7 + 2x &= x - 3 \\ 7 + x &= -3 \\ x &= -10 \end{aligned}$
The number is $-10$.

**26.** $\begin{aligned} \dfrac{7x + 5}{3} &= x + 3 \\ 3\left(\dfrac{7x + 5}{3}\right) &= 3(x + 3) \\ 7x + 5 &= 3x + 9 \\ 4x + 5 &= 9 \\ 4x &= 4 \\ x &= 1 \end{aligned}$

**27.** Let $x = $ the width and $3x - 2 = $ the length.
$\begin{aligned} 2L + 2W &= P \\ 2(3x - 2) + 2x &= 28 \\ 6x - 4 + 2x &= 28 \\ 8x - 4 &= 28 \\ 8x &= 32 \\ x &= 4 \end{aligned}$
$3x - 2 = 3(4) - 2 = 10$
The width is 4 feet and the length is 10 feet.

**28.** $x < 5$, $(-\infty, 5)$

**29.**
$$F = \frac{9}{5}C + 32$$
$$F - 32 = \frac{9}{5}C$$
$$\frac{5}{9}(F - 32) = C$$
$$\frac{5F - 160}{9} = C$$

**30. a.** $x = -1$ is a vertical line and the slope is undefined.

   **b.** $y = 7$ is a horizontal line and the slope is zero.

**31.** $2 < x \le 4$

**32.** $m = \dfrac{y_2 - y_1}{x_2 - x_1} = \dfrac{2}{20} = \dfrac{1}{10} \cdot 100\% = 10\%$

**33.** $3x + y = 12$

   **a.** $(0, \ )$: $3(0) + y = 12$
   $$y = 12, \ (0, 12)$$

   **b.** $(\ , 6)$: $3x + 6 = 12$
   $$3x = 6$$
   $$x = 2, \ (2, 6)$$

   **c.** $(-1, \ )$: $3(-1) + y = 12$
   $$-3 + y = 12$$
   $$y = 15, \ (-1, 15)$$

**34.** $\begin{cases} 3x + 2y = -8 \\ 2x - 6y = -9 \end{cases}$
   Multiply the first equation by 3 and add.
   $$9x + 6y = -24$$
   $$\underline{2x - 6y = -9}$$
   $$11x \quad\quad = -33$$
   $$x = -3$$
   Replace $x$ with $-3$ in the first equation.

$$3(-3) + 2y = -8$$
$$-9 + 2y = -8$$
$$2y = 1$$
$$y = \frac{1}{2}$$
The solution to the system is $\left(-3, \dfrac{1}{2}\right)$.

**35.** $2x + y = 5$

| $x$ | $y$ |
|-----|-----|
| 0 | 5 |
| $\dfrac{5}{2}$ | 0 |

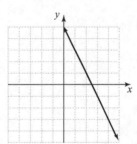

**36.** $\begin{cases} x = -3y + 3 \\ 2x + 9y = 5 \end{cases}$

Replace $x$ with $-3y + 3$ in the second equation.
$$2(-3y + 3) + 9y = 5$$
$$-6y + 6 + 9y = 5$$
$$3y + 6 = 5$$
$$3y = -1$$
$$y = -\frac{1}{3}$$

Replace $y$ with $-\dfrac{1}{3}$ in the first equation.

$$x = -3\left(-\frac{1}{3}\right) + 3 = 1 + 3 = 4$$

The solution to the system is $\left(4, -\dfrac{1}{3}\right)$.

**37.**

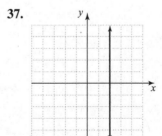

**38. a.** $(-5)^2 = (-5)(-5) = 25$

   **b.** $-5^2 = -(5)(5) = -25$

   **c.** $2 \cdot 5^2 = 2 \cdot 5 \cdot 5 = 50$

**39.** $x = 5$ is a vertical line and the slope is undefined.

**40.** $\dfrac{(z^2)^3 \cdot z^7}{z^9} = \dfrac{z^6 \cdot z^7}{z^9} = z^{6+7-9} = z^4$

**41.** $x + y < 7$
Test $(0, 0)$
$0 + 0 \overset{?}{<} 7$
True
Shade below.

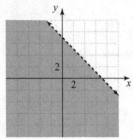

**42.** $(5y^2 - 6) - (y^2 + 2) = 5y^2 - 6 - y^2 - 2 = 4y^2 - 8$

**43.** $(2x^2)(-3x^5) = (2 \cdot -3)(x^2 \cdot x^5) = -6x^{2+5} = -6x^7$

**44.** $-x^2$

   **a.** $-(2)^2 = -4$

   **b.** $-(-2)^2 = -4$

**45.** $(-2x^2 + 5x - 1) + (-2x^2 + x + 3)$
$= -2x^2 + 5x - 1 - 2x^2 + x + 3$
$= -4x^2 + 6x + 2$

**46.** $(10x^2 - 3)(10x^2 + 3) = (10x^2)^2 - 3^2$
$\qquad\qquad\qquad\qquad = 100x^4 - 9$

**47.** $(2x - y)^2 = (2x)^2 - 2(2x)(y) + (y)^2$
$\qquad\qquad = 4x^2 - 4xy + y^2$

**48.** $(10x^2 + 3)^2 = (10x^2)^2 + 2(10x^2)(3) + 3^2$
$\qquad\qquad\quad = 100x^4 + 60x^2 + 9$

**49.** $\dfrac{6m^2 + 2m}{2m} = \dfrac{6m^2}{2m} + \dfrac{2m}{2m} = 3m + 1$

**50. a.** $5^{-1} = \dfrac{1}{5}$

   **b.** $7^{-2} = \dfrac{1}{7^2} = \dfrac{1}{49}$

# Chapter 6

**1. a.** $36 = 2 \cdot 2 \cdot 3 \cdot 3 = 2^2 \cdot 3^2$

$42 = 2 \cdot 3 \cdot 7$

GCF $= 2 \cdot 3 = 6$

**b.** $35 = 5 \cdot 7$

$44 = 2 \cdot 2 \cdot 11$

GCF $= 1$

**c.** $12 = 2 \cdot 2 \cdot 3 = 2^2 \cdot 3$

$16 = 2 \cdot 2 \cdot 2 \cdot 2 = 2^4$

$40 = 2 \cdot 2 \cdot 2 \cdot 5 = 2^3 \cdot 5$

GCF $= 2^2 = 4$

**2. a.** The GCF is $y^4$ since 4 is the smallest exponent to which $y$ is raised.

**b.** The GCF is $x^1$ or $x$, since 1 is the smallest exponent on $x$.

**3. a.** $5y^4 = 5 \cdot y^4$

$15y^2 = 3 \cdot 5 \cdot y^2$

$-20y^3 = -1 \cdot 2 \cdot 2 \cdot 5 \cdot y^3$

GCF $= 5 \cdot y^2 = 5y^2$

**b.** $4x^2 = 2 \cdot 2 \cdot x^2$

$x^3 = x^3$

$3x^8 = 3 \cdot x^8$

GCF $= x^2$

**c.** The GCF of $a^4$, $a^3$, and $a^2$ is $a^2$.

The GCF of $b^2$, $b^5$, and $b^3$ is $b^2$.

Thus, the GCF of $a^4 b^2$, $a^3 b^5$, and $a^2 b^3$ is $a^2 b^2$.

**4. a.** $4t + 12; \text{GCF} = 4$

$4t + 12 = 4 \cdot t + 4 \cdot 3 = 4(t + 3)$

**b.** $y^8 + y^4; \text{GCF} = y^4$

$y^8 + y^4 = y^4 \cdot y^4 + y^4 \cdot 1 = y^4(y^4 + 1)$

**5.** $-8b^6 + 16b^4 - 8b^2$

$= -8b^2(b^4) - 8b^2(-2b^2) - 8b^2(1)$

$= -8b^2(b^4 - 2b^2 + 1)$ or $8b^2(-b^4 + 2b^2 - 1)$

**6.** $5x^4 - 20x = 5x(x^3 - 4)$

**7.** $\dfrac{5}{9}z^5 + \dfrac{1}{9}z^4 - \dfrac{2}{9}z^3 = \dfrac{1}{9}z^3(5z^2 + z - 2)$

**8.** $8a^2 b^4 - 20a^3 b^3 + 12ab^3 = 4ab^3(2ab - 5a^2 + 3)$

**9.** $8(y - 2) + x(y - 2) = (y - 2)(8 + x)$

**10.** $7xy^3(p + q) - (p + q) = 7xy^3(p + q) - 1(p + q)$

$= (p + q)(7xy^3 - 1)$

**11.** $xy + 3y + 4x + 12 = (xy + 3y) + (4x + 12)$

$= y(x + 3) + 4(x + 3)$

$= (x + 3)(y + 4)$

Check: $(x + 3)(y + 4) = xy + 3y + 4x + 12$

**12.** $40x^3 - 24x^2 + 15x - 9 = 8x^2(5x - 3) + 3(5x - 3)$

$= (5x - 3)(8x^2 + 3)$

**13.** $2xy + 3y^2 - 2x - 3y = (2xy + 3y^2) + (-2x - 3y)$

$= y(2x + 3y) - 1(2x + 3y)$

$= (2x + 3y)(y - 1)$

**14.** $7a^3 + 5a^2 + 7a + 5 = (7a^3 + 5a^2) + (7a + 5)$

$= a^2(7a + 5) + 1(7a + 5)$

$= (7a + 5)(a^2 + 1)$

**15.** $4xy + 15 - 12x - 5y = 4xy - 12x - 5y + 15$

$= (4xy - 12x) + (-5y + 15)$

$= 4x(y - 3) - 5(y - 3)$

$= (y - 3)(4x - 5)$

**16.** $9y - 18 + y^3 - 4y^2 = 9(y - 2) + y^2(y - 4)$

There is no common binomial factor, so it cannot be factored by grouping.

**17.** $3xy - 3ay - 6ax + 6a^2 = 3(xy - ay - 2ax + 2a^2)$

$= 3[y(x - a) - 2a(x - a)]$

$= 3(x - a)(y - 2a)$

**Vocabulary, Readiness & Video Check 6.1**

1. Since $5 \cdot 4 = 20$, the numbers 5 and 4 are called <u>factors</u> of 20.

2. The <u>greatest common factor</u> of a list of integers is the largest integer that is a factor of all the integers in the list.

3. The greatest common factor of a list of common variables raised to powers is the variable raised to the <u>least</u> exponent in the list.

4. The process of writing a polynomial as a product is called <u>factoring</u>.

5. $7(x + 3) + y(x + 3)$ is a sum, not a product. The statement is false.

6. $3x^3 + 6x + x^2 + 2 = 3x(x^2 + 2) + (x^2 + 2)$
$$= (x^2 + 2)(3x + 1)$$
The statement is false.

7. The GCF of a list of numbers is the largest number that is a factor of all numbers in the list.

8. The GCF of common variable factors is the variable raised to the smallest exponent.

9. When factoring out a GCF, the number of terms in the other factor should have the same number of terms as your original polynomial.

10. Look for a GCF other than 1 or −1; if you have a simplified four-term polynomial.

**Exercise Set 6.1**

1. $32 = 2 \cdot 2 \cdot 2 \cdot 2 \cdot 2 = 2^5$
$36 = 2 \cdot 2 \cdot 3 \cdot 3 = 2^2 \cdot 3^2$
$GCF = 2 \cdot 2 = 4$

3. $18 = 2 \cdot 3 \cdot 3 = 2 \cdot 3^2$
$42 = 2 \cdot 3 \cdot 7$
$84 = 2 \cdot 2 \cdot 3 \cdot 7 = 2^2 \cdot 3 \cdot 7$
$GCF = 2 \cdot 3 = 6$

5. $24 = 2 \cdot 2 \cdot 2 \cdot 3 = 2^3 \cdot 3$
$14 = 2 \cdot 7$
$21 = 3 \cdot 7$
$GCF = 1$

7. The GCF of $y^2$, $y^4$, and $y^7$ is $y^2$.

9. The GCF of $z^7$, $z^9$, and $z^{11}$ is $z^7$.

11. The GCF of $x^{10}$, $x$, and $x^3$ is $x$.
The GCF of $y^2$, $y^2$, and $y^3$ is $y^2$.
Thus the GCF of $x^{10}y^2$, $xy^2$, and $x^3y^3$ is $xy^2$.

13. $14x = 2 \cdot 7 \cdot x$
$21 = 3 \cdot 7$
$GCF = 7$

15. $12y^4 = 2 \cdot 2 \cdot 3 \cdot y^4$
$20y^3 = 2 \cdot 2 \cdot 5 \cdot y^3$
$GCF = 2 \cdot 2 \cdot y^3 = 4y^3$

17. $-10x^2 = -1 \cdot 2 \cdot 5 \cdot x^2$
$15x^3 = 3 \cdot 5 \cdot x^3$
$GCF = 5 \cdot x^2 = 5x^2$

19. $12x^3 = 2 \cdot 2 \cdot 3 \cdot x^3$
$-6x^4 = -1 \cdot 2 \cdot 3 \cdot x^4$
$3x^5 = 3 \cdot x^5$
$GCF = 3 \cdot x^3 = 3x^3$

21. $-18x^2y = -1 \cdot 2 \cdot 3 \cdot 3 \cdot x^2 \cdot y$
$9x^3y^3 = 3 \cdot 3 \cdot x^3 \cdot y^3$
$36x^3y = 2 \cdot 2 \cdot 3 \cdot 3 \cdot x^3 \cdot y$
$GCF = 3 \cdot 3 \cdot x^2 \cdot y = 9x^2y$

23. $20a^6b^2c^8 = 2 \cdot 2 \cdot 5 \cdot a^6 \cdot b^2 \cdot c^8$
$50a^7b = 2 \cdot 5 \cdot 5 \cdot a^7 \cdot b$
$GCF = 2 \cdot 5 \cdot a^6 \cdot b = 10a^6b$

25. $3a + 6 = 3(a + 2)$

27. $30x - 15 = 15(2x - 1)$

29. $x^3 + 5x^2 = x^2(x + 5)$

31. $6y^4 + 2y^3 = 2y^3(3y + 1)$

33. $4x - 8y + 4 = 4(x - 2y + 1)$

35. $6x^3 - 9x^2 + 12x = 3x(2x^2 - 3x + 4)$

**37.** $a^7b^6 - a^3b^2 + a^2b^5 - a^2b^2$
$= a^2b^2(a^5b^4 - a + b^3 - 1)$

**39.** $8x^5 + 16x^4 - 20x^3 + 12 = 4(2x^5 + 4x^4 - 5x^3 + 3)$

**41.** $\dfrac{1}{3}x^4 + \dfrac{2}{3}x^3 - \dfrac{4}{3}x^5 + \dfrac{1}{3}x$
$= \dfrac{1}{3}x(x^3 + 2x^2 - 4x^4 + 1)$

**43.** $y(x^2 + 2) + 3(x^2 + 2) = (x^2 + 2)(y + 3)$

**45.** $z(y + 4) - 3(y + 4) = (y + 4)(z - 3)$

**47.** $r(z^2 - 6) + (z^2 - 6) = r(z^2 - 6) + 1(z^2 - 6)$
$= (z^2 - 6)(r + 1)$

**49.** $-2x - 14 = -2(x + 7)$

**51.** $-2x^5 + x^7 = -x^5(2 - x^2)$

**53.** $-6a^4 + 9a^3 - 3a^2 = -3a^2(2a^2 - 3a + 1)$

**55.** $x^3 + 2x^2 + 5x + 10 = x^2(x + 2) + 5(x + 2)$
$= (x + 2)(x^2 + 5)$

**57.** $5x + 15 + xy + 3y = 5(x + 3) + y(x + 3)$
$= (x + 3)(5 + y)$

**59.** $6x^3 - 4x^2 + 15x - 10 = 2x^2(3x - 2) + 5(3x - 2)$
$= (3x - 2)(2x^2 + 5)$

**61.** $5m^3 + 6mn + 5m^2 + 6n$
$= m(5m^2 + 6n) + 1(5m^2 + 6n)$
$= (5m^2 + 6n)(m + 1)$

**63.** $2y - 8 + xy - 4x = 2(y - 4) + x(y - 4)$
$= (y - 4)(2 + x)$

**65.** $2x^3 - x^2 + 8x - 4 = x^2(2x - 1) + 4(2x - 1)$
$= (2x - 1)(x^2 + 4)$

**67.** $3x - 3 + x^3 - 4x^2 = 3(x - 1) + x^2(x - 4)$
The polynomial is not factorable by grouping.

**69.** $4x^2 - 8xy - 3x + 6y = 4x(x - 2y) - 3(x - 2y)$
$= (x - 2y)(4x - 3)$

**71.** $5q^2 - 4pq - 5q + 4p = q(5q - 4p) - 1(5q - 4p)$
$= (5q - 4p)(q - 1)$

**73.** $2x^4 + 5x^3 + 2x^2 + 5x = x(2x^3 + 5x^2 + 2x + 5)$
$= x[x^2(2x + 5) + 1(2x + 5)]$
$= x(2x + 5)(x^2 + 1)$

**75.** $12x^2y - 42x^2 - 4y + 14$
$= 2(6x^2y - 21x^2 - 2y + 7)$
$= 2[3x^2(2y - 7) - 1(2y - 7)]$
$= 2(2y - 7)(3x^2 - 1)$

**77.** $32xy^2 - 18x^2 = 2x(16y - 9x)$

**79.** $y(x + 2) - 3(x + 2) = (x + 2)(y - 3)$

**81.** $14x^3y + 7x^2y - 7xy = 7xy(2x^2 + x - 1)$

**83.** $28x^3 - 7x^2 + 12x - 3 = 7x^2(4x - 1) + 3(4x - 1)$
$= (4x - 1)(7x^2 + 3)$

**85.** $-40x^8y^6 - 16x^9y^5 = -8x^8y^5(5y + 2x)$

**87.** $6a^2 + 9ab^2 + 6ab + 9b^3$
$= 3(2a^2 + 3ab^2 + 2ab + 3b^3)$
$= 3[a(2a + 3b^2) + b(2a + 3b^2)]$
$= 3(2a + 3b^2)(a + b)$

**89.** $(x + 2)(x + 5) = x^2 + 5x + 2x + 10 = x^2 + 7x + 10$

**91.** $(b + 1)(b - 4) = b^2 - 4b + b - 4 = b^2 - 3b - 4$

| | Two Numbers | Their Product | Their Sum |
|---|---|---|---|
| **93.** | 2, 6 | 12 | 8 |
| **95.** | −1, −8 | 8 | −9 |
| **97.** | −2, 5 | −10 | 3 |

**99.** **a.** $8 \cdot a - 24 = 8a - 24$

    **b.** $8(a - 3) = 8a - 24$

    **c.** $4(2a - 12) = 8a - 48$

    **d.** $8 \cdot a - 2 \cdot 12 = 8a - 24$

The answer is b.

**101.** $(a+6)(a+2)$ is factored.

**103.** $5(2y+z)-b(2y+z)$ is not factored.

**105.** answers may vary

**107.** answers may vary

**109. a.** $-20x^2+300x+120$
$=-20(6)^2+300(6)+120$
$=-720+1800+120$
$=1200$
There were 1200 million single digital downloads in 2010.

**b.** Let $x=2014-2004=10$.
$-20x^2+300x+120$
$=-20(10)^2+300(10)+120$
$=-2000+3000+120$
$=1120$
There were 1120 million single downloads in 2009.

**c.** $-20x^2+300x+120$
$=-20\cdot x^2+(-20)(-15x)+(-20)(-6)$
$=-20(x^2-15x-6)$

**111.** Subtract the area of the inner rectangle from the area of the outer rectangle.
Outer rectangle: $A=l\cdot w$
$$A=12x\cdot x^2=12x^3$$
Inner rectangle: $A=l\cdot w$
$$A=2\cdot x=2x$$
The area of the shaded region is given by the expression $12x^3-2x=2x(6x^2-1)$.

**113.** Area $=4n^4-24n=4n(n^3-6)$
Since the width is $4n$ units, the length is $(n^3-6)$ units.

**115.** $x^{2n}+2x^n+3x^n+6=x^n(x^n+2)+3(x^n+2)$
$$=(x^n+2)(x^n+3)$$

**117.** $3x^{2n}+21x^n-5x^n-35$
$=3x^n(x^n+7)-5(x^n+7)$
$=(x^n+7)(3x^n-5)$

**Section 6.2 Practice**

**1.**

| Positive Factors of 6 | Sum of Factors |
|---|---|
| 1, 6 | 7 |
| 2, 3 | 5 |

$x^2+5x+6=(x+2)(x+3)$

**2.**

| Negative Factors of 70 | Sum of Factors |
|---|---|
| −1, −70 | −71 |
| −2, −35 | −37 |
| −5, −14 | −19 |
| −7, −10 | −17 |

$x^2-17x+70=(x-7)(x-10)$

**3.**

| Factors of −14 | Sum of Factors |
|---|---|
| −1, 14 | 13 |
| 1, −14 | −13 |
| −2, 7 | 5 |
| 2, −7 | −5 |

$x^2+5x-14=(x-2)(x+7)$

**4.** The first term of each binomial is $p$. Then look for two numbers whose product is −63 and whose sum is −2.
$p^2-2p-63=(p-9)(p+7)$

**5.** The first term of each binomial is $b$. Then look for two numbers whose product is 1 and whose sum is 5. There are no such numbers.
$b^2+5b+1$ is a prime polynomial.

**6.** The first term of each polynomial is $x$. Then look for two terms whose product is $12y^2$ and whose sum is $7y$.
$x^2+7xy+12y^2=(x+3y)(x+4y)$

**7.** The first term of each polynomial is $x^2$. Then look for two numbers whose product is 12 and whose sum is 13.
$x^4+13x^2+12=(x^2+1)(x^2+12)$

**8.** $48 - 14x + x^2 = x^2 - 14x + 48$

The first term of each binomial is $x$. Then look for two factors whose product is 48 and whose sum is $-14$.

$x^2 - 14x + 48 = (x - 6)(x - 8)$

**9.** $4x^2 - 24x + 36 = 4(x^2 - 6x + 9)$

The first term of each binomial is $x$. Then look for two factors whose product is 9 and whose sum is $-6$.

$4(x^2 - 6x + 9) = 4(x - 3)(x - 3)$ or $4(x - 3)^2$

**10.** $3y^4 - 18y^3 - 21y^2 = 3y^2(y^2 - 6y - 7)$

The first term of each binomial is $y$. Then look for two factors whose product is $-7$ and whose sum is $-6$.

$3y^2(y^2 - 6y - 7) = 3y^2(y - 7)(y + 1)$

**Vocabulary, Readiness & Video Check 6.2**

**1.** The statement is true.

**2.** The statement is true.

**3.** Since $4x - 12 = 4(x - 3)$, the statement is false.

**4.** $(x + 2y)^2 = (x + 2y)(x + 2y) \neq (x + 2y)(x + y)$
The statement is false.

**5.** $x^2 + 9x + 20 = (x + 4)(x \underline{+ 5})$

**6.** $x^2 + 12x + 35 = (x + 5)(x \underline{+ 7})$

**7.** $x^2 - 7x + 12 = (x - 4)(x \underline{- 3})$

**8.** $x^2 - 13x + 22 = (x - 2)(x \underline{- 11})$

**9.** $x^2 + 4x + 4 = (x + 2)(x \underline{+ 2})$

**10.** $x^2 + 10x + 24 = (x + 6)(x \underline{+ 4})$

**11.** 15 is positive, so its factors would have to either be both positive or both negative. Since the factors need to sum to $-8$, both factors must be negative.

**12.** Since the sum of the factors is 3, the factors are $-2$ and 5, $(-2 + 5 = 3)$. If you accidentally choose factors whose sum is $-3$, simply "switch" the signs of the factors.

**Exercise Set 6.2**

**1.** $x^2 + 7x + 6 = (x + 6)(x + 1)$

**3.** $y^2 - 10y + 9 = (y - 9)(y - 1)$

**5.** $x^2 - 6x + 9 = (x - 3)(x - 3)$ or $(x - 3)^2$

**7.** $x^2 - 3x - 18 = (x - 6)(x + 3)$

**9.** $x^2 + 3x - 70 = (x + 10)(x - 7)$

**11.** $x^2 + 5x + 2$ is a prime polynomial.

**13.** $x^2 + 8xy + 15y^2 = (x + 5y)(x + 3y)$

**15.** $a^4 - 2a^2 - 15 = (a^2 - 5)(a^2 + 3)$

**17.** $13 + 14m + m^2 = m^2 + 14m + 13 = (m + 13)(m + 1)$

**19.** $10t - 24 + t^2 = t^2 + 10t - 24 = (t - 2)(t + 12)$

**21.** $a^2 - 10ab + 16b^2 = (a - 2b)(a - 8b)$

**23.** $2z^2 + 20z + 32 = 2(z^2 + 10z + 16)$
$= 2(z + 8)(z + 2)$

**25.** $2x^3 - 18x^2 + 40x = 2x(x^2 - 9x + 20)$
$= 2x(x - 5)(x - 4)$

**27.** $x^2 - 3xy - 4y^2 = (x - 4y)(x + y)$

**29.** $x^2 + 15x + 36 = (x + 12)(x + 3)$

**31.** $x^2 - x - 2 = (x - 2)(x + 1)$

**33.** $r^2 - 16r + 48 = (r - 12)(r - 4)$

**35.** $x^2 + xy - 2y^2 = (x + 2y)(x - y)$

**37.** $3x^2 + 9x - 30 = 3(x^2 + 3x - 10) = 3(x + 5)(x - 2)$

**39.** $3x^2 - 60x + 108 = 3(x^2 - 20x + 36)$
$\qquad\qquad\qquad\quad = 3(x - 18)(x - 2)$

**41.** $x^2 - 18x - 144 = (x - 24)(x + 6)$

**43.** $r^2 - 3r + 6$ is a prime polynomial.

**45.** $x^2 - 8x + 15 = (x - 5)(x - 3)$

**47.** $6x^3 + 54x^2 + 120x = 6x(x^2 + 9x + 20)$
$\qquad\qquad\qquad\qquad = 6x(x + 4)(x + 5)$

**49.** $4x^2y + 4xy - 12y = 4y(x^2 + x - 3)$

**51.** $x^2 - 4x - 21 = (x - 7)(x + 3)$

**53.** $x^2 + 7xy + 10y^2 = (x + 5y)(x + 2y)$

**55.** $64 + 24t + 2t^2 = 2t^2 + 24t + 64$
$\qquad\qquad\qquad = 2(t^2 + 12t + 32)$
$\qquad\qquad\qquad = 2(t + 8)(t + 4)$

**57.** $x^3 - 2x^2 - 24x = x(x^2 - 2x - 24)$
$\qquad\qquad\qquad = x(x - 6)(x + 4)$

**59.** $2t^5 - 14t^4 + 24t^3 = 2t^3(t^2 - 7t + 12)$
$\qquad\qquad\qquad\qquad = 2t^3(t - 4)(t - 3)$

**61.** $5x^3y - 25x^2y^2 - 120xy^3 = 5xy(x^2 - 5xy - 24y^2)$
$\qquad\qquad\qquad\qquad\qquad = 5xy(x - 8y)(x + 3y)$

**63.** $162 - 45m + 3m^2 = 3m^2 - 45m + 162$
$\qquad\qquad\qquad\quad = 3(m^2 - 15m + 54)$
$\qquad\qquad\qquad\quad = 3(m - 9)(m - 6)$

**65.** $-x^2 + 12x - 11 = -1(x^2 - 12x + 11)$
$\qquad\qquad\qquad\quad = -1(x - 11)(x - 1)$

**67.** $\dfrac{1}{2}y^2 - \dfrac{9}{2}y - 11 = \dfrac{1}{2}(y^2 - 9y - 22)$
$\qquad\qquad\qquad\quad = \dfrac{1}{2}(y - 11)(y + 2)$

**69.** $x^3y^2 + x^2y - 20x = x(x^2y^2 + xy - 20)$
$\qquad\qquad\qquad\qquad = x(xy - 4)(xy + 5)$

**71.** $(2x + 1)(x + 5) = 2x^2 + 10x + x + 5$
$\qquad\qquad\qquad\quad = 2x^2 + 11x + 5$

**73.** $(5y - 4)(3y - 1) = 15y^2 - 5y - 12y + 4$
$\qquad\qquad\qquad\quad = 15y^2 - 17y + 4$

**75.** $(a + 3b)(9a - 4b) = 9a^2 - 4ab + 27ab - 12b^2$
$\qquad\qquad\qquad\qquad = 9a^2 + 23ab - 12b^2$

**77.** $(x - 3)(x + 8) = x^2 + 8x - 3x - 3(8) = x^2 + 5x - 24$

**79.** Answers may vary

**81.** $P = 2l + 2w$
$l = x^2 + 10x$ and $w = 4x + 33$, so
$P = 2(x^2 + 10x) + 2(4x + 33)$
$\quad = 2x^2 + 20x + 8x + 66$
$\quad = 2x^2 + 28x + 66$
$\quad = 2(x^2 + 14x + 33)$
$\quad = 2(x + 11)(x + 3)$
The perimeter of the rectangle is given by the polynomial $2x^2 + 28x + 66$ which factors as $2(x + 11)(x + 3)$.

**83.** $-16t^2 + 64t + 80 = -16(t^2 - 4t - 5)$
$\qquad\qquad\qquad\qquad = -16(t - 5)(t + 1)$

**85.** $x^2 + \dfrac{1}{2}x + \dfrac{1}{16} = \left(x + \dfrac{1}{4}\right)\left(x + \dfrac{1}{4}\right)$ or $\left(x + \dfrac{1}{4}\right)^2$

**87.** $z^2(x + 1) - 3z(x + 1) - 70(x + 1)$
$\quad = (x + 1)(z^2 - 3z - 70)$
$\quad = (x + 1)(z - 10)(z + 7)$

**89.** $x^{2n} + 8x^n - 20 = (x^n + 10)(x^n - 2)$

**91.** $c$ must be the product of positive numbers that sum to 6.
$6 = 1 + 5; \ 1 \cdot 5 = 5$
$6 = 2 + 4; \ 2 \cdot 4 = 8$
$6 = 3 + 3; \ 3 \cdot 3 = 9$
$x^2 + 6x + c$ if factorable when $c$ is 5, 8, or 9.

**93.** $c$ must be the product of negative numbers that sum to $-4$.
$-4 = -1 + (-3); \ -1 \cdot -3 = 3$
$-4 = -2 + (-2); \ -2 \cdot -2 = 4$
$y^2 - 4y + c$ if factorable when $c$ is 3 or 4.

**95.** $b$ must be the sum of positive numbers whose product is 15.
$15 = 1 \cdot 15; \ 1 + 15 = 16$
$15 = 3 \cdot 5; \ 3 + 5 = 8$
$x^2 + bx + 15$ is factorable when $b$ is 8 or 16.

**97.** $b$ must be the positive sum of a positive number and a negative number whose product is $-27$.
$-27 = 27 \cdot -1; \ 27 + (-1) = 26$
$-27 = 9 \cdot -3; \ 9 + (-3) = 6$
$m^2 + bm - 27$ is factorable when $b$ is 6 or 26.

## Section 6.3 Practice

**1.** Factors of $2x^2$: $2x^2 = 2x \cdot x$
Factors of 15: $15 = 1 \cdot 15, \ 15 = 3 \cdot 5$
Try possible combinations.
Factored form: $2x^2 + 11x + 15 = (2x + 5)(x + 3)$

**2.** Factors of $15x^2$: $15x^2 = 15x \cdot x, \ 15x^2 = 5x \cdot 3x$
Factors of 8: $8 = -1 \cdot -8, \ 8 = -2 \cdot -4$
Try possible combinations.
Factored form: $15x^2 - 22x + 8 = (5x - 4)(3x - 2)$

**3.** Factors of $4x^2$: $4x^2 = 4x \cdot x, \ 4x^2 = 2x \cdot 2x$
Factors of $-3$: $-3 = -1 \cdot 3, \ -3 = 1 \cdot -3$
Try possible combinations.
Factored form: $4x^2 + 11x - 3 = (4x - 1)(x + 3)$

**4.** Factors of $21x^2$: $21x^2 = 21x \cdot x, \ 21x^2 = 3x \cdot 7x$
Factors of
$-2y^2$: $-2y^2 = -2y \cdot y, \ -2y^2 = 2y \cdot -y$
Try possible combinations.
Factored form:
$21x^2 + 11xy - 2y^2 = (7x - y)(3x + 2y)$

**5.** Factors of $2x^4$: $2x^4 = 2x^2 \cdot x^2$
Factors of $-7$: $-7 = -7 \cdot 1, \ -7 = 7 \cdot -1$
Try possible combinations.
$2x^4 - 5x^2 - 7 = (2x^2 - 7)(x^2 + 1)$

**6.** $3x^3 + 17x^2 + 10x = x(3x^2 + 17x + 10)$
Factors of $3x^2$: $3x^2 = 3x \cdot x$
Factors of 10: $10 = 1 \cdot 10, \ 10 = 2 \cdot 5$
Try possible combinations:
$3x^3 + 17x^2 + 10x = x(3x^2 + 17x + 10)$
$\qquad\qquad\qquad\quad = x(3x + 2)(x + 5)$

**7.** $-8x^2 + 2x + 3 = -1(8x^2 - 2x - 3)$
$\qquad\qquad\qquad = -1(4x - 3)(2x + 1)$

**8.** $x^2 = (x)^2$ and $49 = 7^2$
Is $2 \cdot x \cdot 7 = 14x$ the middle term? Yes.
$x^2 + 14x + 49 = (x + 7)^2$

**9.** $4x^2 = (2x)^2$ and $9y^2 = (3y)^2$
Is $2 \cdot 2x \cdot 3y = 12xy$ the middle term? No.
Try other possibilities.
$4x^2 + 20xy + 9y^2 = (2x + 9y)(2x + y)$

**10.** $36n^4 = (6n^2)^2$ and $1 = 1^2$
Is $2 \cdot 6n^2 \cdot 1 = 12n^2$ the middle term? Yes, the opposite of the middle term.
$36n^4 - 12n^2 + 1 = (6n^2 - 1)^2$

**11.** $12x^3 - 84x^2 + 147x = 3x(4x^2 - 28x + 49)$
$\qquad\qquad\qquad\quad = 3x[(2x)^2 - 2 \cdot 2x \cdot 7 + 7^2]$
$\qquad\qquad\qquad\quad = 3x(2x - 7)^2$

## Vocabulary, Readiness & Video Check 6.3

**1.** A perfect square trinomial is a trinomial that is the square of a binomial.

**2.** The term $25y^2$ written as a square is $\underline{(5y)^2}$.

**3.** The expression $x^2 + 10xy + 25y^2$ is called a perfect square trinomial.

**4.** The factorization $(x + 5y)(x + 5y)$ may also be written as $\underline{(x + 5y)^2}$.

**5.** $2x^2 + 5x + 3$ factors as $(2x + 3)(x + 1)$, which is choice d.

**6.** $7x^2 + 9x + 2$ factors as $(7x + 2)(x + 1)$, which is choice b.

**7.** Consider the factors of the first and last terms and the signs of the trinomial. Continue to check by multiplying until you get the middle term of the trinomial.

**257**

**8.** If the GCF has been factored out, then neither binomial can contain a common factor other than 1 or −1. This helps limit your choice of factors for one or both binomials since you cannot choose factors that would give the terms in either binomial a common factor.

**9.** The first and last terms are squares, $a^2$ and $b^2$, and the middle term is $2 \cdot a \cdot b$ or $-2 \cdot a \cdot b$.

**Exercise Set 6.3**

**1.** $5x^2 + 22x + 8 = (5x + 2)(x + 4)$

**3.** $50x^2 + 15x - 2 = (5x + 2)(10x - 1)$

**5.** $25x^2 - 20x + 4 = (5x - 2)(5x - 2)$

**7.** $2x^2 + 13x + 15 = (2x + 3)(x + 5)$

**9.** $8y^2 - 17y + 9 = (y - 1)(8y - 9)$

**11.** $2x^2 - 9x - 5 = (2x + 1)(x - 5)$

**13.** $20r^2 + 27r - 8 = (4r - 1)(5r + 8)$

**15.** $10x^2 + 31x + 3 = (10x + 1)(x + 3)$

**17.** $2m^2 + 17m + 10$ is prime.

**19.** $6x^2 - 13xy + 5y^2 = (3x - 5y)(2x - y)$

**21.** $15m^2 - 16m - 15 = (3m - 5)(5m + 3)$

**23.** $12x^3 + 11x^2 + 2x = x(12x^2 + 11x + 2)$
$$= x(3x + 2)(4x + 1)$$

**25.** $21b^2 - 48b - 45 = 3(7b^2 - 16b - 15)$
$$= 3(7b + 5)(b - 3)$$

**27.** $7z + 12z^2 - 12 = 12z^2 + 7z - 12 = (3z + 4)(4z - 3)$

**29.** $6x^2y^2 - 2xy^2 - 60y^2 = 2y^2(3x^2 - x - 30)$
$$= 2y^2(3x - 10)(x + 3)$$

**31.** $4x^2 - 8x - 21 = (2x - 7)(2x + 3)$

**33.** $-x^2 + 2x + 24 = -1(x^2 - 2x - 24)$
$$= -1(x - 6)(x + 4)$$

**35.** $4x^3 - 9x^2 - 9x = x(4x^2 - 9x - 9)$
$$= x(4x + 3)(x - 3)$$

**37.** $24x^2 - 58x + 9 = (4x - 9)(6x - 1)$

**39.** $x^2 + 22x + 121 = x^2 + 2 \cdot x \cdot 11 + 11^2 = (x + 11)^2$

**41.** $x^2 - 16x + 64 = x^2 - 2 \cdot x \cdot 8 + 8^2 = (x - 8)^2$

**43.** $16a^2 - 24a + 9 = (4a)^2 - 2 \cdot 4a \cdot 3 + 3^2 = (4a - 3)^2$

**45.** $x^4 + 4x^2 + 4 = (x^2)^2 + 2 \cdot x^2 \cdot 2 + 2^2 = (x^2 + 2)^2$

**47.** $2n^2 - 28n + 98 = 2(n^2 - 14n + 49)$
$$= 2(n^2 - 2 \cdot n \cdot 7 + 7^2)$$
$$= 2(n - 7)^2$$

**49.** $16y^2 + 40y + 25 = (4y)^2 + 2 \cdot 4y \cdot 5 + 5^2$
$$= (4y + 5)^2$$

**51.** $2x^2 - 7x - 99 = (2x + 11)(x - 9)$

**53.** $24x^2 + 41x + 12 = (8x + 3)(3x + 4)$

**55.** $3a^2 + 10ab + 3b^2 = (3a + b)(a + 3b)$

**57.** $-9x + 20 + x^2 = x^2 - 9x + 20 = (x - 4)(x - 5)$

**59.** $p^2 + 12pq + 36q^2 = p^2 + 2 \cdot p \cdot 6q + (6q)^2$
$$= (p + 6q)^2$$

**61.** $x^2y^2 - 10xy + 25 = (xy)^2 - 2 \cdot xy \cdot 5 + 5^2$
$$= (xy - 5)^2$$

**63.** $40a^2b + 9ab - 9b = b(40a^2 + 9a - 9)$
$$= b(8a - 3)(5a + 3)$$

**65.** $30x^3 + 38x^2 + 12x = 2x(15x^2 + 19x + 6)$
$$= 2x(3x + 2)(5x + 3)$$

**67.** $6y^3 - 8y^2 - 30y = 2y(3y^2 - 4y - 15)$
$$= 2y(3y + 5)(y - 3)$$

**69.** $10x^4 + 25x^3y - 15x^2y^2 = 5x^2(2x^2 + 5xy - 3y^2)$
$$= 5x^2(2x - y)(x + 3y)$$

**71.** $-14x^2 + 39x - 10 = -1(14x^2 - 39x + 10)$
$$= -1(2x - 5)(7x - 2)$$

**73.** $16p^4 - 40p^3 + 25p^2 = p^2(16p^2 - 40p + 25)$
$$= p^2[(4p)^2 - 2 \cdot 4p \cdot 5 + 5^2]$$
$$= p^2(4p - 5)^2$$

**75.** $x + 3x^2 - 2 = 3x^2 + x - 2 = (3x - 2)(x + 1)$

**77.** $8x^2 + 6xy - 27y^2 = (4x + 9y)(2x - 3y)$

**79.** $1 + 6x^2 + x^4 = x^4 + 6x^2 + 1$ is prime.

**81.** $9x^2 - 24xy + 16y^2 = (3x)^2 - 2 \cdot 3x \cdot 4y + (4y)^2$
$$= (3x - 4y)^2$$

**83.** $18x^2 - 9x - 14 = (6x - 7)(3x + 2)$

**85.** $-27t + 7t^2 - 4 = 7t^2 - 27t - 4 = (7t + 1)(t - 4)$

**87.** $49p^2 - 7p - 2 = (7p + 1)(7p - 2)$

**89.** $m^3 + 18m^2 + 81m = m(m^2 + 18m + 81)$
$$= m(m^2 + 2 \cdot m \cdot 9 + 9^2)$$
$$= m(m + 9)^2$$

**91.** $5x^2y^2 + 20xy + 1$ is prime.

**93.** $6a^5 + 37a^3b^2 + 6ab^4 = a(6a^4 + 37a^2b^2 + 6b^4)$
$$= a(6a^2 + b^2)(a^2 + 6b^2)$$

**95.** $(x - 2)(x + 2) = x^2 + 2x - 2x - 4 = x^2 - 4$

**97.** $(a + 3)(a^2 - 3a + 9)$
$$= a^3 - 3a^2 + 9a + 3a^2 - 9a + 27$$
$$= a^3 + 27$$

**99.** Look for the tallest bar. The age range is 25–34.

**101.** answers may vary

**103.** no

**105.** answers may vary

**107.** $P = (3x^2 + 1) + (6x + 4) + (x^2 + 15x)$
$$= 3x^2 + 1 + 6x + 4 + x^2 + 15x$$
$$= 4x^2 + 21x + 5$$
$$= (4x + 1)(x + 5)$$

**109.** $4x^2 + 2x + \dfrac{1}{4} = (2x)^2 + 2 \cdot 2x \cdot \dfrac{1}{2} + \left(\dfrac{1}{2}\right)^2$
$$= \left(2x + \dfrac{1}{2}\right)^2$$

**111.** $4x^2(y-1)^2 + 10x(y-1)^2 + 25(y-1)^2$
$$= (y-1)^2(4x^2 + 10x + 25)$$

**113.** $16 = 4^2;\ 2 \cdot x \cdot 4 = 8x;\ 8$

**115.** $(a + b)^2 = a^2 + 2ab + b^2$

**117.** $b = 2$: $3x^2 + 2x - 5 = (3x + 5)(x - 1)$
$b = 14$: $3x^2 + 14x - 5 = (3x - 1)(x + 5)$

**119.** $c = 2$: $5x^2 + 7x + 2 = (5x + 2)(x + 1)$

**121.** $-12x^3y^2 + 3x^2y^2 + 15xy^2$
$$= -3xy^2(4x^2 - x - 5)$$
$$= -3xy^2(4x - 5)(x + 1)$$

**123.** $4x^2(y-1)^2 + 20x(y-1)^2 + 25(y-1)^2$
$$= (y-1)^2(4x^2 + 20x + 25)$$
$$= (y-1)^2[(2x)^2 + 2 \cdot 2x \cdot 5 + 5^2]$$
$$= (y-1)^2(2x + 5)^2$$

**125.** $3x^{2n} + 17x^n + 10 = (3x^n + 2)(x^n + 5)$

**127.** Answers may vary

**Section 6.4 Practice**

**1.**

| Factors of $ac = 60$ | Sum of Factors |
|---|---|
| 1, 60 | 61 |
| 2, 30 | 32 |
| 3, 20 | 23 |
| 4, 15 | 19 |
| 5, 12 | 17 |
| 6, 10 | 16 |

← correct sum
   $b = 61$.

$$5x^2 + 61x + 12 = 5x^2 + 1x + 60x + 12$$
$$= x(5x+1) + 12(5x+1)$$
$$= (5x+1)(x+12)$$

**2.**

| Factors of $ac = 60$ | Sum of Factors |
|---|---|
| $-1, -60$ | $-61$ |
| $-2, -30$ | $-32$ |
| $-3, -20$ | $-23$ |
| $-4, -15$ | $-19$ |
| $-5, -12$ | $-17$ |
| $-6, -10$ | $-60$ |

← Correct sum
   $b = -19$

$$12x^2 - 19x + 5 = 12x^2 - 15x - 4x + 5$$
$$= 3x(4x-5) - 1(4x-5)$$
$$= (4x-5)(3x-1)$$

**3.** $30x^2 - 14x - 4 = 2(15x^2 - 7x - 2)$
Find two numbers whose product is
$ac = 15(-2) = -30$ and whose sum is $b$, $-7$. The
numbers are $-10$ and 3.
$$2(15x^2 - 7x - 2) = 2(15x^2 - 10x + 3x - 2)$$
$$= 2[5x(3x-2) + 1(3x-2)]$$
$$= 2(3x-2)(5x+1)$$

**4.** $40m^4 + 5m^3 - 35m^2 = 5m^2(8m^2 + m - 7)$
Find two numbers whose product is
$ac = 8(-7) = -56$ and whose sum is $b$, 1. The
numbers are 8 and $-7$.
$$5m^2(8m^2 + m - 7) = 5m^2(8m^2 + 8m - 7m - 7)$$
$$= 5m^2[8m(m+1) - 7(m+1)]$$
$$= 5m^2(m+1)(8m-7)$$

**5.** Find two numbers whose product is
$ac = 16 \cdot 9 = 144$ and whose sum is $b$, 24. The
numbers are 12 and 12.
$$16x^2 + 24x + 9 = 16x^2 + 12x + 12x + 9$$
$$= 4x(4x+3) + 3(4x+3)$$
$$= (4x+3)(4x+3)$$
$$= (4x+3)^2$$

**Vocabulary, Readiness & Video Check 6.4**

**1.** $a = 1$, $b = 6$, $c = 8$
   $a \cdot c = 1 \cdot 8 = 8$
   $4 \cdot 2 = 8$ and $4 + 2 = 6$; choice a.

**2.** $a = 1$, $b = 11$, $c = 24$
   $a \cdot c = 1 \cdot 24 = 24$
   $8 \cdot 3 = 24$ and $8 + 3 = 11$; choice c.

**3.** $a = 2$, $b = 13$, $c = 6$
   $a \cdot c = 2 \cdot 6 = 12$
   $12 \cdot 1 = 12$ and $12 + 1 = 13$; choice b.

**4.** $a = 4$, $b = 8$, $c = 3$
   $a \cdot c = 4 \cdot 3 = 12$
   $2 \cdot 6 = 12$ and $2 + 6 = 8$; choice d.

**5.** This gives us a four-term polynomial which may
be factored by grouping.

**Exercise Set 6.4**

**1.** $x^2 + 3x + 2x + 6 = x(x+3) + 2(x+3)$
$$= (x+3)(x+2)$$

**3.** $y^2 + 8y - 2y - 16 = y(y+8) - 2(y+8)$
$$= (y+8)(y-2)$$

**5.** $8x^2 - 5x - 24x + 15 = x(8x-5) - 3(8x-5)$
$$= (8x-5)(x-3)$$

**7.** $5x^4 - 3x^2 + 25x^2 - 15 = x^2(5x^2-3) + 5(5x^2-3)$
$$= (5x^2-3)(x^2+5)$$

**9. a.**    $9 \cdot 2 = 18$; $9 + 2 = 11$; $9, 2$

     **b.**    $11x = 9x + 2x$

     **c.**    $\begin{aligned} 6x^2 + 11x + 3 &= 6x^2 + 9x + 2x + 3 \\ &= 3x(2x+3) + 1(2x+3) \\ &= (3x+1)(2x+3) \end{aligned}$

**11. a.**    $-20 \cdot (-3) = 60$; $-20 + (-3) = -23$; $-20, -3$

     **b.**    $-23x = -20x - 3x$

     **c.**    $\begin{aligned} 15x^2 - 23x + 4 &= 15x^2 - 20x - 3x + 4 \\ &= 5x(3x-4) - 1(3x-4) \\ &= (3x-4)(5x-1) \end{aligned}$

**13.**   $ac = 21 \cdot 2 = 42$; $b = 17$; two numbers: 14, 3
$\begin{aligned} 21y^2 + 17y + 2 &= 21y^2 + 14y + 3y + 2 \\ &= 7y(3y+2) + 1(3y+2) \\ &= (3y+2)(7y+1) \end{aligned}$

**15.**   $ac = 7 \cdot (-11) = -77$; $b = -4$;
two numbers: $-11, 7$
$\begin{aligned} 7x^2 - 4x - 11 &= 7x^2 - 11x + 7x - 11 \\ &= x(7x-11) + 1(7x-11) \\ &= (7x-11)(x+1) \end{aligned}$

**17.**   $ac = 10 \cdot 2 = 20$; $b = -9$; two numbers: $-4, -5$
$\begin{aligned} 10x^2 - 9x + 2 &= 10x^2 - 4x - 5x + 2 \\ &= 2x(5x-2) - 1(5x-2) \\ &= (5x-2)(2x-1) \end{aligned}$

**19.**   $ac = 2 \cdot 5 = 10$; $b = -7$; two numbers: $-5, -2$
$\begin{aligned} 2x^2 - 7x + 5 &= 2x^2 - 5x - 2x + 5 \\ &= x(2x-5) - 1(2x-5) \\ &= (2x-5)(x-1) \end{aligned}$

**21.**   $12x + 4x^2 + 9 = 4x^2 + 12x + 9$
$ac = 4 \cdot 9 = 36$; $b = 12$; two numbers: 6, 6
$\begin{aligned} 4x^2 + 12x + 9 &= 4x^2 + 6x + 6x + 9 \\ &= 2x(2x+3) + 3(2x+3) \\ &= (2x+3)(2x+3) \\ &= (2x+3)^2 \end{aligned}$

**23.**   $ac = 4 \cdot (-21) = -84$; $b = -8$;
two numbers: 6, $-14$
$\begin{aligned} 4x^2 - 8x - 21 &= 4x^2 + 6x - 14x - 21 \\ &= 2x(2x+3) - 7(2x+3) \\ &= (2x+3)(2x-7) \end{aligned}$

**25.**   $ac = 10 \cdot 12 = 120$; $b = -23$;
two numbers: $-8, -15$
$\begin{aligned} 10x^2 - 23x + 12 &= 10x^2 - 8x - 15x + 12 \\ &= 2x(5x-4) - 3(5x-4) \\ &= (5x-4)(2x-3) \end{aligned}$

**27.**   $2x^3 + 13x^2 + 15x = x(2x^2 + 13x + 15)$
$ac = 2 \cdot 15 = 30$; $b = 13$; two numbers: 3, 10
$\begin{aligned} x(2x^2 + 13x + 15) &= x(2x^2 + 3x + 10x + 15) \\ &= x[x(2x+3) + 5(2x+3)] \\ &= x(2x+3)(x+5) \end{aligned}$

**29.**   $16y^2 - 34y + 18 = 2(8y^2 - 17y + 9)$
$ac = 8(9) = 72$; $b = -17$; two numbers: $-9, -8$
$\begin{aligned} 2(8y^2 - 17y + 9) &= 2(8y^2 - 9y - 8y + 9) \\ &= 2[y(8y-9) - 1(8y-9)] \\ &= 2(8y-9)(y-1) \end{aligned}$

**31.**   $-13x + 6 + 6x^2 = 6x^2 - 13x + 6$
$ac = 6 \cdot 6 = 36$; $b = -13$; two numbers: $-9, -4$
$\begin{aligned} 6x^2 - 13x + 6 &= 6x^2 - 9x - 4x + 6 \\ &= 3x(2x-3) - 2(2x-3) \\ &= (2x-3)(3x-2) \end{aligned}$

**33.**   $54a^2 - 9a - 30 = 3(18a^2 - 3a - 10)$
$ac = 18(-10) = -180$; $b = -3$;
two numbers: 12, $-15$
$\begin{aligned} 3(18a^2 - 3a - 10) &= 3(18a^2 + 12a - 15a - 10) \\ &= 3[6a(3a+2) - 5(3a+2)] \\ &= 3(3a+2)(6a-5) \end{aligned}$

**35.**   $20a^3 + 37a^2 + 8a = a(20a^2 + 37a + 8)$
$ac = 20(8) = 160$; $b = 37$; two numbers: 5, 32
$\begin{aligned} a(20a^2 + 37a + 8) &= a(20a^2 + 5a + 32a + 8) \\ &= a[5a(4a+1) + 8(4a+1)] \\ &= a(4a+1)(5a+8) \end{aligned}$

**37.**   $12x^3 - 27x^2 - 27x = 3x(4x^2 - 9x - 9)$
$ac = 4(-9) = -36$; $b = -9$; two numbers: 3, $-12$
$\begin{aligned} 3x(4x^2 - 9x - 9) &= 3x(4x^2 + 3x - 12x - 9) \\ &= 3x[x(4x+3) - 3(4x+3)] \\ &= 3x(4x+3)(x-3) \end{aligned}$

       

**39.** $3x^2y + 4xy^2 + y^3 = y(3x^2 + 4xy + y^2)$
$ac = 3 \cdot 1 = 3;\ b = 4;$ two numbers: 1, 3
$y(3x^2 + 4xy + y^2) = y(3x^2 + xy + 3xy + y^2)$
$\qquad\qquad\qquad\quad = y[x(3x + y) + y(3x + y)]$
$\qquad\qquad\qquad\quad = y(3x + y)(x + y)$

**41.** $ac = 20 \cdot 1 = 20;\ b = 7;$ there are no two numbers.
$20z^2 + 7z + 1$ is prime.

**43.** $5x^2 + 50xy + 125y^2 = 5(x^2 + 10xy + 25y^2)$
$ac = 1 \cdot 25 = 25;\ b = 10;$ two numbers: 5, 5
$5(x^2 + 10xy + 25y^2) = 5(x^2 + 5xy + 5xy + 25y^2)$
$\qquad\qquad\qquad\qquad = 5[x(x + 5y) + 5y(x + 5y)]$
$\qquad\qquad\qquad\qquad = 5(x + 5y)(x + 5y)$
$\qquad\qquad\qquad\qquad = 5(x + 5y)^2$

**45.** $24a^2 - 6ab - 30b^2 = 6(4a^2 - ab - 5b^2)$
$ac = 4 \cdot (-5) = -20;\ b = -1;$ two numbers: 4, $-5$
$6(4a^2 - ab - 5b^2) = 6(4a^2 + 4ab - 5ab - 5b^2)$
$\qquad\qquad\qquad\quad = 6[4a(a + b) - 5b(a + b)]$
$\qquad\qquad\qquad\quad = 6(a + b)(4a - 5b)$

**47.** $15p^4 + 31p^3q + 2p^2q^2 = p^2(15p^2 + 31pq + 2q^2)$
$ac = 15(2) = 30;\ b = 31;$ two numbers: 1, 30
$p^2(15p^2 + 31pq + 2q^2)$
$= p^2(15p^2 + pq + 30pq + 2q^2)$
$= p^2[p(15p + q) + 2q(15p + q)]$
$= p^2(15p + q)(p + 2q)$

**49.** $162a^4 - 72a^2 + 8 = 2(81a^4 - 36a^2 + 4)$
$ac = 81 \cdot 4 = 324;\ b = -36;$
two numbers: $-18, -18$
$2(81a^4 - 36a^2 + 4)$
$= 2(81a^4 - 18a^2 - 18a^2 + 4)$
$= 2[9a^2(9a^2 - 2) - 2(9a^2 - 2)]$
$= 2(9a^2 - 2)(9a^2 - 2)$
$= 2(9a^2 - 2)^2$

**51.** $35 + 12x + x^2 = x^2 + 12x + 35$
$ac = 1 \cdot 35 = 35;\ b = 12;$ two numbers: 5, 7
$x^2 + 12x + 35 = x^2 + 5x + 7x + 35$
$\qquad\qquad\qquad = x(x + 5) + 7(x + 5)$
$\qquad\qquad\qquad = (x + 5)(x + 7)$

**53.** $6 - 11x + 5x^2 = 5x^2 - 11x + 6$
$ac = 5 \cdot 6 = 30;\ b = -11;$ two numbers: $-6, -5$
$5x^2 - 11x + 6 = 5x^2 - 6x - 5x + 6$
$\qquad\qquad\qquad = x(5x - 6) - 1(5x - 6)$
$\qquad\qquad\qquad = (5x - 6)(x - 1)$

**55.** $(x - 2)(x + 2) = x^2 - 2^2 = x^2 - 4$

**57.** $(y + 4)(y + 4) = y^2 + 2 \cdot y \cdot 4 + 4^2 = y^2 + 8y + 16$

**59.** $(9z + 5)(9z - 5) = (9z)^2 - 5^2 = 81z^2 - 25$

**61.** $(x - 3)(x^2 + 3x + 9) = x^3 - 3^3 = x^3 - 27$

**63.** $5(2x^2 + 9x + 9) = 10x^2 + 45x + 45$
$ac = 2 \cdot 9 = 18;\ b = 9;$ two numbers: 3, 6
$5(2x^2 + 9x + 9) = 5(2x^2 + 3x + 6x + 9)$
$\qquad\qquad\qquad\quad = 5[x(2x + 3) + 3(2x + 3)]$
$\qquad\qquad\qquad\quad = 15(2x + 3)(x + 3)$

**65.** $x^{2n} + 2x^n + 3x^n + 6 = x^n(x^n + 2) + 3(x^n + 2)$
$\qquad\qquad\qquad\qquad\qquad = (x^n + 2)(x^n + 3)$

**67.** $ac = 3 \cdot (-35) = -105;\ b = 16;$
two numbers: $-5, 21$
$3x^{2n} + 16x^n - 35 = 3x^{2n} - 5x^n + 21x^n - 35$
$\qquad\qquad\qquad\qquad = x^n(3x^n - 5) + 7(3x^n - 5)$
$\qquad\qquad\qquad\qquad = (3x^n - 5)(x^n + 7)$

**69.** answers may vary

**Section 6.5 Practice**

**1.** $x^2 - 81 = x^2 - 9^2 = (x + 9)(x - 9)$

**2. a.** $9x^2 - 1 = (3x)^2 - 1^2 = (3x + 1)(3x - 1)$

**b.** $36a^2 - 49b^2 = (6a)^2 - (7b)^2$
$\qquad\qquad\qquad = (6a + 7b)(6a - 7b)$

**c.** $p^2 - \dfrac{25}{36} = p^2 - \left(\dfrac{5}{6}\right)^2 = \left(p + \dfrac{5}{6}\right)\left(p - \dfrac{5}{6}\right)$

**3.** $p^4 - q^{10} = (p^2)^2 - (q^5)^2 = (p^2 + q^5)(p^2 - q^5)$

**4. a.** $z^4 - 81 = (z^2)^2 - 9^2$

$= (z^2 + 9)(z^2 - 9)$

$= (z^2 + 9)(z + 3)(z - 3)$

  **b.** $m^2 + 49$ is a prime polynomial.

**5.** $36y^3 - 25y = y(36y^2 - 25)$

$= y[(6y)^2 - 5^2]$

$= y(6y + 5)(6y - 5)$

**6.** $80y^4 - 5 = 5(16y^2 - 1)$

$= 5[(4y)^2 - 1^2]$

$= 5(4y + 1)(4y - 1)$

**7.** $-9x^2 + 100 = -1(9x^2 - 100)$

$= -1[(3x)^2 - 10^2]$

$= -1(3x + 10)(3x - 10)$

or $-9x^2 + 100 = 100 - 9x^2$

$= 10^2 - (3x)^2$

$= (10 + 3x)(10 - 3x)$

**8.** $x^3 + 64 = x^3 + 4^3$

$= (x + 4)(x^2 - x \cdot 4 + 4^2)$

$= (x + 4)(x^2 - 4x + 16)$

**9.** $x^3 - 125 = x^3 - 5^3$

$= (x - 5)(x^2 + x \cdot 5 + 5^2)$

$= (x - 5)(x^2 + 5x + 25)$

**10.** $27y^3 + 1 = (3y)^3 + 1^3$

$= (3y + 1)[(3y)^2 - 3y \cdot 1 + 1^2]$

$= (3y + 1)(9y^2 - 3y + 1)$

**11.** $32x^3 - 500y^3$

$= 4(8x^3 - 125y^3)$

$= 4[(2x)^3 - (5y)^3]$

$= 4(2x - 5y)[(2x)^2 + 2x \cdot 5y + (5y)^2]$

$= 4(2x - 5y)(4x^2 + 10xy + 25y^2)$

**Calculator Explorations**

| $x$ | $x^2 - 2x + 1$ | $x^2 - 2x - 1$ | $(x-1)^2$ |
|---|---|---|---|
| 5 | 16 | 14 | 16 |
| −3 | 16 | 14 | 16 |
| 2.7 | 2.89 | 0.89 | 2.89 |
| −12.1 | 171.61 | 169.61 | 171.61 |
| 0 | 1 | −1 | 1 |

**Vocabulary, Readiness & Video Check 6.5**

**1.** The expression $x^3 - 27$ is called a <u>difference of two cubes</u>.

**2.** The expression $x^2 - 49$ is called a <u>difference of two squares</u>.

**3.** The expression $z^3 + 1$ is called a <u>sum of two cubes</u>.

**4.** The binomial $y^2 + 9$ is prime. The statement is false.

**5.** $49x^2 = (7x)^2$

**6.** $25y^4 = (5y^2)^2$

**7.** $8y^3 = (2y)^3$

**8.** $x^6 = (x^2)^3$

**9.** In order to recognize the binomial as a difference of squares and also to identify the terms to use in the special factoring formula

**10.** A prime polynomial is one that can't be factored further.

**11.** First rewrite the original binomial with terms written as cubes. Answers will then vary depending on your interpretation.

**Exercise Set 6.5**

**1.** $x^2 - 4 = x^2 - 2^2 = (x + 2)(x - 2)$

**3.** $81p^2 - 1 = (9p)^2 - 1^2 = (9p + 1)(9p - 1)$

**5.** $25y^2 - 9 = (5y)^2 - 3^2 = (5y+3)(5y-3)$

**7.** $121m^2 - 100n^2 = (11m)^2 - (10n)^2$
$$= (11m+10n)(11m-10n)$$

**9.** $x^2y^2 - 1 = (xy)^2 - 1^2 = (xy+1)(xy-1)$

**11.** $x^2 - \dfrac{1}{4} = x^2 - \left(\dfrac{1}{2}\right)^2 = \left(x+\dfrac{1}{2}\right)\left(x-\dfrac{1}{2}\right)$

**13.** $-4r^2 + 1 = -1(4r^2 - 1)$
$$= -1[(2r)^2 - 1^2]$$
$$= -1(2r+1)(2r-1)$$

**15.** $16r^2 + 1$ is the sum of two squares, $(4r)^2 + 1^2$, not the difference of two squares. $16r^2 + 1$ is a prime polynomial.

**17.** $-36 + x^2 = -1(36 - x^2)$
$$= -1(6^2 - x^2)$$
$$= -1(6+x)(6-x) \text{ or } (-6+x)(6+x)$$

**19.** $m^4 - 1 = (m^2)^2 - 1^2$
$$= (m^2+1)(m^2-1)$$
$$= (m^2+1)(m+1)(m-1)$$

**21.** $m^4 - n^{18} = (m^2)^2 - (n^9)^2$
$$= (m^2+n^9)(m^2-n^9)$$

**23.** $x^3 + 125 = x^3 + 5^3$
$$= (x+5)(x^2 - x\cdot 5 + 5^2)$$
$$= (x+5)(x^2 - 5x + 25)$$

**25.** $8a^3 - 1 = (2a)^3 - 1^3$
$$= (2a-1)[(2a)^2 + 2a\cdot 1 + 1^2]$$
$$= (2a-1)(4a^2 + 2a + 1)$$

**27.** $m^3 + 27n^3 = m^3 + (3n)^3$
$$= (m+3n)[m^2 - m\cdot 3n + (3n)^2]$$
$$= (m+3n)(m^2 - 3mn + 9n^2)$$

**29.** $5k^3 + 40 = 5(k^3 + 8)$
$$= 5(k^3 + 2^3)$$
$$= 5(k+2)[k^2 - k\cdot 2 + 2^2]$$
$$= 5(k+2)(k^2 - 2k + 4)$$

**31.** $x^3y^3 - 64 = (xy)^3 - 4^3$
$$= (xy-4)[(xy)^2 + xy\cdot 4 + 4^2]$$
$$= (xy-4)(x^2y^2 + 4xy + 16)$$

**33.** $250r^3 - 128t^3 = 2(125r^3 - 64t^3)$
$$= 2[(5r)^3 - (4t)^3]$$
$$= 2(5r-4t)[(5r)^2 + 5r\cdot 4t + (4t)^2]$$
$$= 2(5r-4t)(25r^2 + 20rt + 16t^2)$$

**35.** $r^2 - 64 = r^2 - 8^2 = (r+8)(r-8)$

**37.** $x^2 - 169y^2 = x^2 - (13y)^2 = (x+13y)(x-13y)$

**39.** $27 - t^3 = 3^3 - t^3$
$$= (3-t)(3^2 + 3\cdot t + t^2)$$
$$= (3-t)(9 + 3t + t^2)$$

**41.** $18r^2 - 8 = 2(9r^2 - 4)$
$$= 2[(3r)^2 - 2^2]$$
$$= 2(3r+2)(3r-2)$$

**43.** $9xy^2 - 4x = x(9y^2 - 4)$
$$= x[(3y)^2 - 2^2]$$
$$= x(3y+2)(3y-2)$$

**45.** $8m^3 + 64 = 8(m^3 + 8)$
$$= 8(m^3 + 2^3)$$
$$= 8(m+2)(m^2 - m\cdot 2 + 2^2)$$
$$= 8(m+2)(m^2 - 2m + 4)$$

**47.** $xy^3 - 9xyz^2 = xy(y^2 - 9z^2)$
$$= xy[y^2 - (3z)^2]$$
$$= xy(y+3z)(y-3z)$$

**49.** $36x^2 - 64y^2 = 4(9x^2 - 16y^2)$
$$= 4[(3x)^2 - (4y)^2]$$
$$= 4(3x+4y)(3x-4y)$$

**51.** $144 - 81x^2 = 9(16 - 9x^2)$
$$= 9[4^2 - (3x)^2]$$
$$= 9(4 + 3x)(4 - 3x)$$

**53.** $x^3 y^3 - z^6 = (xy)^3 - (z^2)^3$
$$= (xy - z^2)[(xy)^2 + xy \cdot z^2 + (z^2)^2]$$
$$= (xy - z^2)(x^2 y^2 + xyz^2 + z^4)$$

**55.** $49 - \dfrac{9}{25} m^2 = 7^2 - \left(\dfrac{3}{5} m\right)^2 = \left(7 + \dfrac{3}{5} m\right)\left(7 - \dfrac{3}{5} m\right)$

**57.** $t^3 + 343 = t^3 + 7^3$
$$= (t + 7)(t^2 - t \cdot 7 + 7^2)$$
$$= (t + 7)(t^2 - 7t + 49)$$

**59.** $n^3 - 49n = n(n^2 + 49)$

**61.** $x^6 - 81x^2 = x^2(x^4 - 81)$
$$= x^2[(x^2)^2 - 9^2]$$
$$= x^2(x^2 + 9)(x^2 - 9)$$
$$= x^2(x^2 + 9)(x + 3)(x - 3)$$

**63.** $64p^3 q - 81pq^3 = pq(64p^2 - 81q^2)$
$$= pq[(8p)^2 - (9q)^2]$$
$$= pq(8p + 9q)(8p - 9q)$$

**65.** $27x^2 y^3 + xy^2 = xy^2(27xy + 1)$

**67.** $125a^4 - 64ab^3$
$$= a(125a^3 - 64b^3)$$
$$= a[(5a)^3 - (4b)^3]$$
$$= a(5a - 4b)[(5a)^2 + 5a \cdot 4b + (4b)^2]$$
$$= a(5a - 4b)(25a^2 + 20ab + 16b^2)$$

**69.** $16x^4 - 64x^2 = 16x^2(x^2 - 4)$
$$= 16x^2(x^2 - 2^2)$$
$$= 16x^2(x + 2)(x - 2)$$

**71.** $x - 6 = 0$
$$x - 6 + 6 = 0 + 6$$
$$x = 6$$

**73.** $2m + 4 = 0$
$$2m + 4 - 4 = 0 - 4$$
$$2m = -4$$
$$\frac{2m}{2} = \frac{-4}{2}$$
$$m = -2$$

**75.** $5z - 1 = 0$
$$5z - 1 + 1 = 0 + 1$$
$$5z = 1$$
$$\frac{5z}{5} = \frac{1}{5}$$
$$z = \frac{1}{5}$$

**77.** $(x + 2)^2 - y^2 = (x + 2 + y)(x + 2 - y)$

**79.** $a^2(b - 4) - 16(b - 4) = (b - 4)(a^2 - 16)$
$$= (b - 4)(a^2 - 4^2)$$
$$= (b - 4)(a + 4)(a - 4)$$

**81.** $(x^2 + 6x + 9) - 4y^2 = (x + 3)^2 - 4y^2$
$$= (x + 3)^2 - (2y)^2$$
$$= [(x + 3) + 2y][(x + 3) - 2y]$$
$$= (x + 3 + 2y)(x + 3 - 2y)$$

**83.** $x^{2n} - 100 = (x^n)^2 - 10^2 = (x^n + 10)(x^n - 10)$

**85.** $x + 6$ since
$$(x + 6)(x - 6) = x^2 - 6x + 6x - 36$$
$$= x^2 - 36$$
$$= x^2 - 6^2$$

**87.** answers may vary

**89. a.** $2704 - 16t^2 = 2704 - 16(3)^2$
$$= 2704 - 16 \cdot 9$$
$$= 2704 - 144$$
$$= 2560$$
After 3 seconds, the filter is 2560 feet above the river.

**b.** $2704 - 16t^2 = 2704 - 16(7)^2$
$$= 2704 - 16 \cdot 49$$
$$= 2704 - 784$$
$$= 1920$$
After 7 seconds, the filter is 1920 feet above the river.

**c.** The filter lands in the river when its height is 0 feet.

$$2704 - 16t^2 = 0$$
$$(52 + 4t)(52 - 4t) = 0$$
$$52 + 4t = 0 \quad \text{or} \quad 52 - 4t = 0$$
$$4t = -52 \qquad\qquad -4t = -52$$
$$t = -13 \qquad\qquad\quad t = 13$$

Discard $t = -13$ since time cannot be negative. The filter lands in the river after 13 seconds.

**d.** $2704 - 16t^2 = 16(169 - t^2)$
$$= 16[(13)^2 - t^2]$$
$$= 16(13 - t)(13 + t)$$

**91. a.** Let $t = 3$.
$$1600 - 16t^2 = 1600 - 16(3)^2 = 1456$$
After 3 seconds the height is 1456 feet.

**b.** Let $t = 7$.
$$1600 - 16t^2 = 1600 - 16(7)^2 = 816$$
After 7 seconds the height is 816 feet.

**c.** When it hits the ground, the height is 0.
Let $0 = 1600 - 16t^2$.
$$16t^2 = 1600$$
$$t^2 = 100$$
$$t = \sqrt{100}$$
$$t = 10$$
Thus, it will hit the ground after 10 seconds.

**d.** $1600 - 16t^2 = 16(100 - t^2)$
$$= 16(10^2 - t^2)$$
$$= 16(10 + t)(10 - t)$$

**Integrated Review Practice**

**1.** $6x^2 - 11x + 3$
$ac = 6 \cdot 3 = 18$; $b = -11$; two numbers: $-2, -9$
$6x^2 - 11x + 3 = 6x^2 - 2x - 9x + 3$
$$= 2x(3x - 1) - 3(3x - 1)$$
$$= (3x - 1)(2x - 3)$$

**2.** $3x^3 + x^2 - 12x - 4 = (3x^3 + x^2) + (-12x - 4)$
$$= x^2(3x + 1) - 4(3x + 1)$$
$$= (3x + 1)(x^2 - 4)$$
$$= (3x + 1)(x + 2)(x - 2)$$

**3.** $27x^2 - 3y^2 = 3(9x^2 - y^2)$
$$= 3[(3x)^2 - y^2]$$
$$= 3(3x + y)(3x - y)$$

**4.** $8a^3 + b^3 = (2a)^3 + b^3$
$$= (2a + b)[(2a)^2 - 2a \cdot b + b^2]$$
$$= (2a + b)(4a^2 - 2ab + b^2)$$

**5.** $60x^3y^2 - 66x^2y^2 - 36xy^2$
$$= 6xy^2(10x^2 - 11x - 6)$$
$$= 6xy^2(5x + 2)(2x - 3)$$

**Integrated Review**

**1.** $x^2 + 2xy + y^2 = (x + y)(x + y) = (x + y)^2$

**2.** $x^2 - 2xy + y^2 = (x - y)(x - y) = (x - y)^2$

**3.** $a^2 + 11a - 12 = (a + 12)(a - 1)$

**4.** $a^2 - 11a + 10 = (a - 10)(a - 1)$

**5.** $a^2 - a - 6 = (a - 3)(a + 2)$

**6.** $a^2 - 2a + 1 = (a - 1)(a - 1) = (a - 1)^2$

**7.** $x^2 + 2x + 1 = (x + 1)(x + 1) = (x + 1)^2$

**8.** $x^2 + x - 2 = (x + 2)(x - 1)$

**9.** $x^2 + 4x + 3 = (x + 3)(x + 1)$

**10.** $x^2 + x - 6 = (x + 3)(x - 2)$

**11.** $x^2 + 7x + 12 = (x + 4)(x + 3)$

**12.** $x^2 + x - 12 = (x + 4)(x - 3)$

**13.** $x^2 + 3x - 4 = (x + 4)(x - 1)$

**14.** $x^2 - 7x + 10 = (x - 5)(x - 2)$

**15.** $x^2 + 2x - 15 = (x + 5)(x - 3)$

**16.** $x^2 + 11x + 30 = (x + 6)(x + 5)$

**17.** $x^2 - x - 30 = (x-6)(x+5)$

**18.** $x^2 + 11x + 24 = (x+8)(x+3)$

**19.** $2x^2 - 98 = 2(x^2 - 49)$
$= 2(x^2 - 7^2)$
$= 2(x+7)(x-7)$

**20.** $3x^2 - 75 = 3(x^2 - 25)$
$= 3(x^2 - 5^2)$
$= 3(x+5)(x-5)$

**21.** $x^2 + 3x + xy + 3y = x(x+3) + y(x+3)$
$= (x+3)(x+y)$

**22.** $3y - 21 + xy - 7x = 3(y-7) + x(y-7)$
$= (y-7)(3+x)$

**23.** $x^2 + 6x - 16 = (x+8)(x-2)$

**24.** $x^2 - 3x - 28 = (x-7)(x+4)$

**25.** $4x^3 + 20x^2 - 56x = 4x(x^2 + 5x - 14)$
$= 4x(x+7)(x-2)$

**26.** $6x^3 - 6x^2 - 120x = 6x(x^2 - x - 20)$
$= 6x(x-5)(x+4)$

**27.** $12x^2 + 34x + 24 = 2(6x^2 + 17x + 12)$
$= 2(6x^2 + 9x + 8x + 12)$
$= 2[3x(2x+3) + 4(2x+3)]$
$= 2(2x+3)(3x+4)$

**28.** $8a^2 + 6ab - 5b^2 = 8a^2 + 10ab - 4ab - 5b^2$
$= 2a(4a+5b) - b(4a+5b)$
$= (4a+5b)(2a-b)$

**29.** $4a^2 - b^2 = (2a)^2 - b^2 = (2a+b)(2a-b)$

**30.** $28 - 13x - 6x^2 = 28 - 21x + 8x - 6x^2$
$= 7(4-3x) + 2x(4-3x)$
$= (4-3x)(7+2x)$

**31.** $20 - 3x - 2x^2 = 20 - 8x + 5x - 2x^2$
$= 4(5-2x) + x(5-2x)$
$= (5-2x)(4+x)$

**32.** $x^2 - 2x + 4$ is a prime polynomial.

**33.** $a^2 + a - 3$ is a prime polynomial.

**34.** $6y^2 + y - 15 = 6y^2 + 10y - 9y - 15$
$= 2y(3y+5) - 3(3y+5)$
$= (3y+5)(2y-3)$

**35.** $4x^2 - x - 5 = 4x^2 - 5x + 4x - 5$
$= x(4x-5) + 1(4x-5)$
$= (4x-5)(x+1)$

**36.** $x^2 y - y^3 = y(x^2 - y^2) = y(x-y)(x+y)$

**37.** $4t^2 + 36 = 4(t^2 + 9)$

**38.** $x^2 + x + xy + y = x(x+1) + y(x+1)$
$= (x+1)(x+y)$

**39.** $ax + 2x + a + 2 = x(a+2) + 1(a+2)$
$= (a+2)(x+1)$

**40.** $18x^3 - 63x^2 + 9x = 9x(2x^2 - 7x + 1)$

**41.** $12a^3 - 24a^2 + 4a = 4a(3a^2 - 6a + 1)$

**42.** $x^2 + 14x - 32 = (x+16)(x-2)$

**43.** $x^2 - 14x - 48$ is prime.

**44.** $16a^2 - 56ab + 49b^2 = (4a)^2 - 2(4a)(7b) + (7b)^2$
$= (4a - 7b)^2$

**45.** $25p^2 - 70pq + 49q^2 = (5p)^2 - 2(5p)(7q) + (7q)^2$
$= (5p - 7q)^2$

**46.** $7x^2 + 24xy + 9y^2 = 7x^2 + 3xy + 21xy + 9y^2$
$= x(7x+3y) + 3y(7x+3y)$
$= (7x+3y)(x+3y)$

**47.** $125 - 8y^3 = 5^3 - (2y)^3$
$= (5 - 2y)[5^2 + 5 \cdot 2y + (2y)^2]$
$= (5 - 2y)(25 + 10y + 4y^2)$

    

**48.** $64x^3 + 27 = (4x)^3 + 3^3$
$$= (4x+3)[(4x)^2 - 4x \cdot 3 + 3^2]$$
$$= (4x+3)(16x^2 - 12x + 9)$$

**49.** $-x^2 - x + 30 = -1(x^2 + x - 30) = -(x+6)(x-5)$

**50.** $-x^2 + 6x - 8 = -1(x^2 - 6x + 8) = -(x-2)(x-4)$

**51.** $14 + 5x - x^2 = (7-x)(2+x)$

**52.** $3 - 2x - x^2 = (3+x)(1-x)$

**53.** $3x^4y + 6x^3y - 72x^2y = 3x^2y(x^2 + 2x - 24)$
$$= 3x^2y(x+6)(x-4)$$

**54.** $2x^3y + 8x^2y^2 - 10xy^3 = 2xy(x^2 + 4xy - 5y^2)$
$$= 2xy(x+5y)(x-y)$$

**55.** $5x^3y^2 - 40x^2y^3 + 35xy^4 = 5xy^2 - 8xy + 7y^2)$
$$= 5xy^2(x - 7y)(x - y)$$

**56.** $4x^4y - 8x^3y - 60x^2y = 4x^2y(x^2 - 2x - 15)$
$$= 4x^2y(x-5)(x+3)$$

**57.** $12x^3y + 243xy = 3xy(4x^2 + 81)$

**58.** $6x^3y^2 + 8xy^2 = 2xy^2(3x^2 + 4)$

**59.** $4 - x^2 = 2^2 - x^2 = (2+x)(2-x)$

**60.** $9 - y^2 = 3^2 - y^2 = (3+y)(3-y)$

**61.** $3rs - s + 12r - 4 = s(3r-1) + 4(3r-1)$
$$= (3r-1)(s+4)$$

**62.** $x^3 - 2x^2 + 3x - 6 = x^2(x-2) + 3(x-2)$
$$= (x-2)(x^2+3)$$

**63.** $4x^2 - 8xy - 3x + 6y = 4x(x-2y) - 3(x-2y)$
$$= (x-2y)(4x-3)$$

**64.** $4x^2 - 2xy - 7yz + 14xz$
$$= 2x(2x-y) + 7z(-y+2x)$$
$$= (2x-y)(2x+7z)$$

**65.** $6x^2 + 18xy + 12y^2 = 6(x^2 + 3xy + 2y^2)$
$$= 6(x+2)(x+y)$$

**66.** $12x^2 + 46xy - 8y^2 = 2(6x^2 + 23xy - 4y^2)$
$$= 2(6x^2 + 24xy - xy - 4y^2)$$
$$= 2[6x(x+4y) - y(x+4y)]$$
$$= 2(x+4y)(6x-y)$$

**67.** $xy^2 - 4x + 3y^2 - 12 = x(y^2 - 4) + 3(y^2 - 4)$
$$= (y^2 - 4)(x+3)$$
$$= (y^2 - 2^2)(x+3)$$
$$= (y+2)(y-2)(x+3)$$

**68.** $x^2y^2 - 9x^2 + 3y^2 - 27 = x^2(y^2 - 9) + 3(y^2 - 9)$
$$= (y^2 - 9)(x^2 + 3)$$
$$= (y^2 - 3^2)(x^2 + 3)$$
$$= (y-3)(y+3)(x^2 + 3)$$

**69.** $5(x+y) + x(x+y) = (x+y)(5+x)$

**70.** $7(x-y) + y(x-y) = (x-y)(7+y)$

**71.** $14t^2 - 9t + 1 = 14t^2 - 7t - 2t + 1$
$$= 7t(2t-1) - 1(2t-1)$$
$$= (2t-1)(7t-1)$$

**72.** $3t^2 - 5t + 1$ is a prime polynomial.

**73.** $3x^2 + 2x - 5 = 3x^2 + 5x - 3x - 5$
$$= x(3x+5) - 1(3x+5)$$
$$= (3x+5)(x-1)$$

**74.** $7x^2 + 19x - 6 = 7x^2 + 21x - 2x - 6$
$$= 7x(x+3) - 2(x+3)$$
$$= (x+3)(7x-2)$$

**75.** $x^2 + 9xy - 36y^2 = (x+12y)(x-3y)$

**76.** $3x^2 + 10xy - 8y^2 = 3x^2 - 2xy + 12xy - 8y^2$
$$= x(3x-2y) + 4y(3x-2y)$$
$$= (3x-2y)(x+4y)$$

**77.** $1 - 8ab - 20a^2b^2 = 1 - 10ab + 2ab - 20a^2b^2$
$$= 1(1 - 10ab) + 2ab(1 - 10ab)$$
$$= (1 - 10ab)(1 + 2ab)$$

**78.** $1 - 7ab - 60a^2b^2 = 1 - 12ab + 5ab - 60a^2b^2$
$$= 1(1 - 12ab) + 5ab(1 - 12ab)$$
$$= (1 - 12ab)(1 + 5ab)$$

**79.** $9 - 10x^2 + x^4 = (9 - x^2)(1 - x^2)$
$$= (3^2 - x^2)(1^2 - x^2)$$
$$= (3 + x)(3 - x)(1 + x)(1 - x)$$

**80.** $36 - 13x^2 + x^4 = (9 - x^2)(4 - x^2)$
$$= (3^2 - x^2)(2^2 - x^2)$$
$$= (3 + x)(3 - x)(2 + x)(2 - x)$$

**81.** $x^4 - 14x^2 - 32 = (x^2 + 2)(x^2 - 16)$
$$= (x^2 + 2)(x^2 - 4^2)$$
$$= (x^2 + 2)(x + 4)(x - 4)$$

**82.** $x^4 - 22x^2 - 75 = (x^2 + 3)(x^2 - 25)$
$$= (x^2 + 3)(x^2 - 5^2)$$
$$= (x^2 + 3)(x + 5)(x - 5)$$

**83.** $x^2 - 23x + 120 = (x - 15)(x - 8)$

**84.** $y^2 + 22y + 96 = (y + 16)(y + 6)$

**85.** $6x^3 - 28x^2 + 16x = 2x(3x^2 - 14x + 8)$
$$= 2x(3x - 2)(x - 4)$$

**86.** $6y^3 - 8y^2 - 30y = 2y(3y^2 - 4y - 15)$
$$= 2y(3y + 5)(y - 3)$$

**87.** $27x^3 - 125y^3 = (3x)^3 - (5y)^3$
$$= (3x - 5y)[(3x)^2 + 3x \cdot 5y + (5y)^2]$$
$$= (3x - 5y)(9x^2 + 15xy + 25y^2)$$

**88.** $216y^3 - z^3 = (6y)^3 - z^3$
$$= (6y - z)[(6y)^2 + 6y \cdot z + z^2]$$
$$= (6y - z)(36y^2 + 6yz + z^2)$$

**89.** $x^3y^3 + 8z^3 = (xy)^3 + (2z)^3$
$$= (xy + 2z)[(xy)^2 - xy \cdot 2z + (2z)^2]$$
$$= (xy + 2z)(x^2y^2 - 2xyz + 4z^2)$$

**90.** $27a^3b^3 + 8 = (3ab)^3 + 2^3$
$$= (3ab + 2)[(3ab)^2 - 3ab \cdot 2 + 2^2]$$
$$= (3ab + 2)(9a^2b^2 - 6ab + 4)$$

**91.** $2xy - 72x^3y = 2xy(1 - 36x^2)$
$$= 2xy[1^2 - (6x)^2]$$
$$= 2xy(1 + 6x)(1 - 6x)$$

**92.** $2x^3 - 18x = 2x(x^2 - 9)$
$$= 2x(x^2 - 3^2)$$
$$= 2x(x + 3)(x - 3)$$

**93.** $x^3 + 6x^2 - 4x - 24 = x^2(x + 6) - 4(x + 6)$
$$= (x + 6)(x^2 - 4)$$
$$= (x + 6)(x^2 - 2^2)$$
$$= (x + 6)(x + 2)(x - 2)$$

**94.** $x^3 - 2x^2 - 36x + 72 = x^2(x - 2) - 36(x - 2)$
$$= (x - 2)(x^2 - 36)$$
$$= (x - 2)(x^2 - 6^2)$$
$$= (x - 2)(x + 6)(x - 6)$$

**95.** $6a^3 + 10a^2 = 2a^2(3a + 5)$

**96.** $4n^2 - 6n = 2n(2n - 3)$

**97.** $a^2(a + 2) + 2(a + 2) = (a + 2)(a^2 + 2)$

**98.** $a - b + x(a - b) = (a - b)(1 + x)$

**99.** $x^3 - 28 + 7x^2 - 4x = x^3 + 7x^2 - 28 - 4x$
$$= x^2(x + 7) - 4(7 + x)$$
$$= (x + 7)(x^2 - 4)$$
$$= (x + 7)(x^2 - 2^2)$$
$$= (x + 7)(x + 2)(x - 2)$$

**100.** $a^3 - 45 - 9a + 5a^2 = a^3 + 5a^2 - 9a - 45$
$$= a^2(a + 5) - 9(a + 5)$$
$$= (a + 5)(a^2 - 9)$$
$$= (a + 5)(a^2 - 3^2)$$
$$= (a + 5)(a + 3)(a - 3)$$

**101.** $(x - y)^2 - z^2 = (x - y + z)(x - y - z)$

    

**102.** $(x+2y)^2 - 9 = (x+2y)^2 - 3^2$
$\qquad\qquad = (x+2y+3)(x+2y-3)$

**103.** $81 - (5x+1)^2 = 9^2 - (5x+1)^2$
$\qquad\qquad = [9+(5x+1)][9-(5x+1)]$
$\qquad\qquad = (9+5x+1)(9-5x-1)$

**104.** $b^2 - (4a+c)^2$
$\quad = [b+(4a+c)][b-(4a+c)]$
$\quad = (b+4a+c)(b-4a-c)$

**105.** answers may vary

**106.** Yes; $9x^2 + 81y^2 = 9(x^2 + 9y^2)$

**107.** $(x+10)(x-7) = (x-7)(x+10)$
$\qquad\qquad\qquad = -1(x+10)(7-x);$
a, c

**108.** $(x-2)(x-5) = (x-5)(x-2) = (5-x)(2-x);$ b, c

## Section 6.6 Practice

**1.** $(x+4)(x-5) = 0$
$\quad x+4 = 0 \quad$ or $\quad x-5 = 0$
$\qquad x = -4 \qquad\qquad x = 5$
Check:
Let $x = -4$.
$\quad (x+4)(x-5) = 0$
$\quad (-4+4)(-4-5) \stackrel{?}{=} 0$
$\qquad\qquad 0(-9) = 0 \quad$ True
Let $x = 5$.
$\quad (x+4)(x-5) = 0$
$\quad (5+4)(5-5) \stackrel{?}{=} 0$
$\qquad\qquad 9(0) = 0 \quad$ True
The solutions are $-4$ and 5.

**2.** $(x-12)(4x+3) = 0$
$\quad x-12 = 0 \quad$ or $\quad 4x+3 = 0$
$\qquad x = 12 \qquad\qquad 4x = -3$
$\qquad\qquad\qquad\qquad\qquad x = -\dfrac{3}{4}$
Check:
Let $x = 12$.
$\quad (x-12)(4x+3) = 0$
$\quad (12-12)(4(12)+3) \stackrel{?}{=} 0$
$\qquad\qquad 0(51) \stackrel{?}{=} 0$
$\qquad\qquad\quad 0 = 0$ True
Let $x = -\dfrac{3}{4}$.

$(x-12)(4x+3) = 0$
$\left(-\dfrac{3}{4}-12\right)\left[4\left(-\dfrac{3}{4}\right)+3\right] \stackrel{?}{=} 0$
$\qquad\left(-\dfrac{3}{4}-12\right)(0) \stackrel{?}{=} 0$
$\qquad\qquad\qquad 0 = 0 \quad$ True

The solutions are 12 and $-\dfrac{3}{4}$.

**3.** $x(7x-6) = 0$
$\quad x = 0 \quad$ or $\quad 7x-6 = 0$
$\qquad\qquad\qquad\qquad 7x = 6$
$\qquad\qquad\qquad\qquad x = \dfrac{6}{7}$
Check:
Let $x = 0$.
$\quad x(7x-6) = 0$
$\quad 0(7 \cdot 0 - 6) \stackrel{?}{=} 0$
$\qquad 0(-6) = 0 \quad$ True
Let $x = \dfrac{6}{7}$.
$\quad x(7x-6) = 0$
$\quad \dfrac{6}{7}\left(7 \cdot \dfrac{6}{7} - 6\right) \stackrel{?}{=} 0$
$\qquad \dfrac{6}{7}(6-6) \stackrel{?}{=} 0$
$\qquad\quad \dfrac{6}{7}(0) = 0 \quad$ True

The solutions are 0 and $\dfrac{6}{7}$.

**4.** $x^2 - 8x - 48 = 0$
$\quad (x+4)(x-12) = 0$
$\quad x+4 = 0 \quad$ or $\quad x-12 = 0$
$\qquad x = -4 \qquad\qquad x = 12$
Check:
Let $x = -4$.
$\qquad x^2 - 8x - 48 = 0$
$\quad (-4)^2 - 8(-4) - 48 \stackrel{?}{=} 0$
$\qquad\quad 16 + 32 - 48 \stackrel{?}{=} 0$
$\qquad\qquad\quad 48 - 48 \stackrel{?}{=} 0$
$\qquad\qquad\qquad\quad 0 = 0 \quad$ True
Let $x = 12$.

$$x^2 - 8x - 48 = 0$$
$$12^2 - 8 \cdot 12 - 48 \stackrel{?}{=} 0$$
$$144 - 96 - 48 \stackrel{?}{=} 0$$
$$48 - 48 \stackrel{?}{=} 0$$
$$0 = 0 \quad \text{True}$$

The solutions are $-4$ and 12.

**5.** 
$$9x^2 - 24x = -16$$
$$9x^2 - 24x + 16 = 0$$
$$(3x - 4)(3x - 4) = 0$$
$$3x - 4 = 0$$
$$3x = 4$$
$$x = \frac{4}{3}$$

The solution is $\frac{4}{3}$.

**6.** 
$$x(3x + 7) = 6$$
$$3x^2 + 7x = 6$$
$$3x^2 + 7x - 6 = 0$$
$$(3x - 2)(x + 3) = 0$$
$$3x - 2 = 0 \quad \text{or} \quad x + 3 = 0$$
$$3x = 2 \qquad\qquad x = -3$$
$$x = \frac{2}{3}$$

The solutions are $\frac{2}{3}$ and $-3$.

**7.** 
$$-3x^2 - 6x + 72 = 0$$
$$-3(x^2 + 2x - 24) = 0$$
$$-3(x + 6)(x - 4) = 0$$
$$x + 6 = 0 \quad \text{or} \quad x - 4 = 0$$
$$x = -6 \qquad\qquad x = 4$$

The solutions are $-6$ and 4.

**8.** 
$$7x^3 - 63x = 0$$
$$7x(x^2 - 9) = 0$$
$$7x(x + 3)(x - 3) = 0$$
$$7x = 0 \quad \text{or} \quad x + 3 = 0 \quad \text{or} \quad x - 3 = 0$$
$$x = 0 \qquad\qquad x = -3 \qquad\qquad x = 3$$

The solutions are 0, $-3$, and 3.

**9.** 
$$(3x - 2)(2x^2 - 13x + 15) = 0$$
$$(3x - 2)(2x - 3)(x - 5) = 0$$
$$3x - 2 = 0 \quad \text{or} \quad 2x - 3 = 0 \quad \text{or} \quad x - 5 = 0$$
$$3x = 2 \qquad\qquad 2x = 3 \qquad\qquad x = 5$$
$$x = \frac{2}{3} \qquad\qquad x = \frac{3}{2}$$

The solutions are $\frac{2}{3}$, $\frac{3}{2}$, and 5.

**10.** 
$$5x^3 + 5x^2 - 30x = 0$$
$$5x(x^2 + x - 6) = 0$$
$$5x(x + 3)(x - 2) = 0$$
$$5x = 0 \quad \text{or} \quad x + 3 = 0 \quad \text{or} \quad x - 2 = 0$$
$$x = 0 \qquad\qquad x = -3 \qquad\qquad x = 2$$

The solutions are 0, $-3$, and 2.

**11.** 
$$y = x^2 - 6x + 8$$
$$0 = x^2 - 6x + 8$$
$$0 = (x - 4)(x - 2)$$
$$x - 4 = 0 \quad \text{or} \quad x - 2 = 0$$
$$x = 4 \qquad\qquad x = 2$$

The $x$-intercepts of the graph of $y = x^2 - 6x + 8$ are (2, 0) and (4, 0).

**Calculator Explorations**

**1.** $-0.9$, 2.2

**2.** $-2.5$, 3.5

**3.** no real solution

**4.** no real solution

**5.** $-1.8$, 2.8

**6.** $-0.9$, 0.3

**Vocabulary, Readiness & Video Check 6.6**

**1.** An equation that can be written in the form $ax^2 + bx + c = 0$, (with $a \neq 0$), is called a <u>quadratic</u> equation.

**2.** If the product of two numbers is 0, then at least one of the numbers must be <u>0</u>.

**3.** The solutions to $(x - 3)(x + 5) = 0$ are <u>3</u>, <u>$-5$</u>.

**4.** If $a \cdot b = 0$, then <u>$a = 0$ or $b = 0$</u>.

    

**5.** One side of the equation must be a factored polynomial and the other side must be zero.

**6.** Because no matter how many factors you have in a multiplication problem, it's still true that for a zero product, at least one of the factors must be zero.

**7.** To find the *x*-intercepts of any graph in two variables we let $y = 0$. Doing this with our quadratic equation gives us an equation $= 0$ which we can try to solve by factoring.

**Exercise Set 6.6**

**1.** $(x - 6)(x - 7) = 0$
$x - 6 = 0$ or $x - 7 = 0$
$x = 6$ 　　　$x = 7$
The solutions are 6 and 7.

**3.** $(x - 2)(x + 1) = 0$
$x - 2 = 0$ or $x + 1 = 0$
$x = 2$ 　　　$x = -1$
The solutions are 2 and −1.

**5.** $(x + 9)(x + 17) = 0$
$x + 9 = 0$ or $x + 17 = 0$
$x = -9$ 　　　$x = -17$
The solutions are −9 and −17.

**7.** $x(x + 6) = 0$
$x = 0$ or $x + 6 = 0$
　　　　　$x = -6$
The solutions are 0 and −6.

**9.** $3x(x - 8) = 0$
$3x = 0$ or $x - 8 = 0$
$x = 0$ 　　　$x = 8$
The solutions are 0 and 8.

**11.** $(2x + 3)(4x - 5) = 0$
$2x + 3 = 0$ or $4x - 5 = 0$
$2x = -3$ 　　　$4x = 5$
$x = -\dfrac{3}{2}$ 　　　$x = \dfrac{5}{4}$
The solutions are $-\dfrac{3}{2}$ and $\dfrac{5}{4}$.

**13.** $(2x - 7)(7x + 2) = 0$
$2x - 7 = 0$ or $7x + 2 = 0$
$2x = 7$ 　　　$7x = -2$
$x = \dfrac{7}{2}$ 　　　$x = -\dfrac{2}{7}$
The solutions are $\dfrac{7}{2}$ and $-\dfrac{2}{7}$.

**15.** $\left(x - \dfrac{1}{2}\right)\left(x + \dfrac{1}{3}\right) = 0$
$x - \dfrac{1}{2} = 0$ or $x + \dfrac{1}{3} = 0$
$x = \dfrac{1}{2}$ 　　　$x = -\dfrac{1}{3}$
The solutions are $\dfrac{1}{2}$ and $-\dfrac{1}{3}$.

**17.** $(x + 0.2)(x + 1.5) = 0$
$x + 0.2 = 0$ or $x + 1.5 = 0$
$x = -0.2$ 　　　$x = -1.5$
The solutions are −0.2 and −1.5

**19.** $x^2 - 13x + 36 = 0$
$(x - 9)(x - 4) = 0$
$x - 9 = 0$ or $x - 4 = 0$
$x = 9$ 　　　$x = 4$
The solutions are 9 and 4.

**21.** $x^2 + 2x - 8 = 0$
$(x + 4)(x - 2) = 0$
$x + 4 = 0$ or $x - 2 = 0$
$x = -4$ 　　　$x = 2$
The solutions are −4 and 2.

**23.** $x^2 - 7x = 0$
$x(x - 7) = 0$
$x = 0$ or $x - 7 = 0$
　　　　　$x = 7$
The solutions are 0 and 7.

**25.** $x^2 - 4x = 32$
$x^2 - 4x - 32 = 0$
$(x - 8)(x + 4) = 0$
$x - 8 = 0$ or $x + 4 = 0$
$x = 8$ 　　　$x = -4$
The solutions are 8 and −4.

**27.**

$$x^2 = 16$$
$$x^2 - 16 = 0$$
$$(x+4)(x-4) = 0$$
$$x+4 = 0 \quad \text{or} \quad x-4 = 0$$
$$x = -4 \qquad\qquad x = 4$$

The solutions are −4 and 4.

**29.**

$$(x+4)(x-9) = 4x$$
$$x^2 - 5x - 36 = 4x$$
$$x^2 - 9x - 36 = 0$$
$$(x-12)(x+3) = 0$$
$$x-12 = 0 \quad \text{or} \quad x+3 = 0$$
$$x = 12 \qquad\qquad x = -3$$

The solutions are 12 and −3.

**31.**

$$x(3x-1) = 14$$
$$3x^2 - x = 14$$
$$3x^2 - x - 14 = 0$$
$$(3x-7)(x+2) = 0$$
$$3x - 7 = 0 \quad \text{or} \quad x+2 = 0$$
$$3x = 7 \qquad\qquad x = -2$$
$$x = \frac{7}{3}$$

The solutions are $\frac{7}{3}$ and −2.

**33.**

$$-3x^2 + 75 = 0$$
$$-3(x^2 - 25) = 0$$
$$-3(x+5)(x-5) = 0$$
$$x+5 = 0 \quad \text{or} \quad x-5 = 0$$
$$x = -5 \qquad\qquad x = 5$$

The solutions are −5 and 5.

**35.**

$$24x^2 + 44x = 8$$
$$24x^2 + 44x - 8 = 0$$
$$4(6x^2 + 11x - 2) = 0$$
$$4(6x-1)(x+2) = 0$$
$$6x - 1 = 0 \quad \text{or} \quad x+2 = 0$$
$$6x = 1 \qquad\qquad x = -2$$
$$x = \frac{1}{6}$$

The solutions are $\frac{1}{6}$ and −2.

**37.**

$$x^3 - 12x^2 + 32x = 0$$
$$x(x^2 - 12x + 32) = 0$$
$$x(x-8)(x-4) = 0$$
$$x = 0 \quad \text{or} \quad x-8 = 0 \quad \text{or} \quad x-4 = 0$$
$$\qquad\qquad x = 8 \qquad\qquad x = 4$$

The solutions are 0, 8, and 4.

**39.**

$$(4x-3)(16x^2 - 24x + 9) = 0$$
$$(4x-3)(4x-3)^2 = 0$$
$$(4x-3)^3 = 0$$
$$4x - 3 = 0$$
$$4x = 3$$
$$x = \frac{3}{4}$$

The solution is $\frac{3}{4}$.

**41.**

$$4x^3 - x = 0$$
$$x(4x^2 - 1) = 0$$
$$x(2x+1)(2x-1) = 0$$
$$x = 0 \quad \text{or} \quad 2x+1 = 0 \quad \text{or} \quad 2x-1 = 0$$
$$\qquad\qquad 2x = -1 \qquad\qquad 2x = 1$$
$$\qquad\qquad x = -\frac{1}{2} \qquad\qquad x = \frac{1}{2}$$

The solutions are 0, $-\frac{1}{2}$, and $\frac{1}{2}$.

**43.**

$$32x^3 - 4x^2 - 6x = 0$$
$$2x(16x^2 - 2x - 3) = 0$$
$$2x(2x-1)(8x+3) = 0$$
$$2x = 0 \quad \text{or} \quad 2x-1 = 0 \quad \text{or} \quad 8x+3 = 0$$
$$x = 0 \qquad\qquad 2x = 1 \qquad\qquad 8x = -3$$
$$\qquad\qquad x = \frac{1}{2} \qquad\qquad x = -\frac{3}{8}$$

The solutions are 0, $\frac{1}{2}$, and $-\frac{3}{8}$.

**45.**

$$(x+3)(x-2) = 0$$
$$x+3 = 0 \quad \text{or} \quad x-2 = 0$$
$$x = -3 \qquad\qquad x = 2$$

The solutions are −3 and 2.

**47.**

$$x^2 + 20x = 0$$
$$x(x+20) = 0$$
$$x = 0 \quad \text{or} \quad x+20 = 0$$
$$\qquad\qquad x = -20$$

The solutions are 0 and −20.

**49.** $4(x-7)=6$

$4x-28=6$

$4x=34$

$x=\dfrac{34}{4}$

$x=\dfrac{17}{2}$

The solution is $\dfrac{17}{2}$.

**51.** $4y^2-1=0$

$(2y+1)(2y-1)=0$

$2y+1=0$　or　$2y-1=0$

$2y=-1$　　　　$2y=1$

$y=-\dfrac{1}{2}$　　　$y=\dfrac{1}{2}$

The solutions are $-\dfrac{1}{2}$ and $\dfrac{1}{2}$.

**53.** $(2x+3)(2x^2-5x-3)=0$

$(2x+3)(2x+1)(x-3)=0$

$2x+3=0$　or　$2x+1=0$　or　$x-3=0$

$2x=-3$　　　　$2x=-1$　　　$x=3$

$x=-\dfrac{3}{2}$　　　$x=-\dfrac{1}{2}$

The solutions are $-\dfrac{3}{2}$, $-\dfrac{1}{2}$, and 3.

**55.** $x^2-15=-2x$

$x^2+2x-15=0$

$(x+5)(x-3)=0$

$x+5=0$　or　$x-3=0$

$x=-5$　　　$x=3$

The solutions are $-5$ and 3.

**57.** $30x^2-11x-30=0$

$(6x+5)(5x-6)=0$

$6x+5=0$　or　$5x-6=0$

$6x=-5$　　　$5x=6$

$x=-\dfrac{5}{6}$　　　$x=\dfrac{6}{5}$

The solutions are $-\dfrac{5}{6}$ and $\dfrac{6}{5}$.

**59.** $5x^2-6x-8=0$

$(5x+4)(x-2)=0$

$5x+4=0$　or　$x-2=0$

$5x=-4$　　　$x=2$

$x=-\dfrac{4}{5}$

The solutions are $-\dfrac{4}{5}$ and 2.

**61.** $6y^2-22y-40=0$

$2(3y^2-11y-20)=0$

$2(3y+4)(y-5)=0$

$3y+4=0$　or　$y-5=0$

$3y=-4$　　　$y=5$

$y=-\dfrac{4}{3}$

The solutions are $-\dfrac{4}{3}$ and 5.

**63.** $(y-2)(y+3)=6$

$y^2+y-6=6$

$y^2+y-12=0$

$(y+4)(y-3)=0$

$y+4=0$　or　$y-3=0$

$y=-4$　　　$y=3$

The solutions are $-4$ and 3.

**65.** $3x^3+19x^2-72x=0$

$x(3x^2+19x-72)=0$

$x(3x-8)(x+9)=0$

$x=0$　or　$3x-8=0$　or　$x+9=0$

$3x=8$　　　　$x=-9$

$x=\dfrac{8}{3}$

The solutions are 0, $\dfrac{8}{3}$, and $-9$.

**67.** $x^2+14x+49=0$

$(x+7)^2=0$

$x+7=0$

$x=-7$

The solution is $-7$.

**69.** $12y = 8y^2$

$0 = 8y^2 - 12y$

$0 = 4y(2y - 3)$

$4y = 0$    or    $2y - 3 = 0$

  $y = 0$           $2y = 3$

                    $y = \dfrac{3}{2}$

The solutions are 0 and $\dfrac{3}{2}$.

**71.**        $7x^3 - 7x = 0$

         $7x(x^2 - 1) = 0$

   $7x(x + 1)(x - 1) = 0$

$7x = 0$   or   $x + 1 = 0$   or   $x - 1 = 0$

 $x = 0$         $x = -1$          $x = 1$

The solutions are 0, $-1$, and 1.

**73.**    $3x^2 + 8x - 11 = 13 - 6x$

     $3x^2 + 14x - 24 = 0$

     $(3x - 4)(x + 6) = 0$

$3x - 4 = 0$    or    $x + 6 = 0$

    $3x = 4$           $x = -6$

      $x = \dfrac{4}{3}$

The solutions are $\dfrac{4}{3}$ and $-6$.

**75.**     $3x^2 - 20x = -4x^2 - 7x - 6$

     $7x^2 - 13x + 6 = 0$

     $(7x - 6)(x - 1) = 0$

$7x - 6 = 0$    or    $x - 1 = 0$

    $7x = 6$          $x = 1$

      $x = \dfrac{6}{7}$

The solutions are $\dfrac{6}{7}$ and 1.

**77.** Let $y = 0$ and solve for $x$.

    $y = (3x + 4)(x - 1)$

    $0 = (3x + 4)(x - 1)$

$3x + 4 = 0$     or    $x - 1 = 0$

   $3x = -4$          $x = 1$

     $x = -\dfrac{4}{3}$

The intercepts are $\left(-\dfrac{4}{3}, 0\right)$ and $(1, 0)$.

**79.** Let $y = 0$ and solve for $x$.

    $y = x^2 - 3x - 10$

    $0 = x^2 - 3x - 10$

    $0 = (x - 5)(x + 2)$

$x - 5 = 0$    or    $x + 2 = 0$

    $x = 5$          $x = -2$

The $x$-intercepts are $(5, 0)$ and $(-2, 0)$.

**81.** Let $y = 0$ and solve for $x$.

    $y = 2x^2 + 11x - 6$

    $0 = 2x^2 + 11x - 6$

    $0 = (2x - 1)(x + 6)$

$2x - 1 = 0$    or    $x + 6 = 0$

    $2x = 1$          $x = -6$

     $x = \dfrac{1}{2}$

The $x$-intercepts are $\left(\dfrac{1}{2}, 0\right)$ and $(-6, 0)$.

**83.** e; $x$-intercepts are $(-2, 0)$, $(1, 0)$

**85.** b; $x$-intercepts are $(0, 0)$, $(-3, 0)$

**87.** c; $y = 2x^2 - 8 = 2(x - 2)(x + 2)$

    $x$-intercepts are $(2, 0)$, $(-2, 0)$.

**89.** $\dfrac{3}{5} + \dfrac{4}{9} = \dfrac{3 \cdot 9}{5 \cdot 9} + \dfrac{4 \cdot 5}{9 \cdot 5}$

          $= \dfrac{27}{45} + \dfrac{20}{45}$

          $= \dfrac{27 + 20}{45}$

          $= \dfrac{47}{45}$

**91.** $\dfrac{7}{10} - \dfrac{5}{12} = \dfrac{7 \cdot 6}{10 \cdot 6} - \dfrac{5 \cdot 5}{12 \cdot 5}$

            $= \dfrac{42}{60} - \dfrac{25}{60}$

            $= \dfrac{42 - 25}{60}$

            $= \dfrac{17}{60}$

**93.** $\dfrac{7}{8} \div \dfrac{7}{15} = \dfrac{7}{8} \cdot \dfrac{15}{7} = \dfrac{15}{8}$

**95.** $\dfrac{4}{5} \cdot \dfrac{7}{8} = \dfrac{4 \cdot 7}{5 \cdot 8} = \dfrac{4 \cdot 7}{5 \cdot 2 \cdot 4} = \dfrac{7}{10}$

     

**97.** Didn't write the equation in standard form; standard form should be:

$$x(x-2) = 8$$
$$x^2 - 2x = 8$$
$$x^2 - 2x - 8 = 0$$
$$(x-4)(x+2) = 0$$
$$x - 4 = 0 \quad \text{or} \quad x + 2 = 0$$
$$x = 4 \qquad\qquad x = -2$$

**99.** Answers may vary. Possible answer: If the solutions are $x = 6$ and $x = -1$, then, by the zero factor property,

$$x = 6 \quad \text{or} \quad x = -1$$
$$x - 6 = 0 \qquad x + 1 = 0$$
$$(x-6)(x+1) = 0$$

**101.** Answers may vary. Possible answer: If the solutions are $x = 5$ and $x = 7$, then, by the zero factor property,

$$x = 5 \quad \text{or} \quad x = 7$$
$$x - 5 = 0 \qquad x - 7 = 0$$
$$(x-5)(x-7) = 0$$
$$x^2 - 7x - 5x + 35 = 0$$
$$x^2 - 12x + 35 = 0$$

**103.** $y = -16x^2 + 20x + 300$

   **a.**

| time $x$ | 0 | 1 | 2 | 3 | 4 | 5 | 6 |
|---|---|---|---|---|---|---|---|
| height $y$ | 300 | 304 | 276 | 216 | 124 | 0 | −156 |

   **b.** The compass strikes the ground after 5 seconds, when the height, $y$, is zero feet.

   **c.** The maximum height was approximately 304 feet.

   **d.**

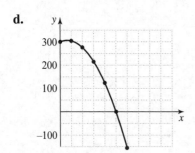

**105.** $(x-3)(3x+4) = (x+2)(x-6)$

$$3x^2 - 5x - 12 = x^2 - 4x - 12$$
$$2x^2 - x = 0$$
$$x(2x-1) = 0$$
$$2x - 1 = 0 \quad \text{or} \quad x = 0$$
$$x = \frac{1}{2}$$

The solutions are $\dfrac{1}{2}$ and 0.

**107.**  $(2x-3)(x+8) = (x-6)(x+4)$

$2x^2 + 13x - 24 = x^2 - 2x - 24$

$x^2 + 15x = 0$

$x(x+15) = 0$

$x + 15 = 0 \quad \text{or} \quad x = 0$

$x = -15$

The solutions are $-15$ and $0$.

**Section 6.7 Practice**

**1.** Find $t$ when $h = 0$.

$h = -16t^2 + 64$

$0 = -16t^2 + 64$

$0 = -16(t^2 - 4)$

$0 = -16(t-2)(t+2)$

$t - 2 = 0 \quad \text{or} \quad t + 2 = 0$

$t = 2 \qquad\qquad t = -2$

Since time cannot be negative, the diver will reach the pool in 2 seconds.

**2.** Let $x$ = the number.

$x^2 - 8x = 48$

$x^2 - 8x - 48 = 0$

$(x-12)(x+4) = 0$

$x - 12 = 0 \quad \text{or} \quad x + 4 = 0$

$x = 12 \qquad\qquad x = -4$

There are two numbers. They are $-4$ and $12$.

**3.** Let $x$ = height, then $3x - 1$ = base.

$A = \frac{1}{2}bh$

$210 = \frac{1}{2}(3x-1)(x)$

$420 = (3x-1)(x)$

$420 = 3x^2 - x$

$0 = 3x^2 - x - 420$

$0 = (3x+35)(x-12)$

$3x + 35 = 0 \qquad \text{or} \quad x - 12 = 0$

$x = -\frac{35}{3} \qquad\qquad x = 12$

Since height cannot be negative, the height is 12 feet and the base is $3(12) - 1 = 35$ feet.

**4.** Let $x$ = first integer, then
$x + 1$ = next consecutive integer.

$x(x+1) = x + (x+1) + 41$

$x^2 + x = 2x + 42$

$x^2 - x - 42 = 0$

$(x-7)(x+6) = 0$

$x - 7 = 0 \quad \text{or} \quad x + 6 = 0$

$x = 7 \qquad\qquad x = -6$

The numbers are 7 and 8 or $-6$ and $-5$.

**5.** Let $x$ = first leg, then $2x - 1$ = second leg, and $2x + 1$ = hypotenuse.

$x^2 + (2x-1)^2 = (2x+1)^2$

$x^2 + 4x^2 - 4x + 1 = 4x^2 + 4x + 1$

$x^2 - 8x = 0$

$x(x-8) = 0$

$x = 0 \quad \text{or} \quad x - 8 = 0$

$\qquad\qquad x = 8$

Since the length cannot be 0, the legs have lengths 8 units and $2(8) - 1 = 15$ units and the hypotenuse has length $2(8) + 1 = 17$ units.

**Vocabulary, Readiness & Video Check 6.7**

**1.** In applications, the context of the problem needs to be considered. Each exercise resulted in both a positive and a negative solution, and a negative solution is not appropriate for any of the problems.

**Exercise Set 6.7**

**1.** Let $x$ = the width, then $x + 4$ = the length.

**3.** Let $x$ = the first odd integer, then
$x + 2$ = the next consecutive odd integer.

**5.** Let $x$ = the base, then $4x + 1$ = the height.

**7.** Let $x$ = the length of one side.

$A = x^2$

$121 = x^2$

$0 = x^2 - 121$

$0 = x^2 - 11^2$

$0 = (x+11)(x-11)$

$x + 11 = 0 \qquad \text{or} \quad x - 11 = 0$

$x = -11 \qquad\qquad x = 11$

Since the length cannot be negative, the sides are 11 units long.

**9.** The perimeter is the sum of the lengths of the sides.

$$120 = (x+5)+(x^2-3x)+(3x-8)+(x+3)$$
$$120 = x+5+x^2-3x+3x-8+x+3$$
$$120 = x^2+2x$$
$$0 = x^2+2x-120$$

$$x^2+2x-120 = 0$$
$$(x+12)(x-10) = 0$$

$$x+12=0 \quad \text{or} \quad x-10=0$$
$$x=-12 \qquad\qquad x=10$$

Since the dimensions cannot be negative, the lengths of the sides are:

$10+5 = 15$ cm, $10^2-3(10) = 70$ cm,

$3(10)-8 = 22$ cm, and $10+3 = 13$ cm.

**11.** $x+5 =$ the base and $x-5 =$ the height.

$$A = bh$$
$$96 = (x+5)(x-5)$$
$$96 = x^2-25$$
$$0 = x^2-121$$

$$x^2-121 = 0$$
$$(x+11)(x-11) = 0$$

$$x+11=0 \quad \text{or} \quad x-11=0$$
$$x=-11 \qquad\qquad x=11$$

Since the dimensions cannot be negative, $x = 11$. The base is $11 + 5 = 16$ miles, and the height is $11 - 5 = 6$ miles.

**13.** Find $t$ when $h = 0$.

$$h = -16t^2+64t+80$$
$$0 = -16t^2+64t+80$$
$$0 = -16(t^2-4t-5)$$
$$0 = -16(t-5)(t+1)$$

$$t-5=0 \quad \text{or} \quad t+1=0$$
$$t=5 \qquad\qquad t=-1$$

Since the time $t$ cannot be negative, the object hits the ground after 5 seconds.

**15.** Let $x =$ the length then $2x - 7 =$ the width.

$$A = lw$$
$$30 = (x)(2x-7)$$
$$30 = 2x^2-7x$$
$$0 = 2x^2-7x-30$$
$$0 = (2x+5)(x-6)$$

$$2x+5=0 \quad \text{or} \quad x-6=0$$
$$x=-\frac{5}{2} \qquad\qquad x=6$$

Since the dimensions cannot be negative, the length is 6 cm and the width is $2(6) - 7 = 5$ cm.

**17.** Let $n = 12$.

$$D = \frac{1}{2}n(n-3)$$
$$D = \frac{1}{2}\cdot 12(12-3) = 6(9) = 54$$

A polygon with 12 sides has 54 diagonals.

**19.** Let $D = 35$ and solve for $n$.

$$D = \frac{1}{2}n(n-3)$$
$$35 = \frac{1}{2}n(n-3)$$
$$70 = n^2-3n$$
$$0 = n^2-3n-70$$
$$0 = (n-10)(n+7)$$

$$n-10=0 \quad \text{or} \quad n+7=0$$
$$n=10 \qquad\qquad n=-7$$

The polygon has 10 sides.

**21.** Let $x =$ the unknown number.

$$x+x^2 = 132$$
$$x^2+x-132 = 0$$
$$(x+12)(x-11) = 0$$

$$x+12=0 \quad \text{or} \quad x-11=0$$
$$x=-12 \qquad\qquad x=11$$

The two numbers are $-12$ and $11$.

**23.** Let $x =$ the first room number, then $x + 1 =$ next room number.

$$x(x+1) = 210$$
$$x^2+x = 210$$
$$x^2+x-210 = 0$$
$$(x-14)(x+15) = 0$$

$$x-14=0 \quad \text{or} \quad x+15=0$$
$$x=14 \qquad\qquad x=-15$$

Since the room number is not negative, the room numbers are 14 and 15.

**25.** Let $x =$ hypotenuse, then $x - 1 =$ height.

$$a^2+b^2 = c^2$$
$$5^2+(x-1)^2 = x^2$$
$$25+x^2-2x+1 = x^2$$
$$26-2x = 0$$
$$26 = 2x$$
$$13 = x$$

The length of the ladder is 13 feet.

**27.** Let $x$ = the length of a side of the original square. Then $x + 3$ = the length of a side of the larger square.

$$64 = (x+3)^2$$
$$64 = x^2 + 6x + 9$$
$$0 = x^2 + 6x - 55$$
$$0 = (x+11)(x-5)$$
$$x+11 = 0 \quad \text{or} \quad x-5 = 0$$
$$x = -11 \qquad\qquad x = 5$$

Since the length cannot be negative, the sides of the original square are 5 inches long.

**29.** Let $x$ = the length of the shorter leg. Then $x + 4$ = the length of the longer leg and $x + 8$ = the length of the hypotenuse. By the Pythagorean theorem,

$$x^2 + (x+4)^2 = (x+8)^2$$
$$x^2 + x^2 + 8x + 16 = x^2 + 16x + 64$$
$$x^2 - 8x - 48 = 0$$
$$(x-12)(x+4) = 0$$
$$x-12 = 0 \quad \text{or} \quad x+4 = 0$$
$$x = 12 \qquad\qquad x = -4$$

Since the length cannot be negative, the sides of the triangle are 12 mm, $12 + 4 = 16$ mm, and $12 + 8 = 20$ mm.

**31.** Let $x$ = the height of the triangle, then $2x$ = the base.

$$A = \frac{1}{2}bh$$
$$100 = \frac{1}{2}(2x)(x)$$
$$100 = x^2$$
$$0 = x^2 - 100$$
$$0 = (x+10)(x-10)$$
$$x+10 = 0 \quad \text{or} \quad x-10 = 0$$
$$x = -10 \qquad\qquad x = 10$$

Since the height cannot be negative, the height of the triangle is 10 km.

**33.** Let $x$ = the length of the shorter leg, then $x + 12$ = the length of the longer leg and $2x - 12$ = the length of the hypotenuse. By the Pythagorean theorem,

$$x^2 + (x+12)^2 = (2x-12)^2$$
$$x^2 + x^2 + 24x + 144 = 4x^2 - 48x + 144$$
$$0 = 2x^2 - 72x$$
$$0 = 2x(x-36)$$

$$2x = 0 \quad \text{or} \quad x-36 = 0$$
$$x = 0 \qquad\qquad x = 36$$

Since the length cannot be zero feet, the shorter leg is 36 feet long.

**35.** Find $t$ when $h = 0$.

$$h = -16t^2 + 1444$$
$$0 = -16t^2 + 1444$$
$$0 = -4(4t^2 - 361)$$
$$0 = -4(2t-19)(2t+19)$$
$$2t-19 = 0 \quad \text{or} \quad 2t+19 = 0$$
$$t = \frac{19}{2} \qquad\qquad t = -\frac{19}{2}$$

Since time cannot be negative, the object reaches the ground in $\frac{19}{2} = 9.5$ seconds.

**37.** Let $P = 100$ and $A = 144$.

$$A = P(1+r)^2$$
$$144 = 100(1+r)^2$$
$$144 = 100 + 200r + 100r^2$$
$$0 = 100r^2 + 200r - 44$$
$$0 = 4(25r^2 + 50r - 11)$$
$$0 = 4(5r-1)(5r+11)$$
$$5r-1 = 0 \quad \text{or} \quad 5r+11 = 0$$
$$5r = 1 \qquad\qquad 5r = -11$$
$$r = \frac{1}{5} \qquad\qquad r = -\frac{11}{5}$$
$$r = 0.2 \qquad\qquad r = -2.2$$

Since the interest rate cannot be negative $r = 0.2$ and the rate is 20%.

**39.** Let $x$ = the length and $x - 7$ = the width.

$$A = lw$$
$$120 = (x-7)(x)$$
$$120 = x^2 - 7x$$
$$0 = x^2 - 7x - 120$$
$$0 = (x+8)(x-15)$$
$$x+8 = 0 \quad \text{or} \quad x-15 = 0$$
$$x = -8 \qquad\qquad x = 15$$

Since the length cannot be negative, the length is 15 miles. The width is $15 - 7 = 8$ miles.

**41.** Let $C = 9500$.

$$C = x^2 - 15x + 50$$
$$9500 = x^2 - 15x + 50$$
$$0 = x^2 - 15x - 9450$$
$$0 = (x+90)(x-105)$$

$x + 90 = 0$ or $x - 105 = 0$

$x = -90$ $x = 105$

Since the number of units cannot be negative the solution is 105 units.

**43.** From the graph, there were approximately 2 million visitors to Glacier National Park in 2009.

**45.** From the graph, there were approximately 1.9 million visitors to Glacier National Park in 2005.

**47.** From the graph, the lines intersect at approximately 2003.

**49.** answers may vary

**51.** $\dfrac{20}{35} = \dfrac{2 \cdot 2 \cdot 5}{5 \cdot 7} = \dfrac{4}{7}$

**53.** $\dfrac{27}{18} = \dfrac{3 \cdot 3 \cdot 3}{2 \cdot 3 \cdot 3} = \dfrac{3}{2}$

**55.** $\dfrac{14}{42} = \dfrac{2 \cdot 7}{2 \cdot 3 \cdot 7} = \dfrac{1}{3}$

**57.** Let $x$ = the rate (in mph) of the slower boat, then $x + 7$ = the rate (in mph) of the faster boat. After one hour, the slower boat has traveled $x$ miles and the faster boat has traveled $x + 7$ miles. By the Pythagorean theorem,

$x^2 + (x + 7)^2 = 17^2$

$x^2 + x^2 + 14x + 49 = 289$

$2x^2 + 14x + 49 = 289$

$2x^2 + 14x - 240 = 0$

$2(x^2 + 7x - 120) = 0$

$2(x + 15)(x - 8) = 0$

$x + 15 = 0$ or $x - 8 = 0$

$x = -15$ $x = 8$

Since the rate cannot be negative, the slower boat travels at 8 mph. The faster boat travels at 8 + 7 = 15 mph.

**59.** Let $x$ = the first number, then $20 - x$ = the other number.

$x^2 + (20 - x)^2 = 218$

$x^2 + 400 - 40x + x^2 = 218$

$2x^2 - 40x + 400 = 218$

$2x^2 - 40x + 182 = 0$

$2(x^2 - 20x + 91) = 0$

$2(x - 13)(x - 7) = 0$

$x - 13 = 0$ or $x - 7 = 0$

$x = 13$ $x = 7$

The numbers are 13 and 7.

**61.** Pool: width = $x$ and length = $x + 6$

Total Area: width = $x + 8$ and length = $x + 14$

Total area = 576 + Pool area

$(x + 14)(x + 8) = 576 + (x + 6)(x)$

$x^2 + 22x + 112 = 576 + x^2 + 6x$

$16x + 112 = 576$

$16x = 464$

$x = 29$

$x + 6 = 29 + 6 = 35$

The pool has length 35 meters and width 29 meters.

**63.** answers may vary

**Chapter 6 Vocabulary Check**

**1.** An equation that can be written in the form $ax^2 + bx + c = 0$ (with $a$ not 0) is called a quadratic equation.

**2.** Factoring is the process of writing an expression as a product.

**3.** The greatest common factor of a list of terms is the product of all common factors.

**4.** A trinomial that is the square of some binomial is called a perfect square trinomial.

**5.** The expression $a^2 - b^2$ is called a difference of two squares.

**6.** The expression $a^3 - b^3$ is called a difference of two cubes.

**7.** The expression $a^3 + b^3$ is called a sum of two cubes.

8. By the zero factor property, if the product of two numbers is 0, then at least one of the numbers must be <u>0</u>.

9. In a right triangle, the side opposite the right angle is called the <u>hypotenuse</u>.

10. In a right triangle, each side adjacent to the right angle is called a <u>leg</u>.

11. The Pythagorean theorem states that $(\text{leg})^2 + (\text{leg})^2 = (\underline{\text{hypotenuse}})^2$.

## Chapter 6 Review

1. $6x^2 - 15x = 3x(2x - 5)$

2. $2x^3 y + 6x^2 y^2 + 8xy^3 = 2xy(x^2 + 3xy + 4y^2)$

3. $20x^2 + 12x = 4x(5x + 3)$

4. $6x^2 y^2 - 3xy^3 = 3xy^2(2x - y)$

5. $3x(2x + 3) - 5(2x + 3) = (2x + 3)(3x - 5)$

6. $5x(x + 1) - (x + 1) = (x + 1)(5x - 1)$

7. $3x^2 - 3x + 2x - 2 = 3x(x-1) + 2(x-1)$
$= (x-1)(3x+2)$

8. $3a^2 + 9ab + 3b^2 + ab = 3a(a + 3b) + b(3b + a)$
$= (a + 3b)(3a + b)$

9. $10a^2 + 5ab + 7b^2 + 14ab$
$= 5a(2a + b) + 7b(b + 2a)$
$= (2a + b)(5a + 7b)$

10. $6x^2 + 10x - 3x - 5 = 2x(3x + 5) - 1(3x + 5)$
$= (3x + 5)(2x - 1)$

11. $x^2 + 6x + 8 = (x + 4)(x + 2)$

12. $x^2 - 11x + 24 = (x - 8)(x - 3)$

13. $x^2 + x + 2$ is prime.

14. $x^2 - x + 2$ is prime.

15. $x^2 + 4xy - 12y^2 = (x + 6y)(x - 2y)$

16. $x^2 + 8xy + 15y^2 = (x + 5y)(x + 3y)$

17. $72 - 18x - 2x^2 = 2(36 - 9x - x^2)$
$= 2(3 - x)(12 + x)$
or
$72 - 18x - 2x^2 = -2x^2 - 18x + 72$
$= -2(x^2 + 9x - 36)$
$= -2(x - 3)(x + 12)$

18. $32 + 12x - 4x^2 = 4(8 + 3x - x^2)$
or
$32 + 12x - 4x^2 = -4x^2 + 12x + 32$
$= -4(x^2 - 3x - 8)$

19. $10a^3 - 110a^2 + 100a = 10a(a^2 - 11a + 10a)$
$= 10a(a - 1)(a - 10)$

20. $5y^3 - 50y^2 + 120y = 5y(y^2 - 10y + 24)$
$= 5y(y - 6)(y - 4)$

21. To factor $x^2 + 2x - 48$, think of two numbers whose product is <u>–48</u> and whose sum is <u>2</u>.

22. The first step in factoring $3x^2 + 15x + 30$ is to factor out the GCF, 3.

23. Factors of $2x^2$: $2x \cdot x$
Factors of 6: $6 = 1 \cdot 6$, $6 = 2 \cdot 3$
$2x^2 + 13x + 6 = (2x + 1)(x + 6)$

24. Factors of $4x^2$: $4x^2 = 4x \cdot x$, $4x^2 = 2x \cdot 2x$
Factors of $-3$: $-3 = -1 \cdot 3$, $-3 = 1 \cdot -3$
$4x^2 + 4x - 3 = (2x + 3)(2x - 1)$

25. Factors of $6x^2$: $6x^2 = 6x \cdot x$, $6x^2 = 3x \cdot 2x$
Factors of $-4y^2$: $-4y^2 = -4y \cdot y$,
$-4y^2 = 4y \cdot -y$, $-4y^2 = -2y \cdot 2y$
$6x^2 + 5xy - 4y^2 = (3x + 4y)(2x - y)$

**26.** $18 \cdot -20y^2 = -360y^2$

$15y \cdot -24y = -360y^2$

$15y + (-24y) = -9y$

$18x^2 - 9xy - 20y^2 = 18x^2 + 15xy - 24xy - 20y^2$
$\qquad = 3x(6x + 5y) - 4y(6x + 5y)$
$\qquad = (6x + 5y)(3x - 4y)$

**27.** $10y^3 + 25y^2 - 60y = 5y(2y^2 + 5y - 12)$

$2 \cdot -12 = -24$

$-3 \cdot 8 = -24$

$-3 + 8 = 5$

$10y^3 + 25y^2 - 60y = 5y(2y^2 + 5y - 12)$
$\qquad = 5y(2y^2 - 3y + 8y - 12)$
$\qquad = 5y[y(2y - 3) + 4(2y - 3)]$
$\qquad = 5y(2y - 3)(y + 4)$

**28.** $60y^3 - 39y^2 + 6y = 3y(20y^2 - 13y + 2)$

$20 \cdot 2 = 40$

$-5 \cdot -8 = 40$

$-5 + (-8) = -13$

$60y^3 - 39y^2 + 6y = 3y(20y^2 - 13y + 2)$
$\qquad = 3y(20y^2 - 5y - 8y + 2)$
$\qquad = 3y[5y(4y - 1) - 2(4y - 1)]$
$\qquad = 3y(4y - 1)(5y - 2)$

**29.** $18x^2 - 60x + 50 = 2(9x^2 - 30x + 25)$
$\qquad = 2[(3x)^2 - 2 \cdot 3x \cdot 5 + 5^2]$
$\qquad = 2(3x - 5)^2$

**30.** $4x^2 - 28xy + 49y^2 = [(2x)^2 - 2 \cdot 2x \cdot 7y + (7y)^2]$
$\qquad = (2x - 7y)^2$

**31.** $4x^2 - 9 = (2x)^2 - 3^2 = (2x + 3)(2x - 3)$

**32.** $9t^2 - 25s^2 = (3t)^2 - (5s)^2 = (3t + 5s)(3t - 5s)$

**33.** $16x^2 + y^2$ is a prime polynomial.

**34.** $x^3 - 8y^3 = x^3 - (2y)^3$
$\qquad = (x - 2y)[x^2 + x \cdot 2y + (2y)^2]$
$\qquad = (x - 2y)(x^2 + 2xy + 4y^2)$

**35.** $8x^3 + 27 = (2x)^3 + 3^3$
$\qquad = (2x + 3)[(2x)^2 - 2x \cdot 3 + 3^2]$
$\qquad = (2x + 3)(4x^2 - 6x + 9)$

**36.** $2x^3 + 8x = 2x(x^2 + 4)$

**37.** $54 - 2x^3y^3 = 2(27 - x^3y^3)$
$\qquad = 2[3^3 - (xy)^3]$
$\qquad = 2(3 - xy)[3^2 + 3 \cdot xy + (xy)^2]$
$\qquad = 2(3 - xy)(9 + 3xy + x^2y^2)$

**38.** $9x^2 - 4y^2 = (3x)^2 - (2y)^2 = (3x - 2y)(3x + 2y)$

**39.** $16x^4 - 1 = (4x^2)^2 - 1^2$
$\qquad = (4x^2 + 1)(4x^2 - 1)$
$\qquad = (4x^2 + 1)[(2x)^2 - 1^2]$
$\qquad = (4x^2 + 1)(2x + 1)(2x - 1)$

**40.** $x^4 + 16$ is a prime polynomial.

**41.** $(x + 6)(x - 2) = 0$

$x + 6 = 0$　or　$x - 2 = 0$

$\quad x = -6$　　　　$x = 2$

The solutions are $-6$ and $2$.

**42.** $3x(x + 1)(7x - 2) = 0$

$3x = 0$　or　$x + 1 = 0$　or　$7x - 2 = 0$

$x = 0$　　　　$x = -1$　　　　$7x = 2$

$\qquad\qquad\qquad\qquad\qquad\qquad x = \dfrac{2}{7}$

The solutions are $0$, $-1$, and $\dfrac{2}{7}$.

**43.** $4(5x + 1)(x + 3) = 0$

$5x + 1 = 0$　or　$x + 3 = 0$

$\quad 5x = -1$　　　　$x = -3$

$\quad x = -\dfrac{1}{5}$

The solutions are $-\dfrac{1}{5}$ and $-3$.

**44.** $x^2 + 8x + 7 = 0$

$(x + 7)(x + 1) = 0$

$x + 7 = 0$　or　$x + 1 = 0$

$\quad x = -7$　　　　$x = -1$

The solutions are $-7$ and $-1$.

**45.**    $x^2 - 2x - 24 = 0$

      $(x-6)(x+4) = 0$

      $x - 6 = 0$   or   $x + 4 = 0$

         $x = 6$          $x = -4$

      The solutions are 6 and –4.

**46.**        $x^2 + 10x = -25$

      $x^2 + 10x + 25 = 0$

      $(x+5)(x+5) = 0$

      $x + 5 = 0$   or   $x + 5 = 0$

         $x = -5$         $x = -5$

      The solution is –5.

**47.**        $x(x-10) = -16$

         $x^2 - 10x = -16$

    $x^2 - 10x + 16 = 0$

      $(x-8)(x-2) = 0$

      $x - 8 = 0$   or   $x - 2 = 0$

         $x = 8$          $x = 2$

      The solutions are 8 and 2.

**48.**    $(3x-1)(9x^2 - 6x + 1) = 0$

      $(3x-1)(3x-1)(3x-1) = 0$

      $3x - 1 = 0$   or   $3x - 1 = 0$   or   $3x - 1 = 0$

         $3x = 1$          $3x = 1$          $3x = 1$

         $x = \dfrac{1}{3}$       $x = \dfrac{1}{3}$       $x = \dfrac{1}{3}$

      The solution is $\dfrac{1}{3}$.

**49.**          $56x^2 - 5x - 6 = 0$

      $56x^2 + 16x - 21x - 6 = 0$

      $8x(7x+2) - 3(7x+2) = 0$

          $(7x+2)(8x-3) = 0$

      $7x + 2 = 0$    or    $8x - 3 = 0$

        $7x = -2$         $8x = 3$

        $x = -\dfrac{2}{7}$       $x = \dfrac{3}{8}$

      The solutions are $-\dfrac{2}{7}$ and $\dfrac{3}{8}$.

**50.**    $20x^2 - 7x - 6 = 0$

      $(4x-3)(5x+2) = 0$

      $4x - 3 = 0$   or   $5x + 2 = 0$

         $4x = 3$           $5x = -2$

        $x = \dfrac{3}{4}$         $x = -\dfrac{2}{5}$

      The solutions are $\dfrac{3}{4}$ and $-\dfrac{2}{5}$.

**51.**    $5(3x+2) = 4$

      $15x + 10 = 4$

        $15x = -6$

          $x = -\dfrac{6}{15} = -\dfrac{2}{5}$

      The solution is $-\dfrac{2}{5}$.

**52.**    $6x^2 - 3x + 8 = 0$

      The equation has no real solution.

**53.**    $12 - 5t = -3$

        $-5t = -15$

          $t = 3$

      The solution is 3.

**54.**    $5x^3 + 20x^2 + 20x = 0$

        $5x(x^2 + 4x + 4) = 0$

        $5x(x+2)(x+2) = 0$

      $x + 2 = 0$    or    $5x = 0$

        $x = -2$         $x = 0$

      The solutions are –2 and 0.

**55.**    $4t^3 - 5t^2 - 21t = 0$

      $t(4t^2 - 5t - 21) = 0$

      $t(4t+7)(t-3) = 0$

      $t = 0$   or   $4t + 7 = 0$     or   $t - 3 = 0$

                      $4t = -7$           $t = 3$

                        $t = -\dfrac{7}{4}$

      The solutions are 0, $-\dfrac{7}{4}$, and 3.

**56.**    Answers may vary. Possible answer:

      $(x-4)(x-5) = 0$

        $x^2 - 9x + 20 = 0$

**57. a.** $7 \neq 2 \cdot 5$

**b.** $10 = 2 \cdot 5$
$$P = 2l + 2w$$
$$= 2(10) + 2(5)$$
$$= 20 + 10$$
$$= 30 \neq 24$$

**c.** $8 = 2 \cdot 4$
$$P = 2l + 2w = 2(8) + 2(4) = 16 + 8 = 24$$

**d.** $10 \neq 2 \cdot 2$

Choice **c** gives the correct dimensions.

**58. a.** $3 \cdot 8 + 1 = 25 \neq 10$

**b.** $3 \cdot 4 + 1 = 13$
$$A = lw = 13(4) = 52 \neq 80$$

**c.** $3 \cdot 4 + 1 = 13 \neq 20$

**d.** $3 \cdot 5 + 1 = 16$
$$A = lw = 5(16) = 80$$

Choice **d** gives the correct dimensions.

**59.**
$$x^2 = 81$$
$$x^2 - 81 = 0$$
$$(x - 9)(x + 9) = 0$$
$$x - 9 = 0 \quad \text{or} \quad x + 9 = 0$$
$$x = 9 \qquad\qquad x = -9$$
Since length is not negative, the length of the side is 9 units.

**60.** $(2x + 3) + (3x + 1) + (x^2 - 3x) + (x + 3) = 47$
$$x^2 + 3x + 7 = 47$$
$$x^2 + 3x - 40 = 0$$
$$(x - 5)(x + 8) = 0$$
$$x - 5 = 0 \quad \text{or} \quad x + 8 = 0$$
$$x = 5 \qquad\qquad x = -8$$
Length is not negative, so $x = 5$. The lengths are:
$x + 3 = 5 + 3 = 8$ units
$2x + 3 = 2(5) + 3 = 13$ units
$3x + 1 = 3(5) + 1 = 16$ units
$x^2 - 3x = 5^2 - 3(5) = 10$ units

**61.** Let $x =$ the width of the flag. Then
$2x - 15 =$ the length of the flag.
$$A = lw$$
$$500 = (2x - 15)(x)$$
$$500 = 2x^2 - 15x$$
$$0 = 2x^2 - 15x - 500$$
$$0 = (2x + 25)(x - 20)$$
$$2x + 25 = 0$$
$$2x = -25 \quad \text{or} \quad x - 20 = 0$$
$$x = -\frac{25}{2} \qquad\qquad x = 20$$
Since the dimensions cannot be negative, the width is 20 inches and the length is $2(20) - 15 = 25$ inches.

**62.** Let $x =$ the height of the sail, then
$4x =$ the base of the sail.
$$A = \frac{1}{2}bh$$
$$162 = \frac{1}{2}(4x)(x)$$
$$162 = 2x^2$$
$$0 = 2x^2 - 162$$
$$0 = 2(x^2 - 81)$$
$$0 = 2(x + 9)(x - 9)$$
$$x + 9 = 0 \quad \text{or} \quad x - 9 = 0$$
$$x = -9 \qquad\qquad x = 9$$
Since the dimensions cannot be negative, the height is 9 yards and the base is $4 \cdot 9 = 36$ yards.

**63.** Let $x =$ the first integer. Then
$x + 1 =$ the next consecutive integer.
$$x(x + 1) = 380$$
$$x^2 + x = 380$$
$$x^2 + x - 380 = 0$$
$$(x + 20)(x - 19) = 0$$
$$x + 20 = 0 \quad \text{or} \quad x - 19 = 0$$
$$x = -20 \qquad\qquad x = 19$$
The integers are 19 and 20.

**64.** Let $x$ be the first positive even integer. Then
$x + 2$ is the next consecutive even integer.
$$x(x + 2) = 440$$
$$x^2 + 2x = 440$$
$$x^2 + 2x - 440 = 0$$
$$(x + 22)(x - 20) = 0$$
$$x + 22 = 0 \quad \text{or} \quad x - 20 = 0$$
$$x = -22 \qquad\qquad x = 20$$
Discard $x = -22$ since it is not positive. The integers are 20 and 20 + 2 = 22.

**65. a.** Let $h = 2800$ and solve for $t$.

$$h = -16t^2 + 440t$$
$$2800 = -16t^2 + 440t$$
$$0 = -16t^2 + 440t - 2800$$
$$0 = -8(2t^2 - 55t + 350)$$
$$0 = -8(2t - 35)(t - 10)$$
$$2t - 35 = 0 \quad \text{or} \quad t - 10 = 0$$
$$2t = 35 \qquad\qquad t = 10$$
$$t = \frac{35}{2}$$
$$t = 17.5$$

The solutions are 17.5 sec and 10 sec. There are two answers because the rocket reaches a height of 2800 feet on its way up and on its way back down.

**b.** Let $h = 0$ and solve for $t$.

$$h = -16t^2 + 440t$$
$$0 = -16t^2 + 440t$$
$$0 = -8t(2t - 55)$$
$$-8t = 0 \quad \text{or} \quad 2t - 55 = 0$$
$$t = 0 \qquad\qquad 2t = 55$$
$$t = \frac{55}{2}$$
$$t = 27.5$$

$t = 0$ is when the rocket is launched, so it reaches the ground again after 27.5 seconds.

**66.** Let $x$ = the length of the longer leg. Then $x + 8$ = the length of the hypotenuse and $x - 8$ = the length of the shorter leg.

$$(x+8)^2 = (x-8)^2 + x^2$$
$$x^2 + 16x + 64 = x^2 - 16x + 64 + x^2$$
$$0 = x^2 - 32x$$
$$0 = x(x - 32)$$
$$x = 0 \quad \text{or} \quad x - 32 = 0$$
$$x = 32$$

The longer leg is 32 centimeters.

**67.** $7x - 63 = 7(x - 9)$

**68.** $11x(4x - 3) - 6(4x - 3) = (4x - 3)(11x - 6)$

**69.** $m^2 - \dfrac{4}{25} = m^2 - \left(\dfrac{2}{5}\right)^2 = \left(m + \dfrac{2}{5}\right)\left(m - \dfrac{2}{5}\right)$

**70.** $3x^3 - 4x^2 + 6x - 8 = x^2(3x - 4) + 2(3x - 4)$
$$= (3x - 4)(x^2 + 2)$$

**71.** $xy + 2x - y - 2 = x(y + 2) - 1(y + 2)$
$$= (y + 2)(x - 1)$$

**72.** $2x^2 + 2x - 24 = 2(x^2 + x - 12) = 2(x + 4)(x - 3)$

**73.** $3x^3 - 30x^2 + 27x = 3x(x^2 - 10x + 9)$
$$= 3x(x - 9)(x - 1)$$

**74.** $4x^2 - 81 = (2x)^2 - 9^2 = (2x + 9)(2x - 9)$

**75.** $2x^2 - 18 = 2(x^2 - 9)$
$$= 2(x^2 - 3^2)$$
$$= 2(x + 3)(x - 3)$$

**76.** $16x^2 - 24x + 9 = (4x)^2 - 2 \cdot 4x \cdot 3 + 3^2$
$$= (4x - 3)^2$$

**77.** $5x^2 + 20x + 20 = 5(x^2 + 4x + 4)$
$$= 5(x^2 + 2 \cdot x \cdot 2 + 2^2)$$
$$= 5(x + 2)^2$$

**78.** $2x^2 + 5x - 12 = (2x - 3)(x + 4)$

**79.** $4x^2y - 6xy^2 = 2xy(2x - 3y)$

**80.** $125x^3 + 27 = (5x)^3 + 3^3$
$$= (5x + 3)[(5x)^2 - 5x \cdot 3 + 3^2]$$
$$= (5x + 3)(25x^2 - 15x + 9)$$

**81.** $24x^2 - 3x - 18 = 3(8x^2 - x - 6)$

**82.** $(x + 7)^2 - y^2 = [(x + 7) + y][(x + 7) - y]$
$$= (x + 7 + y)(x + 7 - y)$$

**83.** $x^2(x + 3) - 4(x + 3) = (x + 3)(x^2 - 4)$
$$= (x + 3)(x^2 - 2^2)$$
$$= (x + 3)(x - 2)(x + 2)$$

**84.** $54a^3b - 2b = 2b(27a^3 - 1)$
$$= 2b[(3a)^3 - 1^3]$$
$$= 2b(3a - 1)[(3a)^2 + 3a \cdot 1 + 1^2]$$
$$= 2b(3a - 1)(9a^2 + 3a + 1)$$

**85.** $(x^2 - 2) + (x^2 - 4x) + (3x^2 - 5x)$
$= x^2 + x^2 + 3x^2 - 4x - 5x - 2$
$= 5x^2 - 9x - 2$
$= (5x + 1)(x - 2)$

**86.** $2(2x^2 + 3) + 2(6x^2 - 14x)$
$= 4x^2 + 6 + 12x^2 - 28x$
$= 16x^2 - 28x + 6$
$= 2(8x^2 - 14x + 3)$
$= 2(4x - 1)(2x - 3)$

**87.** $2x^2 - x - 28 = 0$
$(2x + 7)(x - 4) = 0$
$2x + 7 = 0 \quad$ or $\quad x - 4 = 0$
$x = -\dfrac{7}{2} \qquad\qquad x = 4$

The solutions are $-\dfrac{7}{2}$ and 4.

**88.** $x^2 - 2x = 15$
$x^2 - 2x - 15 = 0$
$(x + 3)(x - 5) = 0$
$x + 3 = 0 \quad$ or $\quad x - 5 = 0$
$x = -3 \qquad\qquad x = 5$
The solutions are $-3$ and 5.

**89.** $2x(x + 7)(x + 4) = 0$
$2x = 0 \quad$ or $\quad x + 7 = 0 \quad$ or $\quad x + 4 = 0$
$x = 0 \qquad\qquad x = -7 \qquad\qquad x = -4$
The solutions are 0, $-7$, and $-4$.

**90.** $x(x - 5) = -6$
$x^2 - 5x = -6$
$x^2 - 5x + 6 = 0$
$(x - 3)(x - 2) = 0$
$x - 3 = 0 \quad$ or $\quad x - 2 = 0$
$x = 3 \qquad\qquad x = 2$
The solutions are 3 and 2.

**91.** $x^2 = 16x$
$x^2 - 16x = 0$
$x(x - 16) = 0$
$x = 0 \quad$ or $\quad x - 16 = 0$
$\qquad\qquad\qquad x = 16$
The solutions are 0 and 16.

**92.** $(x^2 + 3) + (4x + 5) + 2x = 48$
$x^2 + 6x + 8 = 48$
$x^2 + 6x - 40 = 0$
$(x - 4)(x + 10) = 0$
$x - 4 = 0 \quad$ or $\quad x + 10 = 0$
$x = 4 \qquad\qquad x = -10$
Since the length cannot be negative, $x = 4$.
The lengths are:
$x^2 + 3 = 4^2 + 3 = 19$ inches
$4x + 5 = 4 \cdot 4 + 5 = 21$ inches
$2x = 2 \cdot 4 = 8$ inches

**93.** Let $x =$ length, then $x - 4 =$ width.
$\quad A = lw$
$12 = x(x - 4)$
$12 = x^2 - 4x$
$0 = x^2 - 4x - 12$
$0 = (x - 6)(x + 2)$
$x - 6 = 0 \quad$ or $\quad x + 2 = 0$
$x = 6 \qquad\qquad x = -2$
Since length cannot be negative, the length is 6
inches and the width is $6 - 4 = 2$ inches.

**94.** Find $t$ when $h = 0$.
$h = -16t^2 + 729$
$0 = -16t^2 + 729$
$0 = -(16t^2 - 729)$
$0 = -[(4t)^2 - (27)^2]$
$0 = -(4t + 27)(4t - 27)$
$4t + 27 = 0 \qquad$ or $\quad 4t - 27 = 0$
$t = -\dfrac{27}{4} \qquad\qquad t = \dfrac{27}{4}$

Since time cannot be negative, the object reaches
the ground in $\dfrac{27}{4} = 6.75$ seconds.

**95.** Area of large figure $-$ Area of circle
$= [(6x)(5x) - 2x^2] - \pi x^2$
$= 30x^2 - 2x^2 - \pi x^2$
$= 28x^2 - \pi x^2$
$= x^2(28 - \pi)$

## Chapter 6 Test

**1.** $x^2 + 11x + 28 = (x + 7)(x + 4)$

**2.** $49 - m^2 = (7^2 - m^2) = (7 - m)(7 + m)$

**3.** $y^2 + 22y + 121 = y^2 + 2 \cdot y \cdot 11 + 11^2 = (y+11)^2$

**4.** $4(a+3) - y(a+3) = (a+3)(4-y)$

**5.** $x^2 + 4$ is the sum of two perfect squares (not the difference). The polynomial is prime.

**6.** $y^2 - 8y - 48 = (y-12)(y+4)$

**7.** $x^2 + x - 10$ is a prime polynomial.

**8.** $9x^3 + 39x^2 + 12x = 3x(3x^2 + 13x + 4)$
$$= 3x(3x+1)(x+4)$$

**9.** $3a^2 + 3ab - 7a - 7b = 3a(a+b) - 7(a+b)$
$$= (a+b)(3a-7)$$

**10.** $3x^2 - 5x + 2 = (3x-2)(x-1)$

**11.** $x^2 + 14xy + 24y^2 = (x+12y)(x+2y)$

**12.** $180 - 5x^2 = 5(36 - x^2)$
$$= 5(6^2 - x^2)$$
$$= 5(6+x)(6-x)$$

**13.** $6t^2 - t - 5 = (6t+5)(t-1)$

**14.** $xy^2 - 7y^2 - 4x + 28 = y^2(x-7) - 4(x-7)$
$$= (x-7)(y^2 - 4)$$
$$= (x-7)(y^2 - 2^2)$$
$$= (x-7)(y+2)(y-2)$$

**15.** $x - x^5 = x(1 - x^4)$
$$= x[1 - (x^2)^2]$$
$$= x(1+x^2)(1-x^2)$$
$$= x(1+x^2)(1^2 - x^2)$$
$$= x(1+x^2)(1+x)(1-x)$$

**16.** $-xy^3 - x^3y = xy(y^2 + x^2)$

**17.** $64x^3 - 1 = (4x)^3 - 1^3$
$$= (4x-1)[(4x)^2 + 4x \cdot 1 + 1^2]$$
$$= (4x-1)(16x^2 + 4x + 1)$$

**18.** $8y^3 - 64 = 8(y^3 - 8)$
$$= 8(y^3 - 2^3)$$
$$= 8(y-2)(y^2 + y \cdot 2 + 2^2)$$
$$= 8(y-2)(y^2 + 2y + 4)$$

**19.** $(x-3)(x+9) = 0$
$x - 3 = 0$　or　$x + 9 = 0$
$x = 3$　　　　　$x = -9$
The solutions are 3 and −9.

**20.**　　　$x^2 + 5x = 14$
$x^2 + 5x - 14 = 0$
$(x+7)(x-2) = 0$
$x + 7 = 0$　or　$x - 2 = 0$
$x = -7$　　　　$x = 2$
The solutions are −7 and 2.

**21.**　　　$x(x+6) = 7$
$x^2 + 6x = 7$
$x^2 + 6x - 7 = 0$
$(x+7)(x-1) = 0$
$x + 7 = 0$　or　$x - 1 = 0$
$x = -7$　　　　$x = 1$
The solutions are −7 and 1.

**22.** $3x(2x-3)(3x+4) = 0$
$3x = 0$　or　$2x - 3 = 0$　or　$3x + 4 = 0$
$x = 0$　　　　$2x = 3$　　　　　$3x = -4$
$$x = \frac{3}{2} \qquad\qquad x = -\frac{4}{3}$$
The solutions are $0$, $\dfrac{3}{2}$, and $-\dfrac{4}{3}$.

**23.**　　　$5t^3 - 45t = 0$
$5t(t^2 - 9) = 0$
$5t(t+3)(t-3) = 0$
$5t = 0$　or　$t + 3 = 0$　or　$t - 3 = 0$
$t = 0$　　　　$t = -3$　　　　　$t = 3$
The solutions are 0, −3, and 3.

**24.** $t^2 - 2t - 15 = 0$
$(t-5)(t+3) = 0$
$t - 5 = 0$　or　$t + 3 = 0$
$t = 5$　　　　$t = -3$
The solutions are 5 and −3.

　　　　　　**287**

**25.**
$$6x^2 = 15x$$
$$6x^2 - 15x = 0$$
$$3x(2x - 5) = 0$$
$$3x = 0 \quad \text{or} \quad 2x - 5 = 0$$
$$x = 0 \qquad\qquad 2x = 5$$
$$x = \frac{5}{2}$$

The solutions are 0 and $\frac{5}{2}$.

**26.** Let $x$ = the altitude of the triangle, then $x + 9$ = the base.
$$A = \frac{1}{2}bh$$
$$68 = \frac{1}{2}(x + 9)(x)$$
$$136 = x^2 + 9x$$
$$0 = x^2 + 9x - 136$$
$$0 = (x + 17)(x - 8)$$
$$x + 17 = 0 \quad \text{or} \quad x - 8 = 0$$
$$x = -17 \qquad\qquad x = 8$$
Since the length of the base cannot be negative, the base is $8 + 9 = 17$ feet.

**27.** Let $x$ = the first number, then $17 - x$ = the other number.
$$x^2 + (17 - x)^2 = 145$$
$$x^2 + 289 - 34x + x^2 = 145$$
$$2x^2 - 34x + 144 = 0$$
$$2(x^2 - 17x + 72) = 0$$
$$2(x - 9)(x - 8) = 0$$
$$x - 9 = 0 \quad \text{or} \quad x - 8 = 0$$
$$x = 9 \qquad\qquad x = 8$$
The numbers are 8 and 9.

**28.** Find $t$ when $h = 0$.
$$h = -16t^2 + 784$$
$$0 = -16t^2 + 784$$
$$0 = -16(t^2 - 49)$$
$$0 = -16(t + 7)(t - 7)$$
$$t + 7 = 0 \quad \text{or} \quad t - 7 = 0$$
$$t = -7 \qquad\qquad t = 7$$
Since the time cannot be negative, the object reaches the ground after 7 seconds.

**29.** Let $x$ = length of the shorter leg, then $x + 10$ = length of hypotenuse, and $x + 5$ = length of longer leg.
$$x^2 + (x + 5)^2 = (x + 10)^2$$
$$x^2 + x^2 + 10x + 25 = x^2 + 20x + 100$$
$$x^2 - 10x - 75 = 0$$
$$(x + 5)(x - 15) = 0$$
$$x + 5 = 0 \quad \text{or} \quad x - 15 = 0$$
$$x = -5 \qquad\qquad x = 15$$
Since length cannot be negative, the lengths of the triangle sides are:
shorter leg = 15 cm, longer leg = 20 cm, hypotenuse = 25 cm.

## Chapter 6 Cumulative Review

**1. a.** $9 \le 11$

**b.** $8 > 1$

**c.** $3 \ne 4$

**2. a.** $|-5| > |-3|$

**b.** $|0| < |-2|$

**3. a.** $\dfrac{42}{49} = \dfrac{6 \cdot 7}{7 \cdot 7} = \dfrac{6}{7}$

**b.** $\dfrac{11}{27} = \dfrac{11}{3 \cdot 3 \cdot 3} = \dfrac{11}{27}$

**c.** $\dfrac{88}{20} = \dfrac{4 \cdot 22}{4 \cdot 5} = \dfrac{22}{5}$

**4.** Let $x = 20$ and $y = 10$.
$$\frac{x}{y} + 5x = \frac{20}{10} + 5(20) = 2 + 100 = 102$$

**5.** $\dfrac{8 + 2 \cdot 3}{2^2 - 1} = \dfrac{8 + 6}{4 - 1} = \dfrac{14}{3}$

**6.** Let $x = -20$ and $y = 10$.
$$\frac{x}{y} + 5x = \frac{-20}{10} + 5(-20) = -2 - 100 = -102$$

**7. a.** $3 + (-7) + (-8) = 3 + (-15) = -12$

**b.** $[7 + (-10)] + \left[-2 + |-4|\right] = -3 + (-2 + 4)$
$$= -3 + 2$$
$$= -1$$

**8.** Let $x = -20$ and $y = -10$.

$$\frac{x}{y} + 5x = \frac{-20}{-10} + 5(-20) = 2 - 100 = -98$$

**9. a.** $(-8)(4) = -32$

    **b.** $14(-1) = -14$

    **c.** $(-9)(-10) = 90$

**10.** $5 - 2(3x - 7) = 5 - 6x + 14 = -6x + 19$

**11. a.** $7x - 3x = (7 - 3)x = 4x$

    **b.** $10y^2 + y^2 = (10 + 1)y^2 = 11y^2$

    **c.** $8x^2 + 2x - 3x = 8x^2 + (2 - 3)x = 8x^2 - x$

    **d.** $9n^2 - 5n^2 + n^2 = (9 - 5 + 1)n^2 = 5n^2$

**12.** $0.8y + 0.2(y - 1) = 1.8$
$$0.8y + 0.2y - 0.2 = 1.8$$
$$1.0y - 0.2 = 1.8$$
$$y = 2.0$$

**13.** $3 - x = 7$
$$3 - 3 - x = 7 - 3$$
$$-x = 4$$
$$x = -4$$

**14.** $\dfrac{x}{-7} = -4$
$$-7\left(\frac{x}{-7}\right) = -7(-4)$$
$$x = 28$$

**15.** $-3x = 33$
$$\frac{-3x}{-3} = \frac{33}{-3}$$
$$x = -11$$

**16.** $-\dfrac{2}{3}x = -22$
$$\left(-\frac{3}{2}\right)\left(-\frac{2}{3}\right)x = \left(-\frac{3}{2}\right)(-22)$$
$$x = 33$$

**17.** $8(2 - t) = -5t$
$$16 - 8t = -5t$$
$$16 - 8t + 5t = -5t + 5t$$
$$16 - 3t = 0$$
$$16 - 16 - 3t = -16$$
$$-3t = -16$$
$$\frac{-3t}{-3} = \frac{-16}{-3}$$
$$t = \frac{16}{3}$$

**18.** $-z = \dfrac{7z + 3}{5}$
$$5(-z) = 5\left(\frac{7z + 3}{5}\right)$$
$$-5z = 7z + 3$$
$$-5z - 7z = 7z - 7z + 3$$
$$-12z = 3$$
$$\frac{-12z}{-12} = \frac{3}{-12}$$
$$z = -\frac{1}{4}$$

**19.** Let $x =$ the length of the shorter piece and $3x =$ the length of the longer piece.
$$x + 3x = 48$$
$$4x = 48$$
$$x = 12$$
$$3x = 3(12) = 36$$
The pieces are 12 inches and 36 inches in length.

**20.** $3x + 9 \le 5(x - 1)$
$$3x + 9 \le 5x - 5$$
$$-2x + 9 \le -5$$
$$2x \le -14$$
$$\frac{-2x}{-2} \ge \frac{-14}{-2}$$
$$x \ge 7, [7, \infty)$$

**21.** $y = -\dfrac{1}{3}x + 2$

| $x$ | $y$ |
|---|---|
| 0 | 2 |
| −3 | 3 |
| 3 | 1 |

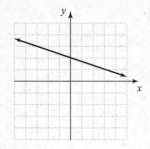

**22.** $-7x - 8y = -9$
$(-1, 2)$: $-7(-1) - 8(2) \overset{?}{=} -9$
$\qquad\qquad 7 - 16 \overset{?}{=} -9$
$\qquad\qquad\qquad -9 = -9$  True
$(-1, 2)$ is a solution of the equation.

**23.** $3x - 4y = 4$
$\qquad -4y = -3x + 4$
$\qquad\quad y = \dfrac{3}{4}x - 1$
$\qquad\quad y = mx + b$
slope $= \dfrac{3}{4}$; $y$-intercept $= (0, -1)$

**24.** $(5, -6)$ and $(5, 2)$
$m = \dfrac{y_2 - y_1}{x_2 - x_1} = \dfrac{2 - (-6)}{5 - 5} = \dfrac{8}{0}$
The slope is undefined.

**25. a.** If $x = 5$, $2x^3 = 2(5)^3 = 2(125) = 250$.

**b.** If $x = -3$, $\dfrac{9}{x^2} = \dfrac{9}{(-3)^2} = \dfrac{9}{9} = 1$.

**26.** $7x - 3y = 2$
$\qquad -3y = -7x + 2$
$\qquad\quad y = \dfrac{-7x}{-3} + \dfrac{2}{-3}$
$\qquad\quad y = \dfrac{7}{3}x - \dfrac{2}{3}$
$\qquad\quad y = mx + b$
slope $= \dfrac{7}{3}$; $y$-intercept $= \left(0, -\dfrac{2}{3}\right)$

**27. a.** $-3x^2$ has degree 2.

**b.** $5x^3yz$ has degree $3 + 1 + 1 = 5$.

**c.** 2 has degree 0.

**28.** Vertical line has equation $x = c$.
Point $(0, 7)$
$x = 0$

**29.** $(2x^3 + 8x^2 - 6x) - (2x^3 - x^2 + 1)$
$= 2x^3 + 8x^2 - 6x - 2x^3 + x^2 - 1$
$= 9x^2 - 6x - 1$

**30.** $m = 4$, $b = \dfrac{1}{2}$
$\qquad y = mx + b$
$\qquad y = 4x + \dfrac{1}{2}$
$\qquad 2y = 8x + 1$
$8x - 2y = -1$

**31.** $(3x + 2)(2x - 5)$
$= 3x(2x) + 3x(-5) + 2(2x) + 2(-5)$
$= 6x^2 - 15x + 4x - 10$
$= 6x^2 - 11x - 10$

**32.** $(-4, 0)$ and $(6, -1)$
$m = \dfrac{y_2 - y_1}{x_2 - x_1} = \dfrac{-1 - 0}{6 - (-4)} = -\dfrac{1}{10}$
$m = -\dfrac{1}{10}$, point $(-4, 0)$
$\qquad y - y_1 = m(x - x_1)$
$\qquad y - 0 = -\dfrac{1}{10}[x - (-4)]$
$\qquad\quad y = -\dfrac{1}{10}x - \dfrac{4}{10}$
$\qquad 10y = -x - 4$
$x + 10y = -4$

**33.** $(3y + 1)^2 = (3y)^2 + 2(3y)(1) + 1^2 = 9y^2 + 6y + 1$

**34.** $\begin{cases} -x + 3y = 18 \\ -3x + 2y = 19 \end{cases}$
Multiply the first equation by $-3$.
$\quad 3x - 9y = -54$
$\underline{-3x + 2y = 19}$
$\qquad -7y = -35$
$\qquad\quad y = 5$
Substitute 5 for $y$ in the first equation.
$-x + 3(5) = 18$
$\quad -x + 15 = 18$
$\qquad\quad -x = 3$
$\qquad\quad\; x = -3$
The solution to the system is $(-3, 5)$.

**35. a.** $3^{-2} = \dfrac{1}{3^2} = \dfrac{1}{9}$

**b.** $2x^{-3} = \dfrac{2}{x^3}$

**c.** $2^{-1} + 4^{-1} = \dfrac{1}{2} + \dfrac{1}{4}$
$= \dfrac{1 \cdot 2}{2 \cdot 2} + \dfrac{1}{4}$
$= \dfrac{2}{4} + \dfrac{1}{4}$
$= \dfrac{2+1}{4}$
$= \dfrac{3}{4}$

**d.** $(-2)^{-4} = \dfrac{1}{(-2)^4} = \dfrac{1}{16}$

**e.** $y^{-4} = \dfrac{1}{y^4}$

**36.** $\dfrac{(5a^7)^2}{a^5} = \dfrac{5^2 a^{14}}{a^5} = 25a^{14-5} = 25a^9$

**37. a.** $367,000,000 = 3.67 \times 10^8$

**b.** $0.000003 = 3.0 \times 10^{-6}$

**c.** $20,520,000,000 = 2.052 \times 10^{10}$

**d.** $0.00085 = 8.5 \times 10^{-4}$

**38.** $(3x - 7y)^2 = (3x)^2 - 2(3x)(7y) + (7y)^2$
$= 9x^2 - 42xy + 49y^2$

**39.**
$$\begin{array}{r} x+4 \\ x+3 \overline{\smash{\big)}\, x^2 + 7x + 12} \\ \underline{x^2 + 3x} \phantom{aaaaaa} \\ 4x + 12 \\ \underline{4x + 12} \\ 0 \end{array}$$

$\dfrac{x^2 + 7x + 12}{x + 3} = x + 4$

**40.** $\dfrac{(xy)^{-3}}{(x^5 y^6)^3} = \dfrac{x^{-3} y^{-3}}{x^{15} y^{18}}$
$= x^{-3-15} y^{-3-18}$
$= x^{-18} y^{-21}$
$= \dfrac{1}{x^{18} y^{21}}$

**41. a.** $x^3,\ x^7,\ x^5$: GCF $= x^3$

**b.** $y,\ y^4,\ y^7$: GCF $= y$

**42.** $z^3 + 7z + z^2 + 7 = z(z^2 + 7) + 1(z^2 + 7)$
$= (z^2 + 7)(z + 1)$

**43.** $x^2 + 7x + 12 = (x + 4)(x + 3)$

**44.** $2x^3 + 2x^2 - 84x = 2x(x^2 + x - 42)$
$= 2x(x + 7)(x - 6)$

**45.** $8x^2 - 22x + 5 = 8x^2 - 20x - 2x + 5$
$= 4x(2x - 5) - 1(2x - 5)$
$= (2x - 5)(4x - 1)$

**46.** $-4x^2 - 23x + 6 = -1(4x^2 + 23x - 6)$
$= -(4x^2 - x + 24x - 6)$
$= -[x(4x - 1) + 6(4x - 1)]$
$= -(4x - 1)(x + 6)$

**47.** $25a^2 - 9b^2 = (5a)^2 - (3b)^2 = (5a + 3b)(5a - 3b)$

**48.** $9xy^2 - 16x = x(9y^2 - 16)$
$= x[(3y)^2 - 4^2]$
$= x(3y + 4)(3y - 4)$

**49.** $(x - 3)(x + 1) = 0$
$x - 3 = 0 \quad \text{or} \quad x + 1 = 0$
$\phantom{x - 3} x = 3 \phantom{aaaaaa} x = -1$
The solutions are 3 and $-1$.

**50.** $\phantom{aaa} x^2 - 13x = -36$
$x^2 - 13x + 36 = 0$
$(x - 9)(x - 4) = 0$
$x - 9 = 0 \quad \text{or} \quad x - 4 = 0$
$\phantom{aa} x = 9 \phantom{aaaaaaa} x = 4$
The solutions are 9 and 4.

# Chapter 7

**1. a.** Replace each $x$ in the expression with 3 and then simplify.
$$\frac{x+6}{3x-2} = \frac{3+6}{3(3)-2} = \frac{9}{9-2} = \frac{9}{7}$$

**b.** Replace each $x$ in the expression with $-3$ and then simplify.
$$\frac{x+6}{3x-2} = \frac{-3+6}{3(-3)-2} = \frac{3}{-9-2} = \frac{3}{-11} \text{ or } -\frac{3}{11}$$

**2. a.** The denominator of $\dfrac{x}{x+6}$ is 0 when
$x + 6 = 0$ or when $x = -6$. Thus, when
$x = -6$, the expression $\dfrac{x}{x+6}$ is undefined.

**b.** The denominator of $\dfrac{x^4 - 3x^2 + 7x}{7}$ is never
0, so there are no values of $x$ for which this expression is undefined.

**c.** Set the denominator equal to zero.
$$x^2 + 6x + 8 = 0$$
$$(x+2)(x+4) = 0$$
$$x + 2 = 0 \quad \text{or} \quad x + 4 = 0$$
$$x = -2 \qquad\qquad x = -4$$
Thus, when $x = -2$ or $x = -4$, the
denominator $x^2 + 6x + 8$ is 0. So the
rational expression $\dfrac{x^2 - 5}{x^2 + 6x + 8}$ is undefined
when $x = -2$ or when $x = -4$.

**d.** No matter which real number $x$ is replaced
by, the denominator $x^4 + 5$ does not equal
0, so there are no real numbers for which
this expression is undefined.

**3.** Factor the numerator and denominator, then look
for common factors.
$$\frac{x^6 - x^5}{6x - 6} = \frac{x^5(x-1)}{6(x-1)} = \frac{x^5}{6}$$

**4.** Factor the numerator and denominator, then look
for common factors.
$$\frac{x^2 + 5x + 4}{x^2 + 2x - 8} = \frac{(x+1)(x+4)}{(x-2)(x+4)} = \frac{x+1}{x-2}$$

**5.** Factor the numerator and denominator, then look
for common factors.
$$\frac{x^3 + 9x^2}{x^2 + 18x + 81} = \frac{x^2(x+9)}{(x+9)(x+9)} = \frac{x^2}{x+9}$$

**6.** Factor the numerator and denominator, then look
for common factors.
$$\frac{x-7}{x^2 - 49} = \frac{x-7}{(x+7)(x-7)} = \frac{1}{x+7}$$

**7. a.** Note that $s - t$ and $t - s$ are opposites. In
other words, $t - s = -1(s - t)$.
$$\frac{s-t}{t-s} = \frac{1 \cdot (s-t)}{-1 \cdot (s-t)} = \frac{1}{-1} = -1$$

**b.** By the commutative property of addition,
$d + 2c = 2c + d$.
$$\frac{2c+d}{d+2c} = \frac{2c+d}{2c+d} = 1$$

**8.** $\dfrac{2x^2 - 5x - 12}{16 - x^2} = \dfrac{(x-4)(2x+3)}{(4-x)(4+x)}$
$$= \frac{(x-4)(2x+3)}{(-1)(x-4)(4+x)}$$
$$= \frac{2x+3}{(-1)(4+x)}$$
$$= -\frac{2x+3}{x+4} \text{ or } \frac{-2x-3}{x+4}$$

**9.** $-\dfrac{x+3}{6x-11} = \dfrac{-(x+3)}{6x-11} = \dfrac{-x-3}{6x-11}$
Also,
$$-\frac{x+3}{6x-11} = \frac{x+3}{-(6x-11)} = \frac{x+3}{-6x+11} \text{ or } \frac{x+3}{11-6x}$$
Thus, some equivalent forms of $-\dfrac{x+3}{6x-11}$ are
$$\frac{-(x+3)}{6x-11}, \frac{-x-3}{6x-11}, \frac{x+3}{-(6x-11)}, \frac{x+3}{-6x+11}, \text{ and}$$
$$\frac{x+3}{11-6x}.$$

**Vocabulary, Readiness & Video Check 7.1**

1. A <u>rational expression</u> is an expression that can be written in the form $\dfrac{P}{Q}$ where $P$ and $Q$ are polynomials and $Q \neq 0$.

2. The expression $\dfrac{x+3}{3+x}$ simplifies to <u>1</u>.

3. The expression $\dfrac{x-3}{3-x}$ simplifies to <u>−1</u>.

4. A rational expression is undefined for values that make the denominator <u>0</u>.

5. The expression $\dfrac{7x}{x-2}$ is undefined for $x = $ <u>2</u>.

6. The process of writing a rational expression in lowest terms is called <u>simplifying</u>.

7. For a rational expression, $-\dfrac{a}{b} = \dfrac{-a}{\underline{b}} = \dfrac{a}{\underline{-b}}$.

8. replacement values for variables; by evaluating the expression for different replacement values—variables are replaced with these values and the expression is simplified

9. Rational expressions are fractions and are therefore undefined if the denominator is zero; if a denominator contains variables, set it equal to zero and solve.

10. Although $x$ is a factor in the numerator, it is not a factor in the denominator—factor means write as a product and the denominator is a difference, not a product.

11. You would need to write parentheses around the numerator or denominator if it had more than one term because the negative sign needs to apply to the entire numerator or denominator.

**Exercise Set 7.1**

1. $\dfrac{x+5}{x+2} = \dfrac{2+5}{2+2} = \dfrac{7}{4}$

3. $\dfrac{4z-1}{z-2} = \dfrac{4(-5)-1}{-5-2} = \dfrac{-20-1}{-7} = \dfrac{-21}{-7} = 3$

5. $\dfrac{y^3}{y^2-1} = \dfrac{(-2)^3}{(-2)^2-1} = \dfrac{-8}{4-1} = \dfrac{-8}{3} = -\dfrac{8}{3}$

7. $\dfrac{x^2+8x+2}{x^2-x-6} = \dfrac{2^2+8(2)+2}{2^2-2-6}$

$= \dfrac{4+16+2}{4-8}$

$= \dfrac{22}{-4}$

$= \dfrac{11 \cdot 2}{-2 \cdot 2}$

$= -\dfrac{11}{2}$

9. $2x = 0$

$x = 0$

The expression is undefined when $x = 0$.

11. $x + 2 = 0$

$x = -2$

The expression is undefined when $x = -2$.

13. $2x - 5 = 0$

$2x = 5$

$x = \dfrac{5}{2}$

The expression is undefined when $x = \dfrac{5}{2}$.

15. The denominator is never zero, so there are no values for which $\dfrac{x^2-5x-2}{4}$ is undefined.

17. $x^2 - 5x - 6 = 0$

$(x+1)(x-6) = 0$

$x+1 = 0$   or   $x-6 = 0$

$x = -1$            $x = 6$

The expression is undefined when $x = -1, 6$.

19. The denominator is never zero, so there are no values for which $\dfrac{9x^3+4}{x^2+36}$ is undefined.

**21.** $3x^2 + 13x + 14 = 0$

$(x+2)(3x+7) = 0$

$x+2 = 0 \quad \text{or} \quad 3x+7 = 0$

$x = -2 \qquad\qquad 3x = -7$

$$x = -\frac{7}{3}$$

The expression is undefined when $x = -2, -\dfrac{7}{3}$.

**23.** $-\dfrac{x-10}{x+8} = \dfrac{-(x-10)}{x+8} = \dfrac{-x+10}{x+8}$ or $\dfrac{10-x}{x+8}$

$-\dfrac{x-10}{x+8} = \dfrac{x-10}{-(x+8)} = \dfrac{x-10}{-x-8}$

**25.** $-\dfrac{5y-3}{y-12} = \dfrac{-(5y-3)}{y-12} = \dfrac{-5y+3}{y-12}$ or $\dfrac{3-5y}{y-12}$

$-\dfrac{5y-3}{y-12} = \dfrac{5y-3}{-(y-12)} = \dfrac{5y-3}{-y+12}$ or $\dfrac{5y-3}{12-y}$

**27.** $\dfrac{x+7}{7+x} = \dfrac{x+7}{x+7} = 1$

**29.** $\dfrac{x-7}{7-x} = \dfrac{x-7}{-1(x-7)} = \dfrac{1}{-1} = -1$

**31.** $\dfrac{2}{8x+16} = \dfrac{2}{8(x+2)} = \dfrac{2(1)}{2(4)(x+2)} = \dfrac{1}{4(x+2)}$

**33.** $\dfrac{x-2}{x^2-4} = \dfrac{x-2}{(x+2)(x-2)} = \dfrac{1}{x+2}$

**35.** $\dfrac{2x-10}{3x-30} = \dfrac{2(x-5)}{3(x-10)}$

The numerator and denominator have no common factors, so the expression cannot be simplified.

**37.** $\dfrac{-5a-5b}{a+b} = \dfrac{-5(a+b)}{a+b} = -5$

**39.** $\dfrac{7x+35}{x^2+5x} = \dfrac{7(x+5)}{x(x+5)} = \dfrac{7}{x}$

**41.** $\dfrac{x+5}{x^2-4x-45} = \dfrac{x+5}{(x+5)(x-9)} = \dfrac{1}{x-9}$

**43.** $\dfrac{5x^2+11x+2}{x+2} = \dfrac{(5x+1)(x+2)}{x+2} = 5x+1$

**45.** $\dfrac{x^3+7x^2}{x^2+5x-14} = \dfrac{x^2(x+7)}{(x+7)(x-2)} = \dfrac{x^2}{x-2}$

**47.** $\dfrac{14x^2-21x}{2x-3} = \dfrac{7x(2x-3)}{2x-3} = 7x$

**49.** $\dfrac{x^2+7x+10}{x^2-3x-10} = \dfrac{(x+2)(x+5)}{(x+2)(x-5)} = \dfrac{x+5}{x-5}$

**51.** $\dfrac{3x^2+7x+2}{3x^2+13x+4} = \dfrac{(3x+1)(x+2)}{(3x+1)(x+4)} = \dfrac{x+2}{x+4}$

**53.** $\dfrac{2x^2-8}{4x-8} = \dfrac{2(x^2-4)}{4(x-2)} = \dfrac{2(x+2)(x-2)}{2\cdot2(x-2)} = \dfrac{x+2}{2}$

**55.** $\dfrac{4-x^2}{x-2} = \dfrac{(-1)(x^2-4)}{x-2}$

$= -\dfrac{(x+2)(x-2)}{x-2}$

$= -(x+2)$ or $-x-2$

**57.** $\dfrac{x^2-1}{x^2-2x+1} = \dfrac{(x+1)(x-1)}{(x-1)(x-1)} = \dfrac{x+1}{x-1}$

**59.** $\dfrac{m^2-6m+9}{m^2-m-6} = \dfrac{(m-3)(m-3)}{(m+2)(m-3)} = \dfrac{m-3}{m+2}$

**61.** $\dfrac{11x^2-22x^3}{6x-12x^2} = \dfrac{11x^2(1-2x)}{6x(1-2x)} = \dfrac{11x}{6}$

**63.** $\dfrac{x^2+xy+2x+2y}{x+2} = \dfrac{x(x+y)+2(x+y)}{x+2}$

$= \dfrac{(x+y)(x+2)}{x+2}$

$= x+y$

**65.** $\dfrac{5x+15-xy-3y}{2x+6} = \dfrac{5(x+3)-y(x+3)}{2(x+3)}$

$= \dfrac{(x+3)(5-y)}{2(x+3)}$

$= \dfrac{5-y}{2}$

**67.** $\dfrac{x^3+8}{x+2} = \dfrac{(x+2)(x^2-2x+4)}{x+2} = x^2-2x+4$

**69.** $\dfrac{x^3-1}{1-x} = \dfrac{(x-1)(x^2+x+1)}{-1(x-1)}$

$\qquad = -1(x^2+x+1)$

$\qquad = -x^2-x-1$

**71.** $\dfrac{2xy+5x-2y-5}{3xy+4x-3y-4} = \dfrac{x(2y+5)-1(2y+5)}{x(3y+4)-1(3y+4)}$

$\qquad = \dfrac{(2y+5)(x-1)}{(3y+4)(x-1)}$

$\qquad = \dfrac{2y+5}{3y+4}$

**73.** $\dfrac{9-x^2}{x-3} = \dfrac{-1(x^2-9)}{x-3}$

$\qquad = -\dfrac{(x+3)(x-3)}{x-3}$

$\qquad = -(x+3) \text{ or } -x-3$

The given answer is correct.

**75.** $\dfrac{7-34x-5x^2}{25x^2-1} = \dfrac{-1(5x^2+34x-7)}{(5x+1)(5x-1)}$

$\qquad = -\dfrac{(x+7)(5x-1)}{(5x+1)(5x-1)}$

$\qquad = -\dfrac{x+7}{5x+1}$

$\qquad = \dfrac{x+7}{-(5x+1)} \text{ or } \dfrac{x+7}{-5x-1}$

The given answer is correct.

**77.** $\dfrac{1}{3} \cdot \dfrac{9}{11} = \dfrac{1 \cdot 9}{3 \cdot 11} = \dfrac{3 \cdot 3}{3 \cdot 11} = \dfrac{3}{11}$

**79.** $\dfrac{1}{3} \div \dfrac{1}{4} = \dfrac{1}{3} \cdot \dfrac{4}{1} = \dfrac{4}{3}$

**81.** $\dfrac{13}{20} \div \dfrac{2}{9} = \dfrac{13}{20} \cdot \dfrac{9}{2} = \dfrac{13 \cdot 9}{20 \cdot 2} = \dfrac{117}{40}$

**83.** $\dfrac{5a-15}{5} = \dfrac{5(a-3)}{5} = a-3$

The statement is correct.

**85.** $\dfrac{1+2}{1+3} = \dfrac{3}{4}$

The statement is incorrect.

**87.** $\dfrac{x}{x+7}$ cannot be simplified.

**89.** Since $5-x = -1(x-5)$, $\dfrac{5-x}{x-5}$ can be simplified.

**91.** answers may vary

**93.** answers may vary

**95. a.** $R = \dfrac{150x^2}{x^2+3}$

$\qquad = \dfrac{150(1)^2}{1^2+3}$

$\qquad = \dfrac{150}{4}$

$\qquad = \$37.5 \text{ million}$

**b.** $R = \dfrac{150x^2}{x^2+3}$

$\qquad = \dfrac{150(2)^2}{2^2+3}$

$\qquad = \dfrac{600}{7}$

$\qquad \approx \$85.7 \text{ million}$

**c.** $85.7 - 37.5 = \$48.2 \text{ million}$

**97.** Let $D = 1000$ and $A = 8$.

$C = \dfrac{DA}{A+12} = \dfrac{1000(8)}{8+12} = \dfrac{8000}{20} = 400$

The child should receive 400 mg.

**99.** $C = \dfrac{100W}{L}$; $W = 5, L = 6.4$

$C = \dfrac{100(5)}{6.4} = \dfrac{500}{6.4} = 78.125$

The skull is medium.

**101.** Use $S = \dfrac{h+d+2t+3r}{b}$ with $h = 183, d = 39$,

$t = 1, r = 42$, and $b = 587$.

$S = \dfrac{183+39+2(1)+3(42)}{587} = \dfrac{350}{587} \approx 0.5963$

Pujols' slugging percentage was about 59.6%.

**103.** $y = \dfrac{x^2 - 25}{x+5} = \dfrac{(x+5)(x-5)}{x+5} = x-5,\ x \neq -5$

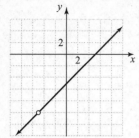

**105.** $y = \dfrac{x^2 + x - 12}{x+4} = \dfrac{(x+4)(x-3)}{x+4} = x-3,\ x \neq -4$

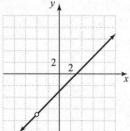

**Section 7.2 Practice**

**1. a.** $\dfrac{4a}{5} \cdot \dfrac{3}{b^2} = \dfrac{4a \cdot 3}{5 \cdot b^2} = \dfrac{12a}{5b^2}$

   **b.** $\dfrac{-3p^4}{q^2} \cdot \dfrac{2q^3}{9p^4} = \dfrac{-3p^4 \cdot 2q^3}{q^2 \cdot 9p^4}$

$$= \dfrac{-1 \cdot 3 \cdot p^4 \cdot 2 \cdot q \cdot q^2}{q^2 \cdot 3 \cdot 3 \cdot p^4}$$

$$= -\dfrac{2q}{3}$$

**2.** $\dfrac{x^2 - x}{5x} \cdot \dfrac{15}{x^2 - 1} = \dfrac{x(x-1)}{5x} \cdot \dfrac{3 \cdot 5}{(x+1)(x-1)}$

$$= \dfrac{x(x-1) \cdot 3 \cdot 5}{5x \cdot (x+1)(x-1)}$$

$$= \dfrac{3}{x+1}$$

**3.** $\dfrac{6 - 3x}{6x + 6x^2} \cdot \dfrac{3x^2 - 2x - 5}{x^2 - 4}$

$$= \dfrac{3(2-x)}{2 \cdot 3 \cdot x(1+x)} \cdot \dfrac{(x+1)(3x-5)}{(x+2)(x-2)}$$

$$= \dfrac{3(2-x)(x+1)(3x-5)}{2 \cdot 3x(1+x)(x+2)(x-2)}$$

$$= \dfrac{-1(x-2)(x+1)(3x-5)}{2x(x+1)(x+2)(x-2)}$$

$$= -\dfrac{3x-5}{2x(x+2)}$$

**4.** $\dfrac{5a^3 b^2}{24} \div \dfrac{10a^5}{6} = \dfrac{5a^3 b^2}{24} \cdot \dfrac{6}{10a^5}$

$$= \dfrac{5a^3 b^2 \cdot 6}{4 \cdot 6 \cdot 2 \cdot 5 \cdot a^2 \cdot a^3}$$

$$= \dfrac{b^2}{8a^2}$$

**5.** $\dfrac{(x-5)^2}{3} \div \dfrac{4x - 20}{9} = \dfrac{(x-5)(x-5)}{3} \cdot \dfrac{9}{4x - 20}$

$$= \dfrac{(x-5)(x-5) \cdot 3 \cdot 3}{3 \cdot 4(x-5)}$$

$$= \dfrac{3(x-5)}{4}$$

**6.** $\dfrac{10x - 2}{x^2 - 9} \div \dfrac{5x^2 - x}{x+3} = \dfrac{10x - 2}{x^2 - 9} \cdot \dfrac{x+3}{5x^2 - x}$

$$= \dfrac{2(5x-1)(x+3)}{(x+3)(x-3) \cdot x(5x-1)}$$

$$= \dfrac{2}{x(x-3)}$$

**7.** $\dfrac{3x^2 - 11x - 4}{2x - 8} \div \dfrac{9x + 3}{6} = \dfrac{3x^2 - 11x - 4}{2x - 8} \cdot \dfrac{6}{9x + 3}$

$$= \dfrac{(3x+1)(x-4) \cdot 2 \cdot 3}{2(x-4) \cdot 3(3x+1)}$$

$$= \dfrac{1}{1} \text{ or } 1$$

**8. a.** $\dfrac{y+9}{8x} \cdot \dfrac{y+9}{2x} = \dfrac{(y+9) \cdot (y+9)}{8x \cdot 2x} = \dfrac{(y+9)^2}{16x^2}$

**b.**
$$\frac{y+9}{8x} \div \frac{y+9}{2} = \frac{y+9}{8x} \cdot \frac{2}{y+9}$$
$$= \frac{(y+9)\cdot 2}{2\cdot 4\cdot x\cdot (y+9)}$$
$$= \frac{1}{4x}$$

**c.**
$$\frac{35x-7x^2}{x^2-25} \cdot \frac{x^2+3x-10}{x^2+4x}$$
$$= \frac{7x(5-x)}{(x+5)(x-5)} \cdot \frac{(x-2)(x+5)}{x(x+4)}$$
$$= \frac{7x\cdot(-1)(x-5)\cdot(x-2)(x+5)}{(x+5)(x-5)\cdot x(x+4)}$$
$$= -\frac{7(x-2)}{x+4}$$

**9.** $288 \text{ sq in.} = \frac{288 \text{ sq in.}}{1} \cdot \frac{1 \text{ sq ft}}{144 \text{ sq in.}} = 2 \text{ sq ft}$

**10.** $3.5 \text{ sq ft} = \frac{3.5 \text{ sq ft}}{1} \cdot \frac{144 \text{ sq in.}}{1 \text{ sq ft}} = 504 \text{ sq in.}$

**11.** $61,000 \text{ sq yd} = 61,000 \text{ sq yd} \cdot \frac{9 \text{ sq ft}}{1 \text{ sq yd}}$
$$= 549,000 \text{ sq ft}$$

**12.** 102.7 feet/second
$$= \frac{102.7 \text{ feet}}{1 \text{ second}} \cdot \frac{3600 \text{ seconds}}{1 \text{ hour}} \cdot \frac{1 \text{ mile}}{5280 \text{ feet}}$$
$$= \frac{102.7\cdot 3600}{5280} \text{ miles/hour}$$
$$\approx 70.0 \text{ miles/hour}$$

**Vocabulary, Readiness & Video Check 7.2**

**1.** The expressions $\frac{x}{2y}$ and $\frac{2y}{x}$ are called reciprocals.

**2.** $\frac{a}{b} \cdot \frac{c}{d} = \frac{a\cdot c}{b\cdot d}$ or $\frac{ac}{bd}$

**3.** $\frac{a}{b} \div \frac{c}{d} = \frac{a\cdot d}{b\cdot c}$ or $\frac{ad}{bc}$

**4.** $\frac{x}{7} \cdot \frac{x}{6} = \frac{x^2}{42}$

**5.** $\frac{x}{7} \div \frac{x}{6} = \frac{6}{7}$

**6.** Yes, multiplying and simplifying rational expressions often requires polynomial factoring. Example 2 alone involves factoring out a GCF, factoring a trinomial with $a \neq 1$, and factoring a difference of squares.

**7.** Dividing rational expressions is exactly like dividing fractions. Therefore, to divide by a rational expression, multiply by its reciprocal.

**8.** Multiplication and division of rational expressions are performed similarly—both involve multiplication—but there are important differences. Note the operation first to see whether you multiply by the reciprocal or not.

**9.** The units in the unit fraction consist of units converting to original units.

**Exercise Set 7.2**

**1.** $\frac{3x}{y^2} \cdot \frac{7y}{4x} = \frac{3\cdot x\cdot 7\cdot y}{y\cdot y\cdot 4\cdot x} = \frac{21}{4y}$

**3.** $\frac{8x}{2} \cdot \frac{x^5}{4x^2} = \frac{8\cdot x^5}{2\cdot 4x^2} = \frac{2\cdot 4\cdot x\cdot x\cdot x^4}{2\cdot 4\cdot x\cdot x} = x^4$

**5.** $-\frac{5a^2b}{30a^2b^2} \cdot b^3 = \frac{5a^2b\cdot b^3}{30a^2b^2}$
$$= -\frac{5\cdot a^2\cdot b\cdot b\cdot b^2}{5\cdot 6\cdot a^2\cdot b^2}$$
$$= -\frac{b\cdot b}{6}$$
$$= -\frac{b^2}{6}$$

**7.** $\frac{x}{2x-14} \cdot \frac{x^2-7x}{5} = \frac{x\cdot(x^2-7x)}{(2x-14)\cdot 5}$
$$= \frac{x\cdot x(x-7)}{2(x-7)\cdot 5}$$
$$= \frac{x\cdot x}{2\cdot 5}$$
$$= \frac{x^2}{10}$$

**9.** $\dfrac{6x+6}{5} \cdot \dfrac{10}{36x+36} = \dfrac{(6x+6) \cdot 10}{5 \cdot (36x+36)}$

$= \dfrac{6(x+1) \cdot 2 \cdot 5}{5 \cdot 36(x+1)}$

$= \dfrac{6 \cdot 5 \cdot 2 \cdot (x+1)}{6 \cdot 5 \cdot 2 \cdot 3 \cdot (x+1)}$

$= \dfrac{1}{3}$

**11.** $\dfrac{(m+n)^2}{m-n} \cdot \dfrac{m}{m^2+mn} = \dfrac{(m+n)(m+n) \cdot m}{(m-n) \cdot m(m+n)} = \dfrac{m+n}{m-n}$

**13.** $\dfrac{x^2-25}{x^2-3x-10} \cdot \dfrac{x+2}{x} = \dfrac{(x^2-25) \cdot (x+2)}{(x^2-3x-10) \cdot x}$

$= \dfrac{(x-5)(x+5) \cdot (x+2)}{(x-5)(x+2) \cdot x}$

$= \dfrac{x+5}{x}$

**15.** $\dfrac{x^2+6x+8}{x^2+x-20} \cdot \dfrac{x^2+2x-15}{x^2+8x+16}$

$= \dfrac{(x+2)(x+4)}{(x+5)(x-4)} \cdot \dfrac{(x+5)(x-3)}{(x+4)(x+4)}$

$= \dfrac{(x+2)(x+4) \cdot (x+5)(x-3)}{(x+5)(x-4) \cdot (x+4)(x+4)}$

$= \dfrac{(x+2)(x-3)}{(x-4)(x+4)}$

**17.** $\dfrac{5x^7}{2x^5} \div \dfrac{15x}{4x^3} = \dfrac{5x^7}{2x^5} \cdot \dfrac{4x^3}{15x}$

$= \dfrac{5 \cdot x^2 \cdot x^5 \cdot 2 \cdot 2 \cdot x \cdot x^2}{2 \cdot x^5 \cdot 3 \cdot 5 \cdot x}$

$= \dfrac{2x^4}{3}$

**19.** $\dfrac{8x^2}{y^3} \div \dfrac{4x^2y^3}{6} = \dfrac{8x^2}{y^3} \cdot \dfrac{6}{4x^2y^3} = \dfrac{2 \cdot 4 \cdot x^2 \cdot 6}{y^3 \cdot 4x^2y^3} = \dfrac{12}{y^6}$

**21.** $\dfrac{(x-6)(x+4)}{4x} \div \dfrac{2x-12}{8x^2}$

$= \dfrac{(x-6)(x+4)}{4x} \cdot \dfrac{8x^2}{2x-12}$

$= \dfrac{(x-6)(x+4) \cdot 2 \cdot 4 \cdot x \cdot x}{4x \cdot 2(x-6)}$

$= x(x+4)$

**23.** $\dfrac{3x^2}{x^2-1} \div \dfrac{x^5}{(x+1)^2} = \dfrac{3x^2}{x^2-1} \cdot \dfrac{(x+1)^2}{x^5}$

$= \dfrac{3x^2 \cdot (x+1)(x+1)}{(x-1)(x+1) \cdot x^2 \cdot x^3}$

$= \dfrac{3(x+1)}{x^3(x-1)}$

**25.** $\dfrac{m^2-n^2}{m+n} \div \dfrac{m}{m^2+nm} = \dfrac{m^2-n^2}{m+n} \cdot \dfrac{m^2+nm}{m}$

$= \dfrac{(m-n)(m+n) \cdot m(m+n)}{(m+n) \cdot m}$

$= (m-n)(m+n)$

$= m^2-n^2$

**27.** $\dfrac{x+2}{7-x} \div \dfrac{x^2-5x+6}{x^2-9x+14} = \dfrac{x+2}{7-x} \cdot \dfrac{x^2-9x+14}{x^2-5x+6}$

$= \dfrac{(x+2) \cdot (x-7)(x-2)}{-1(x-7) \cdot (x-3)(x-2)}$

$= -\dfrac{x+2}{x-3}$

**29.** $\dfrac{x^2+7x+10}{x-1} \div \dfrac{x^2+2x-15}{x-1}$

$= \dfrac{x^2+7x+10}{x-1} \cdot \dfrac{x-1}{x^2+2x-15}$

$= \dfrac{(x+5)(x+2) \cdot (x-1)}{(x-1) \cdot (x+5)(x-3)}$

$= \dfrac{x+2}{x-3}$

**31.** $\dfrac{5x-10}{12} \div \dfrac{4x-8}{8} = \dfrac{5x-10}{12} \cdot \dfrac{8}{4x-8}$

$= \dfrac{5(x-2) \cdot 2 \cdot 4}{6 \cdot 2 \cdot 4(x-2)}$

$= \dfrac{5}{6}$

**33.** $\dfrac{x^2+5x}{8} \cdot \dfrac{9}{3x+15} = \dfrac{x(x+5) \cdot 3 \cdot 3}{8 \cdot 3(x+5)} = \dfrac{3x}{8}$

**35.**
$$\frac{7}{6p^2+q} \div \frac{14}{18p^2+3q} = \frac{7}{6p^2+q} \cdot \frac{18p^2+3q}{14}$$
$$= \frac{7 \cdot 3(6p^2+q)}{(6p^2+q) \cdot 7 \cdot 2}$$
$$= \frac{3}{2}$$

**37.**
$$\frac{3x+4y}{x^2+4xy+4y^2} \cdot \frac{x+2y}{2} = \frac{(3x+4y) \cdot (x+2y)}{(x+2y)(x+2y) \cdot 2}$$
$$= \frac{3x+4y}{2(x+2y)}$$

**39.**
$$\frac{(x+2)^2}{x-2} \div \frac{x^2-4}{2x-4} = \frac{(x+2)^2}{x-2} \cdot \frac{2x-4}{x^2-4}$$
$$= \frac{(x+2)(x+2) \cdot 2(x-2)}{(x-2) \cdot (x+2)(x-2)}$$
$$= \frac{2(x+2)}{x-2}$$

**41.**
$$\frac{x^2-4}{24x} \div \frac{2-x}{6xy} = \frac{x^2-4}{24x} \cdot \frac{6xy}{2-x}$$
$$= \frac{(x+2)(x-2) \cdot 6x \cdot y}{4 \cdot 6x \cdot (-1)(x-2)}$$
$$= -\frac{y(x+2)}{4}$$

**43.**
$$\frac{a^2+7a+12}{a^2+5a+6} \cdot \frac{a^2+8a+15}{a^2+5a+4}$$
$$= \frac{(a+3)(a+4) \cdot (a+5)(a+3)}{(a+3)(a+2) \cdot (a+4)(a+1)}$$
$$= \frac{(a+5)(a+3)}{(a+2)(a+1)}$$

**45.**
$$\frac{5x-20}{3x^2+x} \cdot \frac{3x^2+13x+4}{x^2-16}$$
$$= \frac{5(x-4)}{x(3x+1)} \cdot \frac{(3x+1)(x+4)}{(x+4)(x-4)}$$
$$= \frac{5(x-4) \cdot (3x+1)(x+4)}{x(3x+1) \cdot (x+4)(x-4)}$$
$$= \frac{5}{x}$$

**47.**
$$\frac{8n^2-18}{2n^2-5n+3} \div \frac{6n^2+7n-3}{n^2-9n+8}$$
$$= \frac{8n^2-18}{2n^2-5n+3} \cdot \frac{n^2-9n+8}{6n^2+7n-3}$$
$$= \frac{2(2n+3)(2n-3) \cdot (n-8)(n-1)}{(n-1)(2n-3) \cdot (2n+3)(3n-1)}$$
$$= \frac{2(n-8)}{3n-1}$$

**49.**
$$\frac{x^2-9}{2x} \div \frac{x+3}{8x^4} = \frac{x^2-9}{2x} \cdot \frac{8x^4}{x+3}$$
$$= \frac{(x+3)(x-3)}{2x} \cdot \frac{8x^4}{x+3}$$
$$= \frac{2x \cdot 4x^3 \cdot (x+3)(x-3)}{2x \cdot (x+3)}$$
$$= 4x^3(x-3)$$

**51.**
$$\frac{a^2+ac+ba+bc}{a-b} \div \frac{a+c}{a+b}$$
$$= \frac{a(a+c)+b(a+c)}{a-b} \cdot \frac{a+b}{a+c}$$
$$= \frac{(a+c)(a+b)}{a-b} \cdot \frac{a+b}{a+c}$$
$$= \frac{(a+c) \cdot (a+b) \cdot (a+b)}{(a-b) \cdot (a+c)}$$
$$= \frac{(a+b)^2}{a-b}$$

**53.**
$$\frac{3x^2+8x+5}{x^2+8x+7} \cdot \frac{x+7}{x^2+4} = \frac{(3x+5)(x+1)}{(x+7)(x+1)} \cdot \frac{x+7}{x^2+4}$$
$$= \frac{(3x+5) \cdot (x+1) \cdot (x+7)}{(x+7) \cdot (x+1) \cdot (x^2+4)}$$
$$= \frac{3x+5}{x^2+4}$$

**55.**
$$\frac{x^3+8}{x^2-2x+4} \cdot \frac{4}{x^2-4}$$
$$= \frac{(x+2)(x^2-2x+4)}{x^2-2x+4} \cdot \frac{4}{(x+2)(x-2)}$$
$$= \frac{4 \cdot (x+2) \cdot (x^2-2x+4)}{(x+2)(x-2)(x^2-2x+4)}$$
$$= \frac{4}{x-2}$$

**57.** $\dfrac{a^2-ab}{6a^2+6ab} \div \dfrac{a^3-b^3}{a^2-b^2}$

$= \dfrac{a^2-ab}{6a^2+6ab} \cdot \dfrac{a^2-b^2}{a^3-b^3}$

$= \dfrac{a(a-b)}{6a(a+b)} \cdot \dfrac{(a-b)(a+b)}{(a-b)(a^2+ab+b^2)}$

$= \dfrac{a\cdot(a-b)\cdot(a-b)\cdot(a+b)}{6\cdot a\cdot(a+b)\cdot(a-b)\cdot(a^2+ab+b^2)}$

$= \dfrac{a-b}{6(a^2+ab+b^2)}$

**59.** 10 square feet

$= \dfrac{10 \text{ square feet}}{1} \cdot \dfrac{144 \text{ square inches}}{1 \text{ square foot}}$

$= 1440$ square inches

**61.** 45 square feet $= \dfrac{45 \text{ square feet}}{1} \cdot \dfrac{1 \text{ square yard}}{9 \text{ square feet}}$

$= 5$ square yards

**63.** 3 cubic yards $= \dfrac{3 \text{ cubic yards}}{1} \cdot \dfrac{27 \text{ cubic feet}}{1 \text{ cubic yard}}$

$= 81$ cubic feet

**65.** $\dfrac{50 \text{ miles}}{1 \text{ hour}} = \dfrac{50 \text{ miles}}{1 \text{ hour}} \cdot \dfrac{5280 \text{ feet}}{1 \text{ mile}} \cdot \dfrac{1 \text{ hour}}{3600 \text{ seconds}}$

$\approx 73$ feet per second

**67.** 6.3 square yards

$= \dfrac{6.3 \text{ square yards}}{1} \cdot \dfrac{9 \text{ square feet}}{1 \text{ square yard}}$

$= 56.7$ square feet

**69.** 133,500 square yards

$= \dfrac{133,500 \text{ square yards}}{1} \cdot \dfrac{9 \text{ square feet}}{1 \text{ square yard}}$

$= 1,201,500$ square feet

**71.** 359.2 feet per second

$= \dfrac{359.2 \text{ feet}}{1 \text{ second}} \cdot \dfrac{3600 \text{ seconds}}{1 \text{ hour}} \cdot \dfrac{1 \text{ mile}}{5280 \text{ feet}}$

$\approx 244.9$ miles per hour

**73.** $\dfrac{1}{5} + \dfrac{4}{5} = \dfrac{5}{5} = 1$

**75.** $\dfrac{9}{9} - \dfrac{19}{9} = -\dfrac{10}{9}$

**77.** $\dfrac{6}{5} + \left(\dfrac{1}{5} - \dfrac{8}{5}\right) = \dfrac{6}{5} + \left(-\dfrac{7}{5}\right) = -\dfrac{1}{5}$

**79.** $x - 2y = 6$

| $x$ | $y$ |
|-----|-----|
| 0 | –3 |
| 6 | 0 |

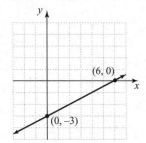

**81.** $\dfrac{4}{a} \cdot \dfrac{1}{b} = \dfrac{4\cdot 1}{a\cdot b} = \dfrac{4}{ab}$

The statement is true.

**83.** $\dfrac{x}{5} \cdot \dfrac{x+3}{4} = \dfrac{x\cdot(x+3)}{5\cdot 4} = \dfrac{x^2+3x}{20}$

The statement is false.

**85.** Area = length · width

$\dfrac{x+5}{9x} \cdot \dfrac{2x}{x^2-25} = \dfrac{(x+5)\cdot 2\cdot x}{9\cdot x\cdot(x+5)(x-5)} = \dfrac{2}{9(x-5)}$

The area of the rectangle is $\dfrac{2}{9(x-5)}$ square feet.

**87.** $\left(\dfrac{x^2-y^2}{x^2+y^2} \div \dfrac{x^2-y^2}{3x}\right) \cdot \dfrac{x^2+y^2}{6}$

$= \dfrac{x^2-y^2}{x^2+y^2} \cdot \dfrac{3x}{x^2-y^2} \cdot \dfrac{x^2+y^2}{6}$

$= \dfrac{(x^2-y^2)\cdot 3x\cdot(x^2+y^2)}{(x^2+y^2)\cdot(x^2-y^2)\cdot 2\cdot 3}$

$= \dfrac{x}{2}$

**89.** $\left(\dfrac{2a+b}{b^2} \cdot \dfrac{3a^2-2ab}{ab+2b^2}\right) \div \dfrac{a^2-3ab+2b^2}{5ab-10b^2}$

$= \dfrac{2a+b}{b^2} \cdot \dfrac{3a^2-2ab}{ab+2b^2} \cdot \dfrac{5ab-10b^2}{a^2-3ab+2b^2}$

$= \dfrac{(2a+b)\cdot(3a^2-2ab)\cdot(5ab-10b^2)}{b^2\cdot(ab+2b^2)\cdot(a^2-3ab+2b^2)}$

$= \dfrac{(2a+b)\cdot a(3a-2b)\cdot 5b(a-2b)}{b^2\cdot b(a+2b)\cdot(a-2b)(a-b)}$

$= \dfrac{5a(2a+b)(3a-2b)}{b^2(a+2b)(a-b)}$

**91.** answers may vary

**Section 7.3 Practice**

**1.** $\dfrac{7a}{4b} + \dfrac{a}{4b} = \dfrac{7a+a}{4b} = \dfrac{8a}{4b} = \dfrac{2a}{b}$

**2.** $\dfrac{3x}{3x-2} - \dfrac{2}{3x-2} = \dfrac{3x-2}{3x-2} = \dfrac{1}{1}$ or 1

**3.** $\dfrac{4x^2+15x}{x+3} - \dfrac{8x+15}{x+3} = \dfrac{(4x^2+15x)-(8x+15)}{x+3}$

$= \dfrac{4x^2+15x-8x-15}{x+3}$

$= \dfrac{4x^2+7x-15}{x+3}$

$= \dfrac{(x+3)(4x-5)}{x+3}$

$= 4x-5$

**4. a.** Find the prime factorization of each denominator.
$14 = 2\cdot 7$
$21 = 3\cdot 7$
The greatest number of times that the factor 2 appears is 1. The greatest number of times that the factor 3 appears is 1. The greatest number of times that the factor 7 appears is 1.
$\text{LCD} = 2^1\cdot 3^1\cdot 7^1 = 42$

**b.** Factor each denominator.
$9y = 3\cdot 3\cdot y = 3^2\cdot y$
$15y^3 = 3\cdot 5\cdot y^3$
The greatest number of times that the factor 3 appears is 2. The greatest number of times that the factor 5 appears is 1. The greatest

number of times that the factor $y$ appears is 3.
$\text{LCD} = 3^2\cdot 5^1\cdot y^3 = 9\cdot 5\cdot y^3 = 45y^3$

**5. a.** The denominators $y-5$ and $y-4$ are completely factored already. The factor $y-5$ appears once and the factor $y-4$ appears once.
$\text{LCD} = (y-5)(y-4)$

**b.** The denominators $a$ and $a+2$ cannot be factored further. The factor $a$ appears once and the factor $a+2$ appears once.
$\text{LCD} = a(a+2)$

**6.** Factor each denominator.
$(2x-1)^2 = (2x-1)^2$
$6x-3 = 3(2x-1)$
The greatest number that the factor $2x-1$ appears in any one denominator is 2. The greatest number of times that the factor 3 appears is 1.
$\text{LCD} = 3(2x-1)^2$

**7.** Factor each denominator.
$x^2+5x+4 = (x+1)(x+4)$
$x^2-16 = (x-4)(x+4)$
$\text{LCD} = (x+1)(x+4)(x-4)$

**8.** The denominators $3-x$ and $x-3$ are opposites. That is, $3-x = -1(x-3)$. Use $x-3$ or $3-x$ as the LCD.
$\text{LCD} = x-3$ or $\text{LCD} = 3-x$

**9. a.** Since $5y(7xy) = 35xy^2$, multiply by 1 in the form of $\dfrac{7xy}{7xy}$.

$\dfrac{3x}{5y} = \dfrac{3x}{5y}\cdot 1 = \dfrac{3x}{5y}\cdot\dfrac{7xy}{7xy} = \dfrac{3x(7xy)}{5y(7xy)} = \dfrac{21x^2y}{35xy^2}$

**b.** First, factor the denominator on the right.
$\dfrac{9x}{4x+7} = \dfrac{}{2(4x+7)}$
To obtain the denominator on the right from the denominator on the left, multiply by 1 in the form of $\dfrac{2}{2}$.

$$\frac{9x}{4x+7} = \frac{9x}{4x+7} \cdot \frac{2}{2}$$

$$= \frac{9x \cdot 2}{(4x+7) \cdot 2}$$

$$= \frac{18x}{2(4x+7)} \text{ or } \frac{18x}{8x+14}$$

**10.** First, factor the denominator $x^2 - 2x - 15$ as $(x+3)(x-5)$. If we multiply the original denominator $(x+3)(x-5)$ by $x-2$, the result is the new denominator $(x-2)(x+3)(x-5)$. Thus, we multiply by 1 in the form $\frac{x-2}{x-2}$.

$$\frac{3}{x^2-2x-15} = \frac{3}{(x+3)(x-5)}$$

$$= \frac{3}{(x+3)(x-5)} \cdot \frac{x-2}{x-2}$$

$$= \frac{3(x-2)}{(x+3)(x-5)(x-2)}$$

$$= \frac{3x-6}{(x-2)(x+30(x-5)}$$

**Vocabulary, Readiness & Video Check 7.3**

**1.** $\dfrac{7}{11} + \dfrac{2}{11} = \dfrac{9}{\underline{11}}$

**2.** $\dfrac{7}{11} - \dfrac{2}{11} = \dfrac{5}{\underline{11}}$

**3.** $\dfrac{a}{b} + \dfrac{c}{b} = \dfrac{a+c}{\underline{b}}$

**4.** $\dfrac{a}{b} - \dfrac{c}{b} = \dfrac{a-c}{\underline{b}}$

**5.** $\dfrac{5}{x} - \dfrac{6+x}{x} = \dfrac{5-(6+x)}{\underline{x}}$

**6.** In order to carry out the subtraction properly—parentheses make sure each term in the numerator is affected by the subtraction and not just the first term.

**7.** We factor denominators into the smallest factors—including coefficients—so we can determine the most number of times each unique factor occurs in any one denominator for the LCD.

**8.** To write an equivalent rational expression, you multiply the <u>numerator</u> of a rational expression by the same expression as the denominator. This means you're multiplying the original rational expression by a factor of <u>one</u> and therefore not changing the <u>value</u> of the original expression.

**Exercise Set 7.3**

**1.** $\dfrac{a+1}{13} + \dfrac{8}{13} = \dfrac{a+1+8}{13} = \dfrac{a+9}{13}$

**3.** $\dfrac{4m}{3n} + \dfrac{5m}{3n} = \dfrac{4m+5m}{3n} = \dfrac{9m}{3n} = \dfrac{3m}{n}$

**5.** $\dfrac{4m}{m-6} - \dfrac{24}{m-6} = \dfrac{4m-24}{m-6} = \dfrac{4(m-6)}{m-6} = 4$

**7.** $\dfrac{9}{3+y} + \dfrac{y+1}{3+y} = \dfrac{9+y+1}{3+y} = \dfrac{10+y}{3+y}$

**9.** $\dfrac{5x^2+4x}{x-1} - \dfrac{6x+3}{x-1} = \dfrac{5x^2+4x-(6x+3)}{x-1}$

$$= \frac{5x^2+4x-6x-3}{x-1}$$

$$= \frac{5x^2-2x-3}{x-1}$$

$$= \frac{(5x+3)(x-1)}{x-1}$$

$$= 5x+3$$

**11.** $\dfrac{4a}{a^2+2a-15} - \dfrac{12}{a^2+2a-15} = \dfrac{4a-12}{a^2+2a-15}$

$$= \frac{4(a-3)}{(a+5)(a-3)}$$

$$= \frac{4}{a+5}$$

**13.** $\dfrac{2x+3}{x^2-x-30} - \dfrac{x-2}{x^2-x-30} = \dfrac{2x+3-(x-2)}{x^2-x-30}$

$$= \frac{2x+3-x+2}{x^2-x-30}$$

$$= \frac{x+5}{x^2-x-30}$$

$$= \frac{x+5}{(x-6)(x+5)}$$

$$= \frac{1}{x-6}$$

**15.** $\dfrac{2x+1}{x-3}+\dfrac{3x+6}{x-3}=\dfrac{2x+1+3x+6}{x-3}=\dfrac{5x+7}{x-3}$

**17.** $\begin{aligned}\dfrac{2x^2}{x-5}-\dfrac{25+x^2}{x-5}&=\dfrac{2x^2-(25+x^2)}{x-5}\\&=\dfrac{2x^2-25-x^2}{x-5}\\&=\dfrac{x^2-25}{x-5}\\&=\dfrac{(x+5)(x-5)}{x-5}\\&=x+5\end{aligned}$

**19.** $\begin{aligned}\dfrac{5x+4}{x-1}-\dfrac{2x+7}{x-1}&=\dfrac{5x+4-(2x+7)}{x-1}\\&=\dfrac{5x+4-2x-7}{x-1}\\&=\dfrac{3x-3}{x-1}\\&=\dfrac{3(x-1)}{x-1}\\&=3\end{aligned}$

**21.** $\begin{aligned}2x&=2\cdot x\\4x^3&=2^2\cdot x^3\\ \text{LCD}&=2^2\cdot x^3=4x^3\end{aligned}$

**23.** $\begin{aligned}8x&=2^3\cdot x\\2x+4&=2(x+2)\\ \text{LCD}&=2^3\cdot x\cdot(x+2)=8x(x+2)\end{aligned}$

**25.** $\begin{aligned}x+3&=x+3\\x-2&=x-2\\ \text{LCD}&=(x+3)(x-2)\end{aligned}$

**27.** $\begin{aligned}x+6&=x+6\\3x+18&=3(x+6)\\ \text{LCD}&=3(x+6)\end{aligned}$

**29.** $\begin{aligned}(x-6)^2&=(x-6)^2\\5x-30&=5(x-6)\\ \text{LCD}&=5(x-6)^2\end{aligned}$

**31.** $\begin{aligned}3x+3&=3\cdot(x+1)\\2x^2+4x+2&=2(x^2+2x+1)=2\cdot(x+1)^2\\ \text{LCD}&=2\cdot3(x+1)^2=6(x+1)^2\end{aligned}$

**33.** $\begin{aligned}x-8&=x-8\\8-x&=-(x-8)\\ \text{LCD}&=x-8\text{ or }8-x\end{aligned}$

**35.** $\begin{aligned}x^2+3x-4&=(x-1)(x+4)\\x^2+2x-3&=(x-1)(x+3)\\ \text{LCD}&=(x-1)(x+4)(x+3)\end{aligned}$

**37.** $\begin{aligned}3x^2+4x+1&=(3x+1)(x+1)\\2x^2-x-1&=(x-1)(2x+1)\\ \text{LCD}&=(3x+1)(x+1)(x-1)(2x+1)\end{aligned}$

**39.** $\begin{aligned}x^2-16&=(x+4)(x-4)\\2x^3-8x^2&=2x^2(x-4)\\ \text{LCD}&=2x^2(x+4)(x-4)\end{aligned}$

**41.** $\dfrac{3}{2x}=\dfrac{3(2x)}{2x(2x)}=\dfrac{6x}{4x^2}$

**43.** $\dfrac{6}{3a}=\dfrac{6(4b^2)}{3a(4b^2)}=\dfrac{24b^2}{12ab^2}$

**45.** $\dfrac{9}{2x+6}=\dfrac{9}{2(x+3)}=\dfrac{9(y)}{2(x+3)(y)}=\dfrac{9y}{2y(x+3)}$

**47.** $\dfrac{9a+2}{5a+10}=\dfrac{9a+2}{5(a+2)}=\dfrac{(9a+2)(b)}{5(a+2)(b)}=\dfrac{9ab+2b}{5b(a+2)}$

**49.** $\begin{aligned}\dfrac{x}{x^3+6x^2+8x}&=\dfrac{x}{x(x+4)(x+2)}\\&=\dfrac{x(x+1)}{x(x+4)(x+2)(x+1)}\\&=\dfrac{x^2+x}{x(x+4)(x+2)(x+1)}\end{aligned}$

**51.** $\dfrac{9y-1}{15x^2-30}=\dfrac{(9y-1)(2)}{(15x^2-30)2}=\dfrac{18y-2}{30x^2-60}$

**53.** $\dfrac{5x}{7}+\dfrac{9x}{7}=\dfrac{5x+9x}{7}=\dfrac{14x}{7}=\dfrac{2x}{1}=2x$

    

**55.** $\dfrac{x+3}{4} \div \dfrac{2x-1}{4} = \dfrac{x+3}{4} \cdot \dfrac{4}{2x-1}$

$= \dfrac{(x+3) \cdot 4}{4 \cdot (2x-1)}$

$= \dfrac{x+3}{2x-1}$

**57.** $\dfrac{x^2}{x-6} - \dfrac{5x+6}{x-6} = \dfrac{x^2 - (5x+6)}{x-6}$

$= \dfrac{x^2 - 5x - 6}{x-6}$

$= \dfrac{(x+1)(x-6)}{x-6}$

$= x+1$

**59.** $\dfrac{-2x}{x^3 - 8x} + \dfrac{3x}{x^3 - 8x} = \dfrac{-2x + 3x}{x^3 - 8x}$

$= \dfrac{x}{x(x^2 - 8)}$

$= \dfrac{1}{x^2 - 8}$

**61.** $\dfrac{12x-6}{x^2 + 3x} \cdot \dfrac{4x^2 + 13x + 3}{4x^2 - 1}$

$= \dfrac{6(2x-1) \cdot (x+3)(4x+1)}{x(x+3) \cdot (2x+1)(2x-1)}$

$= \dfrac{6(4x+1)}{x(2x+1)}$

**63.** LCD = 21

$\dfrac{2}{3} + \dfrac{5}{7} = \dfrac{2(7)}{3(7)} + \dfrac{5(3)}{7(3)} = \dfrac{14}{21} + \dfrac{15}{21} = \dfrac{29}{21}$

**65.** $6 = 2 \cdot 3$

$4 = 2^2$

LCD $= 2^2 \cdot 3 = 12$

$\dfrac{1}{6} - \dfrac{3}{4} = \dfrac{1(2)}{6(2)} - \dfrac{3(3)}{4(3)} = \dfrac{2}{12} - \dfrac{9}{12} = \dfrac{2-9}{12} = -\dfrac{7}{12}$

**67.** $12 = 2 \cdot 2 \cdot 3 = 2^2 \cdot 3$

$20 = 2 \cdot 2 \cdot 5 = 2^2 \cdot 5$

LCD $= 2^2 \cdot 3 \cdot 5 = 60$

$\dfrac{1}{12} + \dfrac{3}{20} = \dfrac{1(5)}{12(5)} + \dfrac{3(3)}{20(3)} = \dfrac{5}{60} + \dfrac{9}{60} = \dfrac{14}{60} = \dfrac{7}{30}$

**69.** $4a - 20 = 4(a - 5)$

$(a-5)^2 = (a-5)^2$

LCD $= 4(a-5)^2$

The correct choice is d.

**71.** answers may vary

**73.** $\dfrac{3}{x} + \dfrac{y}{x} = \dfrac{3+y}{x}$

The correct choice is c.

**75.** $\dfrac{3}{x} \cdot \dfrac{y}{x} = \dfrac{3 \cdot y}{x \cdot x} = \dfrac{3y}{x^2}$

The correct choice is b.

**77.** $\dfrac{5}{2-x} = \dfrac{5(-1)}{(2-x)(-1)} = -\dfrac{5}{x-2}$

**79.** $-\dfrac{7+x}{2-x} = \dfrac{7+x}{(-1)(2-x)} = \dfrac{7+x}{x-2}$

**81.** $P = \dfrac{5}{x-2} + \dfrac{5}{x-2} + \dfrac{5}{x-2} + \dfrac{5}{x-2}$

$= \dfrac{5+5+5+5}{x-2}$

$= \dfrac{20}{x-2}$

The perimeter is $\dfrac{20}{x-2}$ meters.

**83.** answers may vary

**85.** $88 = 2^3 \cdot 11$

$4332 = 2^3 \cdot 3 \cdot 19^2$

LCM $= 2^3 \cdot 3 \cdot 11 \cdot 19^2 = 95,304$

They will align again in 95,304 Earth days.

**87.** answers may vary

**89.** answers may vary

**Section 7.4 Practice**

**1. a.** Since $5 = 5$ and $15 = 3 \cdot 5$, the LCD $= 3 \cdot 5 = 15$.

$$\frac{2x}{5} - \frac{6x}{15} = \frac{2x(3)}{5(3)} - \frac{6x}{15}$$
$$= \frac{6x}{15} - \frac{6x}{15}$$
$$= \frac{6x - 6x}{15}$$
$$= \frac{0}{15}$$
$$= 0$$

**b.** Since $8a = 2^3 \cdot a$ and $12a^2 = 2^2 \cdot 3 \cdot a^2$, the LCD $= 2^3 \cdot 3 \cdot a^2 = 24a^2$.

$$\frac{7}{8a} + \frac{5}{12a^2} = \frac{7(3a)}{8a(3a)} + \frac{5(2)}{12a^2(2)}$$
$$= \frac{21a}{24a^2} + \frac{10}{24a^2}$$
$$= \frac{21a + 10}{24a^2}$$

**2.** Since $x^2 - 25 = (x+5)(x-5)$, the LCD $= (x+5)(x-5)$.

$$\frac{12x}{x^2 - 25} - \frac{6}{x+5} = \frac{12x}{(x+5)(x-5)} - \frac{6(x-5)}{(x+5)(x-5)}$$
$$= \frac{12x - 6(x-5)}{(x+5)(x-5)}$$
$$= \frac{12x - 6x + 30}{(x+5)(x-5)}$$
$$= \frac{6x + 30}{(x+5)(x-5)}$$
$$= \frac{6(x+5)}{(x+5)(x-5)}$$
$$= \frac{6}{x-5}$$

**3.** The LCD is $5y(y + 1)$.

$$\frac{3}{5y} + \frac{2}{y+1} = \frac{3(y+1)}{5y(y+1)} + \frac{2(5y)}{(y+1)(5y)}$$
$$= \frac{3(y+1) + 2(5y)}{5y(y+1)}$$
$$= \frac{3y + 3 + 10y}{5y(y+1)}$$
$$= \frac{13y + 3}{5y(y+1)}$$

**4.** $x - 5$ and $5 - x$ are opposites. Write the denominator $5 - x$ as $-(x - 5)$ and simplify.

$$\frac{6}{x-5} - \frac{7}{5-x} = \frac{6}{x-5} - \frac{7}{-(x-5)}$$
$$= \frac{6}{x-5} - \frac{-7}{x-5}$$
$$= \frac{6 - (-7)}{x-5}$$
$$= \frac{13}{x-5}$$

**5.** Note that 2 is the same as $\frac{2}{1}$. The LCD of $\frac{2}{1}$ and $\frac{b}{b+3}$ is $b + 3$.

$$2 + \frac{b}{b+3} = \frac{2}{1} + \frac{b}{b+3}$$
$$= \frac{2(b+3)}{1(b+3)} + \frac{b}{b+3}$$
$$= \frac{2(b+3) + b}{b+3}$$
$$= \frac{2b + 6 + b}{b+3}$$
$$= \frac{3b + 6}{b+3} \text{ or } \frac{3(b+2)}{b+3}$$

**6.** First, factor the denominators.

$$\frac{5}{2x^2 + 3x} - \frac{3x}{4x+6} = \frac{5}{x(2x+3)} - \frac{3x}{2(2x+3)}$$

The LCD is $2x(2x + 3)$.

$$\frac{5}{2x^2 + 3x} - \frac{3x}{4x+6} = \frac{5(2)}{x(2x+3)(2)} - \frac{3x(x)}{2(2x+3)(x)}$$
$$= \frac{10 - 3x^2}{2x(2x+3)}$$

    

**7.** First, factor the denominators.

$$x^2 + 7x + 12 = (x+4)(x+3)$$

$$x^2 - 9 = (x+3)(x-3)$$

$$\text{LCD} = (x+4)(x+3)(x-3)$$

$$\frac{2x}{x^2+7x+12} + \frac{3x}{x^2-9}$$

$$= \frac{2x}{(x+4)(x+3)} + \frac{3x}{(x+3)(x-3)}$$

$$= \frac{2x(x-3)}{(x+4)(x+3)(x-3)} + \frac{3x(x+4)}{(x+3)(x-3)(x+4)}$$

$$= \frac{2x(x-3)+3x(x+4)}{(x+4)(x+3)(x-3)}$$

$$= \frac{2x^2-6x+3x^2+12x}{(x+4)(x+3)(x-3)}$$

$$= \frac{5x^2+6x}{(x+4)(x+3)(x-3)} \text{ or } \frac{x(5x+6)}{(x+4)(x+3)(x-3)}$$

**Vocabulary, Readiness & Video Check 7.4**

**1.** The first step to perform on $\frac{3}{4} - \frac{y}{4}$ is to subtract the numerators and place the difference over the common denominator; choice d.

**2.** The first step to perform on $\frac{2}{a} \cdot \frac{3}{a+6}$ is to multiply the numerators and multiply the denominators; choice c.

**3.** The first step to perform on $\frac{x+1}{x} \div \frac{x-1}{x}$ is to multiply the first rational expression by the reciprocal of the second rational expression; choice a.

**4.** The first step to perform on $\frac{9}{x-2} - \frac{x}{x+2}$ is to find the LCD and write each expression as an equivalent expression with the LCD as denominator; choice b.

**5.** The problem adds two rational expressions with denominators that are opposites of each other. Recognizing this special case can save you time and effort. If you recognize that one denominator is $-1$ times the other denominator, you may save time.

**Exercise Set 7.4**

**1.** $\text{LCD} = 2 \cdot 3 \cdot x = 6x$

$$\frac{4}{2x} + \frac{9}{3x} = \frac{4(3)}{2x(3)} + \frac{9(2)}{3x(2)}$$

$$= \frac{12}{6x} + \frac{18}{6x}$$

$$= \frac{30}{6x}$$

$$= \frac{5(6)}{6x}$$

$$= \frac{5}{x}$$

**3.** $\text{LCD} = 5b$

$$\frac{15a}{b} - \frac{6b}{5} = \frac{15a(5)}{b(5)} - \frac{6b(b)}{5(b)}$$

$$= \frac{75a}{5b} - \frac{6b^2}{5b}$$

$$= \frac{75a-6b^2}{5b}$$

**5.** $\text{LCD} = 2x^2$

$$\frac{3}{x} + \frac{5}{2x^2} = \frac{3(2x)}{x(2x)} + \frac{5}{2x^2} = \frac{6x}{2x^2} + \frac{5}{2x^2} = \frac{6x+5}{2x^2}$$

**7.** $2x + 2 = 2(x+1)$

$\text{LCD} = 2(x+1)$

$$\frac{6}{x+1} + \frac{10}{2x+2} = \frac{6}{x+1} + \frac{10}{2(x+1)}$$

$$= \frac{6(2)}{(x+1)2} + \frac{10}{2(x+1)}$$

$$= \frac{12}{2(x+1)} + \frac{10}{2(x+1)}$$

$$= \frac{12+10}{2(x+1)}$$

$$= \frac{22}{2(x+1)}$$

$$= \frac{2(11)}{2(x+1)}$$

$$= \frac{11}{x+1}$$

**9.** $x^2 - 4 = (x+2)(x-2)$

LCD $= (x+2)(x-2)$

$$\frac{3}{x+2} - \frac{2x}{x^2-4} = \frac{3(x-2)}{(x+2)(x-2)} - \frac{2x}{(x+2)(x-2)}$$
$$= \frac{3(x-2)-2x}{(x+2)(x-2)}$$
$$= \frac{3x-6-2x}{(x+2)(x-2)}$$
$$= \frac{x-6}{(x+2)(x-2)}$$

**11.** LCD $= 4x(x-2)$

$$\frac{3}{4x} + \frac{8}{x-2} = \frac{3(x-2)}{4x(x-2)} + \frac{8(4x)}{(x-2)(4x)}$$
$$= \frac{3x-6}{4x(x-2)} + \frac{32x}{4x(x-2)}$$
$$= \frac{3x-6+32x}{4x(x-2)}$$
$$= \frac{35x-6}{4x(x-2)}$$

**13.** $3 - x = -(x-3)$

$$\frac{6}{x-3} + \frac{8}{3-x} = \frac{6}{x-3} + \frac{8}{-(x-3)}$$
$$= \frac{6}{x-3} + \frac{-8}{x-3}$$
$$= \frac{6+(-8)}{x-3}$$
$$= -\frac{2}{x-3}$$

**15.** $3 - x = -(x-3)$

$$\frac{9}{x-3} + \frac{9}{3-x} = \frac{9}{x-3} + \frac{9}{-(x-3)}$$
$$= \frac{9}{x-3} + \frac{-9}{x-3}$$
$$= \frac{9+(-9)}{x-3}$$
$$= \frac{0}{x-3}$$
$$= 0$$

**17.** $1 - x^2 = -(x^2-1)$

$$\frac{-8}{x^2-1} - \frac{7}{1-x^2} = \frac{8}{-(x^2-1)} - \frac{7}{1-x^2}$$
$$= \frac{8}{1-x^2} - \frac{7}{1-x^2}$$
$$= \frac{8-7}{1-x^2}$$
$$= \frac{1}{1-x^2} \text{ or } -\frac{1}{x^2-1}$$

**19.** LCD $= x$

$$\frac{5}{x} + 2 = \frac{5}{x} + \frac{2}{1} = \frac{5}{x} + \frac{2(x)}{1(x)} = \frac{5+2x}{x}$$

**21.** LCD $= x - 2$

$$\frac{5}{x-2} + 6 = \frac{5}{x-2} + \frac{6}{1}$$
$$= \frac{5}{x-2} + \frac{6(x-2)}{1(x-2)}$$
$$= \frac{5}{x-2} + \frac{6x-12}{x-2}$$
$$= \frac{5+6x-12}{x-2}$$
$$= \frac{6x-7}{x-2}$$

**23.** LCD $= y + 3$

$$\frac{y+2}{y+3} - 2 = \frac{y+2}{y+3} - \frac{2}{1}$$
$$= \frac{y+2}{y+3} - \frac{2(y+3)}{y+3}$$
$$= \frac{y+2}{y+3} - \frac{2y+6}{y+3}$$
$$= \frac{y+2-(2y+6)}{y+3}$$
$$= \frac{y+2-2y-6}{y+3}$$
$$= \frac{-y-4}{y+3}$$
$$= \frac{-(y+4)}{y+3}$$
$$= -\frac{y+4}{y+3}$$

**25.** LCD $= 4x$

$$\frac{-x+2}{x}-\frac{x-6}{4x}=\frac{(-x+2)(4)}{x(4)}-\frac{x-6}{4x}$$
$$=\frac{4(-x+2)-(x-6)}{4x}$$
$$=\frac{-4x+8-x+6}{4x}$$
$$=\frac{-5x+14}{4x}\text{ or }-\frac{5x-14}{4x}$$

**27.** $\dfrac{5x}{x+2}-\dfrac{3x-4}{x+2}=\dfrac{5x-(3x-4)}{x+2}$
$$=\frac{5x-3x+4}{x+2}$$
$$=\frac{2x+4}{x+2}$$
$$=\frac{2(x+2)}{x+2}$$
$$=2$$

**29.** LCD $= 21$

$$\frac{3x^4}{7}-\frac{4x^2}{21}=\frac{3x^4(3)}{7(3)}-\frac{4x^2}{21}$$
$$=\frac{3(3x^4)-4x^2}{21}$$
$$=\frac{9x^4-4x^2}{21}$$

**31.** LCD $= (x+3)^2$

$$\frac{1}{x+3}-\frac{1}{(x+3)^2}=\frac{1(x+3)}{(x+3)(x+3)}-\frac{1}{(x+3)^2}$$
$$=\frac{x+3}{(x+3)^2}-\frac{1}{(x+3)^2}$$
$$=\frac{x+3-1}{(x+3)^2}$$
$$=\frac{x+2}{(x+3)^2}$$

**33.** LCD $= 5b(b-1)$

$$\frac{4}{5b}+\frac{1}{b-1}=\frac{4(b-1)}{5b(b-1)}+\frac{1(5b)}{(b-1)(5b)}$$
$$=\frac{4b-4}{5b(b-1)}+\frac{5b}{5b(b-1)}$$
$$=\frac{4b-4+5b}{5b(b-1)}$$
$$=\frac{9b-4}{5b(b-1)}$$

**35.** LCD $= m$

$$\frac{2}{m}+1=\frac{2}{m}+\frac{1}{1}=\frac{2}{m}+\frac{1(m)}{1(m)}=\frac{2+m}{m}$$

**37.** LCD $= (x-7)(x-2)$

$$\frac{2x}{x-7}-\frac{x}{x-2}=\frac{2x(x-2)}{(x-7)(x-2)}-\frac{x(x-7)}{(x-2)(x-7)}$$
$$=\frac{2x(x-2)-x(x-7)}{(x-7)(x-2)}$$
$$=\frac{2x^2-4x-x^2+7x}{(x-7)(x-2)}$$
$$=\frac{x^2+3x}{(x-7)(x-2)}\text{ or }\frac{x(x+3)}{(x-7)(x-2)}$$

**39.** $2x-1=-(1-2x)$

$$\frac{6}{1-2x}-\frac{4}{2x-1}=\frac{6}{1-2x}-\frac{4}{-(1-2x)}$$
$$=\frac{6}{1-2x}-\frac{-4}{1-2x}$$
$$=\frac{6-(-4)}{1-2x}$$
$$=\frac{10}{1-2x}$$

**41.** LCD $= (x-1)(x+1)^2$

$$\frac{7}{(x+1)(x-1)}+\frac{8}{(x+1)^2}$$
$$=\frac{7(x+1)}{(x+1)(x-1)(x+1)}+\frac{8(x-1)}{(x+1)^2(x-1)}$$
$$=\frac{7x+7}{(x+1)^2(x-1)}+\frac{8x-8}{(x+1)^2(x-1)}$$
$$=\frac{7x+7+8x-8}{(x+1)^2(x-1)}$$
$$=\frac{15x-1}{(x+1)^2(x-1)}$$

**43.** $x^2 - 1 = (x+1)(x-1)$

$x^2 - 2x + 1 = (x-1)^2$

$\text{LCD} = (x+1)(x-1)^2$

$$\frac{x}{x^2-1} - \frac{2}{x^2-2x+1}$$

$$= \frac{x(x-1)}{(x-1)(x+1)(x-1)} - \frac{2(x+1)}{(x-1)^2(x+1)}$$

$$= \frac{x^2-x}{(x-1)^2(x+1)} - \frac{2x+2}{(x-1)^2(x+1)}$$

$$= \frac{x^2-x-(2x+2)}{(x-1)^2(x+1)}$$

$$= \frac{x^2-x-2x-2}{(x-1)^2(x+1)}$$

$$= \frac{x^2-3x-2}{(x-1)^2(x+1)}$$

**45.** $2a + 6 = 2(a+3)$

$\text{LCD} = 2(a+3)$

$$\frac{3a}{2a+6} - \frac{a-1}{a+3} = \frac{3a}{2(a+3)} - \frac{(a-1)(2)}{(a+3)(2)}$$

$$= \frac{3a}{2(a+3)} - \frac{2a-2}{2(a+3)}$$

$$= \frac{3a-(2a-2)}{2(a+3)}$$

$$= \frac{3a-2a+2}{2(a+3)}$$

$$= \frac{a+2}{2(a+3)}$$

**47.** $\text{LCD} = (2y+3)^2$

$$\frac{y-1}{2y+3} + \frac{3}{(2y+3)^2} = \frac{(y-1)(2y+3)}{(2y+3)(2y+3)} + \frac{3}{(2y+3)^2}$$

$$= \frac{(y-1)(2y+3)+3}{(2y+3)^2}$$

$$= \frac{2y^2+y-3+3}{(2y+3)^2}$$

$$= \frac{2y^2+y}{(2y+3)^2} \text{ or } \frac{y(2y+1)}{(2y+3)^2}$$

**49.** $2 - x = -(x-2)$

$2x - 4 = 2(x-2)$

$\text{LCD} = 2(x-2)$

$$\frac{5}{2-x} + \frac{x}{2x-4} = \frac{5}{-(x-2)} + \frac{x}{2(x-2)}$$

$$= \frac{-5}{x-2} + \frac{x}{2(x-2)}$$

$$= \frac{-5(2)}{(x-2)(2)} + \frac{x}{2(x-2)}$$

$$= \frac{-10}{2(x-2)} + \frac{x}{2(x-2)}$$

$$= \frac{x-10}{2(x-2)}$$

**51.** $x^2 + 6x + 9 = (x+3)^2$

$\text{LCD} = (x+3)^2$

$$\frac{15}{x^2+6x+9} + \frac{2}{x+3} = \frac{15}{(x+3)^2} + \frac{2(x+3)}{(x+3)(x+3)}$$

$$= \frac{15+2(x+3)}{(x+3)^2}$$

$$= \frac{15+2x+6}{(x+3)^2}$$

$$= \frac{2x+21}{(x+3)^2}$$

**53.** $x^2 - 5x - 6 = (x-3)(x-2)$

$\text{LCD} = (x-3)(x-2)$

$$\frac{13}{x^2-5x+6} - \frac{5}{x-3}$$

$$= \frac{13}{(x-3)(x-2)} - \frac{5(x-2)}{(x-3)(x-2)}$$

$$= \frac{13-(5x-10)}{(x-3)(x-2)}$$

$$= \frac{13-5x+10}{(x-3)(x-2)}$$

$$= \frac{-5x+23}{(x-3)(x-2)}$$

**55.** $m^2 - 100 = (m+10)(m-10)$

LCD $= 2(m+10)(m-10)$

$\dfrac{70}{m^2-100} + \dfrac{7}{2(m+10)}$

$= \dfrac{70(2)}{(m+10)(m-10)(2)} + \dfrac{7(m-10)}{2(m+10)(m-10)}$

$= \dfrac{70(2)+7(m-10)}{2(m+10)(m-10)}$

$= \dfrac{140+7m-70}{2(m+10)(m-10)}$

$= \dfrac{7m+70}{2(m+10)(m-10)}$

$= \dfrac{7(m+10)}{2(m+10)(m-10)}$

$= \dfrac{7}{2(m-10)}$

**57.** $x^2 - 5x - 6 = (x-6)(x+1)$

$x^2 - 4x - 5 = (x-5)(x+1)$

LCD $= (x-6)(x+1)(x-5)$

$\dfrac{x+8}{x^2-5x-6} + \dfrac{x+1}{x^2-4x-5}$

$= \dfrac{(x+8)(x-5)}{(x-6)(x+1)(x-5)} + \dfrac{(x+1)(x-6)}{(x-5)(x+1)(x-6)}$

$= \dfrac{x^2+3x-40+x^2-5x-6}{(x-6)(x+1)(x-5)}$

$= \dfrac{2x^2-2x-46}{(x-6)(x+1)(x-5)}$

or $\dfrac{2(x^2-x-23)}{(x-6)(x+1)(x-5)}$

**59.** $4n^2 - 12n + 8 = 4(n-1)(n-2)$

$3n^2 - 6n = 3n(n-2)$

LCD $= 4 \cdot 3n(n-1)(n-2) = 12n(n-1)(n-2)$

$\dfrac{5}{4n^2-12n+8} - \dfrac{3}{3n^2-6n}$

$= \dfrac{5(3n)}{4(n-1)(n-2)(3n)} - \dfrac{3(4)(n-1)}{3n(n-2)(4)(n-1)}$

$= \dfrac{5(3n)-3(4)(n-1)}{12n(n-1)(n-2)}$

$= \dfrac{15n-12n+12}{12n(n-1)(n-2)}$

$= \dfrac{3n+12}{12n(n-1)(n-2)}$

$= \dfrac{3(n+4)}{12n(n-1)(n-2)}$

$= \dfrac{n+4}{4n(n-1)(n-2)}$

**61.** $\dfrac{15x}{x+8} \cdot \dfrac{2x+16}{3x} = \dfrac{15x}{x+8} \cdot \dfrac{2(x+8)}{3x}$

$= \dfrac{2 \cdot 5 \cdot 3x \cdot (x+8)}{3x \cdot (x+8)}$

$= 10$

**63.** $\dfrac{8x+7}{3x+5} - \dfrac{2x-3}{3x+5} = \dfrac{8x+7-(2x-3)}{3x+5}$

$= \dfrac{8x+7-2x+3}{3x+5}$

$= \dfrac{6x+10}{3x+5}$

$= \dfrac{2(3x+5)}{3x+5}$

$= 2$

**65.** $\dfrac{5a+10}{18} \div \dfrac{a^2-4}{10a} = \dfrac{5a+10}{18} \cdot \dfrac{10a}{a^2-4}$

$= \dfrac{5(a+2) \cdot 2 \cdot 5a}{2 \cdot 9 \cdot (a-2)(a+2)}$

$= \dfrac{25a}{9(a-2)}$

**67.** $x^2 - 3x + 2 = (x-2)(x-1)$

LCD $= (x-2)(x-1)$

$$\frac{5}{x^2 - 3x + 2} + \frac{1}{x-2} = \frac{5}{(x-2)(x-1)} + \frac{1}{(x-2)} \cdot \frac{(x-1)}{(x-1)}$$

$$= \frac{5}{(x-2)(x-1)} + \frac{x-1}{(x-2)(x-1)}$$

$$= \frac{5+x-1}{(x-2)(x-1)}$$

$$= \frac{x+4}{(x-2)(x-1)}$$

**69.**
$$3x + 5 = 7$$
$$3x + 5 - 5 = 7 - 5$$
$$3x = 2$$
$$\frac{3x}{3} = \frac{2}{3}$$
$$x = \frac{2}{3}$$

**71.**
$$2x^2 - x - 1 = 0$$
$$(2x+1)(x-1) = 0$$
$$2x + 1 = 0 \quad \text{or} \quad x - 1 = 0$$
$$2x = -1 \qquad\qquad x = 1$$
$$x = -\frac{1}{2}$$

The solutions are $x = -\frac{1}{2}$ and $x = 1$.

**73.**
$$4(x+6) + 3 = -3$$
$$4x + 24 + 3 = -3$$
$$4x + 27 = -3$$
$$4x = -30$$
$$x = \frac{-30}{4} = -\frac{15}{2}$$

**75.** $x^2 - 1 = (x+1)(x-1)$

LCD $= x(x+1)(x-1)$

$$\frac{3}{x} - \frac{2x}{x^2 - 1} + \frac{5}{x+1} = \frac{3(x+1)(x-1)}{x(x+1)(x-1)} - \frac{2x(x)}{(x+1)(x-1)(x)} + \frac{5(x)(x-1)}{(x+1)(x)(x-1)}$$

$$= \frac{3(x+1)(x-1) - 2x(x) + 5x(x-1)}{x(x+1)(x-1)}$$

$$= \frac{3x^2 - 3 - 2x^2 + 5x^2 - 5x}{x(x+1)(x-1)}$$

$$= \frac{6x^2 - 5x - 3}{x(x+1)(x-1)}$$

**77.** $x^2 - 4 = (x+2)(x-2)$

$x^2 - 4x + 4 = (x-2)^2$

$x^2 - x - 6 = (x-3)(x+2)$

$LCD = (x+2)(x-2)^2(x-3)$

$$\frac{5}{x^2-4} + \frac{2}{x^2-4x+4} - \frac{3}{x^2-x-6} = \frac{5(x-2)(x-3)}{(x-2)(x+2)(x-2)(x-3)} + \frac{2(x+2)(x-3)}{(x-2)^2(x+2)(x-3)} - \frac{3(x-2)^2}{(x-3)(x+2)(x-2)^2}$$

$$= \frac{5(x^2-5x+6)}{(x-2)^2(x+2)(x-3)} + \frac{2(x^2-x-6)}{(x-2)^2(x+2)(x-3)} - \frac{3(x^2-4x+4)}{(x-2)^2(x+2)(x-3)}$$

$$= \frac{5x^2-25x+30}{(x-2)^2(x+2)(x-3)} + \frac{2x^2-2x-12}{(x-2)^2(x+2)(x-3)} - \frac{3x^2-12x+12}{(x-2)^2(x+2)(x-3)}$$

$$= \frac{5x^2-25x+30+2x^2-2x-12-3x^2+12x-12}{(x-2)^2(x+2)(x-3)}$$

$$= \frac{4x^2-15x+6}{(x-2)^2(x+2)(x-3)}$$

**79.** $x^2 + 9x + 14 = (x+2)(x+7)$

$x^2 + 10x + 21 = (x+3)(x+7)$

$x^2 + 5x + 6 = (x+2)(x+3)$

$LCD = (x+2)(x+7)(x+3)$

$$\frac{9}{x^2+9x+14} - \frac{3x}{x^2+10x+21} + \frac{x+4}{x^2+5x+6} = \frac{9(x+3)}{(x+2)(x+7)(x+3)} - \frac{3x(x+2)}{(x+3)(x+7)(x+2)} + \frac{(x+4)(x+7)}{(x+2)(x+3)(x+7)}$$

$$= \frac{9(x+3)-3x(x+2)+(x+4)(x+7)}{(x+2)(x+7)(x+3)}$$

$$= \frac{9x+27-3x^2-6x+x^2+11x+28}{(x+2)(x+7)(x+3)}$$

$$= \frac{-2x^2+14x+55}{(x+2)(x+7)(x+3)}$$

**81.** The length of the other board is $\left(\dfrac{3}{x+4} - \dfrac{1}{x-4}\right)$ inches.

$LCD = (x+4)(x-4)$

$$\frac{3}{x+4} - \frac{1}{x-4} = \frac{3(x-4)}{(x+4)(x-4)} - \frac{1(x+4)}{(x-4)(x+4)}$$

$$= \frac{3(x-4)-(x+4)}{(x+4)(x-4)}$$

$$= \frac{3x-12-x-4}{(x+4)(x-4)}$$

$$= \frac{2x-16}{(x+4)(x-4)}$$

The length of the other board is $\dfrac{2x-16}{(x+4)(x-4)}$ inches.

**83.** $1 - \dfrac{G}{P} = \dfrac{1}{1} - \dfrac{G}{P} = \dfrac{1(P)}{1(P)} - \dfrac{G}{P} = \dfrac{P-G}{P}$

**85.** answers may vary

**87.** $90° - \left(\dfrac{40}{x}\right)° = \left(90 - \dfrac{40}{x}\right)°$

LCD $= x$

$\left(90 \cdot \dfrac{x}{x} - \dfrac{40}{x}\right)° = \left(\dfrac{90x}{x} - \dfrac{40}{x}\right)° = \left(\dfrac{90x-40}{x}\right)°$

**89.** answers may vary

## Section 7.5 Practice

**1.** The LCD of 3, 5, and 15 is 15.

$$\dfrac{x}{3} + \dfrac{4}{5} = \dfrac{12}{5}$$

$$15\left(\dfrac{x}{3} + \dfrac{4}{5}\right) = 15\left(\dfrac{2}{15}\right)$$

$$15\left(\dfrac{x}{3}\right) + 15\left(\dfrac{4}{5}\right) = 15\left(\dfrac{2}{15}\right)$$

$$5 \cdot x + 12 = 2$$

$$5x = -10$$

$$x = -2$$

**Check:** $\dfrac{x}{3} + \dfrac{4}{5} = \dfrac{2}{15}$

$$\dfrac{-2}{3} + \dfrac{4}{5} \overset{?}{=} \dfrac{2}{15}$$

$$\dfrac{2}{15} = \dfrac{2}{15} \quad \text{True}$$

This number checks, so the solution is $-2$.

**2.** The LCD of 4, 3, and 12 is 12.

$$\dfrac{x+4}{4} - \dfrac{x-3}{3} = \dfrac{11}{12}$$

$$12\left(\dfrac{x+4}{4} - \dfrac{x-3}{3}\right) = 12\left(\dfrac{11}{12}\right)$$

$$12\left(\dfrac{x+4}{4}\right) - 12\left(\dfrac{x-3}{3}\right) = 12\left(\dfrac{11}{12}\right)$$

$$3(x+4) - 4(x-3) = 11$$

$$3x + 12 - 4x + 12 = 11$$

$$-x + 24 = 11$$

$$-x = -13$$

$$x = 13$$

**Check:** $\dfrac{x+4}{4} - \dfrac{x-3}{3} = \dfrac{11}{12}$

$$\dfrac{13+4}{4} - \dfrac{13-3}{3} \overset{?}{=} \dfrac{11}{12}$$

$$\dfrac{17}{4} - \dfrac{10}{3} \overset{?}{=} \dfrac{11}{12}$$

$$\dfrac{11}{12} = \dfrac{11}{12} \quad \text{True}$$

The solution is 13.

**3.** In this equation, 0 cannot be a solution. The LCD is $x$.

$$8 + \dfrac{7}{x} = x + 2$$

$$x\left(8 + \dfrac{7}{x}\right) = x(x+2)$$

$$x(8) + x\left(\dfrac{7}{x}\right) = x \cdot x + x \cdot 2$$

$$8x + 7 = x^2 + 2x$$

$$0 = x^2 - 6x - 7$$

$$0 = (x+1)(x-7)$$

$$x + 1 = 0 \quad \text{or} \quad x - 7 = 0$$

$$x = -1 \qquad\qquad x = 7$$

Neither $-1$ nor 7 makes the denominator in the original equation equal to 0.

**Check:**

$x = -1$

$$8 + \dfrac{7}{x} = x + 2$$

$$8 + \dfrac{7}{-1} \overset{?}{=} -1 + 2$$

$$8 + (-7) \overset{?}{=} 1$$

$$1 = 1 \quad \text{True}$$

$x = 7$

$$8 + \dfrac{7}{7} \overset{?}{=} 7 + 2$$

$$8 + 1 \overset{?}{=} 9$$

$$9 = 9 \quad \text{True}$$

Both $-1$ and 7 are solutions.

**4.** $x^2 - 5x - 14 = (x+2)(x-7)$

The LCD is $(x+2)(x-7)$.

$$\frac{6x}{x^2-5x-14} - \frac{3}{x+2} = \frac{1}{x-7}$$

$$(x+2)(x-7)\left(\frac{6x}{x^2-5x-14} - \frac{3}{x+2}\right) = (x+2)(x-7)\left(\frac{1}{x-7}\right)$$

$$(x+2)(x-7)\cdot\frac{6x}{x^2-5x-14} - (x+2)(x-7)\cdot\frac{3}{x+2} = (x+2)(x-7)\cdot\frac{1}{x-7}$$

$$6x - 3(x-7) = x+2$$
$$6x - 3x + 21 = x+2$$
$$3x + 21 = x+2$$
$$2x = -19$$
$$x = -\frac{19}{2}$$

Check by replacing $x$ with $-\dfrac{19}{2}$ in the original equation. The solution is $-\dfrac{19}{2}$.

**5.** The LCD is $x - 2$.

$$\frac{7}{x-2} = \frac{3}{x-2} + 4$$

$$(x-2)\left(\frac{7}{x-2}\right) = (x-2)\left(\frac{3}{x-2} + 4\right)$$

$$(x-2)\cdot\frac{7}{x-2} = (x-2)\cdot\frac{3}{x-2} + (x-2)\cdot 4$$

$$7 = 3 + 4x - 8$$
$$7 = 4x - 5$$
$$12 = 4x$$
$$3 = x$$

Check by replacing $x$ with 3 in the original equation. The solution is 3.

**6.** From the denominators in the equation, 5 can't be a solution. The LCD is $x - 5$.

$$x + \frac{x}{x-5} = \frac{5}{x-5} - 7$$

$$(x-5)\left(x + \frac{x}{x-5}\right) = (x-5)\left(\frac{5}{x-5} - 7\right)$$

$$(x-5)(x) + (x-5)\left(\frac{x}{x-5}\right) = (x-5)\left(\frac{5}{x-5}\right) - (x-5)(7)$$

$$x^2 - 5x + x = 5 - 7x + 35$$
$$x^2 - 4x = 40 - 7x$$
$$x^2 + 3x - 40 = 0$$
$$(x+8)(x-5) = 0$$

$$x + 8 = 0 \quad \text{or} \quad x - 5 = 0$$
$$x = -8 \qquad\quad x = 5$$

Since 5 can't be a solution, check by replacing $x$ with $-8$ in the original equation. The only solution is $-8$.

**7.** The LCD is *abx*.

$$\frac{1}{a}+\frac{1}{b}=\frac{1}{x}$$

$$abx\left(\frac{1}{a}+\frac{1}{b}\right)=abx\left(\frac{1}{x}\right)$$

$$abx\left(\frac{1}{a}\right)+abx\left(\frac{1}{b}\right)=abx\cdot\frac{1}{x}$$

$$bx+ax=ab$$

$$ax=ab-bx$$

$$ax=b(a-x)$$

$$\frac{ax}{a-x}=b$$

## Calculator Explorations

**1.** $y_1=\dfrac{x-4}{2}-\dfrac{x-3}{9}$, $y_2=\dfrac{5}{18}$

Use INTERSECT

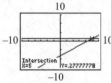

The solution of the equation is 5.

**2.** $y_1=3-\dfrac{6}{x}$, $y_2=x+8$

Use INTERSECT

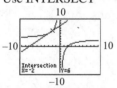

One solution is −3.

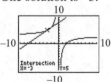

The other solution is −2.

**3.** $y_1=\dfrac{2x}{x-4}$, $y_2=\dfrac{8}{x-4}+1$

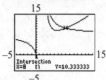

Using TRACE and ZOOM, it is clear that the curves never intersect. The equation has no solution.

**4.** $y_1=x+\dfrac{14}{x-2}$, $y_2=\dfrac{7x}{x-2}+1$

Use INTERSECT

The solution is 8.

## Vocabulary, Readiness & Video Check 7.5

**1.** $4\left(\dfrac{3x}{2}+5\right)=4\left(\dfrac{1}{4}\right)$

$4\left(\dfrac{3x}{2}\right)+4\cdot5=4\left(\dfrac{1}{4}\right)$

$6x+20=1$

The correct choice is c.

**2.** $5x\left(\dfrac{1}{x}-\dfrac{3}{5x}\right)=5x(2)$

$5x\left(\dfrac{1}{x}\right)-5x\left(\dfrac{3}{5x}\right)=5x(2)$

$5-3=10x$

The correct choice is b.

**3.** The LCD of $\dfrac{9}{x}$, $\dfrac{3}{4}$, and $\dfrac{1}{12}=\dfrac{1}{3\cdot4}$ is 12*x*; b.

**4.** The LCD of $\dfrac{8}{3x}$, $\dfrac{1}{x}$, and $\dfrac{7}{9}=\dfrac{7}{3\cdot3}$ is 9*x*; d.

**5.** The LCD of $\dfrac{9}{x-1}$ and $\dfrac{7}{(x-1)^2}$ is $(x-1)^2$; a.

**6.** The LCD of $\dfrac{1}{x-2}$, $\dfrac{3}{x^2-4}=\dfrac{3}{(x+2)(x-2)}$, and

$8=\dfrac{8}{1}$ is $x^2-4$; c.

**7.** These equations are solved in very different ways, so you need to determine the next correct move to make. For a linear equation, you first "move" variables terms on one side and numbers on the other; for a quadratic equation, you first set the equation equal to 0.

8. If there are variables in any denominator, you should first check to see if the proposed solutions make these denominators zero in the original equation, giving you an undefined rational expression. If so, that solution is an extraneous solution and is not a solution to the equation.

9. the steps for solving an equation containing rational expressions; as if it's the only variable in the equation.

**Exercise Set 7.5**

1. The LCD is 5.

$$\frac{x}{5}+3=9$$

$$5\left(\frac{x}{5}+3\right)=5(9)$$

$$5\left(\frac{x}{5}\right)+5(3)=5(9)$$

$$x+15=45$$

$$x=30$$

Check: $\dfrac{x}{5}+3=9$

$$\frac{30}{5}+3 \stackrel{?}{=} 9$$

$$6+3 \stackrel{?}{=} 9$$

$$9=9 \quad \text{True}$$

The solution is 30.

3. The LCD is 12.

$$\frac{x}{2}+\frac{5x}{4}=\frac{x}{12}$$

$$12\left(\frac{x}{2}+\frac{5x}{4}\right)=12\left(\frac{x}{12}\right)$$

$$12\left(\frac{x}{2}\right)+12\left(\frac{5x}{4}\right)=12\left(\frac{x}{12}\right)$$

$$6x+15x=x$$

$$21x=x$$

$$20x=0$$

$$x=0$$

Check: $\dfrac{x}{2}+\dfrac{5x}{4}=\dfrac{x}{12}$

$$\frac{0}{2}+\frac{5\cdot 0}{4} \stackrel{?}{=} \frac{0}{12}$$

$$0+\frac{0}{4} \stackrel{?}{=} 0$$

$$0=0 \quad \text{True}$$

The solution is 0.

5. The LCD is $x$.

$$2-\frac{8}{x}=6$$

$$x\left(2-\frac{8}{x}\right)=x(6)$$

$$x\cdot 2-x\cdot\frac{8}{x}=x\cdot 6$$

$$2x-8=6x$$

$$-8=4x$$

$$-2=x$$

Check: $\quad 2-\dfrac{8}{x}=6$

$$2-\frac{8}{-2} \stackrel{?}{=} 6$$

$$2-(-4) \stackrel{?}{=} 6$$

$$6=6 \quad \text{True}$$

The solution is −2.

7. The LCD is $x$.

$$2+\frac{10}{x}=x+5$$

$$x\left(2+\frac{10}{x}\right)=x(x+5)$$

$$x(2)+x\left(\frac{10}{x}\right)=x(x+5)$$

$$2x+10=x^2+5x$$

$$0=x^2+3x-10$$

$$0=(x+5)(x-2)$$

$$x+5=0 \quad \text{or} \quad x-2=0$$

$$x=-5 \qquad\qquad x=2$$

Check:

$x=-5:\quad 2+\dfrac{10}{x}=x+5$

$$2+\frac{10}{-5} \stackrel{?}{=} -5+5$$

$$2+(-2) \stackrel{?}{=} -5+5$$

$$0=0 \quad \text{True}$$

$x=2:\quad 2+\dfrac{10}{x}=x+5$

$$2+\frac{10}{2} \stackrel{?}{=} 2+5$$

$$2+5 \stackrel{?}{=} 2+5$$

$$7=7 \quad \text{True}$$

Both −5 and 2 are solutions.

**9.** The LCD is 10.

$$\frac{a}{5} = \frac{a-3}{2}$$

$$10\left(\frac{a}{5}\right) = 10\left(\frac{a-3}{2}\right)$$

$$2a = 5(a-3)$$

$$2a = 5a - 15$$

$$-3a = -15$$

$$a = 5$$

Check: $\dfrac{a}{5} = \dfrac{a-3}{2}$

$$\frac{5}{5} \overset{?}{=} \frac{5-3}{2}$$

$$\frac{5}{5} \overset{?}{=} \frac{2}{2}$$

$$1 = 1 \quad \text{True}$$

The solution is 5.

**11.** The LCD is 10.

$$\frac{x-3}{5} + \frac{x-2}{2} = \frac{1}{2}$$

$$10\left(\frac{x-3}{5} + \frac{x-2}{2}\right) = 10\left(\frac{1}{2}\right)$$

$$10\left(\frac{x-3}{5}\right) + 10\left(\frac{x-2}{2}\right) = 10\left(\frac{1}{2}\right)$$

$$2(x-3) + 5(x-2) = 5$$

$$2x - 6 + 5x - 10 = 5$$

$$7x - 16 = 5$$

$$7x = 21$$

$$x = 3$$

Check:

$$\frac{x-3}{5} + \frac{x-2}{2} = \frac{1}{2}$$

$$\frac{3-3}{5} + \frac{3-2}{2} \overset{?}{=} \frac{1}{2}$$

$$\frac{0}{5} + \frac{1}{2} \overset{?}{=} \frac{1}{2}$$

$$0 + \frac{1}{2} \overset{?}{=} \frac{1}{2}$$

$$\frac{1}{2} = \frac{1}{2} \quad \text{True}$$

The solution is 3.

**13.** The LCD is $2a - 5$.

$$\frac{3}{2a-5} = -1$$

$$(2a-5)\left(\frac{3}{2a-5}\right) = (2a-5)(-1)$$

$$3 = -2a + 5$$

$$-2 = -2a$$

$$1 = a$$

Check: $\dfrac{3}{2a-5} = -1$

$$\frac{3}{2(1)-5} \overset{?}{=} -1$$

$$\frac{3}{-3} \overset{?}{=} -1$$

$$-1 = -1 \quad \text{True}$$

The solution is 1.

**15.** The LCD is $y - 4$.

$$\frac{4y}{y-4} + 5 = \frac{5y}{y-4}$$

$$(y-4)\left(\frac{4y}{y-4} + 5\right) = (y-4)\left(\frac{5y}{y-4}\right)$$

$$(y-4)\left(\frac{4y}{y-4}\right) + (y-4)(5) = (y-4)\left(\frac{5y}{y-4}\right)$$

$$4y + 5y - 20 = 5y$$

$$9y - 20 = 5y$$

$$4y - 20 = 0$$

$$4y = 20$$

$$y = 5$$

Check: $\dfrac{4y}{y-4} + 5 = \dfrac{5y}{y-4}$

$$\frac{4(5)}{5-4} + 5 \overset{?}{=} \frac{5(5)}{5-4}$$

$$\frac{20}{1} + 5 \overset{?}{=} \frac{25}{1}$$

$$25 = 25 \quad \text{True}$$

The solution is 5.

**17.** The LCD is $a - 3$.

$$2 + \frac{3}{a-3} = \frac{a}{a-3}$$

$$(a-3)\left(2 + \frac{3}{a-3}\right) = (a-3)\left(\frac{a}{a-3}\right)$$

$$(a-3)(2) + (a-3)\left(\frac{3}{a-3}\right) = a$$

$$2a - 6 + 3 = a$$

$$2a - 3 = a$$

$$-3 = a - 2a$$

$$-3 = -a$$

$$\frac{-3}{-1} = a$$

$$3 = a$$

When $a$ is 3, a denominator equals zero. The equation has no solution.

**19.** $x^2 - 9 = (x+3)(x-3)$

The LCD is $(x + 3)(x - 3)$.

$$\frac{1}{x+3} + \frac{6}{x^2 - 9} = 1$$

$$(x+3)(x-3)\left(\frac{1}{x+3} + \frac{6}{(x+3)(x-3)}\right) = (x+3)(x-3)(1)$$

$$(x+3)(x-3) \cdot \frac{1}{x+3} + (x+3)(x-3) \cdot \frac{6}{(x+3)(x-3)} = (x+3)(x-3) \cdot 1$$

$$x - 3 + 6 = x^2 - 9$$

$$x + 3 = x^2 - 9$$

$$0 = x^2 - x - 12$$

$$0 = (x+3)(x-4)$$

$$x + 3 = 0 \quad \text{or} \quad x - 4 = 0$$
$$x = -3 \qquad\qquad x = 4$$

When $x$ is $-3$, a denominator equals zero. Check $x = 4$.

Check: $\dfrac{1}{x+3} + \dfrac{6}{x^2 - 9} = 1$

$$\frac{1}{4+3} + \frac{6}{4^2 - 9} \stackrel{?}{=} 1$$

$$\frac{1}{7} + \frac{6}{7} \stackrel{?}{=} 1$$

$$1 = 1 \quad \text{True}$$

The solution is 4.

**21.** The LCD is $y + 4$.

$$\frac{2y}{y+4} + \frac{4}{y+4} = 3$$

$$(y+4)\left(\frac{2y}{y+4} + \frac{4}{y+4}\right) = (y+4)(3)$$

$$(y+4)\cdot\frac{2y}{y+4} + (y+4)\cdot\frac{4}{y+4} = (y+4)\cdot 3$$

$$2y + 4 = 3y + 12$$

$$4 = y + 12$$

$$-8 = y$$

Check: $\quad \dfrac{2y}{y+4} + \dfrac{4}{y+4} = 3$

$$\frac{2(-8)}{-8+4} + \frac{4}{-8+4} \stackrel{?}{=} 3$$

$$\frac{-16}{-4} + \frac{4}{-4} \stackrel{?}{=} 3$$

$$4 - 1 \stackrel{?}{=} 3$$

$$3 = 3 \quad \text{True}$$

The solution is $-8$.

**23.** The LCD is $(x + 2)(x - 2)$.

$$\frac{2x}{x+2} - 2 = \frac{x-8}{x-2}$$

$$(x+2)(x-2)\left(\frac{2x}{x+2} - 2\right) = (x+2)(x-2)\left(\frac{x-8}{x-2}\right)$$

$$(x+2)(x-2)\cdot\frac{2x}{x+2} - (x+2)(x-2)(2) = (x+2)(x-2)\cdot\frac{x-8}{x-2}$$

$$2x(x-2) - 2(x^2 - 4) = (x+2)(x-8)$$

$$2x^2 - 4x - 2x^2 + 8 = x^2 - 6x - 16$$

$$-4x + 8 = x^2 - 6x - 16$$

$$0 = x^2 - 2x - 24$$

$$0 = (x+4)(x-6)$$

$$\begin{array}{ccc} x+4=0 & \text{or} & x-6=0 \\ x=-4 & & x=6 \end{array}$$

Check $x = -4$: $\quad \dfrac{2x}{x+2} - 2 = \dfrac{x-8}{x-2}$

$$\frac{2(-4)}{-4+2} - 2 \stackrel{?}{=} \frac{-4-8}{-4-2}$$

$$\frac{-8}{-2} - 2 \stackrel{?}{=} \frac{-12}{-6}$$

$$4 - 2 \stackrel{?}{=} 2$$

$$2 = 2 \quad \text{True}$$

Check $x = 6$: $\dfrac{2x}{x+2} - 2 = \dfrac{x-8}{x-2}$

$$\dfrac{2(6)}{6+2} - 2 \stackrel{?}{=} \dfrac{6-8}{6-2}$$

$$\dfrac{12}{8} - 2 \stackrel{?}{=} \dfrac{-2}{4}$$

$$\dfrac{3}{2} - 2 \stackrel{?}{=} -\dfrac{1}{2}$$

$$-\dfrac{1}{2} = -\dfrac{1}{2} \quad \text{True}$$

The solutions are $-4$ an 6.

**25.** The LCD is $2y$.

$$\dfrac{2}{y} + \dfrac{1}{2} = \dfrac{5}{2y}$$

$$2y\left(\dfrac{2}{y} + \dfrac{1}{2}\right) = 2y\left(\dfrac{5}{2y}\right)$$

$$2y\left(\dfrac{2}{y}\right) + 2y\left(\dfrac{1}{2}\right) = 2y\left(\dfrac{5}{2y}\right)$$

$$4 + y = 5$$

$$y = 1$$

The solution is 1.

**27.** The LCD is $(a-6)(a-1)$.

$$\dfrac{a}{a-6} = \dfrac{-2}{a-1}$$

$$(a-6)(a-1)\left(\dfrac{a}{a-6}\right) = (a-6)(a-1)\left(\dfrac{-2}{a-1}\right)$$

$$a(a-1) = -2(a-6)$$

$$a^2 - a = -2a + 12$$

$$a^2 + a - 12 = 0$$

$$(a+4)(a-3) = 0$$

$$a + 4 = 0 \quad \text{or} \quad a - 3 = 0$$

$$a = -4 \qquad\qquad a = 3$$

The solutions are $-4$ and 3.

**29.** The LCD is $6x$.

$$\dfrac{11}{2x} + \dfrac{2}{3} = \dfrac{7}{2x}$$

$$6x\left(\dfrac{11}{2x} + \dfrac{2}{3}\right) = 6x\left(\dfrac{7}{2x}\right)$$

$$6x \cdot \dfrac{11}{2x} + 6x \cdot \dfrac{2}{3} = 6x \cdot \dfrac{7}{2x}$$

$$33 + 4x = 21$$

$$4x = -12$$

$$x = -3$$

The solution is $-3$.

**31.** The LCD is $(x+2)(x-2)$.

$$\frac{2}{x-2}+1=\frac{x}{x+2}$$

$$(x+2)(x-2)\left(\frac{2}{x-2}+1\right)=(x+2)(x-2)\left(\frac{x}{x+2}\right)$$

$$(x+2)(x-2)\cdot\frac{2}{x-2}+(x+2)(x-2)\cdot 1=(x+2)(x-2)\cdot\frac{x}{x+2}$$

$$2(x+2)+(x+2)(x-2)=x(x-2)$$

$$2x+4+x^2-4=x^2-2x$$

$$x^2+2x=x^2-2x$$

$$2x=-2x$$

$$4x=0$$

$$x=0$$

The solution is 0.

**33.** The LCD is 6.

$$\frac{x+1}{3}-\frac{x-1}{6}=\frac{1}{6}$$

$$6\left(\frac{x+1}{3}-\frac{x-1}{6}\right)=6\left(\frac{1}{6}\right)$$

$$6\left(\frac{x+1}{3}\right)-6\left(\frac{x-1}{6}\right)=6\left(\frac{1}{6}\right)$$

$$2(x+1)-(x-1)=1$$

$$2x+2-x+1=1$$

$$x+3=1$$

$$x=-2$$

The solution is $-2$.

**35.** The LCD is $6(t-4)$.

$$\frac{t}{t-4}=\frac{t+4}{6}$$

$$6(t-4)\left(\frac{t}{t-4}\right)=6(t-4)\left(\frac{t+4}{6}\right)$$

$$6t=(t-4)(t+4)$$

$$6t=t^2-16$$

$$0=t^2-6t-16$$

$$0=(t-8)(t+2)$$

$$t+2=0 \quad \text{or} \quad t-8=0$$

$$t=-2 \qquad\quad t=8$$

The solutions are $-2$ and 8.

**37.** $2y+2=2(y+1)$

$4y+4=2\cdot 2(y+1)$

The LCD is $4(y+1)$.

$$\frac{y}{2y+2}+\frac{2y-16}{4y+4}=\frac{2y-3}{y+1}$$

$$4(y+1)\left(\frac{y}{2(y+1)}+\frac{2y-16}{4(y+1)}\right)=4(y+1)\left(\frac{2y-3}{y+1}\right)$$

$$4(y+1)\left(\frac{y}{2(y+1)}\right)+4(y+1)\left(\frac{2y-16}{4(y+1)}\right)=4(y+1)\left(\frac{2y-3}{y+1}\right)$$

$$2y+2y-16=4(2y-3)$$

$$4y-16=8y-12$$

$$-4y=4$$

$$y=-1$$

In the original equation, $-1$ makes a denominator 0. This equation has no solution.

**39.** $r^2+5r-14=(r+7)(r-2)$

The LCD is $(r+7)(r-2)$.

$$\frac{4r-4}{r^2+5r-14}+\frac{2}{r+7}=\frac{1}{r-2}$$

$$(r+7)(r-2)\left(\frac{4r-4}{(r+7)(r-2)}+\frac{2}{r+7}\right)=(r+7)(r-2)\left(\frac{1}{r-2}\right)$$

$$(r+7)(r-2)\left(\frac{4r-4}{(r+7)(r-2)}\right)+(r+7)(r-2)\left(\frac{2}{r+7}\right)=(r+7)(r-2)\left(\frac{1}{r-2}\right)$$

$$4r-4+2(r-2)=(r+7)(1)$$

$$4r-4+2r-4=r+7$$

$$6r-8=r+7$$

$$5r=15$$

$$r=3$$

The solution is 3.

**41.** $x^2+x-6=(x+3)(x-2)$

The LCD is $(x+3)(x-2)$.

$$\frac{x+1}{x+3}=\frac{x^2-11x}{x^2+x-6}-\frac{x-3}{x-2}$$

$$(x+3)(x-2)\left(\frac{x+1}{x+3}\right)=(x+3)(x-2)\left(\frac{x^2-11x}{(x+3)(x-2)}-\frac{x-3}{x-2}\right)$$

$$(x+3)(x-2)\cdot\frac{x+1}{x+3}=(x+3)(x-2)\cdot\frac{x^2-11x}{(x+3)(x-2)}-(x+3)(x-2)\cdot\frac{x-3}{x-2}$$

$$(x-2)(x+1)=x^2-11x-(x+3)(x-3)$$

$$x^2-x-2=x^2-11x-(x^2-9)$$

$$x^2-x-2=x^2-11x-x^2+9$$

$$x^2-x-2=-11x+9$$

$$x^2+10x-11=0$$

$$(x+11)(x-1)=0$$

$$x+11=0 \quad \text{or} \quad x-1=0$$

$$x=-11 \qquad\qquad x=1$$

The solutions are $-11$ and 1.

**43.** $R = \dfrac{E}{I}$

$I(R) = I\left(\dfrac{E}{I}\right)$

$IR = E$

$I = \dfrac{E}{R}$

**45.** $T = \dfrac{2U}{B+E}$

$(B+E)(T) = (B+E)\left(\dfrac{2U}{B+E}\right)$

$BT + ET = 2U$

$BT = 2U - ET$

$B = \dfrac{2U - ET}{T}$

**47.** $B = \dfrac{705w}{h^2}$

$h^2(B) = h^2\left(\dfrac{705w}{h^2}\right)$

$Bh^2 = 705w$

$\dfrac{Bh^2}{705} = w$

**49.** $N = R + \dfrac{V}{G}$

$G(N) = G\left(R + \dfrac{V}{G}\right)$

$GN = GR + V$

$GN - GR = V$

$G(N - R) = V$

$G = \dfrac{V}{N-R}$

**51.** $\dfrac{C}{\pi r} = 2$

$\pi r\left(\dfrac{C}{\pi r}\right) = \pi r(2)$

$C = 2\pi r$

$\dfrac{C}{2\pi} = \dfrac{2\pi r}{2\pi}$

$\dfrac{C}{2\pi} = r$

**53.** $\dfrac{1}{y} + \dfrac{1}{3} = \dfrac{1}{x}$

$3xy\left(\dfrac{1}{y} + \dfrac{1}{3}\right) = 3xy\left(\dfrac{1}{x}\right)$

$3xy \cdot \dfrac{1}{y} + 3xy \cdot \dfrac{1}{3} = 3xy \cdot \dfrac{1}{x}$

$3x + xy = 3y$

$x(3 + y) = 3y$

$x = \dfrac{3y}{3+y}$

**55.** The reciprocal of $x$ is $\dfrac{1}{x}$.

**57.** The reciprocal of $x$, added to the reciprocal of 2 is $\dfrac{1}{x} + \dfrac{1}{2}$.

**59.** If a tank is filled in 3 hours, then $\dfrac{1}{3}$ of the tank is filled in one hour.

**61.** The graph crosses the $x$-axis at $x = 2$. It crosses the $y$-axis at $y = -2$. The $x$-intercept is $(2, 0)$ and the $y$-intercept is $(0, -2)$.

**63.** The graph crosses the $x$-axis at $x = -4$, $x = -2$ and $x = 3$. It crosses the $y$-axis at $y = 4$. The $x$-intercepts are $(-4, 0)$, $(-2, 0)$ and $(3, 0)$, and the $y$-intercept is $(0, 4)$.

**65.** answers may vary

**67.** expression

$\dfrac{1}{x} + \dfrac{5}{9} = \dfrac{1(9)}{x(9)} + \dfrac{5x}{9x} = \dfrac{5x + 9}{9x}$

**69.** equation

$$\dfrac{5}{x-1} - \dfrac{2}{x} = \dfrac{5}{x(x-1)}$$

$$x(x-1)\left(\dfrac{5}{x-1}\right) - x(x-1)\left(\dfrac{2}{x}\right) = x(x-1)\left(\dfrac{5}{x(x-1)}\right)$$

$$5x - 2(x-1) = 5$$

$$5x - 2x + 2 = 5$$

$$3x = 3$$

$$x = 1$$

1 makes a denominator zero. There is no solution.

**71.**
$$\frac{20x}{3} + \frac{32x}{6} = 180$$
$$6\left(\frac{20x}{3} + \frac{32x}{6}\right) = 6(180)$$
$$6\left(\frac{20x}{3}\right) + 6\left(\frac{32x}{6}\right) = 6(180)$$
$$40x + 32x = 1080$$
$$72x = 1080$$
$$\frac{72x}{72} = \frac{1080}{72}$$
$$x = 15$$
$$\frac{20x}{3} = \frac{20(15)}{3} = 100$$
$$\frac{32x}{6} = \frac{32(15)}{6} = 80$$
The angles are 100° and 80°.

**73.**
$$\frac{150}{x} + \frac{450}{x} = 90$$
$$x\left(\frac{150}{x} + \frac{450}{x}\right) = x(90)$$
$$x\left(\frac{150}{x}\right) + x\left(\frac{450}{x}\right) = x(90)$$
$$150 + 450 = 90x$$
$$600 = 90x$$
$$\frac{600}{90} = \frac{90x}{90}$$
$$\frac{20}{3} = x$$
$$\frac{150}{x} = \frac{150}{\frac{20}{3}} = 150\left(\frac{3}{20}\right) = \frac{45}{2} = 22.5$$
$$\frac{450}{x} = \frac{450}{\frac{20}{3}} = 450\left(\frac{3}{20}\right) = \frac{135}{2} = 67.5$$
The angles are 22.5° and 67.5°.

**75.**

$$\frac{5}{a^2+4a+3}+\frac{2}{a^2+a-6}-\frac{3}{a^2-a-2}=0$$

$$\frac{5}{(a+3)(a+1)}+\frac{2}{(a+3)(a-2)}-\frac{3}{(a-2)(a+1)}=0$$

$$(a+3)(a+1)(a-2)\left(\frac{5}{(a+3)(a+1)}+\frac{2}{(a+3)(a-2)}-\frac{3}{(a-2)(a+1)}\right)=(a+3)(a+1)(a-2)(0)$$

$$(a+3)(a+1)(a-2)\left(\frac{5}{(a+3)(a+1)}\right)+(a+3)(a+1)(a-2)\left(\frac{2}{(a+3)(a-2)}\right)$$

$$-(a+3)(a+1)(a-2)\left(\frac{3}{(a-2)(a+1)}\right)=0$$

$$5(a-2)+2(a+1)-3(a+3)=0$$

$$5a-10+2a+2-3a-9=0$$

$$4a-17=0$$

$$4a=17$$

$$a=\frac{17}{4}$$

The solution is $\frac{17}{4}$.

**Integrated Review**

**1.** expression

$$\frac{1}{x}+\frac{2}{3}=\frac{1(3)}{x(3)}+\frac{2(x)}{3(x)}=\frac{3}{3x}+\frac{2x}{3x}=\frac{3+2x}{3x}$$

**2.** expression

$$\frac{3}{a}+\frac{5}{6}=\frac{3(6)}{a(6)}+\frac{5(a)}{6(a)}=\frac{18}{6a}+\frac{5a}{6a}=\frac{18+5a}{6a}$$

**3.** equation

$$\frac{1}{x}+\frac{2}{3}=\frac{3}{x}$$

$$3x\left(\frac{1}{x}+\frac{2}{3}\right)=3x\left(\frac{3}{x}\right)$$

$$3x\left(\frac{1}{x}\right)+3x\left(\frac{2}{3}\right)=3x\left(\frac{3}{x}\right)$$

$$3+2x=9$$

$$2x=6$$

$$x=3$$

The solution is 3.

**4.** equation

$$\frac{3}{a} + \frac{5}{6} = 1$$

$$6a\left(\frac{3}{a} + \frac{5}{6}\right) = 6a(1)$$

$$6a\left(\frac{3}{a}\right) + 6a\left(\frac{5}{6}\right) = 6a$$

$$18 + 5a = 6a$$

$$18 = a$$

The solution is 18.

**5.** expression

$$\frac{2}{x-1} - \frac{1}{x} = \frac{2(x)}{(x-1)(x)} - \frac{1(x-1)}{x(x-1)}$$

$$= \frac{2x - (x-1)}{x(x-1)}$$

$$= \frac{x+1}{x(x-1)}$$

**6.** expression

$$\frac{4}{x-3} - \frac{1}{x} = \frac{4(x)}{(x-3)(x)} - \frac{1(x-3)}{x(x-3)}$$

$$= \frac{4x - (x-3)}{x(x-3)}$$

$$= \frac{4x - x + 3}{x(x-3)}$$

$$= \frac{3x+3}{x(x-3)}$$

$$= \frac{3(x+1)}{x(x-3)}$$

**7.** equation

$$\frac{2}{x+1} - \frac{1}{x} = 1$$

$$x(x+1)\left(\frac{2}{x+1} - \frac{1}{x}\right) = x(x+1)(1)$$

$$x(x+1)\left(\frac{2}{x+1}\right) - x(x+1)\left(\frac{1}{x}\right) = x(x+1)$$

$$2x - (x+1) = x(x+1)$$

$$2x - x - 1 = x^2 + x$$

$$x - 1 = x^2 + x$$

$$-1 = x^2$$

There is no real number solution.

**8.** equation

$$\frac{4}{x-3} - \frac{1}{x} = \frac{6}{x(x-3)}$$

$$x(x-3)\left(\frac{4}{x-3} - \frac{1}{x}\right) = x(x-3)\left(\frac{6}{x(x-3)}\right)$$

$$x(x-3)\left(\frac{4}{x-3}\right) - x(x-3)\left(\frac{1}{x}\right) = 6$$

$$4x - (x-3) = 6$$

$$4x - x + 3 = 6$$

$$3x + 3 = 6$$

$$3x = 3$$

$$x = 1$$

The solution is 1.

**9.** expression

$$\frac{15x}{x+8} \cdot \frac{2x+16}{3x} = \frac{15x \cdot (2x+16)}{(x+8) \cdot 3x}$$

$$= \frac{3 \cdot 5 \cdot x \cdot 2 \cdot (x+8)}{(x+8) \cdot 3 \cdot x}$$

$$= 5 \cdot 2$$

$$= 10$$

**10.** expression

$$\frac{9z+5}{15} \cdot \frac{5z}{81z^2 - 25} = \frac{(9z+5) \cdot 5z}{15 \cdot (81z^2 - 25)}$$

$$= \frac{(9z+5) \cdot 5 \cdot z}{5 \cdot 3 \cdot (9z+5)(9z-5)}$$

$$= \frac{z}{3(9z-5)}$$

**11.** expression

$$\frac{2x+1}{x-3} + \frac{3x+6}{x-3} = \frac{2x+1+3x+6}{x-3} = \frac{5x+7}{x-3}$$

**12.** expression

$$\frac{4p-3}{2p+7} + \frac{3p+8}{2p+7} = \frac{4p-3+3p+8}{2p+7} = \frac{7p+5}{2p+7}$$

**13.** equation

$$\frac{x+5}{7} = \frac{8}{2}$$

$$14\left(\frac{x+5}{7}\right) = 14\left(\frac{8}{2}\right)$$

$$2(x+5) = 56$$

$$2x + 10 = 56$$

$$2x = 46$$

$$x = 23$$

The solution is 23.

**14.** equation

$$\frac{1}{2} = \frac{x-1}{8}$$

$$8\left(\frac{1}{2}\right) = 8\left(\frac{x-1}{8}\right)$$

$$4 = x-1$$

$$5 = x$$

The solution is 5.

**15.** expression

$$\frac{5a+10}{18} \div \frac{a^2-4}{10a} = \frac{5a+10}{18} \cdot \frac{10a}{a^2-4}$$

$$= \frac{5(a+2)\cdot 2\cdot 5\cdot a}{2\cdot 9(a+2)(a-2)}$$

$$= \frac{5\cdot 5\cdot a}{9(a-2)}$$

$$= \frac{25a}{9(a-2)}$$

**16.** expression

$$\frac{9}{x^2-1} + \frac{12}{3x+3} = \frac{9(3)}{(x+1)(x-1)(3)} + \frac{12(x-1)}{3(x+1)(x-1)}$$

$$= \frac{27+12x-12}{3(x-1)(x+1)}$$

$$= \frac{15+12x}{3(x+1)(x-1)}$$

$$= \frac{3(5+4x)}{3(x+1)(x-1)}$$

$$= \frac{4x+5}{(x+1)(x-1)}$$

**17.** expression

$$\frac{x+2}{3x-1} + \frac{5}{(3x-1)^2} = \frac{(x+2)(3x-1)}{(3x-1)(3x-1)} + \frac{5}{(3x-1)^2}$$

$$= \frac{3x^2+5x-2+5}{(3x-1)^2}$$

$$= \frac{3x^2+5x+3}{(3x-1)^2}$$

**18.** expression

$$\frac{4}{(2x-5)^2} + \frac{x+1}{2x-5} = \frac{4}{(2x-5)^2} + \frac{(x+1)(2x-5)}{(2x-5)(2x-5)}$$

$$= \frac{4+2x^2-3x-5}{(2x-5)^2}$$

$$= \frac{2x^2-3x-1}{(2x-5)^2}$$

**19.** expression

$$\frac{x-7}{x} - \frac{x+2}{5x} = \frac{(x-7)(5)}{x(5)} - \frac{x+2}{5x}$$

$$= \frac{5x-35-x-2}{5x}$$

$$= \frac{4x-37}{5x}$$

**20.** expression

$$\frac{10x-9}{x} - \frac{x-4}{3x} = \frac{(10x-9)(3)}{x(3)} - \frac{x-4}{3x}$$

$$= \frac{30x-27-x+4}{3x}$$

$$= \frac{29x-23}{3x}$$

**21.** equation

$$\frac{3}{x+3} = \frac{5}{x^2-9} - \frac{2}{x-3}$$

$$(x^2-9)\left(\frac{3}{x+3}\right) = (x^2-9)\left(\frac{5}{x^2-9}\right) - (x^2-9)\left(\frac{2}{x-3}\right)$$

$$(x-3)(3) = 5 - (x+3)(2)$$

$$3x-9 = 5-2x-6$$

$$3x-9 = -2x-1$$

$$5x-9 = -1$$

$$5x = 8$$

$$x = \frac{8}{5}$$

The solution is $\frac{8}{5}$.

**22.** equation

$$\frac{9}{x^2-4} + \frac{2}{x+2} = \frac{-1}{x-2}$$

$$(x^2-4)\left(\frac{9}{x^2-4}\right) + (x^2-4)\left(\frac{2}{x+2}\right) = (x^2-4)\left(\frac{-1}{x-2}\right)$$

$$9 + (x-2)(2) = (x+2)(-1)$$

$$9 + 2x - 4 = -x - 2$$

$$2x + 5 = -x - 2$$

$$3x + 5 = -2$$

$$3x = -7$$

$$x = -\frac{7}{3}$$

The solution is $-\frac{7}{3}$.

**23.** answers may vary

**24.** answers may vary

## Section 7.6 Practice

1. Solve the equation as a rational equation.

$$\frac{36}{x} = \frac{4}{11}$$

$$11x \cdot \frac{36}{x} = 11x \cdot \frac{4}{11}$$

$$11 \cdot 36 = x \cdot 4$$

$$396 = 4x$$

$$\frac{396}{4} = \frac{4x}{4}$$

$$99 = x$$

Solve the proportion using cross products.

$$\frac{36}{x} = \frac{4}{11}$$

$$36 \cdot 11 = x \cdot 4$$

$$396 = 4x$$

$$\frac{396}{4} = \frac{4x}{4}$$

$$99 = x$$

**Check:** Both methods give a solution of 99. To check, substitute 99 for $x$ in the original proportion. The solution is 99.

2. $$\frac{3x+2}{9} = \frac{x-1}{2}$$

$$2(3x+2) = 9(x-1)$$

$$6x + 4 = 9x - 9$$

$$6x = 9x - 13$$

$$-3x = -13$$

$$\frac{-3x}{-3} = \frac{-13}{-3}$$

$$x = \frac{13}{3}$$

**Check:** Verify that $\frac{13}{3}$ is the solution.

3. Let $x$ = price of seven 2-liter bottles of Diet Pepsi.

$$\frac{4 \text{ bottles}}{7 \text{ bottles}} = \frac{\text{price of 4 bottles}}{\text{price of 7 bottles}}$$

$$\frac{4}{7} = \frac{5.16}{x}$$

$$4x = 7(5.16)$$

$$4x = 36.12$$

$$x = 9.03$$

**Check:** Verify that 4 bottles is to 7 bottles as $5.16 is to $9.03.
Seven 2-liter bottles of Diet Pepsi cost $9.03.

**4.** Since the triangles are similar, their corresponding sides are in proportion.

$$\frac{20}{8} = \frac{15}{x}$$
$$20x = 8 \cdot 15$$
$$20x = 120$$
$$x = 6$$

**Check:** To check, replace $x$ with 6 in the original proportion and see that a true statement results.
The missing length is 6 meters.

**5.** Let $x$ = the unknown number.

| In words | the quotient of $x$ and 5 | | minus | | $\dfrac{3}{2}$ | | is | | the quotient of $x$ and 10 |
|---|---|---|---|---|---|---|---|---|---|
| | $\downarrow$ | | $\downarrow$ | | $\downarrow$ | | $\downarrow$ | | $\downarrow$ |
| Translate: | $\dfrac{x}{5}$ | | $-$ | | $\dfrac{3}{2}$ | | $=$ | | $\dfrac{x}{10}$ |

The LCD is 10.

$$10\left(\frac{x}{5} - \frac{3}{2}\right) = 10\left(\frac{x}{10}\right)$$
$$10\left(\frac{x}{5}\right) - 10\left(\frac{3}{2}\right) = 10\left(\frac{x}{10}\right)$$
$$2x - 15 = x$$
$$x - 15 = 0$$
$$x = 15$$

**Check:** To check, verify that "the quotient of 15 and 5 minus $\dfrac{3}{2}$ is the quotient of 15 and 10," or $\dfrac{15}{5} - \dfrac{3}{2} = \dfrac{15}{10}$.

**6.** Let $x$ = the time in hours it takes Cindy and Mary to complete the job together. Then
$\dfrac{1}{x}$ = the part of the job they complete in 1 hour.

| | Hours to Complete Total Job | Part of Job Completed in 1 Hour |
|---|---|---|
| Cindy | 3 | $\dfrac{1}{3}$ |
| Mary | 4 | $\dfrac{1}{4}$ |
| Together | $x$ | $\dfrac{1}{x}$ |

The part of the job Cindy completes in 1 hour, added to the part of the job Mary completes in 1 hour is equal to the part of the job they complete together in 1 hour.

$$\frac{1}{3}+\frac{1}{4}=\frac{1}{x}$$

$$12x\left(\frac{1}{3}\right)+12x\left(\frac{1}{4}\right)=12x\left(\frac{1}{x}\right)$$

$$4x+3x=12$$

$$7x=12$$

$$x=\frac{12}{7}\text{ or }1\frac{5}{7}$$

**Check:** The proposed solution is reasonable since $1\frac{5}{7}$ hours is more than half of Cindy's time and less than half of Mary's time. Check $1\frac{5}{7}$ hours in the originally stated problem.

Cindy and Mary can complete the garden planting in $1\frac{5}{7}$ hours.

7. Let $x$ = the speed of the bus. Then since the car's speed is 15 mph faster than that of the bus, the speed of the car is $x + 15$.
Since distance = rate · time, or $d = r \cdot t$, then $t = \frac{d}{r}$.
The bus travels 180 miles in the same time that the car travels 240 miles.

|  | Distance = | Rate · | Time |
|---|---|---|---|
| Bus | 180 | $x$ | $\frac{180}{x}$ |
| Car | 240 | $x+15$ | $\frac{240}{x+15}$ |

Since the car and the bus traveled the same amount of time, $\frac{180}{x}=\frac{240}{x+15}$.

$$\frac{180}{x}=\frac{240}{x+15}$$
$$180(x+15)=240x$$
$$180x+2700=240x$$
$$2700=60x$$
$$45=x$$

The speed of the bus is 45 miles per hour. The speed of the car must then be $x + 15$ or 60 miles per hour.
**Check:** Find the time it takes the car to travel 240 miles and the time it takes the bus to travel 180 miles.

Car: $t=\frac{d}{r}=\frac{240}{60}=4$ hours

Bus: $t=\frac{d}{r}=\frac{180}{45}=4$ hours

Since the times are the same, the proposed solution is correct. The speed of the bus is 45 miles per hour and the speed of the car is 60 miles per hour.

**Vocabulary, Readiness & Video Check 7.6**

1. If both people work together, they can complete the job in less time than either person working alone. That is, in less than 5 hours; choice c.

2. If both inlet pipes are on, they can fill the pond in less time than either pipe alone. That is, in less than 25 hours; choice a.

3. A number: $x$

   The reciprocal of the number: $\frac{1}{x}$

   The reciprocal of the number, decreased by 3: $\frac{1}{x}-3$

4. A number: $y$

   The reciprocal of the number: $\frac{1}{y}$

   The reciprocal of the number, increased by 2: $\frac{1}{y}+2$

5. A number: $z$
   The sum of the number and 5: $z + 5$
   The reciprocal of the sum of the number and 5: $\frac{1}{z+5}$

6. A number: $x$
   The difference of the number and 1: $x - 1$
   The reciprocal of the difference of the number and 1: $\frac{1}{x-1}$

7. A number: $y$
   Twice the number: $2y$

   Eleven divided by twice the number: $\frac{11}{2y}$

**8.** A number: $z$

Triple the number: $3z$

Negative 10 divided by triple the number: $\dfrac{-10}{3y}$

**9.** No; proportions are actually equations containing rational expressions, so they can also be solved by using the steps to solve those equations.

**10.** There are also many ways to set up an incorrect proportion, so just checking your solution in your proportion isn't enough. You need to determine if your solution is reasonable from the relationships given in the problem.

**11.** divided by, quotient

**12.** Two machines (or people) take different amounts of time to complete the task, one faster and one slower than the other. When working together, they will complete the task in less time than the faster machine, so your answer must be less than the time of the faster machine.

**13.**

| | $d$ | $=$ | $r$ | $\cdot$ | $t$ |
|---|---|---|---|---|---|
| car | 325 | | $x+7$ | | $\dfrac{325}{x+7}$ |
| motorcycle | 290 | | $x$ | | $\dfrac{290}{x}$ |

$$\dfrac{325}{x+7} = \dfrac{290}{x}$$

**Exercise Set 7.6**

**1.** $\dfrac{2}{3} = \dfrac{x}{6}$

$12 = 3x$

$4 = x$

**3.** $\dfrac{x}{10} = \dfrac{5}{9}$

$9x = 50$

$x = \dfrac{50}{9}$

**5.** $\dfrac{x+1}{2x+3} = \dfrac{2}{3}$

$3(x+1) = 2(2x+3)$

$3x+3 = 4x+6$

$3 = x+6$

$-3 = x$

**7.** $\dfrac{9}{5} = \dfrac{12}{3x+2}$

$9(3x+2) = 5(12)$

$27x+18 = 60$

$27x = 42$

$x = \dfrac{42}{27} = \dfrac{14}{9}$

**9.** Let $x =$ the elephant's weight on Pluto.

$\dfrac{100}{3} = \dfrac{4100}{x}$

$100x = 3(4100)$

$100x = 12{,}300$

$x = 123$

The elephant's weight is 123 pounds.

**11.** Let $x =$ the number of calories in 43.2 grams.

$\dfrac{110}{28.8} = \dfrac{x}{43.2}$

$110(43.2) = 28.8x$

$4752 = 28.8x$

$165 = x$

There are 165 calories in 43.2 grams.

**13.** $\dfrac{16}{10} = \dfrac{34}{y}$

$16y = 340$

$y = 21.25$

**15.** $\dfrac{28}{20} = \dfrac{8}{y}$

$28y = 160$

$y = \dfrac{160}{28} = \dfrac{40}{7}$

$y = 5\dfrac{5}{7}$ feet

**17.** $3 \cdot \dfrac{1}{x} = 9 \cdot \dfrac{1}{6}$

$\dfrac{3}{x} = \dfrac{9}{6}$

$6x\left(\dfrac{3}{x}\right) = 6x\left(\dfrac{9}{6}\right)$

$18 = 9x$

$x = 2$

The unknown number is 2.

**19.** $\dfrac{3+2x}{x+1} = \dfrac{3}{2}$

$2(x+1)\left(\dfrac{3+2x}{x+1}\right) = 2(x+1)\left(\dfrac{3}{2}\right)$

$2(3+2x) = 3(x+1)$

$6+4x = 3x+3$

$x = -3$

The unknown number is −3.

**21.** Let $x$ be the number of hours for the two surveyors to survey the roadbed together.

| | Hours to Complete Total Job | Part of Job Completed in 1 Hour |
|---|---|---|
| Experienced | 4 | $\dfrac{1}{4}$ |
| Apprentice | 5 | $\dfrac{1}{5}$ |
| Together | $x$ | $\dfrac{1}{x}$ |

$\dfrac{1}{4} + \dfrac{1}{5} = \dfrac{1}{x}$

$20x\left(\dfrac{1}{4}\right) + 20x\left(\dfrac{1}{5}\right) = 20x\left(\dfrac{1}{x}\right)$

$5x + 4x = 20$

$9x = 20$

$x = \dfrac{20}{9}$ or $2\dfrac{2}{9}$

The experienced surveyor and apprentice surveyor, working together, can survey the road in $2\dfrac{2}{9}$ hours.

**23.** Let $x$ be the number of minutes it takes the belts working together.

| | Minutes to Complete Total Job | Part of Job Completed in 1 Minute |
|---|---|---|
| Larger belt | 2 | $\dfrac{1}{2}$ |
| Smaller belt | 6 | $\dfrac{1}{6}$ |
| Both belts | $x$ | $\dfrac{1}{x}$ |

$\dfrac{1}{2} + \dfrac{1}{6} = \dfrac{1}{x}$

$6x\left(\dfrac{1}{2}\right) + 6x\left(\dfrac{1}{6}\right) = 6x\left(\dfrac{1}{x}\right)$

$3x + x = 6$

$4x = 6$

$x = \dfrac{6}{4} = \dfrac{3}{2} = 1\dfrac{1}{2}$

Both belts together can move the cans to the storage area in $1\dfrac{1}{2}$ minutes.

**25.** Let $r$ be the jogger's rate. Then, since distance = rate · time, or $d = r \cdot t$, then $t = \dfrac{d}{r}$.

| | Distance = | Rate · | Time |
|---|---|---|---|
| Trip to Park | 12 | $r$ | $\dfrac{12}{r}$ |
| Return Trip | 18 | $r$ | $\dfrac{18}{r}$ |

Since her time on the return trip is 1 hour longer than on the trip to the park, $\dfrac{18}{r} = \dfrac{12}{r} + 1$.

$r\left(\dfrac{18}{r}\right) = r\left(\dfrac{12}{r}\right) + r(1)$

$18 = 12 + r$

$6 = r$

She jogs at 6 miles per hour.

**27.** Let $r$ be his speed on the first portion. Then his speed on the cooldown portion is $r - 2$.

|  | Distance = | Rate · | Time |
|---|---|---|---|
| 1st portion | 20 | $r$ | $\frac{20}{r}$ |
| Cooldown portion | 16 | $r - 2$ | $\frac{16}{r-2}$ |

$$\frac{20}{r} = \frac{16}{r-2}$$
$$20(r-2) = 16r$$
$$20r - 40 = 16r$$
$$-40 = -4r$$
$$r = 10$$

and $r - 2 = 10 - 2 = 8$

His speed was 10 miles per hour during the first portion and 8 miles per hour during the cooldown portion.

**29.** Let $x =$ the minimum floor space needed by 40 students.
$$\frac{1}{9} = \frac{40}{x}$$
$$1x = 9(40)$$
$$x = 360$$
40 students need 360 square feet.

**31.**
$$\frac{1}{4} = \frac{x}{8}$$
$$8\left(\frac{1}{4}\right) = 8\left(\frac{x}{8}\right)$$
$$2 = x$$
The unknown number is 2.

**33.** Let $x$ be the amount of time it takes Marcus and Tony working together.

|  | Hours to Complete Total Job | Part of Job Completed in 1 Hour |
|---|---|---|
| Marcus | 6 | $\frac{1}{6}$ |
| Tony | 4 | $\frac{1}{4}$ |
| Together | $x$ | $\frac{1}{x}$ |

$$\frac{1}{6} + \frac{1}{4} = \frac{1}{x}$$
$$12x\left(\frac{1}{6}\right) + 12x\left(\frac{1}{4}\right) = 12x\left(\frac{1}{x}\right)$$
$$2x + 3x = 12$$
$$5x = 12$$
$$x = \frac{12}{5} = 2\frac{2}{5}$$
$$45\left(\frac{12}{5}\right) = 108$$

Together Marcus and Tony work for $2\frac{2}{5}$ hours at \$45 per hour. The labor estimate should be \$108.

**35.** Let $w$ be the speed of the wind.

|  | Distance = | Rate · | Time |
|---|---|---|---|
| With wind | 400 | $230 + w$ | $\frac{400}{230+w}$ |
| Against wind | 336 | $230 - w$ | $\frac{336}{230-w}$ |

Since the time with the wind is the same as the time against the wind, $\frac{336}{230-w} = \frac{400}{230+w}$.
$$\frac{336}{230-w} = \frac{400}{230+w}$$
$$336(230 + w) = 400(230 - w)$$
$$77{,}280 + 336w = 92{,}000 - 400w$$
$$736w = 14{,}720$$
$$w = 20$$
The speed of the wind is 20 miles per hour.

**37.**
$$\frac{y}{25} = \frac{3}{2}$$
$$y \cdot 2 = 25 \cdot 3$$
$$y \cdot 2 = 75$$
$$y = \frac{75}{2}$$
$$y = 37\frac{1}{2}$$

The unknown length is $37\frac{1}{2}$ feet.

**39.** Let $x$ be the speed of the slower train. In 3.5 hours, the slower train travels $3.5x$ miles and the faster train travels $3.5(x + 10)$ miles.

$$3.5x + 3.5(x+10) = 322$$
$$3.5x + 3.5x + 35 = 322$$
$$7x + 35 = 322$$
$$7x = 287$$
$$x = 41$$

The slower train travels 41 mph and the faster train travels $41 + 10 = 51$ mph.

**41.**
$$\frac{2}{x-3} - \frac{4}{x+3} = 8 \cdot \frac{1}{x^2 - 9}$$
$$(x-3)(x+3)\left(\frac{2}{x-3} - \frac{4}{x+3}\right) = (x-3)(x+3)\left(\frac{8}{x^2-9}\right)$$
$$(x-3)(x+3)\left(\frac{2}{x-3}\right) - (x-3)(x+3)\left(\frac{4}{x+3}\right) = 8$$
$$2(x+3) - 4(x-3) = 8$$
$$2x + 6 - 4x + 12 = 8$$
$$-2x = -10$$
$$x = 5$$

The unknown number is 5.

**43.** Let $r$ be the rate of the plane in still air.

| | Distance = | Rate | · | Time |
|---|---|---|---|---|
| With wind | 630 | $r + 35$ | | $\frac{630}{r+35}$ |
| Against wind | 455 | $r - 35$ | | $\frac{455}{r-35}$ |

$$\frac{630}{r+35} = \frac{455}{r-35}$$
$$630(r-35) = 455(r+35)$$
$$630r - 22{,}050 = 455r + 15{,}925$$
$$175r = 37{,}975$$
$$r = 217$$

The speed in still air is 217 mph.

**45.** Let $x =$ the number of gallons of water needed.

$$\frac{8}{2} = \frac{36}{x}$$
$$8x = 2(36)$$
$$8x = 72$$
$$x = 9$$

Nine gallons of water are needed for the entire box.

**47.**

| | r × t = d | | |
|---|---|---|---|
| With wind | $16+x$ | $\dfrac{48}{16+x}$ | 48 |
| Into Wind | $16-x$ | $\dfrac{16}{16-x}$ | 16 |

Since the times are the same, $\dfrac{48}{16+x}=\dfrac{16}{16-x}$.

$$\dfrac{48}{16+x}=\dfrac{16}{16-x}$$
$$48(16-x)=16(16+x)$$
$$768-48x=256+16x$$
$$512=64x$$
$$8=x$$

The rate of the wind is 8 miles per hour.

**49.** Let $x$ be the rate of the slower hiker. Then the rate of the faster hiker is $x + 1.1$. In 2 hours, the slower hiker walks $2x$ miles, while the faster hiker walks $2(x + 1.1)$ miles.
$$2x+2(x+1.1)=11$$
$$2x+2x+2.2=11$$
$$4x+2.2=11$$
$$4x=8.8$$
$$x=2.2$$
$x + 1.1 = 2.2 + 1.1 = 3.3$
The hikers walk 2.2 miles per hour and 3.3 miles per hour.

**51.** Let $x$ be the amount of time it takes the second worker to do the job alone.

| | Hours to Complete Total Job | Part of Job Completed in 1 Hour |
|---|---|---|
| Custodian | 3 | $\dfrac{1}{3}$ |
| 2nd Worker | $x$ | $\dfrac{1}{x}$ |
| Together | $1\frac{1}{2}$ or $\frac{3}{2}$ | $\dfrac{2}{3}$ |

$$\dfrac{1}{3}+\dfrac{1}{x}=\dfrac{2}{3}$$
$$3x\left(\dfrac{1}{3}\right)+3x\left(\dfrac{1}{x}\right)=3x\left(\dfrac{2}{3}\right)$$
$$x+3=2x$$
$$3=x$$

It takes the second worker 3 hours to do the job alone.

**53.** Let $x$ be the missing dimension.
$$\dfrac{x}{8}=\dfrac{20}{6}$$
$$6x=8\cdot20$$
$$x=\dfrac{160}{6}$$
$$x=\dfrac{80}{3}=26\dfrac{2}{3}$$
The side is $26\dfrac{2}{3}$ feet long.

**55.**
$$\dfrac{3}{2}=\dfrac{324}{x}$$
$$3\cdot x=2\cdot324$$
$$3x=648$$
$$x=\dfrac{648}{3}=216$$
There should be 216 other nuts in the can.

**57.** Let $t$ be the time in hours that the jet plane travels.

| | distance | rate | time |
|---|---|---|---|
| jet plane | $500t$ | 500 | $t$ |
| propeller plane | $200(t + 2)$ | 200 | $t + 2$ |

$$500t=200(t+2)$$
$$500t=200t+400$$
$$300t=400$$
$$t=\dfrac{400}{300}$$
$$t=\dfrac{4}{3}$$

$$\text{distance}=500t=500\left(\dfrac{4}{3}\right)=666\dfrac{2}{3}$$

The planes are $666\dfrac{2}{3}$ miles from the starting point.

**59.** Let $x$ be the time that it takes the third pipe to fill the pool alone.
$$\dfrac{1}{20}+\dfrac{1}{15}+\dfrac{1}{x}=\dfrac{1}{6}$$
$$60x\left(\dfrac{1}{20}+\dfrac{1}{15}+\dfrac{1}{x}\right)=60x\left(\dfrac{1}{6}\right)$$
$$3x+4x+60=10x$$
$$7x+60=10x$$
$$60=3x$$
$$20=x$$

It will take the third pump 20 hours to do the job alone.

**61.** Let $r$ be the motorcycle's speed.

|  | distance | rate | time |
|---|---|---|---|
| car | 280 | $r + 10$ | $\frac{280}{r+10}$ |
| motorcycle | 240 | $r$ | $\frac{240}{r}$ |

$$\frac{280}{r+10} = \frac{240}{r}$$
$$280r = 240(r+10)$$
$$280r = 240r + 2400$$
$$40r = 2400$$
$$r = 60$$
$$r + 10 = 60 + 10 = 70$$

The motorcycle's speed was 60 miles per hour and the car's speed was 70 miles per hour.

**63.** Let $x$ be the time for the third cook to prepare the same number of pies.
$$\frac{1}{6} + \frac{1}{7} + \frac{1}{x} = \frac{1}{2}$$
$$42x\left(\frac{1}{6} + \frac{1}{7} + \frac{1}{x}\right) = 42x\left(\frac{1}{2}\right)$$
$$7x + 6x + 42 = 21x$$
$$13x + 42 = 21x$$
$$42 = 8x$$
$$\frac{42}{8} = x$$

$$\frac{42}{8} = \frac{21}{4} = 5\frac{1}{4}$$

It will take the third cook $5\frac{1}{4}$ hours to prepare the pies working alone.

**65.** Let $x$ be the number.
$$\frac{x}{3} - 1 = \frac{5}{3}$$
$$3\left(\frac{x}{3} - 1\right) = 3\left(\frac{5}{3}\right)$$
$$x - 3 = 5$$
$$x = 8$$
The number is 8.

**67.** Let $x$ be the speed of the second car.

|  | distance | rate | time |
|---|---|---|---|
| first car | 224 | $x + 14$ | $\frac{224}{x+14}$ |
| second car | 175 | $x$ | $\frac{175}{x}$ |

$$\frac{224}{x+14} = \frac{175}{x}$$
$$224x = 175(x+14)$$
$$224x = 175x + 2450$$
$$49x = 2450$$
$$x = 50$$
$$x + 14 = 50 + 14 = 64$$

The speed of the first car is 64 miles per hour and the speed of the second car is 50 miles per hour.

**69.** Let $x$ be the speed of the plane in still air.

|  | distance | rate | time |
|---|---|---|---|
| with wind | 2160 | $x + 30$ | $\frac{2160}{x+30}$ |
| against wind | 1920 | $x - 30$ | $\frac{1920}{x-30}$ |

$$\frac{2160}{x+30} = \frac{1920}{x-30}$$
$$2160(x-30) = 1920(x+30)$$
$$2160x - 64,800 = 1920x + 57,600$$
$$240x = 122,400$$
$$x = 510$$

The speed of the plane in still air is 510 miles per hour.

**71.** $\frac{9}{12} = \frac{3.75}{x}$
$$9x = 45$$
$$x = 5$$
The missing length is 5.

**73.** $\frac{16}{24} = \frac{9}{x}$
$$16x = 216$$
$$x = 13.5$$
The missing length is 13.5.

**75.** $(-2, 5), (4, -3)$

$$m = \frac{-3-5}{4-(-2)} = \frac{-8}{6} = -\frac{4}{3}$$

Since the slope is negative, the line moves downward.

**77.** $(-3, -6), (1, 5)$

$$m = \frac{5-(-6)}{1-(-3)} = \frac{11}{4}$$

Since the slope is positive, the line moves upward.

**79.** $(3, 7), (3, -2)$

$$m = \frac{-2-7}{3-3} = \frac{-9}{0}$$

The slope is undefined. Since the slope is undefined, the line is vertical.

**81.** The capacity in 2010 was approximately 40,200 megawatts.

**83.** The capacity in 2010 was approximately 40,200 megawatts, or 40.2(1000 megawatts).
40.2(560,000) = 22,512,000
In 2010, the number of megawatts generated from wind would serve the electricity needs of 22,512,000 people.

**85.** Yes, since each side of the equation is one quotient.

**87.** Let $x$ be the number of minutes it takes the first pump to fill the tank. Then it takes $3x$ minutes for the second pump to fill the tank.

|  | Minutes to Complete Total Job | Part of Job Completed in 1 Minute |
|---|---|---|
| 1st Pump | $x$ | $\frac{1}{x}$ |
| 2nd Pump | $3x$ | $\frac{1}{3x}$ |
| Together | 21 | $\frac{1}{21}$ |

$$\frac{1}{x} + \frac{1}{3x} = \frac{1}{21}$$

$$21x\left(\frac{1}{x}\right) + 21x\left(\frac{1}{3x}\right) = 21x\left(\frac{1}{21}\right)$$

$$21 + 7 = x$$

$$28 = x$$

$3x + 3(28) = 84$

The 1st pump takes 28 minutes and the 2nd takes 84 minutes.

**89.** none; answers may vary

**91.** answers may vary

**93.** $D = RT$

$$\frac{D}{T} = \frac{RT}{T}$$

$$\frac{D}{T} = R \text{ or } R = \frac{D}{T}$$

**Section 7.7 Practice**

**1.** Use the ordered pair (2, 10) in the direct variation equation $y = kx$.

$$y = kx$$
$$10 = k \cdot 2$$
$$\frac{10}{2} = \frac{k \cdot 2}{2}$$
$$5 = k$$

Since $k = 5$, the equation is $y = 5x$. To check, see that each given $y$ is 5 times the given $x$.

**2.** Since $y$ varies directly as $x$, the relationship is of the form $y = kx$. Let $y = 12$ and $x = 48$.

$$12 = k \cdot 48$$
$$\frac{12}{48} = \frac{k \cdot 48}{48}$$
$$\frac{1}{4} = k$$

The constant of variation is $\frac{1}{4}$ and the equation is $y = \frac{1}{4}x$. Now replace $x$ with 20.

$$y = \frac{1}{4}x$$
$$y = \frac{1}{4} \cdot 20$$
$$y = 5$$

Thus, when $x$ is 20, $y$ is 5.

**3.** The constant of variation is the same as the slope of the line.

$$\text{slope} = \frac{6-0}{8-0} = \frac{6}{8} = \frac{3}{4}$$

Thus, $k = \frac{3}{4}$ and the variation equation is

$$y = \frac{3}{4}x.$$

4. Use the ordered pair (2, 4) in the inverse

variation equation $y = \dfrac{k}{x}$.

$$y = \frac{k}{x}$$
$$4 = \frac{k}{2}$$
$$2 \cdot 4 = 2 \cdot \frac{k}{2}$$
$$8 = k$$

Since $k = 8$, the equation is $y = \dfrac{8}{x}$.

5. Since $y$ varies inversely as $x$, the constant of variation is the product of the given $x$ and $y$.
$k = xy = (42)(0.05) = 2.1$
The constant of variation is 2.1 and the equation is $y = \dfrac{2.1}{x}$. Now replace $x$ with 70.

$$y = \frac{2.1}{x}$$
$$y = \frac{2.1}{70}$$
$$y = 0.03$$

Thus, when $x$ is 70, $y$ is 0.03.

6. Since the area $A$ varies directly as the square of one of its legs $x$, the equation is $A = kx^2$.
Let $A = 32$ and $x = 8$.

$$A = k \cdot x^2$$
$$32 = k \cdot 8^2$$
$$32 = k \cdot 64$$
$$\frac{1}{2} = k$$

The formula for the area of an isosceles right triangle is then $A = \dfrac{1}{2}x^2$ where $x$ is the length of one leg. Substitute 3.6 for $x$.

$$A = \frac{1}{2}x^2$$
$$A = \frac{1}{2}(3.6)^2$$
$$A = 6.48$$

The area of an isosceles right triangle whose legs measure 3.6 units is 6.48 square units.

7. Since the volume of gas varies inversely with pressure, the equation is $V = \dfrac{k}{P}$.
To find $k$, let $V = 50$ and $P = 20$.

$$V = \frac{k}{P}$$
$$50 = \frac{k}{20}$$
$$1000 = k$$

The equation of variation is $V = \dfrac{1000}{P}$.

Now let $P = 40$.

$$V = \frac{1000}{P}$$
$$V = \frac{1000}{40}$$
$$V = 25$$

At a pressure of 40 atmospheres, the volume of the oxygen is 25 ml.

**Vocabulary, Readiness & Video Check 7.7**

1. $y = \dfrac{k}{x}$, where $k$ is a constant is inverse variation.

2. $y = kx$, where $k$ is a constant is direct variation.

3. $y = 5x$ is direct variation.

4. $y = \dfrac{5}{x}$ is inverse variation.

5. $y = \dfrac{7}{x^2}$ is inverse variation.

6. $y = 6.5x^4$ is direct variation.

7. $y = \dfrac{11}{x}$ is inverse variation.

8. $y = 18x$ is direct variation.

9. $y = 12x^2$ is direct variation.

10. $y = \dfrac{20}{x^3}$ is inverse variation.

11. linear; slope

**12.** With inverse variation, we know that $y = \dfrac{k}{x}$ or

$yx = k$, so we just need to multiply the values of $x$ and $y$ together to find $k$.

**13.** No; the direct relationship is the power of $x$ times a constant, and the inverse relationship is the reciprocal of the power of $x$ times a constant.

**14.** This is a direct variation problem, $y = kx$ (with $k$ positive), so as either amount ($y$ or $x$) increases the other amount also increases. We're asked to find the new distance given a weight increase, so we know that our answer will also show a distance increase.

**Exercise Set 7.7**

**1.** $y = kx$

$3 = k(6)$

$\dfrac{3}{6} = k$

$\dfrac{1}{2} = k$

$y = \dfrac{1}{2}x$

**3.** $y = kx$

$-12 = k(-2)$

$6 = k$

$y = 6x$

**5.** $k = \text{slope} = \dfrac{3-0}{1-0} = \dfrac{3}{1} = 3$

$y = 3x$

**7.** $k = \text{slope} = \dfrac{2-0}{3-0} = \dfrac{2}{3}$

$y = \dfrac{2}{3}x$

**9.** $y = \dfrac{k}{x}$

$7 = \dfrac{k}{1}$

$7 = k$

$y = \dfrac{7}{x}$

**11.** $y = \dfrac{k}{x}$

$0.05 = \dfrac{k}{10}$

$0.5 = k$

$y = \dfrac{0.5}{x}$

**13.** $y = kx$

**15.** $h = \dfrac{k}{t}$

**17.** $z = kx^2$

**19.** $y = \dfrac{k}{z^3}$

**21.** $x = \dfrac{k}{\sqrt{y}}$

**23.** $y = kx$

$20 = k(5)$

$4 = k$

$y = 4x$

$y = 4(10)$

$y = 40$

**25.** $y = \dfrac{k}{x}$

$5 = \dfrac{k}{60}$

$300 = k$

$y = \dfrac{300}{x}$

$y = \dfrac{300}{100}$

$y = 3$

**27.** $z = kx^2$

$96 = k(4)^2$

$96 = 16k$

$6 = k$

$z = 6x^2$

$z = 6(3)^2$

$z = 6(9)$

$z = 54$

**29.** $a = \dfrac{k}{b^3}$

$\dfrac{3}{2} = \dfrac{k}{2^3}$

$\dfrac{3}{2} = \dfrac{k}{8}$

$2k = 24$

$k = 12$

$a = \dfrac{12}{b^3}$

$a = \dfrac{12}{3^3}$

$a = \dfrac{12}{27}$

$a = \dfrac{4}{9}$

**31.** $p = kh$

$112.50 = k(18)$

$6.25 = k$

$p = 6.25h$

$p = 6.25(10)$

$p = 62.5$

Your pay is $62.50 for 10 hours.

**33.** $x = \dfrac{k}{n}$

$9.00 = \dfrac{k}{5000}$

$45,000 = k$

$c = \dfrac{45,000}{n}$

$c = \dfrac{45,000}{7500}$

$c = 6$

The cost is $6.00 per headphone to manufacture 7500 headphones.

**35.** $d = kw$

$4 = k(60)$

$\dfrac{4}{60} = k$

$\dfrac{1}{15} = k$

$d = \dfrac{1}{15}w$

$d = \dfrac{1}{15}(80)$

$d = \dfrac{80}{15}$

$d = 5\dfrac{1}{3}$

The spring stretches $5\dfrac{1}{3}$ inches with an 80-pound weight.

**37.** $w = \dfrac{k}{d^2}$

$180 = \dfrac{k}{4000^2}$

$180 = \dfrac{k}{16,000,000}$

$2,880,000,000 = k$

$w = \dfrac{2,880,000,000}{d^2}$

$w = \dfrac{2,880,000,000}{4010^2}$

$w = \dfrac{2,880,000,000}{16,080,100}$

$w \approx 179.1$

His weight 10 miles above Earth's surface is 179.1 pounds.

**39.** $d = kt^2$

$64 = k(2)^2$

$64 = 4k$

$16 = k$

$d = 16t^2$

$d = 16(10)^2$

$d = 16(100)$

$d = 1600$

He falls 1600 feet in 10 seconds.

**41.** $\dfrac{\frac{3}{4}+\frac{1}{4}}{\frac{3}{8}+\frac{13}{8}} = \dfrac{\frac{3+1}{4}}{\frac{3+13}{8}} = \dfrac{\frac{4}{4}}{\frac{16}{8}} = \dfrac{1}{2}$

**43.** $\dfrac{\frac{2}{5}+\frac{1}{5}}{\frac{7}{10}+\frac{7}{10}} = \dfrac{\frac{2+1}{5}}{\frac{7+7}{10}}$

$= \dfrac{\frac{3}{5}}{\frac{14}{10}}$

$= \dfrac{3}{5} \div \dfrac{14}{10}$

$= \dfrac{3}{5} \cdot \dfrac{10}{14}$

$= \dfrac{3 \cdot 2 \cdot 5}{5 \cdot 2 \cdot 7}$

$= \dfrac{3}{7}$

**45.** $y = kx$
If $x$ is tripled, $y$ is also tripled.

**47.** $P = k\sqrt{l}$
The result of quadrupling $l$ is $4l$.
$\sqrt{4l} = \sqrt{4} \cdot \sqrt{l} = 2\sqrt{l}$
Thus, when the length of the pendulum is quadrupled, the period is doubled.

**Section 7.8 Practice**

**1.** $\dfrac{\frac{3}{4}}{\frac{6}{11}} = \dfrac{3}{4} \div \dfrac{6}{11} = \dfrac{3}{4} \cdot \dfrac{11}{6} = \dfrac{3 \cdot 11}{4 \cdot 2 \cdot 3} = \dfrac{11}{8}$

**2.** $\dfrac{\frac{3}{4}+\frac{2}{3}}{\frac{3}{4}-\frac{1}{5}} = \dfrac{\frac{3(3)}{4(3)}+\frac{2(4)}{3(4)}}{\frac{3(5)}{4(5)}-\frac{1(4)}{5(4)}}$

$= \dfrac{\frac{9}{12}+\frac{8}{12}}{\frac{15}{20}-\frac{4}{20}}$

$= \dfrac{\frac{17}{12}}{\frac{11}{20}}$

$= \dfrac{17}{12} \cdot \dfrac{20}{11}$

$= \dfrac{17 \cdot 4 \cdot 5}{3 \cdot 4 \cdot 11}$

$= \dfrac{85}{33}$

**3.** $\dfrac{\frac{4}{x}-\frac{1}{2}}{\frac{1}{5}-\frac{x}{10}} = \dfrac{\frac{8}{2x}-\frac{x}{2x}}{\frac{2}{10}-\frac{x}{10}}$

$= \dfrac{\frac{8-x}{2x}}{\frac{2-x}{10}}$

$= \dfrac{8-x}{2x} \cdot \dfrac{10}{2-x}$

$= \dfrac{2 \cdot 5(8-x)}{2 \cdot x(2-x)}$

$= \dfrac{5(8-x)}{x(2-x)}$

**4.** The LCD of $\dfrac{3}{4}, \dfrac{2}{3}, \dfrac{3}{4}$, and $\dfrac{1}{5}$ is 60.

$\dfrac{\frac{3}{4}+\frac{2}{3}}{\frac{3}{4}-\frac{1}{5}} = \dfrac{60\left(\frac{3}{4}+\frac{2}{3}\right)}{60\left(\frac{3}{4}-\frac{1}{5}\right)}$

$= \dfrac{60\left(\frac{3}{4}\right)+60\left(\frac{2}{3}\right)}{60\left(\frac{3}{4}\right)-60\left(\frac{1}{5}\right)}$

$= \dfrac{45+40}{45-12}$

$= \dfrac{85}{33}$

**5.** The LCD of $\dfrac{a-b}{b}, \dfrac{a}{b}$, and $\dfrac{4}{1}$ is $b$.

$\dfrac{\frac{a-b}{b}}{\frac{a}{b}+4} = \dfrac{b\left(\frac{a-b}{b}\right)}{b\left(\frac{a}{b}+4\right)} = \dfrac{b\left(\frac{a-b}{b}\right)}{b\left(\frac{a}{b}\right)+b(4)} = \dfrac{a-b}{a+4b}$

**6.** The LCD of $\dfrac{4}{3b}, \dfrac{b}{a}, \dfrac{a}{3}$, and $\dfrac{b}{1}$ is $3ab$.

$\dfrac{\frac{4}{3b}+\frac{b}{a}}{\frac{a}{3}-b} = \dfrac{3ab\left(\frac{4}{3b}+\frac{b}{a}\right)}{3ab\left(\frac{a}{3}-b\right)}$

$= \dfrac{3ab\left(\frac{4}{3b}\right)+3ab\left(\frac{b}{a}\right)}{3ab\left(\frac{a}{3}\right)-3ab(b)}$

$= \dfrac{4a+3b^2}{a^2b-3ab^2}$ or $\dfrac{4a+3b^2}{ab(a-3b)}$

**Vocabulary, Readiness & Video Check 7.8**

**1.** $\dfrac{\frac{y}{2}}{\frac{5x}{2}} = \dfrac{2\left(\frac{y}{2}\right)}{2\left(\frac{5x}{2}\right)} = \dfrac{y}{5x}$

**2.** $\dfrac{\frac{10}{x}}{\frac{z}{x}} = \dfrac{x\left(\frac{10}{x}\right)}{x\left(\frac{z}{x}\right)} = \dfrac{10}{z}$

**3.** $\dfrac{\frac{3}{x}}{\frac{5}{x^2}} = \dfrac{x^2\left(\frac{3}{x}\right)}{x^2\left(\frac{5}{x^2}\right)} = \dfrac{3x}{5}$

**4.** $\dfrac{\frac{a}{10}}{\frac{b}{20}} = \dfrac{20\left(\frac{a}{10}\right)}{20\left(\frac{b}{20}\right)} = \dfrac{2a}{b}$

**5.** a single fraction in the numerator and the denominator

**6.** In Method 2, you find the LCD of all fractions in the complex fraction, then multiply both the numerator and the denominator so that both will no longer contain fractions; in Method 1, you find the LCD of the fractions only in the numerator or only in the denominator in order to get a single fraction in the numerator and/or denominator.

**Exercise Set 7.8**

**1.** $\dfrac{\frac{1}{2}}{\frac{3}{4}} = \dfrac{1}{2} \cdot \dfrac{4}{3} = \dfrac{1 \cdot 2 \cdot 2}{2 \cdot 3} = \dfrac{2}{3}$

**3.** $\dfrac{-\frac{4x}{9}}{-\frac{2x}{3}} = -\dfrac{4x}{9} \cdot -\dfrac{3}{2x} = \dfrac{2 \cdot 2 \cdot 3 \cdot x}{3 \cdot 3 \cdot 2 \cdot x} = \dfrac{2}{3}$

**5.** $\dfrac{\frac{1+x}{6}}{\frac{1+x}{3}} = \dfrac{1+x}{6} \cdot \dfrac{3}{1+x} = \dfrac{3 \cdot (1+x)}{2 \cdot 3 \cdot (1+x)} = \dfrac{1}{2}$

**7.** $\dfrac{\frac{1}{2}+\frac{2}{3}}{\frac{5}{9}-\frac{5}{6}} = \dfrac{\frac{1}{2} \cdot \frac{3}{3}+\frac{2}{3} \cdot \frac{2}{2}}{\frac{5}{9} \cdot \frac{2}{2}-\frac{5}{6} \cdot \frac{3}{3}}$

$= \dfrac{\frac{3}{6}+\frac{4}{6}}{\frac{10}{18}-\frac{15}{18}}$

$= \dfrac{\frac{7}{6}}{-\frac{5}{18}}$

$= \dfrac{7}{6} \cdot -\dfrac{18}{5}$

$= -\dfrac{7 \cdot 3 \cdot 6}{6 \cdot 5}$

$= -\dfrac{21}{5}$

**9.** $\dfrac{2+\frac{7}{10}}{1+\frac{3}{5}} = \dfrac{10\left(2+\frac{7}{10}\right)}{10\left(1+\frac{3}{5}\right)}$

$= \dfrac{10(2)+10\left(\frac{7}{10}\right)}{10(1)+10\left(\frac{3}{5}\right)}$

$= \dfrac{20+7}{10+6}$

$= \dfrac{27}{16}$

**11.** $\dfrac{\frac{1}{3}}{\frac{1}{2}-\frac{1}{4}} = \dfrac{12\left(\frac{1}{3}\right)}{12\left(\frac{1}{2}-\frac{1}{4}\right)} = \dfrac{12\left(\frac{1}{3}\right)}{12\left(\frac{1}{2}\right)-12\left(\frac{1}{4}\right)} = \dfrac{4}{6-3} = \dfrac{4}{3}$

**13.** $\dfrac{-\frac{2}{9}}{-\frac{14}{3}} = -\dfrac{2}{9} \cdot -\dfrac{3}{14} = \dfrac{2 \cdot 3}{3 \cdot 3 \cdot 2 \cdot 7} = \dfrac{1}{21}$

**15.** $\dfrac{-\frac{5}{12x^2}}{\frac{25}{16x^3}} = -\dfrac{5}{12x^2} \cdot \dfrac{16x^3}{25} = -\dfrac{5 \cdot 4 \cdot 4 \cdot x^2 \cdot x}{4 \cdot 3 \cdot x^2 \cdot 5 \cdot 5} = -\dfrac{4x}{15}$

**17.** $\dfrac{\frac{m}{n}-1}{\frac{m}{n}+1} = \dfrac{n\left(\frac{m}{n}-1\right)}{n\left(\frac{m}{n}+1\right)} = \dfrac{n\left(\frac{m}{n}\right)-n(1)}{n\left(\frac{m}{n}\right)+n(1)} = \dfrac{m-n}{m+n}$

**19.** $\dfrac{\frac{1}{5}-\frac{1}{x}}{\frac{7}{10}+\frac{1}{x^2}} = \dfrac{10x^2\left(\frac{1}{5}-\frac{1}{x}\right)}{10x^2\left(\frac{7}{10}+\frac{1}{x^2}\right)}$

$= \dfrac{10x^2\left(\frac{1}{5}\right)-10x^2\left(\frac{1}{x}\right)}{10x^2\left(\frac{7}{10}\right)+10x^2\left(\frac{1}{x^2}\right)}$

$= \dfrac{2x^2-10x}{7x^2+10}$

$= \dfrac{2x(x-5)}{7x^2+10}$

**21.** $\dfrac{1+\frac{1}{y-2}}{y+\frac{1}{y-2}} = \dfrac{(y-2)\left(1+\frac{1}{y-2}\right)}{(y-2)\left(y+\frac{1}{y-2}\right)}$

$= \dfrac{(y-2)(1)+(y-2)\left(\frac{1}{y-2}\right)}{(y-2)(y)+(y-2)\left(\frac{1}{y-2}\right)}$

$= \dfrac{y-2+1}{y^2-2y+1}$

$= \dfrac{y-1}{(y-1)^2}$

$= \dfrac{1}{y-1}$

**23.** $\dfrac{\frac{4y-8}{16}}{\frac{6y-12}{4}} = \dfrac{4y-8}{16}\cdot\dfrac{4}{6y-12} = \dfrac{4(y-2)\cdot 4}{4\cdot 4\cdot 6(y-2)} = \dfrac{1}{6}$

**25.** $\dfrac{\frac{x}{y}+1}{\frac{x}{y}-1} = \dfrac{y\left(\frac{x}{y}+1\right)}{y\left(\frac{x}{y}-1\right)} = \dfrac{y\left(\frac{x}{y}\right)+y(1)}{y\left(\frac{x}{y}\right)-y(1)} = \dfrac{x+y}{x-y}$

**27.** $\dfrac{1}{2+\frac{1}{3}} = \dfrac{3(1)}{3\left(2+\frac{1}{3}\right)} = \dfrac{3(1)}{3(2)+3\left(\frac{1}{3}\right)} = \dfrac{3}{6+1} = \dfrac{3}{7}$

**29.** $\dfrac{\frac{ax+ab}{x^2-b^2}}{\frac{x+b}{x-b}} = \dfrac{ax+ab}{x^2-b^2}\cdot\dfrac{x-b}{x+b}$

$= \dfrac{a(x+b)\cdot(x-b)}{(x+b)(x-b)\cdot(x+b)}$

$= \dfrac{a}{x+b}$

**31.** $\dfrac{\frac{-3+y}{4}}{\frac{8+y}{28}} = \dfrac{-3+y}{4}\cdot\dfrac{28}{8+y} = \dfrac{4\cdot 7\cdot(-3+y)}{4\cdot(8+y)} = \dfrac{7(y-3)}{8+y}$

**33.** $\dfrac{3+\frac{12}{x}}{1-\frac{16}{x^2}} = \dfrac{x^2\left(3+\frac{12}{x}\right)}{x^2\left(1-\frac{16}{x^2}\right)}$

$= \dfrac{x^2(3)+x^2\left(\frac{12}{x}\right)}{x^2(1)-x^2\left(\frac{16}{x^2}\right)}$

$= \dfrac{3x^2+12x}{x^2-16}$

$= \dfrac{3x(x+4)}{(x-4)(x+4)}$

$= \dfrac{3x}{x-4}$

**35.** $\dfrac{\frac{8}{x+4}+2}{\frac{12}{x+4}-2} = \dfrac{(x+4)\left(\frac{8}{x+4}+2\right)}{(x+4)\left(\frac{12}{x+4}-2\right)}$

$= \dfrac{(x+4)\left(\frac{8}{x+4}\right)+(x+4)(2)}{(x+4)\left(\frac{12}{x+4}\right)-(x+4)(2)}$

$= \dfrac{8+2x+8}{12-2x-8}$

$= \dfrac{16+2x}{4-2x}$

$= \dfrac{2(8+x)}{2(2-x)}$

$= \dfrac{8+x}{2-x}$

$= -\dfrac{x+8}{x-2}$

**37.** $\dfrac{\frac{s}{r}+\frac{r}{s}}{\frac{s}{r}-\frac{r}{s}} = \dfrac{rs\left(\frac{s}{r}+\frac{r}{s}\right)}{rs\left(\frac{s}{r}-\frac{r}{s}\right)} = \dfrac{rs\left(\frac{s}{r}\right)+rs\left(\frac{r}{s}\right)}{rs\left(\frac{s}{r}\right)-rs\left(\frac{r}{s}\right)} = \dfrac{s^2+r^2}{s^2-r^2}$

**39.**
$$\frac{\frac{6}{x-5}+\frac{x}{x-2}}{\frac{3}{x-6}-\frac{2}{x-5}}=\frac{(x-5)(x-2)(x-6)\left(\frac{6}{x-5}+\frac{x}{x-2}\right)}{(x-5)(x-2)(x-6)\left(\frac{3}{x-6}-\frac{2}{x-5}\right)}$$

$$=\frac{(x-5)(x-2)(x-6\frac{6}{x-5})+(x-5)(x-2)(x-6)\left(\frac{x}{x-2}\right)}{(x-5)(x-2)(x-6)\left(\frac{3}{x-6}\right)-(x-5)(x-2)(x-6)\left(\frac{2}{x-5}\right)}$$

$$=\frac{6(x-2)(x-6)+x(x-5)(x-6)}{3(x-5)(x-2)-2(x-2)(x-6)}$$

$$=\frac{6x^2-48x+72+x^3-11x^2+30x}{3x^2-21x+30-2x^2+16x-24}$$

$$=\frac{x^3-5x^2-18x+72}{x^2-5x+6}$$

$$=\frac{(x-6)(x-3)(x+4)}{(x-2)(x-3)}$$

$$=\frac{(x-6)(x+4)}{x-2}$$

**41.** The longest bar corresponds to Serena Williams, so Serena Williams has won the most prize money in her career.

**43.** $30.5-22.1=8.4$
The approximate spread in lifetime prize money between Lindsay Davenport and Serena Williams is about $8.4 million.

**45.** answers may vary

**47.** $\dfrac{\frac{1}{3}+\frac{3}{4}}{2}=\dfrac{12\left(\frac{1}{3}+\frac{3}{4}\right)}{12(2)}=\dfrac{12\left(\frac{1}{3}\right)+12\left(\frac{3}{4}\right)}{12(2)}=\dfrac{4+9}{24}=\dfrac{13}{24}$

**49.**
$$\frac{1}{\frac{1}{R_1}+\frac{1}{R_2}}=\frac{R_1R_2(1)}{R_1R_2\left(\frac{1}{R_1}+\frac{1}{R_2}\right)}$$

$$=\frac{R_1R_2}{R_1R_2\left(\frac{1}{R_1}\right)+R_1R_2\left(\frac{1}{R_2}\right)}$$

$$=\frac{R_1R_2}{R_2+R_1}$$

**51.** $\dfrac{x^{-1}+2^{-1}}{x^{-2}-4^{-1}} = \dfrac{\frac{1}{x}+\frac{1}{2}}{\frac{1}{x^2}-\frac{1}{4}}$

$= \dfrac{4x^2\left(\frac{1}{x}+\frac{1}{2}\right)}{4x^2\left(\frac{1}{x^2}-\frac{1}{4}\right)}$

$= \dfrac{4x+2x^2}{4-x^2}$

$= \dfrac{2x^2+4x}{-(x^2-4)}$

$= -\dfrac{2x(x+2)}{(x+2)(x-2)}$

$= -\dfrac{2x}{x-2}$ or $\dfrac{2x}{2-x}$

**53.** $\dfrac{y^{-2}}{1-y^{-2}} = \dfrac{\frac{1}{y^2}}{1-\frac{1}{y^2}}$

$= \dfrac{y^2\left(\frac{1}{y^2}\right)}{y^2\left(1-\frac{1}{y^2}\right)}$

$= \dfrac{y^2\left(\frac{1}{y^2}\right)}{y^2(1)-y^2\left(\frac{1}{y^2}\right)}$

$= \dfrac{1}{y^2-1}$

**55.** $t = \dfrac{d}{r}$

$t = \dfrac{\frac{20x}{3}}{\frac{5x}{9}} = \dfrac{20x}{3}\cdot\dfrac{9}{5x} = \dfrac{4\cdot5x\cdot3\cdot3}{3\cdot5x} = 12$ hours

**Chapter 7 Vocabulary Check**

1. A <u>rational expression</u> is an expression that can be written in the form $\dfrac{P}{Q}$, where $P$ and $Q$ are polynomials and $Q$ is not 0.

2. In a <u>complex fraction</u>, the numerator or denominator or both may contain fractions.

3. For a rational expression, $-\dfrac{a}{b} = \dfrac{-a}{\underline{b}} = \dfrac{a}{\underline{-b}}$.

4. A rational expression is undefined when the <u>denominator</u> is 0.

5. The process of writing a rational expression in lowest terms is called <u>simplifying</u>.

6. The expressions $\dfrac{2x}{7}$ and $\dfrac{7}{2x}$ are called <u>reciprocals</u>.

7. The <u>least common denominator</u> of a list of rational expressions is a polynomial of least degree whose factors include all factors of the denominators in the list.

8. A <u>ratio</u> is the quotient of two numbers.

9. $\dfrac{x}{2} = \dfrac{7}{16}$ is an example of a <u>proportion</u>.

10. If $\dfrac{a}{b} = \dfrac{c}{d}$, then $ad$ and $bc$ are called <u>cross products</u>.

11. The equation $y = \dfrac{k}{x}$ is an example of <u>inverse variation</u>.

12. The equation $y = kx$ is an example of <u>direct variation</u>.

**Chapter 7 Review**

1. The rational expression is undefined when
$$x^2-4=0$$
$$(x-2)(x+2)=0$$
$$x-2=0 \quad\text{or}\quad x+2=0$$
$$x=2 \qquad\qquad x=-2$$

2. The rational expression is undefined when
$$4x^2-4x-15=0$$
$$(2x+3)(2x-5)=0$$
$$2x+3=0 \qquad\text{or}\quad 2x-5=0$$
$$2x=-3 \qquad\qquad 2x=5$$
$$x=-\dfrac{3}{2} \qquad\qquad x=\dfrac{5}{2}$$

3. $\dfrac{2-z}{z+5} = \dfrac{2-(-2)}{-2+5} = \dfrac{2+2}{3} = \dfrac{4}{3}$

**4.** $\dfrac{x^2+xy-y^2}{x+y} = \dfrac{5^2+5\cdot 7-7^2}{5+7}$

$\qquad\qquad = \dfrac{25+35-49}{12}$

$\qquad\qquad = \dfrac{11}{12}$

**5.** $\dfrac{2x+6}{x^2+3x} = \dfrac{2(x+3)}{x(x+3)} = \dfrac{2}{x}$

**6.** $\dfrac{3x-12}{x^2-4x} = \dfrac{3(x-4)}{x(x-4)} = \dfrac{3}{x}$

**7.** $\dfrac{x+2}{x^2-3x-10} = \dfrac{x+2}{(x-5)(x+2)} = \dfrac{1}{x-5}$

**8.** $\dfrac{x+4}{x^2+5x+4} = \dfrac{x+4}{(x+1)(x+4)} = \dfrac{1}{x+1}$

**9.** $\dfrac{x^3-4x}{x^2+3x+2} = \dfrac{x(x^2-4)}{(x+2)(x+1)}$

$\qquad\qquad = \dfrac{x(x-2)(x+2)}{(x+2)(x+1)}$

$\qquad\qquad = \dfrac{x(x-2)}{x+1}$

**10.** $\dfrac{5x^2-125}{x^2+2x-15} = \dfrac{5(x^2-25)}{(x-3)(x+5)}$

$\qquad\qquad = \dfrac{5(x-5)(x+5)}{(x-3)(x+5)}$

$\qquad\qquad = \dfrac{5(x-5)}{x-3}$

**11.** $\dfrac{x^2-x-6}{x^2-3x-10} = \dfrac{(x-3)(x+2)}{(x-5)(x+2)} = \dfrac{x-3}{x-5}$

**12.** $\dfrac{x^2-2x}{x^2+2x-8} = \dfrac{x(x-2)}{(x+4)(x-2)} = \dfrac{x}{x+4}$

**13.** $\dfrac{x^2+xa+xb+ab}{x^2-xc+bx-bc} = \dfrac{x(x+a)+b(x+a)}{x(x-c)+b(x-c)}$

$\qquad\qquad = \dfrac{(x+a)(x+b)}{(x-c)(x+b)}$

$\qquad\qquad = \dfrac{x+a}{x-c}$

**14.** $\dfrac{x^2+5x-2x-10}{x^2-3x-2x+6} = \dfrac{x(x+5)-2(x+5)}{x(x-3)-2(x-3)}$

$\qquad\qquad = \dfrac{(x+5)(x-2)}{(x-3)(x-2)}$

$\qquad\qquad = \dfrac{x+5}{x-3}$

**15.** $\dfrac{4-x}{x^3-64} = -\dfrac{x-4}{x^3-64}$

$\qquad\qquad = -\dfrac{x-4}{(x-4)(x^2+4x+16)}$

$\qquad\qquad = -\dfrac{1}{x^2+4x+16}$

**16.** $\dfrac{x^2-4}{x^3+8} = \dfrac{(x+2)(x-2)}{(x+2)(x^2-2x+4)} = \dfrac{x-2}{x^2-2x+4}$

**17.** $\dfrac{15x^3y^2}{z} \cdot \dfrac{z}{5xy^3} = \dfrac{15x^3y^2\cdot z}{z\cdot 5xy^3}$

$\qquad\qquad = \dfrac{3\cdot 5\cdot x^2\cdot x\cdot y^2\cdot z}{z\cdot 5\cdot x\cdot y^2\cdot y}$

$\qquad\qquad = \dfrac{3x^2}{y}$

**18.** $\dfrac{-y^3}{8} \cdot \dfrac{9x^2}{y^3} = -\dfrac{y^3\cdot 9x^2}{8\cdot y^3} = -\dfrac{9x^2}{8}$

**19.** $\dfrac{x^2-9}{x^2-4} \cdot \dfrac{x-2}{x+3} = \dfrac{(x^2-9)\cdot(x-2)}{(x^2-4)\cdot(x+3)}$

$\qquad\qquad = \dfrac{(x-3)(x+3)(x-2)}{(x+2)(x-2)(x+3)}$

$\qquad\qquad = \dfrac{x-3}{x+2}$

**20.** $\dfrac{2x+5}{x-6} \cdot \dfrac{2x}{-x+6} = \dfrac{2x+5}{x-6} \cdot \dfrac{2x}{-(x-6)}$

$\qquad\qquad = \dfrac{2x+5}{x-6} \cdot \dfrac{-2x}{x-6}$

$\qquad\qquad = \dfrac{(2x+5)\cdot(-2x)}{(x-6)\cdot(x-6)}$

$\qquad\qquad = \dfrac{-2x(2x+5)}{(x-6)^2}$

**21.** $\dfrac{x^2-5x-24}{x^2-x-12} \div \dfrac{x^2-10x+16}{x^2+x-6}$

$= \dfrac{x^2-5x-24}{x^2-x-12} \cdot \dfrac{x^2+x-6}{x^2-10x+16}$

$= \dfrac{(x-8)(x+3) \cdot (x+3)(x-2)}{(x-4)(x+3) \cdot (x-8)(x-2)}$

$= \dfrac{x+3}{x-4}$

**22.** $\dfrac{4x+4y}{xy^2} \div \dfrac{3x+3y}{x^2y} = \dfrac{4x+4y}{xy^2} \cdot \dfrac{x^2y}{3x+3y}$

$= \dfrac{4(x+y) \cdot x \cdot x \cdot y}{x \cdot y \cdot y \cdot 3(x+y)}$

$= \dfrac{4x}{3y}$

**23.** $\dfrac{x^2+x-42}{x-3} \cdot \dfrac{(x-3)^2}{x+7}$

$= \dfrac{(x+7)(x-6) \cdot (x-3)(x-3)}{(x-3) \cdot (x+7)}$

$= (x-6)(x-3)$

**24.** $\dfrac{2a+2b}{3} \cdot \dfrac{a-b}{a^2-b^2} = \dfrac{2(a+b) \cdot (a-b)}{3 \cdot (a+b)(a-b)} = \dfrac{2}{3}$

**25.** $\dfrac{2x^2-9x+9}{8x-12} \div \dfrac{x^2-3x}{2x} = \dfrac{2x^2-9x+9}{8x-12} \cdot \dfrac{2x}{x^2-3x}$

$= \dfrac{(2x-3)(x-3) \cdot 2x}{4(2x-3) \cdot x(x-3)}$

$= \dfrac{2}{4}$

$= \dfrac{1}{2}$

**26.** $\dfrac{x^2-y^2}{x^2+xy} \div \dfrac{3x^2-2xy-y^2}{3x^2+6x}$

$= \dfrac{x^2-y^2}{x^2+xy} \cdot \dfrac{3x^2+6x}{3x^2-2xy-y^2}$

$= \dfrac{(x-y)(x+y) \cdot 3x(x+2)}{x(x+y) \cdot (3x+y)(x-y)}$

$= \dfrac{3(x+2)}{3x+y}$

**27.** $\dfrac{x-y}{4} \div \dfrac{y^2-2y-xy+2x}{16x+24}$

$= \dfrac{x-y}{4} \cdot \dfrac{16x+24}{y^2-2y-xy+2x}$

$= \dfrac{x-y}{4} \cdot \dfrac{8(2x+3)}{y(y-2)-x(y-2)}$

$= \dfrac{x-y}{4} \cdot \dfrac{8(2x+3)}{(y-2)(y-x)}$

$= -\dfrac{y-x}{4} \cdot \dfrac{8(2x+3)}{(y-2)(y-x)}$

$= -\dfrac{2 \cdot 4(y-x)(2x+3)}{4(y-2)(y-x)}$

$= -\dfrac{2(2x+3)}{y-2}$

**28.** $\dfrac{5+x}{7} \div \dfrac{xy+5y-3x-15}{7y-35}$

$= \dfrac{5+x}{7} \cdot \dfrac{7y-35}{xy+5y-3x-15}$

$= \dfrac{(5+x) \cdot 7(y-5)}{7 \cdot (x+5)(y-3)}$

$= \dfrac{y-5}{y-3}$

**29.** $\dfrac{x}{x^2+9x+14} + \dfrac{7}{x^2+9x+14} = \dfrac{x+7}{x^2+9x+14}$

$= \dfrac{x+7}{(x+7)(x+2)}$

$= \dfrac{1}{x+2}$

**30.** $\dfrac{x}{x^2+2x-15} + \dfrac{5}{x^2+2x-15} = \dfrac{x+5}{x^2+2x-15}$

$= \dfrac{x+5}{(x+5)(x-3)}$

$= \dfrac{1}{x-3}$

**31.** $\dfrac{4x-5}{3x^2} - \dfrac{2x+5}{3x^2} = \dfrac{4x-5-(2x+5)}{3x^2}$

$= \dfrac{4x-5-2x-5}{3x^2}$

$= \dfrac{2x-10}{3x^2}$

**32.**
$$\frac{9x+7}{6x^2}-\frac{3x+4}{6x^2}=\frac{9x+7-(3x+4)}{6x^2}$$
$$=\frac{9x+7-3x-4}{6x^2}$$
$$=\frac{6x+3}{6x^2}$$
$$=\frac{3(2x+1)}{3\cdot 2x^2}$$
$$=\frac{2x+1}{2x^2}$$

**33.** $2x=2\cdot x$
$7x=7\cdot x$
$\text{LCD}=2\cdot 7\cdot x=14x$

**34.** $x^2-5x-24=(x-8)(x+3)$
$x^2+11x+24=(x+8)(x+3)$
$\text{LCD}=(x-8)(x+3)(x+8)$

**35.** $\dfrac{5}{7x}=\dfrac{5}{7x}\cdot\dfrac{2x^2y}{2x^2y}=\dfrac{5\cdot 2x^2y}{7x\cdot 2x^2y}=\dfrac{10x^2y}{14x^3y}$

**36.** $\dfrac{9}{4y}=\dfrac{9}{4y}\cdot\dfrac{4y^2x}{4y^2x}=\dfrac{9\cdot 4y^2x}{4y\cdot 4y^2x}=\dfrac{36y^2x}{16y^3x}$

**37.**
$$\frac{x+2}{x^2+11x+18}=\frac{x+2}{(x+9)(x+2)}$$
$$=\frac{(x+2)(x-5)}{(x+9)(x+2)(x-5)}$$
$$=\frac{x^2-3x-10}{(x+2)(x-5)(x+9)}$$

**38.**
$$\frac{3x-5}{x^2+4x+4}=\frac{3x-5}{(x+2)^2}$$
$$=\frac{(3x-5)(x+3)}{(x+2)^2(x+3)}$$
$$=\frac{3x^2+4x-15}{(x+2)^2(x+3)}$$

**39.** $\dfrac{4}{5x^2}-\dfrac{6}{y}=\dfrac{4(y)}{5x^2(y)}-\dfrac{6(5x^2)}{y(5x^2)}=\dfrac{4y-30x^2}{5x^2y}$

**40.**
$$\frac{2}{x-3}-\frac{4}{x-1}=\frac{2(x-1)}{(x-3)(x-1)}-\frac{4(x-3)}{(x-1)(x-3)}$$
$$=\frac{2(x-1)-4(x-3)}{(x-3)(x-1)}$$
$$=\frac{2x-2-4x+12}{(x-3)(x-1)}$$
$$=\frac{-2x+10}{(x-3)(x-1)}$$

**41.**
$$\frac{4}{x+3}-2=\frac{4}{x+3}-\frac{2(x+3)}{x+3}$$
$$=\frac{4-2(x+3)}{x+3}$$
$$=\frac{4-2x-6}{x+3}$$
$$=\frac{-2x-2}{x+3}$$

**42.**
$$\frac{3}{x^2+2x-8}+\frac{2}{x^2-3x+2}$$
$$=\frac{3}{(x+4)(x-2)}+\frac{2}{(x-1)(x-2)}$$
$$=\frac{3(x-1)}{(x+4)(x-2)(x-1)}+\frac{2(x+4)}{(x-1)(x-2)(x+4)}$$
$$=\frac{3(x-1)+2(x+4)}{(x+4)(x-2)(x-1)}$$
$$=\frac{3x-3+2x+8}{(x+4)(x-2)(x-1)}$$
$$=\frac{5x+5}{(x+4)(x-2)(x-1)}$$

**43.**
$$\frac{2x-5}{6x+9}-\frac{4}{2x^2+3x}=\frac{2x-5}{3(2x+3)}-\frac{4}{x(2x+3)}$$
$$=\frac{(2x-5)(x)}{3(2x+3)(x)}-\frac{4(3)}{x(2x+3)(3)}$$
$$=\frac{2x^2-5x-12}{3x(2x+3)}$$
$$=\frac{(2x+3)(x-4)}{3x(2x+3)}$$
$$=\frac{x-4}{3x}$$

      

**44.** $\dfrac{x-1}{x^2-2x+1}-\dfrac{x+1}{x-1}=\dfrac{x-1}{(x-1)^2}-\dfrac{x+1}{x-1}$

$$=\dfrac{1}{x-1}-\dfrac{x+1}{x-1}$$

$$=\dfrac{1-(x+1)}{x-1}$$

$$=\dfrac{1-x-1}{x-1}$$

$$=\dfrac{-x}{x-1}$$

$$=-\dfrac{x}{x-1}$$

**45.** $P=2l+2w$

$$P=2\left(\dfrac{x}{8}\right)+2\left(\dfrac{x+2}{4x}\right)$$

$$=\dfrac{x}{4}+\dfrac{2(x+2)}{4x}$$

$$=\dfrac{x\cdot x}{4\cdot x}+\dfrac{2x+4}{4x}$$

$$=\dfrac{x^2+2x+4}{4x}$$

$A=l\cdot w$

$$A=\dfrac{x}{8}\cdot\dfrac{x+2}{4x}=\dfrac{x\cdot(x+2)}{8\cdot 4x}=\dfrac{x+2}{32}$$

The perimeter is $\dfrac{x^2+2x+4}{4x}$ units and the area is $\dfrac{x+2}{32}$ square units.

**46.** $P=\dfrac{3x}{4x-4}+\dfrac{2x}{3x-3}+\dfrac{x}{x-1}$

$$=\dfrac{3x}{4(x-1)}+\dfrac{2x}{3(x-1)}+\dfrac{x}{x-1}$$

$$=\dfrac{3x(3)}{4(x-1)(3)}+\dfrac{2x(4)}{3(x-1)(4)}+\dfrac{x(12)}{(x-1)(12)}$$

$$=\dfrac{9x+8x+12x}{12(x-1)}$$

$$=\dfrac{29x}{12(x-1)}$$

$A=\dfrac{1}{2}\cdot b\cdot h$

$$A=\dfrac{1}{2}\cdot\dfrac{x}{x-1}\cdot\dfrac{6y}{5}=\dfrac{1\cdot x\cdot 2\cdot 3y}{2\cdot(x-1)\cdot 5}=\dfrac{3xy}{5(x-1)}$$

The perimeter is $\dfrac{29x}{12(x-1)}$ units and the area is $\dfrac{3xy}{5(x-1)}$ square units.

**47.**
$$\frac{n}{10} = 9 - \frac{n}{5}$$
$$10\left(\frac{n}{10}\right) = 10\left(9 - \frac{n}{5}\right)$$
$$10\left(\frac{n}{10}\right) = 10(9) - 10\left(\frac{n}{5}\right)$$
$$n = 90 - 2n$$
$$3n = 90$$
$$n = 30$$

**48.**
$$\frac{2}{x+1} - \frac{1}{x-2} = -\frac{1}{2}$$
$$2(x+1)(x-2)\left(\frac{2}{x+1} - \frac{1}{x-2}\right) = 2(x+1)(x-2)\left(-\frac{1}{2}\right)$$
$$2(x+1)(x-2)\left(\frac{2}{x+1}\right) - 2(x+1)(x-2)\left(\frac{1}{x-2}\right) = 2(x+1)(x-2)\left(-\frac{1}{2}\right)$$
$$4(x-2) - 2(x+1) = -(x+1)(x-2)$$
$$4x - 8 - 2x - 2 = -(x^2 - x - 2)$$
$$2x - 10 = -x^2 + x + 2$$
$$x^2 + x - 12 = 0$$
$$(x+4)(x-3) = 0$$
$$x + 4 = 0 \quad \text{or} \quad x - 3 = 0$$
$$x = -4 \qquad\qquad x = 3$$

**49.**
$$\frac{y}{2y+2} + \frac{2y-16}{4y+4} = \frac{y-3}{y+1}$$
$$\frac{y}{2(y+1)} + \frac{2y-16}{4(y+1)} = \frac{y-3}{y+1}$$
$$4(y+1)\left(\frac{y}{2(y+1)} + \frac{2y-16}{4(y+1)}\right) = 4(y+1)\left(\frac{y-3}{y+1}\right)$$
$$4(y+1)\left(\frac{y}{2(y+1)}\right) + 4(y+1)\left(\frac{2y-16}{4(y+1)}\right) = 4(y+1)\left(\frac{y-3}{y+1}\right)$$
$$2y + 2y - 16 = 4(y-3)$$
$$4y - 16 = 4y - 12$$
$$-16 = -12 \quad \text{False}$$

This equation has no solution.

**50.**

$$\frac{2}{x-3} - \frac{4}{x+3} = \frac{8}{x^2-9}$$

$$(x-3)(x+3)\left(\frac{2}{x-3} - \frac{4}{x+3}\right) = (x-3)(x+3)\left(\frac{8}{(x-3)(x+3)}\right)$$

$$(x-3)(x+3)\left(\frac{2}{x-3}\right) - (x-3)(x+3)\left(\frac{4}{x+3}\right) = 8$$

$$2(x+3) - 4(x-3) = 8$$

$$2x + 6 - 4x + 12 = 8$$

$$-2x + 18 = 8$$

$$-2x = -10$$

$$x = 5$$

**51.**

$$\frac{x-3}{x+1} - \frac{x-6}{x+5} = 0$$

$$(x+1)(x+5)\left(\frac{x-3}{x+1} - \frac{x-6}{x+5}\right) = (x+1)(x+5)(0)$$

$$(x+1)(x+5)\left(\frac{x-3}{x+1}\right) - (x+1)(x+5)\left(\frac{x-6}{x+5}\right) = 0$$

$$(x+5)(x-3) - (x+1)(x-6) = 0$$

$$x^2 + 2x - 15 - (x^2 - 5x - 6) = 0$$

$$x^2 + 2x - 15 - x^2 + 5x + 6 = 0$$

$$7x - 9 = 0$$

$$7x = 9$$

$$x = \frac{9}{7}$$

**52.**

$$x + 5 = \frac{6}{x}$$

$$x(x+5) = x\left(\frac{6}{x}\right)$$

$$x^2 + 5x = 6$$

$$x^2 + 5x - 6 = 0$$

$$(x+6)(x-1) = 0$$

$$x + 6 = 0 \quad \text{or} \quad x - 1 = 0$$

$$x = -6 \qquad\quad x = 1$$

**53.** $\dfrac{4A}{5b} = x^2$

$$4A = 5bx^2$$

$$\frac{4A}{5x^2} = \frac{5bx^2}{5x^2}$$

$$\frac{4A}{5x^2} = b$$

**54.**
$$\frac{x}{7} + \frac{y}{8} = 10$$
$$56\left(\frac{x}{7}\right) + 56\left(\frac{y}{8}\right) = 56(10)$$
$$8x + 7y = 560$$
$$7y = 560 - 8x$$
$$y = \frac{560 - 8x}{7}$$

**55.**
$$\frac{x}{2} = \frac{12}{4}$$
$$4x = 24$$
$$x = 6$$

**56.**
$$\frac{20}{1} = \frac{x}{25}$$
$$500 = x$$

**57.**
$$\frac{2}{x-1} = \frac{3}{x+3}$$
$$2(x+3) = 3(x-1)$$
$$2x + 6 = 3x - 3$$
$$6 = x - 3$$
$$9 = x$$

**58.**
$$\frac{4}{y-3} = \frac{2}{y-3}$$
$$4(y-3) = 2(y-3)$$
$$4y - 12 = 2y - 6$$
$$2y - 12 = -6$$
$$2y = 6$$
$$y = 3$$

$y = 3$ doesn't check, so this equation has no solution.

**59.** Let $x$ = the number of parts processed in 45 minutes.
$$\frac{300}{20} = \frac{x}{45}$$
$$13,500 = 20x$$
$$675 = x$$
675 parts can be processed in 45 minutes.

**60.** Let $x$ = the charge for 3 hours.
$$\frac{90.00}{8} = \frac{x}{3}$$
$$270.00 = 8x$$
$$33.75 = x$$
He charges \$33.75 for 3 hours.

**61.**
$$5 \cdot \frac{1}{x} = \frac{3}{2} \cdot \frac{1}{x} + \frac{7}{6}$$
$$\frac{5}{x} = \frac{3}{2x} + \frac{7}{6}$$
$$6x\left(\frac{5}{x}\right) = 6x\left(\frac{3}{2x}\right) + 6x\left(\frac{7}{6}\right)$$
$$30 = 9 + 7x$$
$$21 = 7x$$
$$x = 3$$
The unknown number is 3.

**62.**
$$\frac{1}{x} = \frac{1}{4-x}$$
$$4 - x = x$$
$$4 = 2x$$
$$2 = x$$
The unknown number is 2.

**63.** Let $r$ be the rate of the faster car. Then the rate of the slower car is $r - 10$.

|  | Distance = Rate · Time |  |  |
|---|---|---|---|
| Fast car | 90 | $r$ | $\frac{90}{r}$ |
| Slow car | 60 | $r - 10$ | $\frac{60}{r-10}$ |

$$\frac{90}{r} = \frac{60}{r-10}$$
$$90(r-10) = 60r$$
$$90r - 900 = 60r$$
$$-900 = -30r$$
$$30 = r$$
$r - 10 = 30 - 10 = 20$
The rate of the fast car is 30 miles per hour and the rate of the slower car is 20 miles per hour.

**64.** Let $r$ be the speed of the boat in still water.

|  | Distance = Rate · Time |  |  |
|---|---|---|---|
| Upstream | 48 | $r - 4$ | $\frac{48}{r-4}$ |
| Downstream | 72 | $r + 4$ | $\frac{72}{r+4}$ |

$$\frac{48}{r-4} = \frac{72}{r+4}$$
$$48(r+4) = 72(r-4)$$
$$48r + 192 = 72r - 288$$
$$480 = 24r$$
$$r = 20$$

The speed of the boat in still water is 20 miles per hour.

**65.** Let $x$ be the time it takes Maria working alone.

|         | Hours to Complete Total Job | Part of Job Completed in 1 Hour |
|---------|:---------------------------:|:-------------------------------:|
| Mark    | 7                           | $\frac{1}{7}$                   |
| Maria   | $x$                         | $\frac{1}{x}$                   |
| Together| 5                           | $\frac{1}{5}$                   |

$$\frac{1}{7} + \frac{1}{x} = \frac{1}{5}$$
$$35x\left(\frac{1}{7}\right) + 35x\left(\frac{1}{x}\right) = 35x\left(\frac{1}{5}\right)$$
$$5x + 35 = 7x$$
$$35 = 2x$$
$$x = \frac{35}{2} \text{ or } 17\frac{1}{2}$$

It takes Maria $17\frac{1}{2}$ hours to complete the job alone.

**66.** Let $x$ be the number of days it takes the pipes to fill the pond together.

|         | Days to Complete Total Job | Part of Job Completed in 1 Day |
|---------|:--------------------------:|:------------------------------:|
| Pipe A  | 20                         | $\frac{1}{20}$                 |
| Pipe B  | 15                         | $\frac{1}{15}$                 |
| Together| $x$                        | $\frac{1}{x}$                  |

$$\frac{1}{20} + \frac{1}{25} = \frac{1}{x}$$
$$60x\left(\frac{1}{20}\right) + 60x\left(\frac{1}{15}\right) = 60x\left(\frac{1}{x}\right)$$
$$3x + 4x = 60$$
$$7x = 60$$
$$x = \frac{60}{7} = 8\frac{4}{7}$$

Both pipes fill the pond in $8\frac{4}{7}$ days.

**67.** $\frac{2}{3} = \frac{10}{x}$
$$2x = 30$$
$$x = 15$$
The missing length is 15.

**68.** $\frac{12}{4} = \frac{18}{x}$
$$12x = 72$$
$$x = 6$$
The missing length is 6.

**69.** $y = kx$
$$40 = k(4)$$
$$10 = k$$

$$y = 10x$$
$$y = 10(11)$$
$$y = 110$$

**70.** $y = \frac{k}{x}$
$$4 = \frac{k}{6}$$
$$24 = k$$

$$y = \frac{24}{x}$$
$$y = \frac{24}{48}$$
$$y = \frac{1}{2}$$

**71.**
$$y = \frac{k}{x^3}$$
$$12.5 = \frac{k}{2^3}$$
$$12.5 = \frac{k}{8}$$
$$100 = k$$

$$y = \frac{100}{x^3}$$
$$y = \frac{100}{3^3}$$
$$y = \frac{100}{27}$$

**72.**
$$y = kx^2$$
$$175 = k(5)^2$$
$$175 = 25k$$
$$7 = k$$

$$y = 7x^2$$
$$y = 7(10)^2$$
$$y = 7(100)$$
$$y = 700$$

**73.**
$$c = \frac{k}{a}$$
$$6600 = \frac{k}{3000}$$
$$19,800,000 = k$$

$$c = \frac{19,800,000}{a}$$
$$c = \frac{19,800,000}{5000}$$
$$c = 3960$$
It costs \$3960 to manufacture 5000 ml of medicine.

**74.**
$$d = kw$$
$$8 = k(150)$$
$$\frac{8}{150} = k$$
$$\frac{4}{75} = k$$

$$d = \frac{4}{75}w$$
$$d = \frac{4}{75}(90)$$
$$d = \frac{360}{75}$$
$$d = 4\frac{4}{5}$$
A 90-pound weight would stretch the spring $4\frac{4}{5}$ inches.

**75.** $\dfrac{\frac{5x}{27}}{-\frac{10xy}{21}} = \dfrac{5x}{27} \cdot -\dfrac{21}{10xy} = -\dfrac{5x \cdot 3 \cdot 7}{3 \cdot 9 \cdot 5 \cdot 2 \cdot x \cdot y} = -\dfrac{7}{18y}$

**76.** $\dfrac{\frac{3}{5}+\frac{2}{7}}{\frac{1}{5}+\frac{5}{6}} = \dfrac{\frac{21}{35}+\frac{10}{35}}{\frac{6}{30}+\frac{25}{30}} = \dfrac{\frac{31}{35}}{\frac{31}{30}} = \dfrac{31}{35} \cdot \dfrac{30}{31} = \dfrac{31 \cdot 5 \cdot 6}{5 \cdot 7 \cdot 31} = \dfrac{6}{7}$

**77.** $\dfrac{3-\frac{1}{y}}{2-\frac{1}{y}} = \dfrac{y\left(3-\frac{1}{y}\right)}{y\left(2-\frac{1}{y}\right)} = \dfrac{y(3)-y\left(\frac{1}{y}\right)}{y(2)-y\left(\frac{1}{y}\right)} = \dfrac{3y-1}{2y-1}$

**78.** $\dfrac{\frac{6}{x+2}+4}{\frac{8}{x+2}-4} = \dfrac{(x+2)\left(\frac{6}{x+2}+4\right)}{(x+2)\left(\frac{8}{x+2}-4\right)}$

$$= \dfrac{(x+2)\left(\frac{6}{x+2}\right)+(x+2)(4)}{(x+2)\left(\frac{8}{x+2}\right)-(x+2)(4)}$$

$$= \dfrac{6+4x+8}{8-4x-8}$$

$$= \dfrac{4x+14}{-4x}$$

$$= -\dfrac{2(2x+7)}{2 \cdot 2x}$$

$$= -\dfrac{2x+7}{2x}$$

**79.** $\dfrac{4x+12}{8x^2+24x} = \dfrac{4(x+3)}{2 \cdot 4 \cdot x(x+3)} = \dfrac{1}{2x}$

**80.** $\dfrac{x^3-6x^2+9x}{x^2+4x-21} = \dfrac{x(x-3)^2}{(x+7)(x-3)} = \dfrac{x(x-3)}{x+7}$

    

**81.** $\dfrac{x^2+9x+20}{x^2-25}\cdot\dfrac{x^2-9x+20}{x^2+8x+16}$

$=\dfrac{(x+4)(x+5)\cdot(x-4)(x-5)}{(x+5)(x-5)\cdot(x+4)(x+4)}$

$=\dfrac{x-4}{x+4}$

**82.** $\dfrac{x^2-x-72}{x^2-x-30}\div\dfrac{x^2+6x-27}{x^2-9x+18}$

$=\dfrac{x^2-x-72}{x^2-x-30}\cdot\dfrac{x^2-9x+18}{x^2+6x-27}$

$=\dfrac{(x-9)(x+8)\cdot(x-3)(x-6)}{(x+5)(x-6)\cdot(x+9)(x-3)}$

$=\dfrac{(x-9)(x+8)}{(x+5)(x+9)}$

**83.** $\dfrac{x}{x^2-36}+\dfrac{6}{x^2-36}=\dfrac{x+6}{x^2-36}$

$=\dfrac{x+6}{(x+6)(x-6)}$

$=\dfrac{1}{x-6}$

**84.** $\dfrac{5x-1}{4x}-\dfrac{3x-2}{4x}=\dfrac{5x-1-(3x-2)}{4x}$

$=\dfrac{5x-1-3x+2}{4x}$

$=\dfrac{2x+1}{4x}$

**85.** $\dfrac{4}{3x^2+8x-3}+\dfrac{2}{3x^2-7x+2}$

$=\dfrac{4}{(x+3)(3x-1)}+\dfrac{2}{(x-2)(3x-1)}$

$=\dfrac{4(x-2)}{(x+3)(3x-1)(x-2)}+\dfrac{2(x+3)}{(x-2)(3x-1)(x+3)}$

$=\dfrac{4(x-2)+2(x+3)}{(x+3)(3x-1)(x-2)}$

$=\dfrac{4x-8+2x+6}{(x+3)(3x-1)(x-2)}$

$=\dfrac{6x-2}{(x+3)(3x-1)(x-2)}$

$=\dfrac{2(3x-1)}{(x+3)(3x-1)(x-2)}$

$=\dfrac{2}{(x+3)(x-2)}$

**86.** $\dfrac{3x}{x^2+9x+14}-\dfrac{6x}{x^2+4x-21}$

$=\dfrac{3x}{(x+7)(x+2)}-\dfrac{6x}{(x+7)(x-3)}$

$=\dfrac{3x(x-3)}{(x+7)(x+2)(x-3)}-\dfrac{6x(x+2)}{(x+7)(x-3)(x+2)}$

$=\dfrac{3x(x-3)-6x(x+2)}{(x+7)(x+2)(x-3)}$

$=\dfrac{3x^2-9x-6x^2-12x}{(x+7)(x+2)(x-3)}$

$=\dfrac{-3x^2-21x}{(x+7)(x+2)(x-3)}$

$=\dfrac{-3x(x+7)}{(x+7)(x+2)(x-3)}$

$=-\dfrac{3x}{(x+2)(x-3)}$

**87.** $\dfrac{4}{a-1}+2=\dfrac{3}{a-1}$

$(a-1)\left(\dfrac{4}{a-1}\right)+(a-1)(2)=(a-1)\left(\dfrac{3}{a-1}\right)$

$4+2(a-1)=3$

$4+2a-2=3$

$2+2a=3$

$2a=1$

$a=\dfrac{1}{2}$

**88.** $\dfrac{x}{x+3}+4=\dfrac{x}{x+3}$

$(x+3)\left(\dfrac{x}{x+3}\right)+(x+3)(4)=(x+3)\left(\dfrac{x}{x+3}\right)$

$x+4(x+3)=x$

$x+4x+12=x$

$5x+12=x$

$12=-4x$

$-3=x$

Since $x=-3$ makes a denominator 0, the solution does not check. This equation has no solution.

**89.** $\dfrac{2x}{3}-\dfrac{1}{6}=\dfrac{x}{2}$

$6\left(\dfrac{2x}{3}\right)-6\left(\dfrac{1}{6}\right)=6\left(\dfrac{x}{2}\right)$

$4x-1=3x$

$-1=-x$

$1=x$

The unknown number is 1.

**90.** Let $x$ be the number of days it takes them to paint the house working together.

|  | Days to Complete Total Job | Part of Job Completed in 1 Day |
|---|---|---|
| Mr. Crocker | 3 | $\frac{1}{3}$ |
| Son | 4 | $\frac{1}{4}$ |
| Together | $x$ | $\frac{1}{x}$ |

$$\frac{1}{3} + \frac{1}{4} = \frac{1}{x}$$
$$12x\left(\frac{1}{3}\right) + 12x\left(\frac{1}{4}\right) = 12x\left(\frac{1}{x}\right)$$
$$4x + 3x = 12$$
$$7x = 12$$
$$x = \frac{12}{7} \text{ or } 1\frac{5}{7}$$

Working together, Mr. Crocker and his son can paint the house in $1\frac{5}{7}$ days.

**91.** $\frac{5}{3} = \frac{10}{x}$

$5x = 30$

$x = 6$

The missing length is 6.

**92.** $\frac{6}{18} = \frac{4}{x}$

$6x = 72$

$x = 12$

The missing length is 12.

**93.** $\dfrac{\frac{1}{4}}{\frac{1}{3} + \frac{1}{2}} = \dfrac{12\left(\frac{1}{4}\right)}{12\left(\frac{1}{3} + \frac{1}{2}\right)} = \dfrac{12\left(\frac{1}{4}\right)}{12\left(\frac{1}{3}\right) + 12\left(\frac{1}{2}\right)} = \dfrac{3}{4 + 6} = \dfrac{3}{10}$

**94.** $\dfrac{4 + \frac{2}{x}}{6 + \frac{3}{x}} = \dfrac{x\left(4 + \frac{2}{x}\right)}{x\left(6 + \frac{3}{x}\right)}$

$= \dfrac{x(4) + x\left(\frac{2}{x}\right)}{x(6) + x\left(\frac{3}{x}\right)}$

$= \dfrac{4x + 2}{6x + 3}$

$= \dfrac{2(2x + 1)}{3(2x + 1)}$

$= \dfrac{2}{3}$

**95.** 1.8 square yards

$= \dfrac{1.8 \text{ square yards}}{1} \cdot \dfrac{9 \text{ square feet}}{1 \text{ square yard}}$

$= 16.2$ square feet

**96.** 135 cubic feet

$= \dfrac{135 \text{ cubic feet}}{1} \cdot \dfrac{1 \text{ cubic yard}}{27 \text{ cubic feet}}$

$= 5$ cubic yards

**Chapter 7 Test**

**1.** The rational expression is undefined when
$$x^2 + 4x + 3 = 0$$
$$(x + 3)(x + 1) = 0$$
$$x + 3 = 0 \quad \text{or} \quad x + 1 = 0$$
$$x = -3 \quad\quad\quad x = -1$$

**2. a.** $C = \dfrac{100x + 3000}{x}$

$= \dfrac{100(200) + 3000}{200}$

$= \dfrac{20,000 + 3000}{200}$

$= \dfrac{23,000}{200}$

$= 115$

The average cost per desk is $115.

**b.** $C = \dfrac{100x + 3000}{x}$

$= \dfrac{100(1000) + 3000}{1000}$

$= \dfrac{100,000 + 3000}{1000}$

$= \dfrac{103,000}{1000}$

$= 103$

The average cost per desk is \$103.

**3.** $\dfrac{3x - 6}{5x - 10} = \dfrac{3(x - 2)}{5(x - 2)} = \dfrac{3}{5}$

**4.** $\dfrac{x + 6}{x^2 + 12x + 36} = \dfrac{x + 6}{(x + 6)^2} = \dfrac{1}{x + 6}$

**5.** $\dfrac{x + 3}{x^3 + 27} = \dfrac{x + 3}{(x + 3)(x^2 - 3x + 9)} = \dfrac{1}{x^2 - 3x + 9}$

**6.** $\dfrac{2m^3 - 2m^2 - 12m}{m^2 - 5m + 6} = \dfrac{2m(m^2 - m - 6)}{(m - 3)(m - 2)}$

$= \dfrac{2m(m - 3)(m + 2)}{(m - 3)(m - 2)}$

$= \dfrac{2m(m + 2)}{m - 2}$

**7.** $\dfrac{ay + 3a + 2y + 6}{ay + 3a + 5y + 15} = \dfrac{(y + 3)(a + 2)}{(y + 3)(a + 5)} = \dfrac{a + 2}{a + 5}$

**8.** $\dfrac{y - x}{x^2 - y^2} = \dfrac{-(x - y)}{(x - y)(x + y)} = -\dfrac{1}{x + y}$

**9.** $\dfrac{3}{x - 1} \cdot (5x - 5) = \dfrac{3}{x - 1} \cdot 5(x - 1) = \dfrac{3 \cdot 5(x - 1)}{x - 1} = 15$

**10.** $\dfrac{y^2 - 5y + 6}{2y + 4} \cdot \dfrac{y + 2}{2y - 6} = \dfrac{(y - 3)(y - 2) \cdot (y + 2)}{2(y + 2) \cdot 2(y - 3)}$

$= \dfrac{y - 2}{4}$

**11.** $\dfrac{15x}{2x + 5} - \dfrac{6 - 4x}{2x + 5} = \dfrac{15x - (6 - 4x)}{2x + 5}$

$= \dfrac{15x - 6 + 4x}{2x + 5}$

$= \dfrac{19x - 6}{2x + 5}$

**12.** $\dfrac{5a}{a^2 - a - 6} - \dfrac{2}{a - 3} = \dfrac{5a}{(a - 3)(a + 2)} - \dfrac{2(a + 2)}{(a - 3)(a + 2)}$

$= \dfrac{5a - 2(a + 2)}{(a - 3)(a + 2)}$

$= \dfrac{5a - 2a - 4}{(a - 3)(a + 2)}$

$= \dfrac{3a - 4}{(a - 3)(a + 2)}$

**13.** $\dfrac{6}{x^2 - 1} + \dfrac{3}{x + 1} = \dfrac{6}{(x + 1)(x - 1)} + \dfrac{3(x - 1)}{(x + 1)(x - 1)}$

$= \dfrac{6 + 3x - 3}{(x + 1)(x - 1)}$

$= \dfrac{3x + 3}{(x + 1)(x - 1)}$

$= \dfrac{3(x + 1)}{(x + 1)(x - 1)}$

$= \dfrac{3}{x - 1}$

**14.** $\dfrac{x^2 - 9}{x^2 - 3x} \div \dfrac{xy + 5x + 3y + 15}{2x + 10}$

$= \dfrac{x^2 - 9}{x^2 - 3x} \cdot \dfrac{2x + 10}{xy + 5x + 3y + 15}$

$= \dfrac{(x - 3)(x + 3) \cdot 2(x + 5)}{x(x - 3) \cdot (x + 3)(y + 5)}$

$= \dfrac{2(x + 5)}{x(y + 5)}$

**15.** $\dfrac{x + 2}{x^2 + 11x + 18} + \dfrac{5}{x^2 - 3x - 10}$

$= \dfrac{x + 2}{(x + 9)(x + 2)} + \dfrac{5}{(x - 5)(x + 2)}$

$= \dfrac{(x + 2)(x - 5)}{(x + 9)(x + 2)(x - 5)} + \dfrac{5(x + 9)}{(x - 5)(x + 2)(x + 9)}$

$= \dfrac{(x + 2)(x - 5) + 5(x + 9)}{(x + 9)(x + 2)(x - 5)}$

$= \dfrac{x^2 - 3x - 10 + 5x + 45}{(x + 9)(x + 2)(x - 5)}$

$= \dfrac{x^2 + 2x + 35}{(x + 9)(x + 2)(x - 5)}$

**16.**
$$\frac{4}{y} - \frac{5}{3} = -\frac{1}{5}$$

$$15y\left(\frac{4}{y} - \frac{5}{3}\right) = 15y\left(-\frac{1}{5}\right)$$

$$15y\left(\frac{4}{y}\right) - 15y\left(\frac{5}{3}\right) = 15y\left(-\frac{1}{5}\right)$$

$$60 - 25y = -3y$$

$$60 = 22y$$

$$\frac{60}{22} = y$$

$$y = \frac{30}{11}$$

**17.**
$$\frac{5}{y+1} = \frac{4}{y+2}$$

$$5(y+2) = 4(y+1)$$

$$5y+10 = 4y+4$$

$$y = -6$$

**18.**
$$\frac{a}{a-3} = \frac{3}{a-3} - \frac{3}{2}$$

$$2(a-3)\left(\frac{a}{a-3}\right) = 2(a-3)\left(\frac{3}{a-3} - \frac{3}{2}\right)$$

$$2a = 2(a-3)\left(\frac{3}{a-3}\right) - 2(a-3)\left(\frac{3}{2}\right)$$

$$2a = 6 - 3(a-3)$$

$$2a = 6 - 3a + 9$$

$$2a = 15 - 3a$$

$$5a = 15$$

$$a = 3$$

In the original equation, 3 makes a denominator 0. This equation has no solution.

**19.**
$$x - \frac{14}{x-1} = 4 - \frac{2x}{x-1}$$

$$(x-1)\left(x - \frac{14}{x-1}\right) = (x-1)\left(4 - \frac{2x}{x-1}\right)$$

$$x(x-1) - 14 = 4(x-1) - 2x$$

$$x^2 - x - 14 = 4x - 4 - 2x$$

$$x^2 - x - 14 = 2x - 4$$

$$x^2 - 3x - 10 = 0$$

$$(x-5)(x+2) = 0$$

$$x - 5 = 0 \quad \text{or} \quad x + 2 = 0$$

$$x = 5 \qquad\qquad x = -2$$

**20.**

$$\frac{10}{x^2-25}=\frac{3}{x+5}+\frac{1}{x-5}$$

$$\frac{10}{(x+5)(x-5)}=\frac{3}{x+5}+\frac{1}{x-5}$$

$$(x+5)(x-5)\left(\frac{10}{(x+5)(x-5)}\right)=(x+5)(x-5)\left(\frac{3}{x+5}\right)+(x+5)(x-5)\left(\frac{1}{x-5}\right)$$

$$10=3(x-5)+1(x+5)$$
$$10=3x-15+x+5$$
$$10=4x-10$$
$$20=4x$$
$$5=x$$

In the original equation 5 makes a denominator 0. This equation has no solution.

**21.** $\dfrac{\frac{5x^2}{yz^2}}{\frac{10x}{z^3}}=\dfrac{5x^2}{yz^2}\cdot\dfrac{z^3}{10x}=-\dfrac{5\cdot x\cdot x\cdot z\cdot z^2}{y\cdot z^2\cdot 2\cdot 5\cdot x}=\dfrac{xz}{2y}$

**22.** $\dfrac{5-\frac{1}{y^2}}{\frac{1}{y}+\frac{2}{y^2}}=\dfrac{y^2\left(5-\frac{1}{y^2}\right)}{y^2\left(\frac{1}{y}+\frac{2}{y^2}\right)}$

$$=\dfrac{y^2(5)-y^2\left(\frac{1}{y^2}\right)}{y^2\left(\frac{1}{y}\right)+y^2\left(\frac{2}{y^2}\right)}$$

$$=\dfrac{5y^2-1}{y+2}$$

**23.** $\dfrac{\frac{b}{a}-\frac{a}{b}}{\frac{1}{b}+\frac{1}{a}}=\dfrac{\left(\frac{b}{a}-\frac{a}{b}\right)ab}{\left(\frac{1}{b}+\frac{1}{a}\right)ab}$

$$=\dfrac{b^2-a^2}{a+b}$$

$$=\dfrac{(b-a)(b+a)}{a+b}$$

$$=b-a$$

**24.** $y=kx$
$10=k(15)$
$\dfrac{10}{15}=k$
$\dfrac{2}{3}=k$

$$y = \frac{2}{3}x$$

$$y = \frac{2}{3}(42)$$

$$y = \frac{84}{3}$$

$$y = 28$$

**25.**

$$y = \frac{k}{x^2}$$

$$8 = \frac{k}{5^2}$$

$$8 = \frac{k}{25}$$

$$200 = k$$

$$y = \frac{200}{x^2}$$

$$y = \frac{200}{15^2}$$

$$y = \frac{200}{225}$$

$$y = \frac{8}{9}$$

**26.** Let $x$ = the number of defective bulbs.

$$\frac{85}{3} = \frac{510}{x}$$

$$85x = 1530$$

$$x = 18$$

Expect to find 18 defective bulbs.

**27.**

$$x + 5 \cdot \frac{1}{x} = 6$$

$$x + \frac{5}{x} = 6$$

$$x\left(x + \frac{5}{x}\right) = x(6)$$

$$x(x) + x\left(\frac{5}{x}\right) = x(6)$$

$$x^2 + 5 = 6x$$

$$x^2 - 6x + 5 = 0$$

$$(x - 5)(x - 1) = 0$$

$$x - 5 = 0 \quad \text{or} \quad x - 1 = 0$$

$$x = 5 \qquad\qquad x = 1$$

The unknown number is 5 or 1.

**28.** Let $r$ be the speed of the boat in still water.

| | Distance | = Rate | · Time |
|---|---|---|---|
| Upstream | 14 | $r - 2$ | $\frac{14}{r-2}$ |
| Downstream | 16 | $r + 2$ | $\frac{16}{r+2}$ |

$$\frac{14}{r-2} = \frac{16}{r+2}$$

$$14(r+2) = 16(r-2)$$

$$14r + 28 = 16r - 32$$

$$60 = 2r$$

$$r = 30$$

The speed of the boat in still water is 30 miles per hour.

**29.** Let $x$ be the number of hours it takes to fill the tank using both pipes.

| | Hours to Complete Total Job | Part of Job Completed in 1 Hour |
|---|---|---|
| 1st Pipe | 12 | $\frac{1}{12}$ |
| 2nd Pipe | 15 | $\frac{1}{15}$ |
| Together | $x$ | $\frac{1}{x}$ |

$$\frac{1}{12} + \frac{1}{15} = \frac{1}{x}$$

$$60x\left(\frac{1}{12}\right) + 60x\left(\frac{1}{15}\right) = 60x\left(\frac{1}{x}\right)$$

$$5x + 4x = 60$$

$$9x = 60$$

$$x = \frac{60}{9} = \frac{20}{3} = 6\frac{2}{3}$$

Together, the pipes can fill the tank in

$6\frac{2}{3}$ hours.

**30.**

$$\frac{8}{x} = \frac{10}{15}$$

$$8(15) = 10x$$

$$120 = 10x$$

$$12 = x$$

The missing length is 12.

**Chapter 7 Cumulative Review**

**1. a.** $\dfrac{15}{x} = 4$

  **b.** $12 - 3 = x$

  **c.** $4x + 17 \neq 21$

  **d.** $3x < 48$

**2. a.** $12 - x = -45$

  **b.** $12x = -45$

  **c.** $x - 10 = 2x$

**3.** Let $x$ = the amount invested at 9% for one year.

| | Principal | · Rate = | Interest |
|---|---|---|---|
| 9% | $x$ | 0.09 | $0.09x$ |
| 7% | $20{,}000 - x$ | 0.07 | $0.07(20{,}000 - x)$ |
| Total | 20,000 | | 1550 |

$$0.09x + 0.07(20{,}000 - x) = 1550$$
$$0.09x + 1400 - 0.07x = 1550$$
$$0.02x + 1400 = 1550$$
$$0.02x = 150$$
$$x = 7500$$
$$20{,}000 - x = 20{,}000 - 7500 = 12{,}500$$
He invested \$7500 at 9% and \$12,500 at 7%.

**4.** Let $x$ be the number of bankruptcies in 1994 then $2x - 80{,}000$ is the number in 2002.
$$x + 2x - 80{,}000 = 2{,}290{,}000$$
$$3x - 80{,}000 = 2{,}290{,}000$$
$$3x = 2{,}370{,}000$$
$$x = 790{,}000$$
$$2x - 80{,}000 = 2(790{,}000) - 80{,}000 = 1{,}500{,}000$$
There were 790,000 bankruptcies in 1994 and 1,500,000 in 2002.

**5.** $x - 3y = 6$

| $x$ | $y$ |
|---|---|
| 0 | -2 |
| 6 | 0 |

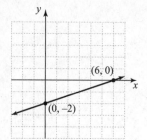

**6.** $7x + 2y = 9$
$$2y = -7x + 9$$
$$y = -\frac{7}{2}x + \frac{9}{2}$$
$$y = mx + b$$
$$m = -\frac{7}{2}$$

**7. a.** $4^2 \cdot 4^5 = 4^{2+5} = 4^7$

  **b.** $x^4 \cdot x^6 = x^{4+6} = x^{10}$

  **c.** $y^3 \cdot y = y^{3+1} = y^4$

  **d.** $y^3 \cdot y^2 \cdot y^7 = y^{3+2+7} = y^{12}$

  **e.** $(-5)^7 \cdot (-5)^8 = (-5)^{7+8} = (-5)^{15}$

  **f.** $a^2 \cdot b^2 = a^2 b^2$

**8. a.** $\dfrac{x^9}{x^7} = x^{9-7} = x^2$

  **b.** $\dfrac{x^{19} y^5}{xy} = x^{19-1} \cdot y^{5-1} = x^{18} y^4$

  **c.** $(x^5 y^2)^3 = x^{5 \cdot 3} y^{2 \cdot 3} = x^{15} y^6$

  **d.** $(-3a^2 b)(5a^3 b) = -15a^{2+3} b^{1+1} = -15a^5 b^2$

**9.** $[(8z + 11) + (9z - 2)] - (5z - 7)$
$$= 8z + 11 + 9z - 2 - 5z + 7$$
$$= 12z + 16$$

**10.** $(x + 1) - (9x^2 - 6x + 2) = x + 1 - 9x^2 + 6x - 2$
$$= -9x^2 + 7x - 1$$

**11.** $(3a+b)^3$

$= (3a+b)(3a+b)^2$

$= (3a+b)[(3a)^2 + 2(3a)(b) + (b)^2]$

$= (3a+b)(9a^2 + 6ab + b^2)$

$= 27a^3 + 18a^2b + 3ab^2 + 9a^2b + 6ab^2 + b^3$

$= 27a^3 + 27a^2b + 9ab^2 + b^3$

**12.** $(2x+1)(5x^2 - x + 2)$

$= 2x(5x^2 - x + 2) + 1(5x^2 - x + 2)$

$= 10x^3 - 2x^2 + 4x + 5x^2 - x + 2$

$= 10x^3 + 3x^2 + 3x + 2$

**13. a.** $(t+2)^2 = (t)^2 + 2(t)(2) + (2)^2 = t^2 + 4t + 4$

   **b.** $(p-q)^2 = (p)^2 - 2(p)(q) + (q)^2$

$= p^2 - 2pq + q^2$

   **c.** $(2x+5)^2 = (2x)^2 + 2(2x)(5) + (5)^2$

$= 4x^2 + 20x + 25$

   **d.** $(x^2 - 7y)^2 = (x^2)^2 - 2(x^2)(7y) + (7y)^2$

$= x^4 - 14x^2y + 49y^2$

**14. a.** $(x+9)^2 = (x)^2 + 2(x)(9) + (9)^2$

$= x^2 + 18x + 81$

   **b.** $(2x+1)(2x-1) = (2x)^2 - (1)^2 = 4x^2 - 1$

   **c.** $8x(x^2+1)(x^2-1) = 8x[(x^2)^2 - (1)^2]$

$= 8x[x^4 - 1]$

$= 8x^5 - 8x$

**15. a.** $\dfrac{1}{x^{-3}} = x^3$

   **b.** $\dfrac{1}{3^{-4}} = 3^4 = 81$

   **c.** $\dfrac{p^{-4}}{q^{-9}} = \dfrac{q^9}{p^4}$

   **d.** $\dfrac{5^{-3}}{2^{-5}} = \dfrac{2^5}{5^3} = \dfrac{32}{125}$

**16. a.** $5^{-3} = \dfrac{1}{5^3} = \dfrac{1}{125}$

   **b.** $\dfrac{9}{x^{-7}} = 9x^7$

   **c.** $\dfrac{11^{-1}}{7^{-2}} = \dfrac{7^2}{11^1} = \dfrac{49}{11}$

**17.**
$$\begin{array}{r} 4x^2 - 4x + 6 \\ 2x+3 \overline{)\, 8x^3 + 4x^2 + 0x + 7} \end{array}$$

$$\begin{array}{r} \underline{8x^3 + 12x^2} \\ -8x^2 + 0x \\ \underline{-8x^2 - 12x} \\ 12x + 7 \\ \underline{12x + 18} \\ -11 \end{array}$$

$$\dfrac{4x^2 + 7 + 8x^3}{2x+3} = 4x^2 - 4x + 6 - \dfrac{11}{2x+3}$$

**18.**
$$\begin{array}{r} 4x^2 + 16x + 55 \\ x-4 \overline{)\, 4x^3 + 0x^2 - 9x + 2} \end{array}$$

$$\begin{array}{r} \underline{4x^3 - 16x^2} \\ 16x^2 - 9x \\ \underline{16x^2 - 64x} \\ 55x + 2 \\ \underline{55x - 220} \\ 222 \end{array}$$

$$\dfrac{4x^3 - 9x + 2}{x-4} = 4x^2 + 16x + 55 + \dfrac{222}{x-4}$$

**19. a.** $28 = 2 \cdot 2 \cdot 7$

$40 = 2 \cdot 2 \cdot 2 \cdot 5$

$\text{GCF} = 2^2 = 4$

   **b.** $55 = 5 \cdot 11$

$21 = 3 \cdot 7$

$\text{GCF} = 1$

   **c.** $15 = 3 \cdot 5$

$18 = 2 \cdot 3 \cdot 3$

$66 = 2 \cdot 3 \cdot 11$

$\text{GCF} = 3$

**20.** $9x^2 = 3 \cdot 3 \cdot x^2$
$6x^3 = 2 \cdot 3 \cdot x^3$
$21x^5 = 3 \cdot 7 \cdot x^5$
$GCF = 3x^2$

**21.** $-9a^5 + 18a^2 - 3a = -3a(3a^4 - 6a + 1)$

**22.** $7x^6 - 7x^5 + 7x^4 = 7x^4(x^2 - x + 1)$

**23.** $3m^2 - 24m - 60 = 3(m^2 - 8m - 20)$
$= 3(m^2 - 10m + 2m - 20)$
$= 3[m(m-10) + 2(m-10)]$
$= 3(m-10)(m+2)$

**24.** $-2a^2 + 10a + 12 = -2(a^2 - 5a - 6)$
$= -2(a+1)(a-6)$

**25.** $3x^2 + 11x + 6 = 3x^2 + 2x + 9x + 6$
$= x(3x+2) + 3(3x+2)$
$= (3x+2)(x+3)$

**26.** $10m^2 - 7m + 1 = 10m^2 - 2m - 5m + 1$
$= 2m(5m-1) - 1(5m-1)$
$= (2m-1)(5m-1)$

**27.** $x^2 + 12x + 36 = x^2 + 2 \cdot x \cdot 6 + 6^2 = (x+6)^2$

**28.** $4x^2 + 12x + 9 = (2x)^2 + 2(2x)(3) + (3)^2$
$= (2x+3)^2$

**29.** $x^2 + 4$ is a prime polynomial.

**30.** $x^2 - 4 = (x)^2 - (2)^2 = (x+2)(x-2)$

**31.** $x^3 + 8 = x^3 + 2^3$
$= (x+2)(x^2 - x \cdot 2 + 2^2)$
$= (x+2)(x^2 - 2x + 4)$

**32.** $27y^3 - 1 = (3y)^3 - (1)^3$
$= (3y-1)[(3y)^2 + 3y(1) + (1)^2]$
$= (3y-1)(9y^2 + 3y + 1)$

**33.** $2x^3 + 3x^2 - 2x - 3 = x^2(2x+3) - 1(2x+3)$
$= (2x+3)(x^2 - 1)$
$= (2x+3)(x^2 - 1^2)$
$= (2x+3)(x+1)(x-1)$

**34.** $3x^3 + 5x^2 - 12x - 20 = x^2(3x+5) - 4(3x+5)$
$= (3x+5)(x^2 - 4)$
$= (3x+5)(x^2 - 2^2)$
$= (3x+5)(x+2)(x-2)$

**35.** $12m^2 - 3n^2 = 3(4m^2 - n^2)$
$= 3[(2m)^2 - (n)^2]$
$= 3(2m+n)(2m-n)$

**36.** $x^5 - x = x(x^4 - 1)$
$= x[(x^2)^2 - 1^2]$
$= x(x^2+1)(x^2-1)$
$= x(x^2+1)(x+1)(x-1)$

**37.** $x(2x-7) = 4$
$2x^2 - 7x = 4$
$2x^2 - 7x - 4 = 0$
$2x^2 - 8x + x - 4 = 0$
$2x(x-4) + 1(x-4) = 0$
$(x-4)(2x+1) = 0$
$2x+1 = 0$ or $x-4 = 0$
$2x = -1$ $\qquad x = 4$
$x = -\dfrac{1}{2}$

**38.** $3x^2 + 5x = 2$
$3x^2 + 5x - 2 = 0$
$3x^2 + 6x - x - 2 = 0$
$3x(x+2) - 1(x+2) = 0$
$(x+2)(3x-1) = 0$
$3x-1 = 0$ or $x+2 = 0$
$3x = 1$ $\qquad x = -2$
$x = \dfrac{1}{3}$

**39.** $y = x^2 - 5x + 4$

$0 = x^2 - 5x + 4$

$0 = (x-4)(x-1)$

$x - 1 = 0 \quad \text{or} \quad x - 4 = 0$

$\quad x = 1 \qquad\qquad x = 4$

The *x*-intercepts are $(1, 0)$ and $(4, 0)$.

**40.** $y = x^2 - x - 6$

$0 = x^2 - x - 6$

$0 = (x-3)(x+2)$

$x + 2 = 0 \quad \text{or} \quad x - 3 = 0$

$\quad x = -2 \qquad\qquad x = 3$

The *x*-intercepts are $(-2, 0)$ and $(3, 0)$.

**41.** Let $x =$ the base and $2x - 2 =$ the height.

$A = \dfrac{1}{2} bh$

$30 = \dfrac{1}{2} x(2x - 2)$

$30 = \dfrac{1}{2} (2x)(x - 1)$

$30 = x(x - 1)$

$30 = x^2 - x$

$0 = x^2 - x - 30$

$0 = (x+5)(x-6)$

$x - 6 = 0 \quad \text{or} \quad x + 5 = 0$

$\quad x = 6 \qquad\qquad x = -5$

Length cannot be negative, so $x = 6$.

$2x - 2 = 2(6) - 2 = 10$

The base is 6 meters and the height is 10 meters.

**42.** Let $x =$ the base and $3x + 5 =$ the height.

$A = bh$

$182 = x(3x + 5)$

$182 = 3x^2 + 5x$

$0 = 3x^2 + 5x - 182$

$0 = 3x^2 + 26x - 21x - 182$

$0 = x(3x + 26) - 7(3x + 26)$

$0 = (x-7)(3x+26)$

$x - 7 = 0 \quad \text{or} \quad 3x + 26 = 0$

$\quad x = 7 \qquad\qquad x = -\dfrac{26}{3}$

Length cannot be negative so $x = 7$.

$3x + 5 = 3(7) + 5 = 26$

The base is 7 ft and the height is 26 ft.

**43.** $\dfrac{5x - 5}{x^3 - x^2} = \dfrac{5(x-1)}{x^2(x-1)} = \dfrac{5}{x^2}$

**44.** $\dfrac{2x^2 - 50}{4x^4 - 20x^3} = \dfrac{2(x^2 - 25)}{4x^3(x-5)}$

$\qquad = \dfrac{2(x+5)(x-5)}{4x^3(x-5)}$

$\qquad = \dfrac{x+5}{2x^3}$

**45.** $\dfrac{6x + 2}{x^2 - 1} \div \dfrac{3x^2 + x}{x - 1} = \dfrac{6x + 2}{x^2 - 1} \cdot \dfrac{x - 1}{3x^2 + x}$

$\qquad = \dfrac{2(3x + 1)}{(x+1)(x-1)} \cdot \dfrac{x - 1}{x(3x + 1)}$

$\qquad = \dfrac{2}{x(x + 1)}$

**46.** $\dfrac{6x^2 - 18x}{3x^2 - 2x} \cdot \dfrac{15x - 10}{x^2 - 10} = \dfrac{6x(x-3) \cdot 5(3x - 2)}{x(3x - 2) \cdot (x+3)(x-3)}$

$\qquad = \dfrac{30}{x + 3}$

**47.** $\dfrac{\frac{x+1}{y}}{\frac{x}{y} + 2} = \dfrac{y\left(\frac{x+1}{y}\right)}{y\left(\frac{x}{y} + 2\right)} = \dfrac{x + 1}{y\left(\frac{x}{y}\right) + 2y} = \dfrac{x + 1}{x + 2y}$

**48.** $\dfrac{\frac{m}{3} + \frac{n}{6}}{\frac{m+n}{12}} = \dfrac{12}{12} \cdot \dfrac{\frac{m}{3} + \frac{n}{6}}{\frac{m+n}{12}}$

$\qquad = \dfrac{12\left(\frac{m}{3}\right) + 12\left(\frac{n}{6}\right)}{12\left(\frac{m+n}{12}\right)}$

$\qquad = \dfrac{4m + 2n}{m + n} \text{ or } \dfrac{2(2m + n)}{m + n}$

# Chapter 8

**1. a.** $\sqrt{\dfrac{4}{81}} = \dfrac{2}{9}$ because $\left(\dfrac{2}{9}\right)^2 = \dfrac{4}{81}$ and $\dfrac{2}{9}$ is positive.

**b.** $-\sqrt{25} = -5$
The negative sign in front of the radical indicates the negative square root of 25.

**c.** $\sqrt{144} = 12$ because $12^2 = 144$ and 12 is positive.

**d.** $\sqrt{0.49} = 0.7$ because $(0.7)^2 = 0.49$ and 0.7 is positive.

**e.** $-\sqrt{1} = -1$
The negative sign in front of the radical indicates the negative square root of 1.

**2. a.** $\sqrt[3]{0} = 0$ because $0^3 = 0$.

**b.** $\sqrt[3]{-64} = -4$ because $(-4)^3 = -64$.

**c.** $\sqrt[3]{\dfrac{1}{8}} = \dfrac{1}{2}$ because $\left(\dfrac{1}{2}\right)^3 = \dfrac{1}{8}$.

**3. a.** $\sqrt[4]{81} = 3$ because $3^4 = 81$ and 3 is positive.

**b.** $\sqrt[5]{100,000} = 10$ because $10^5 = 100,000$.

**c.** $\sqrt[6]{-64}$ is not a real number since the index 6 is even and the radicand $-64$ is negative.

**d.** $\sqrt[3]{-125} = -5$ because $(-5)^3 = -125$.

**4.** $\sqrt{17} \approx 4.123105626$
To three decimal places, $\sqrt{17} = 4.123$.

**5. a.** $\sqrt{x^{10}} = x^5$ because $(x^5)^2 = x^{10}$.

**b.** $\sqrt{y^{14}} = y^7$ because $(y^7)^2 = y^{14}$.

**c.** $\sqrt[3]{125z^9} = 5z^3$ because $(5z^3)^3 = 125z^9$.

**d.** $\sqrt{49x^2} = 7x$ because $(7x)^2 = 49x^2$.

**e.** $\sqrt{\dfrac{z^4}{36}} = \dfrac{z^2}{6}$ because $\left(\dfrac{z^2}{6}\right)^2 = \dfrac{z^4}{36}$.

**f.** $\sqrt[3]{-8a^6b^{12}} = -2a^2b^4$ because $(-2a^2b^4)^3 = -8a^6b^{12}$.

## Calculator Explorations

**1.** $\sqrt{7} \approx 2.646$
This is reasonable; 7 is between 4 and 9 so $\sqrt{7}$ is between $\sqrt{4} = 2$ and $\sqrt{9} = 3$.

**2.** $\sqrt{14} \approx 3.742$
This is reasonable; 14 is between 9 and 16 so $\sqrt{14}$ is between $\sqrt{9} = 3$ and $\sqrt{16} = 4$.

**3.** $\sqrt{11} \approx 3.317$
This is reasonable; 11 is between 9 and 16 so $\sqrt{11}$ is between $\sqrt{9} = 3$ and $\sqrt{16} = 4$.

**4.** $\sqrt{200} \approx 14.142$
This is reasonable; 200 is between 196 and 225 so $\sqrt{200}$ is between $\sqrt{196} = 14$ and $\sqrt{225} = 15$.

**5.** $\sqrt{82} \approx 9.055$
This is reasonable; 82 is between 81 and 100 so $\sqrt{82}$ is between $\sqrt{81} = 9$ and $\sqrt{100} = 10$.

**6.** $\sqrt{46} \approx 6.782$
This is reasonable; 46 is between 36 and 49 so $\sqrt{46}$ is between $\sqrt{36} = 6$ and $\sqrt{49} = 7$.

**7.** $\sqrt[3]{40} \approx 3.420$
This is reasonable; 40 is between 27 and 64 so $\sqrt[3]{40}$ is between $\sqrt[3]{27} = 3$ and $\sqrt[3]{64} = 4$.

**8.** $\sqrt[3]{71} \approx 4.141$
This is reasonable; 71 is between 64 and 125 so $\sqrt[3]{71}$ is between $\sqrt[3]{64} = 4$ and $\sqrt[3]{125} = 5$.

9. $\sqrt[4]{20} \approx 2.115$

This is reasonable; 20 is between 16 and 81 so $\sqrt[4]{20}$ is between $\sqrt[4]{16} = 2$ and $\sqrt[4]{81} = 3$.

10. $\sqrt[4]{15} \approx 1.968$

This is reasonable; 15 is between 1 and 16 so $\sqrt[4]{15}$ is between $\sqrt[4]{1} = 1$ and $\sqrt[4]{16} = 2$.

11. $\sqrt[5]{18} \approx 1.783$

This is reasonable; 18 is between 1 and 32 so $\sqrt[5]{18}$ is between $\sqrt[5]{1} = 1$ and $\sqrt[5]{32} = 2$.

12. $\sqrt[6]{2} \approx 1.122$

This is reasonable; 2 is between 1 and 64 so $\sqrt[6]{2}$ is between $\sqrt[6]{1} = 1$ and $\sqrt[6]{64} = 2$.

## Vocabulary, Readiness & Video Check 8.1

1. In the expression $\sqrt[4]{16}$, the number 4 is called the index, the number 16 is called the radicand, and $\sqrt{\phantom{x}}$ is called the radical sign.

2. The symbol $\sqrt{\phantom{x}}$ is used to denote the positive, or principal, square root.

3. False; $\sqrt{-16}$ is not a real number since the index is even and the radicand is negative.

4. True

5. True

6. True

7. True

8. False; $\sqrt{x^{16}} = x^8$ because $(x^8)^2 = x^{16}$.

9. The radical sign, $\sqrt{\phantom{x}}$, indicates a positive square root only. A negative sign before the radical sign, $-\sqrt{\phantom{x}}$, indicates a negative square root.

10. A square root of a negative number is not a real number, but the cube root of a negative number is a real number.

11. an odd-numbered index

12. Take the two integers that your answer falls between and square them, then check to make sure that the radicand falls between these two squares.

13. Divide the index into each exponent in the radicand—but still check by squaring your answer.

## Exercise Set 8.1

1. $\sqrt{16} = 4$, because $4^2 = 16$ and 4 is positive.

3. $\sqrt{\dfrac{1}{25}} = \dfrac{1}{5}$, because $\left(\dfrac{1}{5}\right)^2 = \dfrac{1}{25}$ and $\dfrac{1}{5}$ is positive.

5. $-\sqrt{100} = -10$, because $10^2 = 100$ and the negative sign indicates the negative square root.

7. $\sqrt{-4}$ is not a real number.

9. $-\sqrt{121} = -11$, because $11^2 = 121$ and the negative sign indicates the negative square root.

11. $\sqrt{\dfrac{9}{25}} = \dfrac{3}{5}$, because $\left(\dfrac{3}{5}\right)^2 = \dfrac{9}{25}$ and $\dfrac{3}{5}$ is positive.

13. $\sqrt{900} = 30$ because $30^2 = 900$ and 30 is positive.

15. $\sqrt{144} = 12$, because $12^2 = 144$ and 12 is positive.

17. $\sqrt{\dfrac{1}{100}} = \dfrac{1}{10}$ because $\left(\dfrac{1}{10}\right)^2 = \dfrac{1}{100}$ and $\dfrac{1}{10}$ is positive.

19. $\sqrt{0.25} = 0.5$ because $(0.5)^2 = 0.25$ and 0.5 is positive.

21. $\sqrt[3]{125} = 5$, because $(5)^3 = 125$.

23. $\sqrt[3]{-64} = -4$, because $(-4)^3 = -64$.

25. $-\sqrt[3]{8} = -2$ because $2^3 = 8$.

**27.** $\sqrt[3]{\dfrac{1}{8}} = \dfrac{1}{2}$, because $\left(\dfrac{1}{2}\right)^3 = \dfrac{1}{8}$.

**29.** $\sqrt[3]{-125} = -5$, because $(-5)^3 = -125$.

**31.** $\sqrt[5]{32} = 2$, because $(2)^5 = 32$.

**33.** $\sqrt{81} = 9$, because $9^2 = 81$ and 9 is positive.

**35.** $\sqrt[4]{-16}$ is not a real number.

**37.** $\sqrt[3]{-\dfrac{27}{64}} = -\dfrac{3}{4}$ because $\left(-\dfrac{3}{4}\right)^3 = -\dfrac{27}{64}$.

**39.** $-\sqrt[4]{625} = -5$, because $\sqrt[4]{625} = 5$.

**41.** $\sqrt[6]{1} = 1$, because $(1)^6 = 1$.

**43.** $\sqrt{7} \approx 2.646$

**45.** $\sqrt{37} \approx 6.083$

**47.** $\sqrt{136} \approx 11.662$

**49.** $\sqrt{2} \approx 1.41$
$90\sqrt{2} \approx 90 \cdot 1.41 = 126.90$
The distance is about 126.90 feet.

**51.** $\sqrt{m^2} = m$, because $m^2 = m^2$.

**53.** $\sqrt{x^4} = x^2$, because $(x^2)^2 = x^4$.

**55.** $\sqrt{9x^8} = 3x^4$, because $(3x^4)^2 = 9x^8$.

**57.** $\sqrt{81x^2} = 9x$, because $(9x)^2 = 81x^2$.

**59.** $\sqrt{a^2 b^4} = ab^2$, because $(ab^2)^2 = a^2 b^4$.

**61.** $\sqrt{16a^6 b^4} = 4a^3 b^2$ because $(4a^3 b^2)^2 = 16a^6 b^4$.

**63.** $\sqrt[3]{a^6 b^{18}} = a^2 b^6$ because $(a^2 b^6)^3 = a^6 b^{18}$.

**65.** $\sqrt[3]{-8x^3 y^{27}} = -2xy^9$ because
$(-2xy^9)^3 = -8x^3 y^{27}$.

**67.** $\sqrt{\dfrac{x^6}{36}} = \dfrac{x^3}{6}$ because $\left(\dfrac{x^3}{6}\right)^2 = \dfrac{x^6}{36}$.

**69.** $\sqrt{\dfrac{25y^2}{9}} = \dfrac{5y}{3}$ because $\left(\dfrac{5y}{3}\right)^2 = \dfrac{25y^2}{9}$.

**71.** $50 = 25 \cdot 2$; 25 is a perfect square

**73.** $32 = 16 \cdot 2$; 16 is a perfect square or $32 = 4 \cdot 8$; 4 is a perfect square

**75.** $28 = 4 \cdot 7$; 4 is a perfect square

**77.** $27 = 9 \cdot 3$; 9 is a perfect square

**79. a.** $\sqrt[7]{-1}$ is a real number because the index is odd.

**b.** $\sqrt[3]{-125}$ is a real number because the index is odd.

**c.** $\sqrt[6]{-128}$ is not a real number because the index is even and the radicand is negative.

**d.** $\sqrt[8]{-1}$ is not a real number because the index is even and the radicand is negative.

**81.** The length of the side is $\sqrt{49}$. Since $7^2 = 49$, $\sqrt{49} = 7$ and the sides of the square have length 7 miles.

**83.** The length of a side is $\sqrt{9.61}$ inches. Since $(3.1)^2 = 9.61$, $\sqrt{9.61} = 3.1$. The length of a side is 3.1 inches.

**85.** $\sqrt{\sqrt{81}} = \sqrt{9} = 3$, since $3^2 = 9$ and $9^2 = 81$.

**87.** $\sqrt{\sqrt{10,000}} = 10$ since $10^2 = 100$ and $100^2 = 10,000$.

**89.** Since $\sqrt{18}$ is between $\sqrt{16}$ and $\sqrt{25}$, then $\sqrt{18}$ is between 4 and 5.

**91.** Since $\sqrt{80}$ is between $\sqrt{64}$ and $\sqrt{81}$, then $\sqrt{80}$ is between 8 and 9.

**93.** $T = 2\pi\sqrt{\dfrac{L}{g}} = 2\pi\sqrt{\dfrac{30}{32}} \approx 2(3.14)(0.968) \approx 6.1$

The period of the pendulum is 6.1 seconds.

**95.** answers may vary

**97.**

| $x$ | $y = \sqrt{x}$ |
|-----|----------------|
| 0 | $\sqrt{0} = 0$ |
| 1 | $\sqrt{1} = 1$ |
| 3 | $\sqrt{3} \approx 1.7$ |
| 4 | $\sqrt{4} = 2$ |
| 9 | $\sqrt{9} = 3$ |

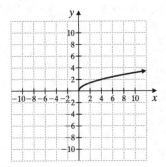

**99.** $\sqrt{x^2} = |x|$

**101.** $\sqrt{(x+2)^2} = |x+2|$

**103.** $y = \sqrt{x-2}$

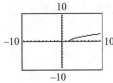

The graph starts at (2, 0).

**105.** $y = \sqrt{x+4}$

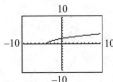

The graph starts at (−4, 0).

## Section 8.2 Practice

**1. a.** $\sqrt{24} = \sqrt{4 \cdot 6} = \sqrt{4} \cdot \sqrt{6} = 2\sqrt{6}$

   **b.** $\sqrt{60} = \sqrt{4 \cdot 15} = \sqrt{4} \cdot \sqrt{15} = 2\sqrt{15}$

   **c.** $\sqrt{42}$ is in simplest form since the radicand 42 contains no perfect square factors other than 1.

   **d.** $\sqrt{300} = \sqrt{100 \cdot 3} = \sqrt{100} \cdot \sqrt{3} = 10\sqrt{3}$

**2.** $5\sqrt{40} = 5 \cdot \sqrt{40}$
$$= 5 \cdot \sqrt{4 \cdot 10}$$
$$= 5 \cdot \sqrt{4} \cdot \sqrt{10}$$
$$= 5 \cdot 2 \cdot \sqrt{10}$$
$$= 10\sqrt{10}$$

**3. a.** $\sqrt{\dfrac{5}{49}} = \dfrac{\sqrt{5}}{\sqrt{49}} = \dfrac{\sqrt{5}}{7}$

   **b.** $\sqrt{\dfrac{9}{100}} = \dfrac{\sqrt{9}}{\sqrt{100}} = \dfrac{3}{10}$

   **c.** $\sqrt{\dfrac{18}{25}} = \dfrac{\sqrt{18}}{\sqrt{25}} = \dfrac{\sqrt{9} \cdot \sqrt{2}}{5} = \dfrac{3\sqrt{2}}{5}$

**4. a.** $\sqrt{x^7} = \sqrt{x^6 \cdot x} = \sqrt{x^6} \cdot \sqrt{x} = x^3\sqrt{x}$

   **b.** $\sqrt{12a^4} = \sqrt{4 \cdot 3 \cdot a^4}$
$$= \sqrt{4a^4 \cdot 3}$$
$$= \sqrt{4a^4} \cdot \sqrt{3}$$
$$= 2a^2\sqrt{3}$$

   **c.** $\sqrt{\dfrac{98}{z^8}} = \dfrac{\sqrt{98}}{\sqrt{z^8}} = \dfrac{\sqrt{49 \cdot 2}}{z^4} = \dfrac{\sqrt{49} \cdot \sqrt{2}}{z^4} = \dfrac{7\sqrt{2}}{z^4}$

   **d.** $\sqrt{\dfrac{11y^9}{49}} = \dfrac{\sqrt{11y^9}}{\sqrt{49}}$
$$= \dfrac{\sqrt{y^8 \cdot 11y}}{7}$$
$$= \dfrac{\sqrt{y^8} \cdot \sqrt{11y}}{7}$$
$$= \dfrac{y^4\sqrt{11y}}{7}$$

**5. a.** $\sqrt[3]{24} = \sqrt[3]{8 \cdot 3} = \sqrt[3]{8} \cdot \sqrt[3]{3} = 2\sqrt[3]{3}$

**b.** The number 38 contains no perfect cube factors, so $\sqrt[3]{38}$ cannot be simplified further.

**c.** $\sqrt[3]{\dfrac{5}{27}} = \dfrac{\sqrt[3]{5}}{\sqrt[3]{27}} = \dfrac{\sqrt[3]{5}}{3}$

**d.** $\sqrt[3]{\dfrac{15}{64}} = \dfrac{\sqrt[3]{15}}{\sqrt[3]{64}} = \dfrac{\sqrt[3]{15}}{4}$

**6. a.** $\sqrt[4]{80} = \sqrt[4]{16 \cdot 5} = \sqrt[4]{16} \cdot \sqrt[4]{5} = 2\sqrt[4]{5}$

**b.** $\sqrt[4]{\dfrac{5}{81}} = \dfrac{\sqrt[4]{5}}{\sqrt[4]{81}} = \dfrac{\sqrt[4]{5}}{3}$

**c.** $\sqrt[5]{96} = \sqrt[5]{32 \cdot 3} = \sqrt[5]{32} \cdot \sqrt[5]{3} = 2\sqrt[5]{3}$

**Vocabulary, Readiness & Video Check 8.2**

**1.** If $\sqrt{a}$ and $\sqrt{b}$ are real numbers, then $\sqrt{a \cdot b} = \underline{\sqrt{a} \cdot \sqrt{b}}$.

**2.** If $\sqrt{a}$ and $\sqrt{b}$ are real numbers, then $\sqrt{\dfrac{a}{b}} = \underline{\dfrac{\sqrt{a}}{\sqrt{b}}}$.

**3.** $\sqrt{16 \cdot 25} = \sqrt{16} \cdot \sqrt{25} = \underline{4} \cdot \underline{5} = \underline{20}$

**4.** $\sqrt{36 \cdot 3} = \sqrt{36} \cdot \sqrt{3} = \underline{6} \cdot \sqrt{3} = \underline{6\sqrt{3}}$

**5.** Factor until you have a product of primes. A repeated prime factor means a perfect square—if more than one factor is repeated you can multiply all the repeated factors together to get one larger perfect square factor.

**6.** In words, the quotient rule for square roots says that the same root of a quotient is equal to the square root of the <u>numerator</u> over the square root of the <u>denominator</u>.

**7.** The power must be 1. Any even power is a perfect square and can be simplified; any higher odd power is the product of an even power times the variable with a power of 1.

**8.** If a factor is repeated the same number of times as the index, then you have a perfect root, and the product rule can be applied.

**Exercise Set 8.2**

**1.** $\sqrt{20} = \sqrt{4 \cdot 5} = \sqrt{4} \cdot \sqrt{5} = 2\sqrt{5}$

**3.** $\sqrt{50} = \sqrt{25 \cdot 2} = \sqrt{25} \cdot \sqrt{2} = 5\sqrt{2}$

**5.** $\sqrt{33}$ can't be simplified.

**7.** $\sqrt{98} = \sqrt{49 \cdot 2} = \sqrt{49} \cdot \sqrt{2} = 7\sqrt{2}$

**9.** $\sqrt{60} = \sqrt{4 \cdot 15} = \sqrt{4} \cdot \sqrt{15} = 2\sqrt{15}$

**11.** $\sqrt{180} = \sqrt{36 \cdot 5} = \sqrt{36} \cdot \sqrt{5} = 6\sqrt{5}$

**13.** $\sqrt{52} = \sqrt{4 \cdot 13} = \sqrt{4} \cdot \sqrt{13} = 2\sqrt{13}$

**15.** $3\sqrt{25} = 3 \cdot \sqrt{25} = 3 \cdot 5 = 15$

**17.** $7\sqrt{63} = 7 \cdot \sqrt{63}$
$= 7 \cdot \sqrt{9 \cdot 7}$
$= 7 \cdot \sqrt{9} \cdot \sqrt{7}$
$= 7 \cdot 3 \cdot \sqrt{7}$
$= 21\sqrt{7}$

**19.** $-5\sqrt{27} = -5 \cdot \sqrt{27}$
$= -5 \cdot \sqrt{9 \cdot 3}$
$= -5 \cdot \sqrt{9} \cdot \sqrt{3}$
$= -5 \cdot 3 \cdot \sqrt{3}$
$= -15\sqrt{3}$

**21.** $\sqrt{\dfrac{8}{25}} = \dfrac{\sqrt{8}}{\sqrt{25}} = \dfrac{\sqrt{4 \cdot 2}}{5} = \dfrac{\sqrt{4} \cdot \sqrt{2}}{5} = \dfrac{2\sqrt{2}}{5}$

**23.** $\sqrt{\dfrac{27}{121}} = \dfrac{\sqrt{27}}{\sqrt{121}} = \dfrac{\sqrt{9 \cdot 3}}{11} = \dfrac{\sqrt{9} \cdot \sqrt{3}}{11} = \dfrac{3\sqrt{3}}{11}$

**25.** $\sqrt{\dfrac{9}{4}} = \dfrac{\sqrt{9}}{\sqrt{4}} = \dfrac{3}{2}$

**27.** $\sqrt{\dfrac{125}{9}} = \dfrac{\sqrt{125}}{\sqrt{9}} = \dfrac{\sqrt{25 \cdot 5}}{3} = \dfrac{\sqrt{25} \cdot \sqrt{5}}{3} = \dfrac{5\sqrt{5}}{3}$

**29.** $\sqrt{\dfrac{11}{36}} = \dfrac{\sqrt{11}}{\sqrt{36}} = \dfrac{\sqrt{11}}{6}$

**31.** $-\sqrt{\dfrac{27}{144}} = -\dfrac{\sqrt{27}}{\sqrt{144}}$

$\qquad\qquad = -\dfrac{\sqrt{9 \cdot 3}}{12}$

$\qquad\qquad = -\dfrac{\sqrt{9} \cdot \sqrt{3}}{12}$

$\qquad\qquad = -\dfrac{3\sqrt{3}}{12}$

$\qquad\qquad = -\dfrac{\sqrt{3}}{4}$

**33.** $\sqrt{x^7} = \sqrt{x^6 \cdot x} = \sqrt{x^6} \cdot \sqrt{x} = x^3\sqrt{x}$

**35.** $\sqrt{x^{13}} = \sqrt{x^{12} \cdot x} = \sqrt{x^{12}} \cdot \sqrt{x} = x^6\sqrt{x}$

**37.** $\sqrt{36a^3} = \sqrt{36a^2 \cdot a} = \sqrt{36a^2} \cdot \sqrt{a} = 6a\sqrt{a}$

**39.** $\sqrt{96x^4} = \sqrt{16x^4 \cdot 6} = \sqrt{16x^4} \cdot \sqrt{6} = 4x^2\sqrt{6}$

**41.** $\sqrt{\dfrac{12}{m^2}} = \dfrac{\sqrt{12}}{\sqrt{m^2}} = \dfrac{\sqrt{4 \cdot 3}}{m} = \dfrac{\sqrt{4} \cdot \sqrt{3}}{m} = \dfrac{2\sqrt{3}}{m}$

**43.** $\sqrt{\dfrac{9x}{y^{10}}} = \dfrac{\sqrt{9x}}{\sqrt{y^{10}}} = \dfrac{\sqrt{9 \cdot x}}{y^5} = \dfrac{\sqrt{9} \cdot \sqrt{x}}{y^5} = \dfrac{3\sqrt{x}}{y^5}$

**45.** $\sqrt{\dfrac{88}{x^{12}}} = \dfrac{\sqrt{88}}{\sqrt{x^{12}}} = \dfrac{\sqrt{4 \cdot 22}}{x^6} = \dfrac{\sqrt{4} \cdot \sqrt{22}}{x^6} = \dfrac{2\sqrt{22}}{x^6}$

**47.** $8\sqrt{4} = 8 \cdot \sqrt{4} = 8 \cdot 2 = 16$

**49.** $\sqrt{\dfrac{36}{121}} = \dfrac{\sqrt{36}}{\sqrt{121}} = \dfrac{6}{11}$

**51.** $\sqrt{175} = \sqrt{25 \cdot 7} = \sqrt{25} \cdot \sqrt{7} = 5\sqrt{7}$

**53.** $\sqrt{\dfrac{20}{9}} = \dfrac{\sqrt{20}}{\sqrt{9}} = \dfrac{\sqrt{4 \cdot 5}}{3} = \dfrac{\sqrt{4} \cdot \sqrt{5}}{3} = \dfrac{2\sqrt{5}}{3}$

**55.** $\sqrt{24m^7} = \sqrt{4m^6 \cdot 6m} = \sqrt{4m^6} \cdot \sqrt{6m} = 2m^3\sqrt{6m}$

**57.** $\sqrt{\dfrac{23y^3}{4x^6}} = \dfrac{\sqrt{23y^3}}{\sqrt{4x^6}}$

$\qquad\qquad = \dfrac{\sqrt{y^2 \cdot 23y}}{2x^3}$

$\qquad\qquad = \dfrac{\sqrt{y^2} \cdot \sqrt{23y}}{2x^3}$

$\qquad\qquad = \dfrac{y\sqrt{23y}}{2x^3}$

**59.** $\sqrt[3]{24} = \sqrt[3]{8 \cdot 3} = \sqrt[3]{8} \cdot \sqrt[3]{3} = 2\sqrt[3]{3}$

**61.** $\sqrt[3]{250} = \sqrt[3]{125 \cdot 2} = \sqrt[3]{125} \cdot \sqrt[3]{2} = 5\sqrt[3]{2}$

**63.** $\sqrt[3]{\dfrac{5}{64}} = \dfrac{\sqrt[3]{5}}{\sqrt[3]{64}} = \dfrac{\sqrt[3]{5}}{4}$

**65.** $\sqrt[3]{\dfrac{23}{8}} = \dfrac{\sqrt[3]{23}}{\sqrt[3]{8}} = \dfrac{\sqrt[3]{23}}{2}$

**67.** $\sqrt[3]{\dfrac{15}{64}} = \dfrac{\sqrt[3]{15}}{\sqrt[3]{64}} = \dfrac{\sqrt[3]{15}}{4}$

**69.** $\sqrt[3]{80} = \sqrt[3]{8 \cdot 10} = \sqrt[3]{8} \cdot \sqrt[3]{10} = 2\sqrt[3]{10}$

**71.** $\sqrt[4]{48} = \sqrt[4]{16 \cdot 3} = \sqrt[4]{16} \cdot \sqrt[4]{3} = 2\sqrt[4]{3}$

**73.** $\sqrt[4]{\dfrac{8}{81}} = \dfrac{\sqrt[4]{8}}{\sqrt[4]{81}} = \dfrac{\sqrt[4]{8}}{3}$

**75.** $\sqrt[5]{96} = \sqrt[5]{32 \cdot 3} = \sqrt[5]{32} \cdot \sqrt[5]{3} = 2\sqrt[5]{3}$

**77.** $\sqrt[5]{\dfrac{5}{32}} = \dfrac{\sqrt[5]{5}}{\sqrt[5]{32}} = \dfrac{\sqrt[5]{5}}{2}$

**79.** $6x + 8x = (6 + 8)x = 14x$

**81.** $(2x+3)(x-5) = 2x^2 - 10x + 3x - 15$

$\qquad\qquad\qquad = 2x^2 - 7x - 15$

**83.** $9y^2 - 9y^2 = 0$

**85.** $\sqrt{x^6 y^3} = \sqrt{x^6 y^2 y} = \sqrt{x^6 y^2} \cdot \sqrt{y} = x^3 y\sqrt{y}$

**87.** $\sqrt{98x^5y^4} = \sqrt{49x^4y^4 \cdot 2x}$
$= \sqrt{49x^4y^4} \cdot \sqrt{2x}$
$= 7x^2y^2\sqrt{2x}$

**89.** $\sqrt[3]{-8x^6} = -2x^2$ because $(-2x^2)^3 = -8x^6$.

**91.** $\sqrt[3]{80} = \sqrt[3]{8 \cdot 10} = \sqrt[3]{8} \cdot \sqrt[3]{10} = 2\sqrt[3]{10}$
The length of each side is $2\sqrt[3]{10}$ inches.

**93.** answers may vary

**95.** Let $A = 120$. The length of a side is $\sqrt{\dfrac{A}{6}}$.

$\sqrt{\dfrac{A}{6}} = \sqrt{\dfrac{120}{6}} = \sqrt{20} = \sqrt{4} \cdot \sqrt{5} = 2\sqrt{5}$

The length of a side is $2\sqrt{5}$ inches.

**97.** Let $A = 30.375$. The length of a side is $\sqrt{\dfrac{A}{6}}$.

$\sqrt{\dfrac{A}{6}} = \sqrt{\dfrac{30.375}{6}} = \sqrt{5.0625} = 2.25$

The length of one side of a Rubik's cube is 2.25 inches.

**99.** Let $n = 1000$
$C = 100\sqrt[3]{n} + 700$
$C = 100\sqrt[3]{1000} + 700$
$= 100(10) + 700$
$= 1700$
The cost is $1700.

**101.** Let $h = 169$ and $w = 64$.
$B = \sqrt{\dfrac{hw}{3600}}$
$B = \sqrt{\dfrac{(169)(64)}{3600}}$
$= \sqrt{\dfrac{10,816}{3600}}$
$= \sqrt{\dfrac{676}{225}}$
$= \dfrac{26}{15}$
$\approx 1.7$
The surface area is about 1.7 sq m.

**Section 8.3 Practice**

**1. a.** $3\sqrt{2} + 5\sqrt{2} = (3+5)\sqrt{2} = 8\sqrt{2}$

**b.** $\sqrt{6} - 8\sqrt{6} = 1\sqrt{6} - 8\sqrt{6} = (1-8)\sqrt{6} = -7\sqrt{6}$

**c.** $6\sqrt[4]{5} - 2\sqrt[4]{5} + 11\sqrt[4]{7} = (6-2)\sqrt[4]{5} + 11\sqrt[4]{7}$
$= 4\sqrt[4]{5} + 11\sqrt[4]{7}$
This expression cannot be simplified further since the radicals are not the same.

**d.** $4\sqrt{13} - 5\sqrt[3]{13}$ cannot be simplified further since the indices are not the same.

**2. a.** $\sqrt{45} + \sqrt{20} = \sqrt{9 \cdot 5} + \sqrt{4 \cdot 5}$
$= \sqrt{9} \cdot \sqrt{5} + \sqrt{4} \cdot \sqrt{5}$
$= 3\sqrt{5} + 2\sqrt{5}$
$= 5\sqrt{5}$

**b.** $\sqrt{36} + 3\sqrt{24} - \sqrt{40} - \sqrt{150}$
$= 6 + 3\sqrt{4 \cdot 6} - \sqrt{4 \cdot 10} - \sqrt{25 \cdot 6}$
$= 6 + 3\sqrt{4} \cdot \sqrt{6} - \sqrt{4} \cdot \sqrt{10} - \sqrt{25} \cdot \sqrt{6}$
$= 6 + 3 \cdot 2\sqrt{6} - 2\sqrt{10} - 5\sqrt{6}$
$= 6 + 6\sqrt{6} - 2\sqrt{10} - 5\sqrt{6}$
$= 6 - 2\sqrt{10} + \sqrt{6}$

**c.** $\sqrt{98} - 5\sqrt{8} = \sqrt{49 \cdot 2} - 5\sqrt{4 \cdot 2}$
$= \sqrt{49} \cdot \sqrt{2} - 5\sqrt{4} \cdot \sqrt{2}$
$= 7\sqrt{2} - 5 \cdot 2\sqrt{2}$
$= 7\sqrt{2} - 10\sqrt{2}$
$= -3\sqrt{2}$

**3.** $\sqrt{x^3} - 8x\sqrt{x} + 3\sqrt{x^2} = \sqrt{x^2 \cdot x} - 8x\sqrt{x} + 3x$
$= \sqrt{x^2} \cdot \sqrt{x} - 8x\sqrt{x} + 3x$
$= x\sqrt{x} - 8x\sqrt{x} + 3x$
$= -7x\sqrt{x} + 3x$

**4.** $4\sqrt[3]{81x^6} - \sqrt[3]{24x^6} = 4\sqrt[3]{27x^6 \cdot 3} - \sqrt[3]{8x^6 \cdot 3}$
$= 4 \cdot \sqrt[3]{27x^6} \cdot \sqrt[3]{3} - \sqrt[3]{8x^6} \cdot \sqrt[3]{3}$
$= 4 \cdot 3x^2 \cdot \sqrt[3]{3} - 2x^2 \cdot \sqrt[3]{3}$
$= 12x^2\sqrt[3]{3} - 2x^2\sqrt[3]{3}$
$= 10x^2\sqrt[3]{3}$

**Vocabulary, Readiness & Video Check 8.3**

1. Radicals that have the same index and same radicand are called <u>like radicals</u>.

2. The expressions $7\sqrt[3]{2x}$ and $-\sqrt[3]{2x}$ are called <u>like radicals</u>.

3. $11\sqrt{2} + 6\sqrt{2} = \underline{17\sqrt{2}}$

4. $\sqrt{5}$ is the same as $\underline{1\sqrt{5}}$.

5. $\sqrt{5} + \sqrt{5} = \underline{2\sqrt{5}}$

6. $9\sqrt{7} - \sqrt{7} = \underline{8\sqrt{7}}$

7. Both like terms and like radicals are combined using the distributive property; also only like (vs. unlike) terms can be combined, as with like radicals.

8. Sometimes you can't see that there are like radicals to combine until you simplify, so you may incorrectly think you cannot add or subtract if you don't simplify first.

**Exercise Set 8.3**

1. $4\sqrt{3} - 8\sqrt{3} = (4-8)\sqrt{3} = -4\sqrt{3}$

3. $3\sqrt{6} + 8\sqrt{6} - 2\sqrt{6} - 5 = (3+8-2)\sqrt{6} - 5$
   $= 9\sqrt{6} - 5$

5. $\sqrt{11} + \sqrt{11} + 11 = 1\sqrt{11} + 1\sqrt{11} + 11$
   $= (1+1)\sqrt{11} + 11$
   $= 2\sqrt{11} + 11$

7. $6\sqrt{5} - 5\sqrt{5} + \sqrt{2} = (6-5)\sqrt{5} + \sqrt{2} = \sqrt{5} + \sqrt{2}$

9. $\sqrt[3]{18} + \sqrt[3]{18} - 4\sqrt[3]{18} = 1\sqrt[3]{18} + 1\sqrt[3]{18} - 4\sqrt[3]{18}$
   $= (1+1-4)\sqrt[3]{18}$
   $= -2\sqrt[3]{18}$

11. $2\sqrt[3]{3} + 5\sqrt[3]{3} - \sqrt{3} = (2+5)\sqrt[3]{3} - \sqrt{3} = 7\sqrt[3]{3} - \sqrt{3}$

13. $2\sqrt[3]{2} - 7\sqrt[3]{2} - 6 = (2-7)\sqrt[3]{2} - 6 = -5\sqrt[3]{2} - 6$

15. $\sqrt{12} + \sqrt{27} = \sqrt{4 \cdot 3} + \sqrt{9 \cdot 3}$
    $= \sqrt{4} \cdot \sqrt{3} + \sqrt{9} \cdot \sqrt{3}$
    $= 2\sqrt{3} + 3\sqrt{3}$
    $= 5\sqrt{3}$

17. $\sqrt{45} + 3\sqrt{20} = \sqrt{9 \cdot 5} + 3\sqrt{4 \cdot 5}$
    $= \sqrt{9} \cdot \sqrt{5} + 3\sqrt{4} \cdot \sqrt{5}$
    $= 3\sqrt{5} + 3(2)\sqrt{5}$
    $= 3\sqrt{5} + 6\sqrt{5}$
    $= 9\sqrt{5}$

19. $2\sqrt{54} - \sqrt{20} + \sqrt{45} - \sqrt{24}$
    $= 2\sqrt{9 \cdot 6} - \sqrt{4 \cdot 5} + \sqrt{9 \cdot 5} - \sqrt{4 \cdot 6}$
    $= 2\sqrt{9} \cdot \sqrt{6} - \sqrt{4} \cdot \sqrt{5} + \sqrt{9} \cdot \sqrt{5} - \sqrt{4} \cdot \sqrt{6}$
    $= 2(3)\sqrt{6} - 2\sqrt{5} + 3\sqrt{5} - 2\sqrt{6}$
    $= 6\sqrt{6} - 2\sqrt{5} + 3\sqrt{5} - 2\sqrt{6}$
    $= 4\sqrt{6} + \sqrt{5}$

21. $4x - 3\sqrt{x^2} + \sqrt{x} = 4x - 3x + \sqrt{x} = x + \sqrt{x}$

23. $\sqrt{25x} + \sqrt{36x} - 11\sqrt{x}$
    $= \sqrt{25} \cdot \sqrt{x} + \sqrt{36} \cdot \sqrt{x} - 11\sqrt{x}$
    $= 5\sqrt{x} + 6\sqrt{x} - 11\sqrt{x}$
    $= 0$

25. $\sqrt{16x} - \sqrt{x^3} = \sqrt{16x} - \sqrt{x^2 \cdot x}$
    $= \sqrt{16} \cdot \sqrt{x} - \sqrt{x^2} \cdot \sqrt{x}$
    $= 4\sqrt{x} - x\sqrt{x}$
    $= (4-x)\sqrt{x}$

27. $12\sqrt{5} - \sqrt{5} - 4\sqrt{5} = (12-1-4)\sqrt{5} = 7\sqrt{5}$

29. $\sqrt{5} + \sqrt[3]{5}$ cannot be simplified.

31. $4 + 8\sqrt{2} - 9 = 8\sqrt{2} + 4 - 9 = 8\sqrt{2} - 5$

33. $8 - \sqrt{2} - 5\sqrt{2} = 8 + (-1-5)\sqrt{2} = 8 - 6\sqrt{2}$

35. $5\sqrt{32} - \sqrt{72} = 5\sqrt{16 \cdot 2} - \sqrt{36 \cdot 2}$
    $= 5\sqrt{16}\sqrt{2} - \sqrt{36}\sqrt{2}$
    $= 5(4)\sqrt{2} - 6\sqrt{2}$
    $= 20\sqrt{2} - 6\sqrt{2}$
    $= 14\sqrt{2}$

**37.** $\sqrt{8}+\sqrt{9}+\sqrt{18}+\sqrt{81}$
$=\sqrt{4\cdot2}+\sqrt{9}+\sqrt{9\cdot2}+\sqrt{81}$
$=\sqrt{4}\cdot\sqrt{2}+3+\sqrt{9}\cdot\sqrt{2}+9$
$=2\sqrt{2}+3+3\sqrt{2}+9$
$=5\sqrt{2}+12$

**39.** $\sqrt{\dfrac{5}{9}}+\sqrt{\dfrac{5}{81}}=\dfrac{\sqrt{5}}{\sqrt{9}}+\dfrac{\sqrt{5}}{\sqrt{81}}$
$=\dfrac{\sqrt{5}}{3}+\dfrac{\sqrt{5}}{9}$
$=\dfrac{3\sqrt{5}}{9}+\dfrac{\sqrt{5}}{9}$
$=\dfrac{3\sqrt{5}+\sqrt{5}}{9}$
$=\dfrac{4\sqrt{5}}{9}$

**41.** $\sqrt{\dfrac{3}{4}}-\sqrt{\dfrac{3}{64}}=\dfrac{\sqrt{3}}{\sqrt{4}}-\dfrac{\sqrt{3}}{\sqrt{64}}$
$=\dfrac{\sqrt{3}}{2}-\dfrac{\sqrt{3}}{8}$
$=\dfrac{4\sqrt{3}}{8}-\dfrac{\sqrt{3}}{8}$
$=\dfrac{4\sqrt{3}-\sqrt{3}}{8}$
$=\dfrac{3\sqrt{3}}{8}$

**43.** $2\sqrt{45}-2\sqrt{20}=2\sqrt{9\cdot5}-2\sqrt{4\cdot5}$
$=2\sqrt{9}\cdot\sqrt{5}-2\sqrt{4}\cdot\sqrt{5}$
$=2(3)\sqrt{5}-2(2)\sqrt{5}$
$=6\sqrt{5}-4\sqrt{5}$
$=2\sqrt{5}$

**45.** $\sqrt{35}-\sqrt{140}=\sqrt{35}-\sqrt{4\cdot35}$
$=\sqrt{35}-\sqrt{4}\cdot\sqrt{35}$
$=\sqrt{35}-2\sqrt{35}$
$=-\sqrt{35}$

**47.** $5\sqrt{2x}+\sqrt{98x}=5\sqrt{2x}+\sqrt{49\cdot2x}$
$=5\sqrt{2x}+\sqrt{49}\cdot\sqrt{2x}$
$=5\sqrt{2x}+7\sqrt{2x}$
$=12\sqrt{2x}$

**49.** $5\sqrt{x}+4\sqrt{4x}-13\sqrt{x}=5\sqrt{x}+4\sqrt{4}\cdot\sqrt{x}-13\sqrt{x}$
$=5\sqrt{x}+4(2)\sqrt{x}-13\sqrt{x}$
$=5\sqrt{x}+8\sqrt{x}-13\sqrt{x}$
$=13\sqrt{x}-13\sqrt{x}$
$=0$

**51.** $\sqrt{3x^3}+3x\sqrt{x}=\sqrt{x^2\cdot3x}+3x\sqrt{x}$
$=\sqrt{x^2}\cdot\sqrt{3x}+3x\sqrt{x}$
$=x\sqrt{3x}+3x\sqrt{x}$

**53.** $\sqrt[3]{81}+\sqrt[3]{24}=\sqrt[3]{27\cdot3}+\sqrt[3]{8\cdot3}$
$=\sqrt[3]{27}\sqrt[3]{3}+\sqrt[3]{8}\sqrt[3]{3}$
$=3\sqrt[3]{3}+2\sqrt[3]{3}$
$=5\sqrt[3]{3}$

**55.** $4\sqrt[3]{9}-\sqrt[3]{243}=4\sqrt[3]{9}-\sqrt[3]{27\cdot9}$
$=4\sqrt[3]{9}-\sqrt[3]{27}\sqrt[3]{9}$
$=4\sqrt[3]{9}-3\sqrt[3]{9}$
$=\sqrt[3]{9}$

**57.** $\sqrt[3]{8}+\sqrt[3]{54}-5=2+\sqrt[3]{27\cdot2}-5$
$=2+\sqrt[3]{27}\cdot\sqrt[3]{2}-5$
$=-3+3\sqrt[3]{2}$

**59.** $\sqrt{32x^2}+\sqrt[3]{32}+\sqrt{4x^2}$
$=\sqrt{16x^2\cdot2}+\sqrt[3]{8\cdot4}+\sqrt{4x^2}$
$=\sqrt{16x^2}\cdot\sqrt{2}+\sqrt[3]{8}\cdot\sqrt[3]{4}+\sqrt{4x^2}$
$=4x\sqrt{2}+2\sqrt[3]{4}+2x$

**61.** $2\sqrt[3]{8x^3}+2\sqrt[3]{16x^3}=2\cdot2x+2\sqrt[3]{8x^3\cdot2}$
$=4x+2\sqrt[3]{8x^3}\sqrt[3]{2}$
$=4x+2\cdot2x\sqrt[3]{2}$
$=4x+4x\sqrt[3]{2}$

**63.** $12\sqrt[3]{y^7}-y^2\sqrt[3]{8y}=12\sqrt[3]{y^6\cdot y}-y^2\sqrt[3]{8}\sqrt[3]{y}$
$=12\sqrt[3]{y^6}\sqrt[3]{y}-2y^2\sqrt[3]{y}$
$=12y^2\sqrt[3]{y}-2y^2\sqrt[3]{y}$
$=(12y^2-2y^2)\sqrt[3]{y}$
$=10y^2\sqrt[3]{y}$

**65.** $\sqrt{40x} + \sqrt[3]{40} - 2\sqrt{10x} - \sqrt[3]{5}$

$= \sqrt{4 \cdot 10x} + \sqrt[3]{8 \cdot 5} - 2\sqrt{10x} - \sqrt[3]{5}$

$= \sqrt{4} \cdot \sqrt{10x} + \sqrt[3]{8} \cdot \sqrt[3]{5} - 2\sqrt{10x} - \sqrt[3]{5}$

$= 2\sqrt{10x} + 2\sqrt[3]{5} - 2\sqrt{10x} - \sqrt[3]{5}$

$= \sqrt[3]{5}$

**67.** $(x+6)^2 = x^2 + 2(6)x + 6^2 = x^2 + 12x + 36$

**69.** $(2x-1)^2 = (2x)^2 + 2(-1)(2x) + (-1)^2$

$\qquad\qquad = 4x^2 - 4x + 1$

**71.** answers may vary

**73.** Let $l = 3\sqrt{5}$ and $w = \sqrt{5}$.

Perimeter $= 2l + 2w$

$\qquad\qquad = 2\left(3\sqrt{5}\right) + 2\left(\sqrt{5}\right)$

$\qquad\qquad = 6\sqrt{5} + 2\sqrt{5}$

$\qquad\qquad = 8\sqrt{5}$ inches

**75.** Let $l = 8$ and $w = 3$.

Area $=$ area of 2 triangles $+$ area of 2 rectangles

$\qquad = 2\left(\dfrac{3\sqrt{27}}{4}\right) + 2lw$

$\qquad = \dfrac{3\sqrt{9} \cdot \sqrt{3}}{2} + 2(8)(3)$

$\qquad = \left(\dfrac{9\sqrt{3}}{2} + 48\right)$ square feet

**77.** The expression can be simplified.

$4\sqrt{2} + 3\sqrt{2} = (4+3)\sqrt{2} = 7\sqrt{2}$

**79.** The expression $6 + 7\sqrt{6}$ cannot be simplified.

**81.** The expression can be simplified.

$\sqrt{7} + \sqrt{7} + \sqrt{7} = (1+1+1)\sqrt{7} = 3\sqrt{7}$

**83.** $\sqrt{\dfrac{x^3}{16}} - x\sqrt{\dfrac{9x}{25}} + \dfrac{\sqrt{81x^3}}{2}$

$= \dfrac{\sqrt{x^2 \cdot x}}{\sqrt{16}} - x\dfrac{\sqrt{9x}}{\sqrt{25}} + \dfrac{\sqrt{81x^2 \cdot x}}{2}$

$= \dfrac{x\sqrt{x}}{4} - x\dfrac{3\sqrt{x}}{5} + \dfrac{9x\sqrt{x}}{2}$

$= \dfrac{5x\sqrt{x}}{4 \cdot 5} - 4x\dfrac{3\sqrt{x}}{4 \cdot 5} + \dfrac{10 \cdot 9x\sqrt{x}}{2 \cdot 10}$

$= \dfrac{5x\sqrt{x} - 12x\sqrt{x} + 90x\sqrt{x}}{20}$

$= \dfrac{83x\sqrt{x}}{20}$

**Section 8.4 Practice**

**1. a.** $\sqrt{11} \cdot \sqrt{7} = \sqrt{11 \cdot 7} = \sqrt{77}$

   **b.** $9\sqrt{10} \cdot 8\sqrt{3} = 9 \cdot 8\sqrt{10 \cdot 3} = 72\sqrt{30}$

   **c.** $\sqrt{5} \cdot \sqrt{10} = \sqrt{5 \cdot 10}$

$\qquad\qquad\qquad = \sqrt{50}$

$\qquad\qquad\qquad = \sqrt{25 \cdot 2}$

$\qquad\qquad\qquad = \sqrt{25} \cdot \sqrt{2}$

$\qquad\qquad\qquad = 5\sqrt{2}$

   **d.** $\sqrt{17} \cdot \sqrt{17} = \sqrt{17 \cdot 17} = \sqrt{289} = 17$

   **e.** $\sqrt{15y} \cdot \sqrt{5y^3} = \sqrt{15y \cdot 5y^3}$

$\qquad\qquad\qquad = \sqrt{75y^4}$

$\qquad\qquad\qquad = \sqrt{25y^4 \cdot 3}$

$\qquad\qquad\qquad = \sqrt{25y^4} \cdot \sqrt{3}$

$\qquad\qquad\qquad = 5y^2\sqrt{3}$

**2.** $\left(2\sqrt{7}\right)^2 = 2^2 \cdot \left(\sqrt{7}\right)^2 = 4 \cdot 7 = 28$

**3.** $\sqrt[3]{10} \cdot \sqrt[3]{50} = \sqrt[3]{10 \cdot 50}$

$\qquad\qquad\qquad = \sqrt[3]{500}$

$\qquad\qquad\qquad = \sqrt[3]{125 \cdot 4}$

$\qquad\qquad\qquad = \sqrt[3]{125} \cdot \sqrt[3]{4}$

$\qquad\qquad\qquad = 5\sqrt[3]{4}$

**4. a.** $\sqrt{3}\left(\sqrt{3}-\sqrt{5}\right)=\sqrt{3}\cdot\sqrt{3}-\sqrt{3}\cdot\sqrt{5}=3-\sqrt{15}$

**b.** $\sqrt{2z}\left(\sqrt{z}+7\sqrt{2}\right)=\sqrt{2z}\cdot\sqrt{z}+\sqrt{2z}\cdot7\sqrt{2}$

$=\sqrt{2z\cdot z}+7\sqrt{2z\cdot2}$

$=\sqrt{2\cdot z^2}+7\sqrt{4\cdot z}$

$=\sqrt{2}\cdot\sqrt{z^2}+7\sqrt{4}\cdot\sqrt{z}$

$=z\sqrt{2}+7\cdot2\cdot\sqrt{z}$

$=z\sqrt{2}+14\sqrt{z}$

**c.** $\left(\sqrt{x}-\sqrt{7}\right)\left(\sqrt{x}+\sqrt{2}\right)$

$=\sqrt{x}\cdot\sqrt{x}+\sqrt{x}\cdot\sqrt{2}-\sqrt{7}\cdot\sqrt{x}-\sqrt{7}\cdot\sqrt{2}$

$=\sqrt{x^2}+\sqrt{2x}-\sqrt{7x}-\sqrt{14}$

$=x+\sqrt{2x}-\sqrt{7x}-\sqrt{14}$

**5. a.** $\left(\sqrt{7}+4\right)\left(\sqrt{7}-4\right)=\left(\sqrt{7}\right)^2-4^2$

$=7-16$

$=-9$

**b.** $\left(\sqrt{3x}-5\right)^2=\left(\sqrt{3x}\right)^2-2\left(\sqrt{3x}\right)(5)+(5)^2$

$=3x-10\sqrt{3x}+25$

**6. a.** $\dfrac{\sqrt{21}}{\sqrt{7}}=\sqrt{\dfrac{21}{7}}=\sqrt{3}$

**b.** $\dfrac{\sqrt{48}}{\sqrt{6}}=\sqrt{\dfrac{48}{6}}=\sqrt{8}=\sqrt{4\cdot2}=\sqrt{4}\cdot\sqrt{2}=2\sqrt{2}$

**c.** $\dfrac{\sqrt{45y^5}}{\sqrt{5y}}=\sqrt{\dfrac{45y^5}{5y}}=\sqrt{9y^4}=3y^2$

**7.** $\dfrac{3\sqrt{625}}{\sqrt[3]{5}}=3\sqrt[3]{\dfrac{625}{5}}=3\sqrt[3]{125}=5$

**8. a.** $\dfrac{4}{\sqrt{5}}=\dfrac{4}{\sqrt{5}}\cdot\dfrac{\sqrt{5}}{\sqrt{5}}=\dfrac{4\cdot\sqrt{5}}{\sqrt{5}\cdot\sqrt{5}}=\dfrac{4\sqrt{5}}{5}$

**b.** $\dfrac{\sqrt{3}}{\sqrt{18}}=\dfrac{\sqrt{3}}{\sqrt{9\cdot2}}$

$=\dfrac{\sqrt{3}}{3\sqrt{2}}$

$=\dfrac{\sqrt{3}}{3\sqrt{2}}\cdot\dfrac{\sqrt{2}}{\sqrt{2}}$

$=\dfrac{\sqrt{3}\cdot\sqrt{2}}{3\sqrt{2}\cdot\sqrt{2}}$

$=\dfrac{\sqrt{6}}{3\cdot2}$

$=\dfrac{\sqrt{6}}{6}$

**c.** $\sqrt{\dfrac{3}{14x}}=\dfrac{\sqrt{3}}{\sqrt{14x}}$

$=\dfrac{\sqrt{3}}{\sqrt{14x}}\cdot\dfrac{\sqrt{14x}}{\sqrt{14x}}$

$=\dfrac{\sqrt{3}\cdot\sqrt{14x}}{\sqrt{14x}\cdot\sqrt{14x}}$

$=\dfrac{\sqrt{42x}}{14x}$

**9. a.** $\dfrac{3}{\sqrt[3]{25}}=\dfrac{3\cdot\sqrt[3]{5}}{\sqrt[3]{25}\cdot\sqrt[3]{5}}=\dfrac{3\sqrt[3]{5}}{\sqrt[3]{125}}=\dfrac{3\sqrt[3]{5}}{5}$

**b.** $\dfrac{\sqrt[3]{6}}{\sqrt[3]{5}}=\dfrac{\sqrt[3]{6}\cdot\sqrt[3]{25}}{\sqrt[3]{5}\cdot\sqrt[3]{25}}=\dfrac{\sqrt[3]{150}}{\sqrt[3]{125}}=\dfrac{\sqrt[3]{150}}{5}$

**10. a.** $\dfrac{4}{1+\sqrt{5}}=\dfrac{4\left(1-\sqrt{5}\right)}{\left(1+\sqrt{5}\right)\left(1-\sqrt{5}\right)}$

$=\dfrac{4\left(1-\sqrt{5}\right)}{1^2-\left(\sqrt{5}\right)^2}$

$=\dfrac{4\left(1-\sqrt{5}\right)}{1-5}$

$=\dfrac{4\left(1-\sqrt{5}\right)}{-4}$

$=-\dfrac{4\left(1-\sqrt{5}\right)}{4}$

$=-1\left(1-\sqrt{5}\right)$

$=-1+\sqrt{5}$

**b.** $\dfrac{\sqrt{3}+2}{\sqrt{3}-1} = \dfrac{\left(\sqrt{3}+2\right)\left(\sqrt{3}+1\right)}{\left(\sqrt{3}-1\right)\left(\sqrt{3}+1\right)}$

$\qquad\qquad = \dfrac{3+\sqrt{3}+2\sqrt{3}+2}{3-1}$

$\qquad\qquad = \dfrac{5+3\sqrt{3}}{2}$

**c.** $\dfrac{8}{5-\sqrt{x}} = \dfrac{8\left(5+\sqrt{x}\right)}{\left(5-\sqrt{x}\right)\left(5+\sqrt{x}\right)} = \dfrac{8\left(5+\sqrt{x}\right)}{25-x}$

**11.** $\dfrac{14-\sqrt{28}}{6} = \dfrac{14-\sqrt{4\cdot 7}}{6}$

$\qquad = \dfrac{14-2\sqrt{7}}{6}$

$\qquad = \dfrac{2\left(7-\sqrt{7}\right)}{2\cdot 3}$

$\qquad = \dfrac{7-\sqrt{7}}{3}$

**Vocabulary, Readiness & Video Check 8.4**

**1.** $\sqrt{7}\cdot\sqrt{3} = \sqrt{21}$

**2.** $\sqrt{10}\cdot\sqrt{10} = \sqrt{100}$ or $10$

**3.** $\dfrac{\sqrt{15}}{\sqrt{3}} = \sqrt{\dfrac{15}{3}}$ or $\sqrt{5}$

**4.** The process of eliminating the radical in the denominator of a radical expression is called rationalizing the denominator.

**5.** The conjugate of $2+\sqrt{3}$ is $2-\sqrt{3}$.

**6.** In each example, the product rule is first used to multiply the radicals and then later used to simplify the radical.

**7.** The square root of a positive number times the square root of the same positive number (or the square root of a positive number squared) is that positive number.

**8.** If you notice that some simplifying can be done to the fraction if both radicands are under one radical.

**9.** To write an equivalent expression without a radical in the denominator.

**10.** Using the FOIL order to multiply, the Outer product and the Inner product are the only terms with radicals and they will subtract out.

**Exercise Set 8.4**

**1.** $\sqrt{8}\cdot\sqrt{2} = \sqrt{8\cdot 2} = \sqrt{16} = 4$

**3.** $\sqrt{10}\cdot\sqrt{5} = \sqrt{10\cdot 5}$

$\qquad = \sqrt{50}$

$\qquad = \sqrt{25\cdot 2}$

$\qquad = \sqrt{25}\cdot\sqrt{2}$

$\qquad = 5\sqrt{2}$

**5.** $\left(\sqrt{6}\right)^2 = \sqrt{6}\cdot\sqrt{6} = 6$

**7.** $\sqrt{2x}\cdot\sqrt{2x} = \left(\sqrt{2x}\right)^2 = 2x$

**9.** $\left(2\sqrt{5}\right)^2 = 2^2\left(\sqrt{5}\right)^2 = 4(5) = 20$

**11.** $\left(6\sqrt{x}\right)^2 = 6^2\left(\sqrt{x}\right)^2 = 36x$

**13.** $\sqrt{3x^5}\cdot\sqrt{6x} = \sqrt{3x^5\cdot 6x}$

$\qquad = \sqrt{18x^6}$

$\qquad = \sqrt{9x^6\cdot 2}$

$\qquad = \sqrt{9x^6}\cdot\sqrt{2}$

$\qquad = 3x^3\sqrt{2}$

**15.** $\sqrt{2xy^2}\cdot\sqrt{8xy} = \sqrt{2xy^2\cdot 8xy}$

$\qquad = \sqrt{16x^2y^3}$

$\qquad = \sqrt{16x^2y^2\cdot y}$

$\qquad = \sqrt{16x^2y^2}\cdot\sqrt{y}$

$\qquad = 4xy\sqrt{y}$

**17.** $\sqrt{6}\left(\sqrt{5}+\sqrt{7}\right) = \sqrt{6}\cdot\sqrt{5}+\sqrt{6}\cdot\sqrt{7} = \sqrt{30}+\sqrt{42}$

**19.** $\sqrt{10}\left(\sqrt{2}+\sqrt{5}\right)=\sqrt{10}\cdot\sqrt{2}+\sqrt{10}\cdot\sqrt{5}$
$$=\sqrt{20}+\sqrt{50}$$
$$=\sqrt{4\cdot5}+\sqrt{25\cdot2}$$
$$=\sqrt{4}\cdot\sqrt{5}+\sqrt{25}\cdot\sqrt{2}$$
$$=2\sqrt{5}+5\sqrt{2}$$

**21.** $\sqrt{7y}\left(\sqrt{y}-2\sqrt{7}\right)=\sqrt{7y}\cdot\sqrt{y}-\sqrt{7y}\cdot2\sqrt{7}$
$$=\sqrt{7y^2}-2\sqrt{49y}$$
$$=\sqrt{y^2}\cdot\sqrt{7}-2\cdot\sqrt{49}\cdot\sqrt{y}$$
$$=y\sqrt{7}-2\cdot7\cdot\sqrt{y}$$
$$=y\sqrt{7}-14\sqrt{y}$$

**23.** $\left(\sqrt{3}+6\right)\left(\sqrt{3}-6\right)=\left(\sqrt{3}\right)^2-6^2=3-36=-33$

**25.** $\left(\sqrt{3}+\sqrt{5}\right)\left(\sqrt{2}-\sqrt{5}\right)$
$$=\sqrt{3}\cdot\sqrt{2}-\sqrt{3}\cdot\sqrt{5}+\sqrt{5}\cdot\sqrt{2}-\left(\sqrt{5}\right)^2$$
$$=\sqrt{6}-\sqrt{15}+\sqrt{10}-5$$

**27.** $\left(2\sqrt{11}+1\right)\left(\sqrt{11}-6\right)$
$$=2\sqrt{11}\cdot\sqrt{11}-2\sqrt{11}\cdot6+1\cdot\sqrt{11}-1\cdot6$$
$$=2\cdot11-12\sqrt{11}+\sqrt{11}-6$$
$$=22-11\sqrt{11}-6$$
$$=16-11\sqrt{11}$$

**29.** $\left(\sqrt{x}+6\right)\left(\sqrt{x}-6\right)=\left(\sqrt{x}\right)^2-(6)^2=x-36$

**31.** $\left(\sqrt{x}-7\right)^2=\left(\sqrt{x}\right)^2-2\cdot\sqrt{x}\cdot7+7^2$
$$=x-14\sqrt{x}+49$$

**33.** $\left(\sqrt{6y}+1\right)^2=\left(\sqrt{6y}\right)^2+2\cdot\sqrt{6y}\cdot1+1^2$
$$=6y+2\sqrt{6y}+1$$

**35.** $\dfrac{\sqrt{32}}{\sqrt{2}}=\sqrt{\dfrac{32}{2}}=\sqrt{16}=4$

**37.** $\dfrac{\sqrt{21}}{\sqrt{3}}=\sqrt{\dfrac{21}{3}}=\sqrt{7}$

**39.** $\dfrac{\sqrt{90}}{\sqrt{5}}=\sqrt{\dfrac{90}{5}}=\sqrt{18}=\sqrt{9\cdot2}=\sqrt{9}\cdot\sqrt{2}=3\sqrt{2}$

**41.** $\dfrac{\sqrt{75y^5}}{\sqrt{3y}}=\sqrt{\dfrac{75y^5}{3y}}=\sqrt{25y^4}=5y^2$

**43.** $\dfrac{\sqrt{150}}{\sqrt{2}}=\sqrt{\dfrac{150}{2}}=\sqrt{75}=\sqrt{25\cdot3}=\sqrt{25}\cdot\sqrt{3}=5\sqrt{3}$

**45.** $\dfrac{\sqrt{72y^5}}{\sqrt{3y^3}}=\sqrt{\dfrac{72y^5}{3y^3}}$
$$=\sqrt{24y^2}$$
$$=\sqrt{4y^2\cdot6}$$
$$=\sqrt{4y^2}\cdot\sqrt{6}$$
$$=2y\sqrt{6}$$

**47.** $\dfrac{\sqrt{24x^3y^4}}{\sqrt{2xy}}=\sqrt{\dfrac{24x^3y^4}{2xy}}$
$$=\sqrt{12x^2y^3}$$
$$=\sqrt{4x^2y^2\cdot3y}$$
$$=\sqrt{4x^2y^2}\cdot\sqrt{3y}$$
$$=2xy\sqrt{3y}$$

**49.** $\dfrac{\sqrt{3}}{\sqrt{5}}=\dfrac{\sqrt{3}\cdot\sqrt{5}}{\sqrt{5}\cdot\sqrt{5}}=\dfrac{\sqrt{15}}{\sqrt{25}}=\dfrac{\sqrt{15}}{5}$

**51.** $\dfrac{7}{\sqrt{2}}=\dfrac{7\cdot\sqrt{2}}{\sqrt{2}\cdot\sqrt{2}}=\dfrac{7\sqrt{2}}{2}$

**53.** $\dfrac{1}{\sqrt{6y}}=\dfrac{1\cdot\sqrt{6y}}{\sqrt{6y}\cdot\sqrt{6y}}=\dfrac{\sqrt{6y}}{6y}$

**55.** $\sqrt{\dfrac{3}{x}}=\dfrac{\sqrt{3}}{\sqrt{x}}=\dfrac{\sqrt{3}\cdot\sqrt{x}}{\sqrt{x}\cdot\sqrt{x}}=\dfrac{\sqrt{3x}}{x}$

**57.** $\sqrt{\dfrac{1}{8}}=\dfrac{\sqrt{1}}{\sqrt{8}}=\dfrac{1}{\sqrt{4}\cdot\sqrt{2}}$
$$=\dfrac{1}{2\sqrt{2}}$$
$$=\dfrac{1\cdot\sqrt{2}}{2\cdot\sqrt{2}\cdot\sqrt{2}}$$
$$=\dfrac{\sqrt{2}}{2\cdot2}$$
$$=\dfrac{\sqrt{2}}{4}$$

**59.** $\sqrt{\dfrac{2}{15}} = \dfrac{\sqrt{2}}{\sqrt{15}} = \dfrac{\sqrt{2} \cdot \sqrt{15}}{\sqrt{15} \cdot \sqrt{15}} = \dfrac{\sqrt{30}}{15}$

**61.** $\dfrac{8y}{\sqrt{5}} = \dfrac{8y \cdot \sqrt{5}}{\sqrt{5} \cdot \sqrt{5}} = \dfrac{8y\sqrt{5}}{5}$

**63.** $\sqrt{\dfrac{y}{12x}} = \dfrac{\sqrt{y}}{\sqrt{12x}}$

$= \dfrac{\sqrt{y}}{\sqrt{4} \cdot \sqrt{3x}}$

$= \dfrac{\sqrt{y}}{2\sqrt{3x}}$

$= \dfrac{\sqrt{y} \cdot \sqrt{3x}}{2 \cdot \sqrt{3x} \cdot \sqrt{3x}}$

$= \dfrac{\sqrt{3xy}}{2 \cdot 3x}$

$= \dfrac{\sqrt{3xy}}{6x}$

**65.** $\dfrac{3}{\sqrt{2}+1} = \dfrac{3 \cdot \left(\sqrt{2}-1\right)}{\left(\sqrt{2}+1\right)\left(\sqrt{2}-1\right)}$

$= \dfrac{3\left(\sqrt{2}-1\right)}{\left(\sqrt{2}\right)^2 - 1^2}$

$= \dfrac{3\left(\sqrt{2}-1\right)}{2-1}$

$= \dfrac{3\left(\sqrt{2}-1\right)}{1}$

$= 3\sqrt{2} - 3$

**67.** $\dfrac{\sqrt{5}+1}{\sqrt{6}-\sqrt{5}} = \dfrac{\left(\sqrt{5}+1\right)\left(\sqrt{6}+\sqrt{5}\right)}{\left(\sqrt{6}-\sqrt{5}\right)\left(\sqrt{6}+\sqrt{5}\right)}$

$= \dfrac{\sqrt{30}+5+\sqrt{6}+\sqrt{5}}{\left(\sqrt{6}\right)^2 - \left(\sqrt{5}\right)^2}$

$= \dfrac{\sqrt{30}+5+\sqrt{6}+\sqrt{5}}{6-5}$

$= \dfrac{\sqrt{30}+5+\sqrt{6}+\sqrt{5}}{1}$

$= \sqrt{30}+5+\sqrt{6}+\sqrt{5}$

**69.** $\dfrac{3}{\sqrt{x}-4} = \dfrac{3\left(\sqrt{x}+4\right)}{\left(\sqrt{x}-4\right)\left(\sqrt{x}+4\right)}$

$= \dfrac{3 \cdot \sqrt{x} + 3 \cdot 4}{\left(\sqrt{x}\right)^2 - 4^2}$

$= \dfrac{3\sqrt{x}-12}{x-16}$

**71.** $\sqrt{\dfrac{3}{20}} = \dfrac{\sqrt{3}}{\sqrt{20}}$

$= \dfrac{\sqrt{3}}{\sqrt{4} \cdot \sqrt{5}}$

$= \dfrac{\sqrt{3}}{2\sqrt{5}}$

$= \dfrac{\sqrt{3} \cdot \sqrt{5}}{2\sqrt{5} \cdot \sqrt{5}}$

$= \dfrac{\sqrt{15}}{2(5)}$

$= \dfrac{\sqrt{15}}{10}$

**73.** $\dfrac{4}{2-\sqrt{5}} = \dfrac{4 \cdot \left(2+\sqrt{5}\right)}{\left(2-\sqrt{5}\right)\left(2+\sqrt{5}\right)}$

$= \dfrac{4\left(2+\sqrt{5}\right)}{2^2 - \left(\sqrt{5}\right)^2}$

$= \dfrac{4\left(2+\sqrt{5}\right)}{4-5}$

$= \dfrac{4\left(2+\sqrt{5}\right)}{-1}$

$= -4\left(2+\sqrt{5}\right)$

$= -8 - 4\sqrt{5}$

**75.** $\dfrac{3x}{\sqrt{2x}} = \dfrac{3x \cdot \sqrt{2x}}{\sqrt{2x} \cdot \sqrt{2x}} = \dfrac{3x\sqrt{2x}}{2x} = \dfrac{3\sqrt{2x}}{2}$

   

**77.** $\dfrac{5}{2+\sqrt{x}} = \dfrac{5\left(2-\sqrt{x}\right)}{\left(2+\sqrt{x}\right)\left(2-\sqrt{x}\right)}$

$\qquad = \dfrac{10-5\sqrt{x}}{2^2-\left(\sqrt{x}\right)^2}$

$\qquad = \dfrac{10-5\sqrt{x}}{4-x}$

**79.** $\dfrac{6+2\sqrt{3}}{2} = \dfrac{2\left(3+\sqrt{3}\right)}{2} = 3+\sqrt{3}$

**81.** $\dfrac{18-12\sqrt{5}}{6} = \dfrac{6\left(3-2\sqrt{5}\right)}{6} = 3-2\sqrt{5}$

**83.** $\dfrac{15\sqrt{3}+5}{5} = \dfrac{5\left(3\sqrt{3}+1\right)}{5} = 3\sqrt{3}+1$

**85.** $\sqrt[3]{12}\cdot\sqrt[3]{4} = \sqrt[3]{12\cdot 4}$

$\qquad = \sqrt[3]{48}$

$\qquad = \sqrt[3]{8\cdot 6}$

$\qquad = \sqrt[3]{8}\cdot\sqrt[3]{6}$

$\qquad = 2\sqrt[3]{6}$

**87.** $2\sqrt[3]{5}\cdot 6\sqrt[3]{2} = 2\cdot 6\cdot\sqrt[3]{5\cdot 2} = 12\sqrt[3]{10}$

**89.** $\sqrt[3]{15}\cdot\sqrt[3]{25} = \sqrt[3]{375} = \sqrt[3]{125\cdot 3} = \sqrt[3]{125}\cdot\sqrt[3]{3} = 5\sqrt[3]{3}$

**91.** $\dfrac{\sqrt[3]{54}}{\sqrt[3]{2}} = \sqrt[3]{\dfrac{54}{2}} = \sqrt[3]{27} = 3$

**93.** $\dfrac{\sqrt[3]{120}}{\sqrt[3]{5}} = \sqrt[3]{\dfrac{120}{5}} = \sqrt[3]{24} = \sqrt[3]{8\cdot 3} = \sqrt[3]{8}\cdot\sqrt[3]{3} = 2\sqrt[3]{3}$

**95.** $\sqrt[3]{\dfrac{5}{4}} = \dfrac{\sqrt[3]{5}}{\sqrt[3]{4}} = \dfrac{\sqrt[3]{5}\cdot\sqrt[3]{2}}{\sqrt[3]{4}\cdot\sqrt[3]{2}} = \dfrac{\sqrt[3]{10}}{\sqrt[3]{8}} = \dfrac{\sqrt[3]{10}}{2}$

**97.** $\dfrac{6}{\sqrt[3]{2}} = \dfrac{6\cdot\sqrt[3]{4}}{\sqrt[3]{2}\cdot\sqrt[3]{4}} = \dfrac{6\sqrt[3]{4}}{\sqrt[3]{8}} = \dfrac{6\sqrt[3]{4}}{2} = 3\sqrt[3]{4}$

**99.** $\sqrt[3]{\dfrac{1}{9}} = \dfrac{\sqrt[3]{1}}{\sqrt[3]{9}} = \dfrac{1\cdot\sqrt[3]{3}}{\sqrt[3]{9}\cdot\sqrt[3]{3}} = \dfrac{\sqrt[3]{3}}{\sqrt[3]{27}} = \dfrac{\sqrt[3]{3}}{3}$

**101.** $\sqrt[3]{\dfrac{2}{9}} = \dfrac{\sqrt[3]{2}}{\sqrt[3]{9}} = \dfrac{\sqrt[3]{2}\cdot\sqrt[3]{3}}{\sqrt[3]{9}\cdot\sqrt[3]{3}} = \dfrac{\sqrt[3]{6}}{\sqrt[3]{27}} = \dfrac{\sqrt[3]{6}}{3}$

**103.** $x+5 = 7^2$

$\qquad x+5 = 49$

$\qquad\quad x = 44$

**105.** $4z^2+6z-12 = (2z)^2$

$\qquad 4z^2+6z-12 = 4z^2$

$\qquad\qquad 6z-12 = 0$

$\qquad\qquad\quad 6z = 12$

$\qquad\qquad\quad\ z = 2$

**107.** $9x^2+5x+4 = (3x+1)^2$

$\qquad 9x^2+5x+4 = 9x^2+6x+1$

$\qquad\qquad 5x+4 = 6x+1$

$\qquad\qquad\quad 4 = x+1$

$\qquad\qquad\quad 3 = x$

**109.** Let $l = 13\sqrt{2}$ and $w = 5\sqrt{6}$.

$\qquad A = lw$

$\qquad = 13\sqrt{2}\cdot 5\sqrt{6}$

$\qquad = 13\cdot 5\cdot\sqrt{2\cdot 6}$

$\qquad = 65\sqrt{12}$

$\qquad = 65\sqrt{4\cdot 3}$

$\qquad = 65\sqrt{4}\cdot\sqrt{3}$

$\qquad = 65(2)\sqrt{3}$

$\qquad = 130\sqrt{3}$ square meters

**111.** $\sqrt{\dfrac{A}{\pi}} = \dfrac{\sqrt{A}}{\sqrt{\pi}} = \dfrac{\sqrt{A}\cdot\sqrt{\pi}}{\sqrt{\pi}\cdot\sqrt{\pi}} = \dfrac{\sqrt{A\pi}}{\pi}$

**113.** True

**115.** False; $\sqrt{3x}\cdot\sqrt{3x} = \left(\sqrt{3x}\right)^2 = 3x$.

**117.** False; $\sqrt{11}+\sqrt{2}$ cannot be simplified further because the radicands are different.

**119.** answers may vary

**121.** answers may vary

**123.** $\dfrac{\sqrt{3}+1}{\sqrt{2}-1} = \dfrac{\left(\sqrt{3}+1\right)\left(\sqrt{3}-1\right)}{\left(\sqrt{2}-1\right)\left(\sqrt{3}-1\right)}$

$\qquad\qquad = \dfrac{\left(\sqrt{3}\right)^2 - 1^2}{\sqrt{2}\cdot\sqrt{3} - \sqrt{2} - \sqrt{3} + 1^2}$

$\qquad\qquad = \dfrac{3-1}{\sqrt{6} - \sqrt{2} - \sqrt{3} + 1}$

$\qquad\qquad = \dfrac{2}{\sqrt{6} - \sqrt{2} - \sqrt{3} + 1}$

**Integrated Review**

**1.** $\sqrt{36} = 6,$ because $6^2 = 36$ and 6 is positive.

**2.** $\sqrt{48} = \sqrt{16\cdot 3} = \sqrt{16}\cdot\sqrt{3} = 4\sqrt{3}$

**3.** $\sqrt{x^4} = x^2,$ because $(x^2)^2 = x^4.$

**4.** $\sqrt{y^7} = \sqrt{y^6\cdot y} = \sqrt{y^6}\sqrt{y} = y^3\sqrt{y}$

**5.** $\sqrt{16x^2} = 4x,$ because $(4x)^2 = 16x^2.$

**6.** $\sqrt{18x^{11}} = \sqrt{9x^{10}\cdot 2x} = \sqrt{9x^{10}}\sqrt{2x} = 3x^5\sqrt{2x}$

**7.** $\sqrt[3]{8} = 2,$ because $(2)^3 = 8.$

**8.** $\sqrt[4]{81} = 3,$ because $(3)^4 = 81.$

**9.** $\sqrt[3]{-27} = -3,$ because $(-3)^3 = -27.$

**10.** $\sqrt{-4}$ is not a real number.

**11.** $\sqrt{\dfrac{11}{9}} = \dfrac{\sqrt{11}}{\sqrt{9}} = \dfrac{\sqrt{11}}{3}$

**12.** $\sqrt[3]{\dfrac{7}{64}} = \dfrac{\sqrt[3]{7}}{\sqrt[3]{64}} = \dfrac{\sqrt[3]{7}}{4}$

**13.** $-\sqrt{16} = -4$ because $4^2 = 16.$

**14.** $-\sqrt{25} = -5$ because $5^2 = 25.$

**15.** $\sqrt{\dfrac{9}{49}} = \dfrac{\sqrt{9}}{\sqrt{49}} = \dfrac{3}{7}$

**16.** $\sqrt{\dfrac{1}{64}} = \dfrac{\sqrt{1}}{\sqrt{64}} = \dfrac{1}{8}$

**17.** $\sqrt{a^8 b^2} = a^4 b$

**18.** $\sqrt{x^{10} y^{20}} = x^5 y^{10}$

**19.** $\sqrt{25m^6} = 5m^3$

**20.** $\sqrt{9n^{16}} = 3n^8$

**21.** $5\sqrt{7} + \sqrt{7} = (5+1)\sqrt{7} = 6\sqrt{7}$

**22.** $\sqrt{50} - \sqrt{8} = \sqrt{25\cdot 2} - \sqrt{4\cdot 2}$

$\qquad\qquad = \sqrt{25}\cdot\sqrt{2} - \sqrt{4}\cdot\sqrt{2}$

$\qquad\qquad = 5\sqrt{2} - 2\sqrt{2}$

$\qquad\qquad = (5-2)\sqrt{2}$

$\qquad\qquad = 3\sqrt{2}$

**23.** $5\sqrt{2} - 5\sqrt{3}$ cannot be simplified.

**24.** $2\sqrt{x} + \sqrt{25x} - \sqrt{36x} + 3x$

$\qquad = 2\sqrt{x} + \sqrt{25\cdot x} - \sqrt{36\cdot x} + 3x$

$\qquad = 2\sqrt{x} + \sqrt{25}\cdot\sqrt{x} - \sqrt{36}\cdot\sqrt{x} + 3x$

$\qquad = 2\sqrt{x} + 5\sqrt{x} - 6\sqrt{x} + 3x$

$\qquad = (2+5-6)\sqrt{x} + 3x$

$\qquad = \sqrt{x} + 3x$

**25.** $\sqrt{2}\cdot\sqrt{15} = \sqrt{2\cdot 15} = \sqrt{30}$

**26.** $\sqrt{3}\cdot\sqrt{3} = \sqrt{3\cdot 3} = \sqrt{9} = 3$

**27.** $\left(2\sqrt{7}\right)^2 = 2^2\left(\sqrt{7}\right)^2 = 4(7) = 28$

**28.** $\left(3\sqrt{5}\right)^2 = 3^2\left(\sqrt{5}\right)^2 = 9(5) = 45$

**29.** $\sqrt{3}\left(\sqrt{11}+1\right) = \sqrt{3}\cdot\sqrt{11} + \sqrt{3}\cdot 1 = \sqrt{33} + \sqrt{3}$

**30.** $\sqrt{6}\left(\sqrt{3}-2\right) = \sqrt{6}\cdot\sqrt{3} - \sqrt{6}\cdot 2$

$\qquad\qquad = \sqrt{18} - 2\sqrt{6}$

$\qquad\qquad = \sqrt{9\cdot 2} - 2\sqrt{6}$

$\qquad\qquad = \sqrt{9}\cdot\sqrt{2} - 2\sqrt{6}$

$\qquad\qquad = 3\sqrt{2} - 2\sqrt{6}$

**31.** $\sqrt{8y} \cdot \sqrt{2y} = \sqrt{8y \cdot 2y} = \sqrt{16y^2} = 4y$

**32.** $\sqrt{15x^2} \cdot \sqrt{3x^2} = \sqrt{15x^2 \cdot 3x^2}$
$$= \sqrt{45x^4}$$
$$= \sqrt{9x^4} \cdot \sqrt{5}$$
$$= 3x^2\sqrt{5}$$

**33.** $\left(\sqrt{x} - 5\right)\left(\sqrt{x} + 2\right) = \sqrt{x^2} + 2\sqrt{x} - 5\sqrt{x} - 10$
$$= x - 3\sqrt{x} - 10$$

**34.** $\left(3 + \sqrt{2}\right)^2 = 3^2 + 2(3)\sqrt{2} + \left(\sqrt{2}\right)^2$
$$= 9 + 6\sqrt{2} + 2$$
$$= 11 + 6\sqrt{2}$$

**35.** $\dfrac{\sqrt{8}}{\sqrt{2}} = \sqrt{\dfrac{8}{2}} = \sqrt{4} = 2$

**36.** $\dfrac{\sqrt{45}}{\sqrt{15}} = \sqrt{\dfrac{45}{15}} = \sqrt{3}$

**37.** $\dfrac{\sqrt{24x^5}}{\sqrt{2x}} = \sqrt{\dfrac{24x^5}{2x}}$
$$= \sqrt{12x^4}$$
$$= \sqrt{4x^4 \cdot 3}$$
$$= \sqrt{4x^4} \cdot \sqrt{3}$$
$$= 2x^2\sqrt{3}$$

**38.** $\dfrac{\sqrt{75a^4b^5}}{\sqrt{5ab}} = \sqrt{\dfrac{75a^4b^5}{5ab}}$
$$= \sqrt{15a^3b^4}$$
$$= \sqrt{a^2b^4 \cdot 15a}$$
$$= \sqrt{a^2b^4} \cdot \sqrt{15a}$$
$$= ab^2\sqrt{15a}$$

**39.** $\sqrt{\dfrac{1}{6}} = \dfrac{\sqrt{1}}{\sqrt{6}} = \dfrac{1}{\sqrt{6}} \cdot \dfrac{\sqrt{6}}{\sqrt{6}} = \dfrac{\sqrt{6}}{6}$

**40.** $\dfrac{x}{\sqrt{20}} = \dfrac{x}{\sqrt{4} \cdot \sqrt{5}}$
$$= \dfrac{x}{2\sqrt{5}}$$
$$= \dfrac{x}{2\sqrt{5}} \cdot \dfrac{\sqrt{5}}{\sqrt{5}}$$
$$= \dfrac{x\sqrt{5}}{2(5)}$$
$$= \dfrac{x\sqrt{5}}{10}$$

**41.** $\dfrac{4}{\sqrt{6} + 1} = \dfrac{4}{\sqrt{6} + 1} \cdot \dfrac{\sqrt{6} - 1}{\sqrt{6} - 1}$
$$= \dfrac{4\left(\sqrt{6} - 1\right)}{6 - 1}$$
$$= \dfrac{4\sqrt{6} - 4}{5}$$

**42.** $\dfrac{\sqrt{2} + 1}{\sqrt{x} - 5} = \dfrac{\sqrt{2} + 1}{\sqrt{x} - 5} \cdot \dfrac{\sqrt{x} + 5}{\sqrt{x} + 5} = \dfrac{\sqrt{2x} + 5\sqrt{2} + \sqrt{x} + 5}{x - 25}$

**Section 8.5 Practice**

**1.** $\sqrt{x - 5} = 2$
$$\left(\sqrt{x - 5}\right)^2 = 2^2$$
$$x - 5 = 4$$
$$x = 9$$
**Check:** $\sqrt{x - 5} = 2$
$$\sqrt{9 - 5} \stackrel{?}{=} 2$$
$$\sqrt{4} \stackrel{?}{=} 2$$
$$2 = 2 \quad \text{True}$$
The solution is 9.

**2.** $\sqrt{x} + 5 = 3$
$$\sqrt{x} = -2$$

Since $\sqrt{x}$ is the principal or nonnegative square root of $x$, $\sqrt{x}$ cannot equal $-2$. The equation has no solution.

**3.** $\sqrt{7x-4} = \sqrt{x}$

$\left(\sqrt{7x-4}\right)^2 = \left(\sqrt{x}\right)^2$

$7x - 4 = x$

$-4 = -6x$

$\dfrac{-4}{-6} = x$

$\dfrac{2}{3} = x$

**Check:** $\sqrt{7x-4} = \sqrt{x}$

$\sqrt{7 \cdot \dfrac{2}{3} - 4} \stackrel{?}{=} \sqrt{\dfrac{2}{3}}$

$\sqrt{\dfrac{14}{3} - 4} \stackrel{?}{=} \sqrt{\dfrac{2}{3}}$

$\sqrt{\dfrac{14}{3} - \dfrac{12}{3}} \stackrel{?}{=} \sqrt{\dfrac{2}{3}}$

$\sqrt{\dfrac{2}{3}} = \sqrt{\dfrac{2}{3}}$   True

The solution is $\dfrac{2}{3}$.

**4.** $\sqrt{16y^2 + 4y - 28} = 4y$

$\left(\sqrt{16y^2 + 4y - 28}\right)^2 = (4y)^2$

$16y^2 + 4y - 28 = 16y^2$

$4y - 28 = 0$

$4y = 28$

$y = 7$

**Check:** $\sqrt{16y^2 + 4y - 28} = 4y$

$\sqrt{16 \cdot 7^2 + 4 \cdot 7 - 28} \stackrel{?}{=} 4 \cdot 7$

$\sqrt{16 \cdot 49 + 28 - 28} \stackrel{?}{=} 28$

$\sqrt{784} \stackrel{?}{=} 28$

$28 = 28$   True

The solution is 7.

**5.** $\sqrt{x+15} - x = -5$

$\sqrt{x+15} = x - 5$

$\left(\sqrt{x+15}\right)^2 = (x-5)^2$

$x + 15 = x^2 - 10x + 25$

$0 = x^2 - 11x + 10$

$0 = (x-1)(x-10)$

$0 = x - 1$   or   $0 = x - 10$

$1 = x$          $10 = x$

**Check:**

Let $x = 1$.

$\sqrt{x+15} - x = -5$

$\sqrt{1+15} - 1 \stackrel{?}{=} -5$

$\sqrt{16} - 1 \stackrel{?}{=} -5$

$4 - 1 \stackrel{?}{=} -5$

$3 \stackrel{?}{=} -5$   False

Let $x = 10$.

$\sqrt{x+15} - x = -5$

$\sqrt{10+15} - 10 \stackrel{?}{=} -5$

$\sqrt{25} - 10 \stackrel{?}{=} -5$

$5 - 10 \stackrel{?}{=} -5$

$-5 = -5$   True

Since replacing $x$ with 1 resulted in a false statement, 1 is an extraneous solution. The only solution is 10.

**6.** $\sqrt{x} - 4 = \sqrt{x-16}$

$\left(\sqrt{x} - 4\right)^2 = \left(\sqrt{x-16}\right)^2$

$x - 8\sqrt{x} + 16 = x - 16$

$-8\sqrt{x} = -32$

$\sqrt{x} = 4$

$x = 16$

Check the proposed solution in the original equation. The solution is 16.

**Vocabulary, Readiness & Video Check 8.5**

**1.** The squaring property can cause extraneous solutions, so you need to check your solutions in the original equation—before the squaring property was applied—to make sure they are actually solutions.

**2.** No; if the first squaring leaves a radical term, this new equation can be thought of as an equation that needs the property applied once.

**Exercise Set 8.5**

**1.** $\sqrt{x} = 9$

$\left(\sqrt{x}\right)^2 = 9^2$

$x = 81$

**3.** $\sqrt{x+5} = 2$

$\left(\sqrt{x+5}\right)^2 = 2^2$

$x + 5 = 4$

$x = -1$

**5.** $\sqrt{x} - 2 = 5$

$\qquad \sqrt{x} = 7$

$\qquad \left(\sqrt{x}\right)^2 = 7^2$

$\qquad \quad x = 49$

**7.** $3\sqrt{x} + 5 = 2$

$\qquad 3\sqrt{x} = -3$

The square root cannot be negative, therefore there is no solution.

**9.** $\sqrt{x} = \sqrt{3x - 8}$

$\left(\sqrt{x}\right)^2 = \left(\sqrt{3x - 8}\right)^2$

$\qquad x = 3x - 8$

$\quad -2x = -8$

$\qquad x = 4$

**11.** $\sqrt{4x - 3} = \sqrt{x + 3}$

$\left(\sqrt{4x - 3}\right)^2 = \left(\sqrt{x + 3}\right)^2$

$\qquad 4x - 3 = x + 3$

$\qquad \quad 3x = 6$

$\qquad \quad x = 2$

**13.** $\sqrt{9x^2 + 2x - 4} = 3x$

$\left(\sqrt{9x^2 + 2x - 4}\right)^2 = (3x)^2$

$\qquad 9x^2 + 2x - 4 = 9x^2$

$\qquad \quad 2x - 4 = 0$

$\qquad \qquad 2x = 4$

$\qquad \qquad x = 2$

**15.** $\sqrt{x} = x - 6$

$\left(\sqrt{x}\right)^2 = (x - 6)^2$

$\qquad x = x^2 - 12x + 36$

$\qquad 0 = x^2 - 13x + 36$

$\qquad 0 = (x - 9)(x - 4)$

$x - 9 = 0 \quad$ or $\quad x - 4 = 0$

$\quad x = 9 \qquad \qquad x = 4$ (extraneous)

**17.** $\sqrt{x + 7} = x + 5$

$\left(\sqrt{x + 7}\right)^2 = (x + 5)^2$

$\qquad x + 7 = x^2 + 10x + 25$

$\qquad \quad 0 = x^2 + 9x + 18$

$\qquad \quad 0 = (x + 3)(x + 6)$

$x + 3 = 0 \quad$ or $\quad x + 6 = 0$

$\quad x = -3 \qquad \qquad x = -6$ (extraneous)

**19.** $\sqrt{3x + 7} - x = 3$

$\qquad \sqrt{3x + 7} = x + 3$

$\left(\sqrt{3x + 7}\right)^2 = (x + 3)^2$

$\qquad 3x + 7 = x^2 + 6x + 9$

$\qquad \quad 0 = x^2 + 3x + 2$

$\qquad \quad 0 = (x + 1)(x + 2)$

$0 = x + 1 \quad$ or $\quad 0 = x + 2$

$-1 = x \qquad \qquad -2 = x$

Both solutions check.

**21.** $\sqrt{16x^2 + 2x + 2} = 4x$

$\left(\sqrt{16x^2 + 2x + 2}\right)^2 = (4x)^2$

$\qquad 16x^2 + 2x + 2 = 16x^2$

$\qquad \qquad 2x + 2 = 0$

$\qquad \qquad \quad 2x = -2$

$\qquad \qquad \quad x = -1$

A check shows that $x = -1$ is an extraneous solution. Therefore, there is no solution.

**23.** $\sqrt{2x^2 + 6x + 9} = 3$

$\left(\sqrt{2x^2 + 6x + 9}\right)^2 = (3)^2$

$\qquad 2x^2 + 6x + 9 = 9$

$\qquad \quad 2x^2 + 6x = 0$

$\qquad \quad 2x(x + 3) = 0$

$2x = 0 \quad$ or $\quad x + 3 = 0$

$\quad x = 0 \qquad \qquad x = -3$

25. $\sqrt{x-7} = \sqrt{x} - 1$

$\left(\sqrt{x-7}\right)^2 = \left(\sqrt{x}-1\right)^2$

$x - 7 = x - 2\sqrt{x} + 1$

$2\sqrt{x} = 8$

$\sqrt{x} = 4$

$\left(\sqrt{x}\right)^2 = (4)^2$

$x = 16$

27. $\sqrt{x} + 2 = \sqrt{x+24}$

$\left(\sqrt{x}+2\right)^2 = \left(\sqrt{x+24}\right)^2$

$x + 4\sqrt{x} + 4 = x + 24$

$4\sqrt{x} = 20$

$\sqrt{x} = 5$

$x = 25$

29. $\sqrt{x+8} = \sqrt{x} + 2$

$\left(\sqrt{x+8}\right)^2 = \left(\sqrt{x}+2\right)^2$

$x + 8 = x + 4\sqrt{x} + 4$

$4 = 4\sqrt{x}$

$1 = \sqrt{x}$

$1^2 = \left(\sqrt{x}\right)^2$

$1 = x$

31. $\sqrt{2x+6} = 4$

$\left(\sqrt{2x+6}\right)^2 = 4^2$

$2x + 6 = 16$

$2x = 10$

$x = 5$

33. $\sqrt{x+6} + 1 = 3$

$\sqrt{x+6} = 2$

$\left(\sqrt{x+6}\right)^2 = 2^2$

$x + 6 = 4$

$x = -2$

35. $\sqrt{x+6} + 5 = 3$

$\sqrt{x+6} = -2$

The square root cannot be negative, therefore there is no solution.

37. $\sqrt{16x^2 - 3x + 6} = 4x$

$\left(\sqrt{16x^2-3x+6}\right)^2 = (4x)^2$

$16x^2 - 3x + 6 = 16x^2$

$-3x + 6 = 0$

$-3x = -6$

$x = 2$

39. $-\sqrt{x} = -6$

$\sqrt{x} = 6$

$\left(\sqrt{x}\right)^2 = 6^2$

$x = 36$

41. $\sqrt{x+9} = \sqrt{x} - 3$

$\left(\sqrt{x+9}\right)^2 = \left(\sqrt{x}-3\right)^2$

$x + 9 = x - 6\sqrt{x} + 9$

$0 = -6\sqrt{x}$

$0 = \sqrt{x}$

$0 = x$

$x = 0$ does not check. The equation has no solution.

43. $\sqrt{2x+1} + 3 = 5$

$\sqrt{2x+1} = 2$

$\left(\sqrt{2x+1}\right)^2 = 2^2$

$2x + 1 = 4$

$2x = 3$

$x = \dfrac{3}{2}$

45. $\sqrt{x} + 3 = 7$

$\sqrt{x} = 4$

$\left(\sqrt{x}\right)^2 = 4^2$

$x = 16$

47. $\sqrt{4x} = \sqrt{2x+6}$

$\left(\sqrt{4x}\right)^2 = \left(\sqrt{2x+6}\right)^2$

$4x = 2x + 6$

$2x = 6$

$x = 3$

**49.**
$$\sqrt{2x+1} = x-7$$
$$\left(\sqrt{2x+1}\right)^2 = (x-7)^2$$
$$2x+1 = x^2 -14x+49$$
$$0 = x^2 -16x+48$$
$$0 = (x-12)(x-4)$$
$$x-12 = 0 \quad \text{or} \quad x-4 = 0$$
$$x = 12 \qquad\qquad x = 4 \text{ (extraneous)}$$

**51.**
$$x = \sqrt{2x-2} +1$$
$$x-1 = \sqrt{2x-2}$$
$$(x-1)^2 = \left(\sqrt{2x-2}\right)^2$$
$$x^2 -2x+1 = 2x-2$$
$$x^2 -4x+3 = 0$$
$$(x-1)(x-3) = 0$$
$$x-1 = 0 \quad \text{or} \quad x-3 = 0$$
$$x = 1 \qquad\qquad x = 3$$

**53.**
$$\sqrt{1-8x} - x = 4$$
$$\sqrt{1-8x} = x+4$$
$$\left(\sqrt{1-8x}\right)^2 = (x+4)^2$$
$$1-8x = x^2 +8x+16$$
$$0 = x^2 +16x+15$$
$$0 = (x+1)(x+15)$$
$$x+1 = 0 \quad \text{or} \quad x+15 = 0$$
$$x = -1 \qquad\qquad x = -15 \text{ (extraneous)}$$

**55.**
$$3x-8 = 19$$
$$3x = 27$$
$$x = 9$$

**57.** Let $x$ = width and $2x$ = length.
$$2(2x+x) = 24$$
$$2(3x) = 24$$
$$6x = 24$$
$$x = 4$$
$$2x = 2(4) = 8$$
The length is 8 inches.

**59.**
$$\sqrt{x-3} +3 = \sqrt{3x+4}$$
$$\left(\sqrt{x-3} +3\right)^2 = \left(\sqrt{3x+4}\right)^2$$
$$(x-3)+6\sqrt{x-3} +9 = 3x+4$$
$$x+6\sqrt{x-3} +6 = 3x+4$$
$$6\sqrt{x-3} = 2x-2$$
$$3\sqrt{x-3} = x-1$$
$$\left(3\sqrt{x-3}\right)^2 = (x-1)^2$$
$$9(x-3) = x^2 -2x+1$$
$$9x-27 = x^2 -2x+1$$
$$0 = x^2 -11x+28$$
$$0 = (x-4)(x-7)$$
$$0 = x-4 \quad \text{or} \quad 0 = x-7$$
$$4 = x \qquad\qquad 7 = x$$
Both solutions check.

**61.** answers may vary

**63.** $b = \sqrt{\dfrac{V}{2}}$

**a.** $b = \sqrt{\dfrac{20}{2}} \approx 3.2$

$b = \sqrt{\dfrac{200}{2}} = 10$

$b = \sqrt{\dfrac{2000}{2}} \approx 31.6$

| $V$ | 20 | 200 | 2000 |
|-----|-----|-----|------|
| $b$ | 3.2 | 10 | 31.6 |

**b.** No; it increases by a factor of $\sqrt{10}$.

**65.** $y_1 = \sqrt{x-2}$, $y_2 = x-5$

The solution is 7.30.

**67.** $y_1 = -\sqrt{x+4},\ y_2 = 5x - 6$

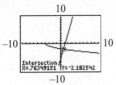

The solution is 0.76.

**Section 8.6 Practice**

1. Use the Pythagorean theorem.
   Let $a = 5$ and $b = 12$.
   $$a^2 + b^2 = c^2$$
   $$5^2 + 12^2 = c^2$$
   $$25 + 144 = c^2$$
   $$169 = c^2$$
   Since $c$ is a length, it is the principal square root of 169.
   $$\sqrt{169} = c$$
   $$13 = c$$
   The hypotenuse has a length of 13 inches.

2. Let $a = 3$ and $c = 7$ with $b$ the unknown leg.
   $$a^2 + b^2 = c^2$$
   $$3^2 + b^2 = 7^2$$
   $$9 + b^2 = 49$$
   $$b^2 = 40$$
   $$b = \sqrt{40} = 2\sqrt{10} \approx 6.32$$
   The exact length of the leg is $2\sqrt{10}$ meters which is approximately 6.32 meters.

3. The points $A$, $B$, and $C$ form a right triangle. The hypotenuse, $\overline{BC}$, has a length of 95 feet and the leg $\overline{AC}$ has a length of 60 feet. In the Pythagorean theorem, let $a = 60$ and $c = 95$ with $b$ the unknown length.
   $$a^2 + b^2 = c^2$$
   $$60^2 + b^2 = 95^2$$
   $$3600 + b^2 = 9025$$
   $$b^2 = 5425$$
   $$b = \sqrt{5425} \approx 74$$
   The length of the bridge is exactly $\sqrt{5425}$ feet which is approximately 74 feet.

4. Use the distance formula with $(x_1,\ y_1) = (-2,\ 5)$ and $(x_2,\ y_2) = (-4,\ -7)$.
   $$d = \sqrt{(x_2 - x_1)^2 + (y_2 - y_1)^2}$$
   $$= \sqrt{[-4 - (-2)]^2 + (-7 - 5)^2}$$
   $$= \sqrt{(-2)^2 + (-12)^2}$$
   $$= \sqrt{4 + 144}$$
   $$= \sqrt{148}$$
   $$= 2\sqrt{37}$$
   The distance is $2\sqrt{37}$ units.

5. Use the formula $v = \sqrt{2gh}$ with $g = 32$ and $h = 12$.
   $$v = \sqrt{2gh} = \sqrt{2 \cdot 32 \cdot 12} = \sqrt{768} = 16\sqrt{3}$$
   The velocity of the object is exactly $16\sqrt{3}$ feet per second, or approximately 27.71 feet per second.

**Vocabulary, Readiness & Video Check 8.6**

1. The Pythagorean theorem applies to right triangles only, and in the formula $a^2 + b^2 = c^2$, $c$ is the length of the hypotenuse.

2. Our answer is a number that when squared equals another value. We're looking for a distance, which must be positive, so our answer must be positive.

3. Be careful of signs because you are subtracting.

4. Both examples ask for an answer rounded to a given place, meaning an estimated answer is expected rather than an exact answer. An exact answer would be given in radical form.

**Exercise Set 8.6**

1. $$a^2 + b^2 = c^2$$
   $$2^2 + 3^2 = c^2$$
   $$4 + 9 = c^2$$
   $$13 = c^2$$
   $$\sqrt{13} = c$$
   The length is $\sqrt{13} \approx 3.61$.

**3.** 
$$a^2 + b^2 = c^2$$
$$3^2 + b^2 = 6^2$$
$$9 + b^2 = 36$$
$$b^2 = 27$$
$$b = \sqrt{27}$$
$$b = 3\sqrt{3}$$
The length is $3\sqrt{3} \approx 5.20$.

**5.** 
$$a^2 + b^2 = c^2$$
$$7^2 + 24^2 = c^2$$
$$49 + 576 = c^2$$
$$625 = c^2$$
$$\sqrt{625} = c$$
$$25 = c$$
The length is 25.

**7.** 
$$a^2 + b^2 = c^2$$
$$a^2 + \left(\sqrt{3}\right)^2 = 5^2$$
$$a^2 + 3 = 25$$
$$a^2 = 22$$
$$a = \sqrt{22}$$
The length is $\sqrt{22} \approx 4.69$.

**9.** 
$$a^2 + b^2 = c^2$$
$$4^2 + b^2 = 13^2$$
$$16 + b^2 = 169$$
$$b^2 = 153$$
$$b = \sqrt{153}$$
$$b = 3\sqrt{17}$$
The length is $3\sqrt{17} \approx 12.37$.

**11.** 
$$a^2 + b^2 = c^2$$
$$4^2 + 5^2 = c^2$$
$$16 + 25 = c^2$$
$$41 = c^2$$
$$\sqrt{41} = c$$
The length is $\sqrt{41} \approx 6.40$.

**13.** 
$$a^2 + b^2 = c^2$$
$$a^2 + 2^2 = 6^2$$
$$a^2 + 4 = 36$$
$$a^2 = 32$$
$$a = \sqrt{32}$$
$$a = 4\sqrt{2}$$
The length is $4\sqrt{2} \approx 5.66$.

**15.** 
$$a^2 + b^2 = c^2$$
$$\left(\sqrt{10}\right)^2 + b^2 = 10^2$$
$$10 + b^2 = 100$$
$$b^2 = 90$$
$$b = \sqrt{90}$$
$$b = 3\sqrt{10}$$
The length is $3\sqrt{10} \approx 9.49$.

**17.** 
$$a^2 + b^2 = c^2$$
$$40^2 + b^2 = 65^2$$
$$1600 + b^2 = 4225$$
$$b^2 = 4225 - 1600$$
$$b^2 = 2625$$
$$\sqrt{b^2} = \sqrt{2625}$$
$$b \approx 51.2 \text{ ft}$$

**19.** 
$$a^2 + b^2 = c^2$$
$$5^2 + 20^2 = c^2$$
$$25 + 400 = c^2$$
$$425 = c^2$$
$$\sqrt{425} = c$$
The length is $\sqrt{425} \approx 20.6$ feet.

**21.** 
$$a^2 + b^2 = c^2$$
$$6^2 + 10^2 = c^2$$
$$36 + 100 = c^2$$
$$136 = c^2$$
$$\sqrt{136} = c$$
The length is $\sqrt{136} \approx 11.7$ feet.

**23.** (3, 6) and (5, 11)

$$d = \sqrt{(x_2 - x_1)^2 + (y_2 - y_1)^2}$$
$$= \sqrt{(5-3)^2 + (11-6)^2}$$
$$= \sqrt{2^2 + 5^2}$$
$$= \sqrt{4 + 25}$$
$$= \sqrt{29}$$

**25.** (−3, 1) and (5, −2)

$$d = \sqrt{(x_2 - x_1)^2 + (y_2 - y_1)^2}$$
$$= \sqrt{[5-(-3)]^2 + (-2-1)^2}$$
$$= \sqrt{(8)^2 + (-3)^2}$$
$$= \sqrt{64 + 9}$$
$$= \sqrt{73}$$

**27.** (3, −2) and (1, −8)

$$d = \sqrt{(x_2 - x_1)^2 + (y_2 - y_1)^2}$$
$$= \sqrt{(1-3)^2 + [-8-(-2)]^2}$$
$$= \sqrt{(-2)^2 + (-6)^2}$$
$$= \sqrt{4 + 36}$$
$$= \sqrt{40}$$
$$= \sqrt{4 \cdot 10}$$
$$= 2\sqrt{10}$$

**29.** $\left(\dfrac{1}{2}, 2\right)$ and (2, −1)

$$d = \sqrt{(x_2 - x_1)^2 + (y_2 - y_1)^2}$$
$$= \sqrt{\left(2 - \frac{1}{2}\right)^2 + (-1-2)^2}$$
$$= \sqrt{\left(\frac{3}{2}\right)^2 + (-3)^2}$$
$$= \sqrt{\frac{9}{4} + 9}$$
$$= \sqrt{\frac{45}{4}}$$
$$= \frac{\sqrt{45}}{\sqrt{4}}$$
$$= \frac{3\sqrt{5}}{2}$$

**31.** (3, −2) and (5, 7)

$$d = \sqrt{(x_2 - x_1)^2 + (y_2 - y_1)^2}$$
$$= \sqrt{(5-3)^2 + [7-(-2)]^2}$$
$$= \sqrt{2^2 + 9^2}$$
$$= \sqrt{4 + 81}$$
$$= \sqrt{85}$$

**33.**
$$b = \sqrt{\frac{3V}{h}}$$
$$6 = \sqrt{\frac{3V}{2}}$$
$$6^2 = \left(\sqrt{\frac{3V}{2}}\right)^2$$
$$36 = \frac{3V}{2}$$
$$24 = V$$
The volume is 24 cubic feet.

**35.** $s = \sqrt{30fd}$
$$s = \sqrt{30(0.35)(280)}$$
$$= \sqrt{2940}$$
$$\approx 54$$
It was moving at 54 mph.

**37.** $v = \sqrt{2.5r}$
$$v = \sqrt{2.5(300)}$$
$$= \sqrt{750}$$
$$\approx 27$$
It can travel at 27 mph.

**39.** $d = 3.5\sqrt{h}$
$$d = 3.5\sqrt{285.4}$$
$$d \approx 59.1$$
You can see a distance of 59.1 kilometers.

**41.** $d = 3.5\sqrt{h}$
$$d = 3.5\sqrt{295.7}$$
$$d \approx 60.2$$
You can see a distance of 60.2 kilometers.

**43.** $2^5 = 2 \cdot 2 \cdot 2 \cdot 2 \cdot 2 = 32$

**45.** $\left(-\dfrac{1}{5}\right)^2 = \left(-\dfrac{1}{5}\right)\left(-\dfrac{1}{5}\right) = \dfrac{1}{25}$

**47.** $x^2 \cdot x^3 = x^{2+3} = x^5$

**49.** $y^3 \cdot y = y^{3+1} = y^4$

**51.** Let $y$ = length of whole base and
$z$ = length of unlabeled section of base.
Find $y$:
$$y^2 + 3^2 = 7^2$$
$$y^2 + 9 = 49$$
$$y^2 = 40$$
$$y = \sqrt{40} = 2\sqrt{10}$$
Find $z$.
$$z^2 + 3^2 = 5^2$$
$$z^2 + 9 = 25$$
$$z^2 = 16$$
$$z = \sqrt{16} = 4$$
Find $x$.
$$x = y - z = 2\sqrt{10} - 4$$

**53.**
$$a^2 + b^2 = c^2$$
$$[60(3)]^2 + [30(3)]^2 = c^2$$
$$180^2 + 90^2 = c^2$$
$$32{,}400 + 8100 = c^2$$
$$40{,}500 = c^2$$
$$\sqrt{40{,}500} = c$$
$$201 \approx c$$
They are about 201 miles apart.

**55.** Answers may vary

**Section 8.7 Practice**

**1. a.** $36^{1/2} = \sqrt{36} = 6$

  **b.** $125^{1/3} = \sqrt[3]{125} = 5$

  **c.** $-\left(\dfrac{1}{81}\right)^{1/4} = -\sqrt[4]{\dfrac{1}{81}} = -\dfrac{1}{3}$

  **d.** $(-1000)^{1/3} = \sqrt[3]{-1000} = -10$

  **e.** $32^{1/5} = \sqrt[5]{32} = 2$

**2. a.** $9^{3/2} = (9^{1/2})^3 = \left(\sqrt{9}\right)^3 = 3^3 = 27$

  **b.** $8^{5/3} = (8^{1/3})^5 = \left(\sqrt[3]{8}\right)^5 = 2^5 = 32$

**c.** $-625^{1/4} = -\sqrt[4]{625} = -5$

**3. a.** $25^{-1/2} = \dfrac{1}{25^{1/2}} = \dfrac{1}{\sqrt{25}} = \dfrac{1}{5}$

  **b.** $1000^{-2/3} = \dfrac{1}{1000^{2/3}}$
$$= \dfrac{1}{\left(\sqrt[3]{1000}\right)^2}$$
$$= \dfrac{1}{10^2}$$
$$= \dfrac{1}{100}$$

  **c.** $-49^{1/2} = -\sqrt{49} = -7$

  **d.** $1024^{-2/5} = \dfrac{1}{1024^{2/5}}$
$$= \dfrac{1}{\left(\sqrt[5]{1024}\right)^2}$$
$$= \dfrac{1}{4^2}$$
$$= \dfrac{1}{16}$$

**4. a.** $6^{3/5} \cdot 6^{7/5} = 6^{(3/5)+(7/5)} = 6^{10/5} = 6^2 = 36$

  **b.** $\dfrac{7^{1/6}}{7^{3/6}} = 7^{(1/6)-(3/6)} = 7^{-2/6} = 7^{-1/3} = \dfrac{1}{7^{1/3}}$

  **c.** $(z^{3/8})^{16} = z^{(3/8)16} = z^6$

  **d.** $\dfrac{a^{3/7}}{a^{-4/7}} = a^{(3/7)-(-4/7)} = a^{7/7} = a^1$ or $a$

  **e.** $\left(\dfrac{x^{5/8}}{y^{2/3}}\right)^{12} = \dfrac{x^{(5/8)12}}{y^{(2/3)12}} = \dfrac{x^{15/2}}{y^8}$

**Vocabulary, Readiness & Video Check 8.7**

**1.** $(-16)^{1/4}$ is not a real number because it is
equivalent to $\sqrt[4]{-16}$ which is not a real number
since it has an even index and a negative
radicand.

**2.** The numerator is the power on the radicand or on the whole expression; the denominator is the index of the radicand.

**3.** A negative fractional exponent will move a base from the numerator to the <u>denominator</u> with the fractional exponent becoming <u>positive</u>.

**4.** Assume you have an expression with fractional exponents. If applying the product rule of exponents, you would <u>add</u> the exponents. If applying the quotient rule of exponents, you would <u>subtract</u> the exponents. If applying the power rule of exponents, you would <u>multiply</u> the exponents.

**Exercise Set 8.7**

**1.** $8^{1/3} = \sqrt[3]{8} = 2$

**3.** $9^{1/2} = \sqrt{9} = 3$

**5.** $16^{3/4} = \left(\sqrt[4]{16}\right)^3 = 2^3 = 8$

**7.** $32^{2/5} = \left(\sqrt[5]{32}\right)^2 = 2^2 = 4$

**9.** $-16^{-1/4} = -\dfrac{1}{16^{1/4}} = -\dfrac{1}{\sqrt[4]{16}} = -\dfrac{1}{2}$

**11.** $16^{-3/2} = \dfrac{1}{16^{3/2}} = \dfrac{1}{\left(\sqrt{16}\right)^3} = \dfrac{1}{4^3} = \dfrac{1}{64}$

**13.** $81^{-3/2} = \dfrac{1}{81^{3/2}} = \dfrac{1}{\left(\sqrt{81}\right)^3} = \dfrac{1}{9^3} = \dfrac{1}{729}$

**15.** $\left(\dfrac{4}{25}\right)^{-1/2} = \dfrac{1}{\left(\frac{4}{25}\right)^{1/2}} = \dfrac{1}{\sqrt{\frac{4}{25}}} = \dfrac{1}{\frac{2}{5}} = \dfrac{5}{2}$

**17.** answers may vary

**19.** $2^{1/3} \cdot 2^{2/3} = 2^{3/3} = 2^1 = 2$

**21.** $\dfrac{4^{3/4}}{4^{1/4}} = 4^{\frac{3}{4}-\frac{1}{4}} = 4^{2/4} = 4^{1/2} = \sqrt{4} = 2$

**23.** $\dfrac{x^{1/6}}{x^{5/6}} = x^{\frac{1}{6}-\frac{5}{6}} = x^{-4/6} = x^{-2/3} = \dfrac{1}{x^{2/3}}$

**25.** $(x^{1/2})^6 = x^{6/2} = x^3$

**27.** answers may vary

**29.** $81^{1/2} = \sqrt{81} = 9$

**31.** $(-8)^{1/3} = \sqrt[3]{-8} = -2$

**33.** $-81^{1/4} = -\left(\sqrt[4]{81}\right) = -(3) = -3$

**35.** $\left(\dfrac{1}{81}\right)^{1/2} = \sqrt{\dfrac{1}{81}} = \dfrac{1}{9}$

**37.** $\left(\dfrac{27}{64}\right)^{1/3} = \dfrac{27^{1/3}}{64^{1/3}} = \dfrac{\sqrt[3]{27}}{\sqrt[3]{64}} = \dfrac{3}{4}$

**39.** $9^{3/2} = \left(\sqrt{9}\right)^3 = (3)^3 = 27$

**41.** $64^{3/2} = \left(\sqrt{64}\right)^3 = (8)^3 = 512$

**43.** $-8^{2/3} = -(8^{2/3}) = -\left(\sqrt[3]{8}\right)^2 = -(2)^2 = -(4) = -4$

**45.** $4^{5/2} = \left(\sqrt{4}\right)^5 = (2)^5 = 32$

**47.** $\left(\dfrac{4}{9}\right)^{3/2} = \dfrac{4^{3/2}}{9^{3/2}} = \dfrac{\left(\sqrt{4}\right)^3}{\left(\sqrt{9}\right)^3} = \dfrac{2^3}{3^3} = \dfrac{8}{27}$

**49.** $\left(\dfrac{1}{81}\right)^{3/4} = \dfrac{1^{3/4}}{81^{3/4}} = \dfrac{\left(\sqrt[4]{1}\right)^3}{\left(\sqrt[4]{81}\right)^3} = \dfrac{1^3}{3^3} = \dfrac{1}{27}$

**51.** $4^{-1/2} = \dfrac{1}{4^{1/2}} = \dfrac{1}{\sqrt{4}} = \dfrac{1}{2}$

**53.** $215^{-1/3} = \dfrac{1}{215^{1/3}} = \dfrac{1}{\sqrt[3]{215}} = \dfrac{1}{5}$

**55.** $625^{-3/4} = \dfrac{1}{625^{3/4}} = \dfrac{1}{\left(\sqrt[4]{625}\right)^3} = \dfrac{1}{5^3} = \dfrac{1}{125}$

**57.** $3^{4/3} \cdot 3^{2/3} = 3^{\frac{4}{3}+\frac{2}{3}} = 3^{6/3} = 3^2 = 9$

**59.** $\dfrac{6^{2/3}}{6^{1/3}} = 6^{\frac{2}{3}-\frac{1}{3}} = 6^{1/3}$

**61.** $(x^{2/3})^9 = x^{\frac{2}{3}\cdot 9} = x^6$

**63.** $\dfrac{6^{1/3}}{6^{-5/3}} = 6^{\frac{1}{3}-\left(-\frac{5}{3}\right)} = 6^{6/3} = 6^2 = 36$

**65.** $\dfrac{3^{-3/5}}{3^{2/5}} = 3^{-\frac{3}{5}-\frac{2}{5}} = 3^{-5/5} = 3^{-1} = \dfrac{1}{3}$

**67.** $\left(\dfrac{x^{1/3}}{y^{3/4}}\right)^2 = \dfrac{(x^{1/3})^2}{(y^{3/4})^2} = \dfrac{x^{2/3}}{y^{3/2}}$

**69.** $\left(\dfrac{x^{2/5}}{y^{3/4}}\right)^8 = \dfrac{(x^{2/5})^8}{(y^{3/4})^8} = \dfrac{x^{16/5}}{y^6}$

**71.** $\begin{cases} x + y < 6 \\ y \ge 2x \end{cases}$

| $x + y < 6$ | $y \ge 2x$ |
|---|---|
| Test $(0, 0)$ | Test $(0, 1)$ |
| ? | ? |
| $0 + 0 < 6$ | $1 \ge 2(0)$ |
| $0 < 6$  True | $1 \ge 0$  True |
| Shade below. | Shade above. |

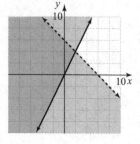

**73.** $\quad x^2 - 4 = 3x$

$x^2 - 3x - 4 = 0$

$(x - 4)(x + 1) = 0$

$x - 4 = 0$   or   $x + 1 = 0$

$x = 4 \qquad\qquad x = -1$

The solutions are $-1$ and $4$.

**75.** $\quad 2x^2 - 5x - 3 = 0$

$2x^2 - 6x + x - 3 = 0$

$2x(x - 3) + 1(x - 3) = 0$

$(2x + 1)(x - 3) = 0$

$2x + 1 = 0$   or   $x - 3 = 0$

$x = -\dfrac{1}{2} \qquad\qquad x = 3$

The solutions are $-\dfrac{1}{2}$ and $3$.

**77.** Let $N = 1.5$ and $P_O = 10,000$.

$P = P_O(1.08)^N = 10,000(1.08)^{1.5} = 11,224$

The population will be 11,224 people.

**79.** $5^{3/4} \approx 3.344$

**81.** $18^{3/5} \approx 5.665$

**Chapter 8 Vocabulary Check**

1. The expressions $5\sqrt{x}$ and $7\sqrt{x}$ are examples of <u>like radicals</u>.

2. In the expression $\sqrt[3]{45}$, the number 3 is the <u>index</u>, the number 45 is the <u>radicand</u>, and $\sqrt{\phantom{x}}$ is called the <u>radical</u> sign.

3. The <u>conjugate</u> of $a + b$ is $a - b$.

4. The <u>principal square root</u> of 25 is 5.

5. The process of eliminating the radical in the denominator of a radical expression is called <u>rationalizing the denominator</u>.

6. The Pythagorean theorem states that for a right triangle, $(\text{leg})^2 + (\text{leg})^2 = (\underline{\text{hypotenuse}})^2$.

**Chapter 8 Review**

1. $\sqrt{81} = 9$, because $9^2 = 81$ and 9 is positive.

2. $-\sqrt{49} = -7$, because $\sqrt{49} = 7$.

3. $\sqrt[3]{27} = 3$, because $(3)^3 = 27$.

4. $\sqrt[4]{16} = 2$, because $2^4 = 16$.

**5.** $-\sqrt{\dfrac{9}{64}} = -\dfrac{3}{8}$, because $\sqrt{\dfrac{9}{64}} = \dfrac{3}{8}$.

**6.** $\sqrt{\dfrac{36}{81}} = \dfrac{6}{9} = \dfrac{2}{3}$, because $\left(\dfrac{6}{9}\right)^2 = \dfrac{36}{81}$.

**7.** $\sqrt[4]{\dfrac{16}{81}} = \dfrac{2}{3}$ because $\left(\dfrac{2}{3}\right)^4 = \dfrac{16}{81}$.

**8.** $\sqrt[3]{-\dfrac{27}{64}} = -\dfrac{3}{4}$ because $\left(-\dfrac{3}{4}\right)^3 = -\dfrac{27}{64}$.

**9.** **c** is not a real number since the index is even and the radicand is negative.

**10.** **a** and **c** are not real numbers since the indices are even and the radicands are odd.

**11.** $\sqrt{x^{12}} = x^6$, because $(x^6)^2 = x^{12}$.

**12.** $\sqrt{x^8} = x^4$, because $(x^4)^2 = x^8$.

**13.** $\sqrt{9x^6} = 3x^3$, because $(3x^3)^2 = 9x^6$.

**14.** $\sqrt{25x^4} = 5x^2$, because $(5x^2)^2 = 25x^4$.

**15.** $\sqrt{\dfrac{16}{y^{10}}} = \dfrac{4}{y^5}$ because $\left(\dfrac{4}{y^5}\right)^2 = \dfrac{16}{y^{10}}$.

**16.** $\sqrt{\dfrac{y^{12}}{49}} = \dfrac{y^6}{7}$ because $\left(\dfrac{y^6}{7}\right)^2 = \dfrac{y^{12}}{49}$.

**17.** $\sqrt{54} = \sqrt{9 \cdot 6} = \sqrt{9} \cdot \sqrt{6} = 3\sqrt{6}$

**18.** $\sqrt{88} = \sqrt{4 \cdot 22} = \sqrt{4} \cdot \sqrt{22} = 2\sqrt{22}$

**19.** $\sqrt{150x^3} = \sqrt{25x^2 \cdot 6x} = \sqrt{25x^2}\sqrt{6x} = 5x\sqrt{6x}$

**20.** $\sqrt{92y^5} = \sqrt{4 \cdot 23 \cdot y^4 \cdot y}$
$= \sqrt{4y^4} \cdot \sqrt{23y}$
$= 2y^2\sqrt{23y}$

**21.** $\sqrt[3]{54} = \sqrt[3]{27 \cdot 2} = \sqrt[3]{27}\sqrt[3]{2} = 3\sqrt[3]{2}$

**22.** $\sqrt[3]{88} = \sqrt[3]{8 \cdot 11} = \sqrt[3]{8} \cdot \sqrt[3]{11} = 2\sqrt[3]{11}$

**23.** $\sqrt[4]{48} = \sqrt[4]{16 \cdot 3} = \sqrt[4]{16}\sqrt[4]{3} = 2\sqrt[4]{3}$

**24.** $\sqrt[4]{162} = \sqrt[4]{81 \cdot 2} = \sqrt[4]{81} \cdot \sqrt[4]{2} = 3\sqrt[4]{2}$

**25.** $\sqrt{\dfrac{18}{25}} = \dfrac{\sqrt{18}}{\sqrt{25}} = \dfrac{\sqrt{9 \cdot 2}}{5} = \dfrac{\sqrt{9} \cdot \sqrt{2}}{5} = \dfrac{3\sqrt{2}}{5}$

**26.** $\sqrt{\dfrac{75}{64}} = \dfrac{\sqrt{75}}{\sqrt{64}} = \dfrac{\sqrt{25 \cdot 3}}{8} = \dfrac{\sqrt{25} \cdot \sqrt{3}}{8} = \dfrac{5\sqrt{3}}{8}$

**27.** $\sqrt{\dfrac{45y^2}{4x^4}} = \dfrac{\sqrt{45y^2}}{\sqrt{4x^4}}$
$= \dfrac{\sqrt{9y^2 \cdot 5}}{2x^2}$
$= \dfrac{\sqrt{9y^2} \cdot \sqrt{5}}{2x^2}$
$= \dfrac{3y\sqrt{5}}{2x^2}$

**28.** $\sqrt{\dfrac{20x^5}{9x^2}} = \sqrt{\dfrac{20x^3}{9}}$
$= \dfrac{\sqrt{4 \cdot 5 \cdot x^2 \cdot x}}{\sqrt{9}}$
$= \dfrac{\sqrt{4x^2} \cdot \sqrt{5x}}{3}$
$= \dfrac{2x\sqrt{5x}}{3}$

**29.** $\sqrt[4]{\dfrac{9}{16}} = \dfrac{\sqrt[4]{9}}{\sqrt[4]{16}} = \dfrac{\sqrt[4]{9}}{2}$

**30.** $\sqrt[3]{\dfrac{40}{27}} = \dfrac{\sqrt[3]{8 \cdot 5}}{\sqrt[3]{27}} = \dfrac{\sqrt[3]{8} \cdot \sqrt[3]{5}}{3} = \dfrac{2\sqrt[3]{5}}{3}$

**31.** $\sqrt[3]{\dfrac{3}{8}} = \dfrac{\sqrt[3]{3}}{\sqrt[3]{8}} = \dfrac{\sqrt[3]{3}}{2}$

**32.** $\sqrt[4]{\dfrac{5}{81}} = \dfrac{\sqrt[4]{5}}{\sqrt[4]{81}} = \dfrac{\sqrt[4]{5}}{3}$

**33.** $3\sqrt[3]{2} + 2\sqrt[3]{3} - 4\sqrt[3]{2} = (3-4)\sqrt[3]{2} + 2\sqrt[3]{3}$
$$= -\sqrt[3]{2} + 2\sqrt[3]{2}$$

**34.** $5\sqrt{2} + 2\sqrt[3]{2} - 8\sqrt{2} = (5-8)\sqrt{2} + 2\sqrt[3]{2}$
$$= -3\sqrt{2} + 2\sqrt[3]{2}$$

**35.** $\sqrt{6} + 2\sqrt[3]{6} - 4\sqrt[3]{6} + 5\sqrt{6} = (1+5)\sqrt{6} + (2-4)\sqrt[3]{6}$
$$= 6\sqrt{6} - 2\sqrt[3]{6}$$

**36.** $3\sqrt{5} - \sqrt[3]{5} - 2\sqrt{5} + 3\sqrt[3]{5} = (3-2)\sqrt{5} + (-1+3)\sqrt[3]{5}$
$$= \sqrt{5} + 2\sqrt[3]{5}$$

**37.** $\sqrt{28x} + \sqrt{63x} + \sqrt[3]{56}$
$$= \sqrt{4 \cdot 7x} + \sqrt{9 \cdot 7x} + \sqrt[3]{8 \cdot 7}$$
$$= \sqrt{4} \cdot \sqrt{7x} + \sqrt{9} \cdot \sqrt{7x} + \sqrt[3]{8} \cdot \sqrt[3]{7}$$
$$= 2\sqrt{7x} + 3\sqrt{7x} + 2\sqrt[3]{7}$$
$$= 5\sqrt{7x} + 2\sqrt[3]{7}$$

**38.** $\sqrt{75y} + \sqrt{48y} - \sqrt[4]{16} = \sqrt{25 \cdot 3y} + \sqrt{16 \cdot 3y} - 2$
$$= \sqrt{25} \cdot \sqrt{3y} + \sqrt{16} \cdot \sqrt{3y} - 2$$
$$= 5\sqrt{3y} + 4\sqrt{3y} - 2$$
$$= 9\sqrt{3y} - 2$$

**39.** $\sqrt{\dfrac{5}{9}} - \sqrt{\dfrac{5}{36}} = \dfrac{\sqrt{5}}{\sqrt{9}} - \dfrac{\sqrt{5}}{\sqrt{36}}$
$$= \dfrac{\sqrt{5}}{3} - \dfrac{\sqrt{5}}{6}$$
$$= \dfrac{2\sqrt{5}}{6} - \dfrac{\sqrt{5}}{6}$$
$$= \dfrac{2\sqrt{5} - \sqrt{5}}{6}$$
$$= \dfrac{\sqrt{5}}{6}$$

**40.** $\sqrt{\dfrac{11}{25}} + \sqrt{\dfrac{11}{16}} = \dfrac{\sqrt{11}}{\sqrt{25}} + \dfrac{\sqrt{11}}{\sqrt{16}}$
$$= \dfrac{\sqrt{11}}{5} + \dfrac{\sqrt{11}}{4}$$
$$= \dfrac{4\sqrt{11}}{20} + \dfrac{5\sqrt{11}}{20}$$
$$= \dfrac{4\sqrt{11} + 5\sqrt{11}}{20}$$
$$= \dfrac{9\sqrt{11}}{20}$$

**41.** $2\sqrt[3]{125} - 5\sqrt[3]{8} = 2(5) - 5(2) = 10 - 10 = 0$

**42.** $3\sqrt[3]{16} - 2\sqrt[3]{2} = 3\sqrt[3]{8 \cdot 2} - 2\sqrt[3]{2}$
$$= 3\sqrt[3]{8} \cdot \sqrt[3]{2} - 2\sqrt[3]{2}$$
$$= 3 \cdot 2\sqrt[3]{2} - 2\sqrt[3]{2}$$
$$= 6\sqrt[3]{2} - 2\sqrt[3]{2}$$
$$= 4\sqrt[3]{2}$$

**43.** $3\sqrt{10} \cdot 2\sqrt{5} = 3 \cdot 2\sqrt{10 \cdot 5}$
$$= 6\sqrt{50}$$
$$= 6\sqrt{25 \cdot 2}$$
$$= 6\sqrt{25}\sqrt{2}$$
$$= 6(5)\sqrt{2}$$
$$= 30\sqrt{2}$$

**44.** $2\sqrt[3]{4} \cdot 5\sqrt[3]{6} = 2 \cdot 5\sqrt[3]{4 \cdot 6}$
$$= 10\sqrt[3]{24}$$
$$= 10\sqrt[3]{8 \cdot 3}$$
$$= 10 \cdot 2\sqrt[3]{3}$$
$$= 20\sqrt[3]{3}$$

**45.** $\sqrt{3}\left(2\sqrt{6} - 3\sqrt{12}\right) = 2\sqrt{18} - 3\sqrt{36}$
$$= 2\sqrt{9 \cdot 2} - 3\sqrt{36}$$
$$= 2\sqrt{9}\sqrt{2} - 3(6)$$
$$= 2(3)\sqrt{2} - 18$$
$$= 6\sqrt{2} - 18$$

**46.** $4\sqrt{5}\left(2\sqrt{10} - 5\sqrt{5}\right) = 8\sqrt{50} - 20\sqrt{25}$
$$= 8\sqrt{25 \cdot 2} - 20(5)$$
$$= 8 \cdot \sqrt{25} \cdot \sqrt{2} - 100$$
$$= 8 \cdot 5 \cdot \sqrt{2} - 100$$
$$= 40\sqrt{2} - 100$$

**47.** $\left(\sqrt{3} + 2\right)\left(\sqrt{6} - 5\right) = \sqrt{18} - 5\sqrt{3} + 2\sqrt{6} - 10$
$$= \sqrt{9 \cdot 2} - 5\sqrt{3} + 2\sqrt{6} - 10$$
$$= \sqrt{9} \cdot \sqrt{2} - 5\sqrt{3} + 2\sqrt{6} - 10$$
$$= 3\sqrt{2} - 5\sqrt{3} + 2\sqrt{6} - 10$$

**48.** $\left(2\sqrt{5} + 1\right)\left(4\sqrt{5} - 3\right) = 8\sqrt{25} - 6\sqrt{5} + 4\sqrt{5} - 3$
$$= 8 \cdot 5 - 2\sqrt{5} - 3$$
$$= 40 - 3 - 2\sqrt{5}$$
$$= 37 - 2\sqrt{5}$$

**49.** $\left(\sqrt{x}-2\right)^2 = \left(\sqrt{x}\right)^2 - 2\cdot\sqrt{x}\cdot 2 + 2^2$
$$= x - 4\sqrt{x} + 4$$

**50.** $\left(\sqrt{y}+4\right)^2 = \left(\sqrt{y}\right)^2 + 2\cdot\sqrt{y}\cdot 4 + 4^2$
$$= y + 8\sqrt{y} + 16$$

**51.** $\dfrac{\sqrt{27}}{\sqrt{3}} = \sqrt{\dfrac{27}{3}} = \sqrt{9} = 3$

**52.** $\dfrac{\sqrt{20}}{\sqrt{5}} = \sqrt{\dfrac{20}{5}} = \sqrt{4} = 2$

**53.** $\dfrac{\sqrt{160}}{\sqrt{8}} = \sqrt{\dfrac{160}{8}} = \sqrt{20} = \sqrt{4\cdot 5} = \sqrt{4}\cdot\sqrt{5} = 2\sqrt{5}$

**54.** $\dfrac{\sqrt{96}}{\sqrt{3}} = \sqrt{\dfrac{96}{3}} = \sqrt{32} = \sqrt{16\cdot 2} = \sqrt{16}\cdot\sqrt{2} = 4\sqrt{2}$

**55.** $\dfrac{\sqrt{30x^6}}{\sqrt{2x^3}} = \sqrt{\dfrac{30x^6}{2x^3}} = \sqrt{15x^3} = \sqrt{x^2\cdot 15x} = x\sqrt{15x}$

**56.** $\dfrac{\sqrt{54x^5y^2}}{\sqrt{3xy^2}} = \sqrt{\dfrac{54x^5y^2}{3xy^2}}$
$$= \sqrt{18x^4}$$
$$= \sqrt{9x^4\cdot 2}$$
$$= \sqrt{9x^4}\cdot\sqrt{2}$$
$$= 3x^2\sqrt{2}$$

**57.** $\dfrac{\sqrt{2}}{\sqrt{11}} = \dfrac{\sqrt{2}\cdot\sqrt{11}}{\sqrt{11}\cdot\sqrt{11}} = \dfrac{\sqrt{22}}{11}$

**58.** $\dfrac{\sqrt{3}}{\sqrt{13}} = \dfrac{\sqrt{3}\cdot\sqrt{13}}{\sqrt{13}\cdot\sqrt{13}} = \dfrac{\sqrt{39}}{13}$

**59.** $\sqrt{\dfrac{5}{6}} = \dfrac{\sqrt{5}}{\sqrt{6}} = \dfrac{\sqrt{5}\cdot\sqrt{6}}{\sqrt{6}\cdot\sqrt{6}} = \dfrac{\sqrt{30}}{\sqrt{36}} = \dfrac{\sqrt{30}}{6}$

**60.** $\sqrt{\dfrac{7}{10}} = \dfrac{\sqrt{7}}{\sqrt{10}} = \dfrac{\sqrt{7}\cdot\sqrt{10}}{\sqrt{10}\cdot\sqrt{10}} = \dfrac{\sqrt{70}}{\sqrt{100}} = \dfrac{\sqrt{70}}{10}$

**61.** $\dfrac{1}{\sqrt{5x}} = \dfrac{1\cdot\sqrt{5x}}{\sqrt{5x}\cdot\sqrt{5x}} = \dfrac{\sqrt{5x}}{5x}$

**62.** $\dfrac{5}{\sqrt{3y}} = \dfrac{5\cdot\sqrt{3y}}{\sqrt{3y}\cdot\sqrt{3y}} = \dfrac{5\sqrt{3y}}{3y}$

**63.** $\sqrt{\dfrac{3}{x}} = \dfrac{\sqrt{3}}{\sqrt{x}} = \dfrac{\sqrt{3}\cdot\sqrt{x}}{\sqrt{x}\cdot\sqrt{x}} = \dfrac{\sqrt{3x}}{x}$

**64.** $\sqrt{\dfrac{6}{y}} = \dfrac{\sqrt{6}}{\sqrt{y}} = \dfrac{\sqrt{6}\cdot\sqrt{y}}{\sqrt{y}\cdot\sqrt{y}} = \dfrac{\sqrt{6y}}{y}$

**65.** $\dfrac{3}{\sqrt{5}-2} = \dfrac{3\cdot\left(\sqrt{5}+2\right)}{\left(\sqrt{5}-2\right)\left(\sqrt{5}+2\right)}$
$$= \dfrac{3\left(\sqrt{5}+2\right)}{\left(\sqrt{5}\right)^2 - 2^2}$$
$$= \dfrac{3\left(\sqrt{5}+2\right)}{5-4}$$
$$= \dfrac{3\left(\sqrt{5}+2\right)}{1}$$
$$= 3\sqrt{5}+6$$

**66.** $\dfrac{8}{\sqrt{10}-3} = \dfrac{8\cdot\left(\sqrt{10}+3\right)}{\left(\sqrt{10}-3\right)\left(\sqrt{10}+3\right)}$
$$= \dfrac{8\left(\sqrt{10}+3\right)}{\left(\sqrt{10}\right)^2 - 3^2}$$
$$= \dfrac{8\left(\sqrt{10}+3\right)}{10-9}$$
$$= \dfrac{8\left(\sqrt{10}+3\right)}{1}$$
$$= 8\sqrt{10}+24$$

**67.** $\dfrac{\sqrt{2}+1}{\sqrt{3}-1} = \dfrac{\left(\sqrt{2}+1\right)\left(\sqrt{3}+1\right)}{\left(\sqrt{3}-1\right)\left(\sqrt{3}+1\right)}$
$$= \dfrac{\sqrt{2}\cdot\sqrt{3}+\sqrt{2}+\sqrt{3}+1^2}{3-1}$$
$$= \dfrac{\sqrt{6}+\sqrt{2}+\sqrt{3}+1}{2}$$

**68.** $\dfrac{\sqrt{3}-2}{\sqrt{5}+2} = \dfrac{\left(\sqrt{3}-2\right)\left(\sqrt{5}-2\right)}{\left(\sqrt{5}+2\right)\left(\sqrt{5}-2\right)}$

$= \dfrac{\sqrt{3}\cdot\sqrt{5}-2\sqrt{3}-2\sqrt{5}+2^2}{5-4}$

$= \sqrt{15}-2\sqrt{3}-2\sqrt{5}+4$

**69.** $\dfrac{10}{\sqrt{x}+5} = \dfrac{10\left(\sqrt{x}-5\right)}{\left(\sqrt{x}+5\right)\left(\sqrt{x}-5\right)} = \dfrac{10\sqrt{x}-50}{x-25}$

**70.** $\dfrac{8}{\sqrt{x}-1} = \dfrac{8\left(\sqrt{x}+1\right)}{\left(\sqrt{x}-1\right)\left(\sqrt{x}+1\right)} = \dfrac{8\sqrt{x}+8}{x-1}$

**71.** $\sqrt[3]{\dfrac{7}{9}} = \dfrac{\sqrt[3]{7}}{\sqrt[3]{9}} = \dfrac{\sqrt[3]{7}\cdot\sqrt[3]{3}}{\sqrt[3]{9}\cdot\sqrt[3]{3}} = \dfrac{\sqrt[3]{21}}{\sqrt[3]{27}} = \dfrac{\sqrt[3]{21}}{3}$

**72.** $\sqrt[3]{\dfrac{3}{4}} = \dfrac{\sqrt[3]{3}}{\sqrt[3]{4}} = \dfrac{\sqrt[3]{3}\cdot\sqrt[3]{2}}{\sqrt[3]{4}\cdot\sqrt[3]{2}} = \dfrac{\sqrt[3]{6}}{\sqrt[3]{8}} = \dfrac{\sqrt[3]{6}}{2}$

**73.** $\sqrt[3]{\dfrac{3}{2}} = \dfrac{\sqrt[3]{3}}{\sqrt[3]{2}} = \dfrac{\sqrt[3]{3}\cdot\sqrt[3]{4}}{\sqrt[3]{2}\cdot\sqrt[3]{4}} = \dfrac{\sqrt[3]{12}}{\sqrt[3]{8}} = \dfrac{\sqrt[3]{12}}{2}$

**74.** $\sqrt[3]{\dfrac{5}{4}} = \dfrac{\sqrt[3]{5}}{\sqrt[3]{4}} = \dfrac{\sqrt[3]{5}\cdot\sqrt[3]{2}}{\sqrt[3]{4}\cdot\sqrt[3]{2}} = \dfrac{\sqrt[3]{10}}{\sqrt[3]{8}} = \dfrac{\sqrt[3]{10}}{2}$

**75.** $\sqrt{2x} = 6$

$\left(\sqrt{2x}\right)^2 = 6^2$

$2x = 36$

$x = 18$

**76.** $\sqrt{x+3} = 4$

$\left(\sqrt{x+3}\right)^2 = 4^2$

$x+3 = 16$

$x = 13$

**77.** $\sqrt{x}+3 = 8$

$\sqrt{x} = 5$

$\left(\sqrt{x}\right)^2 = 5^2$

$x = 25$

**78.** $\sqrt{x}+8 = 3$

$\sqrt{x} = -5$

The square root cannot be negative, therefore there is no solution.

**79.** $\sqrt{2x+1} = x-7$

$\left(\sqrt{2x+1}\right)^2 = (x-7)^2$

$2x+1 = x^2-14x+49$

$0 = x^2-16x+48$

$0 = (x-12)(x-4)$

$x-12 = 0 \quad$ or $\quad x-4 = 0$

$x = 12 \qquad\qquad x = 4 \text{ (extraneous)}$

**80.** $\sqrt{3x+1} = x-1$

$\left(\sqrt{3x+1}\right)^2 = (x-1)^2$

$3x+1 = x^2-2x+1$

$0 = x^2-5x$

$0 = x(x-5)$

$x = 0 \qquad\qquad$ or $\quad x-5 = 0$

$x = 0 \text{ (extraneous)} \qquad x = 5$

**81.** $\sqrt{x}+3 = \sqrt{x+15}$

$\left(\sqrt{x}+3\right)^2 = \left(\sqrt{x+15}\right)^2$

$x+6\sqrt{x}+9 = x+15$

$6\sqrt{x} = 6$

$\sqrt{x} = 1$

$x = 1$

**82.** $\sqrt{x-5} = \sqrt{x}-1$

$\left(\sqrt{x-5}\right)^2 = \left(\sqrt{x}-1\right)^2$

$x-5 = x-2\sqrt{x}+1$

$-6 = -2\sqrt{x}$

$3 = \sqrt{x}$

$9 = x$

**83.** $a^2+b^2 = c^2$

$6^2+9^2 = c^2$

$36+81 = c^2$

$117 = c^2$

$\sqrt{117} = c$

$\sqrt{9\cdot13} = c$

$3\sqrt{13} = c$

The length is $3\sqrt{13} \approx 10.82$.

**84.**
$$a^2 + b^2 = c^2$$
$$5^2 + b^2 = 9^2$$
$$25 + b^2 = 81$$
$$b^2 = 56$$
$$b = \sqrt{56}$$
$$b = \sqrt{4 \cdot 14}$$
$$b = 2\sqrt{14}$$

The length is $2\sqrt{14} \approx 7.48$.

**85.**
$$a^2 + b^2 = c^2$$
$$20^2 + 12^2 = c^2$$
$$400 + 144 = c^2$$
$$544 = c^2$$
$$\sqrt{544} = c$$
$$\sqrt{16 \cdot 34} = c$$
$$4\sqrt{34} = c$$

They are $4\sqrt{34}$ feet apart, approximately 23.32 feet.

**86.**
$$a^2 + b^2 = c^2$$
$$a^2 + 5^2 = 10^2$$
$$a^2 + 25 = 100$$
$$a^2 = 75$$
$$a = \sqrt{75}$$
$$a = \sqrt{25 \cdot 3}$$
$$a = 5\sqrt{3}$$

The length is $5\sqrt{3}$ inches, approximately 8.66 inches.

**87.** $(6, -2)$ and $(-3, 5)$
$$d = \sqrt{(x_2 - x_1)^2 + (y_2 - y_1)^2}$$
$$d = \sqrt{(-3 - 6)^2 + [5 - (-2)]^2}$$
$$d = \sqrt{(-9)^2 + (5 + 2)^2}$$
$$d = \sqrt{81 + 7^2}$$
$$d = \sqrt{81 + 49}$$
$$d = \sqrt{130}$$

**88.** $(2, 8)$ and $(-6, 10)$
$$d = \sqrt{(x_2 - x_1)^2 + (y_2 - y_1)^2}$$
$$\sqrt{(-6 - 2)^2 + (10 - 8)^2} = \sqrt{(-8)^2 + 2^2}$$
$$= \sqrt{64 + 4}$$
$$= \sqrt{68}$$
$$= \sqrt{4 \cdot 17}$$
$$= 2\sqrt{17}$$

**89.** $r = \sqrt{\dfrac{S}{4\pi}}$

$r = \sqrt{\dfrac{72}{4\pi}} \approx 2.4$

The radius is about 2.4 inches.

**90.**
$$r = \sqrt{\frac{S}{4\pi}}$$
$$6 = \sqrt{\frac{S}{4\pi}}$$
$$6^2 = \left(\sqrt{\frac{S}{4\pi}}\right)^2$$
$$36 = \frac{S}{4\pi}$$
$$144\pi = S$$

The surface area is $144\pi$ square inches.

**91.** $\sqrt{a^5} = a^{5/2}$

**92.** $\sqrt[5]{a^3} = a^{3/5}$

**93.** $\sqrt[6]{x^{15}} = x^{15/6} = x^{5/2}$

**94.** $\sqrt[4]{x^{12}} = x^{12/4} = x^3$

**95.** $16^{1/2} = \sqrt{16} = 4$

**96.** $36^{1/2} = \sqrt{36} = 6$

**97.** $(-8)^{1/3} = \sqrt[3]{-8} = -2$

**98.** $(-32)^{1/5} = \sqrt[5]{-32} = -2$

**99.** $-64^{3/2} = -(64^{3/2}) = -\left(\sqrt{64}\right)^3 = -(8)^3 = -512$

**100.** $-8^{2/3} = -\sqrt[3]{8^2} = -\sqrt[3]{64} = -4$

**101.** $\left(\dfrac{16}{81}\right)^{3/4} = \dfrac{16^{3/4}}{81^{3/4}} = \dfrac{\left(\sqrt[4]{16}\right)^3}{\left(\sqrt[4]{81}\right)^3} = \dfrac{2^3}{3^3} = \dfrac{8}{27}$

**102.** $\left(\dfrac{9}{25}\right)^{3/2} = \left(\sqrt{\dfrac{9}{25}}\right)^3 = \left(\dfrac{3}{5}\right)^3 = \dfrac{27}{125}$

**103.** $8^{1/3} \cdot 8^{4/3} = 8^{5/3} = \left(\sqrt[3]{8}\right)^5 = 2^5 = 32$

**104.** $4^{3/2} \cdot 4^{1/2} = 4^{\frac{3}{2}+\frac{1}{2}} = 4^{4/2} = 4^2 = 16$

**105.** $\dfrac{3^{1/6}}{3^{5/6}} = 3^{\frac{1}{6}-\frac{5}{6}} = 3^{-4/6} = 3^{-2/3} = \dfrac{1}{3^{2/3}}$

**106.** $\dfrac{2^{1/4}}{2^{-3/5}} = 2^{\frac{1}{4}-\left(-\frac{3}{5}\right)} = 2^{\frac{1}{4}+\frac{3}{5}} = 2^{\frac{5+12}{20}} = 2^{17/20}$

**107.** $(x^{-1/3})^6 = x^{-\frac{1}{3}\cdot 6} = x^{-2} = \dfrac{1}{x^2}$

**108.** $\left(\dfrac{x^{1/2}}{y^{1/3}}\right)^2 = \dfrac{x^{2/2}}{y^{2/3}} = \dfrac{x}{y^{2/3}}$

**109.** $\sqrt{144} = 12$ because $12^2 = 144$ and 12 is positive.

**110.** $-\sqrt[3]{64} = -\left(\sqrt[3]{64}\right) = -4$ because $4^3 = 64$.

**111.** $\sqrt{16x^{16}} = 4x^8$ because $(4x^8)^2 = 16x^{16}$ and $4x^8$ is positive.

**112.** $\sqrt{4x^{24}} = 2x^{12}$ because $(2x^{12})^2 = 4x^{24}$ and $2x^{12}$ is positive.

**113.** $\sqrt{18x^7} = \sqrt{9x^6 \cdot 2x} = \sqrt{9x^6} \cdot \sqrt{2x} = 3x^3\sqrt{2x}$

**114.** $\sqrt{48y^6} = \sqrt{16y^6 \cdot 3} = \sqrt{16y^6} \cdot \sqrt{3} = 4y^3\sqrt{3}$

**115.** $25^{-1/2} = \dfrac{1}{25^{1/2}} = \dfrac{1}{\sqrt{25}} = \dfrac{1}{5}$

**116.** $64^{-2/3} = \dfrac{1}{64^{2/3}} = \dfrac{1}{\left(\sqrt[3]{64}\right)^2} = \dfrac{1}{4^2} = \dfrac{1}{16}$

**117.** $\sqrt{\dfrac{y^4}{81}} = \dfrac{\sqrt{y^4}}{\sqrt{81}} = \dfrac{y^2}{9}$

**118.** $\sqrt{\dfrac{x^9}{9}} = \dfrac{\sqrt{x^9}}{\sqrt{9}} = \dfrac{\sqrt{x^8 \cdot x}}{3} = \dfrac{x^4\sqrt{x}}{3}$

**119.** $\sqrt{12} + \sqrt{75} = \sqrt{4 \cdot 3} + \sqrt{25 \cdot 3} = 2\sqrt{3} + 5\sqrt{3} = 7\sqrt{3}$

**120.** $\sqrt{63} + \sqrt{28} - \sqrt[3]{27} = \sqrt{9 \cdot 7} + \sqrt{4 \cdot 7} - 3$
$$= 3\sqrt{7} + 2\sqrt{7} - 3$$
$$= 5\sqrt{7} - 3$$

**121.** $\sqrt{\dfrac{3}{16}} - \sqrt{\dfrac{3}{4}} = \dfrac{\sqrt{3}}{\sqrt{16}} - \dfrac{\sqrt{3}}{\sqrt{4}}$
$$= \dfrac{\sqrt{3}}{4} - \dfrac{\sqrt{3}}{2}$$
$$= \dfrac{\sqrt{3}}{4} - \dfrac{2\sqrt{3}}{4}$$
$$= \dfrac{\sqrt{3} - 2\sqrt{3}}{4}$$
$$= -\dfrac{\sqrt{3}}{4}$$

**122.** $\sqrt{45x^3} + x\sqrt{20x} - \sqrt{5x^3}$
$$= \sqrt{9x^2 \cdot 5x} + x\sqrt{4 \cdot 5x} - \sqrt{x^2 \cdot 5x}$$
$$= 3x\sqrt{5x} + x \cdot 2\sqrt{5x} - x\sqrt{5x}$$
$$= 3x\sqrt{5x} + 2x\sqrt{5x} - x\sqrt{5x}$$
$$= 4x\sqrt{5x}$$

**123.** $\sqrt{7} \cdot \sqrt{14} = \sqrt{7 \cdot 14}$
$$= \sqrt{98}$$
$$= \sqrt{49 \cdot 2}$$
$$= \sqrt{49} \cdot \sqrt{2}$$
$$= 7\sqrt{2}$$

**124.** $\sqrt{3}\left(\sqrt{9} - \sqrt{2}\right) = \sqrt{3}\left(3 - \sqrt{2}\right)$
$$= \sqrt{3} \cdot 3 - \sqrt{3} \cdot \sqrt{2}$$
$$= 3\sqrt{3} - \sqrt{6}$$

**125.** $\left(\sqrt{2}+4\right)\left(\sqrt{5}-1\right)=\sqrt{2}\cdot\sqrt{5}-\sqrt{2}+4\sqrt{5}-4$
$$=\sqrt{10}-\sqrt{2}+4\sqrt{5}-4$$

**126.** $\left(\sqrt{x}+3\right)^2=\left(\sqrt{x}\right)^2+2\cdot\sqrt{x}\cdot3+3^2=x+6\sqrt{x}+9$

**127.** $\dfrac{\sqrt{120}}{\sqrt{5}}=\sqrt{\dfrac{120}{5}}=\sqrt{24}=\sqrt{4\cdot6}=2\sqrt{6}$

**128.** $\dfrac{\sqrt{60x^9}}{\sqrt{15x^7}}=\sqrt{\dfrac{60x^9}{15x^7}}=\sqrt{4x^2}=2x$

**129.** $\sqrt{\dfrac{2}{7}}=\dfrac{\sqrt{2}}{\sqrt{7}}=\dfrac{\sqrt{2}\cdot\sqrt{7}}{\sqrt{7}\cdot\sqrt{7}}=\dfrac{\sqrt{14}}{7}$

**130.** $\dfrac{3}{\sqrt{2x}}=\dfrac{3\cdot\sqrt{2x}}{\sqrt{2x}\cdot\sqrt{2x}}=\dfrac{3\sqrt{2x}}{2x}$

**131.** $\dfrac{3}{\sqrt{x}-6}=\dfrac{3\left(\sqrt{x}+6\right)}{\left(\sqrt{x}-6\right)\left(\sqrt{x}+6\right)}=\dfrac{3\sqrt{x}+18}{x-36}$

**132.** $\dfrac{\sqrt{7}-5}{\sqrt{5}+3}=\dfrac{\left(\sqrt{7}-5\right)\left(\sqrt{5}-3\right)}{\left(\sqrt{5}+3\right)\left(\sqrt{5}-3\right)}$
$$=\dfrac{\sqrt{7}\cdot\sqrt{5}-3\sqrt{7}-5\sqrt{5}+5\cdot3}{5-9}$$
$$=\dfrac{\sqrt{35}-3\sqrt{7}-5\sqrt{5}+15}{-4}$$

**133.** $\sqrt{4x}=2$
$$\left(\sqrt{4x}\right)^2=2^2$$
$$4x=4$$
$$x=1$$

**134.** $\sqrt{x-4}=3$
$$\left(\sqrt{x-4}\right)^2=3^2$$
$$x-4=9$$
$$x=13$$

**135.** $\sqrt{4x+8}+6=x$
$$\sqrt{4x+8}=x-6$$
$$\left(\sqrt{4x+8}\right)^2=(x-6)^2$$
$$4x+8=x^2-12x+36$$
$$0=x^2-16x+28$$
$$0=(x-2)(x-14)$$
$0=x-2\qquad$ or $\qquad0=x-14$
$2=x$ (extraneous)$\qquad14=x$

**136.** $\sqrt{x-8}=\sqrt{x}-2$
$$\left(\sqrt{x-8}\right)^2=\left(\sqrt{x}-2\right)^2$$
$$x-8=x-4\sqrt{x}+4$$
$$-12=-4\sqrt{x}$$
$$3=\sqrt{x}$$
$$9=x$$

**137.** The unknown side is the hypotenuse. Let $a=3$ and $b=7$.
$$a^2+b^2=c^2$$
$$3^2+7^2=c^2$$
$$9+49=c^2$$
$$58=c^2$$
$$\sqrt{58}=c$$

The unknown length is $\sqrt{58}$ units or approximately 7.62 units.

**138.** Use the Pythagorean theorem with $a=2$ and $c=6$.
$$a^2+b^2=c^2$$
$$2^2+b^2=6^2$$
$$4+b^2=36$$
$$b^2=32$$
$$b=\sqrt{32}=4\sqrt{2}$$

The length of the rectangle is $4\sqrt{2}$ inches or approximately 5.66 inches.

**139.** Let $x$ be the length of a side of the cube. Since the cube includes only the four walls and roof, there are 5 sides, each of which has area $x^2$ square feet, for a total surface area of $5x^2$ square feet.

$5x^2 = 5120$

$x^2 = 1024$

$x = \sqrt{1024}$

$x = 32$

The length of the sides of the cube is 32 feet.

**Chapter 8 Test**

1. $\sqrt{16} = 4$, because $4^2 = 16$ and 4 is positive.

2. $\sqrt[3]{-125} = -5$, because $(-5)^3 = -125$.

3. $16^{3/4} = \left(\sqrt[4]{16}\right)^3 = 2^3 = 8$

4. $\left(\dfrac{9}{16}\right)^{1/2} = \dfrac{9^{1/2}}{16^{1/2}} = \dfrac{\sqrt{9}}{\sqrt{16}} = \dfrac{3}{4}$

5. $\sqrt[4]{-81}$ is not a real number.

6. $27^{-2/3} = \dfrac{1}{27^{2/3}} = \dfrac{1}{\left(\sqrt[3]{27}\right)^2} = \dfrac{1}{3^2} = \dfrac{1}{9}$

7. $\sqrt{54} = \sqrt{9 \cdot 6} = \sqrt{9} \cdot \sqrt{6} = 3\sqrt{6}$

8. $\sqrt{92} = \sqrt{4 \cdot 23} = \sqrt{4} \cdot \sqrt{23} = 2\sqrt{23}$

9. $\sqrt{3x^6} = \sqrt{x^6} \cdot \sqrt{3} = x^3\sqrt{3}$

10. $\sqrt{8x^4 y^7} = \sqrt{4x^4 y^6 \cdot 2y}$
$= \sqrt{4x^4 y^6}\sqrt{2y}$
$= 2x^2 y^3 \sqrt{2y}$

11. $\sqrt{9x^9} = \sqrt{9x^8}\sqrt{x} = 3x^4\sqrt{x}$

12. $\sqrt[3]{8} = 2$ because $2^3 = 8$.

13. $\sqrt[3]{40} = \sqrt[3]{8 \cdot 5} = \sqrt[3]{8}\sqrt[3]{5} = 2\sqrt[3]{5}$

14. $\sqrt{x^{10}} = x^5$ because $(x^5)^2 = 10$.

15. $\sqrt{y^7} = \sqrt{y^6 \cdot y} = \sqrt{y^6} \cdot \sqrt{y} = y^3\sqrt{y}$

16. $\sqrt{\dfrac{5}{16}} = \dfrac{\sqrt{5}}{\sqrt{16}} = \dfrac{\sqrt{5}}{4}$

17. $\sqrt{\dfrac{y^3}{25}} = \dfrac{\sqrt{y^3}}{\sqrt{25}} = \dfrac{\sqrt{y^2 \cdot y}}{5} = \dfrac{y\sqrt{y}}{5}$

18. $\sqrt[3]{\dfrac{2}{27}} = \dfrac{\sqrt[3]{2}}{\sqrt[3]{27}} = \dfrac{\sqrt[3]{2}}{3}$

19. $3\sqrt{8x} = 3\sqrt{4 \cdot 2x} = 3\sqrt{4}\sqrt{2x} = 3(2)\sqrt{2x} = 6\sqrt{2x}$

20. $\sqrt{13} + \sqrt{13} - 4\sqrt{13} = (1 + 1 - 4)\sqrt{13} = -2\sqrt{13}$

21. $\sqrt{12} - 2\sqrt{75} = \sqrt{4 \cdot 3} - 2\sqrt{25 \cdot 3}$
$= \sqrt{4}\sqrt{3} - 2\sqrt{25}\sqrt{3}$
$= 2\sqrt{3} - 2(5)\sqrt{3}$
$= 2\sqrt{3} - 10\sqrt{3}$
$= -8\sqrt{3}$

22. $\sqrt{2x^2} + \sqrt[3]{54} - x\sqrt{18}$
$= \sqrt{x^2 \cdot 2} + \sqrt[3]{27 \cdot 2} - x\sqrt{9 \cdot 2}$
$= \sqrt{x^2}\sqrt{2} + \sqrt[3]{27}\sqrt[3]{2} - x\sqrt{9}\sqrt{2}$
$= x\sqrt{2} + 3\sqrt[3]{2} - x(3)\sqrt{2}$
$= x\sqrt{2} + 3\sqrt[3]{2} - 3x\sqrt{2}$
$= -2x\sqrt{2} + 3\sqrt[3]{2}$

23. $\sqrt{\dfrac{3}{4}} + \sqrt{\dfrac{3}{25}} = \dfrac{\sqrt{3}}{\sqrt{4}} + \dfrac{\sqrt{3}}{\sqrt{25}}$
$= \dfrac{\sqrt{3}}{2} + \dfrac{\sqrt{3}}{5}$
$= \dfrac{5\sqrt{3}}{10} + \dfrac{2\sqrt{3}}{10}$
$= \dfrac{7\sqrt{3}}{10}$

24. $\sqrt{7} \cdot \sqrt{14} = \sqrt{7 \cdot 14}$
$= \sqrt{98}$
$= \sqrt{49 \cdot 2}$
$= \sqrt{49} \cdot \sqrt{2}$
$= 7\sqrt{2}$

25. $\sqrt{2}\left(\sqrt{6} - \sqrt{5}\right) = \sqrt{2} \cdot \sqrt{6} - \sqrt{2} \cdot \sqrt{5}$
$= \sqrt{12} - \sqrt{10}$
$= \sqrt{4 \cdot 3} - \sqrt{10}$
$= \sqrt{4} \cdot \sqrt{3} - \sqrt{10}$
$= 2\sqrt{3} - \sqrt{10}$

**26.** $\left(\sqrt{x}+2\right)\left(\sqrt{x}-3\right)=\sqrt{x}\cdot\sqrt{x}-3\sqrt{x}+2\sqrt{x}-6$
$$=x-\sqrt{x}-6$$

**27.** $\dfrac{\sqrt{50}}{\sqrt{10}}=\sqrt{\dfrac{50}{10}}=\sqrt{5}$

**28.** $\dfrac{\sqrt{40x^4}}{\sqrt{2x}}=\sqrt{\dfrac{40x^4}{2x}}$
$$=\sqrt{20x^3}$$
$$=\sqrt{4x^2\cdot 5x}$$
$$=\sqrt{4x^2}\cdot\sqrt{5x}$$
$$=2x\sqrt{5x}$$

**29.** $\sqrt{\dfrac{2}{3}}=\dfrac{\sqrt{2}}{\sqrt{3}}=\dfrac{\sqrt{2}\cdot\sqrt{3}}{\sqrt{3}\cdot\sqrt{3}}=\dfrac{\sqrt{6}}{\sqrt{9}}=\dfrac{\sqrt{6}}{3}$

**30.** $\sqrt[3]{\dfrac{5}{9}}=\dfrac{\sqrt[3]{5}}{\sqrt[3]{9}}=\dfrac{\sqrt[3]{5}\cdot\sqrt[3]{3}}{\sqrt[3]{9}\cdot\sqrt[3]{3}}=\dfrac{\sqrt[3]{15}}{\sqrt[3]{27}}=\dfrac{\sqrt[3]{15}}{3}$

**31.** $\sqrt{\dfrac{5}{12x^2}}=\dfrac{\sqrt{5}}{\sqrt{12x^2}}$
$$=\dfrac{\sqrt{5}}{\sqrt{4x^2\cdot 3}}$$
$$=\dfrac{\sqrt{5}}{2x\sqrt{3}}$$
$$=\dfrac{\sqrt{5}\cdot\sqrt{3}}{2x\sqrt{3}\cdot\sqrt{3}}$$
$$=\dfrac{\sqrt{15}}{2x(3)}$$
$$=\dfrac{\sqrt{15}}{6x}$$

**32.** $\dfrac{2\sqrt{3}}{\sqrt{3}-3}=\dfrac{2\sqrt{3}\left(\sqrt{3}+3\right)}{\left(\sqrt{3}-3\right)\left(\sqrt{3}+3\right)}$
$$=\dfrac{2\sqrt{3}\left(\sqrt{3}+3\right)}{3-9}$$
$$=\dfrac{2(3)+6\sqrt{3}}{-6}$$
$$=\dfrac{6+6\sqrt{3}}{-6}$$
$$=\dfrac{6\left(1+\sqrt{3}\right)}{-6}$$
$$=-1\left(1+\sqrt{3}\right)$$
$$=-1-\sqrt{3}$$

**33.** $\sqrt{x}+8=11$
$$\sqrt{x}=3$$
$$\left(\sqrt{x}\right)^2=3^2$$
$$x=9$$

**34.** $\sqrt{3x-6}=\sqrt{x+4}$
$$\left(\sqrt{3x-6}\right)^2=\left(\sqrt{x+4}\right)^2$$
$$3x-6=x+4$$
$$2x=10$$
$$x=5$$

**35.** $\sqrt{2x-2}=x-5$
$$\left(\sqrt{2x-2}\right)^2=(x-5)^2$$
$$2x-2=x^2-10x+25$$
$$0=x^2-12x+27$$
$$0=(x-9)(x-3)$$
$$x-9=0\quad\text{or}\quad x-3=0$$
$$x=9\qquad\qquad x=3\ \text{(extraneous)}$$

**36.** $a^2+b^2=c^2$
$$8^2+b^2=12^2$$
$$64+b^2=144$$
$$b^2=80$$
$$b=\sqrt{80}$$
$$b=\sqrt{16\cdot 5}$$
$$b=4\sqrt{5}$$

The length is $4\sqrt{5}$ inches.

**37.** $(-3, 6)$ and $(-2, 8)$

$$d = \sqrt{(x_2 - x_1)^2 + (y_2 - y_1)^2}$$
$$d = \sqrt{[-2 - (-3)]^2 + (8 - 6)^2}$$
$$d = \sqrt{(-2 + 3)^2 + (2)^2}$$
$$d = \sqrt{1^2 + 4}$$
$$d = \sqrt{1 + 4}$$
$$d = \sqrt{5}$$

**38.** $16^{-3/4} \cdot 16^{-1/4} = 16^{-4/4} = 16^{-1} = \dfrac{1}{16}$

**39.** $\left(\dfrac{x^{2/3}}{y^{2/5}}\right)^5 = \dfrac{x^{10/3}}{y^{10/5}} = \dfrac{x^{10/3}}{y^2}$

## Chapter 8 Cumulative Review

**1. a.** $\dfrac{(-12)(-3) + 3}{-7 - (-2)} = \dfrac{36 + 3}{-7 + 2} = \dfrac{39}{-5} = -\dfrac{39}{5}$

   **b.** $\dfrac{2(-3)^2 - 20}{-5 + 4} = \dfrac{2(9) - 20}{-1}$
$$= -(18 - 20)$$
$$= -(-2)$$
$$= 2$$

**2. a.** $\dfrac{4(-3) - (-6)}{-8 + 4} = \dfrac{-12 + 6}{-4} = \dfrac{-6}{-4} = \dfrac{3}{2}$

   **b.** $\dfrac{3 + (-3)(-2)^3}{-1 - (-4)} = \dfrac{3 + (-3)(-8)}{-1 + 4}$
$$= \dfrac{3 + 24}{3}$$
$$= \dfrac{27}{3}$$
$$= 9$$

**3.** $2x + 3x - 5 + 7 = 10x + 3 - 6x - 4$
$$5x + 2 = 4x - 1$$
$$x + 2 = -1$$
$$x = -3$$

**4.** $6y - 11 + 4 + 2y = 8 + 15y - 8y$
$$8y - 7 = 8 + 7y$$
$$y - 7 = 8$$
$$y = 15$$

**5.** $y = 3x$
$x = -1$: $y = 3(-1) = -3$
$y = 0$: $0 = 3x \Rightarrow x = 0$
$y = -9$: $-9 = 3x \Rightarrow x = -3$

| $x$ | $y$ |
|-----|-----|
| $-1$ | $-3$ |
| $0$ | $0$ |
| $-3$ | $-9$ |

**6.** $2x + y = 6$
$x = 0$: $2(0) + y = 6 \Rightarrow y = 6$
$y = -2$: $2x + (-2) = 6 \Rightarrow 2x = 8 \Rightarrow x = 4$
$x = 3$: $2(3) + y = 6 \Rightarrow 6 + y = 6 \Rightarrow y = 0$

| $x$ | $y$ |
|-----|-----|
| $0$ | $6$ |
| $4$ | $-2$ |
| $3$ | $0$ |

**7.** $m = \dfrac{1}{4},\ b = -3$
$$y = mx + b$$
$$y = \dfrac{1}{4}x - 3$$

**8.** $m = -2, b = 4$
$$y = mx + b$$
$$y = -2x + 4$$
$$2x + y = 4$$

**9.** $y = 5$ is horizontal so a parallel line is also horizontal.
$y = c$
Point $(-2, -3)$
$y = -3$

**10.** $y = m_1 x + b_1$
$y = 2x + 4 \Rightarrow m_1 = 2$

Perpendicular line: $m_2 = -\dfrac{1}{m_1} = -\dfrac{1}{2}$

Point on line 2: $(1, 5)$

$$y = mx + b$$
$$5 = -\frac{1}{2}(1) + b$$
$$\frac{10}{2} + \frac{1}{2} = b$$
$$\frac{11}{2} = b$$

The equation is $y = -\frac{1}{2}x + \frac{11}{2}$.

**11. a.** $y = x$ is a function because its graph is a nonvertical line.

**b.** $y = 2x + 1$ is a function because its graph is a nonvertical line.

**c.** $y = 5$ is a function because its graph is a nonvertical line.

**d.** $x = -1$ is not a function because its graph is a vertical line.

**12. a.** $2x + 3 = y$ or $y = 2x + 3$ is a function because its graph is a nonvertical line.

**b.** $x + 4 = 0$ or $x = -4$ is not a function because its graph is a vertical line.

**c.** $\frac{1}{2}y = 2x$ or $y = 4x$ is a function because its graph is a nonvertical line.

**d.** $y = 0$ is a function because its graph is a nonvertical line.

**13.** $\begin{cases} 2x - 3y = 6 \\ x = 2y \end{cases}$
$(12, 6)$

| $2x - 3y = 6$ | $x = 2y$ |
|---|---|
| $2(12) - 3(6) \stackrel{?}{=} 6$ | $12 \stackrel{?}{=} 2(6)$ |
| $24 - 18 \stackrel{?}{=} 6$ | $12 \stackrel{?}{=} 12$ |
| $6 = 6$  True | $12 = 12$  True |

$(12, 6)$ is a solution.

**14.** $\begin{cases} 2x + y = 4 \\ x + y = 2 \end{cases}$

**a.** $(1, 1)$

| $2x + y = 4$ | $x + y = 2$ |
|---|---|
| $2(1) + (1) \stackrel{?}{=} 4$ | $1 + 1 \stackrel{?}{=} 2$ |
| $2 + 1 \stackrel{?}{=} 4$ | $2 \stackrel{?}{=} 2$ |
| $3 = 4$  False | $2 = 2$  True |

$(1, 1)$ is not a solution.

**b.** $(2, 0)$

| $2x + y = 4$ | $x + y = 2$ |
|---|---|
| $2(2) + (0) \stackrel{?}{=} 4$ | $2 + 0 \stackrel{?}{=} 2$ |
| $4 + 0 \stackrel{?}{=} 4$ | $2 \stackrel{?}{=} 2$ |
| $4 = 4$  True | $2 = 2$  True |

$(2, 0)$ is a solution.

**15.** $\begin{cases} 2x + y = 10 \\ x = y + 2 \end{cases}$

Substitute $y + 2$ for $x$ in the first equation.
$$2(y + 2) + y = 10$$
$$2y + 4 + y = 10$$
$$3y = 6$$
$$y = 2$$

Let $y = 2$ in the second equation.
$$x = (2) + 2$$
$$x = 4$$

The solution of the system is $(4, 2)$.

**16.** $\begin{cases} 3y = x + 10 \\ 2x + 5y = 24 \end{cases}$

Solve the first equation for $x$.
$$3y = x + 10$$
$$3y - 10 = x$$

Substitute $3y - 10$ for $x$ in the second equation.
$$2(3y - 10) + 5y = 24$$
$$6y - 20 + 5y = 24$$
$$11y = 44$$
$$y = 4$$

Let $y = 4$ in the first equation.
$$x = 3(4) - 10$$
$$x = 2$$

The solution of the system is $(2, 4)$.

**17.** $\begin{cases} -x - \dfrac{y}{2} = \dfrac{5}{2} \\ \dfrac{x}{6} - \dfrac{y}{2} = 0 \end{cases}$

Multiply the first equation by 2 and the second equation by 6 to get a simplified system without fractions.
$$\begin{cases} -2x - y = 5 \\ x - 3y = 0 \end{cases}$$

Multiply the second equation by 2 to eliminate $x$.
$$\begin{array}{r} -2x - y = 5 \\ 2x - 6y = 0 \\ \hline -7y = 5 \end{array}$$
$$y = -\frac{5}{7}$$

Now multiply the first simplified equation by $-3$ to eliminate $y$.

$6x + 3y = -15$

$\underline{x - 3y = 0}$

$7x \qquad = -15$

$x = -\dfrac{15}{7}$

The solution of the system is $\left( -\dfrac{15}{7}, -\dfrac{5}{7} \right)$.

**18.** $\begin{cases} \dfrac{x}{2} + y = \dfrac{5}{6} \\ 2x - y = \dfrac{5}{6} \end{cases}$

Multiply both equations by 6.

$3x + 6y = 5$

$\underline{12x - 6y = 5}$

$15x \qquad = 10$

$x = \dfrac{2}{3}$

Let $x = \dfrac{2}{3}$ in the second equation.

$2\left( \dfrac{2}{3} \right) - y = \dfrac{5}{6}$

$\dfrac{4}{3} - y = \dfrac{5}{6}$

$8 - 6y = 5$

$-6y = -3$

$y = \dfrac{1}{2}$

The solution of the system is $\left( \dfrac{2}{3}, \dfrac{1}{2} \right)$.

**19.** Let $x =$ the amount of 25% saline.

No. of liters · Strength = Amt of Saline

| 25% | $x$ | 0.25 | $0.25x$ |
|-----|-----|------|---------|
| 5% | $10 - x$ | 0.05 | $0.05(10 - x)$ |
| 20% | 20 | 0.2 | $0.2(10)$ |

$0.25x + 0.05(10 - x) = 0.2(10)$

$0.25x + 0.5 - 0.05x = 2$

$0.2x + 0.5 = 2$

$0.2x = 1.5$

$x = 7.5$

$10 - x = 10 - 7.5 = 2.5$

Mix 7.5 liters of 25% saline with 2.5 liters of 5% saline.

**20.** Let $x =$ slower speed.

| | $r$ | $\cdot$ $t$ | $=$ $d$ |
|--------|--------|-----|-------------|
| Slower | $x$ | 0.2 | $0.2x$ |
| Faster | $x + 15$ | 0.2 | $0.2(x + 15)$ |

$0.2x + 0.2(x + 15) = 11$

$0.2x + 0.2x + 3 = 11$

$0.4x = 8$

$x = 20$

$x + 15 = 20 + 15 = 35$

The slower streetcar travels at 20 mph.

The faster streetcar travels at 35 mph.

**21.** $\begin{cases} 3x \geq y \\ x + 2y \leq 8 \end{cases}$

$3x \geq y$ $\qquad\qquad$ $x + 2y \leq 8$

Test (0, 1) $\qquad\qquad$ Test (0, 0)

$\qquad ? \qquad\qquad\qquad\qquad ?$

$3(0) \geq 1$ $\qquad\qquad$ $0 + 2(0) \leq 8$

False $\qquad\qquad\qquad$ True

Shade below $\qquad\qquad$ Shade below

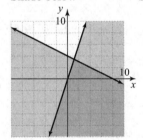

**22.** $\begin{cases} x + y \leq 1 \\ 2x - y \geq 2 \end{cases}$

$x + y \leq 1$ $\qquad\qquad$ $2x - y \geq 2$

Test (0, 0) $\qquad\qquad$ Test (0, 0)

$\qquad ? \qquad\qquad\qquad\qquad ?$

$0 + 0 \leq 1$ $\qquad\qquad$ $2(0) - 0 \geq 2$

True $\qquad\qquad\qquad$ False

Shade below $\qquad\qquad$ Shade below

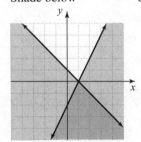

**23.** $-9x^2 + 3xy - 5y^2 + 7xy = -9x^2 - 5y^2 + 10xy$

**24.** $4a^2 + 3a - 2a^2 + 7a - 5 = 2a^2 + 10a - 5$

**25.** $x^2 + 7xy + 6y^2 = (x + 6y)(x + y)$

**26.** $3x^2 + 15x + 18 = 3(x^2 + 5x + 6) = 3(x + 2)(x + 3)$

**27.**
$$\frac{4 - x^2}{3x^2 - 5x - 2} = \frac{-(x^2 - 4)}{3x^2 - 5x - 2}$$
$$= \frac{-(x + 2)(x - 2)}{(3x + 1)(x - 2)}$$
$$= -\frac{x + 2}{3x + 1} \text{ or } \frac{-x - 2}{3x + 1}$$

**28.** $\dfrac{2x^2 + 7x + 3}{x^2 - 9} = \dfrac{(2x + 1)(x + 3)}{(x + 3)(x - 3)} = \dfrac{2x + 1}{x - 3}$

**29.** $\dfrac{3x^3 y^7}{40} \div \dfrac{4x^3}{y^2} = \dfrac{3x^3 y^7}{40} \cdot \dfrac{y^2}{4x^3} = \dfrac{3x^3 y^7 \cdot y^2}{40 \cdot 4x^3} = \dfrac{3y^9}{160}$

**30.**
$$\frac{12x^2 y^3}{5} \div \frac{3y^3}{x} = \frac{12x^2 y^3}{5} \cdot \frac{x}{3y^3}$$
$$= \frac{12x^2 y^3 \cdot x}{5 \cdot 3y^3}$$
$$= \frac{4x^3}{5}$$

**31.** $\dfrac{2y}{2y - 7} - \dfrac{7}{2y - 7} = \dfrac{2y - 7}{2y - 7} = 1$

**32.** $\dfrac{-4x^2}{x + 1} - \dfrac{4x}{x + 1} = \dfrac{-4x^2 - 4x}{x + 1} = -\dfrac{4x(x + 1)}{x + 1} = -4x$

**33.**
$$\frac{2x}{x^2 + 2x + 1} + \frac{x}{x^2 - 1}$$
$$= \frac{2x}{(x + 1)(x + 1)} + \frac{x}{(x + 1)(x - 1)}$$
$$= \frac{2x}{(x + 1)(x + 1)} \cdot \frac{x - 1}{x - 1} + \frac{x}{(x + 1)(x - 1)} \cdot \frac{x + 1}{x + 1}$$
$$= \frac{2x(x - 1) + x(x + 1)}{(x + 1)(x + 1)(x - 1)}$$
$$= \frac{2x^2 - 2x + x^2 + x}{(x + 1)(x + 1)(x - 1)}$$
$$= \frac{3x^2 - x}{(x + 1)(x + 1)(x - 1)}$$
$$= \frac{x(3x - 1)}{(x + 1)^2 (x - 1)}$$

**34.**
$$\frac{3x}{x^2 + 5x + 6} + \frac{1}{x^2 + 2x - 3}$$
$$= \frac{3x}{(x + 2)(x + 3)} + \frac{1}{(x + 3)(x - 1)}$$
$$= \frac{3x}{(x + 2)(x + 3)} \cdot \frac{x - 1}{x - 1} + \frac{1}{(x + 3)(x - 1)} \cdot \frac{x + 2}{x + 2}$$
$$= \frac{3x(x - 1) + 1(x + 2)}{(x + 2)(x + 3)(x - 1)}$$
$$= \frac{3x^2 - 3x + x + 2}{(x + 2)(x + 3)(x - 1)}$$
$$= \frac{3x^2 - 2x + 2}{(x + 2)(x + 3)(x - 1)}$$

**35.**
$$\frac{x}{2} + \frac{8}{3} = \frac{1}{6}$$
$$6\left(\frac{x}{2} + \frac{8}{3}\right) = 6\left(\frac{1}{6}\right)$$
$$6\left(\frac{x}{2}\right) + 6\left(\frac{8}{3}\right) = 1$$
$$3x + 16 = 1$$
$$3x = -15$$
$$x = -5$$

    

**36.**
$$\frac{1}{21}+\frac{x}{7}=\frac{5}{3}$$
$$21\left(\frac{1}{21}+\frac{x}{7}\right)=21\left(\frac{5}{3}\right)$$
$$21\left(\frac{1}{21}\right)+21\left(\frac{x}{7}\right)=35$$
$$1+3x=35$$
$$3x=34$$
$$x=\frac{34}{3}$$

**37.** $\dfrac{2}{3}=\dfrac{10}{x}$
$$2x=30$$
$$x=15$$
The missing length is 15 yards.

**38.** Let $x$ be the missing length.
$$\frac{2}{5}=\frac{5}{x}$$
$$2x=25$$
$$x=\frac{25}{2}$$

The missing length is $\dfrac{25}{2}$ units.

**39.** $\dfrac{\frac{1}{z}-\frac{1}{2}}{\frac{1}{3}-\frac{z}{6}}=\dfrac{6z\left(\frac{1}{z}-\frac{1}{2}\right)}{6z\left(\frac{1}{3}-\frac{z}{6}\right)}$

$$=\frac{6z\left(\frac{1}{z}\right)-6z\left(\frac{1}{2}\right)}{6z\left(\frac{1}{3}\right)-6z\left(\frac{z}{6}\right)}$$

$$=\frac{6-3z}{2z-z^2}$$

$$=\frac{3(2-z)}{z(2-z)}$$

$$=\frac{3}{z}$$

**40.** $\dfrac{x+3}{\frac{1}{x}+\frac{1}{3}}=\dfrac{3x(x+3)}{3x\left(\frac{1}{x}+\frac{1}{3}\right)}$

$$=\frac{3x(x+3)}{3x\left(\frac{1}{x}\right)+3x\left(\frac{1}{3}\right)}$$

$$=\frac{3x(x+3)}{3+x}$$

$$=3x$$

**41. a.** $\sqrt{54}=\sqrt{9\cdot6}=\sqrt{9}\cdot\sqrt{6}=3\sqrt{6}$

**b.** $\sqrt{12}=\sqrt{4\cdot3}=\sqrt{4}\cdot\sqrt{3}=2\sqrt{3}$

**c.** $\sqrt{200}=\sqrt{100\cdot2}=\sqrt{100}\cdot\sqrt{2}=10\sqrt{2}$

**d.** $\sqrt{35}=\sqrt{35}$

**42. a.** $\sqrt{40}=\sqrt{4\cdot10}=\sqrt{4}\cdot\sqrt{10}=2\sqrt{10}$

**b.** $\sqrt{500}=\sqrt{100\cdot5}=\sqrt{100}\cdot\sqrt{5}=10\sqrt{5}$

**c.** $\sqrt{63}=\sqrt{9\cdot7}=\sqrt{9}\cdot\sqrt{7}=3\sqrt{7}$

**d.** $\sqrt{169}=13$

**43. a.** $\left(\sqrt{5}-7\right)\left(\sqrt{5}+7\right)=\left(\sqrt{5}\right)^2-7^2$
$$=5-49$$
$$=-44$$

**b.** $\left(\sqrt{7x}+2\right)^2=\left(\sqrt{7x}\right)^2+2\left(\sqrt{7x}\right)(2)+2^2$
$$=7x+4\sqrt{7x}+4$$

**44. a.** $\left(\sqrt{6}+2\right)^2=\left(\sqrt{6}\right)^2+2\left(\sqrt{6}\right)(2)+2^2$
$$=6+4\sqrt{6}+4$$
$$=10+4\sqrt{6}$$

**b.** $\left(\sqrt{x}+5\right)\left(\sqrt{x}-5\right)=\left(\sqrt{x}\right)^2-5^2=x-25$

**45.** $\sqrt{x}+6=4$
$$\sqrt{x}=-2$$
The square root of a real number cannot be negative. There is no solution.

**46.**
$$\sqrt{x+4}=\sqrt{3x-1}$$
$$\left(\sqrt{x+4}\right)^2=\left(\sqrt{3x-1}\right)^2$$
$$x+4=3x-1$$
$$-2x+4=-1$$
$$-2x=-5$$
$$x=\frac{5}{2}$$

**47.** $a = 6$, $b = 8$

$$c^2 = a^2 + b^2$$
$$c^2 = 6^2 + 8^2$$
$$c^2 = 36 + 64$$
$$c^2 = 100$$
$$\sqrt{c^2} = \sqrt{100}$$
$$c = 10$$

The hypotenuse is 10 inches long.

**48.** $c = 13$, $b = 9$

$$a^2 + b^2 = c^2$$
$$a^2 + 9^2 = 13^2$$
$$a^2 + 81 = 169$$
$$a^2 = 88$$
$$\sqrt{c^2} = \sqrt{88}$$
$$c = \sqrt{4 \cdot 22}$$
$$c = \sqrt{4} \cdot \sqrt{22}$$
$$c = 2\sqrt{22}$$

The other leg is $2\sqrt{22}$ inches long.

**49. a.** $4^{3/2} = \left(\sqrt{4}\right)^3 = 2^3 = 8$

**b.** $27^{2/3} = \left(\sqrt[3]{27}\right)^2 = 3^2 = 9$

**c.** $-16^{3/4} = -\left(\sqrt[4]{16}\right)^3 = -2^3 = -8$

**50. a.** $9^{5/2} = \left(\sqrt{9}\right)^5 = 3^5 = 243$

**b.** $-81^{1/4} = -\left(\sqrt[4]{81}\right) = -3$

**c.** $(-64)^{2/3} = \left(\sqrt[3]{-64}\right)^2 = (-4)^2 = 16$

# Chapter 9

## Section 9.1 Practice

**1.** $x^2 - 16 = 0$

$x^2 = 16$

$x = \sqrt{16}$ or $x = -\sqrt{16}$

$x = 4$ $\qquad$ $x = -4$

Check:

$x^2 - 16 = 0 \qquad\qquad x^2 - 16 = 0$

$4^2 - 16 \stackrel{?}{=} 0 \qquad\quad (-4)^2 - 16 \stackrel{?}{=} 0$

$\qquad 0 = 0$ True $\qquad\qquad 0 = 0$ True

The solutions are 4 and –4.

**2.** $5x^2 = 13$

$x^2 = \dfrac{13}{5}$

$x = \sqrt{\dfrac{13}{5}}$ or $x = -\sqrt{\dfrac{13}{5}}$

$x = \dfrac{\sqrt{5} \cdot \sqrt{13}}{\sqrt{5} \cdot \sqrt{5}} \qquad x = -\dfrac{\sqrt{5} \cdot \sqrt{13}}{\sqrt{5} \cdot \sqrt{5}}$

$x = \dfrac{\sqrt{65}}{5} \qquad\qquad x = -\dfrac{\sqrt{65}}{5}$

The solutions are $-\dfrac{\sqrt{65}}{5}$ and $\dfrac{\sqrt{65}}{5}$.

**3.** $(x-5)^2 = 36$

$x - 5 = \sqrt{36}$ or $x - 5 = -\sqrt{36}$

$x - 5 = 6 \qquad\qquad x - 5 = -6$

$\quad x = 11 \qquad\qquad\quad x = -1$

Check:

$(x-5)^2 \stackrel{?}{=} 36 \qquad\qquad (x-5)^2 = 36$

$(11-5)^2 \stackrel{?}{=} 36 \qquad\quad (-1-5)^2 \stackrel{?}{=} 36$

$\quad 6^2 \stackrel{?}{=} 36 \qquad\qquad\quad (-6)^2 \stackrel{?}{=} 36$

$\qquad 36 = 36$ True $\qquad\qquad 36 = 36$ True

The solutions are –1 and 11.

**4.** $(x+2)^2 = 12$

$x + 2 = \sqrt{12}$ or $x + 2 = -\sqrt{12}$

$x + 2 = 2\sqrt{3} \qquad\qquad x + 2 = -2\sqrt{3}$

$\quad x = -2 + 2\sqrt{3} \qquad\quad x = -2 - 2\sqrt{3}$

The solutions are $-2 \pm 2\sqrt{3}$.

**5.** $(x-8)^2 = -5$

This equation has no real solution because the square root of –5 is not a real number.

**6.** $(3x-5)^2 = 17$

$3x - 5 = \sqrt{17}$ or $3x - 5 = -\sqrt{17}$

$3x = 5 + \sqrt{17} \qquad\qquad 3x = 5 - \sqrt{17}$

$x = \dfrac{5 + \sqrt{17}}{3} \qquad\qquad x = \dfrac{5 - \sqrt{17}}{3}$

The solutions are $\dfrac{5 \pm \sqrt{17}}{3}$.

**7.** Use $h = 16t^2$, where $t$ is the time in seconds and $h$ is the height in feet.

Let $h = 84{,}700$ feet; $84{,}700 = 16t^2$

$84{,}700 = 16t^2$

$5293.75 = t^2$

$\sqrt{5293.75} = t$ or $-\sqrt{5293.75} = t$

$\quad 72.8 \approx t \qquad\qquad\quad -72.8 \approx t$

Reject –72.8 since the time of the fall is not a negative number.

The free fall lasted approximately 72.8 seconds.

## Vocabulary, Readiness & Video Check 9.1

**1.** To solve, $a$ becomes the radicand and the square root of a negative number is not a real number.

**2.** A quadratic equation must be in the form of a variable (or polynomial) squared equal to some non-negative number in order for the property to be used. For Example 6, we use $h = 16t^2$, so this equation can easily be placed in this form. The negative value is rejected because it will not be used in the context of the application.

## Exercise Set 9.1

**1.** $x^2 = 64$

$x = \sqrt{64} = 8$ or $x = -\sqrt{64} = -8$

The solutions are $\pm 8$.

**3.** $x^2 = 21$

$x = \sqrt{21}$ or $x = -\sqrt{21}$

The solutions are $\pm\sqrt{21}$.

Copyright © 2013 Pearson Education, Inc.

**5.** $x^2 = \dfrac{1}{25}$

$x = \sqrt{\dfrac{1}{25}} = \dfrac{1}{5}$ or $x = -\sqrt{\dfrac{1}{25}} = -\dfrac{1}{5}$

The solutions are $\pm\dfrac{1}{5}$.

**7.** $x^2 = -4$

This equation has no real solution because $\sqrt{-4}$ is not a real number.

**9.** $3x^2 = 13$

$x^2 = \dfrac{13}{3}$

$x = \sqrt{\dfrac{13}{3}}$ or $x = -\sqrt{\dfrac{13}{3}}$

$x = \sqrt{\dfrac{13}{3}} \cdot \dfrac{\sqrt{3}}{\sqrt{3}}$      $x = -\sqrt{\dfrac{13}{3}} \cdot \dfrac{\sqrt{3}}{\sqrt{3}}$

$x = \dfrac{\sqrt{39}}{3}$                $x = -\dfrac{\sqrt{39}}{3}$

The solutions are $\pm\dfrac{\sqrt{39}}{3}$.

**11.** $7x^2 = 4$

$x^2 = \dfrac{4}{7}$

$x = \sqrt{\dfrac{4}{7}}$      or   $x = -\sqrt{\dfrac{4}{7}}$

$x = \dfrac{2}{\sqrt{7}} \cdot \dfrac{\sqrt{7}}{\sqrt{7}}$      $x = -\dfrac{2}{\sqrt{7}} \cdot \dfrac{\sqrt{7}}{\sqrt{7}}$

$x = \dfrac{2\sqrt{7}}{7}$            $x = -\dfrac{2\sqrt{7}}{7}$

The solutions are $\pm\dfrac{2\sqrt{7}}{7}$.

**13.** $x^2 - 2 = 0$

$x^2 = 2$

$x = \sqrt{2}$ or $x = -\sqrt{2}$

The solutions are $\pm\sqrt{2}$.

**15.** $2x^2 - 10 = 0$

$2x^2 = 10$

$x^2 = 5$

$x = \sqrt{5}$ or $x = -\sqrt{5}$

The solutions are $\pm\sqrt{5}$.

**17.** $5x^2 + 15 = 0$

$5x^2 = -15$

$x^2 = -3$

This equation has no real solution because the square root of $-3$ is not a real number.

**19.** $(x-5)^2 = 49$

$x - 5 = \sqrt{49}$        or   $x - 5 = -\sqrt{49}$

$x - 5 = 7$                    $x - 5 = -7$

$x = 5 + 7 = 12$          $x = 5 - 7 = -2$

The solutions are $-2$ and $12$.

**21.** $(x+2)^2 = 7$

$x + 2 = \sqrt{7}$        or   $x + 2 = -\sqrt{7}$

$x = -2 + \sqrt{7}$          $x = -2 - \sqrt{7}$

The solutions are $-2 \pm \sqrt{7}$.

**23.** $\left(m - \dfrac{1}{2}\right)^2 = \dfrac{1}{4}$

$m - \dfrac{1}{2} = \sqrt{\dfrac{1}{4}}$     or   $m - \dfrac{1}{2} = -\sqrt{\dfrac{1}{4}}$

$m - \dfrac{1}{2} = \dfrac{1}{2}$            $m - \dfrac{1}{2} = -\dfrac{1}{2}$

$m = \dfrac{1}{2} + \dfrac{1}{2} = 1$        $m = \dfrac{1}{2} - \dfrac{1}{2} = 0$

The solutions are $0$ and $1$.

**25.** $(p+2)^2 = 10$

$p + 2 = \sqrt{10}$        or   $p + 2 = -\sqrt{10}$

$p = -2 + \sqrt{10}$          $p = -2 - \sqrt{10}$

The solutions are $-2 \pm \sqrt{10}$.

**27.** $(3y+2)^2 = 100$

$3y + 2 = \sqrt{100}$     or   $3y + 2 = -\sqrt{100}$

$3y + 2 = 10$                 $3y + 2 = -10$

$3y = -2 + 10$             $3y = -2 - 10$

$3y = 8$                      $3y = -12$

$y = \dfrac{8}{3}$                      $y = -4$

The solutions are $-4$ and $\dfrac{8}{3}$.

**29.** $(z-4)^2 = -9$

This equation has no real solution because $\sqrt{-9}$ is not a real number.

**31.** $(2x-11)^2 = 50$

$2x-11 = \sqrt{50}$ or $2x-11 = -\sqrt{50}$

$2x-11 = 5\sqrt{2}$      $2x-11 = -5\sqrt{2}$

$2x = 11+5\sqrt{2}$      $2x = 11-5\sqrt{2}$

$x = \dfrac{11+5\sqrt{2}}{2}$      $x = \dfrac{11-5\sqrt{2}}{2}$

The solutions are $\dfrac{11\pm5\sqrt{2}}{2}$.

**33.** $(3x-7)^2 = 32$

$3x-7 = \sqrt{32}$ or $3x-7 = -\sqrt{32}$

$3x-7 = 4\sqrt{2}$      $3x-7 = -4\sqrt{2}$

$3x = 7+4\sqrt{2}$      $3x = 7-4\sqrt{2}$

$x = \dfrac{7+4\sqrt{2}}{3}$      $x = \dfrac{7-4\sqrt{2}}{3}$

The solutions are $\dfrac{7\pm4\sqrt{2}}{3}$.

**35.** $(2p-5)^2 = 121$

$2p-5 = \sqrt{121}$ or $2p-5 = -\sqrt{121}$

$2p-5 = 11$      $2p-5 = -11$

$2p = 16$      $2p = -6$

$p = 8$      $p = -3$

The solutions are 8 and $-3$.

**37.** $x^2 - 2 = 0$

$x^2 = 2$

$x = \pm\sqrt{2}$

The solutions are $\pm\sqrt{2}$.

**39.** $(x+6)^2 = 24$

$x+6 = \sqrt{24}$ or $x+6 = -\sqrt{24}$

$x+6 = 2\sqrt{6}$      $x+6 = -2\sqrt{6}$

$x = -6+2\sqrt{6}$      $x = -6-2\sqrt{6}$

The solutions are $-6\pm2\sqrt{6}$.

**41.** $\dfrac{1}{2}n^2 = 5$

$n^2 = 10$

$n = \pm\sqrt{10}$

The solutions are $\pm\sqrt{10}$.

**43.** $(4x-1)^2 = 5$

$4x-1 = \sqrt{5}$ or $4x-1 = -\sqrt{5}$

$4x = 1+\sqrt{5}$      $4x = 1-\sqrt{5}$

$x = \dfrac{1+\sqrt{5}}{4}$      $x = \dfrac{1-\sqrt{5}}{4}$

The solutions are $\dfrac{1\pm\sqrt{5}}{4}$.

**45.** $3z^2 = 36$

$z^2 = 12$

$z = \pm\sqrt{12} = \pm2\sqrt{3}$

The solutions are $\pm2\sqrt{3}$.

**47.** $(8-3x)^2 - 45 = 0$

$(8-3x)^2 = 45$

$8-3x = \sqrt{45}$ or $8-3x = -\sqrt{45}$

$8-3x = 3\sqrt{5}$      $8-3x = -3\sqrt{5}$

$-3x = -8+3\sqrt{5}$      $-3x = -8-3\sqrt{5}$

$x = \dfrac{8-3\sqrt{5}}{3}$      $x = \dfrac{8+3\sqrt{5}}{3}$

The solutions are $\dfrac{8\pm3\sqrt{5}}{3}$.

**49.** Let $A = 20$.

$A = s^2$

$20 = s^2$

$\sqrt{20} = 2\sqrt{5} = s$ or $-\sqrt{20} = -2\sqrt{5} = s$

$4.47 \approx s$      $-4.47 \approx s$

The length of a side is not a negative number so the length is $2\sqrt{5}$ inches or approximately 4.47 inches.

**51.** Let $A = 31{,}329$.

$A = s^2$

$31{,}329 = s^2$

$\sqrt{31{,}329} = s$ or $-\sqrt{31{,}329} = s$

$177 = s$      $-177 = s$

The length of a side is not a negative number, so the length is 177 meters.

**53.** Let $h = 115$.

$$h = 16t^2$$
$$115 = 16t^2$$
$$\frac{115}{16} = t^2$$
$$7.1875 = t^2$$
$$\sqrt{7.1875} = t \quad \text{or} \quad -\sqrt{7.1875} = t$$
$$2.7 \approx t \qquad\qquad -2.7 \approx t$$

The length of the dive is not a negative number so a dive lasts approximately 2.7 seconds.

**55.** Let $h = 4000$.

$$h = 16t^2$$
$$4000 = 16t^2$$
$$\frac{4000}{16} = t^2$$
$$250 = t^2$$
$$\sqrt{250} = t \quad \text{or} \quad -\sqrt{250} = t$$
$$15.8 \approx t \qquad\qquad -15.8 \approx t$$

The length of the fall is not a negative number so the fall would last approximately 15.8 seconds.

**57.** Let $A = 36\pi$.

$$A = \pi r^2$$
$$36\pi = \pi r^2$$
$$36 = r^2$$
$$\sqrt{36} = r \quad \text{or} \quad -\sqrt{36} = r$$
$$6 = r \qquad\qquad -6 = r$$

The radius of the circle is not a negative number so the radius is 6 inches.

**59.** $x^2 + 6x + 9 = x^2 + 2 \cdot x \cdot 3 + 3^2 = (x+3)^2$

**61.** $x^2 - 4x + 4 = x^2 - 2 \cdot x \cdot 2 + 2^2 = (x-2)^2$

**63.** Answers may vary

**65.** $x^2 + 4x + 4 = 16$

$$(x+2)^2 = 16$$
$$x + 2 = \sqrt{16} \quad \text{or} \quad x + 2 = -\sqrt{16}$$
$$x + 2 = 4 \qquad\qquad x + 2 = -4$$
$$x = 2 \qquad\qquad x = -6$$

The solutions are $-6$ and $2$.

**67.** $x^2 + 14x + 49 = 31$

$$(x+7)^2 = 31$$
$$x + 7 = \sqrt{31} \quad \text{or} \quad x + 7 = -\sqrt{31}$$
$$x = -7 + \sqrt{31} \qquad\qquad x = -7 - \sqrt{31}$$

The solutions are $-7 \pm \sqrt{31}$.

**69.** $x^2 = 1.78$

$$x = \sqrt{1.78} \quad \text{or} \quad x = -\sqrt{1.78}$$
$$x \approx 1.33 \qquad\qquad x \approx -1.33$$

The solutions are $\pm 1.33$.

**71.**
$$y = 148(x+1.5)^2 + 589{,}105$$
$$605{,}000 = 148(x+1.5)^2 + 589{,}105$$
$$15{,}895 = 148(x+1.5)^2$$
$$\frac{15{,}895}{148} = (x+1.5)^2$$
$$\sqrt{\frac{15{,}895}{148}} = x + 1.5$$
$$-1.5 + \sqrt{\frac{15{,}895}{148}} = x$$
$$8.9 \approx x$$

or

$$-\sqrt{\frac{15{,}895}{148}} = x + 1.5$$
$$-1.5 - \sqrt{\frac{15{,}895}{148}} = x$$
$$-11.9 \approx x$$

Since the time will not be a negative number in this context, reject the solution of $-11.9$. The solution is $x \approx 9$, so the model predicts that there were 605,000 highway bridges in 2009 $(2000 + 9)$.

**Section 9.2 Practice**

**1.** $x^2 + 2x - 5 = 0$

$$x^2 + 2x = 5$$
$$x^2 + 2x + \left(\frac{2}{2}\right)^2 = 5 + \left(\frac{2}{2}\right)^2$$
$$x^2 + 2x + 1 = 5 + 1$$
$$(x+1)^2 = 6$$
$$x + 1 = \sqrt{6} \quad \text{or} \quad x + 1 = -\sqrt{6}$$
$$x = -1 + \sqrt{6} \qquad\qquad x = -1 - \sqrt{6}$$

The solutions are $-1 \pm \sqrt{6}$.

**2.**

$$x^2 - 8x = -8$$

$$x^2 - 8x + \left(\frac{-8}{2}\right)^2 = -8 + \left(-\frac{8}{2}\right)^2$$

$$x^2 - 8x + 16 = -8 + 16$$

$$(x-4)^2 = 8$$

$$x - 4 = \sqrt{8} \qquad \text{or} \quad x - 4 = -\sqrt{8}$$

$$x - 4 = 2\sqrt{2} \qquad\qquad x - 4 = -2\sqrt{2}$$

$$x = 4 + 2\sqrt{2} \qquad\qquad x = 4 - 2\sqrt{2}$$

The solutions are $4 \pm 2\sqrt{2}$.

**3.**

$$9x^2 - 36x - 13 = 0$$

$$x^2 - 4x - \frac{13}{9} = 0$$

$$x^2 - 4x = \frac{13}{9}$$

$$x^2 - 4x + \left(\frac{-4}{2}\right)^2 = \frac{13}{9} + \left(\frac{-4}{2}\right)^2$$

$$x^2 - 4x + 4 = \frac{13}{9} + 4$$

$$(x-2)^2 = \frac{13}{9} + \frac{36}{9} = \frac{49}{9}$$

$$x - 2 = \sqrt{\frac{49}{9}} \quad \text{or} \quad x - 2 = -\sqrt{\frac{49}{9}}$$

$$x - 2 = \frac{7}{3} \qquad\qquad x - 2 = -\frac{7}{3}$$

$$x = 2 + \frac{7}{3} \qquad\qquad x = 2 - \frac{7}{3}$$

$$x = \frac{13}{3} \qquad\qquad x = -\frac{1}{3}$$

The solutions are $-\frac{1}{3}$ and $\frac{13}{3}$.

**4.**

$$2x^2 + 12x = -20$$

$$x^2 + 6x = -10$$

$$x^2 + 6x + \left(\frac{6}{2}\right)^2 = -10 + \left(\frac{6}{2}\right)^2$$

$$x^2 + 6x + 9 = -10 + 9$$

$$(x+3)^2 = -1$$

There is no real solution since the square root of $-1$ is not a real number.

**5.**

$$2x^2 = 6x - 3$$

$$x^2 = 3x - \frac{3}{2}$$

$$x^2 - 3x = -\frac{3}{2}$$

$$x^2 - 3x + \left(\frac{-3}{2}\right)^2 = -\frac{3}{2} + \left(\frac{-3}{2}\right)^2$$

$$x^2 - 3x + \frac{9}{4} = -\frac{6}{4} + \frac{9}{4}$$

$$\left(x - \frac{3}{2}\right)^2 = \frac{3}{4}$$

$$x - \frac{3}{2} = \sqrt{\frac{3}{4}} \quad \text{or} \quad x - \frac{3}{2} = -\sqrt{\frac{3}{4}}$$

$$x - \frac{3}{2} = \frac{\sqrt{3}}{2} \qquad\qquad x - \frac{3}{2} = -\frac{\sqrt{3}}{2}$$

$$x = \frac{3 + \sqrt{3}}{2} \qquad\qquad x = \frac{3 - \sqrt{3}}{2}$$

The solutions are $\frac{3 \pm \sqrt{3}}{2}$.

**Vocabulary, Readiness & Video Check 9.2**

**1.** By the zero factor property, if the product of two numbers is zero, then at least one of these two numbers must be <u>zero</u>.

**2.** If $a$ is a positive number, and if $x^2 = a$ then $x = \underline{\pm\sqrt{a}}$.

**3.** An equation that can be written in the form $ax^2 + bx + c = 0$ where $a$, $b$, and $c$ are real numbers and $a$ is not zero is called a <u>quadratic equation</u>.

**4.** The process of solving a quadratic equation by writing it in the form $(x+a)^2 = c$ is called <u>completing the square</u>.

**5.** To complete the square on $x^2 + 6x$, add <u>9</u>.

**6.** To complete the square on $x^2 + bx$, add $\underline{\left(\frac{b}{2}\right)^2}$.

**7.** In these examples we are working with equations and whatever is added to one side must also be added to the other to keep equality.

**8.** The coefficient of $y^2$ is 2. The method of completing the square only works when the coefficient of the squared variable is 1, so we must first divide through by 2.

**Exercise Set 9.2**

**1.**
$$x^2 + 8x = -12$$
$$x^2 + 8x + 16 = -12 + 16$$
$$(x+4)^2 = 4$$
$$x + 4 = \sqrt{4} \quad \text{or} \quad x + 4 = -\sqrt{4}$$
$$x = -4 + 2 \qquad\qquad x = -4 - 2$$
$$x = -2 \qquad\qquad\quad x = -6$$
The solutions are $-6$ and $-2$.

**3.**
$$x^2 + 2x - 7 = 0$$
$$x^2 + 2x = 7$$
$$x^2 + 2x + 1 = 7 + 1$$
$$(x+1)^2 = 8$$
$$x + 1 = \sqrt{8} \quad \text{or} \quad x + 1 = -\sqrt{8}$$
$$x + 1 = 2\sqrt{2} \qquad\qquad x + 1 = -2\sqrt{2}$$
$$x = -1 + 2\sqrt{2} \qquad\qquad x = -1 - 2\sqrt{2}$$
The solutions are $-1 \pm 2\sqrt{2}$.

**5.**
$$x^2 - 6x = 0$$
$$x^2 - 6x + 9 = 0 + 9$$
$$(x-3)^2 = 9$$
$$x - 3 = \sqrt{9} \quad \text{or} \quad x - 3 = -\sqrt{9}$$
$$x = 3 + 3 \qquad\qquad x = 3 - 3$$
$$x = 6 \qquad\qquad\quad x = 0$$
The solutions are 0 and 6.

**7.**
$$z^2 + 5z = 7$$
$$z^2 + 5z + \frac{25}{4} = 7 + \frac{25}{4}$$
$$\left(z + \frac{5}{2}\right)^2 = \frac{53}{4}$$
$$z + \frac{5}{2} = \sqrt{\frac{53}{4}} \quad \text{or} \quad z + \frac{5}{2} = -\sqrt{\frac{53}{4}}$$
$$z + \frac{5}{2} = \frac{\sqrt{53}}{2} \qquad\qquad z + \frac{5}{2} = -\frac{\sqrt{53}}{2}$$
$$z = -\frac{5}{2} + \frac{\sqrt{53}}{2} \qquad\qquad z = -\frac{5}{2} - \frac{\sqrt{53}}{2}$$
The solutions are $\dfrac{-5 \pm \sqrt{53}}{2}$.

**9.**
$$x^2 - 2x - 1 = 0$$
$$x^2 - 2x = 1$$
$$x^2 - 2x + 1 = 1 + 1$$
$$(x-1)^2 = 2$$
$$x - 1 = \sqrt{2} \quad \text{or} \quad x - 1 = -\sqrt{2}$$
$$x = 1 + \sqrt{2} \qquad\qquad x = 1 - \sqrt{2}$$
The solutions are $1 \pm \sqrt{2}$.

**11.**
$$y^2 + 5y + 4 = 0$$
$$y^2 + 5y = -4$$
$$y^2 + 5y + \frac{25}{4} = -4 + \frac{25}{4}$$
$$\left(y + \frac{5}{2}\right)^2 = \frac{9}{4}$$
$$y + \frac{5}{2} = \sqrt{\frac{9}{4}} \quad \text{or} \quad y + \frac{5}{2} = -\sqrt{\frac{9}{4}}$$
$$y = -\frac{5}{2} + \frac{3}{2} \qquad\qquad y = -\frac{5}{2} - \frac{3}{2}$$
$$y = -1 \qquad\qquad\qquad y = -4$$
The solutions are $-4$ and $-1$.

**13.**
$$3x^2 - 6x = 24$$
$$x^2 - 2x = 8$$
$$x^2 - 2x + 1 = 8 + 1$$
$$(x-1)^2 = 9$$
$$x - 1 = \sqrt{9} \quad \text{or} \quad x - 1 = -\sqrt{9}$$
$$x = 1 + 3 \qquad\qquad x = 1 - 3$$
$$x = 4 \qquad\qquad\quad x = -2$$
The solutions are $-2$ and 4.

**15.**
$$5x^2 + 10x + 6 = 0$$
$$5x^2 + 10x = -6$$
$$x^2 + 2x = -\frac{6}{5}$$
$$x^2 + 2x + 1 = -\frac{6}{5} + 1$$
$$(x+1)^2 = -\frac{1}{5}$$

This equation has no real solution because $\sqrt{-\dfrac{1}{5}}$ is not a real number.

**17.**
$$2x^2 = 6x + 5$$
$$2x^2 - 6x = 5$$
$$x^2 - 3x = \frac{5}{2}$$
$$x^2 - 3x + \frac{9}{4} = \frac{5}{2} + \frac{9}{4}$$
$$\left(x - \frac{3}{2}\right)^2 = \frac{19}{4}$$
$$x - \frac{3}{2} = \sqrt{\frac{19}{4}} \quad \text{or} \quad x - \frac{3}{2} = -\sqrt{\frac{19}{4}}$$
$$x - \frac{3}{2} = \frac{\sqrt{19}}{2} \qquad x - \frac{3}{2} = -\frac{\sqrt{19}}{2}$$
$$x = \frac{3}{2} + \frac{\sqrt{19}}{2} \qquad x = \frac{3}{2} - \frac{\sqrt{19}}{2}$$

The solutions are $\dfrac{3 \pm \sqrt{19}}{2}$.

**19.**
$$2y^2 + 8y + 5 = 0$$
$$2y^2 + 8y = -5$$
$$y^2 + 4y = -\frac{5}{2}$$
$$y^2 + 4y + 4 = -\frac{5}{2} + 4$$
$$(y + 2)^2 = \frac{3}{2}$$
$$y + 2 = \sqrt{\frac{3}{2}} \quad \text{or} \quad y + 2 = -\sqrt{\frac{3}{2}}$$
$$y + 2 = \frac{\sqrt{3}}{\sqrt{2}} \cdot \frac{\sqrt{2}}{\sqrt{2}} \qquad y + 2 = -\frac{\sqrt{3}}{\sqrt{2}} \cdot \frac{\sqrt{2}}{\sqrt{2}}$$
$$y = -2 + \frac{\sqrt{6}}{2} \qquad y = -2 - \frac{\sqrt{6}}{2}$$

The solutions are $-2 \pm \dfrac{\sqrt{6}}{2}$.

**21.**
$$x^2 + 6x - 25 = 0$$
$$x^2 + 6x = 25$$
$$x^2 + 6x + 9 = 25 + 9$$
$$(x + 3)^2 = 34$$
$$x + 3 = \sqrt{34} \quad \text{or} \quad x + 3 = -\sqrt{34}$$
$$x = -3 + \sqrt{34} \qquad x = -3 - \sqrt{34}$$
The solutions are $-3 \pm \sqrt{34}$.

**23.**
$$x^2 - 3x - 3 = 0$$
$$x^2 - 3x = 3$$
$$x^2 - 3x + \frac{9}{4} = 3 + \frac{9}{4}$$
$$\left(x - \frac{3}{2}\right)^2 = \frac{21}{4}$$
$$x - \frac{3}{2} = \sqrt{\frac{21}{4}} \quad \text{or} \quad x - \frac{3}{2} = -\sqrt{\frac{21}{4}}$$
$$x - \frac{3}{2} = \frac{\sqrt{21}}{2} \qquad x - \frac{3}{2} = -\frac{\sqrt{21}}{2}$$
$$x = \frac{3}{2} + \frac{\sqrt{21}}{2} \qquad x = \frac{3}{2} - \frac{\sqrt{21}}{2}$$

The solutions are $\dfrac{3 \pm \sqrt{21}}{2}$.

**25.**
$$2y^2 - 3y + 1 = 0$$
$$2y^2 - 3y = -1$$
$$y^2 - \frac{3}{2}y = -\frac{1}{2}$$
$$y^2 - \frac{3}{2}y + \frac{9}{16} = -\frac{1}{2} + \frac{9}{16}$$
$$\left(y - \frac{3}{4}\right)^2 = \frac{1}{16}$$
$$y - \frac{3}{4} = \sqrt{\frac{1}{16}} \quad \text{or} \quad y - \frac{3}{4} = -\sqrt{\frac{1}{16}}$$
$$y = \frac{3}{4} + \frac{1}{4} \qquad y = \frac{3}{4} - \frac{1}{4}$$
$$y = 1 \qquad y = \frac{1}{2}$$

The solutions are $\dfrac{1}{2}$ and 1.

**27.**
$$x(x + 3) = 18$$
$$x^2 + 3x = 18$$
$$x^2 + 3x + \frac{9}{4} = 18 + \frac{9}{4}$$
$$\left(x + \frac{3}{2}\right)^2 = \frac{81}{4}$$
$$x + \frac{3}{2} = \sqrt{\frac{81}{4}} \quad \text{or} \quad x + \frac{3}{2} = -\sqrt{\frac{81}{4}}$$
$$x = -\frac{3}{2} + \frac{9}{2} \qquad x = -\frac{3}{2} - \frac{9}{2}$$
$$x = 3 \qquad x = -6$$
The solutions are $-6$ and 3.

**29.** $3z^2 + 6z + 4 = 0$

$$3z^2 + 6z = -4$$

$$z^2 + 2z = -\frac{4}{3}$$

$$z^2 + 2z + 1 = -\frac{4}{3} + 1$$

$$(z+1)^2 = -\frac{1}{3}$$

This equation has no real solution because $\sqrt{-\frac{1}{3}}$ is not a real number.

**31.** $\quad 4x^2 + 16x = 48$

$$x^2 + 4x = 12$$

$$x^2 + 4x + 4 = 12 + 4$$

$$(x+2)^2 = 16$$

$x+2 = \sqrt{16} \quad$ or $\quad x+2 = -\sqrt{16}$

$\quad x = 4 - 2 \qquad\qquad x = -4 - 2$

$\quad x = 2 \qquad\qquad\quad\; x = -6$

The solutions are 2 and $-6$.

**33.** $\dfrac{3}{4} - \sqrt{\dfrac{25}{16}} = \dfrac{3}{4} - \dfrac{5}{4} = -\dfrac{2}{4} = -\dfrac{1}{2}$

**35.** $\dfrac{1}{2} - \sqrt{\dfrac{9}{4}} = \dfrac{1}{2} - \dfrac{3}{2} = -\dfrac{2}{2} = -1$

**37.** $\dfrac{6 + 4\sqrt{5}}{2} = \dfrac{2\left(3 + 2\sqrt{5}\right)}{2} = 3 + 2\sqrt{5}$

**39.** $\dfrac{3 - 9\sqrt{2}}{6} = \dfrac{3\left(1 - 3\sqrt{2}\right)}{3 \cdot 2} = \dfrac{1 - 3\sqrt{2}}{2}$

**41.** answers may vary

**43. a.** $\quad x^2 + 6x + 9 = 11$

$$(x+3)^2 = 11$$

$x+3 = \sqrt{11} \qquad$ or $\quad x+3 = -\sqrt{11}$

$\quad x = -3 + \sqrt{11} \qquad\quad x = -3 - \sqrt{11}$

The solutions are $-3 \pm \sqrt{11}$.

**b.** answers may vary

**45.** $x^2 + kx + 16$

$$\left(\frac{k}{2}\right)^2 = 16$$

$$\frac{k^2}{4} = 16$$

$$k^2 = 64$$

$$k = \pm\sqrt{64}$$

$$k = \pm 8$$

**47.** $\qquad\qquad y = 2.5x^2 + 7.5x + 122$

$$392 = 2.5x^2 + 7.5x + 122$$

$$270 = 2.5x^2 + 7.5x$$

$$108 = x^2 + 3x$$

$$\left(\frac{3}{2}\right)^2 + 108 = x^2 + 3x + \left(\frac{3}{2}\right)^2$$

$$110.25 = (x + 1.5)^2$$

$\sqrt{110.25} = x + 1.5 \quad$ or $\quad -\sqrt{110.25} = x + 1.5$

$\quad 10.5 = x + 1.5 \qquad\qquad -10.5 = x + 1.5$

$\qquad\; 9 = x \qquad\qquad\qquad\;\; -12 = x$

Since the time will not be a negative number in this context, reject the solution $-12$. The solution is $x = 9$, so the model predicts that retail sales from online shopping were \$392 billion in 2011 $(2002 + 9)$.

**49.** $x^2 + 8x = -12$

$y_1 = x^2 + 8x$

$y_2 = -12$

The $x$-coordinates of the intersections, $-6$ and $-2$, are the solutions.

**51.** $2x^2 = 6x + 5$

$y_1 = 2x^2$

$y_2 = 6x + 5$

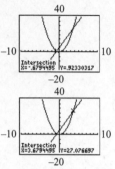

The $x$-coordinates of the intersections, $-0.68$ and $3.68$, are the approximate solutions.

### Section 9.3 Practice

**1.** $5x^2 + x - 2 = 0$

$a = 5,\ b = 1,\ c = -2$

$x = \dfrac{-b \pm \sqrt{b^2 - 4ac}}{2a}$

$x = \dfrac{-1 \pm \sqrt{1^2 - 4(5)(-2)}}{2(5)}$

$= \dfrac{-1 \pm \sqrt{1 + 40}}{10}$

$= \dfrac{-1 \pm \sqrt{41}}{10}$

The solutions are $\dfrac{-1 \pm \sqrt{41}}{10}$.

**2.** $3x^2 + 2x = 8$

$3x^2 + 2x - 8 = 0$

$a = 3,\ b = 2,\ c = -8$

$x = \dfrac{-b \pm \sqrt{b^2 - 4ac}}{2a}$

$x = \dfrac{-2 \pm \sqrt{2^2 - 4(3)(-8)}}{2(3)}$

$= \dfrac{-2 \pm \sqrt{4 + 96}}{6}$

$= \dfrac{-2 \pm \sqrt{100}}{6}$

$= \dfrac{-2 \pm 10}{6}$

$x = \dfrac{-2 + 10}{6} = \dfrac{8}{6} = \dfrac{4}{3}$ or $x = \dfrac{-2 - 10}{6} = \dfrac{-12}{6} = -2$

The solutions are $\dfrac{4}{3}$ and $-2$.

**3.** $3x^2 = 5$

$3x^2 - 5 = 0$

$a = 3,\ b = 0,\ c = -5$

$x = \dfrac{0 \pm \sqrt{0^2 - 4(3)(-5)}}{2(3)}$

$= \dfrac{\pm \sqrt{60}}{6}$

$= \dfrac{\pm 2\sqrt{15}}{6}$

$= \dfrac{\pm \sqrt{15}}{3}$

The solutions are $\pm \dfrac{\sqrt{15}}{3}$.

**4.** $x^2 = 3x - 4$

$x^2 - 3x + 4 = 0$

$a = 1,\ b = -3,\ c = 4$

$x = \dfrac{-(-3) \pm \sqrt{(-3)^2 - 4(1)(4)}}{2(1)}$

$= \dfrac{3 \pm \sqrt{9 - 16}}{2}$

$= \dfrac{3 \pm \sqrt{-7}}{2}$

There is no real number solution because $\sqrt{-7}$ is not a real number.

**5.** $\dfrac{1}{5}x^2 - x = 1$

$\dfrac{1}{5}x^2 - x - 1 = 0$

$x^2 - 5x - 5 = 0$

$a = 1,\ b = -5,\ c = -5$

$x = \dfrac{-(-5) \pm \sqrt{(-5)^2 - 4(1)(-5)}}{2(1)}$

$= \dfrac{5 \pm \sqrt{25 + 20}}{2}$

$= \dfrac{5 \pm \sqrt{45}}{2}$

$= \dfrac{5 \pm 3\sqrt{5}}{2}$

The solutions are $\dfrac{5 \pm 3\sqrt{5}}{2}$.

**6.** The exact solutions are $\dfrac{-1\pm\sqrt{41}}{10}$.

$$\dfrac{-1+\sqrt{41}}{10} \approx 0.5403124237 \approx 0.5$$

$$\dfrac{-1-\sqrt{41}}{10} \approx -0.7403124237 \approx -0.7$$

**7.** $5x^2 + x - 2 = 0$

$a = 5, b = 1, c = -2$

$b^2 - 4ac = 1^2 - 4(5)(-2) = 1 + 40 = 41$

Since $41 > 0$, there are two distinct real solutions.

**8. a.** $x^2 - 10x + 35 = 0$

$a = 1, b = -10, c = 35$

$$\begin{aligned} b^2 - 4ac &= (-10)^2 - 4(1)(35) \\ &= 100 - 140 \\ &= -40 \end{aligned}$$

Since $-40 < 0$, there is no real solution.

**b.** $5x^2 + 3x = 0$

$a = 5, b = 3, c = 0$

$b^2 - 4ac = 3^2 - 4(5)(0) = 9$

Since $9 > 0$, there are two distinct real solutions.

**Vocabulary, Readiness & Video Check 9.3**

**1.** The quadratic formula is $x = \dfrac{-b\pm\sqrt{b^2-4ac}}{2a}$.

**2.** In $x^2 - 6 = 0$, $a = \underline{1}$, $b = \underline{0}$, $c = \underline{-6}$.

**3.** $\dfrac{-1\pm\sqrt{1^2-4(1)(-2)}}{2(1)} = \dfrac{-1\pm\sqrt{1+8}}{2}$

$\qquad\qquad = \dfrac{-1\pm\sqrt{9}}{2}$

$\qquad\qquad = \dfrac{-1\pm 3}{2}$

$\dfrac{-1+3}{2} = \dfrac{2}{2} = 1; \dfrac{-1-3}{2} = \dfrac{-4}{2} = -2$

**4.** $\dfrac{-(-5)\pm\sqrt{(-5)^2-4(2)(3)}}{2(2)} = \dfrac{5\pm\sqrt{25-24}}{4}$

$\qquad\qquad\qquad\qquad = \dfrac{5\pm\sqrt{1}}{4}$

$\qquad\qquad\qquad\qquad = \dfrac{5\pm 1}{4}$

$\dfrac{5+1}{4} = \dfrac{6}{4} = \dfrac{3}{2}; \dfrac{5-1}{4} = \dfrac{4}{4} = 1$

**5.** $\dfrac{-5\pm\sqrt{5^2-4(1)(2)}}{2(1)} = \dfrac{-5\pm\sqrt{25-8}}{2} = \dfrac{-5\pm\sqrt{17}}{2}$

**6.** $\dfrac{-7\pm\sqrt{7^2-4(2)(1)}}{2(2)} = \dfrac{-7\pm\sqrt{49-8}}{4} = \dfrac{-7\pm\sqrt{41}}{4}$

**7. a.** Yes, in order to make sure you have correct values for $a$, $b$, and $c$.

**b.** No; it simplifies calculations, but you would still get a correct answer using fraction values in the formula.

**8.** The exact solution values are keyed into a calculator, and the decimal approximations are then rounded to the requested place value.

**9.** The discriminant is the <u>radicand</u> portion of the quadratic formula and can be used to find out the number of <u>real</u> solutions of a quadratic equation without solving the equation. To use the discriminant, the quadratic equation needs to be written in <u>standard</u> form.

**Exercise Set 9.3**

**1.** $x^2 - 3x + 2 = 0$

$a = 1, b = -3$, and $c = 2$

$x = \dfrac{-(-3)\pm\sqrt{(-3)^2-4(1)(2)}}{2(1)}$

$\quad = \dfrac{3\pm\sqrt{9-8}}{2}$

$\quad = \dfrac{3\pm\sqrt{1}}{2}$

$\quad = \dfrac{3\pm 1}{2}$

$x = \dfrac{3+1}{2} = 2 \quad\text{or}\quad x = \dfrac{3-1}{2} = 1$

The solutions are 1 and 2.

**3.** $3k^2 + 7k + 1 = 0$

$a = 3, b = 7,$ and $c = 1$

$k = \dfrac{-7 \pm \sqrt{7^2 - 4(3)(1)}}{2(3)}$

$= \dfrac{-7 \pm \sqrt{49 - 12}}{6}$

$= \dfrac{-7 \pm \sqrt{37}}{6}$

The solutions are $\dfrac{-7 \pm \sqrt{37}}{6}$.

**5.** $49x^2 - 4 = 0$

$a = 49, b = 0,$ and $c = -4$

$x = \dfrac{-0 \pm \sqrt{0^2 - 4(49)(-4)}}{2(49)}$

$= \dfrac{\pm\sqrt{784}}{98}$

$= \dfrac{\pm 28}{98}$

$= \pm\dfrac{2}{7}$

The solutions are $\pm\dfrac{2}{7}$.

**7.** $5z^2 - 4z + 3 = 0$

$a = 5, b = -4,$ and $c = 3$

$z = \dfrac{-(-4) \pm \sqrt{(-4)^2 - 4(5)(3)}}{2(5)}$

$= \dfrac{4 \pm \sqrt{16 - 60}}{10}$

$= \dfrac{4 \pm \sqrt{-44}}{10}$

There is no real solution because $\sqrt{-44}$ is not a real number.

**9.** $y^2 = 7y + 30$

$y^2 - 7y - 30 = 0$

$a = 1, b = -7,$ and $c = -30$

$y = \dfrac{-(-7) \pm \sqrt{(-7)^2 - 4(1)(-30)}}{2(1)}$

$= \dfrac{7 \pm \sqrt{49 + 120}}{2}$

$= \dfrac{7 \pm \sqrt{169}}{2}$

$= \dfrac{7 \pm 13}{2}$

$y = \dfrac{7 + 13}{2} = 10$   or   $y = \dfrac{7 - 13}{2} = -3$

The solutions are $-3$ and $10$.

**11.** $2x^2 = 10$

$2x^2 - 10 = 0$

$a = 2, b = 0,$ and $c = -10$

$x = \dfrac{-0 \pm \sqrt{0^2 - 4(2)(-10)}}{2(2)}$

$= \dfrac{\pm\sqrt{80}}{4}$

$= \dfrac{\pm 4\sqrt{5}}{4}$

$= \pm\sqrt{5}$

The solutions are $\pm\sqrt{5}$.

**13.** $m^2 - 12 = m$

$m^2 - m - 12 = 0$

$a = 1, b = -1,$ and $c = -12$

$m = \dfrac{-(-1) \pm \sqrt{(-1)^2 - 4(1)(-12)}}{2(1)}$

$= \dfrac{1 \pm \sqrt{1 + 48}}{2}$

$= \dfrac{1 \pm \sqrt{49}}{2}$

$= \dfrac{1 \pm 7}{2}$

$m = \dfrac{1 + 7}{2} = 4$   or   $m = \dfrac{1 - 7}{2} = -3$

The solutions are $-3$ and $4$.

**15.**
$$3 - x^2 = 4x$$
$$-x^2 - 4x + 3 = 0$$
$a = -1, b = -4,$ and $c = 3$
$$x = \frac{-(-4) \pm \sqrt{(-4)^2 - 4(-1)(3)}}{2(-1)}$$
$$= \frac{4 \pm \sqrt{16 + 12}}{-2}$$
$$= \frac{4 \pm \sqrt{28}}{-2}$$
$$= \frac{4 \pm 2\sqrt{7}}{-2}$$
$$= \frac{2(2 \pm \sqrt{7})}{-2}$$
$$= -2 \pm \sqrt{7}$$
The solutions are $-2 \pm \sqrt{7}$.

**17.** $2a^2 - 7a + 3 = 0$
$a = 2, b = -7, c = 3$
$$a = \frac{-(-7) \pm \sqrt{(-7)^2 - 4(2)(3)}}{2(2)}$$
$$= \frac{7 \pm \sqrt{49 - 24}}{4}$$
$$= \frac{7 \pm \sqrt{25}}{4}$$
$$= \frac{7 \pm 5}{4}$$
$$a = \frac{7 + 5}{4} = 3 \quad \text{or} \quad a = \frac{7 - 5}{4} = \frac{1}{2}$$
The solutions are 3 and $\frac{1}{2}$.

**19.** $x^2 - 5x - 2 = 0$
$a = 1, b = -5, c = -2$
$$x = \frac{-(-5) \pm \sqrt{(-5)^2 - 4(1)(-2)}}{2(1)}$$
$$= \frac{5 \pm \sqrt{25 + 8}}{2}$$
$$= \frac{5 \pm \sqrt{33}}{2}$$
The solutions are $\frac{5 \pm \sqrt{33}}{2}$.

**21.** $3x^2 - x - 14 = 0$
$a = 3, b = -1, c = -14$
$$x = \frac{-(-1) \pm \sqrt{(-1)^2 - 4(3)(-14)}}{2(3)}$$
$$= \frac{1 \pm \sqrt{1 + 168}}{6}$$
$$= \frac{1 \pm \sqrt{169}}{6}$$
$$= \frac{1 \pm 13}{6}$$
$$x = \frac{1 + 13}{6} = \frac{7}{3} \quad \text{or} \quad x = \frac{1 - 13}{6} = -2$$
The solutions are $\frac{7}{3}$ and $-2$.

**23.**
$$6x^2 + 9x = 2$$
$$6x^2 + 9x - 2 = 0$$
$a = 6, b = 9,$ and $c = -2$
$$x = \frac{-9 \pm \sqrt{9^2 - 4(6)(-2)}}{2(6)}$$
$$= \frac{-9 \pm \sqrt{81 + 48}}{12}$$
$$= \frac{-9 \pm \sqrt{129}}{12}$$
The solutions are $\frac{-9 \pm \sqrt{129}}{12}$.

**25.**
$$7p^2 + 2 = 8p$$
$$7p^2 - 8p + 2 = 0$$
$a = 7, b = -8,$ and $c = 2$
$$p = \frac{-(-8) \pm \sqrt{(-8)^2 - 4(7)(2)}}{2(7)}$$
$$= \frac{8 \pm \sqrt{64 - 56}}{14}$$
$$= \frac{8 \pm \sqrt{8}}{14}$$
$$= \frac{8 \pm 2\sqrt{2}}{14}$$
$$= \frac{2(4 \pm \sqrt{2})}{2 \cdot 7}$$
$$= \frac{4 \pm \sqrt{2}}{7}$$
The solutions are $\frac{4 \pm \sqrt{2}}{7}$.

**27.** $a^2 - 6a + 2 = 0$

$a = 1$, $b = -6$, and $c = 2$

$$a = \frac{-(-6) \pm \sqrt{(-6)^2 - 4(1)(2)}}{2(1)}$$

$$= \frac{6 \pm \sqrt{36 - 8}}{2}$$

$$= \frac{6 \pm \sqrt{28}}{2}$$

$$= \frac{6 \pm 2\sqrt{7}}{2}$$

$$= \frac{2(3 \pm \sqrt{7})}{2}$$

$$= 3 \pm \sqrt{7}$$

The solutions are $3 \pm \sqrt{7}$.

**29.** $2x^2 - 6x + 3 = 0$

$a = 2$, $b = -6$, and $c = 3$

$$x = \frac{-(-6) \pm \sqrt{(-6)^2 - 4(2)(3)}}{2(2)}$$

$$= \frac{6 \pm \sqrt{36 - 24}}{4}$$

$$= \frac{6 \pm \sqrt{12}}{4}$$

$$= \frac{6 \pm 2\sqrt{3}}{4}$$

$$= \frac{2(3 \pm \sqrt{3})}{2 \cdot 2}$$

$$= \frac{3 \pm \sqrt{3}}{2}$$

The solutions are $\frac{3 \pm \sqrt{3}}{2}$.

**31.** $3x^2 = 1 - 2x$

$3x^2 + 2x - 1 = 0$

$a = 3$, $b = 2$, and $c = -1$

$$x = \frac{-2 \pm \sqrt{2^2 - 4(3)(-1)}}{2(3)}$$

$$= \frac{-2 \pm \sqrt{4 + 12}}{6}$$

$$= \frac{-2 \pm \sqrt{16}}{6}$$

$$= \frac{-2 \pm 4}{6}$$

$$x = \frac{-2 + 4}{6} = \frac{1}{3} \quad \text{or} \quad x = \frac{-2 - 4}{6} = -1$$

The solutions are $-1$ and $\frac{1}{3}$.

**33.** $20y^2 = 3 - 11y$

$20y^2 + 11y - 3 = 0$

$a = 20$, $b = 11$, and $c = -3$

$$y = \frac{-11 \pm \sqrt{11^2 - 4(20)(-3)}}{2(20)}$$

$$= \frac{-11 \pm \sqrt{121 + 240}}{40}$$

$$= \frac{-11 \pm \sqrt{361}}{40}$$

$$= \frac{-11 \pm 19}{40}$$

$$y = \frac{-11 + 19}{40} = \frac{1}{5} \quad \text{or} \quad y = \frac{-11 - 19}{40} = -\frac{3}{4}$$

The solutions are $-\frac{3}{4}$ and $\frac{1}{5}$.

**35.** $x^2 + x + 1 = 0$

$a = 1$, $b = 1$, and $c = 1$

$$x = \frac{-1 \pm \sqrt{1^2 - 4(1)(1)}}{2(1)}$$

$$= \frac{-1 \pm \sqrt{1 - 4}}{2}$$

$$= \frac{-1 \pm \sqrt{-3}}{2}$$

There is no real solution because $\sqrt{-3}$ is not a real number.

**37.**
$$4y^2 = 6y + 1$$
$$4y^2 - 6y - 1 = 0$$
$$a = 4, b = -6, \text{ and } c = -1$$
$$y = \frac{-(-6) \pm \sqrt{(-6)^2 - 4(4)(-1)}}{2(4)}$$
$$= \frac{6 \pm \sqrt{36 + 16}}{8}$$
$$= \frac{6 \pm \sqrt{52}}{8}$$
$$= \frac{6 \pm 2\sqrt{13}}{8}$$
$$= \frac{2(3 \pm \sqrt{13})}{2 \cdot 4}$$
$$= \frac{3 \pm \sqrt{13}}{4}$$

The solutions are $\dfrac{3 \pm \sqrt{13}}{4}$.

**39.**
$$\frac{m^2}{2} = m + \frac{1}{2}$$
$$m^2 = 2m + 1$$
$$m^2 - 2m - 1 = 0$$
$$a = 1, b = -2 \text{ and } c = -1$$
$$m = \frac{-(-2) \pm \sqrt{(-2)^2 - 4(1)(-1)}}{2(1)}$$
$$= \frac{2 \pm \sqrt{4 + 4}}{2}$$
$$= \frac{2 \pm \sqrt{8}}{2}$$
$$= \frac{2 \pm 2\sqrt{2}}{2}$$
$$= \frac{2(1 \pm \sqrt{2})}{2}$$
$$= 1 \pm \sqrt{2}$$

The solutions are $1 \pm \sqrt{2}$.

**41.**
$$3p^2 - \frac{2}{3}p + 1 = 0$$
$$9p^2 - 2p + 3 = 0$$
$$a = 9, b = -2, \text{ and } c = 3$$
$$p = \frac{-(-2) \pm \sqrt{(-2)^2 - 4(9)(3)}}{2(9)}$$
$$= \frac{2 \pm \sqrt{4 - 108}}{18}$$
$$= \frac{2 \pm \sqrt{-104}}{18}$$

There is no real solution because $\sqrt{-104}$ is not a real number.

**43.**
$$4p^2 + \frac{3}{2} = -5p$$
$$8p^2 + 3 = -10p$$
$$8p^2 + 10p + 3 = 0$$
$$a = 8, b = 10, \text{ and } c = 3$$
$$p = \frac{-10 \pm \sqrt{10^2 - 4(8)(3)}}{2(8)}$$
$$= \frac{-10 \pm \sqrt{100 - 96}}{16}$$
$$= \frac{-10 \pm \sqrt{4}}{16}$$
$$= \frac{-10 \pm 2}{16}$$
$$p = \frac{-10 + 2}{16} = -\frac{1}{2} \quad \text{or} \quad p = \frac{-10 - 2}{16} = -\frac{3}{4}$$

The solutions are $-\dfrac{3}{4}$ and $-\dfrac{1}{2}$.

**45.**
$$5x^2 = \frac{7}{2}x + 1$$
$$10x^2 = 7x + 2$$
$$10x^2 - 7x - 2 = 0$$
$$a = 10, b = -7, \text{ and } c = -2$$
$$x = \frac{-(-7) \pm \sqrt{(-7)^2 - 4(10)(-2)}}{2(10)}$$
$$= \frac{7 \pm \sqrt{49 + 80}}{20}$$
$$= \frac{7 \pm \sqrt{129}}{20}$$

The solutions are $\dfrac{7 \pm \sqrt{129}}{20}$.

**47.** $x^2 - \dfrac{11}{2}x - \dfrac{1}{2} = 0$

$2x^2 - 11x - 1 = 0$

$a = 2, b = -11, c = -1$

$x = \dfrac{-(-11) \pm \sqrt{(-11)^2 - 4(2)(-1)}}{2(2)}$

$= \dfrac{11 \pm \sqrt{121 + 8}}{4}$

$= \dfrac{11 \pm \sqrt{129}}{4}$

The solutions are $\dfrac{11 \pm \sqrt{129}}{4}$.

**49.** $5z^2 - 2z = \dfrac{1}{5}$

$25z^2 - 10z = 1$

$25z^2 - 10z - 1 = 0$

$a = 25, b = -10,$ and $c = -1$

$z = \dfrac{-(-10) \pm \sqrt{(-10)^2 - 4(25)(-1)}}{2(25)}$

$= \dfrac{10 \pm \sqrt{100 + 100}}{50}$

$= \dfrac{10 \pm \sqrt{200}}{50}$

$= \dfrac{10 \pm 10\sqrt{2}}{50}$

$= \dfrac{10(1 + \sqrt{2})}{10 \cdot 5}$

$= \dfrac{1 \pm \sqrt{2}}{5}$

The solutions are $\dfrac{1 \pm \sqrt{2}}{5}$.

**51.** $3x^2 = 21$

$3x^2 - 21 = 0$

$a = 3, b = 0, c = -21$

$x = \dfrac{-0 \pm \sqrt{0^2 - 4(3)(-21)}}{2(3)}$

$= \dfrac{\pm\sqrt{252}}{6}$

$= \dfrac{\pm 6\sqrt{7}}{6}$

$= \pm\sqrt{7}$

$\approx \pm 2.6$

The solutions are $\pm\sqrt{7} \approx \pm 2.6$.

**53.** $x^2 + 6x + 1 = 0$

$a = 1, b = 6, c = 1$

$x = \dfrac{-6 \pm \sqrt{6^2 - 4(1)(1)}}{2(1)}$

$= \dfrac{-6 \pm \sqrt{36 - 4}}{2}$

$= \dfrac{-6 \pm \sqrt{32}}{2}$

$= \dfrac{-6 \pm 4\sqrt{2}}{2}$

$= -3 \pm 2\sqrt{2}$

$x = -3 + 2\sqrt{2} \approx 0.2$ or $x = -3 - 2\sqrt{2} \approx -5.8$

The solutions are $-3 \pm 2\sqrt{2}$, approximately $-5.8$ and $-0.2$.

**55.** $x^2 = 9x + 4$

$x^2 - 9x - 4 = 0$

$a = 1, b = -9, c = -4$

$x = \dfrac{-(-9) \pm \sqrt{(-9)^2 - 4(1)(-4)}}{2(1)}$

$= \dfrac{9 \pm \sqrt{81 + 16}}{2}$

$= \dfrac{9 \pm \sqrt{97}}{2}$

$x = \dfrac{9 + \sqrt{97}}{2} \approx 9.4$ or $x = \dfrac{9 - \sqrt{97}}{2} \approx -0.4$

The solutions are $\dfrac{9 \pm \sqrt{97}}{2}$, approximately $9.4$ and $-0.4$.

**57.** $3x^2 - 2x - 2 = 0$

$a = 3, b = -2, c = -2$

$$x = \frac{-(-2) \pm \sqrt{(-2)^2 - 4(3)(-2)}}{2(3)}$$

$$= \frac{2 \pm \sqrt{4 + 24}}{6}$$

$$= \frac{2 \pm \sqrt{28}}{6}$$

$$= \frac{2 \pm 2\sqrt{7}}{6}$$

$$= \frac{1 \pm \sqrt{7}}{3}$$

$x = \frac{1 + \sqrt{7}}{3} \approx 1.2$ or $x = \frac{1 - \sqrt{7}}{3} \approx -0.5$

The solutions are $\frac{1 \pm \sqrt{7}}{3}$, approximately 1.2 and −0.5.

**59.** $x^2 + 3x - 1 = 0$

$a = 1, b = 3, c = -1$

$b^2 - 4ac = 3^2 - 4(1)(-1) = 9 + 4 = 13$

Since the discriminant is a positive number, this equation has two distinct real solutions.

**61.** $3x^2 + x + 5 = 0$

$a = 3, b = 1, c = 5$

$b^2 - 4ac = 1^2 - 4(3)(5) = 1 - 60 = -59$

Since the discriminant is a negative number, this equation has no real solution.

**63.** $4x^2 + 4x = -1$

$4x^2 + 4x + 1 = 0$

$a = 4, b = 4, c = 1$

$b^2 - 4ac = 4^2 - 4(4)(1) = 16 - 16 = 0$

Since the discriminant is 0, this equation has one real solution.

**65.** $9x^2 + 2x = 0$

$a = 9, b = 2, c = 0$

$b^2 - 4ac = 2^2 - 4(9)(0) = 4 - 0 = 4$

Since the discriminant is a positive number this equation has two distinct real solutions.

**67.** $5x^2 + 1 = 0$

$a = 5, b = 0, c = 1$

$b^2 - 4ac = 0^2 - 4(5)(1) = 0 - 20 = -20$

Since the discriminant is a negative number, this equation has no real solution.

**69.** $x^2 + 36 = -12x$

$x^2 + 12x + 36 = 0$

$a = 1, b = 12, c = 36$

$b^2 - 4ac = 12^2 - 4(1)(36) = 144 - 144 = 0$

Since the discriminant is 0, this equation has one real solution.

**71.** $\sqrt{48} = \sqrt{16 \cdot 3} = \sqrt{16} \cdot \sqrt{3} = 4\sqrt{3}$

**73.** $\sqrt{50} = \sqrt{25 \cdot 2} = \sqrt{25} \cdot \sqrt{2} = 5\sqrt{2}$

**75.** Let $x$ = the base and $4x$ = the height.

$$\frac{1}{2}bh = A$$

$$\frac{1}{2}(x)(4x) = 18$$

$$\frac{1}{2}(4x^2) = 18$$

$$2x^2 = 18$$

$$x^2 = 9$$

$$x = \pm\sqrt{9}$$

$$x = \pm 3$$

Since the length can't be negative, base = 3 feet and height = 4(3) = 12 feet.

**77.** $5x^2 + 2 = x; \ a = 5$

$5x^2 - x + 2 = 0$

$a = 5, b = -1, c = 2$

The value of $b$ is −1, choice **c**.

**79.** $7y^2 = 3y; \ b = 3$

$0 = -7y^2 + 3y$

$a = -7, b = 3, c = 0$

The value of $a$ is −7, choice **b**.

**81.** Let $x$ be the width of the chocolate bar. Then the length is $3x - 0.6$.

Area = (length)(width)

$34.65 = (3x - 0.6)(x)$

$34.65 = 3x^2 - 0.6x$

$0 = 3x^2 - 0.6x - 34.65$

$a = 3, b = -0.6, c = -34.65$

$$x = \frac{-b \pm \sqrt{b^2 - 4ac}}{2a}$$

$$x = \frac{-(-0.6) \pm \sqrt{(-0.6)^2 - 4(3)(-34.65)}}{2(3)}$$

$$= \frac{0.6 \pm \sqrt{0.36 + 415.8}}{6}$$

$$= \frac{0.6 \pm \sqrt{416.16}}{6}$$

$$= \frac{0.6 \pm 20.4}{6}$$

$$x = \frac{0.6 + 20.4}{6} = \frac{21}{6} = 3.5$$

or

$$x = \frac{0.6 - 20.4}{6} = \frac{-19.8}{6} = -3.3$$

Since the width is not a negative number, discard the solution $-3.3$.
$3x - 0.6 = 3(3.5) - 0.6 = 10.5 - 0.6 = 9.9$
The width was 3.5 feet and the length was 9.9 feet.

**83.** $x^2 + 3\sqrt{2}x - 5 = 0$

$a = 1,\ b = 3\sqrt{2},\ c = -5$

$$x = \frac{-b \pm \sqrt{b^2 - 4ac}}{2a}$$

$$x = \frac{-3\sqrt{2} \pm \sqrt{\left(3\sqrt{2}\right)^2 - 4(1)(-5)}}{2(1)}$$

$$= \frac{-3\sqrt{2} \pm \sqrt{18 + 20}}{2}$$

$$= \frac{-3\sqrt{2} \pm \sqrt{38}}{2}$$

The solutions are $\dfrac{-3\sqrt{2} \pm \sqrt{38}}{2}$.

**85.** answers may vary

**87.** $7.3z^2 + 5.4z - 1.1 = 0$

$a = 7.3,\ b = 5.4,\ c = -1.1$

$$z = \frac{-b \pm \sqrt{b^2 - 4ac}}{2a}$$

$$z = \frac{-5.4 \pm \sqrt{5.4^2 - 4(7.3)(-1.1)}}{2(7.3)}$$

$$= \frac{-5.4 \pm \sqrt{29.16 + 32.12}}{14.6}$$

$$= \frac{-5.4 \pm \sqrt{61.28}}{14.6}$$

$$z = \frac{-5.4 + \sqrt{61.28}}{14.6} \approx 0.2$$

$$z = \frac{-5.4 - \sqrt{61.28}}{14.6} \approx -0.9$$

The solutions are approximately $-0.9$ and $0.2$.

**89.** Let $h = 30$.

$$y = -16t^2 + 120t + 80$$

$$30 = -16t^2 + 120t + 80$$

$$0 = -16t^2 + 120t + 50$$

$a = -16,\ b = 120,\ \text{and } c = 50$

$$t = \frac{-120 \pm \sqrt{120^2 - 4(-16)(50)}}{2(-16)}$$

$$= \frac{-120 \pm \sqrt{14,400 + 3200}}{-32}$$

$$= \frac{-120 \pm \sqrt{17,600}}{-32}$$

Since the time cannot be negative,

$$t = \frac{-120 - \sqrt{17,600}}{-32} \approx 7.9.$$

The rocket will be 30 feet from the ground after 7.9 seconds.

**91.** $y = 209x^2 - 1917x + 57,131$

Let $y = 71,257$.

$$71,257 = 209x^2 - 1917x + 57,131$$

$$0 = 209x^2 - 1917x - 14,126$$

$a = 209,\ b = -1917,\ c = -14,126$

$$x = \frac{-(-1917) \pm \sqrt{(-1917)^2 - 4(209)(-14,126)}}{2(209)}$$

$$x = \frac{1917 \pm 3935}{418}$$

$$x = \frac{1917 + 3935}{418} = 14 \text{ or } x = \frac{1917 - 3935}{418} \approx -5$$

Since this is a prediction for the future, $x = 14$ and the year is 2014.

**Integrated Review Practice**

**1.** $y^2 - 3y - 4 = 0$

$(y - 4)(y + 1) = 0$

$y - 4 = 0 \quad \text{or} \quad y + 1 = 0$

$y = 4 \qquad\qquad y = -1$

The solutions are 4 and $-1$.

**2.** 
$$(2x+5)^2 = 45$$
$$2x+5 = \pm\sqrt{45}$$
$$2x+5 = \pm 3\sqrt{5}$$
$$2x = -5 \pm 3\sqrt{5}$$
$$x = \frac{-5 \pm 3\sqrt{5}}{2}$$

The solutions are $\dfrac{-5 \pm 3\sqrt{5}}{2}$.

**3.** 
$$x^2 - \frac{5}{2}x = -\frac{3}{2}$$
$$x^2 - \frac{5}{2}x + \frac{3}{2} = 0$$
$$2x^2 - 5x + 3 = 0$$
$$(2x-3)(x-1) = 0$$
$$2x - 3 = 0 \quad \text{or} \quad x - 1 = 0$$
$$2x = 3 \qquad\qquad x = 1$$
$$x = \frac{3}{2}$$

The solutions are $\dfrac{3}{2}$ and 1.

**Integrated Review**

**1.** 
$$5x^2 - 11x + 2 = 0$$
$$(5x-1)(x-2) = 0$$
$$5x - 1 = 0 \quad \text{or} \quad x - 2 = 0$$
$$5x = 1 \qquad\qquad x = 2$$
$$x = \frac{1}{5}$$

The solutions are $\dfrac{1}{5}$ and 2.

**2.** 
$$5x^2 + 13x - 6 = 0$$
$$(5x-2)(x+3) = 0$$
$$5x - 2 = 0 \quad \text{or} \quad x + 3 = 0$$
$$5x = 2 \qquad\qquad x = -3$$
$$x = \frac{2}{5}$$

The solutions are $\dfrac{2}{5}$ and $-3$.

**3.** 
$$x^2 - 1 = 2x$$
$$x^2 - 2x = 1$$
$$x^2 - 2x + 1 = 1 + 1$$
$$(x-1)^2 = 2$$
$$x - 1 = \pm\sqrt{2}$$
$$x = 1 \pm \sqrt{2}$$

The solutions are $1 \pm \sqrt{2}$.

**4.** 
$$x^2 + 7 = 6x$$
$$x^2 - 6x = -7$$
$$x^2 - 6x + 9 = -7 + 9$$
$$(x-3)^2 = 2$$
$$x - 3 = \pm\sqrt{2}$$
$$x = 3 \pm \sqrt{2}$$

The solutions are $3 \pm \sqrt{2}$.

**5.** 
$$a^2 = 20$$
$$a = \pm\sqrt{20}$$
$$= \pm 2\sqrt{5}$$

The solutions are $\pm 2\sqrt{5}$.

**6.** 
$$a^2 = 72$$
$$a = \pm\sqrt{72}$$
$$= \pm 6\sqrt{2}$$

The solutions are $\pm 6\sqrt{2}$.

**7.** 
$$x^2 - x + 4 = 0$$
$$x^2 - x = -4$$
$$x^2 - x + \frac{1}{4} = -4 + \frac{1}{4}$$
$$\left(x - \frac{1}{2}\right)^2 = -\frac{15}{4}$$

There is no real solution.

**8.** 
$$x^2 - 2x + 7 = 0$$
$$x^2 - 2x = -7$$
$$x^2 - 2x + 1 = -7 + 1$$
$$(x-1)^2 = -6$$

There is no real solution.

**9.** $3x^2 - 12x + 12 = 0$

$x^2 - 4x + 4 = 0$

$(x-2)^2 = 0$

$x - 2 = 0$

$x = 2$

The solution is 2.

**10.** $5x^2 - 30x + 45 = 0$

$x^2 - 6x + 9 = 0$

$(x-3)^2 = 0$

$x - 3 = 0$

$x = 3$

The solution is 3.

**11.** $9 - 6p + p^2 = 0$

$(p-3)^2 = 0$

$p - 3 = 0$

$p = 3$

The solution is 3.

**12.** $49 - 28p + 4p^2 = 0$

$(2p-7)^2 = 0$

$2p - 7 = 0$

$2p = 7$

$p = \dfrac{7}{2}$

The solution is $\dfrac{7}{2}$.

**13.** $4y^2 - 16 = 0$

$4y^2 = 16$

$y^2 = 4$

$y = \pm\sqrt{4}$

$y = \pm 2$

The solutions are $\pm 2$.

**14.** $3y^2 - 27 = 0$

$3y^2 = 27$

$y^2 = 9$

$y = \pm\sqrt{9}$

$y = \pm 3$

The solutions are $\pm 3$.

**15.** $x^4 - 3x^3 + 2x^2 = 0$

$x^2(x^2 - 3x + 2) = 0$

$x^2(x-1)(x-2) = 0$

$x^2 = 0$ or $x - 1 = 0$ or $x - 2 = 0$

$x = 0$ $\qquad x = 1$ $\qquad x = 2$

The solutions are 0, 1, and 2.

**16.** $x^3 + 7x^2 + 12x = 0$

$x(x^2 + 7x + 12) = 0$

$x(x+4)(x+3) = 0$

$x = 0$ or $x + 4 = 0$ or $x + 3 = 0$

$\qquad\qquad x = -4 \qquad\quad x = -3$

The solutions are $-4$, $-3$, and 0.

**17.** $(2z+5)^2 = 25$

$2z + 5 = \pm\sqrt{25}$

$2z = -5 \pm 5$

$z = \dfrac{-5 \pm 5}{2}$

$z = \dfrac{-5-5}{2} = -5$ or $z = \dfrac{-5+5}{2} = 0$

The solutions are 0 and $-5$.

**18.** $(3z-4)^2 = 16$

$3z - 4 = \pm\sqrt{16}$

$3z = 4 \pm 4$

$z = \dfrac{4 \pm 4}{3}$

$z = \dfrac{4-4}{3} = 0$ or $z = \dfrac{4+4}{3} = \dfrac{8}{3}$

The solutions are 0 and $\dfrac{8}{3}$.

**19.** $30x = 25x^2 + 2$

$0 = 25x^2 - 30x + 2 = 0$

$a = 25$, $b = -30$, and $c = 2$

$$x = \frac{-(-30) \pm \sqrt{(-30)^2 - 4(25)(2)}}{2(25)}$$

$$= \frac{30 \pm \sqrt{900 - 200}}{50}$$

$$= \frac{30 \pm \sqrt{700}}{50}$$

$$= \frac{30 \pm 10\sqrt{7}}{50}$$

$$= \frac{3 \pm \sqrt{7}}{5}$$

The solutions are $\frac{3 \pm \sqrt{7}}{5}$.

**20.** $12x = 4x^2 + 4$

$0 = 4x^2 - 12x + 4$

$0 = x^2 - 3x + 1$

$a = 1$, $b = -3$, and $c = 1$

$$x = \frac{-(-3) \pm \sqrt{(-3)^2 - 4(1)(1)}}{2(1)}$$

$$= \frac{3 \pm \sqrt{9 - 4}}{2}$$

$$= \frac{3 \pm \sqrt{5}}{2}$$

The solutions are $\frac{3 \pm \sqrt{5}}{2}$.

**21.** $\frac{2}{3}m^2 - \frac{1}{3}m - 1 = 0$

$2m^2 - m - 3 = 0$

$(2m - 3)(m + 1) = 0$

$2m - 3 = 0$  or  $m + 1 = 0$

$2m = 3$        $m = -1$

$m = \frac{3}{2}$

The solutions are $-1$ and $\frac{3}{2}$.

**22.** $\frac{5}{8}m^2 + m - \frac{1}{2} = 0$

$5m^2 + 8m - 4 = 0$

$(5m - 2)(m + 2) = 0$

$5m - 2 = 0$  or  $m + 2 = 0$

$5m = 2$        $m = -2$

$m = \frac{2}{5}$

The solutions are $-2$ and $\frac{2}{5}$.

**23.** $x^2 - \frac{1}{2}x - \frac{1}{5} = 0$

$10x^2 - 5x - 2 = 0$

$a = 10$, $b = -5$, and $c = -2$

$$x = \frac{-(-5) \pm \sqrt{(-5)^2 - 4(10)(-2)}}{2(10)}$$

$$= \frac{5 \pm \sqrt{25 + 80}}{20}$$

$$= \frac{5 \pm \sqrt{105}}{20}$$

The solutions are $\frac{5 \pm \sqrt{105}}{20}$.

**24.** $x^2 + \frac{1}{2}x - \frac{1}{8} = 0$

$8x^2 + 4x - 1 = 0$

$a = 8$, $b = 4$, and $c = -1$

$$x = \frac{-4 \pm \sqrt{4^2 - 4(8)(-1)}}{2(8)}$$

$$= \frac{-4 \pm \sqrt{16 + 32}}{16}$$

$$= \frac{-4 \pm \sqrt{48}}{16}$$

$$= \frac{-4 \pm 4\sqrt{3}}{16}$$

$$= \frac{-1 \pm \sqrt{3}}{4}$$

The solutions are $\frac{-1 \pm \sqrt{3}}{4}$.

**25.** $4x^2 - 27x + 35 = 0$

$(4x - 7)(x - 5) = 0$

$4x - 7 = 0$ or $x - 5 = 0$

$4x = 7$ $\qquad x = 5$

$x = \dfrac{7}{4}$

The solutions are $\dfrac{7}{4}$ and 5.

**26.** $9x^2 - 16x + 7 = 0$

$(9x - 7)(x - 1) = 0$

$9x - 7 = 0$ or $x - 1 = 0$

$9x = 7$ $\qquad x = 1$

$x = \dfrac{7}{9}$

The solutions are $\dfrac{7}{9}$ and 1.

**27.** $(7 - 5x)^2 = 18$

$7 - 5x = \pm\sqrt{18}$

$7 - 5x = \pm 3\sqrt{2}$

$-5x = -7 \pm 3\sqrt{2}$

$\dfrac{-5x}{-5} = \dfrac{-7 \pm 3\sqrt{2}}{-5}$

$x = \dfrac{7 \pm 3\sqrt{2}}{5}$

The solutions are $\dfrac{7 \pm 3\sqrt{2}}{5}$.

**28.** $(5 - 4x)^2 = 75$

$5 - 4x = \pm\sqrt{75}$

$5 - 4x = \pm 5\sqrt{3}$

$-4x = -5 \pm 5\sqrt{3}$

$\dfrac{-4x}{-4} = \dfrac{-5 \pm 5\sqrt{3}}{-4}$

$x = \dfrac{5 \pm 5\sqrt{3}}{4}$

The solutions are $\dfrac{5 \pm 5\sqrt{3}}{4}$.

**29.** $3z^2 - 7z = 12$

$3z^2 - 7z - 12 = 0$

$a = 3$, $b = -7$, and $c = -12$

$z = \dfrac{-(-7) \pm \sqrt{(-7)^2 - 4(3)(-12)}}{2(3)}$

$= \dfrac{7 \pm \sqrt{49 + 144}}{6}$

$= \dfrac{7 \pm \sqrt{193}}{6}$

The solutions are $\dfrac{7 \pm \sqrt{193}}{6}$.

**30.** $6z^2 + 7z = 6$

$6z^2 + 7z - 6 = 0$

$a = 6$, $b = 7$, and $c = -6$

$z = \dfrac{-7 \pm \sqrt{7^2 - 4(6)(-6)}}{2(6)}$

$= \dfrac{-7 \pm \sqrt{49 + 144}}{12}$

$= \dfrac{-7 \pm \sqrt{193}}{12}$

The solutions are $\dfrac{-7 \pm \sqrt{193}}{12}$.

**31.** $x = x^2 - 110$

$0 = x^2 - x - 110$

$0 = (x + 10)(x - 11)$

$x + 10 = 0$ or $x - 11 = 0$

$x = -10$ $\qquad x = 11$

The solutions are $-10$ and 11.

**32.** $x = 56 - x^2$

$x^2 + x - 56 = 0$

$(x + 8)(x - 7) = 0$

$x + 8 = 0$ or $x - 7 = 0$

$x = -8$ $\qquad x = 7$

The solutions are $-8$ and 7.

**33.** $\dfrac{3}{4}x^2 - \dfrac{5}{2}x - 2 = 0$

$3x^2 - 10x - 8 = 0$

$(3x + 2)(x - 4) = 0$

$3x + 2 = 0$ or $x - 4 = 0$

$3x = -2$ $\qquad x = 4$

$x = -\dfrac{2}{3}$

The solutions are $-\dfrac{2}{3}$ and 4.

**34.**
$$x^2 - \frac{6}{5}x - \frac{8}{5} = 0$$
$$5x^2 - 6x - 8 = 0$$
$$(5x + 4)(x - 2) = 0$$
$$5x + 4 = 0 \quad \text{or} \quad x - 2 = 0$$
$$5x = -4 \qquad\qquad x = 2$$
$$x = -\frac{4}{5}$$

The solutions are $-\dfrac{4}{5}$ and 2.

**35.**
$$x^2 - 0.6x + 0.05 = 0$$
$$100x^2 - 60x + 5 = 0$$
$$20x^2 - 12x + 1 = 0$$
$$(10x - 1)(2x - 1) = 0$$
$$10x - 1 = 0 \quad \text{or} \quad 2x - 1 = 0$$
$$10x = 1 \qquad\qquad 2x = 1$$
$$x = \frac{1}{10} = 0.1 \qquad x = \frac{1}{2} = 0.5$$
The solutions are 0.1 and 0.5.

**36.**
$$x^2 - 0.1x + 0.06 = 0$$
$$100x^2 - 10x + 6 = 0$$
$$50x^2 - 5x + 3 = 0$$
$$(5x + 1)(10x - 3) = 0$$
$$5x + 1 = 0 \quad \text{or} \quad 10x - 3 = 0$$
$$5x = -1 \qquad\qquad 10x = 3$$
$$x = -\frac{1}{5} = -0.2 \qquad x = \frac{3}{10} = 0.3$$
The solutions are $-0.2$ and $0.3$.

**37.** $10x^2 - 11x + 2 = 0$
$a = 10$, $b = -11$, and $c = 2$
$$x = \frac{-(-11) \pm \sqrt{(-11)^2 - 4(10)(2)}}{2(10)}$$
$$= \frac{11 \pm \sqrt{121 - 80}}{20}$$
$$= \frac{11 \pm \sqrt{41}}{20}$$

The solutions are $\dfrac{11 \pm \sqrt{41}}{20}$.

**38.** $20x^2 - 11x + 1 = 0$
$a = 20$, $b = -11$, and $c = 1$
$$x = \frac{-(-11) \pm \sqrt{(-11)^2 - 4(20)(1)}}{2(20)}$$
$$= \frac{11 \pm \sqrt{121 - 80}}{40}$$
$$= \frac{11 \pm \sqrt{41}}{40}$$
The solutions are $\dfrac{11 \pm \sqrt{41}}{40}$.

**39.**
$$\frac{1}{2}z^2 - 2z + \frac{3}{4} = 0$$
$$2z^2 - 8z + 3 = 0$$
$a = 2$, $b = -8$, and $c = 3$
$$z = \frac{-(-8) \pm \sqrt{(-8)^2 - 4(2)(3)}}{2(2)}$$
$$= \frac{8 \pm \sqrt{64 - 24}}{4}$$
$$= \frac{8 \pm \sqrt{40}}{4}$$
$$= \frac{8 \pm 2\sqrt{10}}{4}$$
$$= \frac{4 \pm \sqrt{10}}{2}$$

The solutions are $\dfrac{4 \pm \sqrt{10}}{2}$.

**40.**
$$\frac{1}{5}z^2 - \frac{1}{2}z - 2 = 0$$
$$2z^2 - 5z - 20 = 0$$
$a = 2$, $b = -5$, and $c = -20$
$$z = \frac{-(-5) \pm \sqrt{(-5)^2 - 4(2)(-20)}}{2(2)}$$
$$= \frac{5 \pm \sqrt{25 + 160}}{4}$$
$$= \frac{5 \pm \sqrt{185}}{4}$$
The solutions are $\dfrac{5 \pm \sqrt{185}}{4}$.

**41.** answers may vary

**Section 9.4 Practice**

**1. a.** $\sqrt{-36} = \sqrt{-1 \cdot 36} = \sqrt{-1} \cdot \sqrt{36} = i \cdot 6 = 6i$

   **b.** $\sqrt{-15} = \sqrt{-1 \cdot 15} = \sqrt{-1} \cdot \sqrt{15} = i\sqrt{15}$

   **c.** $\sqrt{-48} = \sqrt{-1 \cdot 48}$
$$= \sqrt{-1} \cdot \sqrt{48}$$
$$= i \cdot 4\sqrt{3}$$
$$= 4i\sqrt{3}$$

**2. a.** $6 = 6 + 0i$

   **b.** $0 = 0 + 0i$

   **c.** $\sqrt{24} = 2\sqrt{6} = 2\sqrt{6} + 0i$

   **d.** $\sqrt{-1} = i = 0 + i$

   **e.** $5 + \sqrt{-9} = 5 + \sqrt{-1 \cdot 9}$
$$= 5 + \sqrt{-1} \cdot \sqrt{9}$$
$$= 5 + i \cdot 3$$
$$= 5 + 3i$$

**3. a.** $(4 + 3i) + (-8 - 2i) = [4 + (-8)] + (3i - 2i)$
$$= -4 + i$$

   **b.** $5i + (6 - 9i) = 6 + (5i - 9i) = 6 - 4i$

   **c.** $(3 - 2i) - 4 = (3 - 4) - 2i = -1 - 2i$

**4.** $3i - (13 - 5i) = 3i - 13 + 5i = -13 + 8i$

**5. a.** $2i(3 - 4i) = 2i(3) + 2i(-4i)$
$$= 6i - 8i^2$$
$$= 6i - 8(-1)$$
$$= 6i + 8$$
$$= 8 + 6i$$

   **b.** $(3 + i)(2 - 3i) = 6 - 9i + 2i - 3i^2$
$$= 6 - 7i - 3(-1)$$
$$= 6 - 7i + 3$$
$$= 9 - 7i$$

   **c.** $(5 - 2i)(5 + 2i) = 25 + 10i - 10i - 4i^2$
$$= 25 - 4(-1)$$
$$= 25 + 4$$
$$= 29 + 0i$$

**6.** $\dfrac{3 - i}{2 + 5i} = \dfrac{(3 - i)}{(2 + 5i)} \cdot \dfrac{(2 - 5i)}{(2 - 5i)}$
$$= \frac{6 - 15i - 2i + 5i^2}{4 - 25i^2}$$
$$= \frac{6 - 17i + 5(-1)}{4 - 25(-1)}$$
$$= \frac{6 - 17i - 5}{4 + 25}$$
$$= \frac{1 - 17i}{29}$$
$$= \frac{1}{29} - \frac{17}{29}i$$

**7.** $(x - 3)^2 = -16$
$$x - 3 = \pm\sqrt{-16}$$
$$x - 3 = \pm 4i$$
$$x = 3 \pm 4i$$
The solutions are $3 \pm 4i$.

**8.** $y^2 = 3y - 5$
$$y^2 - 3y + 5 = 0$$
$$a = 1, b = -3, c = 5$$
$$y = \frac{-(-3) \pm \sqrt{(-3)^2 - 4(1)(5)}}{2(1)}$$
$$= \frac{3 \pm \sqrt{9 - 20}}{2}$$
$$= \frac{3 \pm \sqrt{-11}}{2}$$
$$= \frac{3 \pm i\sqrt{11}}{2}$$
The solutions are $\dfrac{3 \pm i\sqrt{11}}{2}$.

**9.** $x^2 + x = -4$
$$x^2 + x + 4 = 0$$
$$a = 1, b = 1, c = 4$$
$$x = \frac{-1 \pm \sqrt{1^2 - 4(1)(4)}}{2(1)}$$
$$= \frac{-1 \pm \sqrt{1 - 16}}{2}$$
$$= \frac{-1 \pm \sqrt{-15}}{2}$$
$$= \frac{-1 \pm i\sqrt{15}}{2}$$
The solutions are $\dfrac{-1 \pm i\sqrt{15}}{2}$.

**Vocabulary, Readiness & Video Check 9.4**

1. A number that can be written in the form $a + bi$ is called a <u>complex</u> number.

2. A complex number that can be written in the form $0 + bi$ is also called a <u>pure imaginary</u> number.

3. A complex number that can be written in the form $a + 0i$ is also called a <u>real</u> number.

4. The <u>conjugate</u> of $a + bi$ is $a - bi$.

5. The form $a + bi$ is called <u>standard</u> form.

6. There are no other differences. The rules and processes for radicals still apply.

7. combining like terms; $i$ is *not* a variable, but a constant, $\sqrt{-1}$

8. The fact that $i^2 = -1$.

9. To write a quotient of complex numbers in standard form $a + bi$, there should be no $i$'s, but only real numbers in the denominator.

10. The discriminant is negative, so the equation has complex, but no real, solutions.

**Exercise Set 9.4**

1. $\sqrt{-9} = \sqrt{-1 \cdot 9} = \sqrt{-1}\sqrt{9} = i \cdot 3 = 3i$

3. $\sqrt{-100} = \sqrt{-1 \cdot 100} = \sqrt{-1}\sqrt{100} = i \cdot 10 = 10i$

5. $\sqrt{-50} = \sqrt{-1 \cdot 25 \cdot 2}$
$= \sqrt{-1}\sqrt{25}\sqrt{2}$
$= i \cdot 5\sqrt{2}$
$= 5i\sqrt{2}$

7. $\sqrt{-63} = \sqrt{-1 \cdot 9 \cdot 7} = \sqrt{-1}\sqrt{9}\sqrt{7} = i \cdot 3\sqrt{7} = 3i\sqrt{7}$

9. $(2-i) + (-5+10i) = 2-5+(-i+10i) = -3+9i$

11. $(-11+3i) - (1-3i) = -11+3i-1+3i$
$= -11-1+(3i+3i)$
$= -12+6i$

13. $(3-4i) - (2-i) = 3-4i-2+i$
$= 3-2+(-4i+i)$
$= 1-3i$

15. $(16+2i) + (-7-6i) = 16+2i-7-6i$
$= 16-7+(2i-6i)$
$= 9-4i$

17. $4i(3-2i) = 12i-8i^2$
$= 12i-8(-1)$
$= 12i+8$
$= 8+12i$

19. $-5i(10+i) = -50i-5i^2$
$= -50i-5(-1)$
$= -50i+5$
$= 5-50i$

21. $(6-2i)(4+i) = 24+6i-8i-2i^2$
$= 24-2i-2(-1)$
$= 24-2i+2$
$= 26-2i$

23. $(3+8i)(3-8i) = 3^2-(8i)^2$
$= 9-64i^2$
$= 9-64(-1)$
$= 9+64$
$= 73$

25. $\dfrac{8-12i}{4} = \dfrac{4(2-3i)}{4} = 2-3i$

27. $\dfrac{7-i}{4-3i} = \dfrac{(7-i)}{(4-3i)} \cdot \dfrac{(4+3i)}{(4+3i)}$
$= \dfrac{28+21i-4i-3i^2}{16-9i^2}$
$= \dfrac{28+17i-3(-1)}{16-9(-1)}$
$= \dfrac{28+17i+3}{16+9}$
$= \dfrac{31+17i}{25}$
$= \dfrac{31}{25} + \dfrac{17}{25}i$

29. $(x+1)^2 = -9$
$x+1 = \pm\sqrt{-9}$
$x+1 = \pm 3i$
$x = -1 \pm 3i$
The solutions are $-1 \pm 3i$.

**31.** $(2z-3)^2 = -12$

$2z - 3 = \pm\sqrt{-12}$

$2z - 3 = \pm\sqrt{-1}\sqrt{4}\sqrt{3}$

$2z - 3 = \pm 2i\sqrt{3}$

$2z = 3 \pm 2i\sqrt{3}$

$z = \dfrac{3 \pm 2i\sqrt{3}}{2}$

The solutions are $\dfrac{3 \pm 2i\sqrt{3}}{2}$.

**33.** $y^2 + 6y + 13 = 0$

$a = 1, b = 6, c = 13$

$y = \dfrac{-6 \pm \sqrt{6^2 - 4(1)(13)}}{2(1)}$

$= \dfrac{-6 \pm \sqrt{36 - 52}}{2}$

$= \dfrac{-6 \pm \sqrt{-16}}{2}$

$= \dfrac{-6 \pm 4i}{2}$

$= \dfrac{2(-3 \pm 2i)}{2}$

$= -3 \pm 2i$

The solutions are $-3 \pm 2i$.

**35.** $4x^2 + 7x + 4 = 0$

$a = 4, b = 7, c = 4$

$x = \dfrac{-7 \pm \sqrt{7^2 - 4(4)(4)}}{2(4)}$

$= \dfrac{-7 \pm \sqrt{49 - 64}}{8}$

$= \dfrac{-7 \pm \sqrt{-15}}{8}$

$= \dfrac{-7 \pm i\sqrt{15}}{8}$

The solutions are $\dfrac{-7 \pm i\sqrt{15}}{8}$.

**37.** $2m^2 - 4m + 5 = 0$

$a = 2, b = -4, c = 5$

$m = \dfrac{-(-4) \pm \sqrt{(-4)^2 - 4(2)(5)}}{2(2)}$

$= \dfrac{4 \pm \sqrt{16 - 40}}{4}$

$= \dfrac{4 \pm \sqrt{-24}}{4}$

$= \dfrac{4 \pm \sqrt{-1 \cdot 4 \cdot 6}}{4}$

$= \dfrac{4 \pm 2i\sqrt{6}}{4}$

$= \dfrac{2\left(2 \pm i\sqrt{6}\right)}{4}$

$= \dfrac{2 \pm i\sqrt{6}}{2}$

The solutions are $\dfrac{2 \pm i\sqrt{6}}{2}$.

**39.** $3 + (12 - 7i) = 3 + 12 - 7i = 15 - 7i$

**41.** $-9i(5i - 7) = -45i^2 + 63i$

$= -45(-1) + 63i$

$= 45 + 63i$

**43.** $(2 - i) - (3 - 4i) = 2 - i - 3 + 4i$

$= 2 - 3 + (-i + 4i)$

$= -1 + 3i$

**45.** $\dfrac{15 + 10i}{5i} = \dfrac{(15 + 10i)}{5i} \cdot \dfrac{(-i)}{(-i)}$

$= \dfrac{-15i - 10i^2}{-5i^2}$

$= \dfrac{-15i - 10(-1)}{-5(-1)}$

$= \dfrac{-15i + 10}{5}$

$= \dfrac{5(-3i + 2)}{5}$

$= -3i + 2$

$= 2 - 3i$

**47.** $-5 + i - (2 + 3i) = -5 + i - 2 - 3i$

$= -5 - 2 + (i - 3i)$

$= -7 - 2i$

**49.** $(4-3i)(4+3i) = (4)^2 - (3i)^2$
$= 16 - 9i^2$
$= 16 - 9(-1)$
$= 16 + 9$
$= 25$

**51.** $\dfrac{4-i}{1+2i} = \dfrac{(4-i)(1-2i)}{(1+2i)(1-2i)}$
$= \dfrac{4-8i-i+2i^2}{(1)^2-(2i)^2}$
$= \dfrac{4-9i+2(-1)}{1-4i^2}$
$= \dfrac{4-9i-2}{1-4(-1)}$
$= \dfrac{2-9i}{5}$
$= \dfrac{2}{5} - \dfrac{9}{5}i$

**53.** $(5+2i)^2 = 5^2 + 2(5)(2i) + (2i)^2$
$= 25 + 20i + 4i^2$
$= 25 + 20i + 4(-1)$
$= 25 + 20i - 4$
$= 21 + 20i$

**55.** $(y-4)^2 = -64$
$y-4 = \pm\sqrt{-64}$
$y-4 = \pm 8i$
$y = 4 \pm 8i$
The solutions are $4 \pm 8i$.

**57.** $4x^2 = -100$
$x^2 = -25$
$x = \pm\sqrt{-25}$
$x = \pm 5i$
The solutions are $\pm 5i$.

**59.** $z^2 + 6z + 10 = 0$
$a = 1, b = 6, c = 10$
$z = \dfrac{-6 \pm \sqrt{6^2 - 4(1)(10)}}{2(1)}$
$= \dfrac{-6 \pm \sqrt{36-40}}{2}$
$= \dfrac{-6 \pm \sqrt{-4}}{2}$
$= \dfrac{-6 \pm 2i}{2}$
$= \dfrac{2(-3 \pm i)}{2}$
$= -3 \pm i$
The solutions are $-3 \pm i$.

**61.** $2a^2 - 5a + 9 = 0$
$a = 2, b = -5, c = 9$
$a = \dfrac{-(-5) \pm \sqrt{(-5)^2 - 4(2)(9)}}{2(2)}$
$= \dfrac{5 \pm \sqrt{25-72}}{4}$
$= \dfrac{5 \pm \sqrt{-47}}{4}$
$= \dfrac{5 \pm i\sqrt{47}}{4}$
The solutions are $\dfrac{5 \pm i\sqrt{47}}{4}$.

**63.** $(2x+8)^2 = -20$
$2x+8 = \pm\sqrt{-20}$
$2x+8 = \pm 2i\sqrt{5}$
$2x = -8 \pm 2i\sqrt{5}$
$x = \dfrac{-8 \pm 2i\sqrt{5}}{2}$
$x = \dfrac{2(-4 \pm i\sqrt{5})}{2}$
$x = -4 \pm i\sqrt{5}$
The solutions are $-4 \pm i\sqrt{5}$.

**65.** $3m^2 + 108 = 0$
$3m^2 = -108$
$m^2 = -36$
$m = \pm\sqrt{-36}$
$m = \pm 6i$
The solutions are $\pm 6i$.

**67.** $x^2 + 14x + 50 = 0$

$a = 1, b = 14, c = 50$

$$x = \frac{-14 \pm \sqrt{14^2 - 4(1)(50)}}{2(1)}$$

$$= \frac{-14 \pm \sqrt{196 - 200}}{2}$$

$$= \frac{-14 \pm \sqrt{-4}}{2}$$

$$= \frac{-14 \pm 2i}{2}$$

$$= \frac{2(-7 \pm i)}{2}$$

$$= -7 \pm i$$

The solutions are $-7 \pm i$.

**69.** $y = -3$

$y = -3$ for all values of $x$.

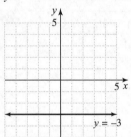

**71.** $y = 3x - 2$

| $x$ | $y$ |
|-----|-----|
| 0 | $-2$ |
| 1 | 1 |

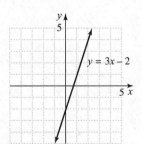

**73.** True

**75.** True

**77.** answers may vary

**Section 9.5 Practice**

**1.** $y = -\dfrac{1}{2}x^2$

| $x$ | $y$ |
|-----|-----|
| 0 | 0 |
| 1 | $-\dfrac{1}{2}$ |
| 2 | $-2$ |
| 4 | $-8$ |
| $-1$ | $-\dfrac{1}{2}$ |
| $-2$ | $-2$ |
| $-4$ | $-8$ |

**2.** $y = x^2 + 1$

$y$-intercept: $x = 0, \ y = 0^2 + 1 = 1$

$x$-intercept: $y = 0$,

$\quad 0 = x^2 + 1$

$\quad -1 = x^2$

No $x$-intercept

| $x$ | $y$ |
|-----|-----|
| 0 | 1 |
| 1 | 2 |
| 2 | 5 |
| 3 | 10 |
| $-1$ | 2 |
| $-2$ | 5 |
| $-3$ | 10 |

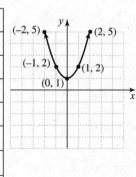

**3.** $y = x^2 - 3x - 4$

$a = 1, b = -3, c = -4$

Vertex:

$$x = \frac{-b}{2a} = \frac{-(-3)}{2(1)} = \frac{3}{2}$$

$$y = x^2 - 3x - 4$$
$$= \left(\frac{3}{2}\right)^2 - 3\left(\frac{3}{2}\right) - 4$$
$$= \frac{9}{4} - \frac{9}{2} - 4$$
$$= \frac{9}{4} - \frac{18}{4} - \frac{16}{4}$$
$$= -\frac{25}{4}$$

The vertex is $\left(\frac{3}{2}, -\frac{25}{4}\right)$.

$x$-intercepts: $y = 0$,
$$0 = x^2 - 3x - 4$$
$$0 = (x + 1)(x - 4)$$
$$x + 1 = 0 \quad \text{or} \quad x - 4 = 0$$
$$x = -1 \qquad x = 4$$

$y$-intercept: $x = 0$,
$$y = x^2 - 3x - 4 = 0^2 - 3(0) - 4 = -4$$

| $x$ | $y$ |
|-----|-----|
| $\frac{3}{2}$ | $-\frac{25}{4}$ |
| $-1$ | $0$ |
| $4$ | $0$ |
| $0$ | $-4$ |
| $1$ | $-6$ |
| $2$ | $-6$ |
| $3$ | $-4$ |

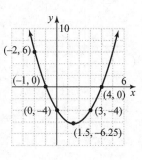

**4.** $y = x^2 + 4x - 7$

$a = 1$, $b = 4$, $c = -7$

vertex:
$$x = \frac{-b}{2a} = \frac{-4}{2(1)} = \frac{-4}{2} = -2$$

$$y = x^2 + 4x - 7$$
$$= (-2)^2 + 4(-2) - 7$$
$$= 4 - 8 - 7$$
$$= -11$$

The vertex is $(-2, -11)$.

$x$-intercepts: $y = 0$,
$$0 = x^2 + 4x - 7$$

$$x = \frac{-4 \pm \sqrt{4^2 - 4(1)(-7)}}{2(1)}$$
$$= \frac{-4 \pm \sqrt{16 + 28}}{2}$$
$$= \frac{-4 \pm \sqrt{44}}{2}$$
$$= \frac{-4 \pm 2\sqrt{11}}{2}$$
$$= -2 \pm \sqrt{11}$$

$y$-intercept: $x = 0$,
$$y = x^2 + 4x - 7 = 0^2 + 4(0) - 7 = -7$$

| $x$ | $y$ |
|-----|-----|
| $-2$ | $-11$ |
| $-2 + \sqrt{11}$ | $0$ |
| $-2 - \sqrt{11}$ | $0$ |
| $0$ | $-7$ |
| $1$ | $-2$ |
| $2$ | $5$ |
| $3$ | $14$ |

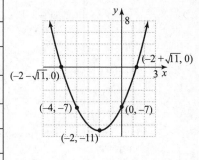

**Graphing Calculator Explorations**

**1.** $x^2 - 7x - 3 = 0$

$y_1 = x^2 - 7x - 3$

$y_2 = 0$

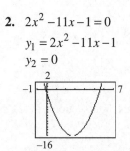

The $x$-coordinates of the intersections, $-0.41$ and $7.41$, are the solutions.

**2.** $2x^2 - 11x - 1 = 0$

$y_1 = 2x^2 - 11x - 1$

$y_2 = 0$

The $x$-coordinates of the intersections, $-0.09$ and $5.59$ are the solutions.

3. $-1.7x^2 + 5.6x - 3.7 = 0$

   $y_1 = -1.7x^2 + 5.6x - 3.7$
   $y_2 = 0$

   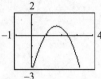

   The *x*-coordinates of the intersections, 0.91 and 2.38, are the solutions.

4. $-5.8x^2 + 2.3x - 3.9 = 0$

   $y_1 = -5.8x^2 + 2.3x - 3.9$
   $y_2 = 0$

   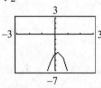

   There are no *x*-intercepts so there are no real solutions.

5. $5.8x^2 - 2.6x - 1.9 = 0$

   $y_1 = 5.8x^2 - 2.6x - 1.9$
   $y_2 = 0$

   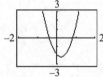

   The *x*-coordinates of the intersections, −0.39 and 0.84, are the solutions.

6. $7.5x^2 - 3.7x - 1.1 = 0$

   $y_1 = 7.5x^2 - 3.7x - 1.1$
   $y_2 = 0$

   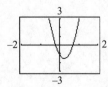

   The *x*-coordinates of the intersections, −0.21 and 0.70 are the solutions.

## Vocabulary, Readiness & Video Check 9.5

1. If a parabola opens upward, the lowest point is called the vertex; if a parabola opens downward, the highest point is called the vertex. If a graph can be folded along a line such that the two sides coincide or form mirror images of each other, we say the graph is symmetric with respect to that line and that line is the axis of symmetry.

2. The vertex; it is a very useful point to plot since it is the highest or lowest point on your graph.

3. For example, if the vertex is in quadrant III or IV and the parabola opens downward, then there won't be any *x*-intercepts, and there's no need to let $y = 0$ and solve the equation for *x*.

## Exercise Set 9.5

1. $y = 2x^2$

| x | y |
|----|----|
| −2 | 8 |
| −1 | 2 |
| 0 | 0 |
| 1 | 2 |
| 2 | 8 |

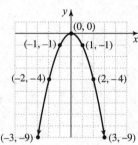

3. $y = -x^2$

| x | y |
|----|----|
| −2 | −4 |
| −1 | −1 |
| 0 | 0 |
| 1 | −1 |
| 2 | −4 |

　　　　Copyright © 2013 Pearson Education, Inc.

**5.** $y = x^2 - 1$

$y$-intercept: $x = 0$, $y = 0^2 - 1 = -1$, $(0, -1)$

vertex: $(0, -1)$

$x$-intercepts: $y = 0$,

$0 = x^2 - 1 = (x+1)(x-1)$

$x + 1 = 0$    or    $x - 1 = 0$

$x = -1$        $x = 1$

$(-1, 0)$ and $(1, 0)$

| $x$ | $y$ |
|-----|-----|
| $-2$ | $3$ |
| $-1$ | $0$ |
| $0$ | $-1$ |
| $1$ | $0$ |
| $2$ | $3$ |

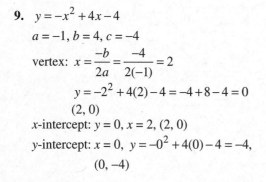

**7.** $y = x^2 + 4$

$y$-intercept: $x = 0$, $y = 0^2 + 4 = 4$, $(0, 4)$

vertex: $(0, 4)$

$x$-intercepts: $y = 0$,

$0 = x^2 + 4$

$-4 = x^2$

There are no $x$-intercepts because there is no real solution to this equation.

| $x$ | $y$ |
|-----|-----|
| $-2$ | $8$ |
| $-1$ | $5$ |
| $0$ | $4$ |
| $1$ | $1$ |
| $2$ | $8$ |

**9.** $y = -x^2 + 4x - 4$

$a = -1$, $b = 4$, $c = -4$

vertex: $x = \dfrac{-b}{2a} = \dfrac{-4}{2(-1)} = 2$

$y = -2^2 + 4(2) - 4 = -4 + 8 - 4 = 0$

$(2, 0)$

$x$-intercept: $y = 0$, $x = 2$, $(2, 0)$

$y$-intercept: $x = 0$, $y = -0^2 + 4(0) - 4 = -4$,

$(0, -4)$

| $x$ | $y$ |
|-----|-----|
| $2$ | $0$ |
| $0$ | $-4$ |
| $1$ | $-1$ |
| $3$ | $-1$ |
| $4$ | $-4$ |

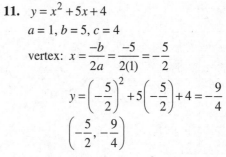

**11.** $y = x^2 + 5x + 4$

$a = 1$, $b = 5$, $c = 4$

vertex: $x = \dfrac{-b}{2a} = \dfrac{-5}{2(1)} = -\dfrac{5}{2}$

$y = \left(-\dfrac{5}{2}\right)^2 + 5\left(-\dfrac{5}{2}\right) + 4 = -\dfrac{9}{4}$

$\left(-\dfrac{5}{2}, -\dfrac{9}{4}\right)$

$y$-intercept: $x = 0$, $y = 0^2 + 5(0) + 4 = 4$, $(0, 4)$

$x$-intercepts: $y = 0$,

$0 = x^2 + 5x + 4$

$0 = (x+4)(x+1)$

$x = -4$    or    $x = -1$

$(-4, 0)$ and $(-1, 0)$

| $x$ | $y$ |
|-----|-----|
| $-5$ | $4$ |
| $-4$ | $0$ |
| $-\dfrac{5}{2}$ | $-\dfrac{9}{4}$ |
| $-1$ | $0$ |
| $0$ | $4$ |

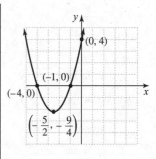

**13.** $y = x^2 - 4x + 5$

$a = 1$, $b = -4$, $c = 5$

vertex: $x = \dfrac{-b}{2a} = \dfrac{-(-4)}{2(1)} = \dfrac{4}{2} = 2$

$y = 2^2 - 4(2) + 5 = 4 - 8 + 5 = 1$

$(2, 1)$

$x$-intercept: $y = 0$, $0 = x^2 - 4x + 5$

$$x = \frac{-(-4) \pm \sqrt{(-4)^2 - 4(1)(5)}}{2(1)}$$

$$= \frac{4 \pm \sqrt{16 - 20}}{2}$$

$$= \frac{4 \pm \sqrt{-4}}{2}$$

There are no $x$-intercepts.

$y$-intercept: $x = 0$, $y = 0^2 - 4(0) + 5 = 5$, $(0, 5)$

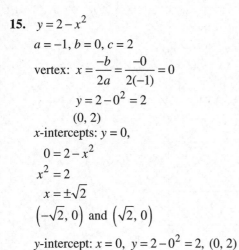

| $x$ | $y$ |
|-----|-----|
| 2 | 1 |
| 0 | 5 |
| 1 | 2 |
| −1 | 10 |
| 3 | 2 |

**15.** $y = 2 - x^2$

$a = -1$, $b = 0$, $c = 2$

vertex: $x = \dfrac{-b}{2a} = \dfrac{-0}{2(-1)} = 0$

$y = 2 - 0^2 = 2$

$(0, 2)$

$x$-intercepts: $y = 0$,

$0 = 2 - x^2$

$x^2 = 2$

$x = \pm\sqrt{2}$

$\left(-\sqrt{2}, 0\right)$ and $\left(\sqrt{2}, 0\right)$

$y$-intercept: $x = 0$, $y = 2 - 0^2 = 2$, $(0, 2)$

| $x$ | $y$ |
|-----|-----|
| 0 | 2 |
| $\sqrt{2}$ | 0 |
| $-\sqrt{2}$ | 0 |
| 0 | 2 |
| 1 | 1 |

**17.** $y = \dfrac{1}{3}x^2$

vertex: $(0, 0)$

$y$-intercept: $x = 0$, $y = \dfrac{1}{3} \cdot 0^2 = 0$, $(0, 0)$

$x$-intercepts: $y = 0$

$0 = \dfrac{1}{3}x^2$

$0 = x^2$

$0 = x$

$(0, 0)$

| $x$ | $y$ |
|-----|-----|
| −6 | 12 |
| −3 | 3 |
| 0 | 0 |
| 3 | 3 |
| 6 | 12 |

**19.** $y = x^2 + 6x$

$a = 1$, $b = 6$, $c = 0$

vertex: $x = \dfrac{-b}{2a} = \dfrac{-6}{2(1)} = -3$

$y = (-3)^2 + 6(-3) = -9$

$(-3, -9)$

$y$-intercept: $x = 0$, $y = 0^2 + 6(0) = 0$, $(0, 0)$

$x$-intercepts: $y = 0$,

$0 = x^2 + 6x$

$0 = x(x + 6)$

$x = -6$ or $x = 0$

$(-6, 0)$ and $(0, 0)$

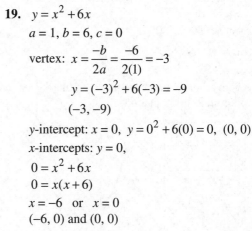

| $x$ | $y$ |
|-----|-----|
| −7 | 7 |
| −6 | 0 |
| −3 | −9 |
| 0 | 0 |
| 1 | 7 |

**21.** $y = x^2 + 2x - 8$

$a = 1, b = 2, c = -8$

vertex: $x = \dfrac{-b}{2a} = \dfrac{-2}{2(1)} = -1$

$\qquad y = (-1)^2 + 2(-1) - 8 = -9$

$\qquad (-1, -9)$

$y$-intercept: $x = 0$, $y = 0^2 + 2(0) - 8 = -8, (0, -8)$

$x$-intercepts: $y = 0$,

$0 = x^2 + 2x - 8$

$0 = (x+4)(x-2)$

$x = -4 \quad$ or $\quad x = 2$

$(-4, 0)$ and $(2, 0)$

| $x$ | $y$ |
|-----|-----|
| $-4$ | $0$ |
| $-2$ | $-8$ |
| $-1$ | $-9$ |
| $0$ | $-8$ |
| $2$ | $0$ |

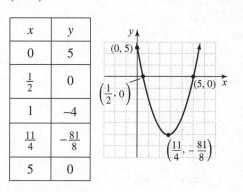

**23.** $y = -\dfrac{1}{2}x^2$

vertex: $(0, 0)$

$y$-intercept: $x = 0$, $y = -\dfrac{1}{2} \cdot 0^2 = 0$, $(0, 0)$

$x$-intercepts: $y = 0$

$0 = -\dfrac{1}{2}x^2$

$0 = x^2$

$0 = x$

$(0, 0)$

| $x$ | $y$ |
|-----|-----|
| $-4$ | $-8$ |
| $-2$ | $-2$ |
| $0$ | $0$ |
| $2$ | $-2$ |
| $4$ | $-8$ |

**25.** $y = 2x^2 - 11x + 5$

$a = 2, b = -11, c = 5$

vertex: $x = \dfrac{-b}{2a} = \dfrac{-(-11)}{2(2)} = \dfrac{11}{4}$

$\qquad y = 2\left(\dfrac{11}{4}\right)^2 - 11\left(\dfrac{11}{4}\right) + 5 = -\dfrac{81}{8}$

$\qquad \left(\dfrac{11}{4}, -\dfrac{81}{8}\right)$

$y$-intercept: $x = 0$, $y = 2 \cdot 0^2 - 11 \cdot 0 + 5 = 5$, $(0, 5)$

$x$-intercepts: $y = 0$

$0 = 2x^2 - 11x + 5$

$0 = (2x - 1)(x - 5)$

$x = \dfrac{1}{2} \quad$ or $\quad x = 5$

$\left(\dfrac{1}{2}, 0\right)$ and $(5, 0)$

| $x$ | $y$ |
|-----|-----|
| $0$ | $5$ |
| $\dfrac{1}{2}$ | $0$ |
| $1$ | $-4$ |
| $\dfrac{11}{4}$ | $-\dfrac{81}{8}$ |
| $5$ | $0$ |

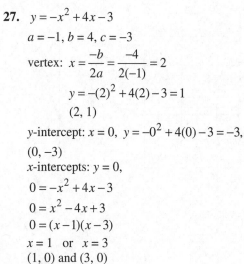

**27.** $y = -x^2 + 4x - 3$

$a = -1, b = 4, c = -3$

vertex: $x = \dfrac{-b}{2a} = \dfrac{-4}{2(-1)} = 2$

$\qquad y = -(2)^2 + 4(2) - 3 = 1$

$\qquad (2, 1)$

$y$-intercept: $x = 0$, $y = -0^2 + 4(0) - 3 = -3$,

$(0, -3)$

$x$-intercepts: $y = 0$,

$0 = -x^2 + 4x - 3$

$0 = x^2 - 4x + 3$

$0 = (x - 1)(x - 3)$

$x = 1 \quad$ or $\quad x = 3$

$(1, 0)$ and $(3, 0)$

| $x$ | $y$ |
|-----|-----|
| 0 | $-3$ |
| 1 | 0 |
| 2 | 1 |
| 3 | 0 |
| 4 | $-3$ |

**29.** $y = x^2 + 2x - 2$

$a = 1, b = 2, c = -2$

vertex: $x = \dfrac{-b}{2a} = \dfrac{-2}{2(1)} = -1$

$\qquad y = (-1)^2 + 2(-1) - 2 = -3$

$\qquad (-1, -3)$

$y$-intercept: $x = 0,\ y = 0^2 + 2(0) - 2 = -2,\ (0, -2)$

$x$-intercepts: $y = 0,$

$0 = x^2 + 2x - 2$

$x = \dfrac{-2 \pm \sqrt{2^2 - 4(1)(-2)}}{2(1)}$

$\quad = \dfrac{-2 \pm \sqrt{12}}{2}$

$\quad = \dfrac{-2 \pm 2\sqrt{3}}{2}$

$\quad = -1 \pm \sqrt{3}$

$\left(-1 - \sqrt{3},\ 0\right)$ and $\left(-1 + \sqrt{3},\ 0\right)$

| $x$ | $y$ |
|-----|-----|
| $-1 - \sqrt{3}$ | 0 |
| $-2$ | $-2$ |
| $-1$ | $-3$ |
| 0 | $-2$ |
| $-1 + \sqrt{3}$ | 0 |

**31.** $y = x^2 - 3x + 1$

$a = 1, b = -3, c = 1$

vertex: $x = \dfrac{-b}{2a} = \dfrac{-(-3)}{2(1)} = \dfrac{3}{2} = 1\dfrac{1}{2}$

$\qquad y = \left(\dfrac{3}{2}\right)^2 - 3\left(\dfrac{3}{2}\right) + 1 = -\dfrac{5}{4} = -1\dfrac{1}{4}$

$\qquad \left(1\dfrac{1}{2},\ -1\dfrac{1}{4}\right)$

$y$-intercept: $x = 0,\ y = 0^2 - 3(0) + 1 = 1,\ (0, 1)$

$x$-intercepts: $y = 0,$

$0 = x^2 - 3x + 1$

$x = \dfrac{-(-3) \pm \sqrt{(-3)^2 - 4(1)(1)}}{2(1)} = \dfrac{3 \pm \sqrt{5}}{2}$

$\left(\dfrac{3 - \sqrt{5}}{2},\ 0\right)$ and $\left(\dfrac{3 + \sqrt{5}}{2},\ 0\right)$

| $x$ | $y$ |
|-----|-----|
| $-1$ | 5 |
| 0 | 1 |
| $\dfrac{3 - \sqrt{5}}{2}$ | 0 |
| $1\dfrac{1}{2}$ | $-1\dfrac{1}{4}$ |
| $\dfrac{3 + \sqrt{5}}{2}$ | 0 |

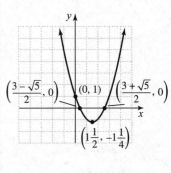

**33.** $\dfrac{\frac{1}{7}}{\frac{2}{5}} = \dfrac{1}{7} \div \dfrac{2}{5} = \dfrac{1}{7} \cdot \dfrac{5}{2} = \dfrac{5}{14}$

**35.** $\dfrac{\frac{1}{x}}{\frac{2}{x^2}} = \dfrac{1}{x} \div \dfrac{2}{x^2} = \dfrac{1}{x} \cdot \dfrac{x^2}{2} = \dfrac{x}{2}$

**37.** $\dfrac{2x}{1 - \frac{1}{x}} = \dfrac{2x}{\frac{x-1}{x}} = 2x \div \dfrac{x-1}{x} = \dfrac{2x}{1} \cdot \dfrac{x}{x-1} = \dfrac{2x^2}{x-1}$

**39.** $\dfrac{\frac{a-b}{2b}}{\frac{b-a}{8b^2}} = \dfrac{a-b}{2b} \div \dfrac{b-a}{8b^2}$

$\qquad = \dfrac{a-b}{2b} \cdot \dfrac{8b^2}{b-a}$

$\qquad = \dfrac{a-b}{2b} \cdot \dfrac{8b^2}{-1(a-b)}$

$\qquad = -4b$

**41.** domain: $(-\infty, \infty)$; range: $(-\infty, 3]$

**43.** domain: $(-\infty, \infty)$; range: $(-\infty, 1]$

**45. a.** The maximum height is about 256 feet.

     **b.** The fireball is at its maximum height after about 4 seconds.

     **c.** The fireball returns to the ground after about 8 seconds.

**47. C**

**49. A**

**Chapter 9 Vocabulary Check**

**1.** If $x^2 = a$, then $x = \sqrt{a}$ or $x = -\sqrt{a}$. This property is called the <u>square root</u> property.

**2.** A number that can be written in the form $a + bi$ is called a <u>complex</u> number.

**3.** The formula $\dfrac{-b}{2a}$ where $y = ax^2 + bx + c$ is called the <u>vertex</u> formula.

**4.** A complex number that can be written in the form $0 + bi$ is also called a <u>pure imaginary</u> number.

**5.** The <u>conjugate</u> of $2 + 3i$ is $2 - 3i$.

**6.** $\sqrt{-1} = \underline{i}$.

**7.** The process of solving a quadratic equation by writing it in the form $(x + a)^2 = c$ is called <u>completing the square</u>.

**8.** The formula $x = \dfrac{-b \pm \sqrt{b^2 - 4ac}}{2a}$ is called the <u>quadratic</u> formula.

**Chapter 9 Review**

**1.** $x^2 = 36$

$\qquad x = \pm\sqrt{36}$

$\qquad x = \pm 6$

The solutions are $\pm 6$.

**2.** $x^2 = 81$

$\qquad x = \pm\sqrt{81}$

$\qquad x = \pm 9$

The solutions are $\pm 9$.

**3.** $k^2 = 50$

$\qquad k = \pm\sqrt{50}$

$\qquad k = \pm 5\sqrt{2}$

The solutions are $\pm 5\sqrt{2}$.

**4.** $k^2 = 45$

$\qquad k = \pm\sqrt{45}$

$\qquad k = \pm 3\sqrt{5}$

The solutions are $\pm 3\sqrt{5}$.

**5.** $(x-11)^2 = 49$

$\qquad x - 11 = \pm\sqrt{49}$

$\qquad x - 11 = \pm 7$

$\qquad\quad x = 11 \pm 7$

$x = 11 - 7 = 4$   or   $x = 11 + 7 = 18$

The solutions are 4 and 18.

**6.** $(x+3)^2 = 100$

$\qquad x + 3 = \pm\sqrt{100}$

$\qquad x + 3 = \pm 10$

$\qquad\quad x = -3 \pm 10$

$x = -3 - 10 = -13$   or   $x = -3 + 10 = 7$

The solutions are $-13$ and 7.

**7.** $(4p+5)^2 = 41$

$\qquad 4p + 5 = \pm\sqrt{41}$

$\qquad\quad 4p = -5 \pm \sqrt{41}$

$\qquad\quad p = \dfrac{-5 \pm \sqrt{41}}{4}$

The solutions are $\dfrac{-5 \pm \sqrt{41}}{4}$.

    

**8.** $(3p+7)^2 = 37$

$3p+7 = \pm\sqrt{37}$

$3p = -7 \pm \sqrt{37}$

$p = \dfrac{-7 \pm \sqrt{37}}{3}$

The solutions are $-\dfrac{7 \pm \sqrt{37}}{3}$.

**9.** Let $h = 100$.

$16t^2 = h$

$16t^2 = 100$

$t^2 = \dfrac{100}{16}$

$t = \pm\sqrt{\dfrac{100}{16}} = \pm\dfrac{10}{4} = \pm 2.5$

The length of time is not a negative number so the dive lasted 2.5 seconds.

**10.** Let $h = 5 \cdot 5280 = 26,400$

$16t^2 = h$

$16t^2 = 26,400$

$t^2 = \dfrac{26,400}{16} = 1650$

$t = \pm\sqrt{1650} \approx \pm 40.6$

The length of time is not a negative number so the fall lasted 40.6 seconds.

**11.** $x^2 - 9x = -8$

$x^2 - 9x + \left(-\dfrac{9}{2}\right)^2 = -8 + \left(-\dfrac{9}{2}\right)^2$

$x^2 - 9x + \dfrac{81}{4} = -\dfrac{32}{4} + \dfrac{81}{4}$

$\left(x - \dfrac{9}{2}\right)^2 = \dfrac{49}{4}$

$x - \dfrac{9}{2} = \sqrt{\dfrac{49}{4}}$   or   $x - \dfrac{9}{2} = -\sqrt{\dfrac{49}{4}}$

$x = \dfrac{9}{2} + \dfrac{7}{2}$         $x = \dfrac{9}{2} - \dfrac{7}{2}$

$x = \dfrac{16}{2} = 8$         $x = \dfrac{2}{2} = 1$

The solutions are 8 and 1.

**12.** $x^2 + 8x = 20$

$x^2 + 8x + \left(\dfrac{8}{2}\right)^2 = 20 + \left(\dfrac{8}{2}\right)^2$

$x^2 + 8x + 16 = 20 + 16$

$(x+4)^2 = 36$

$x + 4 = \sqrt{36}$   or   $x + 4 = -\sqrt{36}$

$x = 6 - 4$         $x = -6 - 4$

$x = 2$         $x = -10$

The solutions are 2 and $-10$.

**13.** $x^2 + 4x = 1$

$x^2 + 4x + \left(\dfrac{4}{2}\right)^2 = 1 + \left(\dfrac{4}{2}\right)^2$

$x^2 + 4x + 4 = 1 + 4$

$(x+2)^2 = 5$

$x + 2 = \sqrt{5}$         or   $x + 2 = -\sqrt{5}$

$x = -2 + \sqrt{5}$         $x = -2 - \sqrt{5}$

The solutions are $-2 \pm \sqrt{5}$.

**14.** $x^2 - 8x = 3$

$x^2 - 8x + \left(\dfrac{-8}{2}\right)^2 = 3 + \left(\dfrac{-8}{2}\right)^2$

$x^2 - 8x + 16 = 3 + 16$

$(x-4)^2 = 19$

$x - 4 = \sqrt{19}$   or   $x - 4 = -\sqrt{19}$

$x = 4 + \sqrt{19}$         $x = 4 - \sqrt{19}$

The solutions are $4 \pm \sqrt{19}$.

**15.** $x^2 - 6x + 7 = 0$

$x^2 - 6x = -7$

$x^2 - 6x + \left(\dfrac{-6}{2}\right)^2 = -7 + \left(\dfrac{-6}{2}\right)^2$

$x^2 - 6x + 9 = -7 + 9$

$(x-3)^2 = 2$

$x - 3 = \sqrt{2}$   or   $x - 3 = -\sqrt{2}$

$x = 3 + \sqrt{2}$         $x = 3 - \sqrt{2}$

The solutions are $3 \pm \sqrt{2}$.

**16.**
$$x^2 + 6x + 7 = 0$$
$$x^2 + 6x = -7$$
$$x^2 + 6x + \left(\frac{6}{2}\right)^2 = -7 + \left(\frac{6}{2}\right)^2$$
$$x^2 + 6x + 9 = -7 + 9$$
$$(x+3)^2 = 2$$
$$x + 3 = \sqrt{2} \qquad \text{or} \quad x + 3 = -\sqrt{2}$$
$$x = -3 + \sqrt{2} \qquad\qquad x = -3 - \sqrt{2}$$

The solutions are $-3 \pm \sqrt{2}$.

**17.**
$$2y^2 + y - 1 = 0$$
$$y^2 + \frac{1}{2}y - \frac{1}{2} = 0$$
$$y^2 + \frac{1}{2}y = \frac{1}{2}$$
$$y^2 + \frac{1}{2}y + \left(\frac{1}{4}\right)^2 = \frac{1}{2} + \left(\frac{1}{4}\right)^2$$
$$y^2 + \frac{1}{2}y + \frac{1}{16} = \frac{1}{2} + \frac{1}{16}$$
$$\left(y + \frac{1}{4}\right)^2 = \frac{9}{16}$$

$$y + \frac{1}{4} = \sqrt{\frac{9}{16}} \qquad \text{or} \quad y + \frac{1}{4} = -\sqrt{\frac{9}{16}}$$
$$y = -\frac{1}{4} + \frac{3}{4} \qquad\qquad y = -\frac{1}{4} - \frac{3}{4}$$
$$y = \frac{1}{2} \qquad\qquad\qquad y = -1$$

The solutions are $\frac{1}{2}$ and $-1$.

**18.**
$$y^2 + 3y - 1 = 0$$
$$y^2 + 3y = 1$$
$$y^2 + 3y + \left(\frac{3}{2}\right)^2 = 1 + \left(\frac{3}{2}\right)^2$$
$$y^2 + 3y + \frac{9}{4} = 1 + \frac{9}{4}$$
$$\left(y + \frac{3}{2}\right)^2 = \frac{13}{4}$$

$$y + \frac{3}{2} = \sqrt{\frac{13}{4}} \qquad \text{or} \quad y + \frac{3}{2} = -\sqrt{\frac{13}{4}}$$
$$y + \frac{3}{2} = \frac{\sqrt{13}}{2} \qquad\qquad y + \frac{3}{2} = -\frac{\sqrt{13}}{2}$$
$$y = -\frac{3}{2} + \frac{\sqrt{13}}{2} \qquad\qquad y = -\frac{3}{2} - \frac{\sqrt{13}}{2}$$

The solutions are $\dfrac{-3 \pm \sqrt{13}}{2}$.

**19.** $9x^2 + 30x + 25 = 0$
$a = 9$, $b = 30$, and $c = 25$
$$x = \frac{-30 \pm \sqrt{30^2 - 4(9)(25)}}{2(9)}$$
$$= \frac{-30 \pm \sqrt{900 - 900}}{18}$$
$$= \frac{-30 \pm \sqrt{0}}{18}$$
$$= -\frac{5}{3}$$

The solution is $-\dfrac{5}{3}$.

**20.** $16x^2 - 72x + 81 = 0$
$a = 16$, $b = -72$, and $c = 81$
$$x = \frac{-(-72) \pm \sqrt{(-72)^2 - 4(16)(81)}}{2(16)}$$
$$= \frac{72 \pm \sqrt{5184 - 5184}}{32}$$
$$= \frac{72 \pm \sqrt{0}}{32}$$
$$= \frac{9}{4}$$

The solution is $\dfrac{9}{4}$.

**443**

**21.**  $7x^2 = 35$

$x^2 = 5$

$x^2 - 5 = 0$

$a = 1, b = 0, c = -5$

$x = \dfrac{-0 \pm \sqrt{0^2 - 4(1)(-5)}}{2(1)}$

$= \pm \dfrac{\sqrt{20}}{2}$

$= \pm \dfrac{2\sqrt{5}}{2}$

$= \pm\sqrt{5}$

The solutions are $\pm\sqrt{5}$.

**22.**  $11x^2 = 33$

$x^2 = 3$

$x^2 - 3 = 0$

$a = 1, b = 0, c = -3$

$x = \dfrac{-0 \pm \sqrt{0^2 - 4(1)(-3)}}{2(1)}$

$= \pm \dfrac{\sqrt{12}}{2}$

$= \pm \dfrac{2\sqrt{3}}{2}$

$= \pm\sqrt{3}$

The solutions are $\pm\sqrt{3}$.

**23.**  $x^2 - 10x + 7 = 0$

$a = 1, b = -10,$ and $c = 7$

$x = \dfrac{-(-10) \pm \sqrt{(-10)^2 - 4(1)(7)}}{2(1)}$

$= \dfrac{10 \pm \sqrt{100 - 28}}{2}$

$= \dfrac{10 \pm \sqrt{72}}{2}$

$= \dfrac{10 \pm 6\sqrt{2}}{2}$

$= 5 \pm 3\sqrt{2}$

The solutions are $5 \pm 3\sqrt{2}$.

**24.**  $x^2 + 4x - 7 = 0$

$a = 1, b = 4,$ and $c = -7$

$x = \dfrac{-4 \pm \sqrt{4^2 - 4(1)(-7)}}{2(1)}$

$= \dfrac{-4 \pm \sqrt{16 + 28}}{2}$

$= \dfrac{-4 \pm \sqrt{44}}{2}$

$= \dfrac{-4 \pm 2\sqrt{11}}{2}$

$= -2 \pm \sqrt{11}$

The solutions are $-2 \pm \sqrt{11}$.

**25.**  $3x^2 + x - 1 = 0$

$a = 3, b = 1, c = -1$

$x = \dfrac{-1 \pm \sqrt{1^2 - 4(3)(-1)}}{2(3)}$

$= \dfrac{-1 \pm \sqrt{1 + 12}}{6}$

$= \dfrac{-1 \pm \sqrt{13}}{6}$

The solutions are $\dfrac{-1 \pm \sqrt{13}}{6}$.

**26.**  $x^2 + 3x - 1 = 0$

$a = 1, b = 3,$ and $c = -1$

$x = \dfrac{-3 \pm \sqrt{3^2 - 4(1)(-1)}}{2(1)}$

$= \dfrac{-3 \pm \sqrt{9 + 4}}{2}$

$= \dfrac{-3 \pm \sqrt{13}}{2}$

The solutions are $\dfrac{-3 \pm \sqrt{13}}{2}$.

**27.**  $2x^2 + x + 5 = 0$

$a = 2, b = 1,$ and $c = 5$

$x = \dfrac{-1 \pm \sqrt{1^2 - 4(2)(5)}}{2(2)}$

$= \dfrac{-1 \pm \sqrt{1 - 40}}{4}$

$= \dfrac{-1 \pm \sqrt{-39}}{4}$

There is no real solution because $\sqrt{-39}$ is not a real number.

**28.** $7x^2 - 3x + 1 = 0$

$a = 7$, $b = -3$ and $c = 1$

$$x = \frac{-(-3) \pm \sqrt{(-3)^2 - 4(7)(1)}}{2(7)}$$

$$= \frac{3 \pm \sqrt{9 - 28}}{14}$$

$$= \frac{3 \pm \sqrt{-19}}{14}$$

There is no real solution because $\sqrt{-19}$ is not a real number.

**29.** From Exercise 25, the exact solutions are $\frac{-1 \pm \sqrt{13}}{6}$.

$$\frac{-1 + \sqrt{13}}{6} \approx 0.4$$

$$\frac{-1 - \sqrt{13}}{6} \approx -0.8$$

**30.** From Exercise 26, the exact solutions are $\frac{-3 \pm \sqrt{13}}{2}$.

$$\frac{-3 + \sqrt{13}}{2} \approx 0.3$$

$$\frac{-3 - \sqrt{13}}{2} \approx -3.3$$

**31.** $x^2 - 7x - 1 = 0$

$a = 1$, $b = -7$, $c = -1$

$b^2 - 4ac = (-7)^2 - 4(1)(-1) = 49 + 4 = 53$

Since the discriminant is a positive number, this equation has two distinct real solutions.

**32.** $x^2 + 6x = 5$

$x^2 + 6x - 5 = 0$

$a = 1$, $b = 6$, $c = -5$

$b^2 - 4ac = 6^2 - 4(1)(-5) = 36 + 20 = 56$

Since the discriminant is a positive number, this equation has two distinct real solutions.

**33.** $x^2 + x + 5 = 0$

$a = 1$, $b = 1$, $c = 5$

$b^2 - 4ac = 1^2 - 4(1)(5) = 1 - 20 = -19$

Since the discriminant is a negative number, this equation has no real solution.

**34.** $5x^2 + 4 = 0$

$a = 5$, $b = 0$, $c = 4$

$b^2 - 4ac = 0^2 - 4(5)(4) = 0 - 80 = -80$

Since the discriminant is a negative number, this equation has no real solution.

**35.** $9x^2 + 1 = 6x$

$9x^2 - 6x + 1 = 0$

$a = 9$, $b = -6$, $c = 1$

$b^2 - 4ac = (-6)^2 - 4(9)(1) = 36 - 36 = 0$

Since the discriminant is 0, this equation has one real solution.

**36.** $x^2 + 25 = 10x$

$x^2 - 10x + 25 = 0$

$a = 1$, $b = -10$, $c = 25$

$b^2 - 4ac = (-10)^2 - 4(1)(25) = 100 - 100 = 0$

Since the discriminant is 0, this equation has one real solution.

**37.** $y = 8x^2 + 109x + 700$

$2590 = 8x^2 + 109x + 700$

$0 = 8x^2 + 109x - 1890$

$a = 8$, $b = 109$, $c = -1890$

$$x = \frac{-b \pm \sqrt{b^2 - 4ac}}{2a}$$

$$x = \frac{-109 \pm \sqrt{109^2 - 4(8)(-1890)}}{2(8)}$$

$$= \frac{-109 \pm \sqrt{11,881 + 60,480}}{16}$$

$$= \frac{-109 \pm \sqrt{72,361}}{16}$$

$$= \frac{-109 \pm 269}{16}$$

$$x = \frac{-109 + 269}{16} = \frac{160}{16} = 10$$

or

$$x = \frac{-109 - 269}{16} = \frac{-378}{16} = -23.625$$

Since time will not be a negative number in this context, discard the solution $-23.625$. The solution is $x = 10$, so the model predicts that the price of platinum will be $2590 per ounce in 2013 (2003 + 10).

**38.**
$$y = 26x^2 + 115x + 493$$
$$3077 = 26x^2 + 115x + 493$$
$$0 = 26x^2 + 115x - 2584$$
$$a = 26, \ b = 115, \ c = -2584$$
$$x = \frac{-b \pm \sqrt{b^2 - 4ac}}{2a}$$
$$x = \frac{-115 \pm \sqrt{115^2 - 4(26)(-2584)}}{2(26)}$$
$$= \frac{-115 \pm \sqrt{13,225 + 268,736}}{52}$$
$$= \frac{-115 \pm \sqrt{281,961}}{52}$$
$$= \frac{-115 \pm 531}{52}$$
$$x = \frac{-115 + 531}{52} = \frac{416}{52} = 8$$
or
$$x = \frac{-115 - 531}{52} = \frac{-646}{52} \approx -12.4$$

Since time will not be a negative number in this context, discard the solution $-12.4$. The solution is $x = 8$, so the model predicts that the price of silver was 3077 cents per ounce in 2011 $(2003 + 8)$.

**39.** $\sqrt{-144} = \sqrt{-1 \cdot 144} = \sqrt{-1} \cdot \sqrt{144} = i \cdot 12 = 12i$

**40.** $\sqrt{-36} = \sqrt{36 \cdot i} = \sqrt{36} \cdot \sqrt{-1} = 6i$

**41.** $\sqrt{-108} = \sqrt{-1 \cdot 36 \cdot 3}$
$$= \sqrt{-1} \cdot \sqrt{36} \cdot \sqrt{3}$$
$$= i \cdot 6\sqrt{3}$$
$$= 6i\sqrt{3}$$

**42.** $\sqrt{-500} = \sqrt{100 \cdot 5 \cdot -1}$
$$= \sqrt{100} \cdot \sqrt{-1} \cdot \sqrt{5}$$
$$= 10i\sqrt{5}$$

**43.** $2i(3 - 5i) = (2i)(3) - (2i)(5i)$
$$= 6i - 10i^2$$
$$= 6i - 10(-1)$$
$$= 10 + 6i$$

**44.** $i(-7 - i) = (i)(-7) - (i)(i)$
$$= -7i - i^2$$
$$= -7i - (-1)$$
$$= 1 - 7i$$

**45.** $(7 - i) + (14 - 9i) = 7 - i + 14 - 9i$
$$= 7 + 14 + (-i - 9i)$$
$$= 21 - 10i$$

**46.** $(10 - 4i) + (9 - 21i) = 10 - 4i + 9 - 21i$
$$= 10 + 9 + (-4i - 21i)$$
$$= 19 - 25i$$

**47.** $3 - (11 + 2i) = 3 - 11 - 2i = -8 - 2i$

**48.** $(-4 - 3i) + 5i = -4 - 3i + 5i = -4 + 2i$

**49.** $(2 - 3i)(3 - 2i) = 6 - 4i - 9i + 6i^2$
$$= 6 - 13i + 6(-1)$$
$$= 6 - 13i - 6$$
$$= -13i$$

**50.** $(2 + 5i)(5 - i) = 10 - 2i + 25i - 5i^2$
$$= 10 - 2i + 25i - 5(-1)$$
$$= 10 + 23i + 5$$
$$= 15 + 23i$$

**51.** $(3 - 4i)(3 + 4i) = (3)^2 - (4i)^2$
$$= 9 - 16i^2$$
$$= 9 - 16(-1)$$
$$= 9 + 16$$
$$= 25$$

**52.** $(7 - 2i)(7 - 2i) = 49 - 14i - 14i + 4i^2$
$$= 49 - 28i + 4(-1)$$
$$= 49 - 28i - 4$$
$$= 49 - 4 - 28i$$
$$= 45 - 28i$$

**53.** $\dfrac{2 - 6i}{4i} = \dfrac{(2 - 6i)(-i)}{4i(-i)}$
$$= \frac{-2i + 6i^2}{-4i^2}$$
$$= \frac{-2i + 6(-1)}{-4(-1)}$$
$$= \frac{-2i - 6}{4}$$
$$= \frac{2(-i - 3)}{4}$$
$$= -\frac{i}{2} - \frac{3}{2}$$
$$= -\frac{3}{2} - \frac{1}{2}i$$

Copyright © 2013 Pearson Education, Inc.

**54.** $\dfrac{5-i}{2i} = \dfrac{(5-i)(-i)}{2i(-i)}$

$\phantom{\dfrac{5-i}{2i}} = \dfrac{-5i + i^2}{-2i^2}$

$\phantom{\dfrac{5-i}{2i}} = \dfrac{-5i + (-1)}{-2(-1)}$

$\phantom{\dfrac{5-i}{2i}} = \dfrac{-1 - 5i}{2}$

$\phantom{\dfrac{5-i}{2i}} = -\dfrac{1}{2} - \dfrac{5}{2}i$

**55.** $\dfrac{4-i}{1+2i} = \dfrac{(4-i)(1-2i)}{(1+2i)(1-2i)}$

$\phantom{\dfrac{4-i}{1+2i}} = \dfrac{4(1) + 4(-2i) - i(1) - i(-2i)}{(1)^2 - (2i)^2}$

$\phantom{\dfrac{4-i}{1+2i}} = \dfrac{4 - 8i - i + 2i^2}{1 - 4i^2}$

$\phantom{\dfrac{4-i}{1+2i}} = \dfrac{4 - 9i + 2(-1)}{1 - 4(-1)}$

$\phantom{\dfrac{4-i}{1+2i}} = \dfrac{4 - 9i - 2}{1 + 4}$

$\phantom{\dfrac{4-i}{1+2i}} = \dfrac{2 - 9i}{5}$

$\phantom{\dfrac{4-i}{1+2i}} = \dfrac{2}{5} - \dfrac{9}{5}i$

**56.** $\dfrac{1+3i}{2-7i} = \dfrac{1+3i}{2-7i} \cdot \dfrac{2+7i}{2+7i}$

$\phantom{\dfrac{1+3i}{2-7i}} = \dfrac{1(2) + 1(7i) + 3i(2) + 3i(7i)}{2^2 - (7i)^2}$

$\phantom{\dfrac{1+3i}{2-7i}} = \dfrac{2 + 7i + 6i + 21i^2}{4 - 49i^2}$

$\phantom{\dfrac{1+3i}{2-7i}} = \dfrac{2 + 13i + 21(-1)}{4 - 49(-1)}$

$\phantom{\dfrac{1+3i}{2-7i}} = \dfrac{2 + 13i - 21}{4 + 49}$

$\phantom{\dfrac{1+3i}{2-7i}} = \dfrac{-19 + 13i}{53}$

$\phantom{\dfrac{1+3i}{2-7i}} = -\dfrac{19}{53} + \dfrac{13}{53}i$

**57.** $3x^2 = -48$

$x^2 = -16$

$x = \pm\sqrt{-16}$

$x = \pm 4i$

The solutions are $\pm 4i$.

**58.** $5x^2 = -125$

$x^2 = -25$

$x = \pm\sqrt{-25}$

$x = \pm 5i$

The solutions are $\pm 5i$.

**59.** $x^2 - 4x + 13 = 0$

$a = 1,\ b = -4,\ c = 13$

$x = \dfrac{-(-4) \pm \sqrt{(-4)^2 - 4(1)(13)}}{2(1)}$

$\phantom{x} = \dfrac{4 \pm \sqrt{16 - 52}}{2}$

$\phantom{x} = \dfrac{4 \pm \sqrt{-36}}{2}$

$\phantom{x} = \dfrac{4 \pm 6i}{2}$

$\phantom{x} = \dfrac{2(2 \pm 3i)}{2}$

$\phantom{x} = 2 \pm 3i$

The solutions are $2 \pm 3i$.

**60.** $x^2 + 4x + 11 = 0$

$a = 1,\ b = 4,\ c = 11$

$x = \dfrac{-4 \pm \sqrt{4^2 - 4(1)(11)}}{2(1)}$

$\phantom{x} = \dfrac{-4 \pm \sqrt{16 - 44}}{2}$

$\phantom{x} = \dfrac{-4 \pm \sqrt{-28}}{2}$

$\phantom{x} = \dfrac{-4 \pm \sqrt{4 \cdot 7 \cdot -1}}{2}$

$\phantom{x} = \dfrac{-4 \pm 2i\sqrt{7}}{2}$

$\phantom{x} = \dfrac{2\left(-2 \pm i\sqrt{7}\right)}{2}$

$\phantom{x} = -2 \pm i\sqrt{7}$

The solutions are $-2 \pm i\sqrt{7}$.

**61.** $y = 5x^2$

The vertex, $x$-intercept, and $y$-intercept are the same, (0, 0).

| $x$ | $y$ |
|----|----|
| $-2$ | 20 |
| $-1$ | 5 |
| 0 | 0 |
| 1 | 5 |
| 2 | 20 |

**62.** $y = -\dfrac{1}{2}x^2$

The vertex, $x$-intercept, and $y$-intercept are the same, (0, 0).

| $x$ | $y$ |
|----|----|
| $-2$ | $-2$ |
| $-1$ | $-\frac{1}{2}$ |
| 0 | 0 |
| 1 | $-\frac{1}{2}$ |
| 2 | $-2$ |

**63.** $y = x^2 - 25$

$a = 1, b = 0, c = -25$

vertex: $x = \dfrac{-b}{2a} = \dfrac{-0}{2(1)} = 0$,  $y = 0^2 - 25 = -25$,

   (0, $-25$)

$y$-intercept: $x = 0$,  $y = 0^2 - 25 = -25$, (0, $-25$)

$x$-intercept: $y = 0$,

   $0 = x^2 - 25$

   $25 = x^2$

   $\pm 5 = x$

   ($-5$, 0), (5, 0)

| $x$ | $y$ |
|----|----|
| 0 | $-25$ |
| $-5$ | 0 |
| 5 | 0 |
| $-1$ | $-24$ |
| 1 | $-24$ |

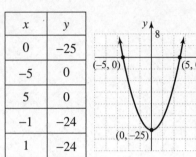

**64.** $y = x^2 - 36$

$a = 1, b = 0, c = -36$

vertex: $x = \dfrac{-b}{2a} = \dfrac{-0}{2(1)} = 0$,  $y = 0^2 - 36 = -36$,

   (0, $-36$)

$y$-intercept: (0, $-36$)

$x$-intercept: $y = 0$,

   $0 = x^2 - 36$

   $36 = x^2$

   $\pm 6 = x$

   ($-6$, 0), (6, 0)

| $x$ | $y$ |
|----|----|
| 0 | $-36$ |
| $-6$ | 0 |
| 6 | 0 |
| $-3$ | $-27$ |
| 3 | $-27$ |

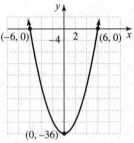

**65.** $y = x^2 + 3$

$a = 1, b = 0, c = 3$

vertex: $x = \dfrac{-b}{2a} = \dfrac{-0}{2(1)} = 0$,  $y = 0^2 + 3 = 0$, (0, 3)

$y$-intercept: (0, 3)

$x$-intercept: $y = 0$,

   $0 = x^2 + 3$

   $-3 = x^2$

There are no $x$-intercepts.

| $x$ | $y$ |
|---|---|
| $-3$ | $12$ |
| $-1$ | $4$ |
| $0$ | $3$ |
| $1$ | $4$ |
| $3$ | $12$ |

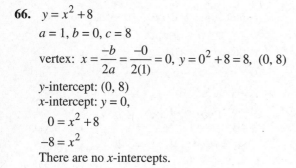

**66.** $y = x^2 + 8$

$a = 1$, $b = 0$, $c = 8$

vertex: $x = \dfrac{-b}{2a} = \dfrac{-0}{2(1)} = 0$, $y = 0^2 + 8 = 8$, $(0, 8)$

$y$-intercept: $(0, 8)$
$x$-intercept: $y = 0$,

$0 = x^2 + 8$

$-8 = x^2$

There are no $x$-intercepts.

| $x$ | $y$ |
|---|---|
| $-2$ | $12$ |
| $-1$ | $9$ |
| $0$ | $8$ |
| $1$ | $9$ |
| $2$ | $12$ |

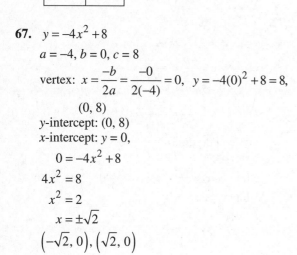

**67.** $y = -4x^2 + 8$

$a = -4$, $b = 0$, $c = 8$

vertex: $x = \dfrac{-b}{2a} = \dfrac{-0}{2(-4)} = 0$, $y = -4(0)^2 + 8 = 8$,

$\qquad (0, 8)$

$y$-intercept: $(0, 8)$
$x$-intercept: $y = 0$,

$0 = -4x^2 + 8$

$4x^2 = 8$

$x^2 = 2$

$x = \pm\sqrt{2}$

$\left(-\sqrt{2}, 0\right), \left(\sqrt{2}, 0\right)$

| $x$ | $y$ |
|---|---|
| $-2$ | $-8$ |
| $-1$ | $4$ |
| $0$ | $8$ |
| $1$ | $4$ |
| $2$ | $-8$ |

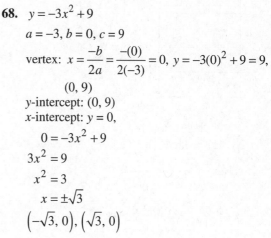

**68.** $y = -3x^2 + 9$

$a = -3$, $b = 0$, $c = 9$

vertex: $x = \dfrac{-b}{2a} = \dfrac{-(0)}{2(-3)} = 0$, $y = -3(0)^2 + 9 = 9$,

$\qquad (0, 9)$

$y$-intercept: $(0, 9)$
$x$-intercept: $y = 0$,

$0 = -3x^2 + 9$

$3x^2 = 9$

$x^2 = 3$

$x = \pm\sqrt{3}$

$\left(-\sqrt{3}, 0\right), \left(\sqrt{3}, 0\right)$

| $x$ | $y$ |
|---|---|
| $-\sqrt{3}$ | $0$ |
| $-1$ | $6$ |
| $0$ | $9$ |
| $1$ | $6$ |
| $\sqrt{3}$ | $0$ |

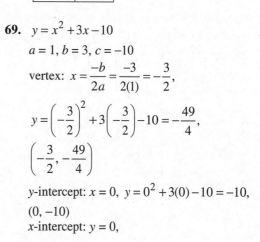

**69.** $y = x^2 + 3x - 10$

$a = 1$, $b = 3$, $c = -10$

vertex: $x = \dfrac{-b}{2a} = \dfrac{-3}{2(1)} = -\dfrac{3}{2}$,

$y = \left(-\dfrac{3}{2}\right)^2 + 3\left(-\dfrac{3}{2}\right) - 10 = -\dfrac{49}{4}$,

$\left(-\dfrac{3}{2}, -\dfrac{49}{4}\right)$

$y$-intercept: $x = 0$, $y = 0^2 + 3(0) - 10 = -10$,

$(0, -10)$
$x$-intercept: $y = 0$,

$0 = x^2 + 3x - 10$

$0 = (x+5)(x-2)$

$x + 5 = 0$ or $x - 2 = 0$

$x = -5$ $\qquad$ $x = 2$

$(-5, 0), (2, 0)$

| $x$ | $y$ |
|---|---|
| $-5$ | $0$ |
| $2$ | $0$ |
| $-\dfrac{3}{2}$ | $-\dfrac{49}{4}$ |
| $0$ | $-10$ |
| $1$ | $-6$ |

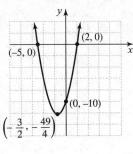

**70.** $y = x^2 + 3x - 4$

$a = 1, b = 3, c = -4$

vertex: $x = \dfrac{-b}{2a} = \dfrac{-3}{2(1)} = -\dfrac{3}{2}$,

$y = \left(-\dfrac{3}{2}\right)^2 + 3\left(-\dfrac{3}{2}\right) - 4 = -\dfrac{25}{4}$,

$\left(-\dfrac{3}{2}, -\dfrac{25}{4}\right)$

*y*-intercept: $x = 0$, $y = 0^2 + 3(0) - 4 = -4$, $(0, -4)$

*x*-intercept: $y = 0$,

$0 = x^2 + 3x - 4$

$0 = (x+4)(x-1)$

$x + 4 = 0$ or $x - 1 = 0$

$x = -4$ $\qquad$ $x = 1$

$(-4, 0), (1, 0)$

| $x$ | $y$ |
|---|---|
| $-\dfrac{3}{2}$ | $-\dfrac{25}{4}$ |
| $-4$ | $0$ |
| $1$ | $0$ |
| $0$ | $-4$ |
| $-1$ | $-6$ |

**71.** $y = -x^2 - 5x - 6$

$a = -1, b = -5, c = -6$

vertex: $x = \dfrac{-b}{2a} = \dfrac{-(-5)}{2(-1)} = -\dfrac{5}{2}$,

$y = -\left(-\dfrac{5}{2}\right)^2 - 5\left(-\dfrac{5}{2}\right) - 6 = \dfrac{1}{4}$,

$\left(-\dfrac{5}{2}, \dfrac{1}{4}\right)$

*y*-intercept: $x = 0$, $y = -0^2 - 5(0) - 6 = -6$,

$\qquad (0, -6)$

*x*-intercept: $y = 0$,

$0 = -x^2 - 5x - 6$

$0 = x^2 + 5x + 6$

$0 = (x+3)(x+2)$

$x + 3 = 0$ or $x + 2 = 0$

$x = -3$ $\qquad$ $x = -2$

$(-3, 0), (-2, 0)$

| $x$ | $y$ |
|---|---|
| $-\dfrac{5}{2}$ | $\dfrac{1}{4}$ |
| $0$ | $-6$ |
| $-2$ | $0$ |
| $-3$ | $0$ |
| $-1$ | $-2$ |

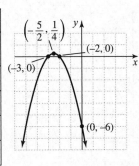

**72.** $y = 3x^2 - x - 2$

$a = 3, b = -1, c = -2$

vertex: $x = \dfrac{-b}{2a} = \dfrac{-(-1)}{2(3)} = \dfrac{1}{6}$

$y = 3\left(\dfrac{1}{6}\right)^2 - \left(\dfrac{1}{6}\right) - 2 = -\dfrac{25}{12}$

$\left(\dfrac{1}{6}, -\dfrac{25}{12}\right)$

*y*-intercept: $x = 0$, $y = 3 \cdot 0^2 - 0 - 2 = -2$ $(0, -2)$

*x*-intercepts: $y = 0$,

$0 = 3x^2 - x - 2$

$0 = (3x+2)(x-1)$

$x = -\dfrac{2}{3}$ or $x = 1$

$\left(-\dfrac{2}{3}, 0\right), (1, 0)$

| $x$ | $y$ |
|-----|-----|
| $-\frac{2}{3}$ | $0$ |
| $0$ | $-2$ |
| $\frac{1}{6}$ | $-\frac{25}{12}$ |
| $1$ | $0$ |

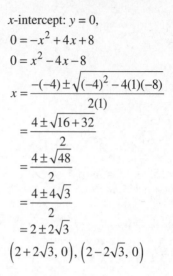

**73.** $y = 2x^2 - 11x - 6$

$a = 2, b = -11, c = -6$

vertex: $x = \dfrac{-b}{2a} = \dfrac{-(-11)}{2(2)} = \dfrac{11}{4}$

$y = 2\left(\dfrac{11}{4}\right)^2 - 11\left(\dfrac{11}{4}\right) - 6 = -\dfrac{169}{8}$.

$\left(\dfrac{11}{4}, -\dfrac{169}{8}\right)$

$y$-intercept: $x = 0, \ y = 2 \cdot 0^2 - 11 \cdot 0 - 6 = -6$,

$(0, -6)$

$x$-intercepts: $y = 0$,

$0 = 2x^2 - 11x - 6$

$0 = (2x + 1)(x - 6)$

$x = -\dfrac{1}{2}$ or $x = 6$

$\left(-\dfrac{1}{2}, 0\right), \ (6, 0)$

| $x$ | $y$ |
|-----|-----|
| $-\frac{1}{2}$ | $0$ |
| $0$ | $-6$ |
| $\frac{11}{4}$ | $-\frac{169}{8}$ |
| $\frac{11}{2}$ | $-6$ |
| $6$ | $0$ |

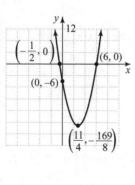

**74.** $y = -x^2 + 4x + 8$

$a = -1, b = 4, c = 8$

vertex: $x = \dfrac{-b}{2a} = \dfrac{-4}{2(-1)} = 2$,

$y = -2^2 + 4(2) + 8 = 12, \ (2, 12)$

$y$-intercept: $x = 0, \ y = -0^2 + 4(0) + 8 = 8, \ (0, 8)$

$x$-intercept: $y = 0$,

$0 = -x^2 + 4x + 8$

$0 = x^2 - 4x - 8$

$x = \dfrac{-(-4) \pm \sqrt{(-4)^2 - 4(1)(-8)}}{2(1)}$

$= \dfrac{4 \pm \sqrt{16 + 32}}{2}$

$= \dfrac{4 \pm \sqrt{48}}{2}$

$= \dfrac{4 \pm 4\sqrt{3}}{2}$

$= 2 \pm 2\sqrt{3}$

$\left(2 + 2\sqrt{3}, 0\right), \left(2 - 2\sqrt{3}, 0\right)$

| $x$ | $y$ |
|-----|-----|
| $2$ | $12$ |
| $0$ | $8$ |
| $2 + 2\sqrt{3}$ | $0$ |
| $2 - 2\sqrt{3}$ | $0$ |
| $1$ | $11$ |

**75.** A

**76.** D

**77.** B

**78.** C

**79.** The equation has one solution because the graph intersects the $x$-axis at one point.

**80.** The equation has two solutions because the graph intersects the $x$-axis at two points.

**81.** The equation has no real solution because the graph does not intersect the $x$-axis.

**82.** The equation has two solutions because the graph intersects the $x$-axis at two points.

**83.** $x^2 = 49$

$x = \pm\sqrt{49}$

$x = \pm 7$

The solutions are $\pm 7$.

**84.** $y^2 = 75$

$y = \pm\sqrt{75}$

$y = \pm 5\sqrt{3}$

The solutions are $\pm 5\sqrt{3}$.

**85.** $(x-7)^2 = 64$

$x - 7 = \pm\sqrt{64}$

$x = 7 \pm 8$

$x = 7 + 8 = 15$ or $x = 7 - 8 = -1$

The solutions are 15 and $-1$.

**86.** $x^2 + 4x = 6$

$x^2 + 4x + \left(\dfrac{4}{2}\right)^2 = 6 + \left(\dfrac{4}{2}\right)^2$

$x^2 + 4x + 4 = 6 + 4$

$(x+2)^2 = 10$

$x + 2 = \sqrt{10}$ or $x + 2 = -\sqrt{10}$

$x = -2 + \sqrt{10}$     $x = -2 - \sqrt{10}$

The solutions are $-2 \pm \sqrt{10}$.

**87.** $3x^2 + x = 2$

$x^2 + \dfrac{1}{3}x = \dfrac{2}{3}$

$x^2 + \dfrac{1}{3}x + \left(\dfrac{1}{6}\right)^2 = \dfrac{2}{3} + \left(\dfrac{1}{6}\right)^2$

$\left(x + \dfrac{1}{6}\right)^2 = \dfrac{25}{36}$

$x + \dfrac{1}{6} = \sqrt{\dfrac{25}{36}}$ or $x + \dfrac{1}{6} = -\sqrt{\dfrac{25}{36}}$

$x = -\dfrac{1}{6} + \dfrac{5}{6}$     $x = -\dfrac{1}{6} - \dfrac{5}{6}$

$x = \dfrac{2}{3}$     $x = -1$

The solutions are $-1$ and $\dfrac{2}{3}$.

**88.** $4x^2 - x - 2 = 0$

$4x^2 - x = 2$

$x^2 - \dfrac{1}{4}x = \dfrac{1}{2}$

$x^2 - \dfrac{1}{4}x + \left(\dfrac{-1}{8}\right)^2 = \dfrac{1}{2} + \left(\dfrac{-1}{8}\right)^2$

$\left(x - \dfrac{1}{8}\right)^2 = \dfrac{33}{64}$

$x - \dfrac{1}{8} = \sqrt{\dfrac{33}{64}}$ or $x - \dfrac{1}{8} = -\sqrt{\dfrac{33}{64}}$

$x - \dfrac{1}{8} = \dfrac{\sqrt{33}}{8}$     $x - \dfrac{1}{8} = -\dfrac{\sqrt{33}}{8}$

$x = \dfrac{1}{8} + \dfrac{\sqrt{33}}{8}$     $x = \dfrac{1}{8} - \dfrac{\sqrt{33}}{8}$

The solutions are $\dfrac{1 \pm \sqrt{33}}{8}$.

**89.** $4x^2 - 3x - 2 = 0$

$a = 4,\ b = -3,\ c = -2$

$x = \dfrac{-(-3) \pm \sqrt{(-3)^2 - 4(4)(-2)}}{2(4)}$

$= \dfrac{3 \pm \sqrt{9 + 32}}{8}$

$= \dfrac{3 \pm \sqrt{41}}{8}$

The solutions are $\dfrac{3 \pm \sqrt{41}}{8}$.

**90.** $5x^2 + x - 2 = 0$

$a = 5,\ b = 1,\ c = -2$

$x = \dfrac{-1 \pm \sqrt{1^2 - 4(5)(-2)}}{2(5)}$

$= \dfrac{-1 \pm \sqrt{1 + 40}}{10}$

$= \dfrac{-1 \pm \sqrt{41}}{10}$

The solutions are $\dfrac{-1 \pm \sqrt{41}}{10}$.

**91.** $4x^2 + 12x + 9 = 0$

$a = 4, b = 12, c = 9$

$x = \dfrac{-12 \pm \sqrt{12^2 - 4(4)(9)}}{2(4)}$

$= \dfrac{-12 \pm \sqrt{144 - 144}}{8}$

$= -\dfrac{3}{2}$

The solution is $-\dfrac{3}{2}$.

**92.** $2x^2 + x + 4 = 0$

$a = 2, b = 1, c = 4$

$x = \dfrac{-1 \pm \sqrt{1^2 - 4(2)(4)}}{2(2)}$

$= \dfrac{-1 \pm \sqrt{1 - 32}}{4}$

$= \dfrac{-1 \pm \sqrt{-31}}{4}$

$= \dfrac{-1 \pm i\sqrt{31}}{4}$

The solutions are $\dfrac{-1 \pm i\sqrt{31}}{4}$.

**93.** $y = 4 - x^2$

$a = -1, b = 0, c = 4$

vertex: $x = \dfrac{-b}{2a} = \dfrac{-0}{2(-1)} = 0$, $y = 4 - 0^2 = 4$,

$\quad$ (0, 4)

$y$-intercept: $x = 0$, $y = 4 - 0^2 = 4$, (0, 4)

$x$-intercept: $y = 0$,

$\quad 0 = 4 - x^2$

$\quad x^2 = 4$

$\quad x = \pm 2$

(−2, 0), (2, 0)

| $x$ | $y$ |
|-----|-----|
| 0 | 4 |
| −2 | 0 |
| 2 | 0 |
| −1 | 3 |
| 1 | 3 |

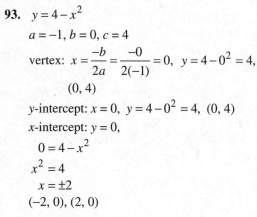

**94.** $y = x^2 + 4$

$a = 1, b = 0, c = 4$

vertex: $x = \dfrac{-b}{2a} = \dfrac{-0}{2(1)} = 0$, $y = 0^2 + 4 = 4$, (0, 4)

$y$-intercept: $x = 0$, $y = 0^2 + 4 = 4$, (0, 4)

$x$-intercept: $y = 0$,

$\quad 0 = x^2 + 4$

$\quad -4 = x^2$

There are no $x$-intercepts.

| $x$ | $y$ |
|-----|-----|
| 0 | 4 |
| −1 | 5 |
| 1 | 5 |
| −2 | 8 |
| 2 | 8 |

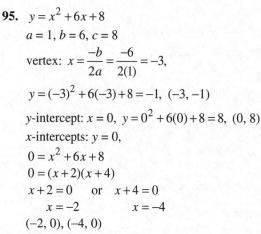

**95.** $y = x^2 + 6x + 8$

$a = 1, b = 6, c = 8$

vertex: $x = \dfrac{-b}{2a} = \dfrac{-6}{2(1)} = -3$,

$y = (-3)^2 + 6(-3) + 8 = -1$, (−3, −1)

$y$-intercept: $x = 0$, $y = 0^2 + 6(0) + 8 = 8$, (0, 8)

$x$-intercepts: $y = 0$,

$\quad 0 = x^2 + 6x + 8$

$\quad 0 = (x + 2)(x + 4)$

$\quad x + 2 = 0 \quad$ or $\quad x + 4 = 0$

$\quad\quad x = -2 \quad\quad\quad\quad x = -4$

(−2, 0), (−4, 0)

| $x$ | $y$ |
|-----|-----|
| −3 | −1 |
| 0 | 8 |
| −2 | 0 |
| −4 | 0 |
| −1 | 3 |

**96.** $y = x^2 - 2x - 4$

$a = 1, b = -2, c = -4$

vertex: $x = \dfrac{-b}{2a} = \dfrac{-(-2)}{2(1)} = 1,$

$y = 1^2 - 2(1) - 4 = -5, \ (1, -5)$

$y$-intercept: $x = 0, \ y = 0^2 - 2(0) - 4 = -4, \ (0, -4)$

$x$-intercepts $y = 0,$

$0 = x^2 - 2x - 4$

$x = \dfrac{-(-2) \pm \sqrt{(-2)^2 - 4(1)(-4)}}{2(1)}$

$= \dfrac{2 \pm \sqrt{4 + 16}}{2}$

$= \dfrac{2 \pm \sqrt{20}}{2}$

$= \dfrac{2 \pm 2\sqrt{5}}{2}$

$= 1 \pm \sqrt{5}$

$\left(1 - \sqrt{5}, 0\right), \left(1 + \sqrt{5}, 0\right)$

| $x$ | $y$ |
|-----------|-----|
| 1 | $-5$ |
| 0 | $-4$ |
| $1 - \sqrt{5}$ | 0 |
| $1 + \sqrt{5}$ | 0 |
| $-1$ | $-1$ |

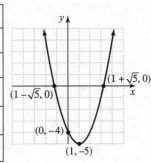

**Chapter 9 Test**

**1.** $5k^2 = 80$

$k^2 = 16$

$k = \pm\sqrt{16}$

$k = \pm 4$

The solutions are $\pm 4$.

**2.** $(3m - 5)^2 = 8$

$3 - 5 = \pm\sqrt{8}$

$3m - 5 = \pm 2\sqrt{2}$

$3m = 5 \pm 2\sqrt{2}$

$m = \dfrac{5 \pm 2\sqrt{2}}{3}$

The solutions are $\dfrac{5 \pm 2\sqrt{2}}{3}$.

**3.** $x^2 - 26x + 160 = 0$

$x^2 - 26x = -160$

$x^2 - 26x + 169 = -160 + 169$

$(x - 13)^2 = 9$

$x - 13 = \pm\sqrt{9}$

$x - 13 = \pm 3$

$x = 13 - 3 \quad \text{or} \quad x = 13 + 3$

$x = 10 \qquad\qquad x = 16$

The solutions are 10 and 16.

**4.** $3x^2 + 12x - 4 = 0$

$x^2 + 4x - \dfrac{4}{3} = 0$

$x^2 + 4x = \dfrac{4}{3}$

$x^2 + 4x + 4 = \dfrac{4}{3} + 4$

$(x + 2)^2 = \dfrac{16}{3}$

$x + 2 = \pm\sqrt{\dfrac{16}{3}}$

$x = -2 \pm \dfrac{4}{\sqrt{3}}$

$x = -2 \pm \dfrac{4\sqrt{3}}{3}$

$x = \dfrac{-6 \pm 4\sqrt{3}}{3}$

The solutions are $\dfrac{-6 \pm 4\sqrt{3}}{3}$.

**5.** $x^2 - 3x - 10 = 0$

$a = 1, b = -3,$ and $c = -10$

$$x = \frac{-(-3) \pm \sqrt{(-3)^2 - 4(1)(-10)}}{2(1)}$$

$$= \frac{3 \pm \sqrt{9 + 40}}{2}$$

$$= \frac{3 \pm \sqrt{49}}{2}$$

$$= \frac{3 \pm 7}{2}$$

$$x = \frac{3 - 7}{2} = -2 \quad \text{or} \quad x = \frac{3 + 7}{2} = 5$$

The solutions are $-2$ and $5$.

**6.** $p^2 - \dfrac{5}{3}p - \dfrac{1}{3} = 0$

$3p^2 - 5p - 1 = 0$

$a = 3, b = -5,$ and $c = -1$

$$p = \frac{-(-5) \pm \sqrt{(-5)^2 - 4(3)(-1)}}{2(3)}$$

$$= \frac{5 \pm \sqrt{25 + 12}}{6}$$

$$= \frac{5 \pm \sqrt{37}}{6}$$

The solutions are $\dfrac{5 \pm \sqrt{37}}{6}$.

**7.** $(3x - 5)(x + 2) = -6$

$3x^2 + x - 10 = -6$

$3x^2 + x - 4 = 0$

$(3x + 4)(x - 1) = 0$

$3x + 4 = 0 \quad \text{or} \quad x - 1 = 0$

$\qquad x = -\dfrac{4}{3} \qquad\qquad x = 1$

The solutions are $-\dfrac{4}{3}$ and $1$.

**8.** $(3x - 1)^2 = 16$

$3x - 1 = \pm\sqrt{16}$

$3x - 1 = \pm 4$

$3x = 1 \pm 4$

$x = \dfrac{1 \pm 4}{3}$

$x = \dfrac{1 - 4}{3} = -1 \quad \text{or} \quad x = \dfrac{1 + 4}{3} = \dfrac{5}{3}$

The solutions are $-1$ and $\dfrac{5}{3}$.

**9.** $3x^2 - 7x - 2 = 0$

$a = 3, b = -7,$ and $c = -2$

$$x = \frac{-(-7) \pm \sqrt{(-7)^2 - 4(3)(-2)}}{2(3)}$$

$$= \frac{7 \pm \sqrt{49 + 24}}{6}$$

$$= \frac{7 \pm \sqrt{73}}{6}$$

The solutions are $\dfrac{7 \pm \sqrt{73}}{6}$.

**10.** $x^2 - 4x + 5 = 0$

$a = 1, b = -4,$ and $c = 5$

$$x = \frac{-(-4) \pm \sqrt{(-4)^2 - 4(1)(5)}}{2(1)}$$

$$= \frac{4 \pm \sqrt{16 - 20}}{2}$$

$$= \frac{4 \pm \sqrt{-4}}{2}$$

$$= \frac{4 \pm 2i}{2}$$

$$= 2 \pm i$$

The solutions are $2 \pm i$.

**11.** $3x^2 - 7x + 2 = 0$

$(3x - 1)(x - 2) = 0$

$3x - 1 = 0 \quad \text{or} \quad x - 2 = 0$

$\qquad x = \dfrac{1}{3} \qquad\qquad x = 2$

The solutions are $\dfrac{1}{3}$ and $2$.

      

**12.** $2x^2 - 6x + 1 = 0$

$a = 1$, $b = -6$, and $c = 1$

$x = \dfrac{-(-6) \pm \sqrt{(-6)^2 - 4(2)(1)}}{2(2)}$

$= \dfrac{6 \pm \sqrt{36 - 8}}{4}$

$= \dfrac{6 \pm \sqrt{28}}{4}$

$= \dfrac{6 \pm 2\sqrt{7}}{4}$

$= \dfrac{3 \pm \sqrt{7}}{2}$

The solutions are $\dfrac{3 \pm \sqrt{7}}{2}$.

**13.**
$9x^3 = x$

$9x^3 - x = 0$

$x(9x^2 - 1) = 0$

$x(3x + 1)(3x - 1) = 0$

$x = 0$ or $3x + 1 = 0$ or $3x - 1 = 0$

$x = 0 \qquad\qquad x = -\dfrac{1}{3} \qquad\qquad x = \dfrac{1}{3}$

The solutions are 0 and $\pm\dfrac{1}{3}$.

**14.** $\sqrt{-25} = i\sqrt{25} = 5i = 0 + 5i$

**15.** $\sqrt{-200} = i\sqrt{100 \cdot 2}$

$= i\sqrt{100} \cdot \sqrt{2}$

$= 10i\sqrt{2}$

$= 0 + 10i\sqrt{2}$

**16.** $(3 + 2i) + (5 - i) = 3 + 2i + 5 - i = 8 + i$

**17.** $(3 + 2i) - (3 - 2i) = 3 + 2i - 3 + 2i = 4i = 0 + 4i$

**18.** $(3 + 2i)(3 - 2i) = (3)^2 - (2i)^2$

$= 9 - 4i^2$

$= 9 - 4(-1)$

$= 9 + 4$

$= 13$

$= 13 + 0i$

**19.** $\dfrac{3 - i}{1 + 2i} = \dfrac{3 - i}{1 + 2i} \cdot \dfrac{1 - 2i}{1 - 2i}$

$= \dfrac{3 - 6i - i + 2i^2}{1^2 - (2i)^2}$

$= \dfrac{3 - 7i + 2(-1)}{1 - 4i^2}$

$= \dfrac{3 - 7i - 2}{1 - 4(-1)}$

$= \dfrac{1 - 7i}{1 + 4}$

$= \dfrac{1}{5} - \dfrac{7}{5}i$

**20.** $y = -5x^2$

$a = -5$, $b = 0$, $c = 0$

vertex: $x = \dfrac{-b}{2a} = \dfrac{-0}{2(-5)} = 0$

$y = -5(0)^2 = 0$, $(0, 0)$

$y$-intercept: $x = 0$, $y = 0$, $(0, 0)$

$x$-intercept: $y = 0$, $x = 0$, $(0, 0)$

| $x$ | $y$ |
|-----|-----|
| $-2$ | $-20$ |
| $-1$ | $-5$ |
| $0$ | $0$ |
| $1$ | $-5$ |
| $2$ | $-20$ |

**21.** $y = x^2 - 4$

$a = 1$, $b = 0$, $c = -4$

vertex: $x = \dfrac{-b}{2a} = \dfrac{-0}{2(1)} = 0$, $y = 0^2 - 4 = -4$,

$(0, -4)$

$y$-intercept: $x = 0$, $y = 0^2 - 4 = -4$, $(0, -4)$

$x$-intercept: $y = 0$,

$0 = x^2 - 4$

$4 = x^2$

$\pm 2 = x$

$(2, 0)$, $(-2, 0)$

| $x$ | $y$ |
|---|---|
| $-2$ | $0$ |
| $-1$ | $-3$ |
| $0$ | $-4$ |
| $1$ | $-3$ |
| $2$ | $0$ |

**22.** $y = x^2 - 7x + 10$

$a = 1,\ b = -7,\ c = 10$

vertex: $x = \dfrac{-b}{2a} = \dfrac{-(-7)}{2(1)} = \dfrac{7}{2}$

$y = \left(\dfrac{7}{2}\right)^2 - 7\left(\dfrac{7}{2}\right) + 10 = -\dfrac{9}{4}$

$\left(\dfrac{7}{2},\ -\dfrac{9}{4}\right)$

$y$-intercept: $x = 0,\ y = 0^2 - 7 \cdot 0 + 10 = 10,\ (0, 10)$

$x$-intercepts: $y = 0$,

$0 = x^2 - 7x + 10$

$0 = (x-2)(x-5)$

$x = 2$ or $x = 5$

$(2, 0),\ (5, 0)$

| $x$ | $y$ |
|---|---|
| $0$ | $10$ |
| $2$ | $0$ |
| $\dfrac{7}{2}$ | $-\dfrac{9}{4}$ |
| $5$ | $0$ |
| $7$ | $10$ |

**23.** $y = 2x^2 + 4x - 1$

$a = 2,\ b = 4,\ c = -1$

vertex: $x = \dfrac{-b}{2a} = \dfrac{-4}{2(2)} = -1,$

$y = 2(-1)^2 + 4(-1) - 1 = -3,\ (-1, -3)$

$x$-intercept: $y = 0$,

$0 = 2x^2 + 4x - 1$

$x = \dfrac{-4 \pm \sqrt{4^2 - 4(2)(-1)}}{2(2)}$

$= \dfrac{-4 \pm \sqrt{16 + 8}}{4}$

$= \dfrac{-4 \pm \sqrt{24}}{4}$

$= \dfrac{-4 \pm 2\sqrt{6}}{4}$

$= \dfrac{-2 \pm \sqrt{6}}{2}$

$\left(\dfrac{-2 - \sqrt{6}}{2},\ 0\right),\ \left(\dfrac{-2 + \sqrt{6}}{2},\ 0\right)$

$y$-intercept: $x = 0,\ y = -1,\ (0, -1)$

| $x$ | $y$ |
|---|---|
| $-2$ | $-1$ |
| $-1$ | $-3$ |
| $0$ | $-1$ |
| $1$ | $5$ |
| $2$ | $15$ |

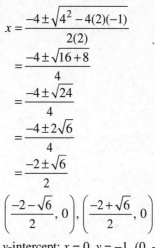

**24.** Let $x =$ the length of the base;
then $4x =$ the height.

$A = \dfrac{1}{2}bh$

$18 = \dfrac{1}{2}x(4x)$

$18 = 2x^2$

$9 = x^2$

$\pm 3 = x$

The length cannot be negative, so the base is 3 feet and the height is $4(3) = 12$ feet.

**25.** $d = \dfrac{n^2 - 3n}{2};\ d = 9$

$9 = \dfrac{n^2 - 3n}{2}$

$18 = n^2 - 3n$

$0 = n^2 - 3n - 18$

$0 = (n-6)(n+3)$

$n - 6 = 0$ or $n + 3 = 0$

$n = 6$        $n = -3$

The number of sides cannot be negative, so $n = 6$ sides.

        

**26.** Let $h = 120.75$.

$$16t^2 = h$$

$$16t^2 = 120.75$$

$$t^2 = \frac{120.75}{16}$$

$$t = \pm\sqrt{\frac{120.75}{16}} \approx \pm 2.7$$

The length of time is not a negative number so the dive lasted 2.7 seconds.

**Chapter 9 Cumulative Review**

**1. a.**   $\dfrac{x-y}{12+x} = \dfrac{2-(-5)}{12+2} = \dfrac{7}{14} = \dfrac{1}{2}$

   **b.**   $x^2 - 3y = (2)^2 - 3(-5) = 4 + 15 = 19$

**2. a.**   $\dfrac{x-y}{7-x} = \dfrac{(-4)-7}{7-(-4)} = \dfrac{-11}{11} = -1$

   **b.**   $x^2 + 2y = (-4)^2 + 2(7) = 16 + 14 = 30$

**3. a.**   $2x + 3x + 5 + 2 = 5x + 7$

   **b.**   $-5a - 3 + a + 2 = -4a - 1$

   **c.**   $4y - 3y^2 = 4y - 3y^2$

   **d.**   $2.3x + 5x - 6 = 7.3x - 6$

   **e.**   $-\dfrac{1}{2}b + b = \dfrac{1}{2}b$

**4. a.**   $4x - 3 + 7 - 5x = -x + 4$

   **b.**   $-6y + 3y - 8 + 8y = 5y - 8$

   **c.**   $2 + 8.1a + a - 6 = 9.1a - 4$

   **d.**   $2x^2 - 2x = 2x^2 - 2x$

**5. a.**   $x$-intercept: $(-3, 0)$
   $y$-intercept: $(0, 2)$

   **b.**   $x$-intercepts: $(-4, 0), (-1, 0)$
   $y$-intercept: $(0, 1)$

   **c.**   $x$-intercept: $(0, 0)$
   $y$-intercept: $(0, 0)$

   **d.**   $x$-intercept: $(2, 0)$
   $y$-intercept: none

   **e.**   $x$-intercepts: $(-1, 0), (3, 0)$
   $y$-intercepts: $(0, 2), (0, -1)$

**6. a.**   $x$-intercept: $(4, 0)$
   $y$-intercept: $(0, 1)$

   **b.**   $x$-intercepts: $(-2, 0), (0, 0), (3, 0)$
   $y$-intercept: $(0, 0)$

   **c.**   $x$-intercept: none
   $y$-intercept: $(0, -3)$

   **d.**   $x$-intercepts: $(-3, 0), (3, 0)$
   $y$-intercepts: $(0, -3), (0, 3)$

**7.**   $y = -\dfrac{1}{5}x + 1: m_1 = -\dfrac{1}{5}$

$$2x + 10y = 3$$

$$y = -\dfrac{1}{5}x + \dfrac{3}{10}: m_2 = -\dfrac{1}{5}$$

$m_1 = m_2$

They are parallel.

**8.**   $y = 3x + 7: m_1 = 3$

$$x + 3y = -15$$

$$y = -\dfrac{1}{3}x - 5: m_2 = -\dfrac{1}{3}$$

$m_1 \cdot m_2 = -1$

They are perpendicular.

**9.**   $\begin{cases} 2x + y = 7 \\ 2y = -4x \end{cases}$

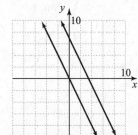

There is no solution, the lines are parallel.

**10.** $\begin{cases} y = x+2 \\ 2x+y = 5 \end{cases}$

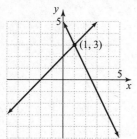

The solution is (1, 3).

**11.** $\begin{cases} 7x-3y = -14 \\ -3x+y = 6 \end{cases}$

Multiply the second equation by 3.

$$\begin{aligned} 7x-3y &= -14 \\ -9x+3y &= 18 \\ \hline -2x\phantom{+3y} &= 4 \\ x &= -2 \end{aligned}$$

Let $x = -2$ in the second equation.

$$\begin{aligned} -3(-2)+y &= 6 \\ 6+y &= 6 \\ y &= 0 \end{aligned}$$

The solution of the system is (–2, 0).

**12.** $\begin{cases} 5x+y = 3 \\ y = -5x \end{cases}$

Substitute $-5x$ for $y$ in the first equation.

$$\begin{aligned} 5x+(-5x) &= 3 \\ 0 &= 3 \end{aligned}$$

There is no solution.

**13.** $\begin{cases} 3x-2y = 2 \\ -9x+6y = -6 \end{cases}$

Multiply the first equation by 3.

$$\begin{aligned} 9x-6y &= 6 \\ -9x+6y &= -6 \\ \hline 0 &= 0 \end{aligned}$$

The system has an infinite number of solutions.

**14.** $\begin{cases} -2x+y = 7 \\ 6x-3y = -21 \end{cases}$

Multiply the first equation by 3.

$$\begin{aligned} -6x+3y &= 21 \\ 6x-3y &= -21 \\ \hline 0 &= 0 \end{aligned}$$

The system has an infinite number of solutions.

**15.** Let $x$ = the rate of Albert.

| | Rate | · Time | = Distance |
|---|---|---|---|
| Albert | $x$ | 2 | $2x$ |
| Louis | $x+1$ | 2 | $2(x+1)$ |
| Total | | | 15 |

$$\begin{aligned} 2x+2(x+1) &= 15 \\ 2x+2x+2 &= 15 \\ 4x+2 &= 15 \\ 4x &= 13 \\ x &= 3.25 \end{aligned}$$

$x+1 = 3.25+1 = 4.25$
Albert: 3.25 mph; Louis: 4.25 mph

**16.** Let $x$ = the number of dimes, and
$15-x$ = the number of quarters.

| | No. of Coins | · Value | = Amt. of Money |
|---|---|---|---|
| Dimes | $x$ | 0.1 | $0.1x$ |
| Quarters | $15-x$ | 0.25 | $0.25(15-x)$ |
| Total | 15 | | 2.85 |

$$\begin{aligned} 0.1x+0.25(15-x) &= 2.85 \\ 0.1x+3.75-0.25x &= 2.85 \\ -0.15x+3.75 &= 2.85 \\ -0.15x &= -0.9 \\ x &= 6 \end{aligned}$$

$15-x = 15-6 = 9$
There are 6 dimes and 9 quarters in the purse.

**17.** $\begin{cases} -3x+4y < 12 \\ x \geq 2 \end{cases}$

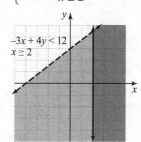

**18.** $\begin{cases} 2x - y \le 6 \\ \quad y \ge 2 \end{cases}$

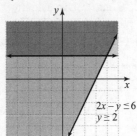

**19. a.** $x^7 \cdot x^4 = x^{7+4} = x^{11}$

**b.** $\left(\dfrac{t}{2}\right)^4 = \dfrac{t^4}{2^4} = \dfrac{t^4}{16}$

**c.** $(9y^5)^2 = 9^2 y^{5 \cdot 2} = 81y^{10}$

**20. a.** $\left(\dfrac{-6x}{y^3}\right)^3 = \dfrac{(-6)^3 x^3}{y^9} = \dfrac{-216x^3}{y^9}$

**b.** $\dfrac{a^2 b^7}{(2b^2)^5} = \dfrac{a^2 b^7}{2^5 b^{10}}$

$\qquad = \dfrac{a^2}{32} \cdot b^{7-10}$

$\qquad = \dfrac{a^2}{32} \cdot b^{-3}$

$\qquad = \dfrac{a^2}{32b^3}$

**c.** $\dfrac{(3y)^2}{y^2} = \dfrac{3^2 y^2}{y^2} = 9y^{2-2} = 9$

**d.** $\dfrac{(x^2 y^4)^2}{xy^3} = \dfrac{x^4 y^8}{xy^3} = x^{4-1} y^{8-3} = x^3 y^5$

**21.** $(5x-1)(2x^2 + 15x + 18) = 0$
$(5x-1)(2x+3)(x+6) = 0$
$5x - 1 = 0 \quad$ or $\quad 2x + 3 = 0 \quad$ or $\quad x + 6 = 0$
$\quad x = \dfrac{1}{5} \qquad\qquad x = -\dfrac{3}{2} \qquad\qquad x = -6$

The solutions are $\dfrac{1}{5}, -\dfrac{3}{2},$ and $-6$.

**22.** $(x+1)(2x^2 - 3x - 5) = 0$
$(x+1)(2x-5)(x+1) = 0$
$x + 1 = 0 \quad$ or $\quad 2x - 5 = 0$
$\quad x = -1 \qquad\qquad x = \dfrac{5}{2}$

The solutions are $-1$ and $\dfrac{5}{2}$.

**23.** $\dfrac{45}{x} = \dfrac{5}{7}$
$45 \cdot 7 = 5x$
$315 = 5x$
$63 = x$

**24.** $\dfrac{2x+7}{3} = \dfrac{x-6}{2}$
$6\left(\dfrac{2x+7}{3}\right) = 6\left(\dfrac{x-6}{2}\right)$
$2(2x+7) = 3(x-6)$
$4x + 14 = 3x - 18$
$x + 14 = -18$
$x = -32$

**25. a.** $\sqrt[4]{16} = \sqrt[4]{2^4} = 2$

**b.** $\sqrt[5]{-32} = \sqrt[5]{(-2)^5} = -2$

**c.** $-\sqrt[3]{8} = -\sqrt[3]{(2)^3} = -2$

**d.** $\sqrt[4]{-81}$ is not a real number.

**26. a.** $\sqrt[3]{27} = \sqrt[3]{3^3} = 3$

**b.** $-\sqrt[4]{256} = -\sqrt[4]{4^4} = -4$

**c.** $\sqrt[3]{-125} = \sqrt[3]{(-5)^3} = -5$

**d.** $\sqrt[5]{1} = \sqrt[5]{1^5} = 1$

**27. a.** $\sqrt{\dfrac{25}{36}} = \dfrac{\sqrt{25}}{\sqrt{36}} = \dfrac{5}{6}$

**b.** $\sqrt{\dfrac{3}{64}} = \dfrac{\sqrt{3}}{\sqrt{64}} = \dfrac{\sqrt{3}}{8}$

c. $\sqrt{\dfrac{40}{81}} = \dfrac{\sqrt{40}}{\sqrt{81}} = \dfrac{\sqrt{4 \cdot 10}}{9} = \dfrac{2\sqrt{10}}{9}$

**28. a.** $\sqrt{\dfrac{4}{25}} = \dfrac{\sqrt{4}}{\sqrt{25}} = \dfrac{2}{5}$

    **b.** $\sqrt{\dfrac{16}{121}} = \dfrac{\sqrt{16}}{\sqrt{121}} = \dfrac{4}{11}$

    **c.** $\sqrt{\dfrac{2}{49}} = \dfrac{\sqrt{2}}{\sqrt{49}} = \dfrac{\sqrt{2}}{7}$

**29. a.** $\sqrt{50} + \sqrt{8} = \sqrt{25 \cdot 2} + \sqrt{4 \cdot 2}$
$$= 5\sqrt{2} + 2\sqrt{2}$$
$$= 7\sqrt{2}$$

    **b.** $7\sqrt{12} - \sqrt{75} = 7\sqrt{4 \cdot 3} - \sqrt{25 \cdot 3}$
$$= 7 \cdot 2\sqrt{3} - 5\sqrt{3}$$
$$= 14\sqrt{3} - 5\sqrt{3}$$
$$= 9\sqrt{3}$$

    **c.** $\sqrt{25} - \sqrt{27} - 2\sqrt{18} - \sqrt{16}$
$$= 5 - \sqrt{9 \cdot 3} - 2\sqrt{9 \cdot 2} - 4$$
$$= 1 - 3\sqrt{3} - 2 \cdot 3\sqrt{2}$$
$$= 1 - 3\sqrt{3} - 6\sqrt{2}$$

**30. a.** $\sqrt{80} + \sqrt{20} = \sqrt{16 \cdot 5} + \sqrt{4 \cdot 5}$
$$= 4\sqrt{5} + 2\sqrt{5}$$
$$= 6\sqrt{5}$$

    **b.** $2\sqrt{98} - 2\sqrt{18} = 2\sqrt{49 \cdot 2} - 2\sqrt{9 \cdot 2}$
$$= 2 \cdot 7\sqrt{2} - 2 \cdot 3\sqrt{2}$$
$$= 14\sqrt{2} - 6\sqrt{2}$$
$$= 8\sqrt{2}$$

    **c.** $\sqrt{32} + \sqrt{121} - \sqrt{12} = \sqrt{16 \cdot 2} + 11 - \sqrt{4 \cdot 3}$
$$= 11 + 4\sqrt{2} - 2\sqrt{3}$$

**31. a.** $\sqrt{7} \cdot \sqrt{3} = \sqrt{7 \cdot 3} = \sqrt{21}$

    **b.** $\sqrt{3} \cdot \sqrt{3} = 3$

    **c.** $\sqrt{3} \cdot \sqrt{15} = \sqrt{3 \cdot 15} = \sqrt{9 \cdot 5} = 3\sqrt{5}$

    **d.** $2\sqrt{3} \cdot 5\sqrt{2} = 10\sqrt{3 \cdot 2} = 10\sqrt{6}$

    **e.** $\sqrt{2x^3} \cdot \sqrt{6x} = \sqrt{12x^4} = \sqrt{4x^4 \cdot 3} = 2x^2\sqrt{3}$

**32. a.** $\sqrt{2} \cdot \sqrt{5} = \sqrt{2 \cdot 5} = \sqrt{10}$

    **b.** $\sqrt{56} \cdot \sqrt{7} = \sqrt{56 \cdot 7}$
$$= \sqrt{392}$$
$$= \sqrt{196 \cdot 2}$$
$$= 14\sqrt{2}$$

    **c.** $\left(4\sqrt{3}\right)^2 = 4^2\left(\sqrt{3}\right)^2 = 16 \cdot 3 = 48$

    **d.** $3\sqrt{8} \cdot 7\sqrt{2} = 21\sqrt{8 \cdot 2} = 21\sqrt{16} = 21 \cdot 4 = 84$

**33.** $\sqrt{x} = \sqrt{5x - 2}$
$$\left(\sqrt{x}\right)^2 = \left(\sqrt{5x - 2}\right)^2$$
$$x = 5x - 2$$
$$0 = 4x - 2$$
$$2 = 4x$$
$$\frac{2}{4} = x$$
$$\frac{1}{2} = x$$

**34.** $\sqrt{x - 4} + 7 = 2$
$$\sqrt{x - 4} = -5$$
The square root of a real number cannot be negative. There is no solution.

**35.** $a^2 + b^2 = c^2$
$$\left(\overline{PQ}\right)^2 + \left(\overline{QR}\right)^2 = \left(\overline{PR}\right)^2$$
$$\left(\overline{PQ}\right)^2 + (240)^2 = (320)^2$$
$$\left(\overline{PQ}\right)^2 + 57,600 = 102,400$$
$$\left(\overline{PQ}\right)^2 = 44,800$$
$$\overline{PQ} = \sqrt{44,800} \approx 212$$
The distance is approximately 212 feet.

**36.** $(-7, 4)$ and $(2, 5)$
$$d = \sqrt{(x_2 - x_1)^2 + (y_2 - y_1)^2}$$
$$d = \sqrt{(2 - (-7))^2 + (5 - 4)^2}$$
$$= \sqrt{9^2 + 1^2}$$
$$= \sqrt{81 + 1}$$
$$= \sqrt{82}$$

**37. a.** $25^{1/2} = \sqrt{25} = 5$

   **b.** $8^{1/3} = \sqrt[3]{8} = 2$

   **c.** $-16^{1/4} = -\sqrt[4]{16} = -2$

   **d.** $(-27)^{1/3} = \sqrt[3]{-27} = -3$

   **e.** $\left(\dfrac{1}{9}\right)^{1/2} = \sqrt{\dfrac{1}{9}} = \dfrac{1}{3}$

**38. a.** $-49^{1/2} = -\sqrt{49} = -7$

   **b.** $256^{1/4} = \sqrt[4]{256} = \sqrt[4]{(4)^4} = 4$

   **c.** $(-64)^{1/3} = \sqrt[3]{-64} = -4$

   **d.** $\left(\dfrac{25}{36}\right)^{1/2} = \sqrt{\dfrac{25}{36}} = \dfrac{5}{6}$

   **e.** $(32)^{1/5} = \sqrt[5]{32} = \sqrt[5]{(2)^5} = 2$

**39.** $2x^2 = 7$

$$x^2 = \frac{7}{2}$$

$$x = \pm\sqrt{\frac{7}{2}}$$

$$x = \pm\frac{\sqrt{7}}{\sqrt{2}} \cdot \frac{\sqrt{2}}{\sqrt{2}} = \pm\frac{\sqrt{14}}{2}$$

**40.** $(x-4)^2 = 3$

$$x - 4 = \pm\sqrt{3}$$

$$x = 4 \pm \sqrt{3}$$

**41.** $x^2 - 10x = -14$

$$x^2 - 10x + 25 = -14 + 25$$

$$(x-5)^2 = 11$$

$$x - 5 = \pm\sqrt{11}$$

$$x = 5 \pm \sqrt{11}$$

**42.** $x^2 + 4x = 8$

$$x^2 + 4x + 4 = 8 + 4$$

$$(x+2)^2 = 12$$

$$x + 2 = \pm\sqrt{12}$$

$$x = -2 \pm \sqrt{4 \cdot 3}$$

$$x = -2 \pm 2\sqrt{3}$$

**43.** $2x^2 - 9x = 5$

$$2x^2 - 9x - 5 = 0$$

$a = 2$, $b = -9$, and $c = -5$

$$x = \frac{-(-9) \pm \sqrt{(-9)^2 - 4(2)(-5)}}{2(2)}$$

$$= \frac{9 \pm \sqrt{81 + 40}}{4}$$

$$= \frac{9 \pm \sqrt{121}}{4}$$

$$= \frac{9 \pm 11}{4}$$

$$x = \frac{9+11}{4} = 5 \quad \text{or} \quad x = \frac{9-11}{4} = -\frac{1}{2}$$

The solutions are $5$ and $-\dfrac{1}{2}$.

**44.** $2x^2 + 5x = 7$

$$2x^2 + 5x - 7 = 0$$

$a = 2$, $b = 5$, and $c = -7$

$$x = \frac{-5 \pm \sqrt{5^2 - 4(2)(-7)}}{2(2)}$$

$$= \frac{-5 \pm \sqrt{25 + 56}}{4}$$

$$= \frac{-5 \pm \sqrt{81}}{4}$$

$$= \frac{-5 \pm 9}{4}$$

$$x = \frac{-5+9}{4} = 1 \quad \text{or} \quad x = \frac{-5-9}{4} = -\frac{7}{2}$$

The solutions are $1$ and $-\dfrac{7}{2}$.

**45. a.** $\sqrt{-4} = i\sqrt{4} = 2i$

   **b.** $\sqrt{-11} = i\sqrt{11}$

   **c.** $\sqrt{-20} = i\sqrt{20} = i\sqrt{4 \cdot 5} = 2i\sqrt{5}$

**46. a.** $\sqrt{-7} = i\sqrt{7}$

   **b.** $\sqrt{-16} = i\sqrt{16} = 4i$

   **c.** $\sqrt{-27} = i\sqrt{27} = i\sqrt{9 \cdot 3} = 3i\sqrt{3}$

**47.** $y = x^2 - 4$

*y*-intercept: $x = 0$, $y = 0^2 - 4 = -4$, $(0, -4)$

vertex: $(0, -4)$

*x*-intercepts: $y = 0$,

$0 = x^2 - 4$

$0 = (x + 2)(x - 2)$

$x + 2 = 0$   or   $x - 2 = 0$

   $x = -2$         $x = 2$

$(-2, 0)$ and $(2, 0)$

| x | y |
|----|----|
| −2 | 0 |
| −1 | −3 |
| 0 | −4 |
| 1 | −3 |
| 2 | 0 |

$y = x^2 - 4$

**48.** $y = x^2 + 2x + 3$

$a = 1, b = 2, c = 3$

vertex: $x = \dfrac{-b}{2a} = \dfrac{-2}{2(1)} = -1$

         $y = (-1)^2 + 2(-1) + 3 = 2$

         $(-1, 2)$

*y*-intercept: $x = 0$, $y = 0^2 + 2(0) + 3 = 3$, $(0, 3)$

*x*-intercepts: $y = 0$, $0 = x^2 + 2x + 3$

$x = \dfrac{-2 \pm \sqrt{2^2 - 4(1)(3)}}{2(1)}$

$= \dfrac{-2 \pm \sqrt{4 - 12}}{2}$

$= \dfrac{-2 \pm \sqrt{-8}}{2}$

There are no real solutions to the equation, so there are no *x*-intercepts.

| x | y |
|----|----|
| −3 | 6 |
| −2 | 3 |
| −1 | 2 |
| 0 | 3 |
| 1 | 6 |

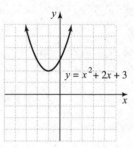

$y = x^2 + 2x + 3$

     

# Appendix A

**Exercise Set A.3**

**1.** $90° − 19° = 71°$

**3.** $90° − 70.8° = 19.2°$

**5.** $90° − 11\dfrac{1}{4}° = 78\dfrac{3}{4}°$

**7.** $180° − 150° = 30°$

**9.** $180° − 30.2° = 149.8°$

**11.** $180° − 79\dfrac{1}{2}° = 100\dfrac{1}{2}°$

**13.** $m\angle 1 = 110°$
$m\angle 2 = 180° − 110° = 70°$
$m\angle 3 = m\angle 2 = 70°$
$m\angle 4 = m\angle 2 = 70°$
$m\angle 5 = m\angle 1 = 110°$
$m\angle 6 = m\angle 4 = 70°$
$m\angle 7 = m\angle 5 = 110°$

**15.** $180° − 11° − 79° = 90°$

**17.** $180° − 25° − 65° = 90°$

**19.** $180° − 30° − 60° = 90°$

**21.** $90° − 45° = 45°$
$45°, 90°$

**23.** $90° − 17° = 73°$
$73°, 90°$

**25.** $90° − 39\dfrac{3}{4}° = 50\dfrac{1}{4}°$
$50\dfrac{1}{4}°, 90°$

**27.** $\dfrac{12}{4} = \dfrac{18}{x}$
$4x\left(\dfrac{12}{4}\right) = 4x\left(\dfrac{18}{x}\right)$
$12x = 72$
$x = 6$

**29.** $\dfrac{6}{9} = \dfrac{3}{x}$
$9x\left(\dfrac{6}{9}\right) = 9x\left(\dfrac{3}{x}\right)$
$6x = 27$
$x = 4.5$

**31.** $a^2 + b^2 = c^2$
$6^2 + 8^2 = c^2$
$36 + 64 = c^2$
$100 = c^2$
$10 = c$

**33.** $a^2 + b^2 = c^2$
$5^2 + b^2 = 13^2$
$25 + b^2 = 169$
$b^2 = 144$
$b = 12$

**Exercise Set A.4**

**1.** Volume $= lwh = 6(4)(3) = 72$ cu in.
Surface area $= 2lh + 2wh + 2lw$
$\qquad = 2(6)(3) + 2(4)(3) + 2(6)(4)$
$\qquad = 36 + 24 + 48$
$\qquad = 108$ sq in.

**3.** Volume $= s^3 = 8^3 = 512$ cu cm
Surface area $= 6s^2 = 6(8)^2 = 384$ sq cm

**5.** Volume $= \dfrac{1}{3}\pi r^2 h = \dfrac{1}{3}\pi(2)^2(3) = 4\pi$ cu yd
$\qquad \approx 4\left(\dfrac{22}{7}\right) = 12\dfrac{4}{7}$ cu yd
Surface area $= \pi r\sqrt{r^2 + h^2} + \pi r^2$
$\qquad = \pi(2)\sqrt{2^2 + 3^2} + \pi 2^2$
$\qquad = \left(2\pi\sqrt{13} + 4\pi\right)$ sq yd
$\qquad \approx 2\sqrt{13}(3.14) + 4(3.14)$
$\qquad = 35.20$ sq yd

**7.** $\text{Volume} = \frac{4}{3}\pi r^3 = \frac{4}{3}\pi(5)^3 = \frac{500}{3}\pi$ cu in.

$\approx \frac{500}{3}\left(\frac{22}{7}\right) = 523\frac{17}{21}$ cu in.

$\text{Surface area} = 4\pi r^2 = 4\pi(5)^2 = 100\pi$ sq in.

$\approx 100\left(\frac{22}{7}\right) = 314\frac{2}{7}$ sq in.

**9.** $\text{Volume} = \frac{1}{3}s^2h = \frac{1}{3}(6)^2(4) = 48$ cu cm

$\text{Surface area} = B + \frac{1}{2}pl$

$= 36 + \frac{1}{2}(24)(5)$

$= 96$ sq cm

**11.** $\text{Volume} = s^3 = \left(1\frac{1}{3}\right)^3 = 2\frac{10}{27}$ cu in.

**13.** $\text{Surface area} = 2lh + 2wh + 2lw$

$= 2(2)(1.4) + 2(3)(1.4) + 2(2)(3)$

$= 5.6 + 8.4 + 12$

$= 26$ sq ft

**15.** $\text{Volume} = \frac{1}{3}s^2h = \frac{1}{3}(5)^2(1.3) = 10\frac{5}{6}$ cu in.

**17.** $\text{Volume} = \frac{1}{3}s^2h = \frac{1}{3}(12)^2(20) = 960$ cu cm

**19.** $\text{Surface area} = 4\pi r^2 = 4\pi(7)^2 = 196\pi$ sq in.

**21.** $\text{Volume} = lwh = 2\left(2\frac{1}{2}\right)\left(1\frac{1}{2}\right) = 7\frac{1}{2}$ cu ft

**23.** $\text{Volume} = \frac{1}{3}\pi r^2h$

$\approx \frac{1}{3}\left(\frac{22}{7}\right)(2)^2(3) = 12\frac{4}{7}$ cu cm

    

# Appendix B

**1.** $\dfrac{30}{10} = \dfrac{15}{y}$

$30y = 150$

$y = 5$

**3.** $\dfrac{8}{15} = \dfrac{z}{6}$

$48 = 15z$

$\dfrac{48}{15} = z$

$z = \dfrac{16}{5}$ or 3.2

**5.** $\dfrac{-3.5}{12.5} = \dfrac{-7}{n}$

$-3.5n = -87.5$

$n = 25$

**7.** $\dfrac{n}{0.6} = \dfrac{0.05}{12}$

$12n = 0.030$

$n = \dfrac{0.030}{12}$

$n = 0.0025$

**9.** $\dfrac{8}{\frac{2}{3}} = \dfrac{24}{n}$

$8n = 16$

$n = 2$

**11.** $\dfrac{7}{9} = \dfrac{35}{3x}$

$21x = 315$

$x = 15$

**13.** $\dfrac{7x}{18} = \dfrac{5}{3}$

$21x = 90$

$x = \dfrac{90}{21}$

$x = \dfrac{30}{7}$

**15.** $\dfrac{11}{7} = \dfrac{4}{x+1}$

$11(x+1) = 28$

$11x + 11 = 28$

$11x = 17$

$x = \dfrac{17}{11}$

**17.** $\dfrac{x-3}{2x+1} = \dfrac{4}{9}$

$9(x-3) = 4(2x+1)$

$9x - 27 = 8x + 4$

$x - 27 = 4$

$x = 31$

**19.** $\dfrac{2x+1}{4} = \dfrac{6x-1}{5}$

$5(2x+1) = 4(6x-1)$

$10x + 5 = 24x - 4$

$5 = 14x - 4$

$9 = 14x$

$\dfrac{9}{14} = x$

**21.** Let $x$ be the number of applications in the 14-oz bottle.

$\dfrac{3}{4} = \dfrac{14}{x}$

$3x = 56$

$x = \dfrac{56}{3} \approx 18.7$

There should be 18 full applications in the 14-oz bottle.

**23.** Let $x$ be the number of weeks that 8 reams last.

$\dfrac{5}{3} = \dfrac{8}{x}$

$5x = 24$

$x = \dfrac{24}{5} = 4.8$

A case of paper will last about 5 weeks.

**25.** Let $x$ be the number of servings for 4 cups of milk.

$$\frac{1\frac{1}{2}}{4} = \frac{4}{x}$$

$$\frac{3}{2}x = 16$$

$$x = 16 \cdot \frac{2}{3} = \frac{32}{3} = 10\frac{2}{3}$$

Ming can make $10\frac{2}{3}$ servings of pancakes.

**27.** Let $x$ be the number of calories in a 24-oz size.

$$\frac{16}{80} = \frac{24}{x}$$

$$16x = 1920$$

$$x = 120$$

There are 120 calories in the 24-oz size.

**29. a.** Let $x$ be the number of teaspoons needed to treat 450 square feet.

$$\frac{25}{1} = \frac{450}{x}$$

$$25x = 450$$

$$x = 18$$

18 teaspoons of granules are needed to treat 450 square feet.

**b.** 18 tsp = 6(3 tsp) = 6(1 tbsp) = 6 tbsp
6 tablespoons of granules must be used.

**31.** Let $x$ be the number of people that 3750 square feet of lawn provides oxygen for.

$$\frac{625}{1} = \frac{3750}{x}$$

$$625x = 3750$$

$$x = 6$$

3750 square feet of lawn provides oxygen for 6 people.

**33.** Let $x$ be the approximate height of the Statue of Liberty.

$$\frac{2}{5\frac{1}{3}} = \frac{42}{x}$$

$$2x = 224$$

$$x = 112$$

The approximate height of the Statue of Liberty is 112 feet. This is a difference of 11 inches from the actual height of 111 feet, 1 inch.

**35.** Let $x$ be the amount of cholesterol in 5 ounces of lobster.

$$\frac{3.5}{72} = \frac{5}{x}$$

$$3.5x = 360$$

$$x = \frac{360}{3.5} \approx 102.9$$

There are about 102.9 milligrams of cholesterol in 5 ounces of lobster.

**37.** Let $x$ be the number of visits in which medication is prescribed.

$$\frac{10}{7} = \frac{620}{x}$$

$$10x = 4340$$

$$x = 434$$

Out of 620 emergency room visits for injury, medication would be prescribed in 434 of the visits.

**39.** Let $x$ be the number of people expected to have worked in the restaurant industry.

$$\frac{3}{1} = \frac{84}{x}$$

$$3x = 84$$

$$x = 28$$

In an office of 84 people, 28 people are likely to have worked in the restaurant industry.

    

# Appendix C

**1.**   9.076
    + 8.004
    17.080

**3.**   27.004
    − 14.200
    12.804

**5.**   107.92
    +   3.04
    110.96

**7.**   10.0
    − 7.6
    2.4

**9.**   126.32
    − 97.89
    28.43

**11.**   3.25
    ×   70
    227.50

**13.**  3)8.1   (quotient 2.7)
       6
       2 1
       2 1
       0

**15.**   55.4050
    −  6.1711
    49.2339

**17.**  0.75)60  becomes  75)6000   (quotient 80)
           600
           00
           00

**19.**  7.612 ÷ 100 = 0.07612

**21.**  2.7)12.312  becomes  27)123.12   (quotient 4.56)
            108
            15 1
            13 5
            1 62
            1 62
            0

**23.**   569.20
       71.25
    +   8.01
    648.46

**25.**   768.00
    −   0.17
    767.83

**27.**   12.000
    +  0.062
    12.062

**29.**   76.00
    −  14.52
    61.48

**31.**  0.43)3.311  becomes  43)331.1   (quotient 7.7)
            301
            30 1
            30 1
            0

**33.**   762.12
       89.70
    +  11.55
    863.37

**35.**   23.400
    −  0.821
    22.579

**37.**   476.12
    − 112.97
    363.15

**39.**   0.007
    + 7.000
    7.007

# Appendix D

**1.** 21, 28, 16, 42, 38

$$\bar{x} = \frac{21+28+16+42+38}{5} = \frac{145}{5} = 29$$

16, 21, 28, 38, 42
median = 28
no mode

**3.** 7.6, 8.2, 8.2, 9.6, 5.7, 9.1

$$\bar{x} = \frac{7.6+8.2+8.2+9.6+5.7+9.1}{6} = \frac{48.4}{6} = 8.1$$

5.7, 7.6, 8.2, 8.2, 9.1, 9.6

$$\text{median} = \frac{8.2+8.2}{2} = 8.2$$

mode = 8.2

**5.** 0.2, 0.3, 0.5, 0.6, 0.6, 0.9, 0.2, 0.7, 1.1

$$\bar{x} = \frac{0.2+0.3+0.5+0.6+0.6+0.9+0.2+0.7+1.1}{9}$$

$$= \frac{5.1}{9}$$

$$= 0.6$$

median = 0.6
mode = 0.2 and 0.6

**7.** 231, 543, 601, 293, 588, 109, 334, 268

$$\bar{x} = \frac{231+543+601+293+588+109+334+268}{8}$$

$$= \frac{2967}{8}$$

$$= 370.9$$

109, 231, 268, 293, 334, 543, 588, 601

$$\text{median} = \frac{293+334}{2} = 313.5$$

no mode

**9.** 1454, 1250, 1136, 1127, 1107

$$\bar{x} = \frac{1454+1250+1136+1127+1107}{5}$$

$$= \frac{6074}{5}$$

$$= 1214.8 \text{ feet}$$

**11.** 1454, 1250, 1136, 1127, 1107, 1046, 1023, 1002

$$\text{median} = \frac{1127+1107}{2} = 1117 \text{ feet}$$

**13.** $\bar{x} = \dfrac{7.8+6.9+7.5+4.7+6.9+7.0}{6}$

$$= \frac{40.8}{6}$$

$$= 6.8 \text{ seconds}$$

**15.** 4.7, 6.9, 6.9, 7.0, 7.5, 7.8
mode = 6.9

**17.** 74, 77, 85, 86, 91, 95

$$\text{median} = \frac{85+86}{2} = 85.5$$

**19.** Sum = 78 + 80 + 66 + 68 + 71 + 64 + 82 + 71
  + 70 + 65 + 70 + 75 + 77 + 86 + 72
  = 1095

$$\bar{x} = \frac{1095}{15} = 73$$

**21.** 64, 65, 66, 68, 70, 70, 71, 71, 72, 75, 77, 78, 80, 82, 86
mode = 70 and 71

**23.** 64, 65, 66, 68, 70, 70, 71, 71, 72, 75, 77, 78, 80, 82, 86
$\qquad\qquad\qquad\qquad\qquad\qquad\qquad\uparrow$
$\qquad\qquad\qquad\qquad\qquad\qquad$ mean = 73
9 rates were lower than the mean.

**25.** __, __, 16, 18, __ ;
Since the mode is 21, at least two of the missing numbers must be 21. The mean is 20. Let the one unknown number be $x$.

$$\bar{x} = \frac{21+21+16+18+x}{5} = 20$$

$$\frac{76+x}{5} = 20$$

$$76+x = 100$$

$$x = 24$$

The missing numbers are 21, 21, 24.

# Practice Final Exam

**1.** $-3^4 = -(3^4) = -81$

**2.** $4^{-3} = \dfrac{1}{4^3} = \dfrac{1}{64}$

**3.** $\begin{aligned} 6[5 + 2(3-8) - 3] &= 6[5 + 2(-5) - 3] \\ &= 6[5 + (-10) - 3] \\ &= 6[-5 - 3] \\ &= 6[-8] \\ &= -48 \end{aligned}$

**4.** $\begin{aligned} &5x^3 + x^2 + 5x - 2 - (8x^3 - 4x^2 + x - 7) \\ &= 5x^3 + x^2 + 5x - 2 - 8x^3 + 4x^2 - x + 7 \\ &= 5x^3 - 8x^3 + x^2 + 4x^2 + 5x - x - 2 + 7 \\ &= -3x^3 + 5x^2 + 4x + 5 \end{aligned}$

**5.** $\begin{aligned} (4x - 2)^2 &= (4x)^2 - 2(4x)(2) + 2^2 \\ &= 16x^2 - 16x + 4 \end{aligned}$

**6.** $\begin{aligned} &(3x + 7)(x^2 + 5x + 2) \\ &= 3x(x^2 + 5x + 2) + 7(x^2 + 5x + 2) \\ &= 3x^3 + 15x^2 + 6x + 7x^2 + 35x + 14 \\ &= 3x^3 + 22x^2 + 41x + 14 \end{aligned}$

**7.** $y^2 - 8y - 48 = (y - 12)(y + 4)$

**8.** $\begin{aligned} 9x^3 + 39x^2 + 12x &= 3x(3x^2 + 13x + 4) \\ &= 3x(3x + 1)(x + 4) \end{aligned}$

**9.** $\begin{aligned} 180 - 5x^2 &= 5(36 - x^2) \\ &= 5(6^2 - x^2) \\ &= 5(6 + x)(6 - x) \end{aligned}$

**10.** $\begin{aligned} 3a^2 + 3ab - 7a - 7b &= 3a(a + b) - 7(a + b) \\ &= (a + b)(3a - 7) \end{aligned}$

**11.** $\begin{aligned} 8y^3 - 64 &= 8(y^3 - 8) \\ &= 8(y^3 - 2^3) \\ &= 8(y - 2)(y^2 + y \cdot 2 + 2^2) \\ &= 8(y - 2)(y^2 + 2y + 4) \end{aligned}$

**12.** $\left(\dfrac{x^2 y^3}{x^3 y^{-4}}\right)^2 = \left(\dfrac{y^{3-(-4)}}{x^{3-2}}\right)^2 = \left(\dfrac{y^7}{x^1}\right)^2 = \dfrac{y^{7 \cdot 2}}{x^{1 \cdot 2}} = \dfrac{y^{14}}{x^2}$

**13.** $\dfrac{5 - \dfrac{1}{y^2}}{\dfrac{1}{y} + \dfrac{2}{y^2}} = \dfrac{y^2\left(5 - \dfrac{1}{y^2}\right)}{y^2\left(\dfrac{1}{y} + \dfrac{2}{y^2}\right)} = \dfrac{5y^2 - 1}{y + 2}$

**14.** $\begin{aligned} &\dfrac{x^2 - 9}{x^2 - 3x} \div \dfrac{xy + 5x + 3y + 15}{2x + 10} \\ &= \dfrac{x^2 - 9}{x^2 - 3x} \cdot \dfrac{2x + 10}{xy + 5x + 3y + 15} \\ &= \dfrac{(x + 3)(x - 3) \cdot 2(x + 5)}{x(x - 3) \cdot (y + 5)(x + 3)} \\ &= \dfrac{2(x + 5)}{x(y + 5)} \end{aligned}$

**15.** $a^2 - a - 6 = (a - 3)(a + 2)$

$\text{LCD} = (a - 3)(a + 2)$

$\begin{aligned} &\dfrac{5a}{a^2 - a - 6} - \dfrac{2}{a - 3} \\ &= \dfrac{5a}{(a - 3)(a + 2)} - \dfrac{2 \cdot (a + 2)}{(a - 3) \cdot (a + 2)} \\ &= \dfrac{5a - 2(a + 2)}{(a - 3)(a + 2)} \\ &= \dfrac{5a - 2a - 4}{(a - 3)(a + 2)} \\ &= \dfrac{3a - 4}{(a - 3)(a + 2)} \end{aligned}$

**16.** $\begin{aligned} 4(n - 5) &= -(4 - 2n) \\ 4n - 20 &= -4 + 2n \\ 2n - 20 &= -4 \\ 2n &= 16 \\ n &= 8 \end{aligned}$

**17.** $\begin{aligned} x(x + 6) &= 7 \\ x^2 + 6x &= 7 \\ x^2 + 6x - 7 &= 0 \\ (x + 7)(x - 1) &= 0 \end{aligned}$

$\begin{array}{lll} x + 7 = 0 & \text{or} & x - 1 = 0 \\ \quad x = -7 & & \quad x = 1 \end{array}$

**18.** $3x - 5 \geq 7x + 3$

$-5 \geq 4x + 3$

$-8 \geq 4x$

$-2 \geq x$

$x \leq -2$

$(-\infty, -2]$

**19.** $2x^2 - 6x + 1 = 0$

$a = 2, b = -6, c = 1$

$x = \dfrac{-b \pm \sqrt{b^2 - 4ac}}{2a}$

$x = \dfrac{-(-6) \pm \sqrt{(-6)^2 - 4(2)(1)}}{2(2)}$

$= \dfrac{6 \pm \sqrt{36 - 8}}{4}$

$= \dfrac{6 \pm \sqrt{28}}{4}$

$= \dfrac{6 \pm 2\sqrt{7}}{4}$

$= \dfrac{3 \pm \sqrt{7}}{2}$

**20.** $\dfrac{4}{y} - \dfrac{5}{3} = -\dfrac{1}{5}$

$15y\left(\dfrac{4}{y}\right) - 15y\left(\dfrac{5}{3}\right) = 15y\left(-\dfrac{1}{5}\right)$

$60 - 25y = -3y$

$60 = 22y$

$\dfrac{60}{22} = y$

$\dfrac{30}{11} = y$

**21.** $\dfrac{5}{y+1} = \dfrac{4}{y+2}$

$5(y+2) = 4(y+1)$

$5y + 10 = 4y + 4$

$y + 10 = 4$

$y = -6$

**22.** $\dfrac{a}{a-3} = \dfrac{3}{a-3} - \dfrac{3}{2}$

$2(a-3)\left(\dfrac{a}{a-3}\right) = 2(a-3)\left(\dfrac{3}{a-3}\right) - 2(a-3)\left(\dfrac{3}{2}\right)$

$2a = 6 - 3(a-3)$

$2a = 6 - 3a + 9$

$2a = 15 - 3a$

$5a = 15$

$a = 3$

In the original equation, $a = 3$ makes a denominator 0, so it is extraneous. The equation has no solution.

**23.** $\sqrt{2x-2} = x - 5$

$\left(\sqrt{2x-2}\right)^2 = (x-5)^2$

$2x - 2 = x^2 - 10x + 25$

$0 = x^2 - 12x + 27$

$0 = (x-9)(x-3)$

$x - 9 = 0$ or $x - 3 = 0$

$x = 9$ $\qquad$ $x = 3$

$x = 3$ is extraneous. The only solution is $x = 9$.

**24.** $5x - 7y = 10$

| $x$ | $y$ |
|-----|-----|
| 2 | 0 |
| 0 | $-\dfrac{10}{7}$ |

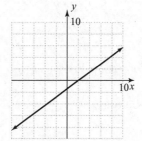

**25.** $x - 3 = 0$

$x = 3$

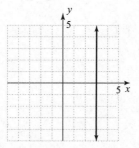

**26.** $y > -4x$

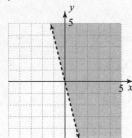

**27.** $m = \dfrac{y_2 - y_1}{x_2 - x_1}$

$m = \dfrac{2 - (-5)}{-1 - 6} = \dfrac{2 + 5}{-7} = \dfrac{7}{-7} = -1$

**28.** $-3x + y = 5$

$\quad\quad y = 3x + 5$

$m = 3$

**29.** $(x_1, y_1) = (2, -5), (x_2, y_2) = (1, 3)$

$m = \dfrac{y_2 - y_1}{x_2 - x_1} = \dfrac{3 - (-5)}{1 - 2} = \dfrac{3 + 5}{-1} = \dfrac{8}{-1} = -8$

$y - y_1 = m(x - x_1)$

$y - (-5) = -8(x - 2)$

$y + 5 = -8x + 16$

$8x + y = 11$

**30.** A line parallel to $x = 7$ is a vertical line. The vertical line through $(-5, -1)$ has equation $x = -5$.

**31.** $\begin{cases} 3x - 2y = -14 \\ \quad\quad y = x + 5 \end{cases}$

Substitute $x + 5$ for $y$ in the first equation.

$3x - 2y = -14$

$3x - 2(x + 5) = -14$

$3x - 2x - 10 = -14$

$\quad\quad\quad x = -4$

Substitute $-4$ for $x$ in the second equation.

$y = x + 5$

$y = -4 + 5$

$y = 1$

The solution is $(-4, 1)$.

**32.** $\begin{cases} 4x - 6y = 7 \\ -2x + 3y = 0 \end{cases}$

Multiply the second equation by 2 and add the result to the first equation.

$4x - 6y = 7$

$\underline{-4x + 6y = 0}$

$\quad\quad\quad 0 = 7$

The statement $0 = 7$ is false, so the system has no solution.

**33.** $h(x) = x^3 - x$

**a.** $h(-1) = (-1)^3 - (-1) = -1 + 1 = 0$

**b.** $h(0) = 0^3 - 0 = 0 - 0 = 0$

**c.** $h(4) = 4^3 - 4 = 64 - 4 = 60$

**34. a.** The graph crosses the $x$-axis at 0 and 4. Therefore, the $x$-intercepts are $(0, 0)$ and $(4, 0)$. The graph crosses the $y$-axis at 0. The $y$-intercept is $(0, 0)$.

**b.** Domain: $(-\infty, \infty)$

Range: $(-\infty, 4]$

**35.** $\sqrt{16} = 4$ because $4^2 = 16$ and 4 is positive.

**36.** $27^{-2/3} = \dfrac{1}{27^{2/3}} = \dfrac{1}{\left(\sqrt[3]{27}\right)^2} = \dfrac{1}{3^2} = \dfrac{1}{9}$

**37.** $\left(\dfrac{9}{16}\right)^{1/2} = \dfrac{9^{1/2}}{16^{1/2}} = \dfrac{\sqrt{9}}{\sqrt{16}} = \dfrac{3}{4}$

**38.** $\sqrt{54} = \sqrt{9 \cdot 6} = \sqrt{9} \cdot \sqrt{6} = 3\sqrt{6}$

**39.** $\sqrt{9x^9} = \sqrt{9x^8 \cdot x} = \sqrt{9x^8} \cdot \sqrt{x} = 3x^4\sqrt{x}$

**40.** $\sqrt{12} - 2\sqrt{75} = \sqrt{4 \cdot 3} - 2\sqrt{25 \cdot 3}$

$= 2\sqrt{3} - 2 \cdot 5\sqrt{3}$

$= 2\sqrt{3} - 10\sqrt{3}$

$= -8\sqrt{3}$

**41.** $\dfrac{\sqrt{40x^4}}{\sqrt{2x}} = \sqrt{\dfrac{40x^4}{2x}}$

$\qquad = \sqrt{20x^3}$

$\qquad = \sqrt{4x^2 \cdot 5x}$

$\qquad = 2x\sqrt{5x}$

**42.** $\sqrt{2}\left(\sqrt{6} - \sqrt{5}\right) = \sqrt{2} \cdot \sqrt{6} - \sqrt{2} \cdot \sqrt{5}$

$\qquad\qquad\qquad = \sqrt{12} - \sqrt{10}$

$\qquad\qquad\qquad = \sqrt{4 \cdot 3} - \sqrt{10}$

$\qquad\qquad\qquad = 2\sqrt{3} - \sqrt{10}$

**43.** $\sqrt{\dfrac{5}{12x^2}} = \dfrac{\sqrt{5}}{\sqrt{12x^2}}$

$\qquad\quad = \dfrac{\sqrt{5}}{\sqrt{4x^2 \cdot 3}}$

$\qquad\quad = \dfrac{\sqrt{5}}{2x\sqrt{3}}$

$\qquad\quad = \dfrac{\sqrt{5} \cdot \sqrt{3}}{2x\sqrt{3} \cdot \sqrt{3}}$

$\qquad\quad = \dfrac{\sqrt{15}}{6x}$

**44.** $\dfrac{2\sqrt{3}}{\sqrt{3}-3} = \dfrac{2\sqrt{3}\left(\sqrt{3}+3\right)}{\left(\sqrt{3}-3\right)\left(\sqrt{3}+3\right)}$

$\qquad\quad = \dfrac{2 \cdot 3 + 6\sqrt{3}}{3-9}$

$\qquad\quad = \dfrac{6 + 6\sqrt{3}}{-6}$

$\qquad\quad = \dfrac{6\left(1+\sqrt{3}\right)}{-6}$

$\qquad\quad = -\left(1+\sqrt{3}\right)$

$\qquad\quad = -1 - \sqrt{3}$

**45.** Let $x$ be the number.

$x + 5 \cdot \dfrac{1}{x} = 6$

$x\left(x + \dfrac{5}{x}\right) = x \cdot 6$

$x^2 + 5 = 6x$

$x^2 - 6x + 5 = 0$

$(x-1)(x-5) = 0$

$x - 1 = 0 \quad \text{or} \quad x - 5 = 0$

$\quad x = 1 \qquad\qquad x = 5$

The number is 1 or 5.

**46.** Let $x$ be the smaller area code. Then the other area code is $2x$.

$x + 2x = 1203$

$3x = 1203$

$x = 401$

$2x = 2(401) = 802$

The area codes are 401 (Rhode Island) and 802 (Vermont).

**47.** Let $x$ be the number of hours since the trains left. One train will have traveled $50x$ miles and the other will have traveled $64x$ miles.

$50x + 64x = 285$

$114x = 285$

$x = \dfrac{285}{114} = \dfrac{5}{2}$ or $2\dfrac{1}{2}$

The trains are 285 miles apart after $2\dfrac{1}{2}$ hours.

**48.** Let $x$ be the amount of 12% solution.

| Amount | Percent | Total Saline |
|--------|---------|--------------|
| $x$ | $12\% = 0.12$ | $0.12x$ |
| 80 | $22\% = 0.22$ | $0.22 \cdot 80 = 17.6$ |
| $80 + x$ | $16\% = 0.16$ | $0.16(80 + x)$ $= 12.8 + 0.16x$ |

$0.12x + 17.6 = 12.8 + 0.16x$

$4.8 = 0.04x$

$120 = x$

120 cc of 12% saline solution should be added.